Neuroactive Steroids in Brain Function, Behavior and Neuropsychiatric Disorders

Michael S. Ritsner • Abraham Weizman
Editors

Neuroactive Steroids in Brain Function, Behavior and Neuropsychiatric Disorders

Novel Strategies for Research and Treatment

Springer

Editors:
Michael S. Ritsner, M.D., Ph.D.
Associate Professor of Psychiatry and Head of Cognitive & Psychobiology Research Laboratory
The Rappaport Faculty of Medicine
Technion - Israel Institute of Technology
Haifa and Director
Acute Department
Sha'ar Menashe Mental Health Center
Hadera, Israel

Abraham Weizman, M.D.
Professor of Psychiatry
Director of Felsenstein Medical Research Center
Sackler Faculty of Medicine
Tel Aviv University, and Director of Research Unit
Geha Mental Health Center
Petah Tikva, Israel

ISBN 978-1-4020-6853-9 ISBN 978-1-4020-6854-6 (eBook)

Library of Congress Control Number: 2008920052

Printed on acid-free paper

9 8 7 6 5 4 3 2 1

springer.com

Foreword

Over the past few decades the involvement of neurosteroids in brain function and mental health has attracted much interest, not only from a neuroscience perspective, but also from clinical and therapeutic aspects. In recent years, much of the research has focused on the multifunctional position of neurosteroids in the nervous system. The areas that have been explored, in particular, are the cellular and molecular mechanisms involved in the activity of neurosteroids in the brain, the role of neurosteroids in brain development, neurodegeneration and neuroplasticity, and their pharmacological properties. The large increase in the number of publications highlights this amplified interest in neurosteroids. Over the last 7 years at least 400 papers have been published describing the putative role of neurosteroids in the modulation of basic brain function and in the etiology and treatment of psychiatric and neurological disorders.

Contributors to this book are amongst the most active basic researchers and clinicians in the field and provide new perspectives not only clarifying some of the ongoing controversies, but also proposing diverse aspects and new insights. The book is organized in two major sections: "Neuroactive steroids and brain function', and 'Neuroactive steroids and neuropsychiatric disorders". Though the selection of topics has been influenced by the current state of the art and the issues that require further elaboration, one should not assume that these are the only significant issues at this time. As often happens in publications composed of contributions by multiple authors from diverse orientations and academic backgrounds, differences in approaches and opinions are inevitable, as is some overlap. Indeed we consider such diversity to be one of this book's strengths. We also believe this book to be the first of its kind to go beyond the neuropsychiatric disorders and delve into the neurobiological basis for clinical symptomatology, psychopathology, emotional regulation, cognitive functioning, and novel pharmacological strategies.

We sincerely hope that this book will broaden the knowledge and deepen the understanding of the complex role of neurosteroids in the regulation of brain functions and in the pathophysiology and pharmacotherapy of neuropsychiatric disorders, and that it will be of interest to a broad spectrum of readers including clinicians, researchers and students, in the fields of neuroscience and mental health.

2007

Michael S. Ritsner
Abraham Weizman

Contents

Contributors

Lotta Andréen, M.D., Ph.D.
Umeå Neurosteroid Research Center, Department of Clinical Sciences, University of Umeå, Sweden, lotta.andreen@lvn.se

Torbjörn Bäckström, M.D., Ph.D.
Professor, Umeå Neurosteroid Research Center, Department of Clinical Sciences, University of Umeå, Norrlands University Hospital, Sweden, Torbjorn.Backstrom@obgyn.umu.se

Glen B. Baker, Ph.D., D.Sc.
Professor, Department of Psychiatry, Mackenzie Centre, University of Alberta, Edmonton, Alberta, Canada, glen.baker@ualberta.ca

Juan Beauquis
Laboratory of Neuroendocrine Biochemistry, Instituto de Biologia y Medicina Experimental, Buenos Aires, Argentina, and Department of Biochemistry, Faculty of Medicine, University of Buenos Aires, Buenos Aires, Argentina, jbeauquis@dna.uba.ar

Vita Birzniece, M.D., Ph.D.
Umeå Neurosteroid Research Center, Department of Clinical Sciences, University of Umeå, Norrlands University Hospital, Sweden, Vita.Birzniece@obgyn.umu.se

Marie Bixo, M.D., Ph.D.
Associate professor, Umeå Neurosteroid Research Center, Department of Clinical Sciences, University of Umeå, Norrlands University Hospital, Sweden, Marie.Bixo@Obgyn.umu.se

Inger Björn, M.D., Ph.D.
Umeå Neurosteroid Research Center, Department of Clinical Sciences, University of Umeå, Norrlands University Hospital, Sweden, Inger.Bjorn@obgyn.umu.se

Stefan R. Bornstein, M.D., Ph.D.
Professor of Medicine, Director and Chair,
Department of Medicine III, University of Dresden, Dresden,
Germany, stefan.bornstein@uniklinikum-dresden.de

Thomas Borowski, Ph.D.
Assistant Professor, Department of Psychology & Neuroscience, Claremont,
California, USA, thomas_borowski@pitzer.edu

Frank van Broekhoven, M.D., Ph.D.s
Department of Psychiatry, Radboud University Nijmegen Medical Center,
Nijmegen, The Netherlands, F.vanBroekhoven@psy.umcn.nl.

Mike B. Calford, Ph.D.
Professor, Pro Vice-Chancellor, Health, University of Newcastle, New South
Wales 2308, Australia, Mike.Calford@newcastle.edu.au

Patrick S. Calhoun, Ph.D.
Director, Health Services Research Core, VA Mid-Atlantic Mental Illness
Research, Education & Clinical Center (MIRECC), Director, Psychology
Post-Doctoral Training, Durham VA Medical Center,
Assistant Clinical Professor, Department of Psychiatry and Behavioral Sciences,
Duke University Medical Center, Durham, North Carolina, USA,
Patrick.calhoun@duke.edu

Carlo Caltagirone, M.D.
IRCCS Santa Lucia Foundation, Department of Clinical and Behavioural
Neurology, Rome, and Department of Neuroscience, Tor Vergata University,
Rome, Italy, c.caltegirone@hsantolucia.it

Victor V. Chaban, Ph.D.
Assistant Professor, Department of Biomedical Sciences,
Charles Drew University of Medicine and Science,
David Geffen School of Medicine at UCLA, Los Angeles, California,
USA, victorchaban@cdrewu.edu

Ioannis Charalampopoulos, Ph.D.
Research Fellow, Department of Pharmacology, School of Medicine,
University of Crete, Ioannis.charalampopoulos@ki.se

Elias Castanas, M.D., Ph.D.
Professor, Department of Experimental Endocrinology, School of Medicine,
University of Crete, castanas@med.uoc.gr

Glenn H. Dillon, Ph.D.
Professor and Vice President for Research, Department of Pharmacology and Neuroscience, University of North Texas Health Science Center, USA

Akiko Dohi
Graduate Student, Department of Pharmacology and Neuroscience, University of North Texas Health Science Center, Fort Worth, Texas, USA

Bernardo Dubrovsky
Research Scientist, McGill University, Montreal, QC, Canada, bernardo.dubrovsky@mcgill.ca

Guillen Fernández, M.D., Ph.D.
Professor, F.C. Donders Centre for Cognitive Neuroimaging, Radboud University Nijmegen, The Netherlands, guillen.fernandez@fcdonders.ru.nl

Cheryl A. Frye, Ph.D.
Professor, Department of Psychology, The University at Albany-SUNY, Albany, New York, USA, cafrye@albany.edu

Luis Miguel Garcia-Segura
Instituto Cajal, C.S.I.C., Spain, lmgs@cajal.csic.es

Lisa R. Gerak
Assistant Professor, Department of Pharmacology, The University of Texas Health Science Center at San Antonio, Texas, USA

Anatoly Gibel, M.D.
Senior Psychiatrists, Acute Department, Sha'ar Menashe Mental Health Center, Israel, gibel@shaar-menashe.org.il

Brett C. Ginsburg
Assistant Professor, Division of Alcohol and Drug Addiction, Department of Psychiatry, The University of Texas Health Science Center at San Antonio, Texas, USA, ginsburg@uthscsa.edu

Samantha S. Gizerian, Ph.D.
Assistant Professor, Department of Biomedical Sciences, Charles R. Drew University of Medicine and Science, Los Angeles, California, USA, sgizerian@cdrewu.edu

Pavel Golubchik, M.D.
Child and Adolescent Outpatient Clinic, Geha Mental Health Center,
Petah Tiqva, Israel, pgolubchik@clalit.org.il

Achille Gravanis, Pharm.D., Ph.D.
Professor, Department of Pharmacology, School of Medicine,
University of Crete, gravanis@med.uoc.gr

David Haage, Ph.D.
Umeå Neurosteroid Research Center, Department of Clinical Sciences,
University of Umeå, Norrlands University Hospital, Sweden,
David.Haage@obgyn.umu.se

Tibor Hajszan, M.D., Ph.D.
Institute of Biophysics, Biological Research Center,
Szeged, Hungary, hajszan@brc.hu

Yasushi Hojo
Department of Biophysics and Life Sciences, Graduate School of Arts and
Sciences, The University of Tokyo, Japan

Françoise Homo-Delarche
CNRS UMR 7059, Université Paris, Diderot, Paris, France,
fhomodel@wanadoo.fr

Zsofia Hoyk, Ph.D.
Institute of Biophysics, Biological Research Center, Szeged,
Hungary, zsofi@brc.hu

Bryan Hyler
The Charles Drew University School of Medicine, Los Angeles, California, USA,
hylermed@sbcglobal.net

Hirotaka Ishii
Department of Biophysics and Life Sciences, Graduate School of Arts and
Sciences, The University of Tokyo, Japan

Inga-Maj Johansson, Ph.D.
Associate professor, Umeå Neurosteroid Research Center, Department of Clinical
Sciences, University of Umeå, Norrlands University Hospital, Sweden,
Inga-Maj.Johansson@obgyn.umu.se

Kristiina Kask, M.D, Ph.D.s
Department of Women's and Children's Health, Academic Hospital, Uppsala,
Sweden, Kristiina.Kask@kbh.uu.se

Suguru Kawato
Professor, Department of Biophysics and Life Sciences, Graduate School of Arts and Sciences, University of Tokyo, Japan, kawato@phys.c.u-tokyo.ac.jp

Erin M. MacKenzie, B.Sc.
Department of Psychiatry, Mackenzie Centre, University of Alberta, Edmonton, Alberta, Canada, emm3@ualberta.ca

Jason D. Kilts, Ph.D.
Senior Research Associate, Department of Psychiatry and Behavioral Sciences, Duke University Medical Center, Durham VA Medical Center, Durham, North Carolina, USA, kilts001@mc.duke.edu

Tetsuya Kimoto
Department of Biophysics and Life Sciences, Graduate School of Arts and Sciences, The University of Tokyo, Japan

Alexander W. Krug, M.D.
Research scientist, Medizinische Klinik III, Medizinische Fakultät Carl Gustav Carus Universität Dresden, Germany, alexander.krug@uniklinikum-dresden.de

Sonsoles de Lacalle, M.D., Ph.D.
Associate Professor and Chair, Department of Biomedical Sciences, The Charles Drew University, Los Angeles, California, USA, sdelaca@cdrewu.edu

Charlotte Lindblad, Ph.D.s,
Umeå Neurosteroid Research Center, Department of Clinical Sciences, University of Umeå, Norrlands University Hospital, Sweden, Charlotte.Lindblad@obgyn.umu.se

Ron Loewenthal, M.D.
Tissue-Typing Laboratory and Bone Marrow Registry, The Safra Children's Hospital, The Chaim Sheba Medical Center, Tel Hashomer, The Sackler Faculty of Medicine, Tel Aviv University, Israel

Magnus Löfgren, Ph.D.s
Umeå Neurosteroid Research Center, Department of Clinical Sciences, University of Umeå, Norrlands University Hospital, Sweden, Magnus.Lofgren@obgyn.umu.se

Per Lundgren, Ph.D.
Umeå Neurosteroid Research Center, Department of Clinical Sciences, University of Umeå, Norrlands University Hospital, Sweden, per.lundgren@obgyn.umu.se

Rachel Maayan, Ph.D.
Laboratory of Biological Psychiatry, Felsenstein Medical Research Center, Sackler Faculty of Medicine, Tel Aviv University, Israel

Lance R. McMahon
Assistant Professor, Department of Pharmacology, The University of Texas Health Science Center at San Antonio, Texas, USA

Roger D. Madison, Ph.D.
Associate Research Professor, Department of Experimental Surgery, Associate Medical Research Professor, Department of Neurobiology, Duke University Medical Center, Research Career Scientist, Durham VA Medical Center, Durham, North Carolina, USA, madis001@mc.duke.edu

Andrew N. Margioris, M.D.
Professor, Department of Clinical Chemistry, School of Medicine, University of Crete, andym@med.uoc.gr

Christine E. Marx's, M.D., M,A.
Associate Professor, Department of Psychiatry and Behavioral Sciences, Duke University Medical Center, Director, Clinical Interventions Core; Director, Psychiatry Fellowship Program, VA Mid-Atlantic Mental Illness Research, Education, and Clinical Center (MIRECC), Durham VA Medical Center, Durham, North Carolina, USA, marx001@mc.duke.edu

Roberto C. Melcangi, Ph.D.
Professor, Department of Endocrinology, Laboratory of Neuroendocrinology, University of Milan, Italy, roberto.melcangi@unimi.it

Jean-Michel Le Melledo, M.D.
Professor, Department of Psychiatry, Mackenzie Centre, University of Alberta, Edmonton, Alberta, Canada, jean-michel.lemelledo@ualberta.ca

Flavia di Michele, M.D., Ph.D.
Mental Health Service and Department of Neuroscience, Tor Vergata University, Rome, Italy

Hideo Mukai
Department of Biophysics and Life Sciences, Graduate School of Arts and Sciences, The University of Tokyo, Japan

Gen Murakami
Department of Biophysics and Life Sciences, Graduate School of Arts and Sciences, The University of Tokyo, Japan

Jennifer C. Naylor, Ph.D.
Research Associate, Duke University Medical Center, Durham VA Medical Center, North Carolina, USA, naylorjc@duke.edu

Alejandro F. De Nicola
Director, Instituto de Biologia y Medicina Experimental, Buenos Aires, Argentina, denicola@dna.uba.ar

Sigrid Nyberg, Ph.D.
Umeå Neurosteroid Research Center, Department of Clinical Sciences, University of Umeå, Norrlands University Hospital, Sweden, sigrid.nyberg@vll.se

Mari Ogiue-Ikeda
Department of Biophysics and Life Sciences, Graduate School of Arts and Sciences, The University of Tokyo, Japan

Arpad Parducz, Ph.D., D.Sc.
Institute of Biophysics, Biological Research Center, Szeged, Hungary, parducz@brc.hu

Victoria M. Payne, M.D., M.S.
Psychiatry Research Fellow, VA Mid-Atlantic Mental Illness Research, Education, and Clinical Center (MIRECC), Durham VA Medical Center, Durham, North Carolina, USA, payne@biac.duke.edu

Luciana Pietranera
Laboratory of Neuroendocrine Biochemistry, Instituto de Biologia y Medicina Experimental, Buenos Aires, Argentina, and Department of Biochemistry, Faculty of Medicine, University of Buenos Aires, Buenos Aires, Argentina, lpietra@dna.uba.ar

Gianna Ragagnin, Ph.D.
Umeå Neurosteroid Research Center, Department of Clinical Sciences, University of Umeå, Norrlands University Hospital, Sweden, Gianna.Ragagnin@obgyn.umu.se

Mozibur Rahman, M.D., Ph.D.
Umeå Neurosteroid Research Center, Department of Clinical Sciences, University of Umeå, Norrlands University Hospital, Sweden, mozibur.rahman@obgyn.umu.se

Yael Ratner, M.D.
Senior Psychiatrists, Acute Department, Sha'ar Menashe Mental Health Center, Israel, yaelratner@gmail.com

Madeline E. Rhodes
Psychology Department, Smith College, Northampton, Massachusetts, USA, mrhodes@email.smith.edu

Michael S. Ritsner, M.D., Ph.D.
Associate Professor of psychiatry, and Head, Cognitive and Psychobiology Research Laboratory, The Rappaport Faculty of Medicine, Technion, Israel Institute of Technology, Haifa Director, Acute Department, Sha'ar Menashe Mental Health Center, Hadera, Israel, ritsner@shaar-menashe.org.il; mritsner@gmail.com

John D. Roache
Professor, Division of Alcohol and Drug Addiction Department of Psychiatry, The University of Texas Health Science Center, San Antonio, Texas, USA

Yuri B. Saalmann, Ph.D.
Research Scientist, Cognitive Neuroscience Laboratory, Department of Psychology, Royal Holloway, University of London, Yuri.Saalmann@rhul.ac.uk

Colin J. Saldanha Ph.D.
Department of Biological Sciences, Lehigh University, Bethlehem, Pennsylvania, USA, colin@lehigh.edu

Flavia E. Saravia
Laboratory of Neuroendocrine Biochemistry, Instituto de Biologia y Medicina Experimental, Buenos Aires, and Department of Biochemistry, Faculty of Medicine, University of Buenos Aires, Buenos Aires, Argentina, fsaravia@dna.uba.ar

Barney A. Schlinger
Professor and Vice-Chair, Department of Physiological Science, UCLA, Los Angeles, California, USA, schlinge@lifesci.ucla.edu

Meharvan Singh, Ph.D.
Associate Professor, Department of Pharmacology and Neuroscience, University of North Texas Health Science Center, USA, msingh@hsc.unt.edu

Gianfranco Spalletta, M.D., Ph.D.
IRCCS Santa Lucia Foundation, Department of Clinical and Behavioural Neurology, Rome, and Department of Neuroscience, Tor Vergata University, Rome, Italy, g.spalletta@hsantalucia.it

Daniel Stein, M.D.
Pediatric Psychosomatic Department, Safra Children's Hospital, The Chaim Sheba Medical Center, Tel Hashomer, Ramat Gan, Israel, dstein@netvision.net.il

Jessica Strömberg, Ph.D.
Umeå Neurosteroid Research Center, Department of Clinical Sciences, University of Umeå, Norrlands University Hospital, Sweden, Jessica, Stromberg@obgyn.umu.se

Inger Sundström-Poromaa, M.D., Ph.D.
Professor, Department Women's and Children's Health, Academic Hospital, Uppsala, Sweden, inger.sundstrom@kbh.uu.se

Jennifer L. Strauss, Ph.D.
Assistant Professor, Department of Psychiatry and Behavioral Sciences, Duke University Medical Center, Health Scientist, Durham VA Medical Center, Durham, North Carolina, USA, jennifer.strauss@duke.edu

Rael D. Strous, M.D.
Beer Yaakov Mental Health Center, Sackler Faculty of Medicine, Tel Aviv University, Israel, raels@post.tau.ac.il

Erika Timby, M.D., Ph.D.
Student, Umeå neurosteroid Research Center, Department of Clinical Sciences, Umeå University, Sweden, Erika.Timby@vll.se

Christos Tsatsanis, Ph.D.
Assistant Professor, Department of Clinical Chemistry, School of Medicine, University of Crete, tsatsani@med.uoc.gr

Sahruh Turkmen, M.D., Ph.D.
Umeå Neurosteroid Research Center, Department of Clinical Sciences, University of Umeå, Norrlands University Hospital, Sweden, sahruhturkmen@hotmail.com

Ming-De Wang, M.D., Ph.D.
Associate Professor, Umeå Neurosteroid Research Center, Department of Clinical Sciences, University of Umeå, Norrlands University Hospital, Sweden, mingde.wang@obgyn.umu.se

Abraham Weizman, M.D.
Professor of Psychiatry and Director of Felsenstein Medical Research Center, Sackler Faculty of Medicine, Tel Aviv University and Director of Research Unit, Geha Mental Health Center, Petah Tikva, Israel, aweizman@clalit.org.il

Guido van Wingen, Ph.D., F.C.
Donders Centre for Cognitive Neuroimaging, Radboud University Nijmegen, The Netherlands, Guido.vanWingen@fcdonders.ru.nl

Christian G. Ziegler, Ph.D.
Research Scientist, Medizinische Klinik III, Medizinische Fakultät Carl Gustav Carus, Universität Dresden, Dresden, Germany

Part I
Neuroactive Steroids and Brain Functions

Chapter 1
Neurosteroid: Molecular Mechanisms of Action on the $GABA_A$ Receptor

Ming-De Wang*,†, Mozibur Rahman†, Jessica Strömberg, Per Lundgren, David Haage, Inga-Maj Johansson, and Torbjörn Bäckström

Abstract Neurosteroids represent a class of endogenous steroids that are synthesized in the brain, the adrenals and the gonads and have potent and selective effects on the gamma-aminobutyric acid type A ($GABA_A$ receptor). 3α-Hydroxy A-ring-reduced metabolites of progesterone, deoxycorticosterone and testosterone enhance the the Cl^- flux through $GABA_A$ receptor conductance at nanomolar concentrations in a non-genomic (rapid and direct) manner. Studies on the $GABA_A$ receptors have shown that allopregnanolone (3α-OH-5α-pregnan-20-one), 5α-androstane-3α,17α-diol (Adiol) and (3α 5α3, 21-dihydroxypregnan-20-one (3α5αTHDOC) enhance the GABA mediated Cl^- currents acting on a site (or sites) distinct from the GABA, benzodiazepine, barbiturate and picrotoxin binding sites. This modulation site (or sites) has a well-defined structure–activity relationship with a 3α-hydroxy and a 20-ketone configuration in the pregnane molecule required for agonist action. However, the neurosteroid pregnenolone sulfate (PS) is a non-competitive $GABA_A$ receptor antagonist and inhibits GABA-activated Cl^- currents in an activation-dependent manner. 3β-hydroxy A-ring reduced pregnane steroids are pregnenolone sulfate-like $GABA_A$ receptors antagonists and inhibit the $GABA_A$ receptor's function and its potentiation induced by their 3α-diasteromers in a non-competitive manner. The specificity of neurosteroid action on the $GABA_A$ receptor results from a variety of molecular mechanisms, including receptor subunit composition, receptor activation–deactivation, and steroid concentration. Here, we will review the GABA-modulatory actions of the neurosteroids. The molecular mechanisms underpinning the non-genomic effect of agonist and antagonist neurosteroids will be discussed with particular emphasis being given to the role of $GABA_A$ receptor isoforms.

Keywords Neurosteroid, GABA, allopregnanolone, pregnenolone sulfate, agonist, antagonist

* Umeå Neurosteroid Research Center, Department of Clinical Science, Section of Obstetric & Gynecology, Umeå University, 901 85 Umeå, Sweden

† Ming-De Wang and Mozibur Rahman contributed equally in this work.

M.S. Ritsner and A. Weizman (eds.), *Neuroactive Steroids in Brain Function, Behavior and Neuropsychiatric Disorders*

Abbreviations CNS central nervous system; GABA γ-amino butyric acid; NMDA *N*-methyl-D-aspartate; PMDD premenstrual dystrophic disorder; PKC protein kinase C; TBPS *t*-butyl bicyclophosphorothionate; P450scc P-450 side chain cleavage; $P450_{21}$ steroid 21 hydroxylase; 3α-HSD 3α-hydroxysteroid dehydrogenase; 3β-HSD 3β-hydroxysteroid dehydrogenase; StAR steroidogenic acute-regulatory; 3α-OH 3α-hydroxyl; 3β-OH 3β-hydroxyl; 3α5αp 3α-OH-5α-pregnan-20-one, allopregnanolone; 3α5βP 3α-OH-5β-pregnan-20-one, pregnanolone; 3α5αTHDOC 3α, 5α-tetrahydrodeoxy-corticosterone; 3α5βTHDOC 3α, 5β-tetrahydrodeoxy-corticosterone; PS pregnenolone sulfate; DHEAS dehydroepiandrosterone sulfate; 3β5αP 3β-OH-5α -pregnan-20-one, isoallopregnanolone; 3β5βP 3β-OH-5β -pregnan-20-one, epipregnanolone; 3β5βTHDOC 5β-pregnan-3β, 21-diol-20-one; 3α-adiol/3α5αADL 5α-androstane-3α,17β-diol; DOC deoxycorticosterone; sIPSC spontaneous inhibitory post-synaptic current; DGC dentate gyrus granule cells; CGC cerebellar granule cells; MPN medial preoptic nucleus; EC_{50} concentration of drug that produces 50% of maximum response

1.1 The Metabolites of Sex and Stress Hormones are Neuroactive

The sex hormones, estrogen, progesterone and testosterone, influence a variety of behaviors in vertebrates in addition to their functional roles in the reproductive system.[1] In adult women, estradiol is mainly produced in the granulose cell from the developing follicle and the corpus luteum in the ovary. Progesterone is synthesized mainly in the granulose cells of corpus luteum, the placenta and the adrenal glands.[2] Sex hormones act through genomic mechanisms through the intracellular receptors that are located in the nucleus or cytoplasm.[3] They act as ligand-activated transcription factors in the regulation of gene expression (Fig. 1.1). However, metabolites of progesterone and several stress hormones act on the membrane bound receptor via a fast non-genomic mechanism. The receptor binding to DNA and RNA synthesis is thus not required (Fig. 1.1).[1–3] While the genomic action of sex hormones requires a time period from minutes to hours and limited by the rate of protein biosynthesis,[4] the modulatory effect on the membrane receptor is fast occurring event and requires only milliseconds to seconds.[4] Today it is known that metabolites of sex and stress hormones act non-genomically in the CNS and alter neuronal excitability.[5–7] The term "neurosteroid" was introduced to describe these steroid metabolites that modulate neuronal activity.[6] The 3αOH A-ring reduced metabolites of progesterone and deoxycorticosterone, allopregnanolone (3α5αP) and 3α, 5α-tetrahydrodeoxycorticosterone (3α5αTHDOC) were first shown to modulate neuronal excitability by interaction with the $GABA_A$ receptor.[5] Several other neurotransmitters like the NMDA,[8] the nicotinic,[9] the muscarinic,[10] the serotonergic[11] and the adrenergic system[12] are also targets for neurosteroids (Fig. 1.1). The functional modulation of

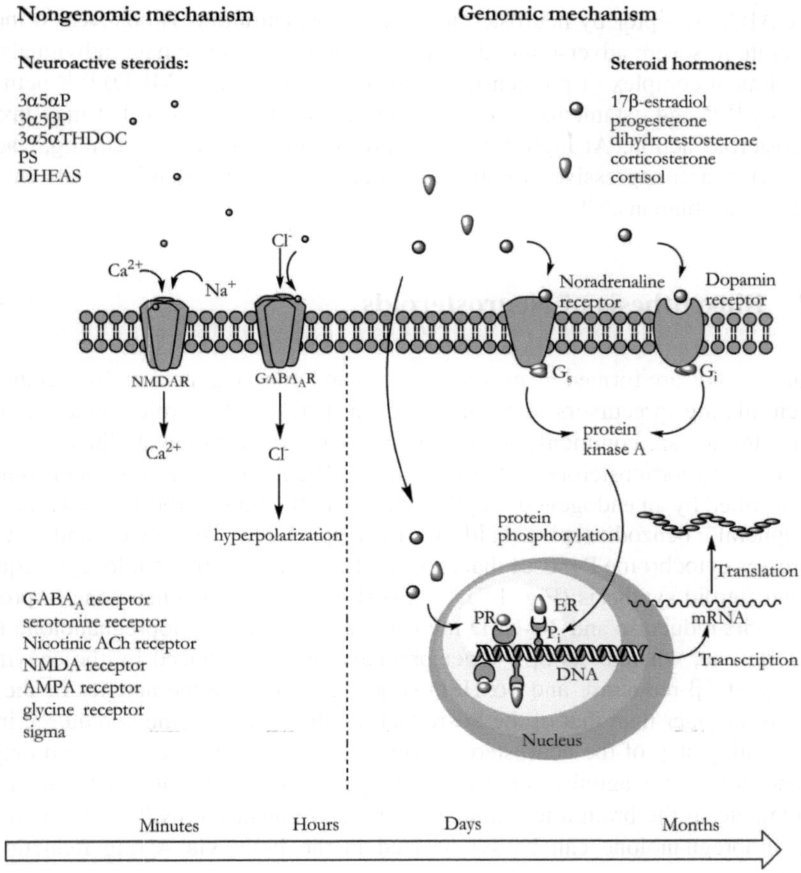

Fig. 1.1 The genomic and non-genomic site of action of steroid hormones and their metabolites. At the left the non genomic mechanism is presented. Neuroactive steroids are exemplified as well as different membrane receptors which they target. The reaction time for neurosteroids are fast which is indicated by the time line. An example is when the $GABA_A$-receptor agonist steroid 3α5αP (Allopregnanolone) binds to the receptor enhancing the GABA effect which results in an increase in the chloride ion influx into the cell. The increase in intracellular chloride ions leads to a hyperpolarization. Such action can be as fast as in the millisecond interval

The genomic mechanism and an example of steroid hormones are presented at the right side. Classical steroid hormones penetrate the cell membrane. In the cytosol steroids bind to their specific receptors estrogen receptor (ER) and progesterone receptors (PR), respectively. The steroid–receptor complex translocates into the nucleus and binds to specific response elements of the DNA. This activates the transcription of mRNA which later can be translated into proteins. The action of the estrogen receptor can be regulated by phosphorylation by protein kinase A. The activation or inactivation of protein kinase A is in this example regulated by a stimulating G-protein coupled to the adrenergic (α2) receptor or by an inactivating G-protein coupled to the dopamine (D1-like) receptor. That occurs when dopamine or noradrenalin binds to their receptors. The genomic action is slow and long lasting the $GABA_A$ receptor by neurosteroids at low concentrations is believed to induce moderate to severe adverse mood changes in up to 20% of female individuals.[13, 14] The clinical complex of premenstrual dystrophic disorders (PMDD),[15, 16] petit mal epilepsy,[17, 18] and catamenial epilepsy[19] are among the disorders that may involve neurosteroid action. At higher doses, neurosteroids may affect learning,[20] act as anxiolytic, anti-aggressive, sedative/anesthetic, and anti-epileptic agents in both animals and humans.[6, 21–23]

1.2 Biosynthesis of Neurosteroids

Neurosteroids are formed *de novo* in neurons and glia, or generated by metabolism of circulating precursors that originate in peripheral steroidogenic organs.[24] Neurosteroids are commonly referred to as A-ring reduced metabolites of progesterone, deoxycorticosterone and testosterone.[24] The brain synthesis of neurosteroids is controlled by an endogenous peptide-diazepam binding inhibitor, a ligand for the "peripheral" benzodiazepine binding site (Fig. 1.1).[25] Astrocytes and neurons expresses cytochrome P450scc that convert cholesterol to pregnenolone, is required for the steroid synthesis (Fig. 1.2).[26, 27] 3β-HSD converts pregnenolone to progesterone. 5α-reductase and 3α-HSD involve the synthesis of allopregnanolone from progesterone, whereas its 5β-isomer pregnanolone is produced by the enzymatic activity of 5β-reductase and 3α-HSD (Fig. 1.2). Because the activity of the 3α-HSD is stronger than that of the 5α-reductase, the latter enzyme is required in the rate-limiting step of the neurosteroid synthesis.[28, 29] Allopregnanolone and pregnenolone are potent agonists on the $GABA_A$ receptor.[5, 6, 30] Allopregnanolone can accumulate in the brain after adrenalectomy and gonadectomy.[31–33] This indicates that allopregnanolone can be synthesized in the brain via A-ring reduction of progesterone.

Pregnenolone sulfate (PS) and dehydroepiandrosterone sulfate (DHEAS) are naturally occurring neurosteroids that inhibit the $GABA_A$ receptor function.[6] PS is synthesized from pregnenolone by the enzyme sulfotransferase (Fig. 1.2). Conversion from DHEA to DHEAS is also mediated by sulfotransferase. DHEA is metabolized from pregnenolone by cytochrome $P450_{C17}$.[34] On the other hand, 3β-HSD is essential for the synthesis of 3β-OH steroids, i.e., epipregnanolone and isoallopregnanolone[35](Fig. 1.3). PS, DHEAS and 3β-OH steroids act as $GABA_A$ receptor antagonists[36, 37] and can be measured from human blood samples.[38]

Stress hormones, 3α5α-THDOC and 3α5βTHDOC are also potent modulators of the $GABA_A$ receptor.[5, 39–41] Both 3α5αTHDOC and 3α5βTHDOC have significant sedative effects *in vivo*. 3α5αTHDOC is a metabolite of the mineralocorticoid DOC and is responsible for the sedative and anti-seizure activity of DOC in animal models.[42] DOC can be metabolized from progesterone and this conversion is mediated by $P450_{21}$.[43] The conversion of deoxycorticosterone into 3α5αTHDOC occurs both in peripheral tissues and in the brain.[44]

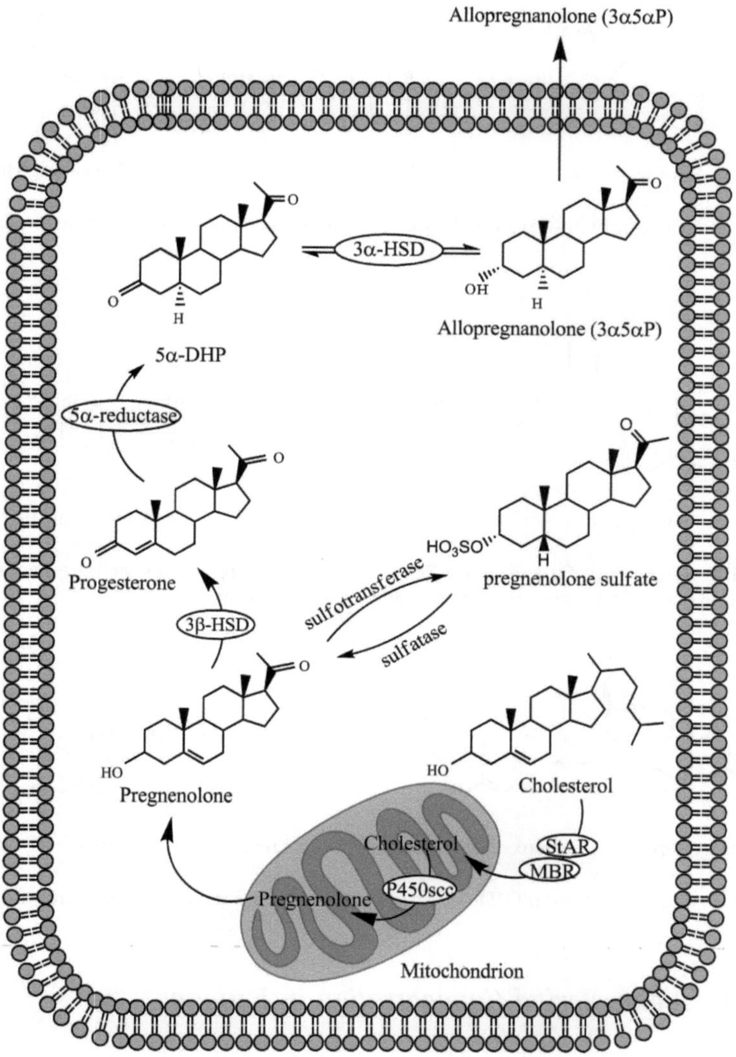

Fig. 1.2 Bio-synthesis of allopregnanolone and pregnenolone sulfate (PS) from cholesterol within the neuron or glial cell. Enzymes involved are P450 side-chain cleavage (P450scc), 3α-hydroxysteroid dehydrogenase (3α-HSD), 3β-hydroxysteroid dehydrogenase (3β-HSD). Transport of cholesterol across the mitochondrial membrane is enhanced by the steroidogenic acute-regulatory (StAR) protein and the mitochondrial benzodiazepine receptor (MBR). 5α-DHP represents 5α-dihydroprogesterone

The A-ring reduced metabolites of testosterone, 3α5α-adiol, acts as a GABA$_A$ receptor agonist.[45] The synthesis of 3α5α-THDOC and 3α5α-adiol is shown in Figs. 1.4 and 1.5.

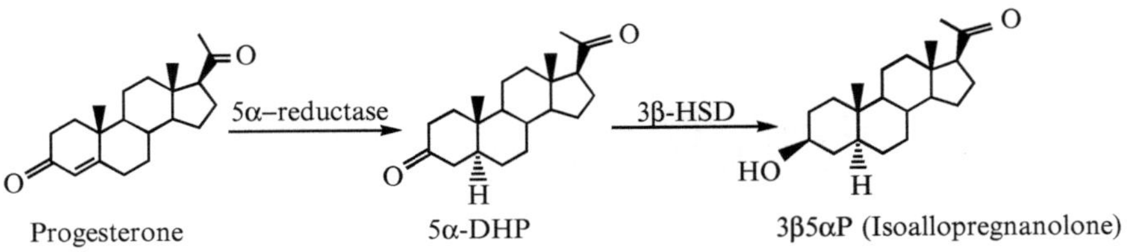

Fig. 1.3 Synthesis of 3β-OH steroids isoallopregnanolone. Enzymes involved are indicated at the arrows

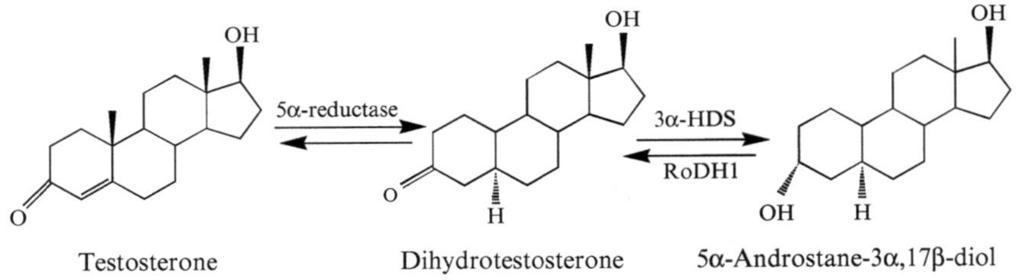

Fig. 1.4 Synthesis of 5α-androstane 3α,17β-diol. RoDH1 represents the rat retinol dehydrogenase 1 enzyme.[287] Enzymes involved are indicated at the arrows

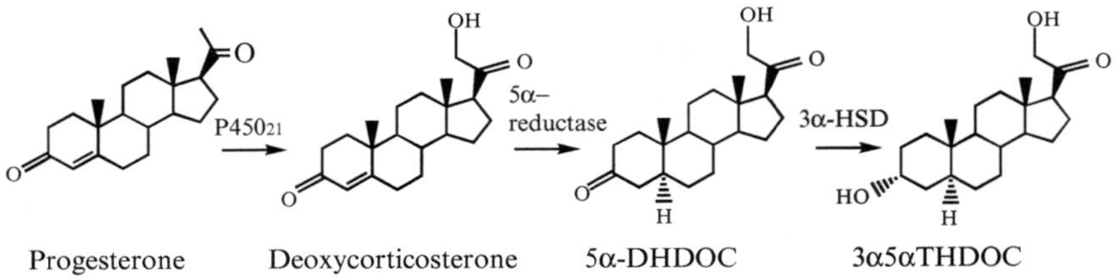

Fig. 1.5 Synthesis of 3α5αTHDOC. Enzymes involved are indicated at the arrows

1.2.1 Neurosteriod Concentration in Human

The physiological concentration of the neurosteriods depends on the type of sampled tissue, stress, and fluctuation of precursor hormones in the body. Table 1.1 shows the normal range of neurosteriods in the human plasma and/brain tissue.

1.3 The GABA$_A$ Receptor

GABA mediates most of the inhibitory neurotransmission in the mammalian brain. GABA is synthesized by two isoforms of glutamic acid decarboxylase (GAD).[46–48] GABA-mediated inhibition is crucially in both short-term and long-term regulation of

Table 1.1 Neurosteroid concentration in human plasma and brain (mean ± SEM)

	Plasma (nM)		Brain (nM)	
Steroids	Follicular	Luteal	Postmenopausal	Luteal
Progesterone	5.0 ± 0.50 [288]	34.7 ± 2.40	65 [289]	137 [289]
3α-OH-5β-pregnan-20-one	0.6 ± 0.00 [290]	1.1 ± 0.50	–	114 [289]
3α-OH-5α-pregnan-20-one	0.53 ± 0.19 [291]	2.14 ± 0.37	47 [289]	66 [289]
3β-OH-5α-pregnan-20-one	0.29 ± 0.14 [291]	1.23 ± 0.28	–	–
3β-OH-5β-pregnan-20-one	0.09 ± 0.08 [291]	0.26 ± 0.13	–	–
Pregnenolone sulfate	11.2 ± 0.6 [288]	15.2 ± 0.8 [289]	–	100 [292]
Pregnenolone	*2.19 [293]	–	–	–
3α5α-Androstane-3α,17β-diol	*0.475 [293]	–	–	–

Superscript numerals denote references
* Concentration in adult men

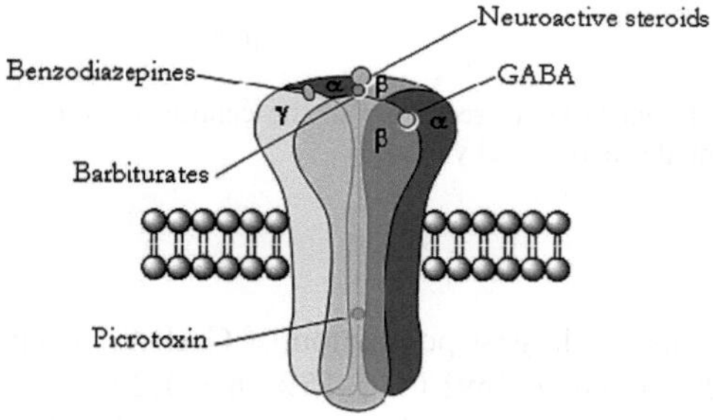

Fig. 1.6 Presentation of the $GABA_A$ receptor with putative binding sites for neuroactive steroids, barbiturates, benzodiazepines, and picrotoxin. At the right, the transmembrane domains of five subunits are shown. Note that the second transmembrane domain of each subunit forms the chloride ion channel lining

neuronal excitability. It has been estimated that approximately 33% of the synapses in the mammalian cerebral cortex are GABAergic.[49] Three different types of receptors, $GABA_A$, $GABA_B$ and $GABA_C$ receptors can be identified in the CNS.[50] $GABA_A$ receptors are members of the ligand gated ion channel superfamily[51] which is coupled to the Cl^- channel. The metabotropic $GABA_B$ receptor is coupled to certain K^+ and Ca^{2+} channels via a GTP binding protein (G-protein) and/or other messengers.[52] The $GABA_A$ receptor is bicuculline sensitive and the $GABA_B$ receptor is baclofen sensitive. The third type of GABA receptor, insensitive to both bicuculline and baclofen, is designated as the $GABA_C$ receptor.[53] The $GABA_C$ responses are also of the fast type associated with the opening of an Cl^- channel; they are unaffected by typical modulators of $GABA_A$ receptor such as benzodiazepines and barbiturates.[50, 54]

To date, 16 isoforms of the $GABA_A$ receptor have been identified.[55] These comprise α1–6, β1–3, γ1–3, δ, ε, π and θ. The $GABA_A$ receptor is pentameric proteins[56] of five different subunits containing 2α/2β/1γ or δ/ε/π/θ subunit variants (Fig. 1.6).[57–59] Immunological, pharmacological, and functional analysis give the evidence that the

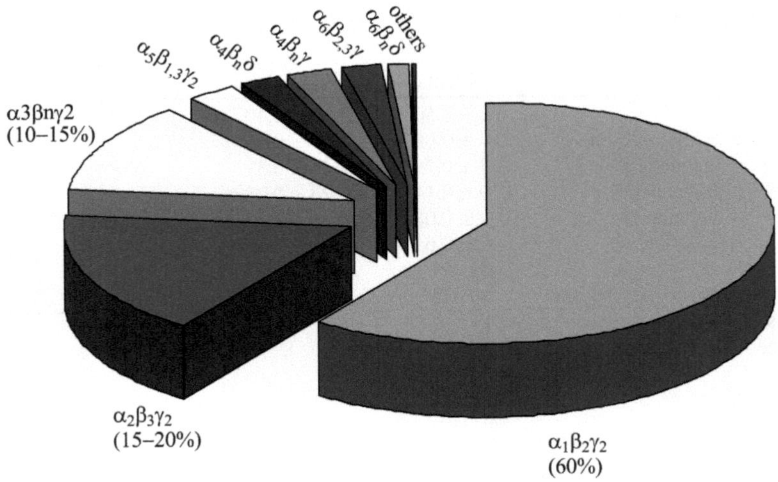

Fig. 1.7 The proportions of different $GABA_A$ receptors in the central nervous system (CNS) (based on the report by Mohler et al.[61] and Fritschy et al.[62])

α1β2γ2 combination represents the largest population of $GABA_A$ receptor (~60%) followed by α2β3γ2 (~15–20%) and α3βnγ2 (~10–15%, n = 1,2 or 3; Fig. 1.7).[60–63] Receptors containing the α4-, α5-, and α6-subnit, as well as the β1-, γ1–3, δ-, π- and θ-subunit, form a minor receptor population. The α4βnδ, α4βnγ and the α6β2/3γ2 receptor – each of these is less than 5% of all the $GABA_A$ receptor quantity (Fig. 1.7). α6βnδ has a small population in the cerebellum and α6β2/3γ2 receptors are located exclusively in the cerebellum. Another $GABA_A$ receptor subunit namely, ρ-subunits are expressed primarily in the retina, forming $GABA_C$ receptor.[64] Structurally, the $GABA_C$ receptor differs from classical $GABA_A$ receptors in that the Cl^- channel comprises a homopentamer of ρ-subunit rather than heteropentamers.[64]

The expression of $GABA_A$ receptor subtypes in the adult brain exhibits a remarkable regional and neuronal specificity which suggests that individual subtypes are present in distinct neuronal circuits. The α1β2γ2 receptor is present in most brain areas and it is localized to interneurons in the hippocampus and cortex (layer I–IV), and cerebral Purkinje cells.[65] The α2β3γ2 receptor is present in cerebral cortex (layer I–IV), hippocampal formation, amygdale, striatum, olfactory bulb, hypothalamus, superior colliculi, and motor nuclei.[62] The α3βnγ2, α3γ2 and α3θ receptors are abundant in the cerebral cortex (layers V–VI), amygdala, olfactory bulb, thalamic reticular and intralaminar nuclei, superior colliculus, brainstem, spinal cord, and locus coeruleus. The α4βnδ (n = 1, 2 or 3) receptor is presented in the dentate gyrus and thalamus. The α5β3γ2 receptor is localized in hippocampal pyramidal cells, deep cortical layers, amygdala, olfactory bulb, hypothalamus, superior colliculus, superior olivary nucleus, spinal trigeminal nucleus and spinal cord. The α6β2/3γ2, α6β2/3δ and α6β2/3γ2 receptors are found mainly in the cerebellum and dorsal cochlear nucleus.[62]

1.3.1 Pharmacological Properties of the $GABA_A$ Receptors

The $GABA_A$ receptor can be modulated by a number of therapeutic agents, including benzodiazepines,[66, 67] barbiturates,[68] anaesthetics, ethanol,[69] zinc[70] and neurosteroids.[71, 72] Pharmacological analysis of $GABA_A$ receptors revealed that the α-subunit determine the GABA affinity,[73] α- and γ-subunit regulate benzodiazepine site pharmacology.[73–76] In fact, the binding pocket of benzodiazepines is located between the α- and γ-subunits.[77] The α1-, α2-, α3-, and α5-containing $GABA_A$ receptors are diazepam-sensitive receptors, whereas the α4- and α6-subunit containing $GABA_A$ receptors are insensitive to diazepam.[78, 79] The α1-, α2-, α3-, and α5-subunits are distinguished further by their affinity to zolpidem ($\alpha 1 > \alpha 2 = \alpha 3 >> \alpha 5$) and various β-carbolines ($\alpha 1 > \alpha 2 = \alpha 3$).[80, 81] Neurosteroids are allosteric modulators of recombinant and native $GABA_A$ receptors.[82] Neurochemical, electrophysiological, and behavioral evidence accumulated over the last two decades suggest that the $GABA_A$ receptor also mediate certain acute and chronic actions of ethanol.[83–85] Ethanol exhibits an array of central depressant actions, including anxiolytic, anticonvulsant, sedative-hypnotic, muscle relaxant, and general anesthetic effects, in a dose-dependent manner.[83, 86] However, some recent attempts have failed to reproduce these effect by ethanol.[87, 88]

Bicuculline, Zn^{2+}, TBPS, picrotoxin and pregnenolone sulfate (PS) are the antagonists of $GABA_A$ receptor. Bicuculline is the competitive blocker of GABA-site.[89, 90] Picrotoxin, Zn^{2+} and PS inhibit the $GABA_A$ receptor through direct and indirect block of the Cl^- channel.[37, 90, 91] Experimental convulsants like pentylenetetrazol and the cage convulsant TBPS act as Cl^- channel blockers on the $GABA_A$ receptor.[92]

1.3.2 Pharmacological Properties of the $GABA_A$ Receptor Depends on the Subunit Composition

There is evidence that many subunits are related to several behavioral effects. Transgenic mouse models enable exciting new perspectives for the development of the pharmacology of modulating agents of $GABA_A$ receptors and their effect on different subunits. Benzodiazepines have dual properties on the $GABA_A$ function-sedative and anxiolytic. A histidine residue at α1, α2, α3, and α5 subunits is crucial for benzodiazepines binding. The sedative effect of benzodiazepines is mediated via the α1 subunit containing $GABA_A$ receptors.[93] In mice, replacing histidine by arginine at position 101 at α1 subunit leads to abolition of the motor component of sedation, amnesic and anticonvulsant effects whereas anxiolytic and muscle relaxant effects of diazepam are retained.[93, 94] In contrast, the α2 subunit appears to be a major determinant for the anxiolytic and muscle relaxant effects of benzodiazepines.[95, 96] Agonistic activity at the α3 subunit also mediate the anxiolytic activity of benzodiazepines, although it occurred only at high receptor occupancy.[97] Recently, it has been shown that the α5 subunit is important for sedative tolerance development to benzodiazepines and for acquisition and expression of associative memory and spatial learning.[98–101] In addition, the α4 subunit is also implicated in the regulation of anxiety.[102] A concentration-dependent decrease of the α4 subunit is seen after 4-day application of allopregnanolone to developing neuronal cells,[103] whereas in the hippocampus and cerebellum, an increase in this subunit can be detected after withdrawal from chronic progesterone (or allopregnanolone) exposure and after short-term treatment.[102, 104–106] The α6 subunit is highly sensitive to pentobarbital effects[107] and neuroactive steroids.[108] In recent years, there has been an accumulating amount of evidence for the involvement of $GABA_A$ receptors in the action of anesthetics. *In vitro* studied on recombinant receptors, the intravenous anesthetic etomidate shows $GABA_A$ subtype selectivity for β2- and β3-containing receptors.[109] A recent study shows that β3-containing receptors are the primary mediators of the anesthetic effects whereas β2, probably in combination with α1, and γ subunits mediates the sedating effects of etomidate.[110] The γ-subunit is thought to confer sensitivity of the receptor to benzodiazepines.[111, 112] The γ1- and γ3-containing receptors have a 10–30-fold lower affinity for flunitrazepam than do receptors containing γ2-subunit.[112–115] On the other hand, absence of γ subunit is essential for zinc sensitivity.[91] The γ2-subunit is also involved in anxiety regulation, and it is changed during hormone treatment and pregnancy.[104, 116–119] The γ2-subunit is essential for clustering of $GABA_A$ receptors and gephyrin and synaptic localization.[117, 120 62, 121] The δ-subunit is responsible for tonic conductance and important for neuroactive steroid modulation on $GABA_A$ receptor.[122] Receptor knockout studies have revealed that the absence of the δ-subunit decreases the sensitivity to neuroactive steroids such as pregnanolone and alphaxalone, thereby influencing the duration of anesthesia and the anxiolytic effect of those steroids.[123] The ε-subunit reduces neuroactive steroid and anesthetic modulation.[108, 124, 125]

Functionally, distinct subunit-specific properties have been identified in both recombinant and native receptors, supporting the concept that $GABA_A$ receptor heterogeneity is a major facet determining the functional properties of GABAergic inhibitory circuits.[126, 127] In particular, the type of α-subunit determines the kinetics of receptor deactivation,[128–130] and the presence of the δ-subunit results in markedly increased agonist affinity and apparent lack of desensitization.[131–133]

1.3.3 Synaptic and Extrasynaptic $GABA_A$ Receptors

The $GABA_A$ receptor, like other receptors in the brain can be synaptic or extrasynaptic. Receptors located at the synaptic junctions are known as "synaptic" whereas the receptors over the neuronal surface outside the synaptic cleft are known as the "extra-synaptic".[134] The synaptic receptors contain γ-subunits and can be rapidly expressed on the neuronal membrane. It is sensitive to both benzodiazepines and neurosteroids. On the other hand, the extrasynaptic receptors contain δ-subunits instead of γ-subunits and are usually difficult to express on the neuronal membrane. δ-subunits are highly sensitive to GABA and neorsteroids but not to benzodiazepines. The expression of α1β2/3γ2, α2β2/3γ2 and α3β2/3γ2 receptors are the predominant synaptic receptors.[135] Receptors that contain α4-, α5- or α6-subunits, α6βnδ, α4βnδ and α5βnγ2 receptors, are predominantly or exclusively extrasynaptic (Fig. 1.8).[62, 136] In the synapse, each vesicle is thought to release several thousands of GABA molecules into the synaptic cleft and leading to a high concentration of GABA (0.3–1.0 mM) in a time span of 10–100 ms.[137, 138] There are only a few number of receptors, from ten to a few hundreds, clustered opposite the release site.[139–141] Synaptically released GABA acting on postsynaptic $GABA_A$ receptors producing "phasic" inhibition, whereas "tonic" inhibition results from the continuous activation of extrasynaptic receptors by ambient GABA.[135] The main feature of

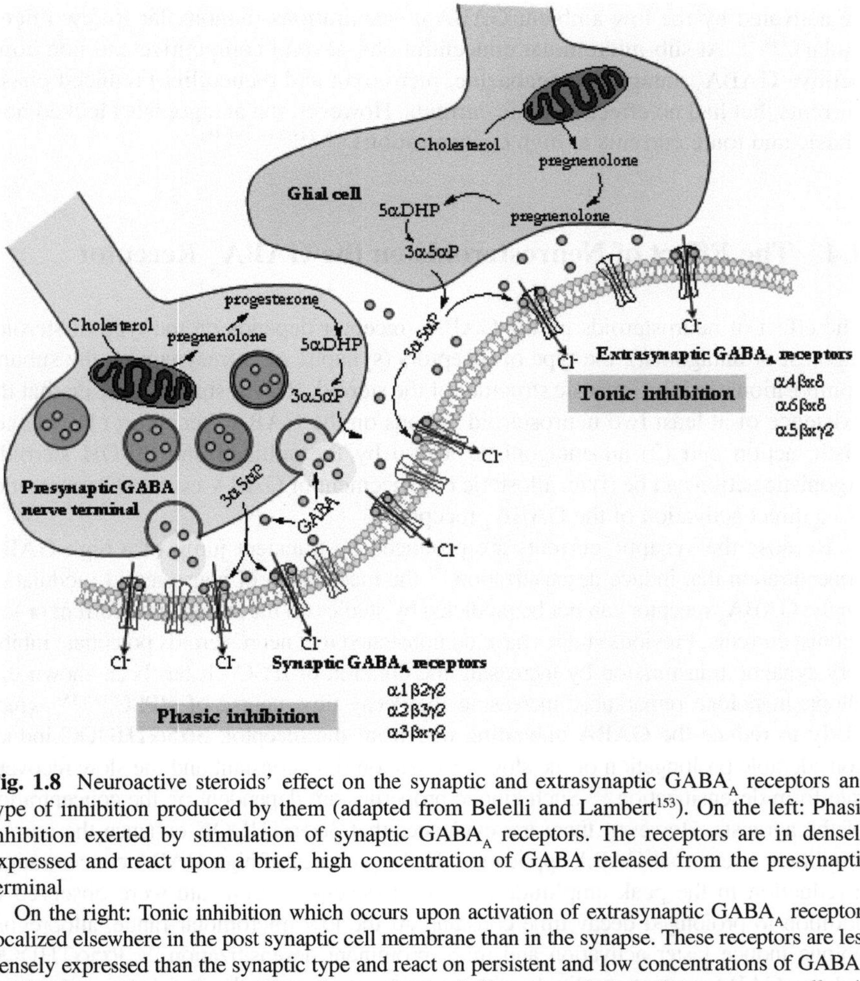

Fig. 1.8 Neuroactive steroids' effect on the synaptic and extrasynaptic $GABA_A$ receptors and type of inhibition produced by them (adapted from Belelli and Lambert[153]). On the left: Phasic inhibition exerted by stimulation of synaptic $GABA_A$ receptors. The receptors are in densely expressed and react upon a brief, high concentration of GABA released from the presynaptic terminal

On the right: Tonic inhibition which occurs upon activation of extrasynaptic $GABA_A$ receptors localized elsewhere in the post synaptic cell membrane than in the synapse. These receptors are less densely expressed than the synaptic type and react on persistent and low concentrations of GABA

This figure shows also that there is a synthesis of allopregnanolone in the neuron as well as in the glial cell. Allopregnanolone interacts with synaptic and extrasynaptic $GABA_A$ receptors phasic inhibition is the rapid synchronous opening of a relatively small number of channels that are clustered at the synaptic junction, whereas tonic inhibition results from random, temporally dispersed activation of receptors that are distributed (albeit in a potentially non-uniform manner) over the neuronal surface. Although both modes of action clearly impact upon neuronal information processing at the cellular and network level, the extent to which each receptor pool influences brain excitability in normal and diseased states is not known. GABA-mediated tonic conductance is found in granule cells of the dentate gyrus,[135, 142] thalamocortical relay neurons of the ventral basal complex,[143] layer V pyramidal neurons in the somatosensory cortex,[144] CA1 pyramidal cells,[145] and certain inhibitory interneurons in the CA1 region of the hippocampus.[146] Unlike receptors mediating phasic current, tonically active $GABA_A$ receptors show unusual high GABA affinity[147] and be activated by the low ambient GABA concentrations (nanomolar to few micromolar).[148, 149] At sub-micromolar concentrations, several competitive and non-competitive $GABA_A$ antagonists (gabazine, picrotoxin and bicuculline) reduced phasic currents, but had no effect on tonic currents. However, the antagonists blocked both phasic and tonic currents at high concentrations.[135, 145, 146, 150, 151]

1.4 The Effect of Neurosteroids on the $GABA_A$ Receptor

The effect of neurosteroids on the $GABA_A$ receptor depends on the type of steroids (agonist or antagonist), the type of receptors (synaptic of extrasynaptic), the subunit compositions and the intrinsic structure of the steroid. Recent studies indicate that the existence of at least two neurosteroid actions on the $GABA_A$ receptor: (1) An agonistic action and (2) an antagonistic action by the sulfated and 3β-OH steroids. Agonistic action can be (i) an allosteric enhancement of GABA-evoked Cl^- current or (ii) a direct activation of the $GABA_A$ receptor.

Because the synaptic currents are produced by transient jump of a high GABA concentration that induce desensitization,[152] the mechanism of neurosteroid modulation on the $GABA_A$ receptor can not be predicted by studies on the steady state current or low agonist currents. Previous studies have demonstrated that neurosteroids potentiate inhibitory synaptic transmission by increasing the duration of IPSC.[5] It has been shown that allopregnanolone remarkably increasing the decay time course of sIPSC,[153, 154] which likely to reduce the GABA unbinding rate from the receptor. 3α5αTHDOC induce considerable prolongation of the slow deactivation time constant and the slow recovery rate from desensitization.[155] Such kinetic properties are dependent on the concentration of the agonists. However, there are evidences that 3α5αTHDOC enhance the current amplitude in external GABA application. In the presence of high GABA concentration, a reduction in the peak amplitude and faster desensitization rate were observed in addition to prolonged decay time constant. At the low micromolar range, allopregnanolone induce faster activation and more prominent desensitization.[154] 3α5αTHDOC reduces GABA current amplitude without apparent effect on the fast deactivation component and its contribution to the peak amplitude.[122, 155, 156] There are evidences of the developmental changes of neurosteroid modulation of the $GABA_A$ receptor. The effect of 3α5αTHDOC on the slow deactivation of GABA response in cerebellar neurons is greater in younger rats than adult rats.[155] More studies revealed the possibility that the δ-subunit containing $GABA_A$ receptor may related to the developmental changes of the neuronal plasticity.[157] In summary, we believe that the prolongation of slow deactivation and altered receptor kinetics in terms of entry and exit from the desensitized state, are essential to the allosteric modulation of the $GABA_A$ receptor by neurosteoids.

Agonist neurosteroids affect not only the time course of sIPSC but also the frequency of the sIPSCs. Such increase in frequency is due to the presynaptic effect of neurosteroids[158,159] by increasing the presynaptic Ca^{2+} permeability. The effect of neurosteroids on the presynaptic neurons is believed to be mediated by the $GABA_A$ receptor.

1.4.1 Synaptic Effect of Neurosteroids

The brief inhibitory response of neurosteroids by activating the postsynaptic $GABA_A$ receptor is a phasic response. Synaptic $GABA_A$ receptors are ternary complexes that commonly incorporate the γ2-subunit in combination with one α (mainly α1/2/3) and one β2/3-subunit. However, these receptor isoforms can also be located extra-synaptically.[135] The kinetic of agonist steroids at synaptic $GABA_A$ receptor has been studied thoroughly by measuring the sIPSC from neurons in brain slice. Neurosteoids have little effect on the onset time and peak amplitude of the sIPSC. Agonist neurosteroids prolong the decay time constant of IPSC.[5,155,160] However, this effect is neuron specific. In hippocampal CA1 neurons, cerebellar granule cells and Purkinje neurons, neurosteroids prolong the sIPSC at relatively low concentration (in the nM range).[156,161,162] On the other hand, micromolar concentrations are required to produce equivalent responses in oxytocin neurons of hypothalamus.[163,164] Moreover, in the preoptic cells in the hypothalamus, 100 nM allopregnanolone prolong the spontaneous current.[160,165] This indicates that the neurons in the same brain region can show heterogeneity. In addition, the effect of 3α5αTHDOC on GABA-binding kinetic is more profound in the hippocampal CA3 and subiculam than that in CA1 and entorhinal cortex.[166]

At higher concentrations (>10 μM) which can occur in the brain during parturition,[167] neurosteroids activate the $GABA_A$ receptor directly[5] in a similar pattern as barbiturates by interacting with different sites on $GABA_A$ receptor.[168] This "GABA-mimetic" effect of neurosteroid is sufficient to suppress the excitatory neurotransmission.[169]

1.4.2 Extrasynaptic Effect of Neurosteroids

The response of neurosteroids at relatively low concentrations is mediated by the activation of extrasynaptic $GABA_A$ receptors containing α4, α6 and δ-subunits (Fig. 1.8). Extrasynaptic receptors identified at the granule cells of the dentate gyrus and cerebellum, and the relay neurons of the thalamus, are distinct from the synaptic receptors. Extrasynaptic conductance can have a considerable influence on neuronal excitability.[170] Extrasynaptic receptors exhibit both a high GABA affinity and reduced receptor desensitization in the continued presence of the agonist.[62] Such properties render these receptors ideally suited to sense the low ambient concentrations (~0.5–1 μM) of the extrasynaptic neurotransmitters.[171] Extrasynaptic receptors containing the δ-subunit are highly sensitive to neurosteroids in certain brain region.[172] At low "physiological" concentrations (10–100 nM), 3α,5α-THDOC selectively enhance the tonic conductance, with little or no effect on the phasic conductance in mouse DGCs and CGCs.[122, 135, 153] Tonic inhibition is reduced in the δ-subunit "knock-out" mice, and the residual tonic current was insensitive to 3α5α-THDOC.[122, 123] However, the extrasynaptic effects of neurosteroids also reveal regional difference in the CNS. 250 nM 3α5αTHDOC has no effect on the tonic inhibition in ventrobasalis complex of the thalamus.[143] On the other hand, the tonic conductance of hippocampal CA1 neurons expressing $GABA_A$ receptor with α5-subunit, is affected by 3α5αTHDOC (≥100 nM).[122, 135, 153] It is suggested that the modulation of tonic currents is influenced by local neurosteroid metabolism. The inhibition of metabolism can greatly enhance the response of the tonic current in dentate granule cells to endogenous neurosteroids, but not on the synthetic metabolically stable ganaxalone.[173] Neurosteroid metabolism reveals regional specificity that also contributes to the regional specificity of the tonic $GABA_A$ receptor-mediated currents. In summary, evidence is emerging that the $GABA_A$ receptor mediated tonic conductance present in some neurons may have a considerable influence on neuronal signaling and network activity.[174–176] The high sensitivity of neurosteroids in extrasynaptic receptor may represent an important target for neurosteroids.

1.4.3 $GABA_A$ Receptor Subunit Dependence of Neurosteroid Action

The effect of neurosteroids on the $GABA_A$ receptor can be attributable to variations in the receptor subunit composition. The action of allopregnanolone was not influenced by the α-subunit, when co-expressed with β1 and γ2L subunits in *Xenopus* oocytes.[177] The GABA responses at α1β1γ2 and α3β1γ2 receptor are enhanced by low concentration of allopregnanolone (≥3 nM), whereas several folds higher concentrations are needed to obtain the equivalent response on α2-,α4-, α5- or α6 subunit containing receptors. Likewise, the subtypes of the β-subunit (1–3) have little action on the neurosteroid effect.[108, 178, 179] The presence of a γ-subunit is not a prerequisite for the neurosteroid activity. In fact, the efficacy of allopregnanolone action at the binary α1β1 receptor is higher than that at the ternary α1β1γ2 receptors.[108, 180] Given the γ-subunit have little or no effect on the maximal GABA-modulatory effect of allopregnanolone, it significantly influences the potency of the steroid with "physiological concentrations" (3–30 nM).[108] However, the inhibition of $GABA_A$ receptor by pregnenolone sulfate did not vary between binary and ternary receptor.[181] The potencies and efficacies of PS to inhibit GABA saturating concentration at the α1β2γ2L and α1β2 receptor were identical.[181] On the other hand, δ-subunit when co expressed with α4- and β3- subunits, a receptor thought to be naturally present in the thalamus shows high steroid sensitivity compare to γ-subunit containing receptor.[108, 124] Receptors incorporating the ε-subunit are reported to be insensitive to the modulatory actions of the pregnane steroids, not the direct GABA-mimetic effect.[82]

1.5 Structure–Activity Relationship

The structure–activity relationship of the neurosteroid has been summarized in a number of articles.[182–185] The systemic investigation of several isomer of $GABA_A$ receptor active neurosteroids revealed a couple crucial structure variation, namely the geometry between ring A/B; a hydrogen bond donator in C3 position; a hydrogen-bond acceptor in C20 position and/or a flexible bond at C17 position.[186–188] The α- and β-configuration refer to constituents below and above the plane of the steroid backbone. Variation of the core and substitution in certain positions of the steroid molecules has crucial consequence on the effect of neurosteroid.

1.5.1 Backbone Structure of Neurosteroids

Anesthetic steroids typically have a saturated backbone of four rings, although this is not an absolute requirement for activity. The four rings form a rigid framework for positioning the hydrogen bonding groups in three-dimensional spaces. The presence of hydrogen bond donor in the α-configuration at C3 and β-configuration at C17 are critical for the sedative action of agonist neurosteroids.[186, 189, 190] These groups are important for the binding of neurosteroids to a variety of proteins by means of hydrogen-binding with polar or charged residues.[191, 192] Replacing the hydrogen of hydroxyl with methyl, thus eliminating the ability of the steroid to donate a hydrogen bond in this region, results in dramatic reduction in its potency.[193]

1.5.2 Steroid-Mimetic Core

Replacing the steroid skeleton with an alicyclic framework like 2-cyclohexylideneperhydro-4,7-methanoindene derivatives give a compound that showed weak potentiating activity.[194] When the carbonitrile is replaced with an acetyl group the derivatives become inactive.[195]

1.5.3 Replacing C5-Position

The configuration of C5-reduction is important to potency. Even if steroids with either 5α or 5β conformations are active, spatial difference in this position may affect the pharmacology of the neurosteroids. Allopregnanolone with 5α-reduction is generally more potent than its 5β-isomer pregnanolone as $GABA_A$ receptor agonist both *in vivo* and *in vitro*. Studies with 3α5α-THDOC and its stereoisomer 3α5β-THDOC revealed important differences in potency, efficacy and regional selectivity at the $GABA_A$ receptor in favor of 5α-reduction.[196, 197] 5α-steroids, but not 5β-steroids, show a high degree of enantioselectivity/enantiospecificy in their action as modulator of the $GABA_A$ receptor and as anesthetics.[198] The efficacy of 3β5βP to inhibit $GABA_A$ receptor is significantly different form 3β5αP.[199, 200] 5β-reduced antagonist neurosteroids cause significant higher inhibition than the 5α-isomers. Although model has been proposed to account for steroids binding at the same site by invoking conformational changes,[186, 195] it is possible that they occupy different sites favored by their respective stereochemistries. Removal of angular methyl group of 5α-steroids, i.e., 19-nor derivatives has variable effect on the potentiating activity of 5α- and 5β-steroids, whereas incorporation of a methyl group at C5 eliminate or diminishes activity.[201]

1.5.4 Replacing C11-Position

The very potent but opposing effects of picomolar concentrations of cortisol and cortisone on GABA receptors in the intestine, with cortisol enhancing and cortisone reducing GABA responses, is very interesting both from a structure–activity and a physiological point of view. Cortisol and cortisone differ in structure only by the level of oxidation at carbon 11, cortisol being the 11β-hydroxy compound and cortisone the 11-oxo compound.[202] 11-ketones like alphaxalone and 11α-amines i.e., minaxalone retain good activity and this is one of the few cases where polar groups are tolerated on the α-face of the steroid without serious loss potentiating activity.[195] However in Cl^- uptake studies in cortical synaptosomes the 3α5α metabolite of cortisol and cortisone are antagonistic but enhance the positive allosteric effect of allopregnanolone.[203]

1.5.5 Replacing C20-Position

Pregnan-20-one is easily metabolized to pregnan-20-ols. In the presence of GABA, allopregnanolone produces a complete inhibition of $[^{35}S]$TBPS binding to a preparation of rat cortical membranes, whereas the reduced metabolize of this steroid 5α-pregnan-3α,20α diol causes only a partial displacement.[204] Reduction of the carbon at C20 position on the potency and the magnitude of the GABA enhancing effect were determined. Reduction of the ketone at the 20 position resulted in a decrease of potency and efficacy to 20α-or 20β-ol reduces the affinity and efficacy of pregnan steroids.[205] However, a ketone is the preferred 20-substituent for pregnanes for good its efficacy.[195, 205, 206]

1.5.6 Replacing C21-Position

A variety of constituents can be accommodated at the C21-position, i.e., halides,[207] alcohols,[208] thiosulfate[209] and heterocycles,[210] generally without serious loss of activity. 3α5αTHDOC is a potent and efficacious $GABA_A$ receptor modulator, whereas the 21-OH analogue of pregnenolone is not. It rather antagonizes the effect of 3α5αP.[195, 208]

1.5.7 Steroid Nucleus and Benz[e]indines

Andros-5-enes and pregn-5-enes with polar functional groups at the 17β- and C20 positions respectively retain good inhibitory activity.[195] Certain benz[e]indines are tricyclic steroid analogues in which the steroid-A ring is opened and partially removed. They have steroid like effect on neurons.[195, 211] At sub-micromolar concentrations, benz[e]indenes like anesthetic steroids, augment the effects of GABA, while at higher concentrations these agents directly open chloride channels in the absence of GABA.[195, 211] The GABA potentiating effects of benz[e]indenes also depend on the GABA concentration, with prominent augmentation of responses to low (sub-EC_{50}) GABA concentrations and little effect on peak responses at saturating GABA levels. The effect of benz[e]indene by introducing the C1 and C19 atoms[212] and altering the configuration at C5[213] showed that the potentiating activity is retained in these compounds. However benz[e]indene with 11 ketone loses its potentiating activity.[214]

1.5.8 Enantiomers

Enantiomers are the mirror image of a compound. Certain steroids and benz[e]indenes augment inhibitory synaptic transmission enantioselectively and provide strong support for the hypothesis that steroids act at specific sites on synaptic $GABA_A$ receptors rather than via alteration of membrane lipids.[215] 3α5αP and 3α5βP is potent enhancer of $GABA_A$ receptor. The non natural enantiomers of these two steroids are reported to be less potent.[215] Enatiomers of a weak potentiator androstane are significantly more active modulator of the $GABA_A$ receptor. Conversion of the C17 carbonyl group of androstane into a spiro-epoxide group yields steroids with activities comparable to 3a5αP and 3a5βP.[215] Moreover, over the range of 0.1–10 μM, the (+)-enantiomers of 3α-hydroxy-5α-androstane-17β-carbonitrile (ACN), and the benz[e]indene are more potent and effective than their (–)-enantiomeric counterparts in enhancing $GABA_A$ receptor-mediated evoked synaptic currents[216–218] and their concentration for direct effect are not physiologically relevant. However, the excellent enantioselectivity observed in the above mentioned 5α-steroids is not retained for 5β-steroids.[219] In addition, the action of PS on $GABA_A$ receptor is not enantioselective.[220]

1.5.9 Androstane Steroids

Like pregnan steroids, metabolites of testosterone such as androstane-diol are also modulators of GABA. It was found that the systemic 3α-androstane-diol administration conditions a place preference more effectively than does systemic administration of dihydrotestosterone or testosterone.[221] Additionally, plasma concentration of 3α-androstane-diol is increased compare to dihydrotestosterone and testosterone. However, the potency and efficacy effect of 3α-androstane steroid is lower than those of 3α-pregnan steroids.[199]

1.5.10 C6 and C7 Positions

Structural modifications of C6 and C7 with methyl group at the axial or the equatorial position also affect the modulation of the $GABA_A$ receptor. 6β-methyl substitutions make the 3α5αP and 3α5β steroids as potent as their parent steroids. On the other hand, 6α, 7α or 7β-methyl constituents results in reduced potency.[222]

1.5.11 Aza- and Oxa-Steroids

17a-Aza-D-homosteroids retain excellent activity relative to the neurosteroid allopregnanolone and 3α5αTHDOC.[198] The preferred side-chain conformation has the carbonyl oxygen atom on the β-face like the ketone of the pregnanes. An isoteric replacement of the C6 atom of the steroids with an oxygen atom yielded oxasteroids which are 100-fold less active. This indicates that hydrophobicity in this region is crucial for retaining of activity.[223]

1.5.12 D-ring Epoxides and Oxetanes

D-ring epoxides and oxetanes are the conformationally constrained steroids which are near perpendicular to the plane of D-ring above C17. A series of 13,24-cyclo-18,21-dinorcholane analogues of 3α5α steroids which contain a C-24 carbonyl group have activities comparable to parent steroids.[224] The fifth ring present in these analogues provides an invariant framework upon which hydrogen bond accepting groups can be positioned to prove for the receptor hydrogen-bond interactions from many different directions and varying distances in or above the plane of C, D-rings.

1.6 Antagonist Neurosteroids on the $GABA_A$ Receptor

Neurosteroids can both enhance and inhibit GABAergic neurotransmission.[36,165, 197, 199] It has been shown earlier that 3β-hydroxy A-ring reduced pregnane steroids (3β-OH steroids) and pregnenolone sulfate (PS) inhibit $GABA_A$ receptor-coupled Cl^- channels.[36, 181, 200, 225] At concentrations that had little effect on GABAergic synaptic currents, 3β-OH steroids and low concentrations of pregnenolone sulfate significantly reversed the potentiation of synaptic currents induced by 3α-OH A-ring reduced steroids.[36] Although antagonist steroids reduced the potentiation induced by 3α-OH steroids, they acted non-competitively with respect to agonist steroids and inhibited the largest degrees of potentiation most effectively.[36] 3β-OH steroids co-applied with GABA alone significantly inhibited GABA responses at concentrations $\geq EC_{50}$.[36, 225–227] This direct, non-competitive effect of 3β-OH steroids and pregnenolone sulfate on the GABA response was sufficient to account for the apparent antagonism of agonist steroids.[199] In summary, both PS and 3β-OH steroids inhibit $GABA_A$ receptors more effectively under conditions that promote agonist binding or channel opening.[36] It was suggested that the interaction between antagonist and agonist steroids was due to a use-dependent action of antagonist steroids at recombinant and synaptic $GABA_A$ receptors.[36]

1.6.1 Mechanism of Pregnenolone Sulfate and other Sulfated Steroids

Sulfated steroids like pregnenolone sulfate and dehydroepiandrosterone sulfate can produce profound effects on behavior. PS, an abundant neurosteroid, enhances learning[228, 229] while antagonizing the impairment of learning and memory produced by ethanol and scopolamine.[230] PS may play a role in cognition and have been reported as negative modulators of $GABA_A$ receptors based on electrophysiological studies and GABA mediated $^{36}Cl^-$ uptake by rat brain synaptosomes.[231, 232] Structure–activity relationships for $GABA_A$ receptor modulation are different for sulfated inhibitory steroids vs nonsulfated potentiating steroids.[233] Potentiation by nonsulfated steroids requires 3α-stereochemistry. Pregnenolone by itself is inactive at the $GABA_A$ receptor. In contrast, both 3α and 3β isomers of PS is inhibitory. Although the addition of a negatively charged sulfate group or hemisuccinate group at the C3 position converts the neurosteroid from being potentiating to inhibitory.[233] A negatively charged group at C3 is not absolutely essential for inhibition since the nonsulfated neurosteroid DHEA is also inhibitory, however less potent than its sulfated derivative DHEAS.[233, 234] Steroids such as 11-keto derivative of PS are of particular interest, as it can behave as a positive, negative, or neutral modulator on the $GABA_A$ receptor. It was suggested that 11-keto pregnenolone sulfate (11-keto PS) exerts a dual action on distinct positive and negative steroid modulation sites associated with the $GABA_A$ receptor.[233]

Consistent with the idea that pregnenolone sulfate induces activation-dependent inhibition on the $GABA_A$ receptor,[36] another report revealed that the inhibition of basal inhibitory postsynaptic currents (IPSCs) by antagonist steroids was correlated with the basal decay time of the IPSCs.[37] Analysis of single channel behavior in the presence of GABA and PS suggested no difference in the ability of PS to block liganded closed vs liganded open receptors.[235] These results leave open the possibility that PS may prefer liganded over unliganded receptors, consistent with a model of state-dependence to PS actions.

An earlier study observed that pregnenolone sulfate inhibits GABA-gated Cl^- current by enhancing receptor desensitization and stabilizing desensitized state with prolonged application of low-affinity GABA agonists to nucleated membrane patched.[236] Note that the promotion of desensitization is selective for prolonged GABA applications.[236] It is well-known that the time course of GABA-mediated IPSCs is influenced strongly by the kinetics of the $GABA_A$ receptor desensitization.[152, 237] Desensitized states are thought to buffer receptors in bound conformations that make it possible for channels to re-open before GABA unbinds. The fast phase of desensitization limits the open probability of the channels, influences peak synaptic currents, and contributes to the fast component of IPSC decay. The slow component of decay may result from reopening of GABA channels after exit from desensitized states.[152] On the other hand, recombinant $GABA_A$ receptors composed of defined subunit combinations also give rise to currents with complex decay kinetics.[238, 239] In cultured hippocampal neurons, PS decreases the peak current of the inhibitory autapic currents (IACs), enhances the fast and slow phases of deactivation after brief application of a few milliseconds. After longer application (~100ms or more), PS enhances both the fast and slow phases of desensitization of the GABA current.[37, 236] Moreover, PS reduces the agonist steroid evoked prolongation of sIPSC and reduces channel-opening frequency.[240] However, recording the sIPSC of neurons of medial preoptic nucleus shows that PS neither affects the peak amplitude nor the decay of sIPSC.[160] Instead, PS reduces the frequency of the sIPSC at higher concentration.[160] The main effect of PS on the sIPSC time course is explained by a simplified model that this substance reduce the rate of desensitization while the agonist steroids assumed to reduce the unbinding rate of GABA from the receptor.[160]

1.6.2 3β-OH Steroid as $GABA_A$ Receptor Antagonist

According to earlier studies, PS,[241] 3β-OH steroids or carboxylated steroids[242] are more effective against GABA responses gated by high concentrations of GABA. In electrophysiological studies, 3β5βP antagonized the 3α5βP-induced enhancement of GABA current.[227, 243] However, 3β5αP, not 3β5βP diminished the inhibitory effect of 3α5αP on population spikes evoked in rat hippocampal CA1 stratum pyramidal.[244] Studies on the chloride uptake into synaptosomes in rat cerebral cortex, in hippocampus and sIPSC in medial preoptic nucleus (MPN) showed that 3β-OH steroids reduced the 3α5αP-enhanced GABA response.[165] In the presence of low concentration GABA, some of the 3β-OH steroids potentiated GABA-evoked chloride ion uptake and prolonged the decay time, whereas the others had little or no effect on GABA stimulated current. Therefore, certain 3β-OH steroids, namely 5β-pregnane-3β,20(S)-diol and 3β-OH-5βpregnane-20-one have both agonistic and antagonist property.[36, 165] Moreover, another study with the Cl- uptake method has shown that 3β-OH-5α-pregnan-20 one is a useful antagonist of 3α5αP-enhanced GABA response.[225] However, in $GABA_A$ receptor expressed in *Xenopus* oocytes showed that 3β-steroids inhibit GABA response at near-saturating concentrations.[199] It is still unclear why the effect of 3β-OH steroids varies with different methods. We suggest that the steroid structure, drug application time and base line GABA concentration may be responsible for the discrepancy.

1.6.3 Mechanism of Action of 3β-Hydroxy Steroids

It is suggested that a homologous mutation of the residue at 2' position closest to the cytoplasmic end of the M_2 helix to *serine* on both the α_1 and the β_2 subunit, α_{1V256S} and β_{2A252S}, reduced the desensitization rate of GABA-activation at saturating doses.[200] In a receptor complex with reduced desensitization components to GABA-activation (e.g., mutant receptors), the PS-inhibition was also greatly reduced.[198] On the other hand, PS has been shown earlier to increase the deactivation rate of the GABA-evoked sIPSCs recorded by patch-clamp techniques at the acutely dissociated neurons from the MPN area of the *rat* brain slice.[160] Recent studies confirmed the findings that PS increased the fast offset rate of GABA-activation.[200] However, the slow component of the offset time course was decreased by PS in a dose-dependent manner. The potencies of 5α-pregnan-3β,20α-diol and 5β-pregnan-3β,20β-diol to influence offset time courses of GABA-activation were significantly lower than PS, in accordance with our earlier findings while PS was significantly more potent to inhibit both peak and steady-state GABA-currents than 3β-OH steroids.[36, 181, 200] PS-inhibition was already seen at low dose of GABA-response ($\leq EC_{20}$),[37] whereas the inhibition by 3β-OH steroids on the current response was first seen at higher end ($>EC_{50}$) of the GABA dose–response curve.[36] Obviously, 5α-pregnan-3β,20α-diol and 5β-pregnan-3β,20β-diol prolonged the fast offset time course of GABA-response, suggesting that the inherent association between 3β-OH steroids and receptor has rather high affinity. However, we cannot fully exclude a more complicated model in which a multivalent interaction between hydrophobic steroids and receptor is employed. Kinetic properties of 3β-OH steroids have also been elucidated in acute dissociate neurons from the MPN area of the hypothalamus.[165] These steroids have no effect on the activation phase or the maximum amplitude of 3α5αP enhanced IPSCs. They rather affect the slow deactivation phase. A recent report shows that 3β-OH steroids exert their effect by reducing the 3α5αP induced prolongation of decay time constant (τ_{decay}).[165]

1.6.4 Structural Modification of 3β-Position

In order to improve the bioavailability of the potentiating steroids, modifications or substitutions of side chains have been tried including the 3β-position. An advantage of incorporating 3β-substituents is that potential hormonal activity resulting from oxidative metabolism is abolished, and oral bioavailability is improved.[189] Along with α in C5, 3β-ethyl, -propyl, -trifluoromethyl and -(benzyloxy)methyl, as well as substituents of the form 3β-XCH_2, where X is Cl, Br, or I or contains unsaturation, show limited efficacy in inhibiting [^{35}S]TBPS binding.[189] An inhibition of [^{35}S]TBPS binding indirectly indicates an agonistic action on the $GABA_A$ receptor. With 5β, all of the 3β-substituted derivatives of pregnanolone inhibit TBPS via a single class of binding sites. In addition, all of the 3-substituted 5β-sterols tested are full inhibitors of [^{35}S] TBPS binding.[189] Electrophysiological measurements using α1β2γ2L receptors expressed in oocytes showed that 3β-methyl- and 3β-(azidomethyl)-3α-hydroxy -5α-pregnan-20-one are potent full efficacy modulators and that 3α-hydroxy-3β-(trifluoromethyl)-5α-pregnan-20-one is a low-efficacy modulator.[189] These results indicate that protection of the OH-group at the 3α position by modification of the 3β-position in 3α5αP and 3α5βP maintains activity at the neuroactive steroid site on the $GABA_A$ receptor. In animal studies, ganaxolone is an orally active anticonvulsant steroid, while the naturally occurring progesterone metabolites 3α5αP and 3α5βP are inactive when administered orally, suggesting that 3β-substitution slows metabolism of the 3-hydroxyl, resulting in orally bioavailable steroid modulators of the $GABA_A$ receptor.[188]

Moreover, another substitution at the 3β position; 3β-(*p*-acetyl phenylethynyl)- 3α-hydroxy-5α-pregnan-20-one (Co 152791) was more potent than 3α5αP both in [^{35}S]TBPS binding in human recombinant receptor and *Xenopus* oocyte expressing $GABA_A$ receptor.[245] Ganaxolone, another 3β-methyl-substituted analog of the endogenous neurosteroid allopregnanolone, has a high-affinity, stereoselective, as positive allosteric modulator of the $GABA_A$ receptor complex and exhibits potent anticonvulsant activity across a range of animal procedures.[173, 246] The profile of anticonvulsant activity obtained for ganaxolone supports clinical evaluation of this drug as an antiepileptic therapy with potential utility in the treatment of generalized absence seizures as well as simple and complex partial seizures.[247]

Screening for the selective antagonists of neuroactive steroid potentiation of GABA responses is going on many labs. Some success has been achieved in this regard. An steroid analogue 3α,5α-17-phenylandrost-16-en-3-ol(17PA) blocked the 3α5αP-enhanced GABA response and direct GABA-mimetic effect of 3α5αP.[197] However, 17PA had no or little effect on 3α5βP steroid. The main drawback of this substance is that the results cannot be reproduced because of difficulty to dissolve and there is no effect up to 5 μM concentration (J.J. Lambert et al., personal communication 2007). Furthermore, this substance demands further evaluation in the animal behavior model.

1.7 Site of Action of the Neurosteroid

Determining how neuroactive steroids interact with the $GABA_A$ receptor is a pre-requisite for understanding their physiological and pathophysiological roles in the brain. Neuroactive steroids bind to $GABA_A$ receptors at a site that is distinct from the recognition sites for GABA, benzodiazepines, and barbiturates. This results in allosteric modulation of GABA binding or channel gating. Studies using $GABA_A$/glycine receptor chimeras suggest an allosteric action of neuroactive steroids at the N-terminal side of the middle of the second transmembrane domain of the $GABA_A$ receptor β1 and/or α2 subunits.[248] Electrophysiological studies have confirmed that neuroactive steroids enhance Cl^- current through increasing both channel frequency and channel open duration.[72, 155, 249] Neuroactive steroids do not require direct aqueous access to the receptor, and membrane accumulation is required for receptor modulation.[250] In a recent study, Hosie et al.[251] identified two discrete binding sites in the receptor's transmembrane domains that mediate the potentiating and direct activation effects of neuroactive steroids. Their potentiating effect of 3α5αTHDOC is mediated by a cavity formed by the α-subunit transmembrane domains. On the other hand, their direct activation of 3α5αTHDOC is mediated by interfacial residues between α and β-subunits and is enhanced by steroid binding to the potentiation site. These profiles indicate that two distinct neuroactive steroid binding sites may exist; αTHr236 and βTyr284 residues in the transmembrane domain initiate direct activation whereas αGln241 and αAsn407 mediate the potentiating response.[251]

It is generally accepted that the sulfate moiety is critical in producing a steroid that blocks rather than potentiates $GABA_A$ receptors.[233] The fact that an anionic group is critical gave suggestion to us that the sulfate might actually interact with residues forming the binding site mediating antagonism. However, no voltage dependency was observed with pregnenolone sulfate-inhibition on the GABA response,[235, 252, 253] which suggests that the charged sulfate moiety does not interact significantly with the membrane field as PS approaches the transition state between unbound and unblocked to bound and blocked.[254] Therefore, the lack of voltage dependence of pregnenolone sulfate-inhibition suggests that it is unlikely that the sulfate moiety interacts with a site deep within the channel.[235] Results from our earlier report suggest that pregnenolone sulfate was a γ-subunit independent inhibitor at $GABA_A$ receptors.[181] However, residues deep in the channel at the 2' position in the M_2 helix of both α_1- and β_2-subunit were critical for pregnenolone sulfate-inhibition.[36, 181, 235] Theoretically, a large, rigid molecule such as a steroid would bind end-on rather than with its long axis across the channel. It is thus possible that either the A-ring (where the sulfate moiety is attached) or the D-ring (the other end of the molecule) would penetrate the Cl^- channel most deeply. The lack of voltage dependence indicates that the A-ring is unlikely to penetrate to the 2' position. The structure of the D-ring of PS is identical to that for many 3β-hydroxysteroids. Since the inhibitory properties of 5α-pregnan-3β,20α-diol and 5β-pregnan-3β,20β-diol were also reduced in the $\alpha_{1V256S}\beta_2\gamma_{2L}$ receptor, we think it is more likely that the uncharged portions of the steroid molecules interact with the 2' residue closest to the cytoplasmic end of the M_2 helix on the α_1-subunit.

The site of action of sulfated neuroactive steroids on $GABA_A$ receptors remains unclear. Based upon the observation that PS reduced the apparent affinity of [^{35}S]-TBPS, Sousa and Ticku[255] suggested that PS and DHEAS might bind at the picrotoxin/cage convulsant site. However, a mutation to the transmembrane M2 channel domain eliminated picrotoxin sensitivity but the inhibitory effects of PS and DHEAS persisted.[256, 257] The absence of voltage sensitivity or alteration of single-channel open time argues against a binding site within the pore. Akk et al.[235] identified a valine residue in the channel domain of the α1 subunit that slowed the development of PS inhibition when mutated to serine, but concluded that this residue is unlikely to be part of the binding site and likely influences PS action indirectly.[235] Using homologous mutation of the residue at 2' position closest to the cytoplasmic end of the M2 helix to *serine* on both α1 and β2 subunit, $\alpha_{1V256}S$ and $\beta_{2A252}S$, it was found that effect of PS is greatly reduced in these two mutant.[181, 200, 235] This suggests that these two amino acids are involved in the PS mediated inhibition. However, it is unclear whether these specific mutations at the cytoplasmic end of M2 helix exert their effect by altering the allosteric mechanism or by directly altering a binding site. A recent study in *Caenorhabditis elegans* identified multiple residues in transmembrane domain 1 (M1), as well as a residue near the extracellular end of the M2 helix, that are critical for low-micromolar inhibition of *C. elegans* $GABA_A$ receptors by PS.[258] This latter residue is of particular interest, as it is a positively charged arginine that could potentially coordinate with the negatively charged sulfate of PS. The *C. elegans* receptor exhibits some pharmacological differences as compared to mammalian $GABA_A$ receptors (e.g., pregnanolone is inhibitory). So, these results may or may not be relevant to mammalian receptors; however, it is notable that an arginine residue is also found in this region of mammalian $GABA_A$ receptor subunits.[256, 258]

Block of responses to high-efficacy agonists by this sulfated steroid is greater than block of responses to partial agonists at saturating concentrations. This is called "activation dependant" or "state dependant inhibition".[36, 37] Picrotoxin, another GABA channel blocker superficially similar to pregnenolone sulfate in its activation dependence but the site of action of pregnenolone sulfate does not require a functional picrotoxin site for inhibition of GABA responses.[257] The GABA receptor antagonist Zn^{2+} also acts in activation dependant manner. However, the Zn^{2+} binding site is located in the interface between the α- and β-subunit and its activity is minimal in ternary αβγ receptor.[91]

1.8 Neurosteroids and Behavior

A number of review articles have discussed the important role of neurosteroids in treating behavioral disorders by interacting with the $GABA_A$ receptor.[188, 259–261] There are a few obstacles preventing the clinical use of endogenous neurosteroids. First of all, naturally occurring neurosteroids such as allopregnanolone have low bioavailability because they are rapidly inactivated and eliminated by glucoronide or sulfate conjugation at the 3α-hydroxy group. The second obstacle is that the

3α-hydroxy group of allopregnanolone may undergo oxidation to a ketone, restoring its activity at nuclear hormone receptors.[262] Ganaxalone (3α-hydroxy-3β-methyl-5α-pregnane-20-one), the 3β-methyl analogue of allopregnanolone, is an example of synthetic neurosteroid congener that overcomes these limitations.[247] Like allopregnanolone, ganaxalone is a positive allosteric modulator of $GABA_A$ receptor.[247]

Neurosteroids modulate anxiety and stress level. After an acute stress stimulus there is a release of progesterone, pregnenolone, allopregnanolone and 3α5α-THDOC in the blood circulation.[263–265] Allopregnanolone and 3α5αTHDOC have been demonstrated to possess potent anxiolytic activity in several animal anxiety models.[266 39, 259, 267] However, the anxiolytic effect of allopregnanolone has not been shown in human.[268] A number of recent reports indicate that allopregnanolone induces aggression and anxiety at low concentrations.[14, 102, 269] A recent clinical study showed that the allopregnanolone and cortisol levels are increased during the examination of Ph.D. students.[270] Therefore, it has been suggested that allopregnanolone has biphasic effects in certain individuals.[269] At low doses it has an adverse, anxiogenic effect, and this effect decreases with increasing doses and the beneficial and calming property occurs.[13, 14] A major concern with potential new anxiolytics is whether they suffer the same drawbacks as classical benzodiazepines. Using the elevated plus-maze paradigm for assessing the anxiolytic activity, selective effects of neurosteroids have been reported which differ from diazepam.[271] Synthetic derivative Co 2–6749 30, which retains a 3β-trifluromethyl group that should block metabolism and enhance oral bioavailability, is selected for clinical development because there is a large separation between anxiolytic-like effects and side effects.[40, 272]

Hippocampus is a key brain area for learning and memory functions.[273] Women with PMDD show often difficulties in concentration and develop fatigue during the luteal phase of the menstrual cycle; this is associated with high circulating levels of allopregnanolone.[274] The enzymes needed for the production of allopregnanolone are present in the hippocampus.[24] Using the Morris water maze paradigm, allopregnanolone was found to inhibit learning.[20] Antagonist neurosteroid 3β-20β-dihydroxy-5α-pregnane, reduces the negative effect of allopregnanolone on the learning in the water maze.[275] PS infused into the basal magnocellular nucleus enhance memory performance, whereas allopregnanolone disrupted memory.[229] Pregnenolone, DHEA and DHEAS increased memory when injected systemically, centrally or into the amygdala.[276–278] There is evidence that the concentration of DHEA and DHEAS are decreased in patients suffering from Alzheimer's disease.[279–281] It is promising that certain neurosteroids should be further explored in the context of prevention and treatment of Alzheimer's disease and mild cognitive impairment.

Niemann–Pick type C (NP–C) disease is a rare but fatal autosomal recessive, childhood-onset, neurodegenerative disorder. This lysosomal lipid storage disorder is characterized by a defect in intracellular cholesterol trafficking, resulting in lysosomal accumulation of unspecified cholesterol and glycolipid.[282] Mice model of this disease have shown that loss of neurosteroidogenic enzyme activity (e.g., P450sec, 3β-HSD, 5α-reductase and 3α HSD) in the early neonatal period resulting prior to the onset of neurological symptoms.[283] NP–C mice treated with allopregnanolone resulted in increasing life-span, substantial delay in onset of neurological symptoms, survival of cerebellar Purkinje and granule cell neurons, and reduction in cholesterol and ganglioside accumulation.[283–286] A synthetic $GABA_A$ neurosteroid, ganaxalone treated NP–C mice also produced beneficial neurological effects. However, these effects were not as robust as those obtained using allopregnanolone.[283] Thus, allopregnanolone may elicit its effects through $GABA_A$ receptors as well as through other mechanisms, possibly through pregnane-X-receptors (PXR). These studies with mouse models suggest that neuroactive steroids may ultimately be useful for the treatment of human diseases.

1.9 Conclusion and Future Directions

In addition to the slow genomic action of the parents steroids, the non-genomic and rapid action of the neuroactive steroids have a significant role in modulation of $GABA_A$ receptor system and shift in mood and behavior. Neuroactive steroids with 3α-OH configuration such as 3α5αP and 3α5αTHDOC are potent and effective modulators of $GABA_A$ receptor in the mammalian CNS and have efficacy in several clinical models. On the other hand, substantial evidences from earlier report show that the 3β-OH steroids, like PS are usually antagonist in nature. OH-group at C3 position and ketone at C20 are critical for neurosteroid activity. Neuroactive steroids activity depends on brain regions and the types of neuron. Locations of the receptors at the synapse (synaptic and extrasynaptic) also have a significant role in their activity since certain extrasynaptic receptors are highly sensitive to neuroactive steroids. Subunit compositions of $GABA_A$ receptor, enzymes related to synthesis and metabolism and phosphorylation mechanism also determine the neuroactive steroids' activity. Further subtlety of neurosteroid sensitivity arises from developmental changes and plasticity due to stress and fluctuation of precursor steroids during ovarian cycle. Recent studies have revealed neurosteroids may influence the expression of certain extrasynaptic receptors. However, the underlying mechanism of this interaction remains to be determined.

Binding site of agonist steroid 3α5αTHDOC has recently been elucidated.[251] This finding will lead the generation of mice expressing neurosteroid insensitive $GABA_A$ receptor isoforms that will be give valuable information about the role of these steroids in human disorders. Moreover, to evaluate the effect of antagonist steroids as potential therapeutics for neurological disorders such as, PMDD and catamenial epilepsy, specific neurosteroid antagonists and the binding site of these steroids will be area for future research interest. Like selective serotonin re-uptake inhibitors (SSRI) as mood stabilizer, regulation of specific neuroactive steroids synthesizing enzymes is another promising strategy in future for the development of novel neuroactive steroid-based treatment for anxiety, epilepsy and depression. However, the main hindrance in this approach is the lack of specificity of the enzyme which may lead to undesirable hormonal imbalance.

Acknowledgements This work was supported by the Swedish Research Council-medicine (project No. 4X-11198; 73p-15450), Insamlingsstiftelsen för medicinsk forskning vid Umeå Universitet, Svenska läkaresällskapet, Socialstyrelsens forskningsfond, Umeå Kvinnokliniken forskningsfond and the EU regional fund's objective 1 program.

References

1. Rupprecht R. Neuroactive steroids: mechanisms of action and neuropsychopharmacological properties. Psychoneuroendocrinology 2003; 28(2):139–168.
2. Baulieu EE, Robel P. Non-genomic mechanisms of action of steroid hormones. Ciba Found Symp 1995; 191:24–37; discussion 37–42.
3. Frye CA, Mermelstein PG, DeBold JF. Evidence for a non-genomic action of progestins on sexual receptivity in hamster ventral tegmental area but not hypothalamus. Brain Res 1992; 578(1–2):87–93.
4. McEwen BS. Non-genomic and genomic effects of steroids on neural activity. Trends Pharmacol Sci 1991; 12(4):141–147.
5. Majewska MD, Harrison NL, Schwartz RD, et al. Steroid hormone metabolites are barbiturate-like modulators of the GABA receptor. Science 1986; 232(4753):1004–1007.
6. Paul SM, Purdy RH. Neuroactive steroids. FASEB J 1992; 6(6):2311–2322.
7. Lambert JJ, Belelli D, Hill-Venning C, et al. Neurosteroids and GABAA receptor function. Trends Pharmacol Sci 1995; 16(9):295–303.
8. Wu FS, Gibbs TT, Farb DH. Pregnenolone sulfate: a positive allosteric modulator at the *N*-methyl-D-aspartate receptor. Mol Pharmacol 1991; 40(3):333–336.
9. Valera S, Ballivet M, Bertrand D. Progesterone modulates a neuronal nicotinic acetylcholine receptor. Proc Natl Acad Sci USA 1992; 89(20):9949–9953.
10. Klangkalya B, Chan A. Structure–activity relationships of steroid hormones on muscarinic receptor binding. J Steroid Biochem 1988; 29(1):111–118.
11. Parry BL. The role of central serotonergic dysfunction in the aetiology of premenstrual dysphoric disorder: therapeutic implications. CNS Drugs 2001; 15(4):277–285.
12. Halbreich U. The etiology, biology, and evolving pathology of premenstrual syndromes. Psychoneuroendocrinology 2003; 28(Suppl 3):55–99.
13. Beauchamp MH, Ormerod BK, Jhamandas K, et al. Neurosteroids and reward: allopregnanolone produces a conditioned place aversion in rats. Pharmacol Biochem Behav 2000; 67(1):29–35.
14. Fish EW, Faccidomo S, DeBold JF, et al. Alcohol, allopregnanolone and aggression in mice. Psychopharmacology (Berl) 2001; 153(4):473–483.
15. Backstrom T, Andreen L, Birzniece V, et al. The role of hormones and hormonal treatments in premenstrual syndrome. CNS Drugs 2003; 17(5):325–342.
16. Sundstrom Poromaa I, Smith S, Gulinello M. GABA receptors, progesterone and premenstrual dysphoric disorder. Arch Womens Ment Health 2003; 6(1):23–41.
17. Banerjee PK, Snead OC, III. Neuroactive steroids exacerbate gamma-hydroxybutyric acid-induced absence seizures in rats. Eur J Pharmacol 1998; 359(1):41–48.
18. Grunewald RA, Aliberti V, Panayiotopoulos CP. Exacerbation of typical absence seizures by progesterone. Seizure 1992; 1(2):137–138.
19. Backstrom T. Epilepsy in women. Oestrogen and progesterone plasma levels. Experientia 1976; 32(2):248–249.
20. Johansson IM, Birzniece V, Lindblad C, et al. Allopregnanolone inhibits learning in the Morris water maze. Brain Res 2002; 934(2):125–131.
21. Backstrom T, Gee KW, Lan N, et al. Steroids in relation to epilepsy and anaesthesia. Ciba Found Symp 1990; 153:225–230; discussion 230–229.
22. Wang M, Backstrom T, Sundstrom I, et al. Neuroactive steroids and central nervous system disorders. Int Rev Neurobiol 2001; 46:421–459.

23. Bjorn I, Bixo M, Nojd KS, et al. The impact of different doses of medroxyprogesterone acetate on mood symptoms in sequential hormonal therapy. Gynecol Endocrinol 2002; 16(1):1–8.
24. Compagnone NA, Mellon SH. Neurosteroids: biosynthesis and function of these novel neuromodulators. Front Neuroendocrinol 2000; 21(1):1–56.
25. Costa E, Cheney DL, Grayson DR, et al. Pharmacology of neurosteroid biosynthesis. Role of the mitochondrial DBI receptor (MDR) complex. Ann NY Acad Sci 1994; 746:223–242.
26. Zwain IH, Yen SS. Neurosteroidogenesis in astrocytes, oligodendrocytes, and neurons of cerebral cortex of rat brain. Endocrinology 1999; 140(8):3843–3852.
27. Patte-Mensah C, Kappes V, Freund-Mercier MJ, et al. Cellular distribution and bioactivity of the key steroidogenic enzyme, cytochrome P450side chain cleavage, in sensory neural pathways. J Neurochem 2003; 86(5):1233–1246.
28. Penning TM, Sharp RB, Krieger NR. Purification and properties of 3 alpha-hydroxysteroid dehydrogenase from rat brain cytosol. Inhibition by nonsteroidal anti-inflammatory drugs and progestins. J Biol Chem 1985; 260(28):15266–15272.
29. Russell DW, Wilson JD. Steroid 5 alpha-reductase: two genes/two enzymes. Annu Rev Biochem 1994; 63:25–61.
30. Zhu D, Wang MD, Backstrom T, et al. Evaluation and comparison of the pharmacokinetic and pharmacodynamic properties of allopregnanolone and pregnanolone at induction of anaesthesia in the male rat. Br J Anaesth 2001; 86(3):403–412.
31. Corpechot C, Young J, Calvel M, et al. Neurosteroids: 3 alpha-hydroxy-5 alpha-pregnan-20-one and its precursors in the brain, plasma, and steroidogenic glands of male and female rats. Endocrinology 1993; 133(3):1003–1009.
32. Purdy RH, Morrow AL, Moore PH, Jr., et al. Stress-induced elevations of gamma-aminobutyric acid type A receptor-active steroids in the rat brain. Proc Natl Acad Sci USA 1991; 88(10):4553–4557.
33. Cheney DL, Uzunov D, Costa E, et al. Gas chromatographic-mass fragmentographic quantitation of 3 alpha-hydroxy-5 alpha-pregnan-20-one (allopregnanolone) and its precursors in blood and brain of adrenalectomized and castrated rats. J Neurosci 1995; 15(6):4641–4650.
34. Mensah-Nyagan AG, Do-Rego JL, Beaujean D, et al. Neurosteroids: expression of steroidogenic enzymes and regulation of steroid biosynthesis in the central nervous system. Pharmacol Rev 1999; 51(1):63–81.
35. Stromstedt M, Warner M, Banner CD, et al. Role of brain cytochrome P450 in regulation of the level of anesthetic steroids in the brain. Mol Pharmacol 1993; 44(5):1077–1083.
36. Wang M, He Y, Eisenman LN, et al. 3beta-hydroxypregnane steroids are pregnenolone sulfate-like GABA(A) receptor antagonists. J Neurosci 2002; 22(9):3366–3375.
37. Eisenman LN, He Y, Fields C, et al. Activation dependent properties of pregnenolone sulfate inhibition of GABAA receptor-mediated current. J Physiol 2003; 550:679–691.
38. Hill M, Bicikova M, Parizek A, et al. Neuroactive steroids, their precursors and polar conjugates during parturition and postpartum in maternal blood: 2. Time profiles of pregnanolone isomers. J Steroid Biochem Mol Biol 2001; 78(1):51–57.
39. Crawley JN, Glowa JR, Majewska MD, et al. Anxiolytic activity of an endogenous adrenal steroid. Brain Res 1986; 398(2):382–385.
40. Gasior M, Carter RB, Witkin JM. Neuroactive steroids: potential therapeutic use in neurological and psychiatric disorders. Trends Pharmacol Sci 1999; 20(3):107–112.
41. Lambert JJ, Harney SC, Belelli D, et al. Neurosteroid modulation of recombinant and synaptic GABAA receptors. Int Rev Neurobiol 2001; 46:177–205.
42. Reddy DS, Rogawski MA. Stress-induced deoxycorticosterone-derived neurosteroids modulate GABA(A) receptor function and seizure susceptibility. J Neurosci 2002; 22(9):3795–3805.
43. Edwards HE, Vimal S, Burnham WM. The acute anticonvulsant effects of deoxycorticosterone in developing rats: role of metabolites and mineralocorticoid-receptor responses. Epilepsia 2005; 46(12):1888–1897.
44. Reddy DS. Is there a physiological role for the neurosteroid THDOC in stress-sensitive conditions? Trends Pharmacol Sci 2003; 24(3):103–106.

45. Frye CA, Van Keuren KR, Erskine MS. Behavioral effects of 3 alpha-androstanediol. I: Modulation of sexual receptivity and promotion of GABA-stimulated chloride flux. Behav Brain Res 1996; 79(1–2):109–118.
46. Erlander MG, Tillakaratne NJ, Feldblum S, et al. Two genes encode distinct glutamate decarboxylases. Neuron 1991; 7(1):91–100.
47. Esclapez M, Tillakaratne NJ, Kaufman DL, et al. Comparative localization of two forms of glutamic acid decarboxylase and their mRNAs in rat brain supports the concept of functional differences between the forms. J Neurosci 1994; 14(3 Pt 2):1834–1855.
48. Soghomonian JJ, Martin DL. Two isoforms of glutamate decarboxylase: why? Trends Pharmacol Sci 1998; 19(12):500–505.
49. Purvez D, Augustine GJ, Fitzpatrick D, et al. Neuroscinece, 3rd ed. Sunderland, MA: Sinauer, 2004, p. 143.
50. Sivilotti L, Nistri A. GABA receptor mechanisms in the central nervous system. Prog Neurobiol 1991; 36(1):35–92.
51. Schofield PR, Darlison MG, Fujita N, et al. Sequence and functional expression of the GABA A receptor shows a ligand-gated receptor super-family. Nature 1987; 328(6127):221–227.
52. Hill DR, Bowery NG. 3H-baclofen and 3H-GABA bind to bicuculline-insensitive GABA B sites in rat brain. Nature 1981; 290(5802):149–152.
53. Drew CA, Johnston GA, Weatherby RP. Bicuculline-insensitive GABA receptors: studies on the binding of (-)-baclofen to rat cerebellar membranes. Neurosci Lett 1984; 52(3):317–321.
54. Bormann J, Feigenspan A. GABAC receptors. Trends Neurosci 1995; 18(12):515–519.
55. Whiting PJ. GABA-A receptor subtypes in the brain: a paradigm for CNS drug discovery? Drug Discov Today 2003; 8(10):445–450.
56. Nayeem N, Green TP, Martin IL, et al. Quaternary structure of the native GABAA receptor determined by electron microscopic image analysis. J Neurochem 1994; 62(2):815–818.
57. Farrar SJ, Whiting PJ, Bonnert TP, et al. Stoichiometry of a ligand-gated ion channel determined by fluorescence energy transfer. J Biol Chem 1999; 274(15):10100–10104.
58. Knight AR, Stephenson FA, Tallman JF, et al. Monospecific antibodies as probes for the stoichiometry of recombinant GABA(A) receptors. Receptors Channels 2000; 7(3):213–226.
59. Klausberger T, Ehya N, Fuchs K, et al. Detection and binding properties of GABA(A) receptor assembly intermediates. J Biol Chem 2001; 276(19):16024–16032.
60. Mohler H, Fritschy JM, Rudolph U. A new benzodiazepine pharmacology. J Pharmacol Exp Ther 2002; 300(1):2–8.
61. Mohler H, Fritschy JM, Crestani F, et al. Specific GABA(A) circuits in brain development and therapy. Biochem Pharmacol 2004; 68(8):1685–1690.
62. Fritschy JM, Brunig I. Formation and plasticity of GABAergic synapses: physiological mechanisms and pathophysiological implications. Pharmacol Ther 2003; 98(3):299–323.
63. Wallner M, Hanchar HJ, Olsen RW. Ethanol enhances alpha 4 beta 3 delta and alpha 6 beta 3 delta gamma-aminobutyric acid type A receptors at low concentrations known to affect humans. Proc Natl Acad Sci USA 2003; 100(25):15218–15223.
64. Bormann J. The "ABC" of GABA receptors. Trends Pharmacol Sci 2000; 21(1):16–19.
65. McKernan RM, Whiting PJ. Which GABAA-receptor subtypes really occur in the brain? Trends Neurosci 1996; 19(4):139–143.
66. Macdonald RL, Olsen RW. GABAA receptor channels. Annu Rev Neurosci 1994; 17:569–602.
67. Sieghart W. GABAA receptors: ligand-gated Cl- ion channels modulated by multiple drug-binding sites. Trends Pharmacol Sci 1992; 13(12):446–450.
68. Smith MC, Riskin BJ. The clinical use of barbiturates in neurological disorders. Drugs 1991; 42(3):365–378.
69. Harris RA, Proctor WR, McQuilkin SJ, et al. Ethanol increases GABAA responses in cells stably transfected with receptor subunits. Alcohol Clin Exp Res 1995; 19(1):226–232.
70. Smart TG. A novel modulatory binding site for zinc on the GABAA receptor complex in cultured rat neurones. J Physiol 1992; 447:587–625.

71. Hawkinson JE, Kimbrough CL, Belelli D, et al. Correlation of neuroactive steroid modulation of [35S]t-butylbicyclophosphorothionate and [3H]flunitrazepam binding and gamma-aminobutyric acid A receptor function. Mol Pharmacol 1994; 46(5):977–985.
72. Puia G, Santi MR, Vicini S, et al. Neurosteroids act on recombinant human GABAA receptors. Neuron 1990; 4(5):759–765.
73. Smith AJ, Alder L, Silk J, et al. Effect of alpha subunit on allosteric modulation of ion channel function in stably expressed human recombinant gamma-aminobutyric acid(A) receptors determined using (36)Cl ion flux. Mol Pharmacol 2001; 59(5):1108–1118.
74. McKernan RM, Wafford K, Quirk K, et al. The pharmacology of the benzodiazepine site of the GABA-A receptor is dependent on the type of gamma-subunit present. J Recept Signal Transduct Res 1995; 15(1–4):173–183.
75. Buhr A, Baur R, Malherbe P, et al. Point mutations of the alpha 1 beta 2 gamma 2 gamma-aminobutyric acid(A) receptor affecting modulation of the channel by ligands of the benzodiazepine binding site. Mol Pharmacol 1996; 49(6):1080–1084.
76. Wingrove PB, Thompson SA, Wafford KA, et al. Key amino acids in the gamma subunit of the gamma-aminobutyric acid A receptor that determine ligand binding and modulation at the benzodiazepine site. Mol Pharmacol 1997; 52(5):874–881.
77. Sigel E. Mapping of the benzodiazepine recognition site on GABA(A) receptors. Curr Top Med Chem 2002; 2(8):833–839.
78. Wingrove PB, Safo P, Wheat L, et al. Mechanism of alpha-subunit selectivity of benzodiazepine pharmacology at gamma-aminobutyric acid type A receptors. Eur J Pharmacol 2002; 437(1–2):31–39.
79. Benson JA, Low K, Keist R, et al. Pharmacology of recombinant gamma-aminobutyric acidA receptors rendered diazepam-insensitive by point-mutated alpha-subunits. FEBS Lett 1998; 431(3):400–404.
80. Sieghart W. Structure and pharmacology of gamma-aminobutyric acid A receptor subtypes. Pharmacol Rev 1995; 47(2):181–234.
81. Mohler H, Fritschy JM, Luscher B, et al. The GABAA receptors. From subunits to diverse functions. Ion Channels 1996; 4:89–113.
82. Lambert JJ, Belelli D, Harney SC, et al. Modulation of native and recombinant GABA(A) receptors by endogenous and synthetic neuroactive steroids. Brain Res Brain Res Rev 2001; 37(1–3):68–80.
83. Deitrich RA, Dunwiddie TV, Harris RA, et al. Mechanism of action of ethanol: initial central nervous system actions. Pharmacol Rev 1989; 41(4):489–537.
84. Ueno S, Harris RA, Messing RO, et al. Alcohol actions on GABA(A) receptors: from protein structure to mouse behavior. Alcohol Clin Exp Res 2001; 25(5 Suppl ISBRA):76S–81S.
85. Harris RA, Valenzuela CF, Brozowski S, et al. Adaptation of gamma-aminobutyric acid type A receptors to alcohol exposure: studies with stably transfected cells. J Pharmacol Exp Ther 1998; 284(1):180–188.
86. Frye GD, Chapin RE, Vogel RA, et al. Effects of acute and chronic 1,3-butanediol treatment on central nervous system function: a comparison with ethanol. J Pharmacol Exp Ther 1981; 216(2):306–314.
87. Carta M, Mameli M, Valenzuela CF. Alcohol enhances GABAergic transmission to cerebellar granule cells via an increase in Golgi cell excitability. J Neurosci 2004; 24(15):3746–3751.
88. Borghese CM, Storustovu S, Ebert B, et al. The delta subunit of gamma-aminobutyric acid type A receptors does not confer sensitivity to low concentrations of ethanol. J Pharmacol Exp Ther 2006; 316(3):1360–1368.
89. Polenzani L, Woodward RM, Miledi R. Expression of mammalian gamma-aminobutyric acid receptors with distinct pharmacology in Xenopus oocytes. Proc Natl Acad Sci USA 1991; 88(10):4318–4322.
90. Krishek BJ, Moss SJ, Smart TG. A functional comparison of the antagonists bicuculline and picrotoxin at recombinant GABAA receptors. Neuropharmacology 1996; 35(9–10): 1289–1298.

91. Hosie AM, Dunne EL, Harvey RJ, et al. Zinc-mediated inhibition of GABA(A) receptors: discrete binding sites underlie subtype specificity. Nat Neurosci 2003; 6(4):362–369.
92. DeLorey TM, Olsen RW. Gamma-aminobutyric acidA receptor structure and function. J Biol Chem 1992; 267(24):16747–16750.
93. Rudolph U, Crestani F, Benke D, et al. Benzodiazepine actions mediated by specific gamma-aminobutyric acid(A) receptor subtypes. Nature 1999; 401(6755):796–800.
94. McKernan RM, Rosahl TW, Reynolds DS et al. Sedative but not anxiolytic properties of benzodiazepines are mediated by the GABA(A) receptor alpha1 subtype. Nat Neurosci 2000; 3(6):587–592.
95. Low K, Crestani F, Keist R, et al. Molecular and neuronal substrate for the selective attenuation of anxiety. Science 2000; 290(5489):131–134.
96. Crestani F, Low K, Keist R, et al. Molecular targets for the myorelaxant action of diazepam. Mol Pharmacol 2001; 59(3):442–445.
97. Dias R, Sheppard WF, Fradley RL, et al. Evidence for a significant role of alpha 3-containing GABAA receptors in mediating the anxiolytic effects of benzodiazepines. J Neurosci 2005; 25(46):10682–10688.
98. Collinson N, Kuenzi FM, Jarolimek W, et al. Enhanced learning and memory and altered GABAergic synaptic transmission in mice lacking the alpha 5 subunit of the GABAA receptor. J Neurosci 2002; 22(13):5572–5580.
99. Crestani F, Keist R, Fritschy JM, et al. Trace fear conditioning involves hippocampal alpha5 GABA(A) receptors. Proc Natl Acad Sci USA 2002; 99(13):8980–8985.
100. van Rijnsoever C, Tauber M, Choulli MK, et al. Requirement of alpha5-GABAA receptors for the development of tolerance to the sedative action of diazepam in mice. J Neurosci 2004; 24(30):6785–6790.
101. Yee BK, Hauser J, Dolgov VV, et al. GABA receptors containing the alpha5 subunit mediate the trace effect in aversive and appetitive conditioning and extinction of conditioned fear. Eur J Neurosci 2004; 20(7):1928–1936.
102. Gulinello M, Gong QH, Li X, et al. Short-term exposure to a neuroactive steroid increases alpha4 GABA(A) receptor subunit levels in association with increased anxiety in the female rat. Brain Res 2001; 910(1–2):55–66.
103. Grobin AC, Morrow AL. 3alpha-hydroxy-5alpha-pregnan-20-one exposure reduces GABA(A) receptor alpha4 subunit mRNA levels. Eur J Pharmacol 2000; 409(2):R1–2.
104. Concas A, Follesa P, Barbaccia ML, et al. Physiological modulation of GABA(A) receptor plasticity by progesterone metabolites. Eur J Pharmacol 1999; 375(1–3):225–235.
105. Follesa P, Concas A, Porcu P, et al. Role of allopregnanolone in regulation of GABA(A) receptor plasticity during long-term exposure to and withdrawal from progesterone. Brain Res Brain Res Rev 2001; 37(1–3):81–90.
106. Smith SS, Gong QH, Hsu FC, et al. GABA(A) receptor alpha4 subunit suppression prevents withdrawal properties of an endogenous steroid. Nature 1998; 392(6679):926–930.
107. Thompson SA, Whiting PJ, Wafford KA. Barbiturate interactions at the human GABAA receptor: dependence on receptor subunit combination. Br J Pharmacol 1996; 117(3): 521–527.
108. Belelli D, Casula A, Ling A, et al. The influence of subunit composition on the interaction of neurosteroids with GABA(A) receptors. Neuropharmacology 2002; 43(4):651–661.
109. Hill-Venning C, Belelli D, Peters JA, et al. Subunit-dependent interaction of the general anaesthetic etomidate with the gamma-aminobutyric acid type A receptor. Br J Pharmacol 1997; 120(5):749–756.
110. Wafford KA, Macaulay AJ, Fradley R, et al. Differentiating the role of gamma-aminobutyric acid type A (GABAA) receptor subtypes. Biochem Soc Trans 2004; 32(Pt 3):553–556.
111. Pritchett DB, Luddens H, Seeburg PH. Type I and type II GABAA-benzodiazepine receptors produced in transfected cells. Science 1989; 245(4924):1389–1392.
112. Ymer S, Draguhn A, Wisden W, et al. Structural and functional characterization of the gamma 1 subunit of GABAA/benzodiazepine receptors. EMBO J 1990; 9(10):3261–3267.

113. Wafford KA, Bain CJ, Whiting PJ, et al. Functional comparison of the role of gamma subunits in recombinant human gamma-aminobutyric acidA/benzodiazepine receptors. Mol Pharmacol 1993; 44(2):437–442.
114. Luddens H, Pritchett DB, Kohler M, et al. Cerebellar GABAA receptor selective for a behavioural alcohol antagonist. Nature 1990; 346(6285):648–651.
115. Hadingham KL, Wafford KA, Thompson SA, et al. Expression and pharmacology of human GABAA receptors containing gamma 3 subunits. Eur J Pharmacol 1995; 291(3):301–309.
116. Crestani F, Lorez M, Baer K, et al. Decreased GABAA-receptor clustering results in enhanced anxiety and a bias for threat cues. Nat Neurosci 1999; 2(9):833–839.
117. Essrich C, Lorez M, Benson JA, et al. Postsynaptic clustering of major GABAA receptor subtypes requires the gamma 2 subunit and gephyrin. Nat Neurosci 1998; 1(7):563–571.
118. Follesa P, Floris S, Tuligi G, et al. Molecular and functional adaptation of the GABA(A) receptor complex during pregnancy and after delivery in the rat brain. Eur J Neurosci 1998; 10(9):2905–2912.
119. Kittler JT, Wang J, Connolly CN, et al. Analysis of GABAA receptor assembly in mammalian cell lines and hippocampal neurons using gamma 2 subunit green fluorescent protein chimeras. Mol Cell Neurosci 2000; 16(4):440–452.
120. Baer K, Essrich C, Benson JA, et al. Postsynaptic clustering of gamma-aminobutyric acid type A receptors by the gamma3 subunit in vivo. Proc Natl Acad Sci USA 1999; 96(22):12860–12865.
121. Schweizer C, Balsiger S, Bluethmann H, et al. The gamma 2 subunit of GABA(A) receptors is required for maintenance of receptors at mature synapses. Mol Cell Neurosci 2003; 24(2):442–450.
122. Stell BM, Brickley SG, Tang CY, et al. Neuroactive steroids reduce neuronal excitability by selectively enhancing tonic inhibition mediated by delta subunit-containing GABAA receptors. Proc Natl Acad Sci USA 2003; 100(24):14439–14444.
123. Mihalek RM, Banerjee PK, Korpi ER, et al. Attenuated sensitivity to neuroactive steroids in gamma-aminobutyrate type A receptor delta subunit knockout mice. Proc Natl Acad Sci USA 1999; 96(22):12905–12910.
124. Davies PA, Hanna MC, Hales TG, et al. Insensitivity to anaesthetic agents conferred by a class of GABA(A) receptor subunit. Nature 1997; 385(6619):820–823.
125. Thompson SA, Bonnert TP, Cagetti E, et al. Overexpression of the GABA(A) receptor epsilon subunit results in insensitivity to anaesthetics. Neuropharmacology 2002; 43(4):662–668.
126. Mohler H, Crestani F, Rudolph U. GABA(A)-receptor subtypes: a new pharmacology. Curr Opin Pharmacol 2001; 1(1):22–25.
127. Sieghart W. Unraveling the function of GABA(A) receptor subtypes. Trends Pharmacol Sci 2000; 21(11):411–413.
128. Verdoorn TA, Draguhn A, Ymer S, et al. Functional properties of recombinant rat GABAA receptors depend upon subunit composition. Neuron 1990; 4(6):919–928.
129. Hutcheon B, Morley P, Poulter MO. Developmental change in GABAA receptor desensitization kinetics and its role in synapse function in rat cortical neurons. J Physiol 2000; 522 Pt 1:3–17.
130. Devor A, Fritschy JM, Yarom Y. Spatial distribution and subunit composition of GABA(A) receptors in the inferior olivary nucleus. J Neurophysiol 2001; 85(4):1686–1696.
131. Burgard EC, Tietz EI, Neelands TR, et al. Properties of recombinant gamma-aminobutyric acid A receptor isoforms containing the alpha 5 subunit subtype. Mol Pharmacol 1996; 50(1):119–127.
132. Fisher JL, Macdonald RL. Single channel properties of recombinant GABAA receptors containing gamma 2 or delta subtypes expressed with alpha 1 and beta 3 subtypes in mouse L929 cells. J Physiol 1997; 505 (Pt 2):283–297.
133. Adkins CE, Pillai GV, Kerby J, et al. alpha4beta3delta GABA(A) receptors characterized by fluorescence resonance energy transfer-derived measurements of membrane potential. J Biol Chem 2001; 276(42):38934–38939.

134. Mody I, Pearce RA. Diversity of inhibitory neurotransmission through GABA(A) receptors. Trends Neurosci 2004; 27(9):569–575.
135. Farrant M, Nusser Z. Variations on an inhibitory theme: phasic and tonic activation of GABA(A) receptors. Nat Rev Neurosci 2005; 6(3):215–229.
136. Nusser Z, Sieghart W, Somogyi P. Segregation of different GABAA receptors to synaptic and extrasynaptic membranes of cerebellar granule cells. J Neurosci 1998; 18(5):1693–1703.
137. Semyanov A, Walker MC, Kullmann DM, et al. Tonically active GABA A receptors: modulating gain and maintaining the tone. Trends Neurosci 2004; 27(5):262–269.
138. Mozrzymas JW, Zarnowska ED, Pytel M, et al. Modulation of GABA(A) receptors by hydrogen ions reveals synaptic GABA transient and a crucial role of the desensitization process. J Neurosci 2003; 23(22):7981–7992.
139. Nusser Z, Cull-Candy S, Farrant M. Differences in synaptic GABA(A) receptor number underlie variation in GABA mini amplitude. Neuron 1997; 19(3):697–709.
140. Mody I, De Koninck Y, Otis TS, et al. Bridging the cleft at GABA synapses in the brain. Trends Neurosci 1994; 17(12):517–525.
141. Brickley SG, Cull-Candy SG, Farrant M. Single-channel properties of synaptic and extrasynaptic GABAA receptors suggest differential targeting of receptor subtypes. J Neurosci 1999; 19(8):2960–2973.
142. Nusser Z, Mody I. Selective modulation of tonic and phasic inhibitions in dentate gyrus granule cells. J Neurophysiol 2002; 87(5):2624–2628.
143. Porcello DM, Huntsman MM, Mihalek RM, et al. Intact synaptic GABAergic inhibition and altered neurosteroid modulation of thalamic relay neurons in mice lacking delta subunit. J Neurophysiol 2003; 89(3):1378–1386.
144. Yamada J, Okabe A, Toyoda H, et al. Cl- uptake promoting depolarizing GABA actions in immature rat neocortical neurones is mediated by NKCC1. J Physiol 2004; 557(Pt 3):829–841.
145. Bai D, Zhu G, Pennefather P, et al. Distinct functional and pharmacological properties of tonic and quantal inhibitory postsynaptic currents mediated by gamma-aminobutyric acid(A) receptors in hippocampal neurons. Mol Pharmacol 2001; 59(4):814–824.
146. Semyanov A, Walker MC, Kullmann DM. GABA uptake regulates cortical excitability via cell type-specific tonic inhibition. Nat Neurosci 2003; 6(5):484–490.
147. Saxena NC, Macdonald RL. Properties of putative cerebellar gamma-aminobutyric acid A receptor isoforms. Mol Pharmacol 1996; 49(3):567–579.
148. Lerma J, Herranz AS, Herreras O, et al. In vivo determination of extracellular concentration of amino acids in the rat hippocampus. A method based on brain dialysis and computerized analysis. Brain Res 1986; 384(1):145–155.
149. Tossman U, Jonsson G, Ungerstedt U. Regional distribution and extracellular levels of amino acids in rat central nervous system. Acta Physiol Scand 1986; 127(4):533–545.
150. Stell BM, Mody I. Receptors with different affinities mediate phasic and tonic GABA(A) conductances in hippocampal neurons. J Neurosci 2002; 22(10):RC223.
151. Yeung JY, Canning KJ, Zhu G, et al. Tonically activated GABAA receptors in hippocampal neurons are high-affinity, low-conductance sensors for extracellular GABA. Mol Pharmacol 2003; 63(1):2–8.
152. Jones MV, Westbrook GL. Desensitized states prolong GABAA channel responses to brief agonist pulses. Neuron 1995; 15(1):181–191.
153. Belelli D, Lambert JJ. Neurosteroids: endogenous regulators of the GABA(A) receptor. Nat Rev Neurosci 2005; 6(7):565–575.
154. Haage D, Johansson S. Neurosteroid modulation of synaptic and GABA-evoked currents in neurons from the rat medial preoptic nucleus. J Neurophysiol 1999; 82(1):143–151.
155. Zhu WJ, Vicini S. Neurosteroid prolongs GABAA channel deactivation by altering kinetics of desensitized states. J Neurosci 1997; 17(11):4022–4031.
156. Cooper EJ, Johnston GA, Edwards FA. Effects of a naturally occurring neurosteroid on GABAA IPSCs during development in rat hippocampal or cerebellar slices. J Physiol 1999; 521 Pt 2:437–449.

157. Shen H, Gong QH, Aoki C, et al. Reversal of neurosteroid effects at alpha4beta2delta GABAA receptors triggers anxiety at puberty. Nat Neurosci 2007; 10(4):469–477.
158. Poisbeau P, Feltz P, Schlichter R. Modulation of GABAA receptor-mediated IPSCs by neuroactive steroids in a rat hypothalamo-hypophyseal coculture model. J Physiol 1997; 500 (Pt 2):475–485.
159. Haage D, Druzin M, Johansson S. Allopregnanolone modulates spontaneous GABA release via presynaptic Cl- permeability in rat preoptic nerve terminals. Brain Res 2002; 958(2):405–413.
160. Haage D, Backstrom T, Johansson S. Interaction between allopregnanolone and pregnenolone sulfate in modulating GABA-mediated synaptic currents in neurons from the rat medial preoptic nucleus. Brain Res 2005; 1033(1):58–67.
161. Harney SC, Frenguelli BG, Lambert JJ. Phosphorylation influences neurosteroid modulation of synaptic GABAA receptors in rat CA1 and dentate gyrus neurones. Neuropharmacology 2003; 45(6):873–883.
162. Vicini S, Losi G, Homanics GE. GABA(A) receptor delta subunit deletion prevents neurosteroid modulation of inhibitory synaptic currents in cerebellar neurons. Neuropharmacology 2002; 43(4):646–650.
163. Brussaard AB, Kits KS, Baker RE, et al. Plasticity in fast synaptic inhibition of adult oxytocin neurons caused by switch in GABA(A) receptor subunit expression. Neuron 1997; 19(5):1103–1114.
164. Koksma JJ, van Kesteren RE, Rosahl TW, et al. Oxytocin regulates neurosteroid modulation of GABA(A) receptors in supraoptic nucleus around parturition. J Neurosci 2003; 23(3):788–797.
165. Stromberg J, Haage D, Taube M, et al. Neurosteroid modulation of allopregnanolone and GABA effect on the GABA-A receptor. Neuroscience 2006; 143(1):73–81.
166. Nguyen Q, Sapp DW, Van Ness PC, et al. Modulation of GABAA receptor binding in human brain by neuroactive steroids: species and brain regional differences. Synapse 1995; 19(2):77–87.
167. Stoffel-Wagner B. Neurosteroid biosynthesis in the human brain and its clinical implications. Ann NY Acad Sci 2003; 1007:64–78.
168. Kerr DI, Ong J. GABA agonists and antagonists. Med Res Rev 1992; 12(6):593–636.
169. Shu HJ, Eisenman LN, Jinadasa D, et al. Slow actions of neuroactive steroids at GABAA receptors. J Neurosci 2004; 24(30):6667–6675.
170. Leroy C, Poisbeau P, Keller AF, et al. Pharmacological plasticity of GABA(A) receptors at dentate gyrus synapses in a rat model of temporal lobe epilepsy. J Physiol 2004; 557(Pt 2):473–487.
171. Kennedy RT, Thompson JE, Vickroy TW. In vivo monitoring of amino acids by direct sampling of brain extracellular fluid at ultralow flow rates and capillary electrophoresis. J Neurosci Methods 2002; 114(1):39–49.
172. Wohlfarth KM, Bianchi MT, Macdonald RL. Enhanced neurosteroid potentiation of ternary GABA(A) receptors containing the delta subunit. J Neurosci 2002; 22(5):1541–1549.
173. Belelli D, Herd MB. The contraceptive agent Provera enhances GABA(A) receptor-mediated inhibitory neurotransmission in the rat hippocampus: evidence for endogenous neurosteroids? J Neurosci 2003; 23(31):10013–10020.
174. Brickley SG, Farrant M, Swanson GT, et al. CNQX increases GABA-mediated synaptic transmission in the cerebellum by an AMPA/kainate receptor-independent mechanism. Neuropharmacology 2001; 41(6):730–736.
175. Mitchell SJ, Silver RA. Shunting inhibition modulates neuronal gain during synaptic excitation. Neuron 2003; 38(3):433–445.
176. Hamann M, Rossi DJ, Attwell D. Tonic and spillover inhibition of granule cells control information flow through cerebellar cortex. Neuron 2002; 33(4):625–633.
177. Belelli D, Herd MB, Mitchell EA, et al. Neuroactive steroids and inhibitory neurotransmission: mechanisms of action and physiological relevance. Neuroscience 2006; 138(3): 821–829.

178. Hadingham KL, Wingrove PB, Wafford KA et al. Role of the beta subunit in determining the pharmacology of human gamma-aminobutyric acid type A receptors. Mol Pharmacol 1993; 44(6):1211–1218.
179. Sanna E, Murgia A, Casula A, et al. Differential subunit dependence of the actions of the general anesthetics alphaxalone and etomidate at gamma-aminobutyric acid type A receptors expressed in Xenopus laevis oocytes. Mol Pharmacol 1997; 51(3):484–490.
180. Maitra R, Reynolds JN. Subunit dependent modulation of GABAA receptor function by neuroactive steroids. Brain Res 1999; 819(1–2):75–82.
181. Wang MD, Rahman M, Zhu D, et al. Pregnenolone sulphate and Zn^{2+} inhibit recombinant rat GABA(A) receptor through different channel property. Acta Physiol (Oxf) 2006; 188(3–4): 153–162.
182. Laubach GD, P'An SY, Rudel HW. Steroid anesthetic agent. Science 1955; 122(3158):78.
183. Gyermek L, Iriarte J, Crabbe P. Steroids. CCCX. Structure–activity relationship of some steroidal hypnotic agents. J Med Chem 1968; 11(1):117–125.
184. Gyermek L, Soyka LF. Steroid anesthetics. Anesthesiology 1975; 42(3):331–344.
185. Sear JW. ORG 21465, a new water-soluble steroid hypnotic: more of the same or something different? Br J Anaesth 1997; 79(4):417–419.
186. Purdy RH, Morrow AL, Blinn JR, et al. Synthesis, metabolism, and pharmacological activity of 3 alpha-hydroxy steroids which potentiate GABA-receptor-mediated chloride ion uptake in rat cerebral cortical synaptoneurosomes. J Med Chem 1990; 33(6):1572–1581.
187. Ragagnin G, Rahman M, Zingmark E, et al. Structure–activity relationship of $GABA_A$-steroids antagonists. 4th International Meeting- Steroids and Nervous system, Torino, Italy 2007, Abstract book, p. 197.
188. Zorumski CF, Mennerick S, Isenberg KE, et al. Potential clinical uses of neuroactive steroids. Curr Opin Investig Drugs 2000; 1(3):360–369.
189. Hogenkamp DJ, Tahir SH, Hawkinson JE, et al. Synthesis and in vitro activity of 3 beta-substituted-3 alpha-hydroxypregnan-20-ones: allosteric modulators of the GABAA receptor. J Med Chem 1997; 40(1):61–72.
190. Hawkinson JE, Kimbrough CL, McCauley LD, et al. The neuroactive steroid 3 alpha-hydroxy-5 beta-pregnan-20-one is a two-component modulator of ligand binding to the GABAA receptor. Eur J Pharmacol 1994; 269(2):157–163.
191. Grishkovskaya I, Avvakumov GV, Sklenar G, et al. Crystal structure of human sex hormone-binding globulin: steroid transport by a laminin G-like domain. EMBO J 2000; 19(4): 504–512.
192. Brzozowski AM, Pike AC, Dauter Z, et al. Molecular basis of agonism and antagonism in the oestrogen receptor. Nature 1997; 389(6652):753–758.
193. Upasani RB, Yang KC, Acosta-Burruel M, et al. 3 alpha-Hydroxy-3 beta-(phenylethynyl)-5 beta-pregnan-20-ones: synthesis and pharmacological activity of neuroactive steroids with high affinity for GABAA receptors. J Med Chem 1997; 40(1):73–84.
194. Burden PM, Allan RD, Hambley T, et al. The synthesis of 2-cyclohexylideneperhydro-4,7-methanoindenes. Non-steroidal analogues of steroidal GABAA receptor modulators. J Chem Soc Perkin Trans I 1998:3163–3167.
195. Hamilton NM. Interaction of steroids with the GABA(A) receptor. Curr Top Med Chem 2002; 2(8):887–902.
196. Gee KW, Lan NC. Gamma-aminobutyric acidA receptor complexes in rat frontal cortex and spinal cord show differential responses to steroid modulation. Mol Pharmacol 1991; 40(6):995–999.
197. Mennerick S, He Y, Jiang X, et al. Selective antagonism of 5alpha-reduced neurosteroid effects at GABA(A) receptors. Mol Pharmacol 2004; 65(5):1191–1197.
198. Covey DF, Han M, Kumar AS, et al. Neurosteroid analogues. 8. Structure–activity studies of N-acylated 17a-aza-D-homosteroid analogues of the anesthetic steroids (3alpha, 5alpha)- and (3alpha,5beta)-3-hydroxypregnan-20-one. J Med Chem 2000; 43(17):3201–3204.

199. Rahman M, Lindblad C, Johansson IM, et al. Neurosteroid modulation of recombinant rat alpha(5)beta(2)gamma(2L) and alpha(1)beta(2)gamma(2L) GABA(A) receptors in Xenopus oocyte. Eur J Pharmacol 2006; 547(1–3):37–44.
200. Wang MD, Rahman M, Zhu D, et al. 3Beta-hydroxysteroids and pregnenolone sulfate inhibit recombinant rat GABA(A) receptor through different channel property. Eur J Pharmacol 2007; 557(2–3):124–131.
201. Han M, Zorumski CF, Covey DF. Neurosteroid analogues. 4. The effect of methyl substitution at the C-5 and C-10 positions of neurosteroids on electrophysiological activity at GABAA receptors. J Med Chem 1996; 39(21):4218–4232.
202. Ong J, Kerr DI, Capper HR, et al. Cortisone: a potent GABAA antagonist in the guinea-pig isolated ileum. J Pharm Pharmacol 1990; 42(9):662–664.
203. Stromberg J, Backstrom T, Lundgren P. Rapid non-genomic effect of glucocorticoid metabolites and neurosteroids on the gamma-aminobutyric acid-A receptor. Eur J Neurosci 2005; 21(8):2083–2088.
204. Gee KW. Steroid modulation of the GABA/benzodiazepine receptor-linked chloride ionophore. Mol Neurobiol 1988; 2(4):291–317.
205. Belelli D, Lambert JJ, Peters JA, et al. Modulation of human recombinant GABAA receptors by pregnanediols. Neuropharmacology 1996; 35(9–10):1223–1231.
206. McCauley LD, Liu V, Chen JS, et al. Selective actions of certain neuroactive pregnanediols at the gamma-aminobutyric acid type A receptor complex in rat brain. Mol Pharmacol 1995; 47(2):354–362.
207. Hill-Venning C, Peters JA, Callachan H, et al. The anaesthetic action and modulation of GABAA receptor activity by the novel water-soluble aminosteroid Org 20599. Neuropharmacology 1996; 35(9–10):1209–1222.
208. Xue BG, Whittemore ER, Park CH, et al. Partial agonism by 3alpha,21-dihydroxy-5beta-pregnan-20-one at the gamma-aminobutyric acidA receptor neurosteroid site. J Pharmacol Exp Ther 1997; 281(3):1095–1101.
209. Fick DB, Hawkinson JE, Acosta-Burruel M, et al. Sodium S-(3 α-hydroxypregnan-20-on-21-yl) thiosulfates: allosteric modulation modulators of GABAA receptor. Book of abstracts, 217th ACS national meeting, Anaheim, CA, United States 1999; March 21–25.
210. Yang J, Kumanovic S, Xia H, et al. Neuroactive steroids: synthesis and SAR of 21-(N-heteroaryl)-3α-hydroxy-3β-methyl-5α-pregnan-20-ones. Book of abstracts, 217th ACS national meeting, Anaheim, CA, United States 1999; March 21–25.
211. Rodgers-Neame NT, Covey DF, Hu Y, et al. Effects of a benz[e]indene on gamma-aminobutyric acid-gated chloride currents in cultured postnatal rat hippocampal neurons. Mol Pharmacol 1992; 42(6):952–957.
212. Hu Y, Zorumski CF, Covey DF. Neurosteroid analogues: structure–activity studies of benz[e]indene modulators of GABAA receptor function. 1. The effect of 6-methyl substitution on the electrophysiological activity of 7-substituted benz[e]indene-3-carbonitriles. J Med Chem 1993; 36(24):3956–3967.
213. Han M, Hu Y, Zorumski CF, et al. Neurosteroid analogues. 3. The synthesis and electrophysiological evaluation of benz[e]indene congeners of neuroactive steroids having the 5 beta-configuration. J Med Chem 1995; 38(22):4548–4556.
214. Hu Y, Zorumski CF, Covey DF. Neurosteroid analogues. 2. Total synthesis and electrophysiological evaluation of benz[e]indene analoguess of the anesthetic steroid alphaxalone. J Org Chem 1995; 50(22):3619–3625.
215. Katona BW, Krishnan K, Cai ZY, et al. Neurosteroid analogues. 12. Potent enhancement of GABA-mediated chloride currents at GABA(A) receptors by ent-androgens. Eur J Med Chem 2007.
216. Wittmer LL, Hu Y, Kalkbrenner M, et al. Enantioselectivity of steroid-induced gamma-aminobutyric acidA receptor modulation and anesthesia. Mol Pharmacol 1996; 50(6):1581–1586.
217. Zorumski CF, Mennerick SJ, Covey DF. Enantioselective modulation of GABAergic synaptic transmission by steroids and benz[e]indenes in hippocampal microcultures. Synapse 1998; 29(2):162–171.

218. Zorumski CF, Wittmer LL, Isenberg KE, et al. Effects of neurosteroid and benz[e]indene enantiomers on GABAA receptors in cultured hippocampal neurons and transfected HEK-293 cells. Neuropharmacology 1996; 35(9–10):1161–1168.
219. Covey DF, Nathan D, Kalkbrenner M, et al. Enantioselectivity of pregnanolone-induced gamma-aminobutyric acid(A) receptor modulation and anesthesia. J Pharmacol Exp Ther 2000; 293(3):1009–1016.
220. Nilsson KR, Zorumski CF, Covey DF. Neurosteroid analogues. 6. The synthesis and GABAA receptor pharmacology of enantiomers of dehydroepiandrosterone sulfate, pregnenolone sulfate, and (3alpha, 5beta)-3-hydroxypregnan-20-one sulfate. J Med Chem 1998; 41(14):2604–2613.
221. Frye CA. The role of neurosteroids and non-genomic effects of progestins and androgens in mediating sexual receptivity of rodents. Brain Res Brain Res Rev 2001; 37(1–3):201–222.
222. Zeng CM, Manion BD, Benz A, et al. Neurosteroid analogues. 10. The effect of methyl group substitution at the C-6 and C-7 positions on the GABA modulatory and anesthetic actions of (3alpha, 5alpha)- and (3alpha,5beta)-3-hydroxypregnan-20-one. J Med Chem 2005; 48(8):3051–3059.
223. Nicoletti D, Ghini AA, Furtmuller R, et al. Synthesis and GABA(A) receptor activity of 6-oxa-analogs of neurosteroids. Steroids 2000; 65(6):349–356.
224. Jiang X, Manion BD, Benz A, et al. Neurosteroid analogues. 9. Conformationally constrained pregnanes: structure–activity studies of 13,24-cyclo-18,21-dinorcholane analogues of the GABA modulatory and anesthetic steroids (3alpha,5alpha)- and (3alpha,5beta)-3-hydroxypregnan-20-one. J Med Chem 2003; 46(25):5334–5348.
225. Lundgren P, Stromberg J, Backstrom T, et al. Allopregnanolone-stimulated GABA-mediated chloride ion flux is inhibited by 3beta-hydroxy-5alpha-pregnan-20-one (isoallopregnanolone). Brain Res 2003; 982(1):45–53.
226. Garrett KM, Gan J. Enhancement of gamma-aminobutyric acidA receptor activity by alpha-chloralose. J Pharmacol Exp Ther 1998; 285(2):680–686.
227. Maitra R, Reynolds JN. Modulation of GABA(A) receptor function by neuroactive steroids: evidence for heterogeneity of steroid sensitivity of recombinant GABA(A) receptor isoforms. Can J Physiol Pharmacol 1998; 76(9):909–920.
228. Flood JF, Morley JE, Roberts E. Pregnenolone sulfate enhances post-training memory processes when injected in very low doses into limbic system structures: the amygdala is by far the most sensitive. Proc Natl Acad Sci USA 1995; 92(23):10806–10810.
229. Mayo W, Dellu F, Robel P, et al. Infusion of neurosteroids into the nucleus basalis magnocellularis affects cognitive processes in the rat. Brain Res 1993; 607(1–2):324–328.
230. Melchior CL, Ritzmann RF. Neurosteroids block the memory-impairing effects of ethanol in mice. Pharmacol Biochem Behav 1996; 53(1):51–56.
231. Majewska MD, Demirgoren S, Spivak CE, et al. The neurosteroid dehydroepiandrosterone sulfate is an allosteric antagonist of the GABAA receptor. Brain Res 1990; 526(1):143–146.
232. Demirgoren S, Majewska MD, Spivak CE, et al. Receptor binding and electrophysiological effects of dehydroepiandrosterone sulfate, an antagonist of the GABAA receptor. Neuroscience 1991; 45(1):127–135.
233. Park-Chung M, Malayev A, Purdy RH, et al. Sulfated and unsulfated steroids modulate gamma-aminobutyric acidA receptor function through distinct sites. Brain Res 1999; 830(1):72–87.
234. Imamura M, Prasad C. Modulation of GABA-gated chloride ion influx in the brain by dehydroepiandrosterone and its metabolites. Biochem Biophys Res Commun 1998; 243(3):771–775.
235. Akk G, Bracamontes J, Steinbach JH. Pregnenolone sulfate block of GABA(A) receptors: mechanism and involvement of a residue in the M2 region of the alpha subunit. J Physiol 2001; 532(Pt 3):673–684.
236. Shen W, Mennerick S, Covey DF, et al. Pregnenolone sulfate modulates inhibitory synaptic transmission by enhancing GABA(A) receptor desensitization. J Neurosci 2000; 20(10):3571–3579.

237. Celentano JJ, Wong RK. Multiphasic desensitization of the GABAA receptor in outside-out patches. Biophys J 1994; 66(4):1039–1050.
238. Lavoie AM, Tingey JJ, Harrison NL, et al. Activation and deactivation rates of recombinant GABA(A) receptor channels are dependent on alpha-subunit isoform. Biophys J 1997; 73(5):2518–2526.
239. Haas KF, Macdonald RL. GABAA receptor subunit gamma2 and delta subtypes confer unique kinetic properties on recombinant GABAA receptor currents in mouse fibroblasts. J Physiol 1999; 514 (Pt 1):27–45.
240. Mienville JM, Vicini S. Pregnenolone sulfate antagonizes GABAA receptor-mediated currents via a reduction of channel opening frequency. Brain Res 1989; 489(1):190–194.
241. Woodward RM, Polenzani L, Miledi R. Effects of steroids on gamma-aminobutyric acid receptors expressed in Xenopus oocytes by poly(A) + RNA from mammalian brain and retina. Mol Pharmacol 1992; 41(1):89–103.
242. Mennerick S, Zeng CM, Benz A, et al. Effects on gamma-aminobutyric acid (GABA)(A) receptors of a neuroactive steroid that negatively modulates glutamate neurotransmission and augments GABA neurotransmission. Mol Pharmacol 2001; 60(4):732–741.
243. Maione S, Berrino L, Vitagliano S, et al. Pregnenolone sulfate increases the convulsant potency of *N*-methyl-D-aspartate in mice. Eur J Pharmacol 1992; 219(3):477–479.
244. Wang MD, Backstrom T, Landgren S. The inhibitory effects of allopregnanolone and pregnanolone on the population spike, evoked in the rat hippocampal CA1 stratum pyramidale in vitro, can be blocked selectively by epiallopregnanolone. Acta Physiol Scand 2000; 169(4):333–341.
245. Hawkinson JE, Acosta-Burruel M, Yang KC, et al. Substituted 3beta-phenylethynyl derivatives of 3alpha-hydroxy-5alpha-pregnan-20-one: remarkably potent neuroactive steroid modulators of gamma-aminobutyric acidA receptors. J Pharmacol Exp Ther 1998; 287(1):198–207.
246. Reddy DS. Pharmacology of endogenous neuroactive steroids. Crit Rev Neurobiol 2003; 15(3–4):197–234.
247. Carter RB, Wood PL, Wieland S, et al. Characterization of the anticonvulsant properties of ganaxolone (CCD 1042; 3alpha-hydroxy-3beta-methyl-5alpha-pregnan-20-one), a selective, high-affinity, steroid modulator of the gamma-aminobutyric acid(A) receptor. J Pharmacol Exp Ther 1997; 280(3):1284–1295.
248. Rick CE, Ye Q, Finn SE, et al. Neurosteroids act on the GABA(A) receptor at sites on the N-terminal side of the middle of TM2. Neuroreport 1998; 9(3):379–383.
249. Callachan H, Cottrell GA, Hather NY, et al. Modulation of the GABAA receptor by progesterone metabolites. Proc R Soc Lond B Biol Sci 1987; 231(1264):359–369.
250. Akk G, Shu HJ, Wang C, et al. Neurosteroid access to the GABAA receptor. J Neurosci 2005; 25(50):11605–11613.
251. Hosie AM, Wilkins ME, da Silva HM, et al. Endogenous neurosteroids regulate GABAA receptors through two discrete transmembrane sites. Nature 2006; 444(7118):486–489.
252. Majewska MD, Schwartz RD. Pregnenolone-sulfate: an endogenous antagonist of the gamma-aminobutyric acid receptor complex in brain? Brain Res 1987; 404(1–2):355–360.
253. Majewska MD, Mienville JM, Vicini S. Neurosteroid pregnenolone sulfate antagonizes electrophysiological responses to GABA in neurons. Neurosci Lett 1988; 90(3):279–284.
254. Woodhull AM. Ionic blockage of sodium channels in nerve. J Gen Physiol 1973; 61(6):687–708.
255. Sousa A, Ticku MK. Interactions of the neurosteroid dehydroepiandrosterone sulfate with the GABA(A) receptor complex reveals that it may act via the picrotoxin site. J Pharmacol Exp Ther 1997; 282(2):827–833.
256. Gibbs TT, Russek SJ, Farb DH. Sulfated steroids as endogenous neuromodulators. Pharmacol Biochem Behav 2006; 84(4):555–567.
257. Shen W, Mennerick S, Zorumski EC, et al. Pregnenolone sulfate and dehydroepiandrosterone sulfate inhibit GABA-gated chloride currents in Xenopus oocytes expressing picrotoxin-insensitive GABA(A) receptors. Neuropharmacology 1999; 38(2):267–271.

258. Wardell B, Marik PS, Piper D, et al. Residues in the first transmembrane domain of the Caenorhabditis elegans GABA(A) receptor confer sensitivity to the neurosteroid pregnenolone sulfate. Br J Pharmacol 2006; 148(2):162–172.
259. Reddy DS, Kulkarni SK. Development of neurosteroid-based novel psychotropic drugs. Prog Med Chem 2000; 37:135–175.
260. Majewska MD. Neurosteroids: endogenous bimodal modulators of the GABAA receptor. Mechanism of action and physiological significance. Prog Neurobiol 1992; 38(4):379–395.
261. Rupprecht R, di Michele F, Hermann B, et al. Neuroactive steroids: molecular mechanisms of action and implications for neuropsychopharmacology. Brain Res Brain Res Rev 2001; 37(1–3):59–67.
262. Rupprecht R, Reul JM, Trapp T, et al. Progesterone receptor-mediated effects of neuroactive steroids. Neuron 1993; 11(3):523–530.
263. Purdy RH, Moore PH, Jr., Morrow AL, et al. Neurosteroids and GABAA receptor function. Adv Biochem Psychopharmacol 1992; 47:87–92.
264. Barbaccia ML, Concas A, Serra M, et al. Stress and neurosteroids in adult and aged rats. Exp Gerontol 1998; 33(7–8):697–712.
265. Serra M, Pisu MG, Littera M, et al. Social isolation-induced decreases in both the abundance of neuroactive steroids and GABA(A) receptor function in rat brain. J Neurochem 2000; 75(2):732–740.
266. Bitran D, Shiekh M, McLeod M. Anxiolytic effect of progesterone is mediated by the neurosteroid allopregnanolone at brain GABAA receptors. J Neuroendocrinol 1995; 7(3):171–177.
267. Wieland S, Belluzzi JD, Stein L, et al. Comparative behavioral characterization of the neuroactive steroids 3 alpha-OH,5 alpha-pregnan-20-one and 3 alpha-OH,5 beta-pregnan-20-one in rodents. Psychopharmacology (Berl) 1995; 118(1):65–71.
268. Wihlback A-C, Sundstrom-Poromaa I, Backstrom T. Action by and sensitivity to neuroactive steroids in menstrual cycle related CNS disorders. Psychopharmacology (Berl) 2006; 186(3):388–401.
269. Miczek KA, Fish EW, De Bold JF. Neurosteroids, GABAA receptors, and escalated aggressive behavior. Horm Behav 2003; 44(3):242–257.
270. Droogleever Fortuyn HA, van Broekhoven F, Span PN, et al. Effects of PhD examination stress on allopregnanolone and cortisol plasma levels and peripheral benzodiazepine receptor density. Psychoneuroendocrinology 2004; 29(10):1341–1344.
271. Rodgers RJ, Johnson NJ. Behaviorally selective effects of neuroactive steroids on plus-maze anxiety in mice. Pharmacol Biochem Behav 1998; 59(1):221–232.
272. Vanover KE, Rosenzweig-Lipson S, Hawkinson JE, et al. Characterization of the anxiolytic properties of a novel neuroactive steroid, Co 2–6749 (GMA-839; WAY-141839; 3alpha, 21-dihydroxy-3beta-trifluoromethyl-19-nor-5beta-pregnan-20-one), a selective modulator of gamma-aminobutyric acid(A) receptors. J Pharmacol Exp Ther 2000; 295(1):337–345.
273. Farr SA, Uezu K, Creonte TA, et al. Modulation of memory processing in the cingulate cortex of mice. Pharmacol Biochem Behav 2000; 65(3):363–368.
274. Sundstro II, Backstrom T. Citalopram increases pregnanolone sensitivity in patients with premenstrual dysphoric disorder. Acta Physiol Scand 1999; 167(2):A6-A7.
275. Turkmen S, Lundgren P, Birzniece V, et al. 3beta-20beta-dihydroxy-5alpha-pregnane (UC1011) antagonism of the GABA potentiation and the learning impairment induced in rats by allopregnanolone. Eur J Neurosci 2004; 20(6):1604–1612.
276. Flood JF, Morley JE, Roberts E. Memory-enhancing effects in male mice of pregnenolone and steroids metabolically derived from it. Proc Natl Acad Sci USA 1992; 89(5):1567–1571.
277. Wolkowitz OM, Reus VI, Roberts E, et al. Antidepressant and cognition-enhancing effects of DHEA in major depression. Ann NY Acad Sci 1995; 774:337–339.
278. Flood JF, Smith GE, Roberts E. Dehydroepiandrosterone and its sulfate enhance memory retention in mice. Brain Res 1988; 447(2):269–278.

279. Sunderland T, Merril CR, Harrington MG, et al. Reduced plasma dehydroepiandrosterone concentrations in Alzheimer's disease. Lancet 1989; 2(8662):570.
280. Nasman B, Olsson T, Backstrom T, et al. Serum dehydroepiandrosterone sulfate in Alzheimer's disease and in multi-infarct dementia. Biol Psychiatry 1991; 30(7):684–690.
281. Hillen T, Lun A, Reischies FM, et al. DHEA-S plasma levels and incidence of Alzheimer's disease. Biol Psychiatry 2000; 47(2):161–163.
282. Sun X, Marks DL, Park WD, et al. Niemann-Pick C variant detection by altered sphingolipid trafficking and correlation with mutations within a specific domain of NPC1. Am J Hum Genet 2001; 68(6):1361–1372.
283. Mellon SH, Gong W, Schonemann MD. Endogenous and synthetic neurosteroids in treatment of Niemann-Pick Type C disease. Brain Res Rev 2007.
284. Ahmad I, Lope-Piedrafita S, Bi X, et al. Allopregnanolone treatment, both as a single injection or repetitively, delays demyelination and enhances survival of Niemann-Pick C mice. J Neurosci Res 2005; 82(6):811–821.
285. Langmade SJ, Gale SE, Frolov A, et al. Pregnane X receptor (PXR) activation: a mechanism for neuroprotection in a mouse model of Niemann-Pick C disease. Proc Natl Acad Sci USA 2006; 103(37):13807–13812.
286. Griffin LD, Gong W, Verot L, et al. Niemann-Pick type C disease involves disrupted neurosteroidogenesis and responds to allopregnanolone. Nat Med 2004; 10(7):704–711.
287. Biswas MG, Russell DW. Expression cloning and characterization of oxidative 17beta- and 3alpha-hydroxysteroid dehydrogenases from rat and human prostate. J Biol Chem 1997; 272(25):15959–15966.
288. Wang M, Seippel L, Purdy RH, et al. Relationship between symptom severity and steroid variation in women with premenstrual syndrome: study on serum pregnenolone, pregnenolone sulfate, 5 alpha-pregnane-3,20-dione and 3 alpha-hydroxy-5 alpha-pregnan-20-one. J Clin Endocrinol Metab 1996; 81(3):1076–1082.
289. Bixo M, Andersson A, Winblad B, et al. Progesterone, 5alpha-pregnane-3,20-dione and 3alpha-hydroxy-5alpha-pregnane-20-one in specific regions of the human female brain in different endocrine states. Brain Res 1997; 764(1–2):173–178.
290. Sundstrom I, Andersson A, Nyberg S, et al. Patients with premenstrual syndrome have a different sensitivity to a neuroactive steroid during the menstrual cycle compared to control subjects. Neuroendocrinology 1998; 67(2):126–138.
291. Havlikova H, Hill M, Kancheva L, et al. Serum profiles of free and conjugated neuroactive pregnanolone isomers in nonpregnant women of fertile age. J Clin Endocrinol Metab 2006; 91(8):3092–3099.
292. Lanthier A, Patwardhan VV. Sex steroids and 5-en-3 beta-hydroxysteroids in specific regions of the human brain and cranial nerves. J Steroid Biochem 1986; 25(3):445–449.
293. Kancheva L, Havlikova H, Hill M, et al. Neuroactive steroids in adult men. 4th International Meeting- Steroids and Nervous system, Torino, Italy 2007, Abstract book, p. 152.

Chapter 2
Steroid Metabolism in Glial Cells

Roberto C. Melcangi[1] and Luis M. Garcia-Segura[2]

Abstract Glial cells exert a fundamental role in the production of neuroactive steroids in central and peripheral nervous system. Steroids so formed, or activated in glia may then act by autocrine pathway on the same cells or by paracrine pathway on different glial types or on neuronal compartment. These mechanisms are then integrated with classical endocrine mechanisms exerted by steroid hormones coming from peripheral glands. The finding that neuroactive steroids are signaling molecules in glia–neuron and glia–glia crosstalk, together with the effects exerted by them in physiological and pathological conditions, strengthens the concept that they exert a key role in regulating nervous function.

Keywords Aromatase, astrocytes, autocrine pathway, cellular crosstalk, endocrine pathway, microglia, neuroendocrine control, neuroprotection, oligodendrocytes, paracrine pathway, pregnenolone, progesterone, Schwann cells, steroidogenesis, testosterone

Abbreviations P450 aro aromatase; P450 scc P450 Side-Chain Clevage; 3α-diol 5α-androstane-3α,17β-diol; AMPA alpha-amino-3-hydroxy-5-methyl-4-isoxazolepropionic acid; CNS central nervous system; DHEA dehydroepiandrosterone; DHP dihydroprogesterone; DHT dihydrotestosterone; DRG dorsal root ganglia; GABA gamma-amino butyric acid; GFAP glial fibrillary acidic protein; 3α-HSD 3α-hydroxysteroid dehydrogenase; 3β-HSD 3β-hydroxy-steroid dehydrogenase; 11β-HSD 11β-hydroxysteroid dehydrogenase; P450 c17 17α-hydroxylase/C17–20-lyase; 17β-HSD 17β-hydroxylase/C17–20-lyase; NMDA *N*-methyl-D-aspartate; PNS peripheral nervous system; PREG pregnenolone; 5α-R 5α-reductase; StAR steroidogenic acute regulatory protein; THP tetrahydroprogesterone; TSPO translocator protein-18 kDa

[1]Department of Endocrinology and Center of Excellence on Neurodegenerative Diseases, University of Milan, Italy

[2]Instituto Cajal, C.S.I.C., Madrid, Spain

M.S. Ritsner and A. Weizman (eds.), *Neuroactive Steroids in Brain Function, Behavior and Neuropsychiatric Disorders*

2.1 Steroid Hormones, Neurosteroids and Neuroactive Steroids

The finding that the nervous system is a target for steroids is a well-ascertained concept. Indeed, several observations have shown that steroids may regulate the development and function of the nervous system and affect mood, behavior and cognition. It has been construed for a long time that steroids exert their actions in the nervous system exclusively by a classical endocrine mechanism involving steroid production by endocrine glands such as the adrenals and gonads, secretion into the blood stream, crossing of the blood–brain barrier and then regulating the central (CNS) and peripheral (PNS) nervous system in various ways. However, further findings have suggested that the effects of steroids in the nervous system are also due to paracrine and/or autocrine mechanisms. This idea is based on observations obtained by Baulieu and coworkers indicating that the synthesis of steroids is not only a peculiarity of peripheral glands but may also occur in the CNS (i.e., the formation of the so-called neurosteroids).[1,2] This term describes those steroids synthesized in the brain directly from cholesterol, which itself is synthesized from glucose within the brain.[3–5] Recent observations indicated that also PNS synthesizes neurosteroids.[6–8] On the other hand, because in many circumstances it is difficult to discriminate whether the steroid effect is due to in situ synthesis (i.e., neurosteroids), to the peripheral hormones (i.e., steroid hormones), or to an enzymatic conversion of steroids in metabolites which are more active and in some cases utilize a different mechanism of action, some investigators in this field use now the term neuroactive steroids. This term was initially introduced by Paul and Purdy[9] to refer to steroids that rapidly alter the excitability of neurons by binding to membrane-bound receptors, but in its present broader usage reefers to steroids that act on the nervous system, regardless of their mechanism of action.

2.2 Synthesis and Steroid Metabolism in Glial Cells

Both neurons and glial cells are involved in the synthesis, metabolism and action of steroids in the nervous system. Concerning glial cells, it has been demonstrated that in the rodent and human CNS, astrocytes and oligodendrocytes express several enzymes involved in the steroidogenesis and consequently these types of glial cells are able to produce different kinds of steroids.[10–16] As shown in Fig. 2.1, cholesterol is converted to pregnenolone (PREG), catalyzed by the cytochrome P450 side-chain cleavage (P450scc), and pregnenolone is converted to progesterone, catalyzed by the 3β-hydroxy-steroid dehydrogenase (3β-HSD). PREG and progesterone can be further converted, via 17-hydroxy-pregnenolone and 17-hydroxy-progesterone, respectively to dehydroepiandrosterone (DHEA) and androstenedione, catalyzed by the 17α-hydroxylase/C17–20-lyase (P450 c17), and from there to androstenediol and testosterone, respectively, catalyzed by 17β-hydroxylase/C17–20-lyase (17β-HSD).

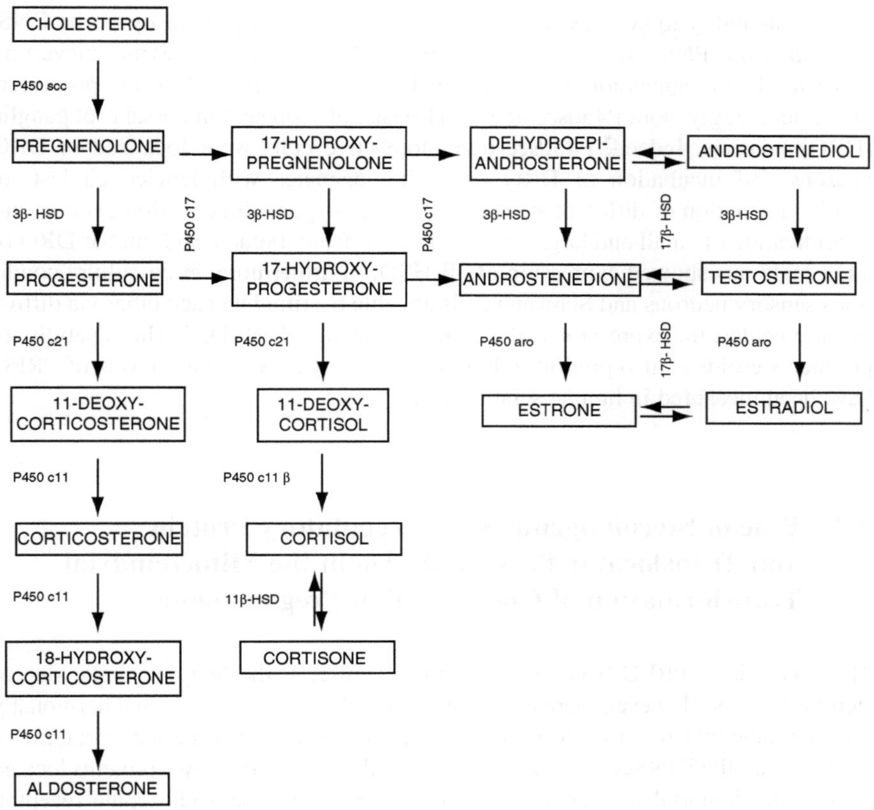

Fig. 2.1 Biosynthesis of steroids in glial cells

Androstenedione and testosterone, can be converted to estrone and estradiol, respectively, catalyzed by aromatase (P450 aro). In addition to these pathways, producing androgens and estrogens, progesterone and its derivative 17-hydroxy-progesterone can be converted to the glucocorticoids corticosterone and cortisol, respectively. Cortisol can be further converted to the glucocorticoid cortisone, catalyzed by 11β-hydroxysteroid dehydrogenase (11β-HSD), and corticosterone via 18-hydroxy-corticosterone to the mineralocorticoid aldosterone.

Astrocytes appear to be the most active steroidogenic cells in the brain, since cultures of these cells produce PREG, progesterone, DHEA, androstenedione, testosterone and estradiol. On the contrary, oligodendrocytes seem to be able to form only PREG and androstenedione. The steroidogenic activity of microglia has not been assessed. However, systemic macrophages that share many properties and a common origin with microglia, express aromatase and produce androgenic and estrogenic derivatives.[17]

The capability to synthesize steroids seems not to be a peculiarity of the CNS, since also the PNS expresses cytochrome P450scc (i.e., the enzyme converting cholesterol to pregnenolone), and 3β-HSD, which converts PREG into progesterone. Interestingly, both P450scc and 3β-HSD are also present in dorsal root ganglia (DRG) neurons. Indeed, P450scc-immunoreactive cells were localized in DRG neurons and incubation of DRG tissue homogenates with labeled cholesterol yielded formation of different steroids including pregnenolone.[18] Moreover, in situ hybridization of small and large sensory neurons from thoracic and lumbar DRG of adult male rats showed expression of 3β-HSD.[19] Furthermore, in co-culture conditions sensory neurons and Schwann cells are able to stimulate each other via diffusible molecules, the expression and activity of enzyme 3β-HSD.[7,20] The capability to produce steroids is also present in human PNS. Indeed, very high levels of PREG have been measured in human sciatic nerves.[19,21]

2.3 Role of Steroidogenic Acute Regulatory Protein and Translocator Protein 18 kDa in the Mitochondrial Transformation of Cholesterol in Pregnenolone

The synthesis of PREG from cholesterol by P450scc is the first enzymatic step of steroidogenesis. However, there is a previous step that is rate limiting and hormonally regulated: the transfer of cholesterol from the outer to the inner mitochondrial membrane, where the P450scc enzyme is located. In this mechanism, two proteins located in the mitochondrial membranes, such as the peripheral benzodiazepine receptor, now re-named as translocator protein-18 kDa (TSPO)[22] and the steroidogenic acute regulatory protein (StAR)[23] take an active role (Fig. 2.2). TSPO was initially described on peripheral tissues as a second binding site for diazepam, which binds with higher affinity to gamma-amino butyric acid Type A ($GABA_A$) receptors on the nervous system. Further studies have then demonstrated that TSPO is pharmacologically and structurally distinct from the central benzodiazepine/$GABA_A$ receptors. TSPO is an 18-kDa peptide located predominantly in the mitochondrial membranes and represents a critical component of the permeability transition pore, a multiprotein complex implicated in the regulation of apoptosis. In addition, TSPO has been related with the regulation of several physiological events, including the control of steroidogenesis. TSPO is expressed in the nervous system and predominantly in glial cells of CNS and PNS. Neural injury and different neuropathological conditions result in the induction of the expression of this molecule. Indeed, TSPO basal expression is upregulated in gliomas,[24,25] in neurodegenerative disorders (e.g., Alzheimer's disease),[26–28] as well as in various forms of brain injury and inflammation[29,30] induced by neurotoxins.[30–33] The induction of TSPO expression after injury in the CNS is mainly restricted to microglia and astrocytes.[34] A very similar effect occurs in the PNS. Indeed, the expression of TSPO in Schwann cells is increased after nerve lesion, and return to normal levels when regeneration is completed.[35]

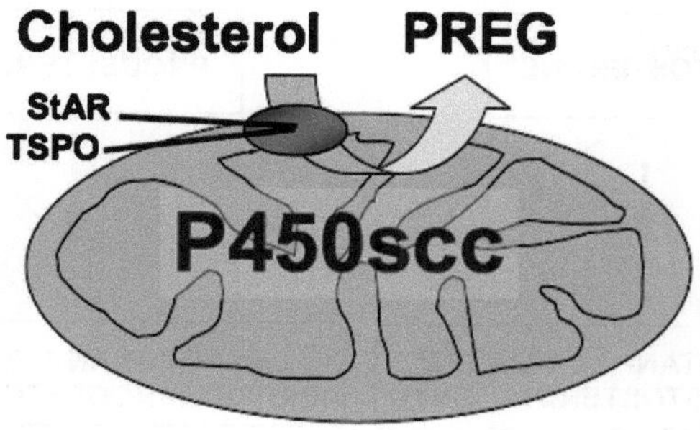

Fig. 2.2 The synthesis of pregnenolone (PREG) from cholesterol by the cholesterol side-chain cleavage enzyme (P450scc) is the first enzymatic step of steroidogenesis. Proteins located in the mitochondrial membranes, such as the translocator protein-18 kDa (TSPO) and the steroidogenic acute regulatory protein (StAR), allow cholesterol to cross the hydrophilic intermembrane space. Both TSPO and StAR expression has been detected in glial cells

StAR is formed as a 37-kDa protein, which is rapidly transported into mitochondria where it is cleaved, generating a mature 30-kDa intramitochondrial StAR protein that is inactive. This molecule appears to be widely distributed throughout the brain, although different levels of expression have been detected between different brain areas. Moreover, StAR expression seems to be restricted to very specific neuronal and astroglial populations in each brain area, although it is predominantly expressed by neurons.[36] In the PNS, StAR is expressed by Schwann cells.[37] Therefore, both neurons and glial cells may participate in the conversion of cholesterol into PREG.

2.4 Synthesis and Metabolism of Progesterone

Glial cells of CNS and PNS may convert not only PREG in progesterone, but also possess several enzymes able to transform native steroids into neuroactive steroids. In particular, glial cells are able to metabolize native steroid hormones into their 5α- and 3α-hydroxy-5α reduced derivatives via the enzymatic complex formed by the 5α-reductase (5α-R) and the 3α-hydroxysteroid dehydrogenase (3α-HSD)[38–41] (Fig. 2.3). This enzymatic complex is very versatile, since every steroid possessing the delta 4-3keto configuration may be first 5α-reduced and subsequently 3α-hydroxylated. Progesterone is a substrate for this enzymatic complex. Namely, in CNS astrocytes and oligodendrocytes convert this steroid by the action of 5α-R, into dihydroprogesterone (DHP) and subsequently, via the action of 3α-HSD, into tetrahydroprogesterone (THP), also known as allopregnanolone. However, among glial cells the conversion of DHP into THP seems to be different. Indeed, observations obtained in culture have indicated that this enzymatic conversion is very high in type 1 astrocytes (which probably correspond in vivo to the protoplasmatic

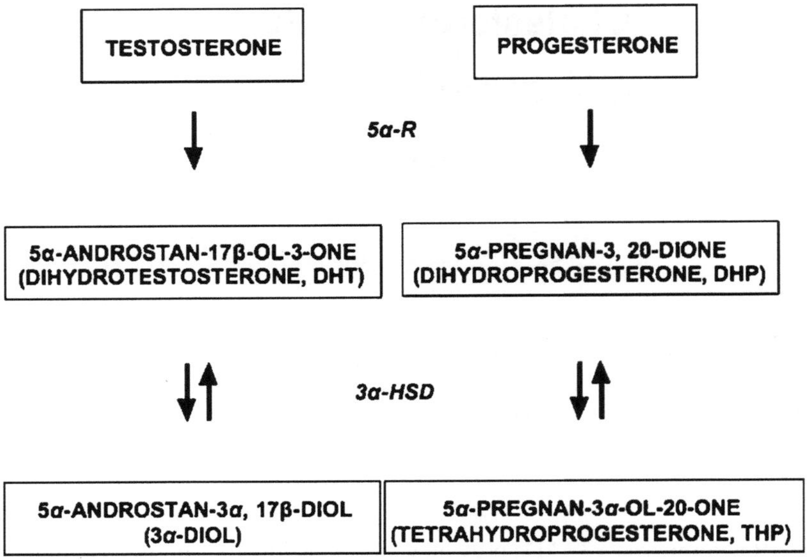

Fig. 2.3 Testosterone and progesterone metabolism by the enzymes 5α-reductase (5α-R) and 3α-hydroxysteroid-dehydrogenase (3α-HSD). Similarly, the 5α-R converts corticosterone to dihydrocorticosterone (DHC) and 11-deoxycorticosterone to dihydrodeoxycorticosterone (DHDOC), and the 3α-HSD further converts dihydrodeoxycorticosterone to tetrahydrodeoxycorticosterone (THDOC)

astrocytes) in comparison to what observed in type 2 astrocytes (which probably correspond in vivo to the fibrous astrocytes) and in oligodendrocytes. However, the conversion of DHP into THP in oligodendrocytes is significantly higher than that present in type 2 astrocytes.[42]

These enzymatic conversions have an important impact on the mechanism of action of progesterone. In fact, DHP, like progesterone itself, acts on a classical steroid receptor, such as the progesterone receptor, while THP acts on a non-classical steroid receptor, like the $GABA_A$ receptor.[43] It is important to recall, that the conversion of DHP into THP, by the enzyme 3α-HSD, is a reversible process.[40] However, 3α-oxidation to yield DHP may be also exerted by 17β-hydroxysteroid dehydrogenase type 10.[44] Interestingly, normal and activated astrocytes, such as those present in Alzheimer type pathology, express this enzyme.[45]

Progesterone metabolism by glial cells is developmentally regulated. Namely, the formation of DHP in fully differentiated oligodendrocytes is fivefold higher than in oligodendrocyte pre-progenitors and in oligodendrocyte progenitors. On the contrary, the formation of THP is higher in oligodendrocyte pre-progenitors and decreases with oligodendrocyte differentiation.[46] These findings emphasize that not only differentiated CNS cells possess the 5α-R/3α-HSD system since considerable enzymatic activities converting progesterone are also present in undifferentiated cells, as shown in studies performed on stem cells originating from the mouse striatum.[47]

The ability to convert steroid hormones into 5α- and 3α-hydroxy-5α-reduced derivatives is also present in myelin forming cells of PNS (i.e., Schwann cells). However, the extent of progesterone metabolism seems to be different respect to what observed in the cellular counterpart of CNS. Therefore, the conversion of progesterone into DHP occurring in Schwann cells is at least four times higher than in oligodendrocytes, while that of DHP into THP is lower than in oligodendrocytes.[48]

2.5 Androgen Metabolism

Testosterone is a further substrate for the action of 5α-R and of 3α-HSD enzymes. As shown in Fig. 2.3, similarly to what mentioned above in case of progesterone, testosterone can be converted into dihydrotestosterone (DHT) by the enzyme 5α-R and then in 5α-androstane-3α, 17β-diol (3α-diol, also known as androstanediol) by the enzyme 3α-HSD. Both enzymatic steps occur in glial cells. However, similarly to what reported in case of progesterone, these enzymatic steps are different depending of the type of glial cells considered. For instance, type 2 astrocytes and oligodendrocytes are more active in converting testosterone into DHT than type 1 astrocytes. Moreover, the conversion of DHT into 3α-diol appears to be mainly localized in type 1 astrocytes.[49] Also in this case the enzymatic conversions of testosterone into its reduced metabolites may have an important impact on the mechanism of action of the steroid. In fact, while DHT, like its precursor testosterone, is able to bind the classical steroid receptor for androgens, 3α-diol is able to bind to a non-classical steroid receptor, like the $GABA_A$ receptor.

Testosterone metabolism by glial cells is also affected by cellular interactions, since the presence of neurons stimulate the activity of both the 5α-R and 3α-HSD enzymes in astrocytes.[50] The nature of these neuronal signals has been also investigated. In mixed glial cultures the formation of DHT is increased by exposure to cAMP analogs, like for instance 8-Br-cAMP.[51] In agreement with this finding, in cultures of glioma cells the stimulation of beta-adrenergic receptor induces together an activation of the cyclic AMP (cAMP)/protein kinase A-mediated signaling pathway a rapid elevation of 5α-R mRNA levels.[52]

The other alternative metabolic pathway of testosterone is its conversion to estradiol that will be described below.

2.6 Estradiol Synthesis

The enzyme aromatase (P450 aro), that is able to convert testosterone into estradiol, is not expressed by glial cells of the CNS of rodents under normal circumstances, since only neurons possess such an activity.[53] However, aromatase is expressed by glial cells in primary cultures of developing zebra finch telencephalon[54] and rodent astrocytic cultures from the cerebral cortex of neonatal rats are able to produce estradiol and estrone.[13,55] In addition to species differences, it is possible that some culture conditions may induce aromatase expression in astrocytes. Indeed, stressful conditions, such as serum deprivation, induce aromatase expression in cultured rat astrocytes.[56] Furthermore, aromatase is expressed by astrocytes in the brain of birds and mammals after brain injury,[57,58] indicating that the enzyme may be induced de novo in these cells under specific circumstances. In contrast to rodents, aromatase is expressed in radial glia in the brain of teleost fish under normal conditions[59] and also in a subpopulation of astrocytes in the human brain.[60] Aromatase expression in glia seems to be regulated by steroid levels. A first demonstration in a vocal fish, the plainfin midshipman, shows in preoptic area and in dimorphic sonic/vocal motor nucleus of testosterone- and estradiol-treated females an increase of aromatase mRNA levels.[61]

2.7 Physiological Implications of Glial Steroidogenesis

Glial cells are not only able to synthesize neurosteroids and to metabolize these or steroid hormones into neuroactive steroids but these cells are also target for their actions. This is not surprising because it has been clearly demonstrated that oligodendrocytes, astrocytes and Schwann cells possess classical intracellular receptors for many families of hormonal steroids (e.g., receptors for glucocorticoids, minerocorticoids, androgens, estrogens and progesterone).[62–67] Moreover, these kinds of glial cells also express non-classical steroid receptors, like for instance ($GABA_A$ and $GABA_B$ receptor), *N*-methyl-D-aspartate (NMDA) receptor, alpha-amino-3-hydroxy-5-methyl-4-isoxazolepropionic acid (AMPA) receptor, and an atypical intracellular receptor like the sigma 1,[68–76] which have been also demonstrated to be a target for neuroactive steroid actions.

Schwann cells, oligodendrocytes, astroglia and microglia are all affected by neuroactive steroids. In particular, neuroactive steroids affect proliferation, gene expression, differentiation, metabolism and morphology of glial cells. For instance, estradiol and progesterone enhance the proliferation of Schwann cells[77] and of oligodendrocyte precursors,[78,79] while decrease in female the astrocyte proliferation after brain injury in the cerebral cortex and in the hippocampus.[80] Estradiol also decreases proliferation and increases cell death in primary cortical astrocytic cultures,[81] while glucocorticoids decrease oligodendrocyte precursor proliferation.[82]

Examples of effects on gene expression are those exerted by progesterone and derivatives (DHP and THP) or testosterone and derivatives (DHT and 3α-diol) on the expression of myelin proteins in Schwann cells and in oligodendrocytes.[62,70,83–87] Moreover, these neuroactive steroids also modulate the expression of transcription factors regulating the expression of myelin proteins in Schwann cells.[88–90]

Effects of neuroactive steroids on glial gene expression and morphology have also been extensively studied in astroglia.[91–99] For instance, corticosterone and dexamethasone inhibit glial fibrillary acidic protein (GFAP) expression in the neonatal and adult rat brain.[100–104] PREG, DHEA and sex steroids affect astroglia cell shape and regulate GFAP in different brain areas.[105–109] However, it is important to highlight that the effect of steroids may be different depending whether they are acting under physiological or pathological conditions. For instance, estradiol increases GFAP expression and promotes the extension of GFAP immunoreactive processes under physiological conditions while the same steroid decreases GFAP and vimentin expression in gliotic injured tissue.[110–113]

Examples of other molecules affected by steroids in astrocytes and microglia are apolipoprotein E,[114] heat shock proteins,[115,116] metalloproteinase-9,[117] aquaporin-4,[118,119] glutamate transporters GLT-1 and GLAST[120] and a variety of growth factors and cytokines, such as transforming growth factor alpha,[121,122] transforming growth factor beta1,[123–126] transforming growth factor beta2,[126] basic fibroblast growth factor,[127,128] insulin-like growth factor I,[129] tumor necrosis factor-alpha,[130] and interleukin-12.[131]

Furthermore, steroids may also affect neuronal function by the modification of glial metabolism. For instance, glucocorticoids are able to influence the metabolism of glutamate in astrocytes and Schwann cells, increasing the expression of both glutamine synthetase and glutamate dehydrogenase in astrocytes[99,132] and the expression of glutamine synthetase and cytosolic aspartate aminotransferase in Schwann cells.[133]

Altogether, these observations indicate that glial steroidogenesis has important functional implications, like, for instance, the regulation of synaptic function and connectivity and the neuroendocrine regulation. Thus, it has been demonstrated that sex steroids affect the synthesis, accumulation and release by hypothalamic astrocytes and tanycytes of different growth factors that regulate neurons synthesizing and secreting the luteinizing hormone releasing hormone (i.e., the hormone deputed to the neuroendocrine control of reproductive functions).

2.8 Therapeutic Implications of Glial Steroidogenesis

The effects exerted by neuroactive steroids on myelin formation and neuronal survival indicate that these molecules may be considered not only as key regulators of nervous function but also as protective agents. Examples supporting this assertion have been obtained in experimental model of Niemann-Pick type C disease (a fatal, autosomal recessive childhood neurogenerative disease),[134] Alzheimer's disease,[135] Parkinson's disease,[136] ischemic[137] or spinal cord injury,[138,139] acquired (after trauma, during aging, diabetic neuropathy, etc.) or inherited peripheral neuropathy, such as Charcot-Marie-Tooth disease.[75,140–142]

A possible therapeutic strategy for neurodegenerative events might consequently imply the treatment with neuroactive steroid themselves, or synthetic ligands for their classical or non-classical steroid receptors. However, also to take in consideration the in situ formation and conversion of steroids at the levels of glial compartment might be extremely important. In fact, as demonstrated in several experimental models, a local increase in the levels of steroids or activation of steroidogenic enzymes may be observed during neurodegenerative events. These findings have suggested that an increase in steroidogenesis is part of an overall mechanism used by the nervous tissue to cope with neurodegenerative conditions. Consequently, modulation of steroidogenesis or of enzymes able to increase neuroactive derivatives (e.g., 5α-R, 3α-HSD, aromatase, etc.) at the level of glial compartment might be also considered as an interesting therapeutic strategy. Indeed, recent observations have indicated that ligands of TSPO, which as mentioned above is involved in the first step of steroidogenesis, exerts protective effects both in CNS and PNS[143,144] and that in situ inhibition of aromatase activity increased neurodegeneration.[145]

2.9 Summary

The observations here reported indicate that glial compartment (i.e., astrocytes, oligodendrocytes, Schwann cells and microglia) exerts a fundamental role in the production of steroids in CNS and PNS. Steroids so formed, or activated in glia compartment may then act by autocrine pathway on the same cells or by paracrine pathway on different glial types or on neuronal compartment. Indeed, classical and non-classical steroid receptors have been reported not only in neuronal but also in glial cells. Paracrine and autocrine effects are then integrated with classical endocrine mechanisms exerted by steroid hormones. Thus, the nervous system is a target for two different pools of steroids, one coming from the peripheral glands and the second one originating directly in the nervous system. The finding that neuroactive steroids may be considered as signaling molecules in glia–neuron and glia–glia crosstalk together with the effects exerted by these molecules in physio-pathological conditions strengthen the concept that they are important key regulators of nervous function.

References

1. Corpechot C, Robel P, Axelson M, et al. Characterization and measurement of dehydroepiandrosterone sulfate in rat brain. Proc Natl Acad Sci USA 1981; 78:4704–4707.
2. Baulieu EE. Neurosteroids: a novel function of the brain. Psychoneuroendocrinology 1998; 23:963–987.
3. Morell P, Jurevics H. Origin of cholesterol in myelin. Neurochem Res 1996; 21:463–470.
4. Turley SD, Burns DK, Dietschy JM. Preferential utilization of newly synthesized cholesterol for brain growth in neonatal lambs. Am J Physiol 1998; 274:1099–1105.
5. Vance JE, Hayashi H, Karten B. Cholesterol homeostasis in neurons and glial cells. Semin Cell Dev Biol 2005; 16:193–212.
6. Koenig HL, Schumacher M, Ferzas B, et al. Progesterone synthesis and myelin formation by Schwann cells. Science 1995; 268:1500–1503.
7. Schumacher M, Guennoun R, Mercier G, et al. Progesterone synthesis and myelin formation in peripheral nerves. Brain Res Rev 2001; 37:343–359.
8. Garcia-Segura LM, Melcangi RC. Steroids and glial cell function. Glia 2006; 54:485–498.
9. Paul SM, Purdy RH. Neuroactive steroids. FASEB J 1992; 6:2311–2322.
10. Hu ZY, Bourreau E, Jung-Testas I, et al. Oligodendrocytes mitochondria convert cholesterol to pregnenolone. Proc Natl Acad Sci USA 1987; 84:8215–8229.

11. Akwa Y, Sananès N, Gouézou M, et al. Astrocytes and neurosteroids: metabolism of pregnenolone and dehydroepiandrosterone. Regulation by cell density. J Cell Biol 1993; 121:135–143.
12. Mensah-Nyagan AG, Do-Rego JL, Beaujean D, et al. Neurosteroids: expression of steroidogenic enzymes and regulation of steroid biosynthesis in the central nervous system. Pharmacol Rev 1999; 51:63–81.
13. Zwain I, Yen SSC. Dehydroepiandrosterone (DHEA): biosynthesis and metabolism in the brain. Endocrinology 1999; 140:880–887.
14. Zwain I, Yen SSC. Neurosteroidogenesis in astrocytes, oligodendrocytes and neurons of cerebral cortex of rat brain. Endocrinology 1999; 140:3843–3852.
15. Mellon SH, Griffin LD, Compagnone NA. Byosynthesis and action of neurosteroids. Brain Res Rev 2001; 37:3–12.
16. Schumacher M, Weill-Engerer S, Liere P, et al. Steroid hormones and neurosteroids in normal and pathological aging of the nervous system. Prog Neurobiol 2003; 71:3–29.
17. Schmidt M, Kreutz M, Loffler G, et al. Conversion of dehydroepiandrosterone to downstream steroid hormones in macrophages. J Endocrinol 2000; 161–169.
18. Patte-Mensah C, Kappes V, Freund-Mercier MJ, et al. Cellular distribution and bioactivity of the key steroidogenic enzyme, cytochrome P450 side-chain cleavage, in sensory neural pathways. J Neurochem 2003; 86:1233–1246.
19. Schumacher M, Guennoun R, Robert F, et al. Local synthesis and dual actions of progesterone in the nervous system: neuroprotection and myelination. Growth Horm IGF Res 2004; 14:518–533.
20. Guennoun R, Schumacher M, Robert F, et al. Neurosteroids: expression of funtional 3β-hydroxysteroid dehydrogenase by rat sensory neurons and Schwann cells. J Neurosci 1997; 9:2236–2247.
21. Morfin R, Young J, Corpéchot C, et al. Neurosteroids: pregnenolone in human sciatic nerves. Proc Natl Acad Sci USA 1992; 89:6790–6793.
22. Papadopoulos V, Baraldi M, Guilarte TR, et al. Translocator protein (18 kDa): new nomenclature for the peripheral-type benzodiazepine receptor based on its structure and molecular function. Trends Pharmacol Sci 2006; 27:402–409.
23. Lavaque E, Sierra A, Azcoitia I, et al. Steroidogenic acute regulatory protein in the brain. Neuroscience. 2006; 138:741–747.
24. Cornu P, Benavides J, Scatton B, et al. Increase in omega 3 (peripheral-type benzodiazepine) binding site densities in different types of human brain tumors. A quantitative autoradiography study. Acta Neurochir 1992; 119:146–152.
25. Miettinen H, Kononen J, Haapasalo H, et al. Expression of peripheral-type benzodiazepine receptor and diazepam binding inhibitor in human astrocytomas: relationship to cell proliferation. Cancer Res 1995; 15:2691–2695.
26. Owen F, Poulter M, Waddington JL, et al. [3H]R05–4864 and [3H]flunitrazepam binding in kainate-lesioned rat striatum and in temporal cortex of brains from patients with senile dementia of the Alzheimer type. Brain Res 1983; 278:373–375.
27. McGeer EG, Singh EA, McGeer PL. Peripheral-type benzodiazepine binding in Alzheimer disease. Alzheim Dis Assoc Disord 1988; 2:331–336.
28. Diorio D, Welner SA, Butterworth RF, et al. Peripheral benzodiazepine binding sites in Alzheimer's disease frontal and temporal cortex. Neurobiol Aging 1991; 12:255–258.
29. Wilms H, Claasen J, Rohl C, et al. Involvement of benzodiazepine receptors in neuroinflammatory and neurodegenerative diseases: evidence from activated microglial cells in vitro. Neurobiol Dis 2003; 14:417–424.
30. Chen MK, Baidoo K, Verina T, et al. Peripheral benzodiazepine receptor imaging in CNS demyelination: functional implications of anatomical and cellular localization. Brain 2004; 127:1379–1392.
31. Kuhlmann AC, Guilarte TR. The peripheral benzodiazepine receptor is a sensitive indicator of domoic acid neurotoxicity. Brain Res 1997; 751:281–288.
32. Kuhlmann AC, Guilarte TR. Regional and temporal expression of the peripheral benzodiazepine receptor in MPTP neurotoxicity. Toxicol Sci 1999; 48:107–116.

33. Kuhlmann AC, Guilarte TR. Cellular and subcellular localization of peripheral benzodiazepine receptors after trimethyltin neurotoxicity. J Neurochem 2000; 74:1694–1704.
34. Kassiou M, Meikle SR, Banati RB. Ligands for peripheral benzodiazepine binding sites in glial cells. Brain Res Brain Res Rev 2005; 48:207–210.
35. Lacor P, Gandolfo P, Tonon MC, et al Regulation of the expression of peripheral benzodiazepine receptors and their endogenous ligands during rat sciatic nerve degeneration and regeneration: a role for PBR in neurosteroidogenesis. Brain Res 1999; 815:70–80.
36. Sierra A, Lavaque E, Perez-Martin M, et al. Steroidogenic acute regulatory protein in the rat brain: cellular distribution, developmental regulation and overexpression after injury. Eur J Neurosci 2003; 18:1458–1467.
37. Benmessahel Y, Troadec JD, Cadepond F, et al. Downregulation of steroidogenic acute regulatory protein (StAR) gene expression by cyclic AMP in cultured Schwann cells. Glia 2004; 45:213–228.
38. Melcangi RC, Magnaghi V, Martini L. Steroid metabolism and effects in central and peripheral glial cells. J Neurobiol 1999; 40:471–483.
39. Melcangi RC, Magnaghi V, Galbiati M, et al. Glial cells: a target for steroid hormones. Prog Brain Res 2001; 132:31–40.
40. Melcangi RC, Magnaghi V, Galbiati M, et al. Formation and effects of neuroactive steroids in the central and peripheral nervous system. Int Rev Neurobiol 2001; 46:145–176.
41. Kiyokage E, Toida K, Suzuki-Yamamoto T, et al. Localization of 5alpha-reductase in the rat main olfactory bulb. J Comp Neurol 2005; 493:381–395.
42. Melcangi RC, Celotti F, Martini L. Progesterone 5α-reduction in neurons, astrocytes and oligodendrocytes. Brain Res 1994; 639:202–206.
43. Belelli D, Lambert, JJ. Neurosteroids: endogenous regulators of the GABA(A) receptor. Nature Rev Neurosci 2005; 6:565–575.
44. He XY, Wegiel J, Yang YZ, et al. Type 10 17beta-hydroxysteroid dehydrogenase catalyzing the oxidation of steroid modulators of aminobutyric acid type A receptors. Mol Cell Endocrinol 2005; 229:111–117.
45. He XY, Wegiel J, Yang SY. Intracellular oxidation of allopregnanolone by human brain type 10 17beta-hydroxysteroid dehydrogenase. Brain Res 2005; 1040:29–35.
46. Gago N, Akwa Y, Sananes N, et al. Progesterone and the oligodendroglial lineage: stage-dependent biosynthesis and metabolism. Glia 2001; 36:295–308.
47. Melcangi RC, Froelichsthal P, Martini L, et al. Steroid metabolizing enzymes in pluripotential progenitor CNS cells: effect of differentiation and maturation. Neuroscience 1996; 72:467–475.
48. Melcangi RC, Magnaghi V, Cavarretta I, et al. Age-induced decrease of glycoprotein P0 and myelin basic protein gene expression in the rat sciatic nerve. Repair by steroid derivatives. Neuroscience 1998; 85:569–578.
49. Melcangi RC, Celotti F, Castano P, et al. Differential localization of the 5α-reductase and the 3α-hydroxysteroid dehydrogenase in neuronal and glial cultures. Endocrinology 1993; 132:1252–1259.
50. Melcangi RC, Celotti F, Martini L. Neurons influence the metabolism of testosterone in cultured astrocytes via humoral signals. Endocrine 1994; 2:709–713.
51. Melcangi RC, Celotti F, Castano P, et al. Intracellular signalling systems controlling the 5α-reductase in glial cell cultures. Brain Res 1992; 585:411–415.
52. Morita K, Arimochi H, Tsuruo Y. Adrenergic activation of 5alpha-reductase gene expression in rat C6 glioma cells: involvement of cyclic AMP/protein kinase A-mediated signaling pathway. J Mol Neurosci 2004; 22:205–212.
53. Negri-Cesi P, Melcangi RC, Celotti F, et al. Aromatase activity in cultured brain cells: difference between neurons and glia. Brain Res 1992; 589:327–332.
54. Schlinger BA, Amur-Umarjee S, Shen P, et al. Neuronal and non-neuronal aromatase in primary cultures of developing zebra finch telencephalon. J Neurosci 1994; 14:7541–7552.
55. Zwain IH, Yen SSC, Cheng CY. Astrocytes cultured in vitro produce estradiol-17β and express aromatase cytochrome P-450 (P-450 AROM) mRNA. Biochim Biophys Acta 1997; 1334:338–348.

56. Azcoitia I, Sierra A, Veiga S, et al. Aromatase expression by reactive astroglia is neuroprotective. Ann NY Acad Sci 2003; 1007:298–305.
57. Garcia-Segura LM, Wozniak A, Azcoitia I, et al. Aromatase expression by astrocytes after brain injury: implications for local estrogen formation in brain repair. Neuroscience 1999; 89:567–578.
58. Peterson RS, Saldanha CJ, Schlinger BA. Rapid upregulation of aromatase mRNA and protein following neural injury in the zebra finch (Taeniopygia guttata). J Neuroendocrinol 2001; 13:317–323.
59. Forlano PM, Deitcher DL, Myers DA, et al. Anatomical distribution and cellular basis for high levels of aromatase activity in the brain of teleost fish: aromatase enzyme and mRNA expression identify glia as source. J Neurosci 2001; 21:8943–8955.
60. Yague JG, Munoz A, de Monasterio-Schrader P, et al. Aromatase expression in the human temporal cortex. Neuroscience 2006; 138:389–401.
61. Forlano PM, Bass AH. Steroid regulation of brain aromatase expression in glia: female preoptic and vocal motor nuclei. J Neurobiol 2005; 65:50–58.
62. Jung-Testas I, Renoir M, Bugnard H, et al. Demonstration of steroid hormone receptors and steroid action in primary cultures of rat glial cells. J Steroid Biochem Mol Biol 1992; 41:621–631.
63. Jung-Testas I, Schumacher M, Robel P, et al. Demonstration of progesterone receptors in rat Schwann cells. J Steroid Biochem Mol Biol 1996; 58:77–82.
64. Langub MC, Watson RE Jr. Estrogen receptor-immunoreactive glia, endothelia, and ependyma in guinea pig preoptic area and median eminence: electron microscopy. Endocrinology 1992; 130:364–372.
65. Wolff, JE, Laterra, J, Goldstein, GW. Steroid inhibition of neural microvessel morphogenesis in vitro: receptor mediation and astroglial dependence. J Neurochem 1992; 58:1023–1032.
66. Magnaghi V, Cavarretta I, Zucchi I, et al. P0 gene expression is modulated by androgens in the sciatic nerve of adult male rats. Mol Brain Res 1999; 70:36–44.
67. Groyer G, Eychenne B, Girard C, et al. Expression and functional state of the corticosteroid receptors and 11 beta-hydroxysteroid dehydrogenase type 2 in Schwann cells. Endocrinology 2006; 147:4339–4350.
68. Bovolin, P, Santi, MR, Puia, G, et al. Expression patterns of γ-aminobutyric acid type A receptor subunit mRNAs in primary cultures of granule neurons and astrocytes from neonatal rat cerebella. Proc Natl Acad Sci USA 1992; 89:9344–9348.
69. Hosli, E, Otten, U, Hosli, L. Expression of GABAA receptors by reactive astrocytes in explant and primary cultures of rat CNS. Int J Dev Neurosci 1997; 15:949–960.
70. Melcangi RC, Magnaghi V, Cavarretta I, et al. Progesterone derivatives are able to influence peripheral myelin protein 22 and P0 gene expression: possible mechanisms of action. J Neurosci Res 1999; 56:349–357.
71. Verkhratsky A, Steinhauser C. Ion channels in glial cells. Brain Res Rev 2000; 32:380–412.
72. Magnaghi V, Ballabio M, Cavarretta ITR, et al. $GABA_B$ receptors in Schwann cells influence proliferation and myelin protein expression. Eur J Neurosci 2004; 19:2641–2649.
73. Palacios G, Muro A, Verdu E, et al. Immunohistochemical localization of the sigma (1) receptor in Schwann cells of rat sciatic nerve. Brain Res. 2004; 1007:65–70.
74. Angehagen M, Ronnback L, Hansson E, et al. Topiramate reduces AMPA-induced Ca(2+) transients and inhibits GluR1 subunit phosphorylation in astrocytes from primary cultures. J Neurochem 2005; 94:1124–1130.
75. Melcangi RC, Cavarretta IT, Ballabio M, et al. Peripheral nerves: a target for the action of neuroactive steroids. Brain Res Rev 2005; 48:328–338.
76. Lalo U, Pankratov Y, Kirchhoff T, et al. NMDA receptors mediate neuron-to-glia signaling in mouse cortical astrocytes. J Neurosci 2006; 26:2673–2683.
77. Svenningsen AF, Kanje M. Estrogen and progesterone stimulate Schwann cell proliferation in a sex- and age-dependent manner. J Neurosci Res 1999; 57:124–130.
78. Marin-Husstege M, Muggirioni M, Raban D, et al. Oligodendrocyte progenitor proliferation and maturation is differentially regulated by male and female sex steroid hormones. Dev Neurosci 2004; 26:245–254.

79. Ghoumari AM, Baulieu EE, Schumacher M. Progesterone increases oligodendroglial cell proliferation in rat cerebellar slice cultures. Neuroscience 2005; 135:47–58.
80. Garcia-Estrada J, Del Rio JA, Luquin S, et al. Gonadal hormones down-regulate reactive gliosis and astrocyte proliferation after a penetrating brain injury. Brain Res 1993; 628:271–278.
81. Zhang L, Li B, Zhao W, et al. Sex-related differences in MAPKs activation in rat astrocytes: effects of estrogen on cell death. Brain Res Mol Brain Res 2002; 103:1–11.
82. Alonso G. Prolonged corticosterone treatment of adult rats inhibits the proliferation of oligodendrocyte progenitors present throughout white and gray matter regions of the brain. Glia 2000; 31:219–231.
83. Désarnaud F, Do Thi AN, Brown AM, et al. Progesterone stimulates the activity of the promoters of peripheral myelin protein-22 and protein zero genes in Schwann cells. J Neurochem 1998; 71:1765–1768.
84. Désarnaud F, Bidichandani S, Patel PI, et al. Glucocorticosteroids stimulate the activity of the promoters of peripheral myelin protein-22 and protein zero genes in Schwann cells. Brain Res 2000; 865:12–16.
85. Melcangi RC, Magnaghi V, Martini L. Aging in peripheral nerves: regulation of myelin protein genes by steroid hormones. Prog Neurobiol 2000; 60:291–308.
86. Ghoumari AM, Ibanez C, El-Etr M, et al. Progesterone and its metabolites increase myelin basic protein expression in organotypic slice cultures of rat cerebellum. J Neurochem 2003; 86:848–859.
87. Ibanez C, Shields SA, El-Etr M, et al. Steroids and the reversal of age-associated changes in myelination and remyelination. Prog Neurobiol 2003; 71:49–56.
88. Guennoun R, Benmessahel Y, Delespierre B, et al. Progesterone stimulates Krox-20 gene expression in Schwann cells. Mol Brain Res 2001; 90:75–82.
89. Mercier G, Turque N, Schumacher M. Early activation of transcription factor expression in Schwann cells by progesterone. Mol Brain Res 2001, 97:137–148.
90. Magnaghi V, Ballabio M, Roglio I, et al. Progesterone derivatives increase expression of Krox-20 and Sox-10 in rat Schwann cells. J Mol Neurosci 2007; 31:149–157.
91. Luquin S, Naftolin F, Garcia-Segura LM. Natural fluctuation and gonadal hormone regulation of astrocyte immunoreactivity in dentate gyrus. J Neurobiol 1993; 24:913–924.
92. Garcia-Segura LM, Chowen JA, Duenas M, et al. Gonadal steroids as promoters of neuroglial plasticity. Psychoneuroendocrinology 1994; 19:445–453.
93. Garcia-Segura LM, Luquin S, Parducz A, et al. Gonadal hormone regulation of glial fibrillary acidic protein immunoreactivity and glial ultrastructure in the rat neuroendocrine hypothalamus. Glia 1994; 10:59–69.
94. Chowen JA, Busiguina S, Garcia-Segura LM. Sexual dimorphism and sex steroid modulation of glial fibrillary acidic protein (GFAP) mRNA and immunoreactive levels in the rat hypothalamus. Neuroscience 1995; 69:519–532.
95. Jones KJ, Coers S, Storer PD, et al. Androgenic regulation of the central glia response following nerve damage. J Neurobiol 1999; 40:560–573.
96. Jordan CL. Glia as mediators of steroid hormone action on the nervous system: an overview. J Neurobiol 1999; 96; 40:434–445.
97. Mong JA, McCarthy MM. Steroid-induced developmental plasticity in hypothalamic astrocytes: implications for synaptic pattering. J Neurobiol 1999; 40:602–619.
98. Nichols NR. Glial responses to steroids as markers of brain aging. J Neurobiol 1999; 40:585–601.
99. Vardimon L, Ben-Dror I, Avisar N, et al. Glucocorticoid control of glial gene expression. J Neurobiol 1999; 40:513–527.
100. Nichols NR, Osterburg HH, Masters JN, et al. Messenger RNA for glial fibrillary acidic protein is decreased in rat brain following acute and chronic corticosterone treatment. Mol Brain Res 1990; 7:1–7.
101. Nichols NR, Masters JN, Finch CE. Changes in gene expression in hippocampus in response to glucocorticoids and stress. Brain Res Bull 1990; 24:659–662.

102. Tsuneishi S, Takada S, Motoike T, et al. Effects of dexamethasone on the expression of myelin basic protein, proteolipid protein, and glial fibrillary acidic protein genes in developing rat brain. Dev Brain Res 1991; 61:117–123.
103. Laping NJ, Nichols NR, Day JR, et al. Corticosterone differentially regulates the bilateral response of astrocyte mRNA in the hippocampus to entorhinal cortex lesions in male rats. Mol Brain Res 1991; 10:291–297.
104. Laping NJ, Teter B, Nichols NR, Glial fibrillary acidic protein: regulation by hormones, cytokines, and growth factors. Brain Pathology 1994; 1:259–275.
105. Tranque PA, Suarez I, Olmos G, et al. Estradiol-induced redistribution of glial fibrillary acidic protein immunoreactivity in the rat brain. Brain Res 1987; 406:348–351.
106. Day JR, Laping NJ, Lampert-Etchells M, et al. Gonadal steroids regulate the expression of glial fibrillary acidic protein in the adult male rat hippocampus. Neuroscience 1993; 55:435–443.
107. Del Cerro S, Garcia-Estrada J, Garcia-Segura LM. Neurosteroids modulate the reaction of astroglia to high extracellular potassium levels. Glia 1996; 18:293–305.
108. Legrand A, Alonso G. Pregnenolone reverses the age-dependent accumulation of glial fibrillary acidic protein within astrocytes of specific regions of the rat brain. Brain Res 1998; 802:125–133.
109. Stone DJ, Song Y, Anderson CP, et al. Bidirectional transcription regulation of glial fibrillary acidic protein by estradiol *in vivo* and *in vitro*. Endocrinology 1998; 139:3202–3209.
110. Garcia-Estrada J, Luquin S, Fernandez A, et al. Dehydroepiandrosterone, pregnenolone and sex steroids down-regulate reactive astroglia in the male rat brain after a penetrating brain injury. Int J Dev Neurosci 1999; 17:145–151.
111. Ciriza I, Azcoitia I, Garcia-Segura LM. Reduced progesterone metabolites protect rat hippocampal neurones from kainic acid excitotoxicity in vivo. J Neuroendocrinol 2004; 16:58–63.
112. Ciriza I, Carrero, P, Azcoitia I, et al. Selective estrogen receptor modulators protect hippocampal neurons from kainic acid excitotoxicity: differences with the effect of estradiol. J Neurobiol 2004; 61:209–221.
113. Hoyk S, Parducz A, Garcia-Segura LM. Dehydroepiandrosterone regulates astroglia reaction to denervation of olfactory glomeruli. Glia 2004; 48:207–216.
114. Stone DJ, Rozovsky I, Morgan TE, et al. Astrocytes and microglia respond to estrogen with increased apoE mRNA *in vivo* and in vitro. Exp Neurol 1997; 143:313–318.
115. Mydlarski MB, Liberman A, Schipper HM. Estrogen induction of glial heat shock proteins: implications for hypothalamic aging. Neurobiol Aging 1995; 16:977–981.
116. Lu A, Ran RQ, Clark J, et al. 17-beta-estradiol induces heat shock proteins in brain arteries and potentiates ischemic heat shock protein induction in glia and neurons. J Cereb Blood Flow Metab 2002; 22:183–195.
117. Vegeto E, Bonincontro C, Pollio G, et al. Estrogen prevents the lipopolysaccharide-induced inflammatory response in microglia. J Neurosci 2001; 21:1809–1818.
118. Gu F, Hata R, Toku K, et al. Testosterone up-regulates aquaporin-4 expression in cultured astrocytes. J Neurosci Res 2003; 72:709–715.
119. Tomas-Camardiel M, Venero JL, Herrera AJ, et al. Blood-brain barrier disruption highly induces aquaporin-4 mRNA and protein in perivascular and parenchymal astrocytes: protective effect by estradiol treatment in ovariectomized animals. J Neurosci Res 2005; 80:235–246.
120. Pawlak J, Brito V, Kuppers E, et al. Regulation of glutamate transporter GLAST and GLT-1 expression in astrocytes by estrogen. Brain Res Mol Brain Res 2005; 138:1–7.
121. Ma YJ, Berg-von der Emde K, Moholt-Siebert M, et al. Region specific regulation of transforming growth factor alpha (TGFα) gene expression in astrocytes of the neuroendocrine brain. J Neurosci 1994; 14:5644–5651.
122. Galbiati M, Martini L, Melcangi RC. Oestrogens, via transforming growth factor a, modulate basic fibroblast growth factor synthesis in hypothalamic astrocytes: in vitro observations. J Neuroendocrinol 2002; 14:829–835.

123. Buchanan CD, Mahesh VB, et al. Estrogen-astrocyte-luteinizing hormone-releasing hormone signaling: a role for transforming growth factor β1. Biol Reprod 2000; 62:1710–1721.
124. Zwain IH, Arroyo A, Amato P, et al. A role for hypothalamic astrocytes in dehydroepiandrosterone and estradiol regulation of gonadotropin-releasing hormone (GnRH) release by GnRH neurons. Neuroendocrinology 2002; 75:375–383.
125. Sortino MA, Chisari M, Merlo S, et al. Glia mediates the neuroprotective action of estradiol on beta-amyloid-induced neuronal death. Endocrinology 2004; 145:5080–5086.
126. Dhandapani KM, Wade FM, Mahesh VB, et al. Astrocyte-derived transforming growth factor-β mediates the neuroprotective effects of 17β-estradiol: involvement of nonclassical genomic signaling pathways. Endocrinology 2005; 146:2749–2759.
127. Flores C, Salmaso N, Cain S, et al. Ovariectomy of adult rats leads to increased expression of astrocytic basic fibroblast growth factor in the ventral tegmental area and in dopaminergic projection regions of the entorhinal and prefrontal cortex. J Neurosci 1999; 19:8665–8673.
128. Magnaghi V, Riva MA, Cavarretta I, et al. Corticosteroids regulate the gene expression of FGF-1 and FGF-2 in cultured rat astrocytes. J Mol Neurosci 2000; 15:11–18.
129. Garcia-Segura, LM, McCarthy, MM. Minireview: role of glia in neuroendocrine function. Endocrinology 2004; 145:1082–1086.
130. Kipper-Galperin M, Galilly R, Danenberg HD, et al. Dehydroepiandrosterone selectively inhibits production of tumor necrosis factor alpha and interleukin-6 in astrocytes. Int J Dev Neurosci 1999; 17:765–775.
131. Soucy G, Boivin G, Labrie F, et al. Estradiol is required for a proper immune response to bacterial and viral pathogens in the female brain. J Immunol 2005; 174:6391–6398.
132. Hardin-Pouzet H, Giraudon P, Belin MF, et al. Glucocorticoid upregulation of glutamate dehydrogenase gene expression in vitro in astrocytes. Mol Brain Res 1996; 37:324–328.
133. Grenier J, Trousson A, Chauchereau A, et al. Differential recruitment of p160 coactivators by glucocorticoid receptor between Schwann cells and astrocytes. Mol Endocrinol 2006; 20:254–267.
134. Griffin LD, Gong W, Verot L, et al. Niemann-pick type C disease involves disrupted neurosteroidogenesis and responds to allopregnanolone. Nat. Med. 2004; 10:704–711.
135. Dudas B, Hanin I, Rose M, et al. Protection against inflammatory neurodegeneration and glial cell death by 7beta-hydroxy epiandrosterone, a novel neurosteroid. Neurobiol Dis 2004; 15:262–268.
136. Morale MC, Serra PA, L'Episcopo F, et al. Estrogen, neuroinflammation and neuroprotection in Parkinson's disease: glia dictates resistance versus vulnerability to neurodegeneration. Neuroscience 2006; 138:869–878.
137. Wise PM, Dubal DB, Rau SW, et al. Are estrogens protective or risk factors in brain injury and neurodegeneration? Reevaluation after the Women's health initiative. Endocr Rev 2005; 26:308–312.
138. Fiore C, Inman DM, Hirose S, et al. Treatment with the neurosteroid dehydroepiandrosterone promotes recovery of motor behavior after moderate contunsive spinal cord injury in the mouse. J Neurosci Res 2004; 75:391–400.
139. Gonzalez SL, Labombarda F, Gonzalez Deniselle MC, et al. Progesterone up-regulates neuronal brain-derived neurotrophic factor expression in the injured spinal cord. Neuroscience 2004; 125:605–614.
140. Sereda MW, Meyer zu Horste G, Suter U, et al. Therapeutic administration of progesterone antagonist in a model of Charcot-Marie-Tooth disease (CMT-1A). Nature Neurosci 2003; 9:1533–1537.
141. Leonelli E, Bianchi R, Cavaletti G, et al. Progesterone and its derivatives are neuroprotective agents in experimental diabetic neuropathy: a multimodal analysis. Neuroscience 2007; 144:1293–1304.
142. Roglio I, Bianchi R, Giatti S, et al. Testosterone derivatives are neuroprotective agents in experimental diabetic neuropathy. Cell Mol Life Sci 2007; 64:1158–1168.

143. Leonelli E, Yague JG, Ballabio M, et al. Ro5-4864, a synthetic ligand of peripheral benzodiazepine receptor, reduces aging-associated myelin degeneration in the sciatic nerve of male rats. Mech Ageing Dev 2005; 126:1159–1163.
144. Veiga S, Azcoitia I, Garcia-Segura LM. Ro5-4864, a peripheral benzodiazepine receptor ligand, reduces reactive gliosis and protects hippocampal hilar neurons from kainic acid excitotoxicity. J Neurosci Res 2005; 80:129–137.
145. Garcia-Segura LM, Veiga S, Sierra A, et al. Aromatase: a neuroprotective enzyme. Prog Neurobiol. 2003; 71:31–41.

Chapter 3
Involvement of Neuroactive Steroids in Hippocampal Disorders: Lessons from Animal Models

Alejandro F. De Nicola[1,2], Luciana Pietranera[1,2], Juan Beauquis[1], Françoise Homo-Delarche[3], and Flavia E. Saravia[1,2]

Abstract Steroids secreted from peripheral endocrine glands and acting on the brain are called neuroactive steroids. Under physiological conditions, neuroactive steroids modulate multiple brain functions, and in aging, trauma or neurodegeneration their role varies from neuroprotective to neurotoxic depending on the chemical properties of the steroid. In this regard, it is known that excess levels of adrenal steroids (gluco and mineralocorticoids) sensitize the hippocampus to the deleterious effects of a pathological environment, whereas the sex hormone estradiol is a powerful hippocampal neuroprotectant. We studied the protective role of estrogens in the ailing hippocampus in animal models of aging and age-associated diseases such as diabetes mellitus and hypertension of genetic (SHR) or mineralocorticoid origin. These models present in common a glucocorticoid or mineralocorticoid overdrive, a process that exacerbates hippocampal neuropathology, according to: (1) decreased proliferation of neuronal progenitors in the subgranular zone (SGZ) of the dentate gyrus (DG); (2) astroglial reactivity, with increased expression of the glial fibrillary acidic protein (GFAP); and (3) decreased neuronal density in the hilus of the DG. These pathological changes were reversible by treatment with estrogens, which ameliorated the hippocampal parameters in middle age mice and models of diabetes and hypertension. Thus, estradiol treatment stimulated progenitor proliferation in the SGZ, normalized the density of GFAP+ astrocytes, and avoided the loss of hilar neurons. Given the important role of the hippocampus in learning, memory and neuroendocrine events, estrogens may offer therapeutic advantages for the treatment of brain disturbances accompanying systemic diseases and aging.

Keywords Aging, dentate gyrus, diabetes mellitus, estrogens, glial fibrillary acidic protein, hippocampus, hypertension, neurogenesis, neuroprotection

---

[1] Laboratory of Neuroendocrine Biochemistry, Instituto de Biologia y Medicina Experimental, Buenos Aires, Argentina

[2] Department of Human Biochemistry, Faculty of Medicine, University of Buenos Aires, Buenos Aires, Argentina

[3] CNRS UMR 7059, Université Paris 7/D, Paris-Diderot, France

M.S. Ritsner and A. Weizman (eds.), *Neuroactive Steroids in Brain Function, Behavior and Neuropsychiatric Disorders*

Abbreviations ACTH adrenocorticotrophin; ANOVA analysis of variance; AVP arginine vasopressin; Bcl_2 β-cell leukemia/lymphoma 2; BrdU bromodeoxyuridine; CHOL cholesterol; CRH corticotrophin-releasing hormone; DCX doublecortin; DG dentate gyrus; DOC deoxycorticosterone; DOCA deoxycorticosterone acetate; E_2 17β-estradiol; ER estrogen receptor; GFAP glial fibrillary acidic protein; GLC granule cell layer; GR glucocorticoid receptor; HPA hypothalamic pituitary adrenal axis; IGF-1 insulin-like growth factor 1; MA middle age; MR mineralocorticoid receptor; NADPH nicotinamide adenine dinucleotide phosphate reduced form; NMDA *n*-methyl-D-aspartic acid; NOD no-obese diabetic mice; POMC propiomelanocortin; PTSD post-traumatic stress disorder; SGZ subgranular zone; SHR spontaneously hypertensive rat; Src sarcoma virus; STZ streptozotocin; SVZ subventricular zone; T_1D type 1 diabetes; WH Woman Health Initiative

3.1 Introduction

3.1.1 Hippocampal Pathology of Aging and Aging-Associated Diseases

Aging is accompanied by pathological changes that preferentially target the hippocampus. Thus, deficits in learning, memory and neurogenesis, changes of neurotransmission, ion channels and electrical activity, altered expression of neuropeptides, growth factors and their receptors, anomalous expression of genes and transcription factors, changes of steroid receptors, vasculopathy, high nitrergic activity, increased oxidative stress, neuronal loss and atrophy, astrogliosis and demyelination account for most abnormalities of the aging hippocampus.[1–4] As stated by several authors,[5–8] a typical feature of aging is the reduction of hippocampal neurogenesis. In the adult, this process is restricted to the subventricular zone (SVZ) and the subgranular zone (SGZ) of the dentate gyrus (DG). Progenitors in the DG proliferate, migrate into the granular cell layer (GCL) and differentiate into mature granule cells.[5,6] Functionally, neurogenesis is associated with learning and memory and acquisition of a fear-conditioned response.[9,10] Since newly-formed neurons rapidly extend their projections to the CA3 pyramidal subfield of the hippocampus, changes of neurogenesis extend beyond the DG, according to Kemperman et al.[7] Thus, the deficient neurogenesis characteristic of aging may upset the whole hippocampal function.

Brain disturbances are also present in age-related diseases such as diabetes mellitus. The various behavioral, neuroendocrine, and neurophysiological abnormalities appearing in uncontrolled type 1 diabetes mellitus (T1D) support the concept of a diabetic encephalopathy.[11–15] At the regional level, diabetes alters, among other structures, glial cells and neurons of the hippocampus. At the neuronal level, hippocampal neurons in T1D are indeed highly vulnerable, as stated by Reagan et al.[16]

Thus, dendritic atrophy, down-regulation of glucocorticoid receptors, altered expression of insulin-growth factor-I (IGF-I) receptors, decreased glucose transporters and susceptibility to apoptosis are present in hippocampal neurons of diabetic animals.[15–19] Moreover, streptozotocin (STZ)-induced TD1 in mice present increased expression of early genes in the pyramidal cell layer and DG, and hyperactivity of NADPH-diaphorase/nitric oxide synthase in the CA3 pyramidal subfield. At the glial cell level, astrogliosis, with increased expression of the glial fibrillary acidic protein (GFAP) is a prominent hippocampal feature of spontaneous and induced models of T1D, such as the non-obese diabetic mice (NOD), and in STZ diabetic rat, respectively.[20,21] In animals with T1D, the increased production of nitric oxide together with the beneficial effects of antioxidants suggests that oxidative stress may play a role in the abnormalities of diabetic encephalopathy.[22,23]

Hypertension, as an age-associated disease, bears a marked neuropathology. Mechanisms responsible for brain changes associated to hypertension include vasculopathy caused by remodeling of the cerebral vessels, vasoconstriction and ischemia.[24] A common model to study brain disturbances of essential hypertension is the spontaneous hypertensive rat (SHR). Substantial evidences for changes of hippocampal structure, neurochemistry and behaviour exist in SHR, compared to their normotensive controls, the Wistar-Kyoto rat (WKY). Hippocampal changes include a reduction of grey matter volume, cytoskeletal breakdown, blood–brain barrier dysfunction, neuronal apoptosis and astrocyte reactivity.[25–28] In another form of experimental hypertension, employing the mineralocorticoid DOCA (deoxycorticosterone acetate)/salt model, brain abnormalities have been mainly described in extrahippocampal regions, such as the hypothalamus and structures of the periventricular and anterior hypothalamic region.[29,30] In contrast to the SHR, there is minimum information on hippocampal involvement in DOCA-treated rats. However, since factors contributing to blood pressure increase may be common to SHR and DOCA/salt hypertension, hippocampal changes are likely to be similar in both models. That a mineralocorticoid overdrive may be operative in these models is suggested from observations that mineralocorticoids play a role in some cases previously diagnosed as essential hypertension.[25]

3.1.2 Adrenal Steroid Overdrive in Aging and Age-Associated Diseases

There are numerous situations commonly associated with disturbances of the hypothalamic–pituitary–adrenal (HPA) axis which have negative consequences upon the hippocampus. In addition to Cushing's syndrome, HPA axis dysfunction is a common disturbance of stress-related diseases, aging, diabetes mellitus, depression, post-traumatic stress disorder (PTSD) and neurodegeneration of the Alzheimer's type. These diseases, with the exception of PTSD, present increased levels of circulating glucocorticoids, a sustained response to stress, flattening of the diurnal rhythm of plasma corticosteroids and ACTH, decreased glucocorticoid negative feedback and hippocampal neuropathology.[31,32] Signs of hippocampal vulnerability include glucocorticoid receptor (GR) down-regulation, atrophy of apical dendrites of pyramidal neurons in the CA3 region, astrogliosis, reduced neurogenesis in the DG, neuronal loss and reduced synthesis of brain-derived neurotrophic factor (BDNF).[33,34] Inappropriate activation of GR by high levels of circulating glucocorticoids induces receptor down-regulation, because GR acts as a natural repressor at the promoter of the GR gene. This event contributes to further dysfunction of the HPA axis.[32]

As already pointed out by the work of Cameron and McKay,[5] Heine et al.,[6] Kemperman et al.[7] and Kuhn et al.,[8] proliferation and migration of newborn cells in the DG is strongly reduced in aging animals. In part, this reduction may be exacerbated by a dysfunction of the HPA axis.[2,35–37] Dysfunction (overdrive) of this system increases the secretion of adrenal steroids, which negatively impact on hippocampal integrity and on cell proliferation and differentiation in the DG.[4,38] Progenitors in the DG express GR and mineralocorticoid receptors (MR).[38] Therefore, hippocampal neurogenesis and HPA axis activity are inversely correlated, suggesting that the low hippocampal neurogenesis of senescent animals may be an index of glucocorticoid-mediated neurotoxicity, as stated by Cameron and McKay[5] and Kemperman et al.[7] Indeed, adrenalectomy of adult or senescent rats stimulates cell proliferation in the DG more so than in young animals.[5]

It is important to consider the factors intrinsic to the glucocorticoid overdrive of aging and diabetes mellitus. For example, in aging rodents, basal levels of circulating glucocorticoids are normal or higher than normal but stress-induced levels of glucocorticoids persist longer than normal.[32,37] The amount of corticosteroid receptors present in the hippocampus of old rats is controversial, although most authors, including ourselves, have reported decreased hippocampal GR and MR, reviewed in Ferrini et al.[2] In cases of GR deficiency, it is hypothesized that diminished steroid feedback enhances the deleterious effects of excess levels of circulating glucocorticoids. This possibility is based on evidence that hippocampal neurons are highly vulnerable to increased glucocorticoid action.[31,32,36] Hippocampal damage releases the inhibitory gabaergic tone exerted upon the hypothalamus, resulting in increased production of hypothalamic neuropeptides. For example, hippocampectomized animals show increased synthesis of corticotrophin-releasing hormone (CRH) and arginine vasopressin (AVP) mRNA and decreased sensitivity to steroid feedback.[39,40] The effect of hippocampectomy is reproduced in other animal models. For example, GR heterozygous mutant mice (GR+/−) show a disinhibited HPA system and a pathological dexamethasone inhibition test.[41]

Although to a lesser extent than hippocampectomy, GR deficiency leads to similar alterations in aging and experimental diabetes mellitus. Typical findings in rodents with T1D induced by STZ or in genetic forms of the disease, include elevated levels of circulating corticosterone, hyperresponse to stress, flattening of circadian rhythms, increased synthesis of hypothalamic neuropeptides, down-regulation of GR and lack of the inhibitory feedback effect of glucocorticoids.[15,20,33] Both aging animals and STZ-induced T1D are useful models to analyze plastic changes of the HPA axis, mainly the possibility that changes are not permanent but reversible by factors modifying the function of this axis.

It is also important to consider that not only glucocorticoid, but also mineralocorticoid overdrive has negative consequences for the brain. Mineralocorticoid actions in the brain include modulation of memory and learning processes, maintenance of ionic balance and membrane excitability, regulation of cardiovascular functions and fluid ingestion, interaction with neuropeptide and neurotransmitters systems as well as a fine control of the HPA axis.[42–45] Part of mineralocorticoid effects are subsequent to hormone binding to MR expressed in anatomically defined regions of the brain. However, non-genomic actions due to direct membrane effects of parent hormones or their reduced metabolites have also been described, particularly with tetrahydro derivatives of the mineralocorticoid deoxycorticosterone (DOC) which bind to the GABA/benzodiazepine receptor.[47,48] It has been suggested that normal activation of MR is neuroprotective, whereas its pathological activation leads to brain damage. In the cardiovascular system, MR has been labeled as a "death receptor" because its hyperactivation leads to a pathological reaction.[49] It remains to be explored if brain MR can be converted into a damaging molecule.

When given to experimental animals at high doses, mineralocorticoids are able to induce hypertension of central origin. In normal rats, van de Berg et al.[50] have shown that activation of central MR but not GR increases systolic blood pressure and that daily warming and stress imposes mineralocorticoid dependency of the pressor response. Adrenal steroids have also a role in SHR, which present abnormalities in the blood pressure response to hormones. For instance, it is known that adrenal-derived corticosterone is essential for the development of hypertension in young SHR, while alterations in corticosteroid negative feedback occur in adult animals.[51,52] Mineralocorticoids appear to participate in the maintenance of hypertension of SHR. In this regard, administration of the MR antagonist RU28318 to SHR induces a long-lasting reduction of blood pressure, whereas infusion of aldosterone increases it in adrenalectomized SHR but not WKY rats.[53,54] Receptor studies have shown increased hippocampal and hypothalamic MR binding capacity in SHR compared to WYK, enhanced activation of MR in heart and kidney and increased heart MR mRNA content in the stroke-prone SHR.[55–57] We observed hypersensitivity of the vasopressin system to high dose mineralocorticoid administration to SHR, suggesting a mineralocorticoid overdrive in these animals.[58] Thus, different experimental paradigms suggest that disturbances of the response to stress, high sensitivity to mineralocorticoids and abnormalities of central and peripheral MR, components of the mineralocorticoid overdrive, contribute to hippocampal neuropathology in SHR and DOCA-salt hypertension.

3.1.3 Neurochemical Basis of Estrogen Neuroprotection in the Hippocampus

In contrast to the neurotoxicity caused by a mineralo or glucocorticoid overdrive, estrogens qualify as "neuroprotectans", as shown by many workers in the field.[59–62] Thus, estrogens prevent cell death, increase neuronal survival and neurite outgrowth, stimulate synaptogenesis and regulate neurotransmission in various experimental situations. McEwen and colleagues first reported that estrogens increase dendritic spine formation and synaptic density in CA1 pyramidal cells, an effect probably mediated by estrogen receptors (ER).[59] In hippocampal neurons in culture, estrogens protect against glutamate toxicity, glucose deprivation, FeSO toxicity and amyloid-β peptide toxicity, the hallmark of Alzheimer's disease, reviewed in Goodman et al.[60] According to Garcia Segura et al.,[63] estradiol interaction with IGF1 is an important event leading to neuroprotection. Some of the estrogen effects could be genomically mediated, after interaction of ligand with ER. Estrogen binding has been reported in the hippocampal pyramidal cells and the hilus of the DG. Of the two isoforms of the estrogen receptor, ERα and ERβ, the β isoform is abundantly expressed in hippocampus, whereas ERα is found in CA1 interneurons and a subset of pyramidal and granule cells.[64]

Estrogens play an important role in the hippocampus, related to limbic-associated functions and learning and memory. One way estrogens modulate these processes is by enhancement of neurogenesis – the proliferation, migration and differentiation of new neurons - in the DG of adult animals.[7,65] In this region, and in the subventricular zone, neurogenesis continues into adulthood.[5] For instance, uptake of the thymidine analog bromodeoxyuridine (BrdU) by proliferating cells of the DG is higher in adult proestrus than in estrus rats, suggesting the participation of endogenous hormones, whereas BrdU+ cells are more abundant in ovariectomized-estradiol (E_2) replaced rats than in ovariectomized-vehicle treated rats.[66] The increase in proliferation is transient, and diminishes in animals subjected to prolonged ovariectomy or chronically overloaded with estrogens.[67,68] A gender difference has been suggested, because in males the response of hippocampal neurogenesis to E_2 is attenuated. In our experience, a ceiling effect probably accounts for the failure of E_2 to increase granule cell proliferation in normal male mice. The effect on neurogenesis seems ER-mediated, since both ER α and ERβ mRNA are found in 80% of proliferating cells of the DG labeled with the Ki67 antibody, and in an important proportion of cells showing a more mature phenotype.[69] A sign of estrogen neuroprotection may be the positive control of neurogenesis. This possibility receives support from published data showing that E_2 enhances neurogenesis under situations presenting a deficient cell proliferation, such as ischemic stroke,[62] diabetes mellitus and aging.[70–72]

It is interesting to consider that generation of new neurons in the DG under the influence of estrogens may be also important for maintenance of hippocampal function and for adequate control of the HPA axis.[73] Thus, in addition to being neuroprotective, estrogens behave as glucocorticoid antagonists, in the sense that under pathological circumstances, estrogens and adrenal corticosteroids may exert opposite effects in the brain. Previous work demonstrated that adrenal status is highly dependent on estrogens. In rodents, there are sex differences in circulating glucocorticoid levels as well as in adrenal size, which is higher in female rats. Estrogens also increase transcortin, the main glucocorticoid transport protein in plasma. Acting upon the anterior pituitary and hypothalamus, estrogens play a facilitatory role on ACTH and corticosterone release during stress. Also, GR are targets of estrogens, as shown by increases in GR mRNA in the amygdala and hypothalamus of estrogen-treated animals.[74] We have previously reported that a prolonged estrogen treatment increases GR binding and immunoreactivity in several brain regions, while it also abolishes GR down-regulation in rats exposed to chronic glucocorticoid treatment.[75,76] Other laboratories have also shown that estrogens and phytoestrogens enhance GR abundance in hippocampus and regulate the HPA axis.[77] Working on the anterior pituitary and the PVN, estrogens down-regulate transcription of the CRH and AVP genes in the PVN and the proopiomelanocortin (POMC) gene in the pituitary corticotroph.[78] Both ER isoforms seem to be involved, since ERβ critically modulates the HPA axis response to stress, whereas ERα may directly influence CRH neurons. As already known, AVP mRNA expression is reduced after DEX treatment, due to glucocorticoid negative feedback on the AVP-producing cells. Because estrogen treatment enhances the DEX effect on AVP mRNA, the sensitizing effect may be due to increased GR levels of estrogen-treated animals.[75] Altogether, our studies raise the possibility that positive modulation of GR by estrogens could reinforce the glucocorticoid feedback mechanism. In human studies, stress-induced glucocorticoid elevations in postmenopausal women are blunted by estrogen replacement.

3.2 Results

3.2.1 Estrogen Effects in the Hippocampus of Middle Age Mice

That estrogens positively control neurogenesis has been mostly informed for young animals,[66, 67] while few studies evaluated a similar role in old animals. In one study[79] 22-month-old ovariectomized rats received prolonged treatment with E_2 valerianate or phytoestrogens from soy bean extract. The authors concluded that cell proliferation in the old brain remains responsive to natural estrogens and phytoestrogens. Although in an indirect manner, a second report demonstrated that infusion of insulin growth factor type I (IGF-1) to 22-month-old rats increases both neurogenesis and blood levels of E_2, suggesting steroid participation in the cell proliferation of this age group.[80] Interestingly, hippocampal neurogenesis starts to decline well before old age. For instance, cell proliferation and migration through the GCL is high in 2-week-old rats, it weakens at 1.5 month of age, and is drastically inhibited in 12-month (middle age) and 24-month (old age) rats.[6] Considering this early decline, studies to elucidate estrogenic effects on the steps leading to neuronal maturation in the DG and on other hippocampal indicators of aging remain an important subject. This issue has clinical relevance, because the Women Health Initiative (WHI) randomized clinical trial claimed that estrogen alone increases the risk of developing mild cognitive impairment,[81] a process considered hippocampal-dependent. The WHI trial has been criticized on the grounds that recruited postmenopausal women several years after cessation of the menstrual period, and at an age span when estrogen responsiveness diminishes.[82] Clinical trials support the effectiveness of hormone-replacement therapy in prevention rather than improvement of mental deterioration.[83,84] Since middle age animals are fully responsive to E_2,[85,86] this age period seems appropriate to counteract the development of age-associated neuropathology. Thus, middle age provides an attractive window of time to explore potential modulation of changes associated with aging, and paradoxically, the literature is not abundant in this period of life.

To fully appreciate hormonal effects in the middle age hippocampus, it seems important to expand the study to other age-sensitive parameters besides neurogenesis. A typical biomarker of the aging brain is the astrocyte hypertrophy, with increased expression of GFAP.[87–89] Estrogens produce a down-regulatory effect on the astrocytosis with high GFAP expression of the brain of very old rats (22–26 months at the time of killing) and young rats receiving castration, traumatic or excitotoxic lesions.[90–92] Another aging marker is the neuronal population of cells in the hilar region, which are markedly lost in old animals.[93] Hilar neurons are extremely vulnerable to excitotoxic or ischemic hippocampal injury.[94] It is unknown if estrogens prevent the hilar neuronal loss during aging, although a rescue effect of acute or chronic E_2 administration follows toxin-induced degeneration of neurons in this region.[93,95]

In order to elucidate if estrogens regulate neurogenesis at middle age, a study was performed in male C57Bl/6 mice divided into a 4-month-old young group and a 10–12 middle age group.[72] In one experiment, we employed a short course of E_2 treatment, in which case a pellet containing 15 µg of E_2 was inserted sc into middle age mice and left in place for 10 days. Mice killed at this time point were used to determine: (a) cell proliferation by measuring incorporation of BrdU into the S phase of the cell cycle; (b) the number of GFAP+ hippocampal astrocytes; and (c) the number of hilar neurons in the DG. A second experiment was designed to study the effects of long E_2 treatment, in which case middle age mice received sc a pellet containing 150 µg E_2 or cholesterol at day 0 and a second 150 µg E_2 pellet or placebo 30 days after the first. In total, mice were exposed to E_2 during 60 days. Mice killed at this time point were used to study: (a) cell proliferation using an antibody against the proliferation marker Ki67, which labels all dividing cells[6]; (b) progenitor survival and migration into the GCL by counting BrdU+ cells labeled 21 days before sacrifice; (c) immunostaining for the neuroblast marker doublecortin (DCX), to label immature neurons which are in the migratory pathway; (d) the number of hippocampal GFAP+ astrocytes; and (e) the number of hilar neurons.

To study cell proliferation in the DG, middle age mice implanted with cholesterol or E_2 pellets during 10 days received BrdU injections 4 and 2 h before sacrifice. Figure 3.1 shows morphological evidence for the pronounced depletion of BrdU+ cells found in the DG of middle age mice (Fig. 3.1A). In contrast, scattered BrdU+ cells were clearly visualized (arrows) in the SGZ of both upper and lower blades of the DG of E_2-treated middle age mice (Fig. 3.1B). Computerized stereology[96] demonstrated that about 250 cells incorporated the thymidine analog in the hippocampus of steroid-naïve middle age mice, whereas the number of BrdU+ cells doubled after E_2 treatment. The SGZ of E_2-treated middle age mice showed clusters of BrdU+ cells suggesting active mitoses; such clusters were never seen in cholesterol-implanted

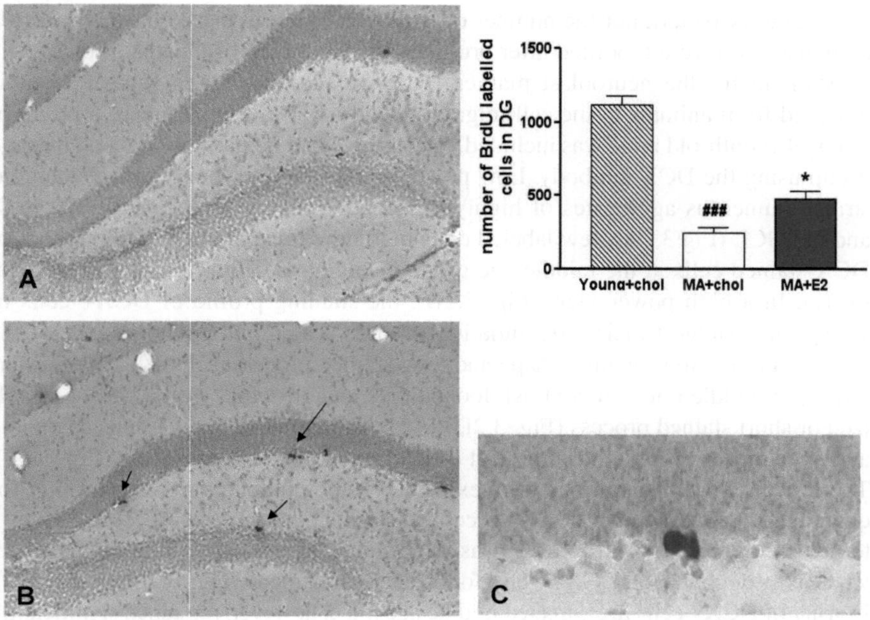

Fig. 3.1 Effects of estradiol on cell proliferation in the hippocampus of middle age (MA) mice. A and B: representative photomicrographs of the dentate gyrus of MA mice receiving cholesterol (A) or short E_2 treatment (B). BrdU+ cells were scarce in a colesterol- treated MA mouse but clearly visible in an E_2-treated MA mouse (arrows in B). C: a mitotic figure in an E_2-treated MA mouse. Quantitative analysis, as shown in the graph, demonstrated a pronounced reduction of BrdU incorporation in untreated MA mice (MA+Chol) compared to young mice (Young + Chol, ### $p<0.001$). After E_2 treatment, the number of BrdU+ cells of MA mice was significantly higher than in cholesterol-treated MA mice ($^*p<0.05$). MA mice treated with E_2 demonstrated a partial recovery respect of young mice. Magnification 40× for A–B, and 400× for C (From Saravia et al.[72]. With permission)

middle age mice. Notwithstanding the significant stimulation of cell proliferation in E_2-treated aging animals, rates for this group were below the ≈1,100 BrdU+ cells per DG reported previously for young mice.[70,71]

In the second experiment, we introduced two variables: (a) E_2 treatment was increased to 150 μg, given at day 0 and again at day 30, for a total exposure of 2 months; and (b) dose of BrdU was raised to 250 mg/kg and given 21 days before killing. BrdU labeling was determined in GCL and in cells in a migratory pathway from the SGZ to the outer third of the GCL. Whereas most BrdU+ cells remained in the SGZ, small quantities distributed along the inner (one third), middle (two thirds) and outer (three thirds) zones of the GCL and in the hilus of the DG. However, no significant differences were obtained between cholesterol-implanted or E_2-implanted middle age mice at any migratory distance considered. A trend existed for the total number of BrdU+ cells being higher in E_2-treated middle mice than in steroid-free mice, but differences were not significant. Therefore, neither the migratory pattern nor the number of BrdU+ cells surviving 21 days after BrdU administration were modified after prolonged E_2 treatment of the aging mice.

Staining for the neuroblast marker DCX was carried out in sets of sections obtained from animals in the cell migration study. For this experiment, a group of young 4-month-old mice was included, because we had no previous data on this age group using the DCX antibody. Low power microscopy of young animals demonstrated numerous aggregates of highly branched DCX+ cells located in the SGZ and the GCL (Fig 3.2A), few labeled cells in middle mice but increased number of DCX-stained cells in the middle age group receiving E_2 (Fig. 3.2C and E, respectively). In a high power view (Fig. 3.2B), the staining profile of DCX+ cells of young mice suggest their differentiation towards a neuronal phenotype, i.e., cells showed a round to triangular shape and long stained processes. DCX+ cells in steroid naïve middle mice, in contrast, looked pale and atrophic, with a hardly visualized or short stained process (Fig. 3.2D). After E_2 treatment, DCX+ cells of middle age mice presented a round soma and a clearly visible neuritic process (Fig. 3.2F). These qualitative observations were expanded using computerized stereology[96] for quantitation of the number of DCX+ cells and the length of their processes. Thus, a tenfold reduction of DCX+ cells was determined in untreated middle age mice, whereas a doubling of DCX+ cells followed E_2 treatment (Fig. 3.2, upper graph). Density of DCX+ cells in E_2-treated middle age mice, however, remained significantly lower than the young group. Concerning the DCX+ cell processes, steroid treatment during 2 months significantly enhanced the length of neurites in middle age mice (Fig. 3.2, bottom graph). The morphological appearance of the DCX + processes in the middle mice exposed to E_2 was closer to the young group, while those corresponding to untreated middle age mice were shorter and convoluted. Thus, these data confirmed that E_2 treatment significantly stimulated cell differentiation, although the morphology and number of DCX+ cells of the hormone-stimulated middle age group was still behind the younger counterpart.

The experiment using the long E_2 treatment also explored if DG cell proliferation remained hormone sensitive after E_2 exposure for 2 months. As the animals were injected with BrdU 21 days before sacrifice, we decided to employ immunocytochemistry for Ki67, a well established biomarker of cell proliferation.[6] Quantitative analysis of the data showed that proliferation rates were much higher in young than in both groups of middle mice. Still, steroid treatment was effective to boost the number of Ki67 positive cells in the DG of middle age mice respect of the steroid-naïve group.

Hilar neuronal number was also modified by E_2 treatment of middle age mice. In untreated middle age mice, fewer cresyl violet-stained neurons were visible in the hilar region compared to the 10-day E_2-treated group. Neuronal density data (per mm^3) confirmed a significant E_2 effect on the number of hilar neurons of middle age mice compared to the absence of hormonal treatment, regardless of the duration of steroid exposure (short or long E_2 treatment). Similar to other parameters described before, middle age groups showed a significant difference respect of the young group. However, the reduction of hilar neuronal density could not be attributable to changes in hilar volume, because measurement of this parameter did not show significant differences between young and middle age mice.

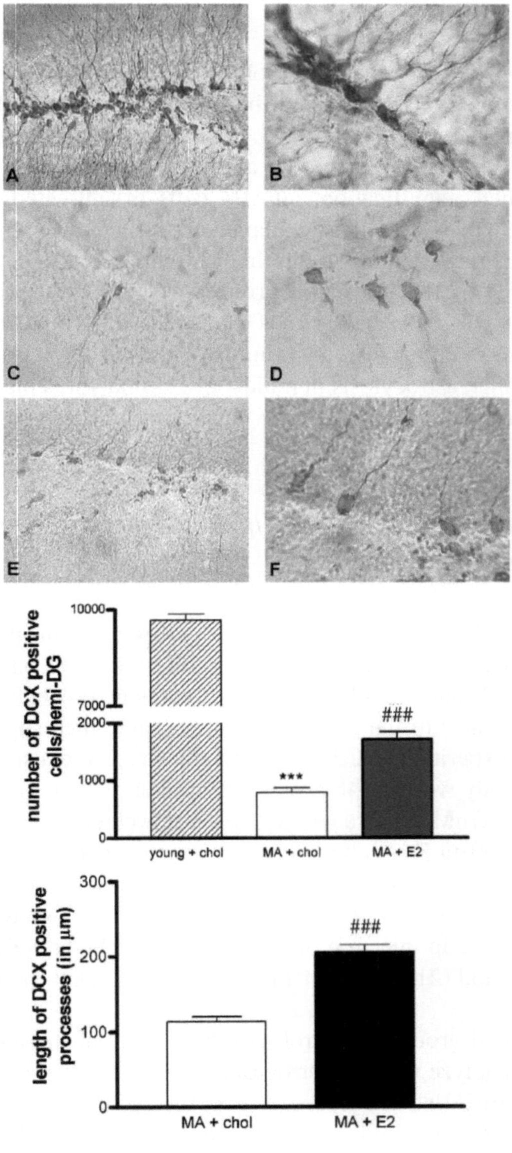

Fig. 3.2 Morphology and quantitative analysis of doublecortin (DCX) + cells (differentiating progenitors) in the dentate gyrus of young mice and middle age (MA) mice with and without E_2 treatment. Numerous DCX+ cells with abundant stained processes characterized the young group, as observed in low (A) and high power (B) views, in sharp contrast with the atrophic appearance of cells of untreated MA mice (C, D). In the E_2-treated MA group, DCX+ cells were more abundant with staining of cell processes (E, F). Magnification in A, C and E: 400×; B, D and F: 1,000×. Upon quantitation, DCX cell density in untreated MA mice (white column in upper graph) was significantly lower than in young group ($^{***}p<0.0001$). E_2-treated MA mice showed increased number of DCX+ cells (dark column in upper graph) and increased length of processes (dark column in lower graph) respect of hormone-free MA mice (### $p<0.0001$ for both cell number and cell processes). Still, DCX+ cell number of E_2-treated MA mice did not reach the level of young mice (stippled column of upper graph) (From Saravia et al.[72]. With permission)

Finally, the response of astrocytes from middle age mice to short and long E_2 treatment was examined in the hilus of the DG. Cholesterol implanted middle age mice showed numerous GFAP+ cells showing a stellate appearance, in contrast to the paucity of GFAP+ cells in middle age mice receiving E_2 treatment during 10 days. At higher magnification, GFAP+ cells of untreated middle age mice showed a more reactive profile with enlarged bodies and deeply stained processes, compared to the E_2-treated aging group. In quantitative analysis, density of GFAP+ cells per unit area of the untreated group was near 2.5-fold higher than the E_2-treated group ($p < 0.02$ and $p < 0.001$ for the short and long E_2 treatment, respectively. Thus, in contrast to the stimulatory effect of E_2 on cell proliferation and hilar cell number, astrocytosis in middle age animals was markedly reduced by hormone treatment, independently of the length of it.

3.2.2 Estrogen Effects in the Hippocampus of Mice with Type I Diabetes

As already discussed, if diabetes mellitus is not properly controlled, it originates an encephalopathy suggesting acceleration of brain aging. We have conclusive evidences that in STZ-induced T1D mice, E_2 treatment prevents the hippocampal abnormalities, similarly to the findings in middle age mice subjected to E_2 treatment. In the T1D experiments, C57Bl/6 male mice received a single ip injection of STZ (200 mg/kg body weight); 48 h after STZ, a marked hyperglycemia developed in treated mice (>15 nM). Ten days after STZ or vehicle injection, a cholesterol pellet containing 200 μg E_2 or cholesterol only was implanted sc under the skin of the neck in the diabetic or controls groups. Ten days after E_2 or vehicle treatment, mice were killed to study cell proliferation in the DG.[70, 71] Two h before killing, all mice received a single ip injection of BrdU at 50 μg/g body weight. It is accepted that at this time period (2 h), BrdU incorporation measures the extent of cell proliferation only.[97]

Four experimental groups – control, control + E_2, diabetic and diabetic + E_2 – were prepared to analyze the number of BrdU-positive cells in the DG. In normoglycemic mice, BrdU labeled cells were observed in the suprapyramidal and infrapyramidal layers at the middle level of the DG (Fig. 3.3, CTRL). Cell counting in both layers of each DG demonstrated numerous BrdU+ cell in control mice, a figure not significantly modified 10 days after E_2 treatment (CTRL + E_2). In contrast, a marked reduction was found in the steroid-naïve diabetic group (Diab), which was restored to control levels after 10 days of E_2 treatment (Diab + E_2). Quantitative analysis is presented in Fig. 3.3, bottom graph. As observed in the photomicrographs (Fig. 3.3), BrdU-labeled cells from vehicle-treated controls, controls plus estrogen and estrogenized diabetic mice showed similar clusters of dark stained nucleus and irregular shape, without morphological differences between untreated and estrogenized groups. The pattern of BrdU staining is representative of immature cells under division as is vastly known from literature reports.

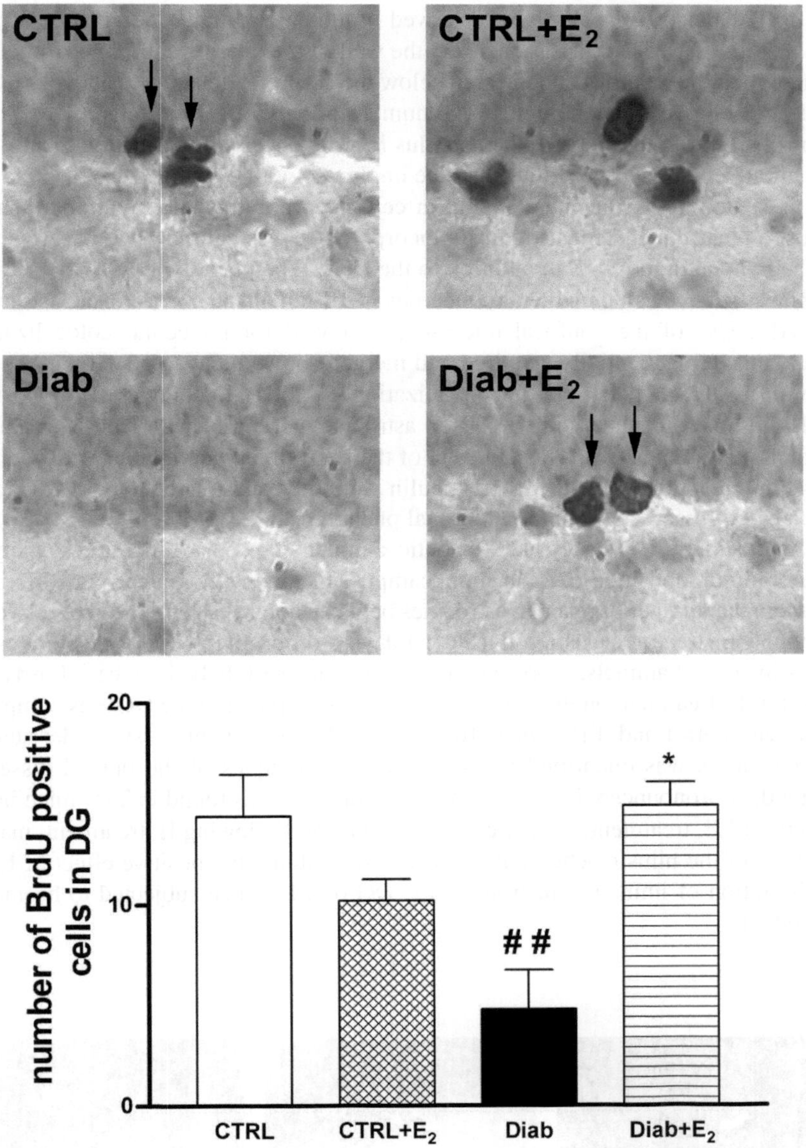

Fig. 3.3 Effects of diabetes and estradiol (E_2) treatment on cell proliferation in the dentate gyrus. Control (CTRL) and diabetic mice (Diab) were implanted with a single pellet containing cholesterol or cholesterol plus 200 μg 17-β E_2 (CTRL + E_2, Diab + E_2). The photomicrograph shows BrdU-positive cells in a CTRL mouse and a CTRL + E_2-treated mouse and a pronounced reduction in a diabetic mouse (Diab). BrdU-positive cells are recovered in a diabetic mouse receiving E_2 (Diab + E_2). Lower graph: Computerized image analysis of the number of BrdU-immunopositive cells in the dentate gyrus. Statistical analysis using ANOVA followed by Bonferroni's post hoc test demonstrated: ## significantly smaller than all other groups ($p < 0.01$) and $^*p < 0.05$ vs Diab group (Modified from Saravia et al.[70]. With permission)

BrdU labeled cells were also observed in the SVZ, another neurogenic area of the mature CNS. These areas enclosed the wall of the lateral ventricle near the fimbria of the hippocampus and the area below the corpus callosum and above the CA1 and CA2 hippocampal subfields. The number of BrdU positive cells in SVZ from control, control plus E_2, and diabetic plus E_2 groups was similar histologically and quantitatively. In contrast, diabetic mice implanted with the cholesterol vehicle pellet presented a significant reduction in cells incorporating the thymidine analog. Thus, a clear enhancement of BrdU incorporation was caused by E_2 treatment of diabetic mice in the SVZ in addition to the CGL. The phenotype of BrdU-positive cells was studied using markers for neurons (β-III-tubulin and Neu-N) and astrocytes (GFAP). Use of the confocal microscope allowed for molecular colocalization. Thus, images sequentially acquired and merged showed that some cells in the DG showed BrdU and β-III-tubulin colocalization (Fig. 3.4). In contrast, BrdU-positive cells did not colocalize with GFAP, an astrocyte marker, or with Neu-N, a marker of more mature neurons. Thus, in spite of the short time interval after BrdU administration, reaction with the β-III-tubulin antibody suggested some BrdU cells already expressed an immature neuronal phenotype.

As stated by Revsin et al.,[20] diabetic animals showed an increase density of GFAP-immunopositive cells in hippocampus. In this case, a clear cut difference exists in the number of reactive astrocytes between control and T1D mice. However, although E_2 treatment lasting during 10 days did not affect GFAP-positive astrocytes in control animals, it prevented the astrocytosis of diabetic mice.[71] Lastly, the effect of E_2 treatment on the number of hilar cells populating the DG was compared between control and T1D mice. To this end, the number of cresyl-violet stained hilar neurons was quantified according to the principles of the optical dissector method. A pronounced decrease in hilar cell number was found in T1D mice in the absence of E_2 treatment, compared to control mice. Following E_2 treatment, mature neurons of the hilus reached normal levels, resembling the positive effect of E_2 on proliferation of immature neurons in the DG of T1D mice subjected to hormonal treatment.

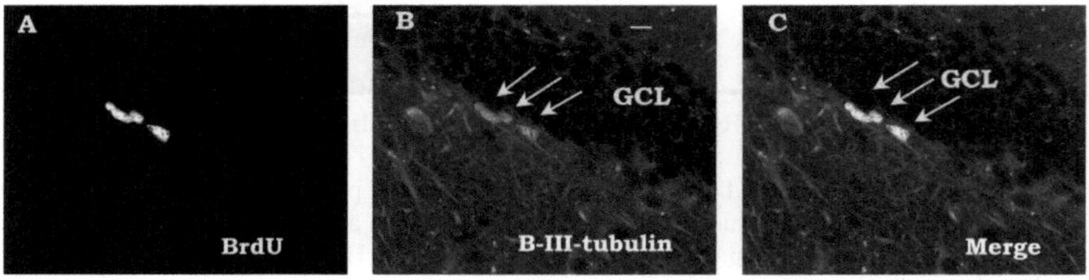

Fig. 3.4 Representative confocal – laser scanning microscopic images corresponding to BrdU-labeled cells in the dentate gyrus from a control mouse (A) and to cells exhibiting an immature neuronal phenotype (β-III-tubulin) (B). Merging showed colocalization of some BrdU-labeled cells with β-III-tubulin (C) indicated by arrows in B and C. Scale bar: 10 µm, GCL: Granular cell layer (From Saravia et al.[71]. With permission)

3.2.3 *Estrogen Effects on the Hippocampus in Hypertensive Rat Models*

In this section, we present data on the effects of E_2 on the hippocampus from rats bearing hypertension of genetic or mineralocorticoid origin. To compare hypertensive animals with previous studies carried out in middle age and T1D mice, we focused the analysis on effects on cell proliferation, astrocytosis and hilar cell number. To study events in genetic hypertension, we employed male 16-week-old normotensive WKY rats and hypertensive SHR showing systolic blood pressure values of 190 ± 5 mm Hg. For the DOCA-salt model, adult Sprague-Dawley rats were divided into four groups, all of which received 1% NaCl as drinking solution for 3 weeks. DOCA + salt treated rats developed hypertension (160 ± 1.2 mm Hg). Rats in group 1 remained untreated (CTL), whereas those of group 2 received the E_2 implant (CTL + E_2, see dose below). Group 3 received 10 mg DOCA/rat in oil every other day for 3 weeks (DOCA), whereas rats in group 4 received DOCA plus the E_2 implant (DOCA + E_2). Half of the animals in both paradigms (i.e., SHR and DOCA) received sc a single pellet of 14 mg E_2 benzoate during the last two weeks of the experiment. For cell proliferation studies, rats received a single ip injection of BrdU at 50 µg/g BW and were killed 2 h later, to measure the extent of cell proliferation (MacMahon and McDermott[97]). To this end, a computerized steorology procedure was used to measure BrdU incorporation throughout the entire DG (Howart and Reed[96]). Separate brain sections were used to determine the density of GFAP + cells using immunocytochemistry and the number of hilar cells after cresyl violet staining.

In control WKY rats, single labeled cells as well as some dividing cells at the middle level of the DG were observed, i.e., a pattern of BrdU staining representative of precursor cells undergoing division. Significantly fewer BrdU-labeled cells were present in the SHR group, the reduction amounting to 50% respect of WKY rats (Fig. 3.5, upper graph). However, E_2 treatment for 2 weeks significantly increased the number of BrdU labeling in SHR but not in WKY rats (Fig. 3.5). In the second series of experiments, BrdU labeling was compared in the DG of four groups of Sprague-Dawley rats: CTL, CTL + E_2, DOCA and DOCA + E_2. Remarkably, the DOCA group showed the lowest number of BrdU + cells, whereas E_2 treatment of this group normalized cell proliferation (Fig. 3.5, lower graph). These results strongly supported a positive effect of E_2 on cell proliferation in both hypertensive rat models.

The second objective of our study was to examine the response of the astrocyte marker GFAP in the *stratum radiatum* below the CA1 and CA3 subfields of the dorsal hippocampus and the hilus of the DG. In the experiment using the SHR/WKY strains, from 120 to 190 GFAP + cells per unit area (mm^2) were measured in the three hippocampal regions of the control WKY rats. In SHR, density of GFAP + cells was twofold higher than WKY for CA1, CA3 and DG.[31] Also, the staining intensity and size of GFAP-labeled cells was appreciably elevated compared to WKY rats. A similar pattern of GFAP expression was found in the DOCA-salt

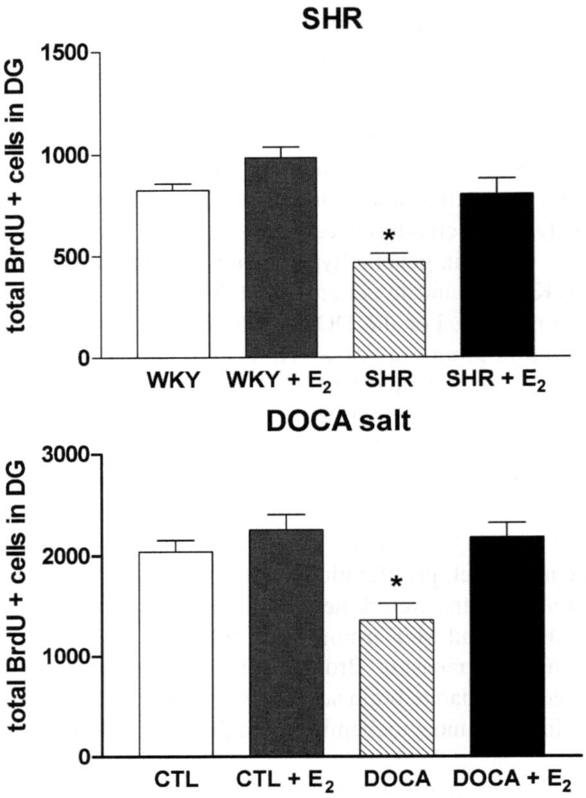

Fig. 3.5 Effects of E_2 on cell proliferation in hypertensive rat models. Upper graph: Number of BrdU + cells in the dentate gyrus of SHR and Wistar-Kyoto rats (WKY) without or with E_2 treatment (WKY + E_2, SHR + E_2). Lower graph: BrdU + cells in control rats (CTL) and deoxycorticosterone acetate (DOCA)-treated rats drinking salt solution, without or with E_2 treatment (CTL + E_2, DOCA + E_2). Estrogen-treated rats received a single E_2 pellet weighing 14 mg under the skin of the neck, which was left in place for 2 weeks. The optical dissector method was employed to determine the number of BrdU + cells in the entire dentate gyrus (Howart and Reed[96]). In both experimental paradigms, untreated hypertensive rats showed the lowest BrdU incorporation ($^*p < 0.01$ vs all other groups in corresponding experiment, by ANOVA followed by Bonferroni's post hoc test). It is observed that E_2 treatment recovered BrdU incorporation both in SHR and DOCA-salt-treated rats (From Pietranera et al.,[30] and unpublished data)

treated rats, in that this group showed a remarkable astrocytosis. Similarly to SHR, GFAP positive cells in DOCA + salt treated animals showed a more reactive profile with enlarged bodies and deeply stained processes, compared to the other groups. E_2 treatment for 2 weeks showed the expected down-regulatory effect on GFAP cell density both in the SHR and DOCA + salt treated rats, whereas the hormone was without effect in normotensive animals of both experimental paradigms (i.e., WKY + E_2, CTL + E_2).

Lastly, to elucidate if changes in the proportion of proliferating cells (immature neurones) and of glial cells were accompanied by changes of mature neurons, the number of cresyl violet-stained hilar neurones was determined. In the SHR/WKY experiment, an approximate 50% reduction in the number of hilar neurons was present in SHR compared to WKY. In the mineralocorticoid-loaded animals, the number of hilar neurones was markedly decreased in the DOCA group respect of CTL, and CTL + E_2 rats, all groups receiving high salt treatment in the drinking water. In both models of hypertension, E_2 treatment restored the number of hilar neurons to the control levels of WKY, for the SHR group, and to normal levels shown by CTL and CTL +E_2 for the DOCA hypertensive group. Therefore, neuronal changes in the hilar region of hypertensive models seemed to be plastic events, sensitive to normalization by treatment with the neuroactive steroid E_2.

3.3 Discussion

3.3.1 Hippocampus, Middle Age and Estradiol

Middle age represent a special period of life useful to detect events in the route to senescence. In this age group, we decided to explore the effectiveness of estrogen treatment to attenuate the hippocampal abnormalities associated with physiological aging. To this end, we approached neuronal parameters, including the different steps of neurogenesis, the number of hilar neurons and one glial cell parameter – GFAP expression by astrocytes – which are reportedly disturbed in senescent animals. The results indicated that out of the three steps of adult neurogenesis (proliferation, migration, and differentiation), E_2 stimulated progenitor proliferation and differentiation. The effects of E_2 on neurogenesis were accompanied by decreased astrocytosis and by increased density of hilar neurons in the DG and GCL.

An important finding was the demonstration that a moderate dose of E_2 given for 10 days doubled cell proliferation in DG, suggesting that this neurogenic zone remains hormone sensitive through middle age. However, estrogen stimulated levels of BrdU incorporation at middle age were still lower than that obtained for young male animals. Therefore, other factors besides steroids may be needed for full recovery of cell proliferation, or recovery of plasticity may be limited in middle age mice. In contrast to the stimulatory effect on cell proliferation, E_2 did not change migration into the GCL. This result was not unexpected, because previous reports suggested that estrogen effects can be transient and do not go beyond the proliferation step[66–68]. In this respect, middle age animals behaved like young animals studied by other investigators.

It is possible that the E_2 effects on cell proliferation at middle age may be of functional significance, because when steroid treatment was prolonged to 2 months, differentiation of cells towards a neuronal lineage was highly stimulated. In our investigation, progenitors were labeled for DCX, a microtubule binding protein transiently expressed by proliferating cells and newly generated neuroblasts.[98] As demonstrated by us and others[72,98,99] the incidence of DCX+ cells was markedly decreased in aging rats and middle age mice. However, E_2 treatment not only increased the density of DCX labeled cells, but also increased staining of the perykarion and neurite processes oriented parallel and into the GCL. From literature reports, it is known that immature granule cells rapidly extend axons into CA3 and become integrated into the hippocampal circuit, where they could influence hippocampal function at early stages of neurogenesis.[7,100] Therefore, after E_2 stimulation of cell differentiation, some missing functions may resume in the hippocampus of middle age mice, since the enlarged DCX+ processes found in steroid-treated middle age mice could reinforce the synaptic interconnections of the hippocampal circuit. Regarding the mechanism for the E_2 effects, it has been demonstrated by Olariu et al.[101] that the reduced cell proliferation in middle age animals is due to loss of granule cell precursors. Therefore, increases of cell proliferation of middle age mice might by due to an expansion of the progenitor precursor pool.

Furthermore, the stimulation of cell proliferation in the DG of middle age mice may be viewed as another example of E_2 neuroprotection. Mechanistically, E_2 effects on the hippocampus might be partly mediated by classical ER. Thus, the presence of mRNAs for the isoforms ERα and ERβ colocalize with markers of cell proliferation and differentiation.[64,69] The block of neurogenesis by administration of an ER antagonist,[79] further indicate a role for the intracellular receptor, although pathways independent of the classical ER have been considered to be part of this event.[102,103] Estrogen also regulate the expression of IGF-I and of brain-derived neurotrophic factor (BDNF) and activate cellular cascades involving IGF-I receptors and BDNF receptors.[79,104,105] These growth factors promote proliferation and neuronal differentiation, and their interaction with E_2 may explain in part the stimulatory effects on neurogenesis. This interaction does not exclude that effects on neurogenesis comprises additional activities of E_2. As it is now accepted, E_2 effects in the brain involves multiple mechanisms, such as prevention of excitotoxicity and inflammation, antioxidant effects, inhibition of apoptosis and stimulation of the anti-apoptotic gene Bcl_2, regulation of cholinergic neurotransmission and dendritic remodeling, among others.[63,102,103,106] Although these effects may be genomically mediated, extranuclear sites of estrogen action involving membrane, synaptic and mitochondrial sites are also likely players. Estrogen effects at extranuclear sites can activate different kinases (phosphatydylinositol 3-kinase, a Ca^{2+} independent protein kinase C isoform, Src kinase, mitogen-activated kinase), regulate Ca^{2+} influx, the ERK 1/2 pathway and glutamate-activated NMDA receptors. Simultaneously, mitochondria play a role in ion homeostasis and cell survival by sequestration of Ca^{2+} and protection in response to E_2 stimulation (reviewed in Morrison et al.[107]). The role of different receptor isoforms besides ERα and ERβ has been discussed by Toran-Allerand.[108] Recently, McEwen and Millner[109] detected extranuclear ERβ immunoreactivity in DCX positive, newly born cells of the DG. Therefore, a relevant endeavour would be to establish a link between nuclear and extranuclear sites with the E_2 stimulation of neurogenesis and other parameters in the brain of aging animals and in age-associated diseases.

Lastly, a role of estrogens on the modulation of the HPA axis may be an additional factor for the enhancement of DG cell proliferation caused by E_2. The hyperactive HPA axis of old animals,[32, 35–37] can be attenuated by E_2 action at the hypothalamic and hippocampal levels.[2,78] Glucocorticoids inhibit hippocampal neurogenesis[4,5,7] and E_2 could enhance neurogenesis by restraining the central stimulation of adrenal glucocorticoid secretion.

An additional neuronal effect resulted from E_2 treatment of middle age mice, associated with the recovery of hilar neurons. Hilar cells are a group of heterogenous neurons, some of them exhibiting a gabaergic phenotype and showing connections with the DG and other hippocampal fields.[110] Their loss is observed after excitotoxic and ischemic injury suggesting an extreme vulnerability to oxidative agents,[93] which are greatly generated during aging. Therefore, we hypothesized that recovery of hilar neurons supports the powerful neuroprotective and antioxidant effects of estrogens in the brain.

It is also feasible to establish a link between the effects of E_2 on neuronal proliferation and differentiation, to reactive astrocytosis. As shown by several authors[87–89] reactive GFAP-expressing astrocytes are a hallmark of a number of brain pathologies, including aging. The E_2 effect on GFAP+ astrocytes in middle age mice was expected on the grounds that this hormone down-regulates the reactive astrocytosis in the brain of old rats, young castrated rats, and animals with traumatic or excitotoxic lesions, a stated by Day et al.,[90] Garcia-Ovejero et al.[91] and Lei et al.[92] These effects may be direct, because astrocytes express ER and an estrogen-response element exists on the 5′ upstream region of the GFAP promoter.[93, 111] The reduced ability to manifest a GFAP reactive phenotype of the astrocytes of E_2-treated middle age mice might prove beneficial for neurogenesis. In this respect, cell proliferation in the granular layer of the DG of 18 month old female mice is 48% higher in a $GFAP^{-/-}$ $Vimentin^{-/-}$ background, suggesting that genetic attenuation of age-associated reactive gliosis stimulates cell proliferation and neurogenesis (Larsson et al.,[112] Pekny and Nilsson[113]). Rozowsky et al.[114] reported that E_2 – even at physiological levels – inhibits GFAP expression in vivo and in vitro in a transcriptionally mediated manner.

A conclusion from this section would be that in middle age mice, from the three basic steps of adult neurogenesis (proliferation, migration, and differentiation), E_2 stimulated progenitor proliferation and differentiation towards a neuronal lineage. This result, in conjunction with recovery from other aging indicators, supports a wide range protection of hippocampal function at middle mice by estrogenic hormones.

3.3.2 Hippocampus, Diabetes and Estradiol

The various alterations of neuronal structure, function and metabolism reported in T1D by several groups,[11–16] were recently shown by us to be accompanied by a pronounced reduction of DG cell proliferation.[70, 71] Additionally, we showed that BrdU incorporation, a measure of cell proliferation in animals killed 2 h after nucleotide administration was diminished in SVZ cells of diabetic mouse brain. Due to the schedule of BrdU administration performed in our experiments with STZ-diabetic mice, i.e., one injection 2h before sacrifice, only some proliferating cells showed colocalization with markers of an immature neuronal phenotype like β-III-tubulin and we were not able to found Neu-N/BrdU or GFAP/BrdU double-labeled cells. Kempermann et al.[7] emphasised in a recent study that cell fate decisions towards neuronal development are made soon after division. In functional terms the impaired neurogenesis in DG may lie beneath the numerous deficits in learning and memory behaviour previously shown in diabetic animals.[13, 115]

Of particular interest, E_2 treatment during 10 days completely restored cell proliferation in DG and SVZ of diabetic mice. It is also worth to point out that E_2 effects were exclusively observed in diabetic but not in normal mice. This finding was not entirely unexpected since it was reported that in normal animals, estrogens did not significantly affect the number of BrdU-immunoreactive cells in rat DG unless the cellular environment is damaged by application of a toxin.[67,116] Diabetic mice also showed abnormal expression of astrocyte markers in hippocampus. Thus, increased number of GFAP+ cells, indicative of astrogliosis, was found in *stratum radiatum* below the CA1 hippocampal subfield of diabetic mice. Both parameters were reverted to normal by the E_2 regime that up-regulated cell proliferation. Furthermore, the stimulation of cell proliferation in DG and SVZ of diabetic mice and the effects on GFAP+ astrocytes and hilar cell number may be viewed as another example of estrogen neuroprotection. We have already discussed the mechanism for E_2 down-regulation of GFAP expression. The reported studies demonstrated that hippocampal neuropathology of uncontrolled diabetes is a reversible condition and sensitive to estrogen treatment. Studies in animal models of TD1 may open up new venues for understanding the beneficial role of steroid hormones in diabetic encephalopathy.

3.3.3 Hippocampus, Hypertension and Estradiol

The results obtained in hypertensive rat models support the hypothesis that the hippocampus is vulnerable to hypertension.[25–29,117] In the current investigation, neurochemical changes of hippocampus were compared in DOCA-salt hypertensive rats, bearing an induced model of hypertension, and in SHR bearing a genetic form of hypertension. Changes in both hypertensive models were remarkably similar, including a decreased cell proliferation in DG, increased number of astrocytes immunopositive for GFAP and a reduced number of hilar neurones in the DG. These similarities suggest that DOCA-salt treated rats and SHR may share etiopathogenic factors. In this way, the effects on neuronal proliferation, hilar neurones and reactive astrocytosis may be linked to vascular damage induced by a mineralocorticoid overdrive. In other words, the pro-inflammatory role of MR could increase cellular vulnerability when combined with ischemic hypertensive injury. A similar mechanism develops for glucocorticoid effects in the hippocampus during hypoxia/ischemia.[118] Notwithstanding the impact of hypertension on these measures, however, adrenocortical hormones of endogenous origin, or excess activity of MR may enhance the hippocampal vulnerability of SHR. SHR are hypersensitive to mineralocorticoids, in that they show enhanced AVP synthesis and c-Fos expression after DOCA treatment (Pietranera et al.[29]). MR also participate in the maintenance of hypertension and vascular lesions of SHR (Rahmouni et al.,[53] Rocha et al.[119]). Interestingly, both MR and GR and highly expressed in the microcirculation of SHR, at least in the mesentery.[120] Based on these premises, it has been hypothesized that SHR bear a form of endocrine hypertension in which activation of MR plays a substantial role.[53,57,119] Needless to say that mineralocorticoids could directly damage the hippocampus of DOCA + salt treated rats, which are hormone overloaded in the context of a high salt intake. Then, salt and high blood pressure may sensitize the hippocampus to the toxic effects of high DOCA treatment, rendering the hippocampal neuropathology of DOCA + salt treated rats a mimic of SHR. In conclusion, abnormalities of glial cells and neurons of the hippocampus are present in animals with genetic or induced forms of hypertension. These similarities suggest common etiopathogenic factors, based on the existence of a mineralocorticoid overdrive. To reinforce this hypothesis, pharmacological modulation of MR and GR using agonists/antagonists is needed to determine if such a modulation correlates with changes of hippocampal neuropathology. A similar experimental approach has been used to identify in different animal models receptor subtypes involved in the central regulation of blood pressure.[50,53,54]

Therefore, the challenge imposed by these data was to establish if changes of glial cells and neurons of SHR and DOCA + salt treated rats were irreversible or plastic events. This problem was solved by treatment with a neuroactive steroid bringing neuroprotection to the hippocampus. In this sense, E_2 played the expected neuroprotective role, supported by previous data in middle age mice and T1D mice, in which estrogenization attenuated the hippocampal neuropathology. In hypertensive rats, 2 weeks of E_2 treatment increased cell proliferation and the number of hilar neurons in DG, at the time that GFAP positive astrocyte density was attenuated. The marked effects of estrogens on hippocampal markers supports that the hormone may play an important role in the susceptibility to hypertensive damage.

3.3.4 Concluding Remarks

It seems relevant now to consider the lessons learned from animal models. The neuroprotective effects of E_2 in the hippocampus of middle age mice, T1D mice and hypertensive rats, encourage the use of estrogens for pharmacological intervention. However, natural estrogens are not free from unwanted side effects in breast and uterus and cannot be used in males. Instead, several estrogen-like neuroactive steroids are available that overcome an undesired "estrogenicity" on peripheral targets. One such compound is 17 α-estradiol, which shows neuroprotective and anti-oxidant effects without binding to ER isoforms.[121] In addition, semisynthetic delta 8-9-dehydro homologs of 17 α-estradiol[122] are non-feminizing estrogens with less effect on uterus and breast, that increase cell proliferation in the DG of ovariectomized mice and show a positive effect on cholineacetyltransferase in brain.[123] These compounds can offer therapeutic advantages to counteract the neuropathology of the hippocampus in aging and age-associated diseases.

Acknowledgements We thank the following institutions for financial support: FONCYT (BID 1728 OC AR PICT 2004 # 25610), the National Research Council of Argentina (CONICET, PIP 5542), and the University of Buenos Aires (M022). The authors deeply acknowledge the assistance of Paulina Roig and Analia Lima for the experiments detailed in this chapter.

References

1. Driscoll I, Sutherland RJ. The aging hippocampus: navigating between rat and human experiments. Rev Neurosci 2005; 16:87–121.
2. Ferrini M, Piroli G, Frontera M, Falbo A, Lima A, De Nicola AF. Estrogens normalize the hypothalamic-pituitary-adrenal axis response to stress and increase glucocorticoid receptor immuno-reactivity in hippocampus of aging male rats. Neuroendocrinology 1999; 69:129–137.
3. McEwen BS. Stress and the aging hippocampus. Front Neuroendocrinol 1999; 20:40–79.
4. Miller DB, O'Callaghan JP Aging, stress and the hippocampus. Ageing Res Rev 2005; 4:123–140.
5. Cameron HA, McKay RD. Restoring production of hippocampal neurons in old age. Nat Neurosci 1999; 2:894–897.
6. Heine VM, Maslam S, Joels M, Lucassen PJ. Prominent decline of newborn cell proliferation, differentiation, and apoptosis in the aging dentate gyrus, in absence of an age-related hypothalamus-pituitary-adrenal axis activation. Neurobiol Aging 2004; 25:361–375.
7. Kempermann G, Gast D, Gage FH. Neuroplasticity in old age: sustained fivefold induction of hippocampal neurogenesis by long-term environmental enrichment. 2000; 52:135–143.
8. Kuhn HG, Dickinson-Anson H, Gage FH. Neurogenesis in the dentate gyrus of the adult rat: age-related decrease of neuronal progenitor proliferation. J Neurosci 1996; 16:2027–2033.
9. Gould E, Tanapat P, Rydel T, Hastings N. Regulation of hippocampal neurogenesis in adulthood. Biol Psychiatry 2000; 48:715–720.
10. Shors TJ, Townsend DA, Zhao M, Kozorovitskiy Y. Neurogenesis may relate to some but not all types of hippocampal-dependent learning. Hippocampus 2002; 12:578–584.
11. Artola A, Kamal A, Ramakers GM, Gardoni F, DiLuca M, Biessels GJ, Cattabeni F, Gispen WH. Synaptic plasticity in the diabetic brain: advanced aging? Prog Brain Res 2002; 138:305–314.
12. Biessels GJ, van der Heide LP, Kamal A, Bleys RL, Gispen WH. Ageing and diabetes: implications for brain function. Eur J Pharmacol 2002; 441:1–14.
13. Gispen WH, Biessels, GJ. Cognition and synaptic plasticity in diabetes mellitus. TINS 2000; 23:542–549.
14. Rowlands NE, Bellush LL. Diabetes mellitus: stress, neurochemistry and behavior. Neurosci Biobehav Rev 1989; 13:199–206.
15. De Nicola AF, Magariños AM, Foglia VG. Neuroendocrine regulation in experimental diabetes (Houssay Lecture). In: Diabetes. Rifkin H, Colwell JA, Taylor SI (eds). Elsevier, Amsterdam, 1991:3–8.
16. Reagan LP, Magariños, AM, McEwen BS. Neurological changes induced by stress in streptozotocin diabetic rats. Ann NY Acad Sci 1999; 893:126–137.
17. Magariños A M, McEwen BS. Experimental diabetes in rats causes hippocampal dendritic and synaptic reorganization and increased glucocorticoid reactivity to stress. PNAS USA 2000; 97:10056–11061.

18. Bestetti G, Rossi GL. Hypothalamic lesions in rats with long-term streptozotocin-induced diabetes mellitus. Acta Neurophatol 1980; 52:119–127.
19. Russel JW, Sullivan KA, Windebank AJ, Herrmann DN, Feldman EL. Neurons undergo apoptosis in animal and cell culture models of diabetes. Neurobiol Dis 1999; 6:347–363.
20. Revsin Y, Saravia F, Roig P, Lima A, de Kloet ER, Homo-Delarche F, De Nicola AF. Neuronal and astroglial alterations in the hippocampus of a mouse model for type 1 diabetes. Brain Res 2005; 1038:22–31.
21. Saravia F, Revsin Y, Gonzalez Deniselle MC, Gonzalez S, Roig P, Lima A, Homo-Delarche F, De Nicola AF. Increased astrocyte reactivity in the hippocampus of murine models of type I diabetes: the nonobese diabetic (NOD) and streptozotocin-treated mice. Brain Res 2002; 957:345–353.
22. McEwen BS, Magariños AM, Reagan LP. Studies of hormone action in the hippocampal formation. Possible relevance to depression and diabetes. J Psychsomatic Res 2002; 53:883–890.
23. Piotrowski P, Wierzbicka K, Smiatek M. Neuronal death in the rat hippocampus in experimental diabetes and cerebral ischemia treated with antioxidants. Folia Neuropathol 2001; 39:147–154.
24. Rigsby CS, Cannady WE, Dorrance AM. Aldosterone: good guy or bad guy in cerebrovascular disease? Trends Endocrinol Metab 200; 16:401–406.
25. Sabbatini M, Strocchi P, Vitaioli L, Amenta P: The hippocampus in spontaneously hypertensive rats: a quantitative microanatomical study. Neuroscience 2000; 100:251–258.
26. Sabbatini M, Catalani A, Consoli C, Marletta N, Tomassoni D and Avola R: The hippocampus in spontaneously hypertensive rats: an animal model of dementia. Mech Ageng Dev 2002; 123:547–559.
27. Tomassoni D, Avola R, Di Tullio A, Sabbatini M, Vitaioli L, Amenta F: Increased expression of glial fibrillary acidic protein in the brain of spontaneously hypertensive rats. Clin Exp Hyp 2004; 26:335–350.
28. Ueno M, Sakamoto H, Tomimoto H, Akiguchi I, Onodera M, Huang C-L, and Kamenishi K: Blood–brain barrier is impaired in the hippocampus of young adult spontaneously hypertensive rats. Acta Neuropathol 2002; 107:532–538.
29. Pietranera L, Saravia F, McEwen BS, Lucas LL, Johnson AK, De Nicola AF: Changes in Fos expression in various brain regions during deoxycorticosterone acetate treatment: relation to salt appetite, vasopressin mRNA and the mineralocorticoid receptor. Neuroendocrinology 2001; 74:396–406.
30. Pietranera L, Saravia F, Gonzalez Deniselle MC, Roig P, Lima A, De Nicola AF. Abnormalities of the hippocampus are similar in deoxycorticosterone acetate-salt hypertensive rats and spontaneously hypertensive rats. J Neuroendocrinol 2006; 18:466–474.
31. De Kloet ER, Joels M, Holsboer F: Stress and the brain: from adaptation to disease. Nat Rev Neurosci 2005; 6:463–475.
32. Sapolsky R, Krey LC, McEwen BS: The neuroendocrinology of stress and aging: The glucocorticoid cascade hypothesis. Endocr Rev 1986; 7:284–304.
33. Tornello S, Orti E, De Nicola AF, Rainbow TC, McEwen BS: Regulation of glucocorticoid receptors in brain by CORT treatment of adrenalectomized rats. Neuroendocrinology 1982; 35:411–417.
34. Watanabe Y, Gould E, McEwen BS: Stress induces atrophy of apical dendrites of hippocampal CA3 pyramidal neurons. Brain Res 1992; 588:341–345.
35. Dalm S, Enthoven L, Meijer OC, van der Mark MH, Karssen AM, de Kloet ER, Oitzl MS. Age-related changes in hypothalamic-pituitary-adrenal axis activity of male C57BL/6J mice. Neuroendocrinology 2005; 81:372–380.
36. Sapolsky RM. Glucocorticoids, stress and their adverse neurological effects: relevance to aging. Exp Gerontol 1999; 34:721–723.
37. Van Eekelen JAM, Rots NY, Sutanto W, de Kloet ER. The effect of aging on stress responsiveness and central corticosteroid receptors in the Brown Norway rat. Neurobiol Aging 1991; 13:159–170.

38. Wong EYH, Herbert J. Roles of mineralocorticoid and glucocorticoid receptors in the regulation of progenitor proliferation in the adult hippocampus. Eur J Neurosci 2005; 22:785–792.
39. Herman JP, Schafer MK, Young EA, Thompson R, Doiuglass J, Akil H, Watson SJ. Evidence for hippocampal regulation of neuroendocrine neurones of the hypothalamic-pituitary-adrenocortical axis. J Neurosci 1989; 9:3071–3082.
40. Magariños AM, Somoza G, De Nicola AF. Glucocorticoid negative feedback and glucocorticoid receptors after hippocampectomy in rats. Horm Metab Res 1987; 19:105–109.
41. Ridder S, Chourbaji S, Hellweg R, Urani A, Zacher C, Schmid W, Zink M, Hortnagl H, Flor H, Henn FA, Schutz G, Gass P. Mice with genetically altered glucocorticoid receptor expression show altered sensitivity for stress-induced depressive reactions. J Neurosci 2005; 25:6243–6250.
42. Joels M, De Kloet ER. Control of neuronal excitability by corticosteroid hormones. Trends Neurosci 1992; 15:25–30.
43. McEwen BS, Lambdin LT, Rainbow TC, De Nicola AF. Aldosterone effects on salt appetite in adrenalectomized rats. Neuroendocrinology 1986; 43:38–43.
44. Mondadori C, Hausler A. Aldosterone receptors are involved in the mediation of the memory-enhancing effects of piracetam. Brain Res 1990; 524:203–207.
45. Ratka A, Sutanto W, Bloemers M, De Kloet ER. On the role of brain mineralocorticoid (type I) and glucocorticoid (type II) receptors in neuroendocrine regulation. Neuroendocrinology 1989; 50:117–123.
46. Ahima R, Krozowski Z, Harlan R. Type I corticosteroid receptor-like immunoreactivity in the rat CNS: distribution and regulation by corticosteroids. J Comp Neurol 1991; 313:522–538.
47. Boldyreff B, Wehling M: Rapid aldosterone actions: from the membrane to signalling cascades to gene transcription and physiological effects. J Steroid Biochem Mol Biol 2003; 85:375–381.
48. Lucas LR, Pompei P, McEwen BS. Effects of deoxycorticosterone acetate and diazepam on neuropeptidergic neurons in rat striatum. Neuroreport 1997; 8:811–816.
49. Funder JW. Aldosterone, mineralocorticoid receptors and vascular inflammation. Mol Cell Endocrinol 2004; 217:263–269.
50. Van den Berg DTWM, De Jong W, De Kloet ER. Mineralocorticoid antagonists inhibits stress-induced blood pressure response after repeated daily warming. Am J Physiol 1994; 267: E921–E926
51. Hashimoto K, Makino S, Hirasawa R, Takao T, Sugawara M, Murakami K, Ono K, Ota Z. Abnormalities in the hypothalamic-pituitary-adrenal axis in spontaneously hypertensive rats during development of hypertension. Endocrinology 1989; 125:1161–1167.
52. Gomez F, Lahmame A. De Kloet ER, Armario A. Hypothalamic-pituitary-adrenal response to chronic stress in five inbred rat strains: differential responses are mainly located at the adrenocortical level. Neuroendocrinology 1996; 63:327–337.
53. Rahmouni K, Barthelmebs M, Grima M, Imbs JL, De Jong W: Involvement of brain mineralocorticoid receptor in salt-enhanced hypertension in spontaneously hypertensive rats. Hypertension 2001; 38:902–906.
54. Kenyon CJ, De Conti GA, Cupolo NA, Morris DJ. The role of aldosterone in the development of hypertension in spontaneously hypertensive rats. Endocrinology 1981; 109:1841–1845.
55. Sutanto W, Oitzl MS, Rots NY, Schobitz B, Van den Berg DT, Van Dijken HH, Mos J, Cools AR, Tilders FJ, Koolhaas JM. Corticosteroid receptor plasticity in the central nervous system of various rat models. Endocr Regul 1992; 26:111–118.
56. Konishi A, Tazawa C, Miki Y, Darnel AD, Suzuki T, Ohta Y, Suzuki T, Tabayashi K, Sasano H. The possible roles of mineralocorticoid receptor and 11 β-hydroxysteroid dehydrogenase type 2 in cardiac fibrosis in the spontaneously hypertensive rat. J Steroid Biochem Mol Biol 2003:85:439–442.
57. Mirshahi M, Nicolas C, Agarwal MK. Enhanced activation of the mineralocorticoid receptor in genetically hypertensive rats. Biochem Biophys Res Commun 1998; 244:120–125.
58. Pietranera, L, Saravia, F, Roig, P, Lima, A, De Nicola, AF. Mineralocorticoid treatment upregulates the hypothalamic vasopressinergic system of spontaneously hypertensive rats. Neuroendocrinology 2004; 80:100–110.

59. Gould E, Wooley CS, Frankfurt M, McEwen BS. Gonadal steroids regulate dendritic spine density in hippocampal pyramidal cells in adulthood. J Neurosci 1990; 10:1286–1291.
60. Goodman Y, Bruce AJ, Cheng B, Mattson MP. Estrogens attenuate and CORT exacerbates excitotoxicity, oxidative injury, and amyloid -peptide toxicity in hippocampal neurons. J Neurochem 1996; 66:1836–1844.
61. McEwen BS: Estrogen actions throughout the brain. Rec Progr Horm Res 2002; 57:357–370.
62. Suzuki S, Gerhold LM, Bottner M, Rau SW, Dela Cruz C, Yang E, Zhu H, Yu J, Cashion AB, Kindy MS, Merchenthaler I, Gage FH, Wise PM. Estradiol enhances neurogenesis following ischemic stroke through estrogen receptors alpha and beta. J Comp Neurol 2007; 500:1064–1075.
63. Garcia-Segura LM, Sanz A, Mendez P. Cross-talk between IGF-I and estradiol in the brain: focus on neuroprotection. Neuroendocrinology 2006; 84(4):275–279.
64. Shughrue PJ, Lane MV, Merchentaler I. Comparative distribution of estrogen receptor α and β mRNA in the rat central nervous system. J Comp Neurol 1997; 388:507–525.
65. Gould E, Tanapat P, Rydel T, Hastings N. Regulation of hippocampal neurogenesis in adulthood. Biol Psychiatry 2000; 48:715–720.
66. Tanapat P, Hastings NB, Gould E. Ovarian steroids influence cell proliferation in the dentate gyrus of the adult female rat in a dose- and time-dependent manner. J Comp Neurol 2005; 481:252–265.
67. Tanapat P, Hastings NB, Reeves AJ, Gould E. Estrogen stimulates a transient increase in the number of new neurons in the dentate gyrus of the adult female rat. J Neurosci 1999; 19:5792–5801.
68. Ormerod BK, Lee TT-L, Galea LAM. Estradiol initially enhances but subsequently suppresses (via adrenal steroids) granule cell proliferation in the dentate gyrus of adult female rats. J Neurobiol 2003; 55:247–260.
69. Isgor C, Watson SJ. Estrogen receptor alpha and beta mRNA expressions by proliferating and differentiating cells in the adult rat dentate gyrus and subventricular zone. Neuroscience 2005; 134:847–856.
70. Saravia F, Revsin Y, Lux-Lantos V, Beauquis J, Homo-Delarche F, De Nicola AF. Oestradiol restores cell proliferation in dentate gryus and subventricular zone of streptozotocin-diabetic mice. J. Neuroendocrinol 2004; 16:704–710.
71. Sararavia FE, Beauquis J, Revsin Y, Homo-Delarche F, de Kloet ER, De Nicola AF. Hippocampal neuropathology of diabetes mellitus is relieved by estrogen treatment. Cell Mol Neurobiol 2006; 26:943–957.
72. Saravia FE, Beauquis J, Pietranera L, De Nicola AF. Neuroprotective effects of estradiol in the hippocampus of middle age mice: involvement of neurons and glial cells. Psychoneuroendocrinology 2007; 32:480–492.
73. De Nicola AF, Saravia FE, Beauquis J, Pietranera L, Ferrini MG. Estrogens and neuroendocrine hypothalamic-pituitary-adrenal axis function. Front Horm Res. 2006; 35:157–168.
74. Pfeiffer A, Lapointe B, Barden N: Hormonal regulation of type II glucocorticoid receptor messenger ribonucleic acid in rat brain. Endocrinology 1991; 129:2166–2174.
75. Ferrini M, De Nicola AF: Estrogens up-regulate type I and type II glucocorticoid receptors in brain regions from ovariectomized rats. Life Sci 1991; 48:2593–2601.
76. Ferrini M, Lima A, De Nicola AF. Estradiol abolishes down-regulation of glucocorticoid receptors in brain. Life Sci 1995; 57:2403–2412.
77. Lephart ED, Galindo E, Bu LH. Stress (hypothalamic-pituitary-adrenal axis) and pain response in male rats exposed lifelong to high vs. low phytoestrogen diets. Neurosci Lett 2003; 342:65–68.
78. Isgor C, Cecchi M, Kabbai M, Akil H, Watson SJ: Estrogen receptor beta in the paraventricular nucleus of hypothalamus regulates the neuroendocrine response to stress and is regulated by CORT. Neuroscience. 2003; 121:837–845.
79. Perez-Martin M, Salazar V, Castillo C, Ariznavarreta C, Azcoitia I, Garcia-Segura LM, Tresguerres JA. Estradiol and soy extract increase the production of new cells in the dentate gyrus of old rats. Exp Gerontol 2005; 40:450–453.

80. Darnaudery M, Perez-Martin M, Belizaire G, Maccari S, Garcia-Segura LM. Insulin-like growth factor 1 reduces age-related disorders induced by prenatal stress in female rats. Neurobiol Aging 2006; 27:119–127.
81. Resnick SM, Maki PM, Rapp SR, Espeland MA, Brunner R, Coker LH, Granek IA, Hogan P, Ockene JK, Shumaker SA. Women's Health Initiative Study of Cognitive Aging Investigators. Effects of combination estrogen plus progestin hormone treatment on cognition and affect. J Clin Endocrinol Metab 2006; 91:1802–1810.
82. Foster TC. Interaction of rapid signal transduction cascades and gene expression in mediating estrogen effects on memory over the life span. Front Neuroendocrinol 2005; 26:51–64.
83. Henderson VW, Paganini-Hill A, Emanuel CK, Dunn ME, Buckwalter JG. Estrogen replacement therapy in older women. Comparisons between Alzheimer's disease cases and nondemented control subjects. Arch Neurol 1994; 51:896–900.
84. Wang PN, Liao SQ, Liu RS, Liu CY, Chao HT, Lu SR, Yu HY, Wang SJ, Liu HC. Effects of estrogen on cognition, mood, and cerebral blood flow in AD: a controlled study. Neurology 2000; 54:2061–2066.
85. Wise PM. Estrogen therapy: does it help or hurt the adult and aging brain? Insights derived from animal models. Neuroscience 2006; 138:831–835.
86. Brewer GJ, Reichensperger JD, Brinton RD. Prevention of age-related dysregulation of calcium dynamics by estrogen in neurons. Neurobiol Aging 2006; 27:306–317.
87. David J-P, Fallet-Bianco C, Vermersch P, Frigard B, Di Menza C, Delacourte A. Viellissement cerebral normal: étude de la réaction gliale. C R Acad Sci Paris 1994; 317:749–753.
88. Goss JR, Finch CE, Morgan DG. Age-related changes in glial fibrillary acidic protein mRNA in the mouse brain. Neurobiol Aging 1991; 12:165–170.
89. Nichols NR, Day JR, Laping NJ, Johnson SA, Finch CE. GFAP mRNA increases with age in rat and human brain. Neurobiol Aging 1993; 14:421–429.
90. Day JR, Laping NJ, Lampert-Etchells M, Brown SA, O'Callaghan JP, McNeill TH, Finch CE. Gonadal steroids regulate the expression of glial fibrillary acidic protein in the adult male rat hippocampus. Neuroscience 1993; 55:435–443.
91. Garcia-Ovejero D, Veiga S, Garcia-Segura LM, Doncarlos LL. Glial expression of estrogen and androgen receptors after rat brain injury. J Comp Neurol 2002; 450:256–71.
92. Lei DL, Long JM, Hengemihle J, O'Neill J, Manaye KF, Ingram DK, Mouton PR. Effects of estrogen and raloxifene on neuroglia number and morphology in the hippocampus of aged female mice. Neuroscience 2003; 121:659–666.
93. Azcoitia I, Sierra A, Garcia-Segura LM. Estradiol prevents kainic acid-induced neuronal loss in the rat dentate gyrus. Neuroreport 1998; 9:3075–3079.
94. Cadacio CL, Milner TA, Gallagher M, Pierce JP. Hilar neuropeptide Y interneuron loss in the aged rat hippocampal formation. Exp Neurol 2003; 183:147–158.
95. Picazo O, Azcoitia I, Garcia-Segura LM. Neuroprotective and neurotoxic effects of estrogens. Brain Res 2003; 990:20–27.
96. Howart C, Reed M. Unbiased stereology. Three-Dimensional Measurement in Microscopy. Oxford: Bios Scientific Publishers, 1998, p. 45.
97. McMahon SS, McDermott KW. Proliferation and migration of glial precursor cells in the developing rat spinal cord. J Neurocytol 2001; 30:821–828.
98. Brown JP, Couillard-Despres S, Cooper-Kuhn CM, Winkler J, Aigner L, Kuhn HG. Transient expression of doublecortin during adult neurogenesis. J Comp Neurol 2003; 467:1–10.
99. McDonald HY, Wojtowicz JM. Dynamics of neurogenesis in the dentate gyrus of adult rats. Neurosci Lett 2005; 385:70–75.
100. Hastings NB, Gould E. Rapid extension of axons into the CA3 region by adult-generated granule cells. J Comp Neurol 1999; 413:146–154.
101. Olariu A, Cleaver KM, Cameron HA. Decreased neurogenesis in aged rats results from loss of granule cell precursors without lengthening of the cell cycle. J Comp Neurol 2007; 501:659–667.
102. Behl, C. Estrogen as a neuroprotective hormone. Nat Rev Neurosci 2002; 3:433–442.

103. McEwen BS, Akama K, Alves S, Brake SG, Bulloch K, Lee S, Li C, Yuen G, Milner, T A. Tracking the estrogen receptor in neurons: implications for estrogen-induced synapse formation. Proc Natl Acad Sci USA 2001; 98:7093–7100.
104. Cardona-Gomez GP, Mendez P, DonCarlos LL, Azcoitia I, Garcia-Segura LM. Interactions of estrogens and insulin-like growth factor-I in the brain: implications for neuroprotection. Brain Res Rev 2001; 37:320–334.
105. Scharfman HE, Maclusky NJ. Similarities between actions of estrogen and BDNF in the hippocampus: coincidence or clue? Trends Neurosci 2005; 28:79–85.
106. Diaz-Brinton R, Chen S, Montoya M, Hsieh D, Minaya J, Kim J, Chu HP. The women's health initiative estrogen replacement therapy is neurotrophic and neuroprotective. Neurobiol Aging 2000; 21:475–496.
107. Morrison JH, Brinton RD, Schmidt PJ, Gore AC. Estrogen, menopause, and the aging brain: how basic neuroscience can inform hormone therapy in women. J Neurosci 2006; 26:10332–10348.
108. Toran-Allerand CD. Estrogen and the brain: beyond ER-alpha and ER-beta. Exp Gerontol 2004; 39:1579–1586.
109. McEwen BS, Milner TA. Hippocampal formation: shedding light on the influence of sex and stress on the brain. Brain Res Rev 2007; 55:343–355.
110. Mody I, Otis TS, Bragin A, Hsu M, Bizsaki G. GABAergic inhibition of granule cells and hilar al synchrony following ischemia-induced hilar neuronal loss. Neuroscience 1995; 69:139–150.
111. Stone DJ, Song Y, Anderson CP, Krohn KK, Finch CE, Rozovsky I. Bidirectional transcription regulation of glial fibrillary acidic protein by estradiol in vivo and in vitro. Endocrinology 1998; 139:3202–3209.
112. Larsson A, Wilhelmsson U, Pekna M, Pekny M. Increased cell proliferation and neurogenesis in the hippocampal dentate gyrus of old GFAP(-/-)Vim(-/-) mice. Neurochem Res 2004; 29:2069–2073.
113. Pekny M, Nilsson M. Astrocyte activation and reactive gliosis. Glia 2005; 50:427–434.
114. Rozovsky I, Wei M, Stone DJ, Zanjani H, Anderson CP, Morgan TE, Finch CE. Estradiol (E2) enhances neurite outgrowth by repressing glial fibrillary acidic protein expression and reorganizing laminin. Endocrinology 2002; 143:636–646.
115. Flood JF, Mooradian AD, Morley JE. Characteristics of learning and memory in streptozocin-induced diabetic mice. Diabetes 1990; 39:1391–1398.
116. Liu Z, Gastard M, Verina T, Bora S, Mouton PR, Koliatsos VE. Estrogens modulate experimentally induced apoptosis of granule cells in the adult hippocampus. J Comp Neurol 2001; 441:1–8.
117. Perfilieva E, Risedal A, Nyberg A, Johansson BB, EWriksson PS. Gender and strain influence on neurogenesis in dentate gyrus of young rats. J Cer Blood Flow Metab 2001; 21:211–217.
118. Krugers HS, Maslam S, Korf J, Joels M. The corticosteroid synthesis inhibitor metyrapone prevents hypoxia/ischemia-induced loss of synaptic function in the rat hippocampus. Stroke 2000; 31:1162–1172.
119. Rocha R, Chander PN, Khanna K, Zuckerman A, Stier CT: Mineralocorticoid blockade reduces vascular injury in stroke-prone hypertensive rats. Hypertension 1998; 31:451–458.
120. Delano FA, Schmid-Schonbein GW. Enhancement of glucocorticoid and mineralocorticoid receptor density in the microcirculation of the spontaneously hypertensive rat. Mirocirculation 2004; 11:69–78.
121. Toran-Allerand CD, Tinnikov AA, Singh RJ, Nethrapalli IS. 17alpha-estradiol: a brain-active estrogen? Endocrinology 2005; 146:3843–3850.
122. Romer W, Oettel M, Menzenbach B, Droescher P, Schwarz S. Novel estrogens and their radical scavenging effects, iron-chelating, and total antioxidative activities: 17 alpha-substituted analogs of delta 9(11)-dehydro-17 beta-estradiol. Steroids 1997; 62:688–694.
123. Honjo H, Iwasa K, Fushiki S, Hosoda T, Tatsumi H, Mihara M, Hirasugi Y, Oida M, Kariya K, Kikuchi N, Kawata M. Estrogen and non-feminizing estrogen for Alzheimer's disease. Endocr J 2003; 50:361–367.

Chapter 4
Estrogen Modulation of Visceral Nociception

Victor V. Chaban

Abstract The incidence of persistent, episodic, or chronic visceral pain disorders such as irritable bowel syndrome and other "functional" syndromes (fibromyalgia, interstitial cystitis, chronic pelvic pain and others) are more prevalent in females. Defining the site(s) and mechanisms through which sex steroid hormones modulate visceral nociception is an important step in understanding the gender differences in pain perception. One such mechanism may be the convergence of nociceptive stimuli and estrogen input on the primary afferent neurons, which innervate viscera. In our study we determined how estrogen interferes with pain transmission at the level of primary sensory neurons. Based on our results, it is likely that estrogen receptors expressed in primary afferent neurons located in the dorsal root ganglion modulate purinergic and opiate receptors-mediating chemical signaling associated with nociception. Nociception is a balance of pro- and anti-nociceptive inputs that is subject to regulation depending on the state of the organism. The modulation of purinergic receptor- mediated increase in intracellular calcium concentration and attenuation of opioid receptors functions by estrogen observed in primary afferent neurons strongly suggest that estrogen modulates visceral pain processing peripherally. Moreover, estrogen appears to have different actions on nociceptive signaling depending on the input acting as anti-nociceptive by inhibiting purinergic receptors-mediating response but also interferes with anti-nociceptive actions of opioids.

Keywords Estradiol, dorsal root ganglion, nociception, uterus, colon

Abbreviations IBS irritable bowel syndrome; DRG dorsal root ganglion; E2 17β-estradiol; ER estrogen receptor; VGCC voltage-gated calcium channel; PGE_2 prostaglandin E2; MOP μ-opioid receptor

Department of Biomedical Sciences, Charles Drew University of Medicine and Science; Department of Neurobiology, David Geffen School of Medicine, University of California, Los Angeles, CA, USA

M.S. Ritsner and A. Weizman (eds.), *Neuroactive Steroids in Brain Function, Behavior and Neuropsychiatric Disorders*

The fact that changes in visceral sensitivity associated with irritable bowel syndrome (IBS), fibromyalgia, interstitial cystitis, chronic pelvic pain are similar in all "functional" disorders as stated by Berkley[1] and others[2–7] suggests a model in which alteration in the central circuits in predisposed individuals are triggered by the similar pathophysiology. Azpiroz et al.[8] and Riley et al.[9] state that the variations of symptoms and pain perception across the menstrual cycle and that sexual intercourse triggers symptoms in a large percentage of female patients with functional disorders suggesting the involvement of sex steroids and the reproductive tract. In our research we hypothesized that 17β-estradiol (E_2) directly modulate the visceral inputs in the primary sensory neurons which can explain the observed clinical and gender differences in visceral hypersensitivity associated with "functional" syndromes and may suggest potential targets for therapeutic prevention.

4.1 Estrogen Action on Primary Sensory Afferent Neurons Correlates with Peripheral Nociception

The cell bodies of visceral primary afferent neurons that transmit information about chemical or mechanical stimulation from the periphery to the spinal cord are located in the dorsal root ganglion (DRG). Nociceptors are small to medium size DRG neurons whose peripheral processes detect potentially damaging physical and chemical stimuli. Based on the published material from the last decade, ATP has emerged as a putative signal for visceral pain. ATP is released by distention of the viscera and tissue damage (for review see Burnstock[10]). Visceral nociceptive C-fibers are activated by ATP and excitatory amino acids that are released by noxious stimuli from cells in target organs (paracrine action) and from afferent terminals themselves (autocrine action as stated by Burnstock,[11] or in sensory ganglia as stated by Matsuka et al.[12]). In our studies we observed that DRG neurons innervating viscera have a greater $[Ca^{2+}]_i$ response to subsequent NMDA stimulation than somatic afferents as stated by McRoberts et al.[13] This observation indicates that these neurons express receptors with higher permeability to Ca^{2+}, which can modulate transduction of nociceptive signals and suggests that visceral afferents are functionally different from somatic afferents. Sensitization of primary afferent neurons may play a role in the enhanced perception of visceral sensations and pain observed in clinical presentations of functional syndromes.

Several lines of evidence indicate that E_2 directly influence the functions of primary afferent neurons. Both subtypes of estrogen receptors (ERα and ERβ) are present in DRG neurons including the small-diameter putative nociceptors.[14] In vitro, the ATP-sensitive DRG neurons respond to E_2[15] which correlated well with the idea that visceral afferents are E_2-sensitive: first, visceral pain is affected by hormonal level in cycling females[9,16,17]; second, there are sex differences in the prevalence of functional disorders involving the viscera[18]; and third, putative visceral afferents[13] fit into the population of DRG neurons that are sensitive to E_2.[19] These data suggest that in addition actions of E_2 in the central nervous system, E_2 can act in the periphery to modulate estrogen receptor (ER) as stated by Aloisi et al.[20] ERs were traditionally envisioned as E_2-activated transcription factors. However, more recent studies indicate that E_2 has a multiplicity of actions: membrane, cytoplasmic and nuclear (reviewed in Nadal et al.[21]). E_2 modulates cellular activity by altering ion channel opening, G-protein signaling, and activation of trophic factor-like signal transduction pathways.[19,22–24] These effects have been ascribed to membrane-associated receptors.[25] Relevant to the present proposal, ERs are distributed in regions of the central and peripheral nervous system that mediate nociception. For example, ERs are expressed in dorsal horn neurons of the spinal cord[26,27] and DRG neurons.[14,28,29] Results from our laboratory[19] and others[22,30] indicate that E_2 acts in neurons to modulate L-type of voltage-gated calcium channel (VGCC).

ATP plays vital roles in cellular metabolism and as an extracellular signal and transduces noxious stimuli by activating purinergic, ATP-gated P2X receptors on primary afferent fibers (reviewed by Dunn et al.[31]). Opening of P2X channels results in membrane depolarization sufficient to trigger action potentials and Ca^{2+} influx through VGCC associated with nociception.[32] According to this theory, the pain of tissue irritation (mechanical distortion or inflammation) is due to ATP activation of high threshold nociceptors in which there is an dramatic increase of intracellular Ca^{2+}.[10,33] The predominant ATP receptor in small diameter nociceptive DRG neurons is the $P2X_3$.[34,35] $P2X_3$-null mice have reduced pain-related behavior in response to noxious stimuli.[36,37] Significantly, inflammation dramatically alters purinoception by causing a severalfold increase in ATP-activated currents, alters the voltage dependence of P2X receptors, and enhances the expression of P2X receptors increasing neuronal hypersensitivity as stated by Xu and Huang.[38]

In our experiments, DRG neurons were tested for ATP-induced $[Ca^{2+}]_i$ flux by fluorescence ratio imaging as described.[39] ATP (10–20 μM) caused transient $[Ca^{2+}]_i$ increase in >50% of small to medium size rat DRG neurons. The response to ATP was fully reversed after a 10–15 min washout. These results indicate that ATP-induced Ca^{2+} flux is dependent on the activation of P2X receptors and plasma membrane Ca^{2+} channels. Both SERM (selective estrogen receptor modulator) tamoxifen, and the pure ER antagonist ICI 182,780 blocked the E_2 attenuation of ATP-induced $[Ca^{2+}]_i$ transients as we showed in Chaban et al.[19] (Fig. 4.1A). Increasingly, many E_2 actions appear to be mediated through a membrane ER.[23,40] Membrane ERs are coded by the same transcripts as those that code intracellular ERα and ERβ based on experiments in transfected cells.[25] Membrane ERs have several features that distinguish them from intracellular receptors: time of activation, dose, and that membrane impermeable constructs mimic the effect of E_2.[41] In our study, the rapid time course and the higher doses required for ER activation suggest that E_2 inhibition of ATP-induced $[Ca^{2+}]_i$ flux was mediated through a membrane associated ER. A membrane impermeable construct, E-6-BSA was used to determine whether the E_2 actions on ATP-induced $[Ca^{2+}]_i$ were mediated at the membrane. E-6-BSA, filtered to remove any potentially unconjugated E_2, mimicked the effect of E_2, indicating that diffusion into DRG neurons was not essential for the E_2 inhibition of ATP-induced $[Ca^{2+}]_i$ transients (Fig. 4.1B). The cellular $[Ca^{2+}]_i$ response initiated by ATP has two components: the initial ion flux is through

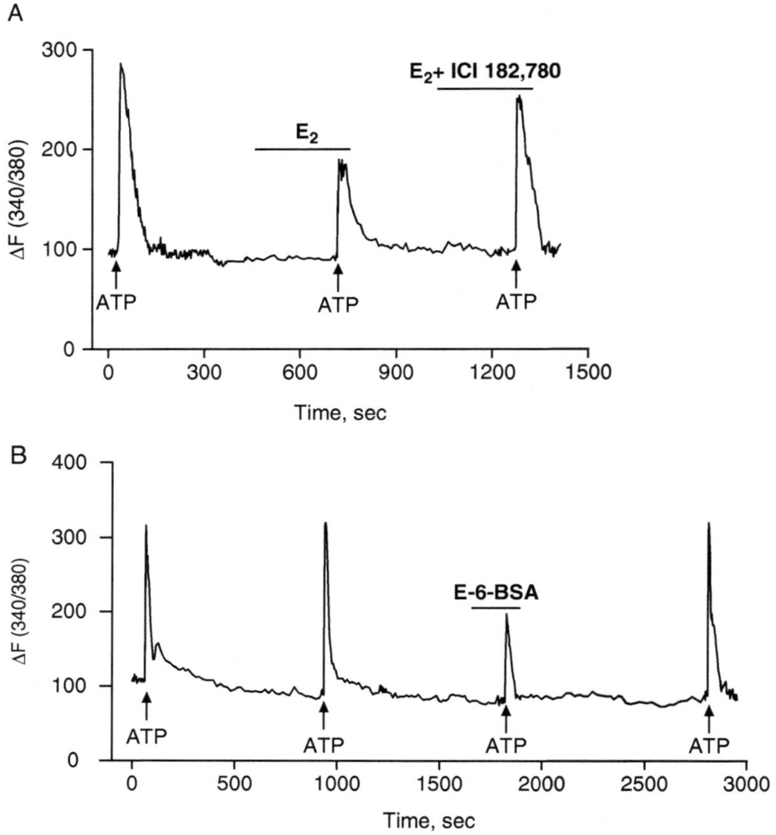

Fig. 4.1 Effect of 17β-estradiol (E_2) on ATP-induced Ca^{2+} signaling in DRG neurons (from Chaban et al.[19]). (**A**). E_2 added 5 min prior to ATP decreased the ATP-induced $[Ca^{2+}]_i$ flux. ER antagonist ICI 182,780 (1 μM) blocked the E_2 response. (**B**) E-6-BSA inhibited the ATP-induced $[Ca^{2+}]_i$ transient similar to E_2 indicating action via a membrane-associated receptor. After wash-out with experimental medium, ATP response on $[Ca^{2+}]_i$ returned to initial (control) amplitude of stimulation

activated P2X channels and the secondary response is the opening of VGCC in response to membrane depolarization as stated by Koshimizu et al.[32] The entire $[Ca^{2+}]_i$ transient was blocked with the purine receptor antagonist PPADS, but the $[Ca^{2+}]_i$ response was partially inhibited by E_2 treatment suggesting that E_2 did not directly antagonize the P2X receptor. DRG neurons express the L-, N- and P-type VGCC.[42] Specific VGCC channel antagonists were used to characterize which VGCC was blocked by E_2. Although ω-conotoxin and ω-agatoxin each attenuated the ATP-induced $[Ca^{2+}]_i$ transient, further treatment with E_2 produced a greater inhibition of $[Ca^{2+}]_i$ suggesting another VGCC. Nifedipine significantly inhibited L-type VGCC and further E_2 treatment did not result in an additional inhibition, suggesting that E_2 acts on L-type VGCC to attenuate ATP-induced $[Ca^{2+}]_i$ flux. This result is consistent with E_2 blockade of L-type Ca^{2+} channels in PC-12 cells,[43] neostriatal and hippocampal neurons.[22,44] Together these results indicate that E_2 attenuated the ATP-induced $[Ca^{2+}]_i$ spike by blocking the L-type VGCC.

In the next set of experiments we investigated whether the rapid E_2 attenuation of ATP-induced increase $[Ca^{2+}]_i$ requires the group II metabotropic glutamate receptor $mGluR_{2/3}$. DRG neurons were stimulated twice, once with ATP for 5 s, and then again in the presence of E_2 or E_2 (100 nM) + LY341495, an $mGluR_{2/3}$ inhibitor. Treatment with E_2 significantly attenuated the ATP-induced $[Ca^{2+}]_i$. Co-application of E_2 and LY341495 blocked the E_2 attenuation of the ATP-induced $[Ca^{2+}]_i$ transient. These data indicate that the rapid action of E_2 in DRG neurons is dependent on the $mGluR_{2/3}$, and demonstrate that membrane estrogen receptor initiated signaling involves an interaction with mGluRs.

In a series of experiments that parallel to those with rat DRGs, we were testing ATP responses on DRG neurons collected from wild type (Wt, C57/B1) and ERα knockout (ERαKO) and ERβ knockout ERβKO mice[45] to verify which ER subtype is required for E_2 attenuation of ATP-induced increase in $[Ca^{2+}]_i$. ATP induced $[Ca^{2+}]_i$ transients in >50% of small diameter DRG neurons from Wt mice. This was similar to the percentage observed in rat-derived DRG. In DRG from Wt mice, a 5 min incubation with E_2 inhibited ATP-induced $[Ca^{2+}]_i$ in ATP-responsive neurons. These results show that Wt mouse and rat DRG respond similarly to ATP and to E_2. In contrast, E_2 did not attenuate the ATP-induced $[Ca^{2+}]_i$ in DRG neurons from ERαKO mice indicating that this E_2 inhibition was dependent on ERα.[15,45]

4.2 Estrogen Modulation of Opioid Receptors Inhibition

One prominent way of E_2 modulation of neuronal excitability is through interactions with opioid systems.[46–48] E_2 modulates both the classic, anti-nociceptive opioids (e.g., enkephalins, β-endorphin)[49,50]; and the pro-nociceptive nociceptin/orphanin FQ systems. Nociceptin is the endogenous ligand for the opioid receptor-like orphanin receptor-1 (ORL_1). In the hypothalamus and limbic system E_2 increases nociceptin mRNA expression[48] ORL_1 activation produces anti-nociceptive effects in the male rat trigeminal system and pro-nociceptive effects in that of the female.[51] Moreover, E_2 reduces the efficacy of μ-opioid receptor (MOP) agonist inhibition of nociceptive response to cervical distension in rats and the response to uterine cervical distension.[52] Additionally, E_2 attenuates the ability of MOP agonists to hyperpolarize hypothalamic neurons.[23] These results are congruent with reports that sex steroids appear to mediate differential responses to opiate analgesics in men and women.[53,54] Studies from our laboratory indicate that in DRG neurons, E_2 modulates DAMGO inhibition of ATP-induced $[Ca^{2+}]_i$ flux and reduces opioid blockade of PGE_2-induced cAMP. These actions are correlated with an E_2-mediated reduction in the number of

MOP bindings sites. In our experiments, animals were treated with estradiol benzoate (EB) and processed for radioreceptor binding by incubating membrane homogenates with 0.1–25 nM of ^{3}H-DAMGO in the presence or absence of 5 μM DAMGO. MOP binding affinity (K_d) was not affected. The number of ^{3}H-DAMGO binding sites (B_{max}) was significantly reduced in EB-treated rats compared with membranes from OVX rats (Fig. 4.2A). This is consistent with demonstrated E_2 action, induction of opioid gene expression resulting in increased endogenous opioid peptides, and release of endogenous opioids. The decrease of membrane MOP binding sites indicates a site of action that explains E_2-induced decreased responsiveness to MOP agonists as stated by Maggi et al.[55]

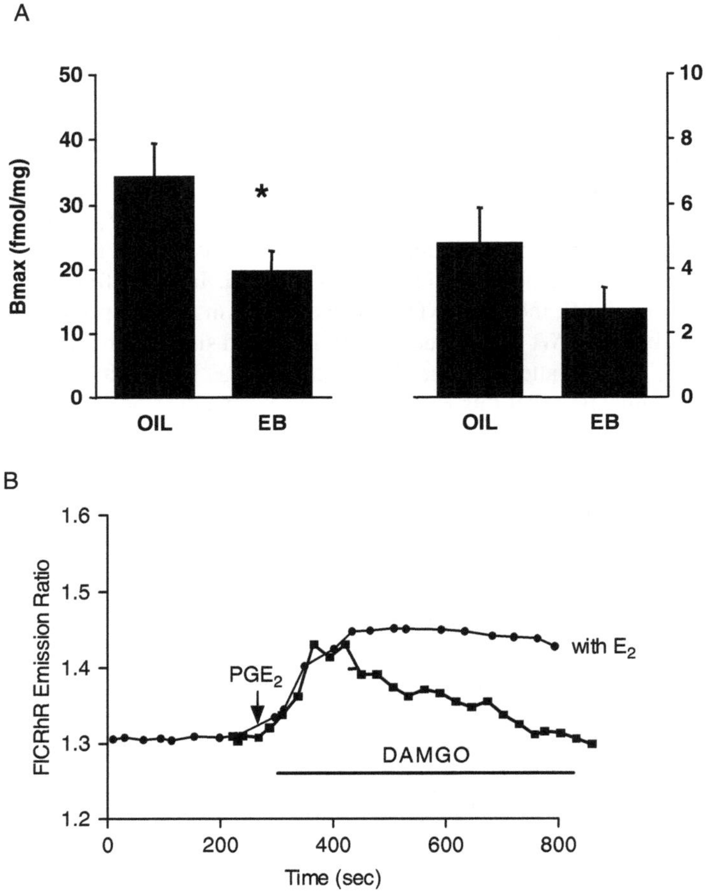

Fig. 4.2 Estradiol attenuates μ-opioid receptor (MOP)-mediated responses in the DRG. (**A**) Effect of E_2 on number of MOP binding sites (B_{max}) and MOP binding affinity (K_d) of DRG membranes. (**B**) E_2 attenuates MOP decrease of PGE_2 -induced $[cAMP]_i$ in DRG neurons. DAMGO inhibited PGE_2-induced increase in $[cAMP]_i$ (lower trace). E_2 treatment for 5 min attenuated the MOP-induced inhibition of PGE_2-induced $[cAMP]_i$ accumulation (upper trace)

The response of primary afferent neurons to pro-inflammatory stimuli has been modeled by exposure to prostaglandin E2 (PGE_2). The FlCRhR probe (recombinant fluorescein- and rhodamine-labeled cAMP-dependent protein kinase) allowed us to image $[cAMP]_i$ in PGE_2-stimulated DRG neurons over time. The time course of the PGE_2-stimulated $[cAMP]_i$ was decreased by the MOP agonist ^{3}H-D-Ala^2, N-Me-Phe^4, glycinol^5-enkephalin (DAMGO, 10 nM, Fig. 4.2B). E_2 (100 nM for 5 min) alone did not increase $[cAMP]_i$ levels but prevented DAMGO attenuation of the PGE_2 effect. These data are consistent with an E_2 blockade of MOP action in hypothalamic neurons as shown by Lagrange et al.[56]

The actions of E_2 on DAMGO inhibition of $[cAMP]_i$ appear to be different from E_2's actions on L-type VGCC and suggest that E_2 actions on nociception are contextual. That is, E_2 actions will be determined by the inputs received by DRG neuronal processes. Several possible mechanisms may account for the E_2 attenuation of opioid effects: E_2 could down-regulate the number of MOP binding sites, reduce the affinity of MOP or decrease the coupling of the MOP to its G-protein (for review see Sinchak and Micevych[48]). These results agree with previously characterized E_2 actions on MOP that indicate that internalization of MOP (e.g., removal from the plasma membrane) is dependent on the release of endogenous opioid peptides.[24,48] This is consistent with demonstrated E_2 action, induction of opioid gene expression resulting in increased endogenous opioid peptides,[49] and release of endogenous opioids.[26,46,47]

4.3 Cross-Sensitization of Visceral DRG Neurons

The inflammatory process produces a number of mediators which activate nociceptors by interacting with ligand-gated ion channels or by sensitizing primary afferents.[57] One mechanism for sensitization involves phosphorylation of ion channels and receptors including P2X receptors.[57,58] P2X receptor function is enhanced by a number of agents including ATP and substance P.[59,60] Inflammation does not change the percentage of total cells responding to ATP but sensitizes the ATP response by increasing the expression of $P2X_2$ and $P2X_3$.[38] Thus, the greater behavioral sensitivity during the inflammation is due to a twofold to threefold increase in ATP responses suggesting that a small amount of ATP would evoke depolarization sufficient to elicit action potentials in DRG neurons as stated by Xu and Huang.[38] This pathological response arises from sensitization of DRG and dorsal horn neurons to external stimuli.[38,61] Sensitization accounts for a dramatically lowered nociceptive threshold to mechanical manipulation of the inflamed area.[62] Within the context of the cross-sensitization hypothesis, inflammation sensitizes non-inflamed viscera that are innervated by the same DRG and/or cross-sensitization occurs as a result of intra-DRG release of sensitizing mediators such as ATP in the DRG.[12] In this model, the reproductive tract inflammation can sensitize DRG neurons innervating the colon, a structure that itself is not affected by inflammation. Recently we found that up to 5–7% of lumbosacral DRG neurons are colon-specific and 5–10% are uterus-specific. Among them up to 5% of labeled DRG neurons innervate both visceral organs in L_1-S_3 levels as we stated in Chaban et al.[63] Compared with control animals, in rats after the uterine inflammation induced with 10% allyl isothiocyanate (mustard oil), endometrial integrity of uterus was significantly disrupted with infiltration of CD45-positive leukocytes. As expected, histological sections from colon showed no signs of inflammation: no CD45+ cells and the integrity of the colon wall was intact.

Sensory neurons can release substance P (SP) intraganglionically during inflammation and experiments have revealed that the SP-containing neural pathway plays an important role in the pathogenesis of IBS and other functional disorders. Extracellular signal-regulated kinase (ERK) is one of the mitogen-activated protein kinases (MAPKs) that is involve in the sensitization of neurons in the DRG. Phosphorylation of ERK (pERK) occurred in DRG neurons that were involved in transmission of various noxious signals and contributes to allodynia that is common in the patients with functional pain-associated diseases.

In our experiments we found that double-labeled DRG neurons (these innervating both uterus and colon) express both pERK. In control rats 5.7% neurons innervating uterus and 3.4% neurons innervating both uterus and colons, were positive for pERK. In rats with induced inflammation in the uterus, the number of pERK positive DRG neurons markedly increased with an average of 24.9% DRG neurons innervating uterus were positive, 20.5% neurons innervating both uterus and colon were positive for pERK. After the uterine inflammation, SP expression in DRG neurons also was significantly increased compared with control rats. Average 43.1% DRG neurons innervating uterus were positive, 33.3% neurons innervating both uterus and colon were positive for SP. Similarly to pERK expression the bigger increase in SP immunoreactivity was seen in DRG neurons innervating both uterus and colon.

In our model, inflammation in the reproductive tract can also cross-sensitize the response to ATP in colonic DRG neurons. Although it is generally accepted that each primary afferent neuron is a single sensory channel (Fig. 4.3A), several studies have challenged that view and our own data using retrograde labeling demonstrate that a population of DRG neuron can innervate different visceral organs: colon and uterus (Fig. 4.3B). This small population of dichotomizing (branching) neurons that innervate the colon and uterus provide an anatomical support for an additional mechanism for sensory convergence in the DRG in sensitization of one viscus by another. The fact that the same primary afferent innervates both the uterus and the colon suggest a plausible mechanism that this population of dichotomizing neurons plays a vital role in the visceral nociceptive transduction and can be modulated by estrogens.

Recent studies demonstrate that E_2 has a significant role in modulating visceral sensitivity, indicating that E_2-induced alterations in sensory processing may underlie sex-based differences in functional pain symptoms as stated by Al-Chaer and Traub.[64] However, reports of E_2 modulation of visceral and somatic nociceptive sensitivity are inconsistent. For example, elevated E_2 levels have been reported to increase the threshold to cutaneous stimuli[65] but decrease the percentage of escape

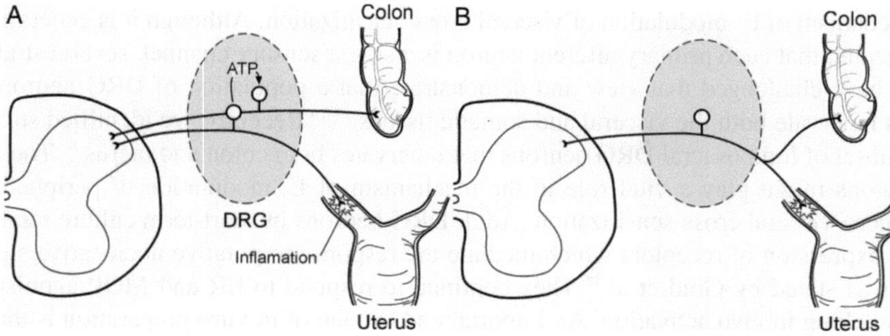

Fig. 4.3 Non-exclusive models of two possibilities for viscero-visceral cross-sensitization in the primary sensory neurons. (**A**) ATP released by a neuron innervating the inflamed uterus acts on a neighboring neuron sensitizing its responses to colonic distention. (**B**) The same neuron innervates the uterus and colon. Uterus inflammation directly sensitizes the neuron to colonic distention

responses to ureteral calculosis.[66,67] However, nociceptive sensitivity appear to increase when E_2 levels are elevated.[68,69] Indeed in most clinical studies, women report more severe pain levels, more frequent pain and longer duration of pain than men as stated by Berkley[1] and Filligim and Edwards.[70] To help resolve these inconsistencies the extensive studies of E_2 actions on the primary afferents will be very useful. Labeled DRGs had responses that suggested functional P2X,[15] NMDA,[13] VR1, mechano- and opioid receptors.[71]

4.4 Conclusion

Despite the broad spectrum of E_2 effects in the nervous system, the mechanisms of E_2 modulation of nociception remains elusive. Several facts have emerged recently, First, both ER subtypes have been reported in small-diameter DRG neurons, a subset of which are nociceptors[28,72] and E_2 modulates signaling in these neurons. Second, our data show that E_2 blocks ATP-induced $[Ca^{2+}]_i$ signaling in DRG neurons through a membrane-associated ERα.[45] Third, this E_2-mediated effect is attenuated by ER-specific antagonists and prevented in animals without a functional ERα.[45] In primary DRG neuronal cultures, E_2 attenuated the ATP-induced $[Ca^{2+}]_i$ responses through L-type VGCC and $mGluR_{2/3}$. E_2 also interfered with the MOP attenuation of PGE_2-induced cAMP rise and decreased the number of MOP-binding sites. Inflammation induced in the uterus increase the number of pERK and SP positive neurons in visceral DRG neurons that innervate both uterus and colon. Together, these data support the idea that E_2 can gate primary afferent response to increase or decrease nociception.

A central (spinal cord) site for the cross-organ sensitization and E_2 action has also been proposed. While it is likely that E_2 modulates several sites of nociception, very little is known to date about the peripheral sites of E_2 action. This is the overarching goal of our investigation to show the potential role of a peripheral site, DRG, in the mechanism of E_2 modulation of visceral cross-sensitization. Although it is generally accepted that each primary afferent neuron is a single sensory channel, several studies have challenged that view and demonstrate that a population of DRG neurons can innervate both the visceral and somatic tissues.[73–76] Recently, we identified such a subset of lumbosacral DRG neurons that innervates both colon and uterus.[63] These neurons might play a vital role in the mechanism of E_2 modulation of peripheral viscero-visceral cross-sensitization. Adult DRG neurons in short-term culture retain the expression of receptors which mediate the response to putative nociceptive signals as stated by Gold et al.[77] They continue to respond to ER and MOP agonists mimicking in vivo activation. An important advantage of in vitro preparation is that these neurons can be studied apart from endogenous signals. A large body of literature supports the idea that E_2 modulates nociceptive responses in functional pain syndromes, however, whether E_2 is pro- or anti-nociceptive remains unresolved due to differences in experimental paradigms. Within the context of our hypothesis E_2 modulation of nociceptive response depends on the type of pain, its durations and the involvement of other anti-nociceptive mechanisms.

From a public health prospective, the outcome of the study of estrogen modulation of visceral nociception at the level of primary sensory neurons will have a substantial impact on the knowledge of nociceptive functional diseases and will help achieve a deeper understanding of gender differences presented in clinical aspects of these symptoms. Only a thorough understanding of the mechanism implicated in these phenomenon can truly contribute to the designed of new and more efficient therapies. Many illnesses affect women and men differently. Some disorders are more common in women, and some express themselves with different symptoms. In calling for greater focus on gender-based biomedical research, future studies will help to identify barriers to the achievement of knowledge about gender differences. Reaching further than the concrete basic science contribution, these studies are a liaison between the basic science work and the clinical aspects and will address a crucial question in women's health. Clinical studies suggest the comorbidity of functional pain syndromes such IBS with psychiatric disorders such as anxiety, depression, and somatoform disorders approaching 40–60%. Nociceptive systems implicated in the etiology of functional disorders all pose health risks. Thus, designing new gender-specific therapies will have a major impact on health-related quality of life in patients with functional pain disorders significantly reducing psychiatric interventions.

Acknowledgements The author would like to thank Drs. Paul Micevych and Andrea Rapkin from University of California, Los Angeles and Dr. John McDonald from Harbor-UCLA Medical Center in Los Angeles for their support and discussions for this study. This research was supported by NIH Grant MD000545.

References

1. Berkley KJ. Sex differences in pain. Behav Brain Sci 1997; 20(3):371–380; discussion 435–513.

2. Lee OY, Mayer EA, Schmulson M, Chang L, et al. Gender-related differences in IBS symptoms. Am J Gastroenterol 2001; 96(7):2184–2193.
3. Toner BB, Akman D. Gender role and irritable bowel syndrome: literature review and hypothesis. Am J Gastroenterol 2000; 95(1):11–16.
4. Mayer EA, Naliboff BD, Chang L. Basic pathophysiologic mechanisms in irritable bowel syndrome. Dig Dis 2001; 19(3):212–218.
5. Mayer EA, Naliboff BD, Chang L, et al. V. Stress and irritable bowel syndrome. Am J Physiol Gastrointest Liver Physiol 2001; 280(4):G519–G524.
6. Fillingim RB, Ness TJ. Sex-related hormonal influences on pain and analgesic responses. Neurosci Biobehav Rev 2000; 24(4):485–501.
7. Mogil JS, Chesler EJ, Wilson SG, et al. Sex differences in thermal nociception and morphine antinociception in rodents depend on genotype. Neurosci Biobehav Rev 2000; 24(3):375–389.
8. Azpiroz F, Dapoigny M, Pace F, et al. Nongastrointestinal disorders in the irritable bowel syndrome. Digestion 2000; 62(1):66–72.
9. Riley JL, 3rd, Robinson ME, Wise EA, et al. Sex differences in the perception of noxious experimental stimuli: a meta-analysis. Pain 1998; 74(2–3):181–187.
10. Burnstock G. Purine-mediated signalling in pain and visceral perception. Trends Pharmacol Sci 2001; 22(4):182–188.
11. Burnstock G. P2X receptors in sensory neurones. Br J Anaesth 2000; 84(4):476–488.
12. Matsuka Y, Neubert JK, Maidment NT, et al. Concurrent release of ATP and substance P within guinea pig trigeminal ganglia in vivo. Brain Res 2001; 915(2):248–255.
13. McRoberts JA, Coutinho SV, Marvizon JC, et al., Role of peripheral *N*-methyl-D-aspartate (NMDA) receptors in visceral nociception in rats. Gastroenterology 2001; 120(7):1737–1748.
14. Papka RE, Storey-Workley M. Estrogen receptor-alpha and -beta coexist in a subpopulation of sensory neurons of female rat dorsal root ganglia. Neurosci Lett 2002; 319(2):71–74.
15. Chaban V, Mayer E, Ennes H, Micevych P. Rapid effect of estrogen on ATP-induced $[Ca^{2+}]_i$ signaling in cultured DRG neurons from wild type and ERKO mice. Gastroenterology 2003; 124(4):A1.
16. Sapsed-Byrne S, Ma D, Ridout D, Holdcroft A. Estrous cycle phase variations in visceromotor and cardiovascular responses to colonic distension in the anesthetized rat. Brain Res 1996; 742(1–2):10–16.
17. Traub RJ, Murphy A. Colonic inflammation induces fos expression in the thoracolumbar spinal cord increasing activity in the spinoparabrachial pathway. Pain 2002; 95(1–2):93–102.
18. Sand P. Chronic pain syndromes of gynecologic origin. J. Reprod. Med. 2004; 49:230–234.
19. Chaban VV, Mayer EA, Ennes HS, Micevych PE. Estradiol inhibits ATP-induced intracellular calcium concentration increase in dorsal root ganglia neurons. Neuroscience 2003; 118(4):941–948.
20. Aloisi AM, Ceccarelli I, Herdegen T. Gonadectomy and persistent pain differently affect hippocampal c-Fos expression in male and female rats. Neurosci Lett 2000; 281(1):29–32.
21. Nadal A, Diaz M, Valverde MA. The estrogen trinity: membrane, cytosolic, and nuclear effects. News Physiol Sci 2001; 16:251–255.
22. Mermelstein PG, Becker JB, Surmeier DJ. Estradiol reduces calcium currents in rat neostriatal neurons via a membrane receptor. J Neurosci 1996; 16(2):595–604.
23. Kelly MJ, Wagner EJ. Estrogen modulation of G-protein-coupled receptors. Trends Endocrinol Metab 1999; 10(9):369–374.
24. Eckersell CB, Popper P, Micevych PE. Estrogen-induced alteration of mu-opioid receptor immunoreactivity in the medial preoptic nucleus and medial amygdala. J Neurosci 1998; 18(10):3967–3976.
25. Razandi M, Pedram A, Greene GL, Levin ER. Cell membrane and nuclear estrogen receptors (ERs) originate from a single transcript: studies of ERalpha and ERbeta expressed in Chinese hamster ovary cells. Mol Endocrinol 1999; 13(2):307–319.
26. Amandusson A, et al. Estrogen-induced alterations of spinal cord enkephalin gene expression. Pain 1999; 83(2):243–248.

27. Williams SJ, Papka RE. Estrogen receptor-immunoreactive neurons are present in the female rat lumbosacral spinal cord. J Neurosci Res 1996; 46(4):492–501.
28. Taleghany N, Sarajari S, DonCarlos LL, et al. Differential expression of estrogen receptor alpha and beta in rat dorsal root ganglion neurons. J Neurosci Res 1999; 57(5):603–615.
29. Papka RE, Storey-Workley M, Shughrue PJ, et al. Estrogen receptor-alpha and beta- immunoreactivity and mRNA in neurons of sensory and autonomic ganglia and spinal cord. Cell Tissue Res 2001; 304(2):193–214.
30. Lee DY, Chai YG, Lee EB, Kim KW, et al. 17Beta-estradiol inhibits high-voltage-activated calcium channel currents in rat sensory neurons via a non-genomic mechanism. Life Sci 2002; 70(17):2047–2059.
31. Dunn PM, Zhong Y, Burnstock G. P2X receptors in peripheral neurons. Prog Neurobiol 2001; 65(2):107–134.
32. Koshimizu TA, Van Goor F, Tomic M, et al. Characterization of calcium signaling by purinergic receptor-channels expressed in excitable cells. Mol Pharmacol 2000; 58(5):936–945.
33. Bodin P, Burnstock G. Purinergic signalling: ATP release. Neurochem Res 2001; 26(8–9):959–969.
34. Ueno S, Tsuda M, Iwanaga T, Inoue K. Cell type-specific ATP-activated responses in rat dorsal root ganglion neurons. Br J Pharmacol 1999; 126(2):429–436.
35. Chen CC, Akopian AN, Sivilotti L, et al. A P2X purinoceptor expressed by a subset of sensory neurons. Nature 1995; 377(6548):428–431.
36. Cockayne DA, Hamilton SG, Zhu QM, Dunn PM, et al. Urinary bladder hyporeflexia and reduced pain-related behaviour in P2X3-deficient mice. Nature 2000; 407(6807):1011–1015.
37. Zhong Y, Dunn PM, Bardini M, et al. Changes in P2X receptor responses of sensory neurons from P2X3-deficient mice. Eur J Neurosci 2001; 14(11):1784–1792.
38. Xu GY, Huang LY. Peripheral inflammation sensitizes P2X receptor-mediated responses in rat dorsal root ganglion neurons. J Neurosci 2002; 22(1):93–102.
39. Chaban VV, McRoberts JA, Ennes HS, Mayer EA. Nitric oxide synthase inhibitors enhance mechanosensitive Ca(2+) influx in cultured dorsal root ganglion neurons. Brain Res 2001; 903(1–2):74–85.
40. Levin ER. Cell localization, physiology, and nongenomic actions of estrogen receptors. J Appl Physiol 2001; 91(4):1860–1867.
41. Norman AW, Mizwicki MT, Norman DP. Steroid-hormone rapid actions, membrane receptors and a conformational ensemble model. Nat Rev Drug Discov 2004; 3(1):27–41.
42. Rusin KI, Moises HC. Mu-Opioid receptor activation reduces multiple components of high-threshold calcium current in rat sensory neurons. J Neurosci 1995; 15(6):4315–4327.
43. Kim YJ, Hur EM, Park TJ, Kim KT Nongenomic inhibition of catecholamine secretion by 17beta-estradiol in PC12 cells. J Neurochem 2000; 74(6):2490–2496.
44. Kurata K, Takebayashi M, Kagaya A, et al. Effect of beta-estradiol on voltage-gated Ca(2+) channels in rat hippocampal neurons: a comparison with dehydroepiandrosterone. Eur J Pharmacol 2001; 416(3):203–212.
45. Chaban VV, Micevych PE. Estrogen receptor-alpha mediates estradiol attenuation of ATP-induced Ca2 + signaling in mouse dorsal root ganglion neurons. J Neurosci Res 2005; 81(1):31–37.
46. Micevych PE, Eckersell CB, Brecha N, Holland KL. Estrogen modulation of opioid and cholecystokinin systems in the limbic-hypothalamic circuit. Brain Res Bull 1997; 44(4):335–343.
47. Micevych P, Sinchak K. Estrogen and endogenous opioids regulate CCK in reproductive circuits. Peptides 2001; 22(8):1235–1244.
48. Sinchak K, Micevych P. Visualizing activation of opioid circuits by internalization of G protein-coupled receptors. Mol. Neurobiol. 2003; 27(2):197–222.
49. Priest CA, Vink KL, Micevych PE. Temporal regulation by estrogen of beta-preprotachykinin mRNA expression in the rat ventromedial nucleus of the hypothalamus. Brain Res Mol Brain Res 1995; 28(1):61–71.
50. Holland KL, Popper P, Micevych PE. Infusion of CCK-A receptor mRNA antisense oligodeoxynucleotides inhibits lordosis behavior. Physiol Behav 1997; 62(3):537–543.

51. Flores CA, et al. Orphanin FQ produces gender-specific modulation of trigeminal nociception: behavioral and electrophysiological observations. Neuroscience 2001; 105(2):489–498.
52. Sandner-Kiesling A and Eisenach JC. Estrogen reduces efficacy of mu- but Not kappa-opioid agonist inhibition in response to uterine cervical distension. Anesthesiology 2002; 96(2):375–379.
53. Miaskowski C. Women and pain. Crit Care Nurs Clin North Am 1997; 9(4): 453–458.
54. Zubieta JK, Smith YR, Bueller JA, et al. Mu-opioid receptor-mediated antinociceptive responses differ in men and women. J Neurosci 2002; 22(12):5100–5107.
55. Maggi R, Ma ZQ, Pimpinelli F, et al. Decrease of the number of opioid receptors and of the responsiveness to morphineduring neuronal differentiation induced by 17beta-estradiol in estrogen receptor-transfected neuroblastoma cells (SK-ER3). Neuroendocrinology 1999; 69(1):54–62.
56. Lagrange AH, Ronnekleiv OK, Kelly MJ. Modulation of G protein-coupled receptors by an estrogen receptor that activates protein kinase A. Mol Pharmacol 1997; 51(4):605–612.
57. Dai Y, Fukuoka T, Wang H, et al. Contribution of sensitized P2X receptors in inflamed tissue to the mechanical hypersensitivity revealed by phosphorylated ERK in DRG neurons. Pain 2004; 108(3):258–266.
58. Gold MS, Levine JD, Correa AM. Modulation of TTX-R INa by PKC and PKA and their role in PGE2-induced sensitization of rat sensory neurons in vitro. J Neurosci 1998; 18(24):10345–10355.
59. Wildman SS, King BF, Burnstock G. Potentiation of ATP-responses at a recombinant P2×2 receptor by neurotransmitters and related substances. Br J Pharmacol 1997; 120(2):221–224.
60. Stoop R, Surprenant A, North RA. Different sensitivities to pH of ATP-induced currents at four cloned P2X receptors. J Neurophysiol 1997; 78(4):1837–1840.
61. Woolf CJ, Doubell TP. The pathophysiology of chronic pain–increased sensitivity to low threshold A beta-fibre inputs. Curr Opin Neurobiol 1994; 4(4):525–534.
62. Reichling DB, Levine JD. The primary afferent nociceptor as pattern generator. Pain 1999; Suppl 6:S103–S109.
63. Chaban V, Christensen A, Wakamatsu M, et al. The same dorsal root ganglion neurons innervate uterus and colon in the rat. Neuroreport 2007; 18(3):209–212.
64. Al-Chaer ED, Traub RJ. Biological basis of visceral pain: recent developments. Pain 2002; 96(3):221–225.
65. Martinez-Gomez M, Cruz Y, Salas M, et al. Assessing pain threshold in the rat: changes with estrus and time of day. Physiol Behav 1994; 55(4):651–657.
66. Giamberardino MA, Affaitati G, Valente R, et al. Changes in visceral pain reactivity as a function of estrous cycle in female rats with artificial ureteral calculosis. Brain Res 1997; 774(1–2):234–238.
67. Bradshaw HB, Berkley KJ. Estrogen replacement reverses ovariectomy-induced vaginal hyperalgesia in the rat. Maturitas 2002; 41(2):157–165.
68. Holdcroft A. Hormones and the gut. Br J Anaesth 2000; 85(1):58–68.
69. Bereiter DA. Sex differences in brainstem neural activation after injury to the TMJ region. Cells Tissues Organs 2001; 169(3):226–237.
70. Fillingim RB, Edwards RR. The association of hormone replacement therapy with experimental pain responses in postmenopausal women. Pain 2001; 92(1–2):229–234.
71. Gschossmann JM, Chaban VV, McRoberts JA, et al. Mechanical activation of dorsal root ganglion cells in vitro: comparison with capsaicin and modulation by kappa-opioids. Brain Res 2000; 856(1–2):101–110.
72. Papka RE, Mowa CN. Estrogen receptors in the spinal cord, sensory ganglia, and pelvic autonomic ganglia. Int Rev Cytol 2003; 231:91–127.
73. Bahr R, Blumberg H, Janig W. Do dichotomizing afferent fibers exist which supply visceral organs as well as somatic structures? A contribution to the problem or referred pain. Neurosci Lett 1981; 24(1):25–28.
74. Devor M, Obermayer M. Membrane differentiation in rat dorsal root ganglia and possible consequences for back pain. Neurosci Lett 1984; 51(3):341–346.
75. Pierau FK, Fellmer G, Taylor DC. Somato-visceral convergence in cat dorsal root ganglion neurones demonstrated by double-labelling with fluorescent tracers. Brain Res 1984; 321(1):63–70.

76. Habler HJ, Janig W, Koltzenburg M. Dichotomizing unmyelinated afferents supplying pelvic viscera and perineum are rare in the sacral segments of the cat. Neurosci Lett 1988; 94(1–2):119–124.
77. Gold MS, Dastmalchi S, Levine JD. Co-expression of nociceptor properties in dorsal root ganglion neurons from the adult rat in vitro. Neuroscience 1996; 71(1):265–275.

Chapter 5
Neuroactive Steroids: Effects on Cognitive Functions

Torbjörn Bäckström[1], Vita Birzniece[1], Guillén Fernández[2], Inga-Maj Johansson[1], Kristiina Kask[3], Charlotte Lindblad[1], Per Lundgren[1], Sigrid Nyberg[1], Gianna Ragagnin[1], Inger Sundström-Poromaa[3], Jessica Strömberg[1], Sahruh Turkmen[1], Ming-De Wang[1], Frank van Broekhoven[4], and Guido van Wingen[2]

Abstract Dementia and Alzheimer's disease afflict approximately 5% of the population over 65, and >20% of the population >80. These disorders are likely to increase substantially given the change in the demographic pattern. The cost for the society and suffering for the families are tremendous. Stress and disorders with increased production of steroids seem to induce cognitive impairment. Likewise drugs that induce high activity in the Gamma Amino Butyric Acid-A ($GABA_A$) receptor seem to be involved in dementia development and cognitive impairment. This review will concentrate on arguments indicating a link between stress disorders-neuroactive steroids active on the $GABA_A$ receptor (GABA-steroids) and permanent cognitive impairment. Chronic long-term exposure by all $GABA_A$ receptor agonists, e.g., benzodiazepines, barbiturates and alcohol, give permanent memory and learning impairment. The sex-steroid medroxyprogesterone, given as postmenopausal hormone therapy, double the dementia frequency in 5 years. The neuroactive steroid allopregnanolone inhibits learning in rat studies. Chronically high cortisol and GABA-steroid levels give irreversible cognitive damages. During stress the production of both cortisol and GABA-steroids increase in parallel. GABA-steroid production occurs in the adrenals and are produced and regulated as cortisol. Patients with advanced Alzheimer's disease have similar cortisol and GABA-steroid response to adrenal stimulation as chronically stressed animals. Patients with mild Alzheimer's disease have a high and non-suppressible production of cortisol and GABA-steroids. Cortisol metabolites increase the effect of allopregnanolone on the $GABA_A$ receptor. Chronic stress and "burn-out syndrome" gives permanent cognitive damages and are frequent in the patient history of patients with Alzheimer's disease. An impaired

[1]Umeå Neurosteroid Research Center, Department of Clinical Science, Norrlands University Hospital, SE 901 85, Umeå, Sweden

[2]F.C. Donders Centre for Cognitive Neuroimaging, Radboud University Nijmegen, Kapittelweg 29, 6525 EN, The Netherlands

[3]Department Women's and Children's Health, Uppsala University, Uppsala, Sweden

[4]Department of Psychiatry, Radboud University Nijmegen Medical Center, Reinier Postlaan 10, 6525 GC, Nijmegen, The Netherlands

M.S. Ritsner and A. Weizman (eds.), *Neuroactive Steroids in Brain Function, Behavior and Neuropsychiatric Disorders*

cholinergic system is involved in Alzheimer's disease. Allopregnanolone hamper memory-related cholinergic action. $GABA_A$ receptor subunit alpha5 is mainly localized in the hippocampus, i.e., in the region for learning and memory. Blockade of the $GABA_A$ receptor subunit alpha5 increased learning and memory. An allopregnanolone antagonist is shown to antagonize the learning and memory disruption of allopregnanolone both in vitro in receptor pharmacological studies and in vivo in studies using the Morris Water Maze model.

Keywords GABA-steroids, $GABA_A$ receptor, dementia, Alzheimer's disease, learning, memory, stress, burn-out-syndrome, allopregnanolone, tetra-hydrodesoxycorticosterone

Abbreviations 5HT 5-hydroxytryptamine = serotonin; AD Alzheimer's disease; Allo-THF allotetrahydrocortisol; CNS central nervous system; DHEAS dihydroepiandroseronesulfate; fMRI functional magnetic resonance imaging; GABA gamma-aminobutyric acid; $GABA_A$ one receptor for GABA; GABA-steroids positive modulators of $GABA_A$ receptor; HT hormone therapy; LTP long-term potentiation; MCI mild cognitive impairment; MPA medroxyprogesterone; MRI magnetic resonance imaging; PET positron emission tomography; PMDD premenstrual dysphoric disorder (PMDD),; RU486 mifepristone; THDOC tetrahydrodesoxycorticosterone; UC1011 3beta-20beta-dihydroxy-5alpha-pregnane; WHI Women's Health Initiative

5.1 Introduction

Dementia and Alzheimer's disease afflict approximately 5% of the population over 65, and >20% of the population >80. These disorders are likely to increase substantially given the change in the demographic pattern. The suffering for the families and cost for the society are tremendous. Chronic stress, "burn-out" syndrome, stress-steroids and sex-steroids are linked to development of dementia.[1–3] In women taking postmenopausal hormone therapy (HT), where the GABA-steroid precursor medroxyprogesterone is given, and the risk for dementia (Alzheimer's disease) doubled after 5 years of treatment.[1] Medroxyprogesterone can induce anesthesia by acting on the $GABA_A$ receptor.[4] Estrogen on the other hand has been claimed to protect against dementia.[5] This was, however, not confirmed in the large Women's Health Initiative (WHI) study but estrogen by is itself did not increase the risk to develop dementia.[6] The medroxyprogesterone exposure during HT treatment resembles the exposure of stress-steroids during chronic stress. $GABA_A$ receptor agonists are known to impair memory and the progesterone metabolite allopregnanolone, is a very potent $GABA_A$ receptor agonist. Allopregnanolone can inhibit learning in rats tested in Morris water maze an accepted model for learning and memory testing.[7] It is well known that other $GABA_A$ receptor agonists, e.g., benzodiazepines,[8] barbiturates[9] and alcohol[10,11] impair memory and learning and increase the risk for permanent damages, although the risk with low and moderate alcohol consumption is under debate.[12]

The interest for the influence of sex and stress steroids in CNS-related diseases has increased noticeably, in particular since it was discovered that steroid metabolites directly and potently modulates neurotransmitter systems.[13] Others and we have shown that, with the exception of estradiol, it is not the classic hormones that achieve these effects but their 3alpha-hydroxy-5alpha /beta metabolites, which have a direct impact on the GABA system via the $GABA_A$ receptor.[14–16] These so called GABA-steroids hereby influence the brain's largest inhibiting system, the GABA system, by enhancing the GABA's own effect on the $GABA_A$ receptor, but also other neurotransmitter systems like the glutamate and serotonin systems.[15] The effects of the GABA-steroids are similar to the effects of barbiturates, benzodiazepines (e.g., Valium) and alcohol, substances also known to target the $GABA_A$ receptor.

5.1.1 *Stress and Sex Steroids and CNS Active Metabolites, GABA-Steroids*

Sex and stress hormones in this context are defined as steroids produced by the adrenals during stress, in the ovary during the menstrual cycle and in the testis, but also directly in the brain in both sexes.[17] GABA-steroids are $GABA_A$ receptor positive modulators of the GABA effect on the $GABA_A$ receptor.[13] Chemically they are 3alpha-hydroxy-5alpha/beta metabolites of progesterone (pregnanolone and allopregnanolone), testosterone (3alpha-5alpha-androstan-diol) and desoxycorticosterone (tetrahydro-desoxycorticosterone, THDOC). The concentration of GABA-steroids in brain and serum varies with the production activity in the adrenals, ovaries and testicles.[18–21] Interestingly, also the classical stress hormones cortisol and corticosterone are metabolized in a similar way and allotetrahydrocortisol (Allo-THF, 3alpha-hydroxy-5alpha-cortisol) interacts with the $GABA_A$ receptor and enhances the effect of the GABA-steroids.[17] Allo-THF enhance the inhibitory effect of allopregnanolone giving an even larger inhibition the neural activity than by allopregnanolone alone.[16] That is during acute and chronic stress the neurons are exposed to a strong enhancing effect of the neural inhibition. During acute stress the GABA-steroids allopregnanolone and THDOC increase in the brain in normal animals. In animals subjected to chronic long-term stress there is a decrease in allopregnanolone concentration in cortex at rest but a higher increase after acute stress than in normal control animals.[22] Patients with Alzheimer's disease show an increased glucocorticoid production compared to healthy elderly control subjects and in addition patients with Alzheimer's disease show an increased 5-alpha reduction. Thus an increased glucocorticoid production is an early feature of Alzheimer's disease with an enhanced metabolism to 5-reduced metabolites of cortisol.[23]

5.1.2 Memory

Memory is a label for a set of diverse cognitive capacities by which humans and animals retain information and reconstruct past experiences. Often the retained knowledge is related to emotions. The hippocampus and amygdala are important brain regions for memory processes, suggesting a powerful relation between emotional situations and strong memories.[24] Spatial memory involves the ability to map the environment. The Morris water maze is the favored test for the study of hippocampal-dependent spatial learning and memory in rats.[25] The Morris water maze consists of a water pool with a hidden escape platform where the subject must learn the location of the platform using distal cues. Performance in the Morris water maze depends on several mechanisms, including attention, learning, memory, vision, and motor coordination.[26]

5.1.3 Dementia

Dementia is a permanent and progressive decline in cognitive function and behavior due to damage or disease in the brain beyond what might be expected from normal aging. Affected areas may be memory, attention, language, and problem solving. Especially in the later stages of the condition, affected persons may be disoriented in time, in place, and in person. Symptoms of dementia can be reversible or irreversible depending upon the etiology of the disease. Less than 10% of cases of dementia are reversed.

Alzheimer's disease (AD) is the most common form of dementia. AD is a well-characterized disease, with clear diagnostic criteria.[27] Mild Cognitive Impairment (MCI) is the first phase of AD but also the preliminary stage of other types of dementia.[28] Although an early intervention is more likely to effectively prevent disease progression, MCI is not yet a single disorder and at present it is hard to differentiate the different groups of patients from each other and predict the ones developing AD. The etiology of AD is still unclear but impaired function in cholinergic and glutamate systems are involved in Alzheimer's disease.[29,30] The role of over activity in the GABA system seems to be involved in AD etiology. Chronic long-term exposure by all $GABA_A$ receptor agonists, e.g., benzodiazepines, barbiturates, alcohol and GABA-steroids, give permanent memory and learning impairment and increased risk of developing dementia.[8–10]

GABA-steroids are also involved in the etiology of dementia. A large clinical trial showed an increased occurrence of Alzheimer's dementia among patients taking postmenopausal estrogen/progestagen replacement therapy compared to placebo already after 4 years of treatment.[1] The GABA-steroid precursor medroxyprogesteron was given as postmenopausal hormone therapy and the dementia frequency after 5 years of treatment was double in the medroxyprogesterone-treated group compared to the placebo-treated group.[1] The hazard ratio was 2.05. This resulted in an additional 23 cases of dementia per 10,000 women per year. Alzheimer's disease was the most common classification of dementia. Medroxyprogesterone, can in high dosages induce anesthesia, probably via metabolism into $GABA_A$ receptor active metabolites.[4] The GABA-steroid levels are similar to those found in mild Alzheimer's disease as mild AD patients have a high and non-suppressible production of cortisol and GABA-steroids.[31] This is one background to the focus of the present review.

5.2 Neural Structures Involved in Memory and Learning

In both humans and animals, the hippocampus is an important brain area for learning and memory functions.[32,33] Hippocampus is especially essential for the formation of long-term declarative memories.[34] The $GABA_A$ receptor alpha5 subunit is almost exclusively expressed in the hippocampus and seems to be involved in decreasing spatial learning (Fig. 5.1). Alpha5 knockout mice had significantly better performance in a water maze model in comparison with wild-type mice. The importance of hippocampus in humans is indicated in studies of licensed London taxi drivers with many years of experience. Positron emission tomography (PET) has been used to examine neural substrates of topographical memory retrieval. Their recall of complex routes around the city resulted in activation of a network of brain regions, including the right hippocampus.[35] In addition, structural MRIs of the brains of taxi drivers show significantly larger posterior hippocampi compared with control subjects.[36] In this study, the length of time spent as a taxi driver positively correlated with the volume of the right posterior hippocampus. Long-term potentiation (LTP) especially in hippocampus has been proposed as an in vitro model for long-term memory. LTP is triggered by Ca^{2+} influx following activation of NMDA receptors and it gives a rapid and lasting increase in synaptic strength.[37] However,

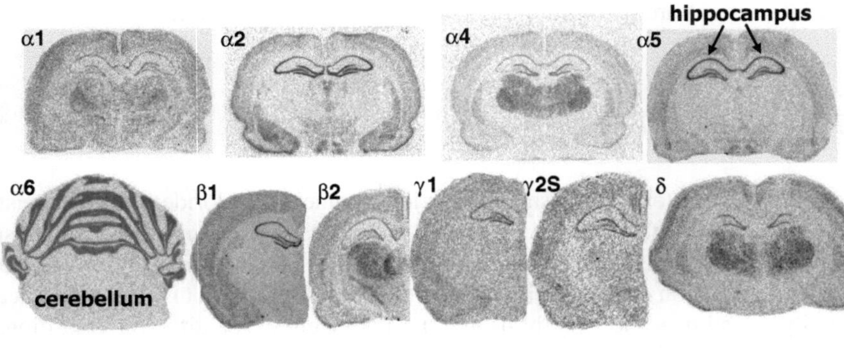

Fig. 5.1 $GABA_A$ receptor subunit localization in the rat brain as identifies by using in situ hybridization. Alpha6 is mainly localized in the cerebellum, and alpha5 in hippocampus (Johansson et al. 2007, unpublished data)

there are controversial data on the LTP requirement in memory formation.[38] The CA1 area of hippocampus exhibit impaired storage and retrieval of spatial memory, but unimpaired non-spatial learning.[39] However, the CA3 region is important for associative memory recall.[40] In hippocampus, the major CNS site for memory and learning, allopregnanolone decreases the neural activity.[14,41] GABA-steroids also impair LPT formation in hippocampus[42–44] and GABA-steroids hamper memory-related cholinergic action in neurons projecting to hippocampus.[45]

5.2.1 Neurotransmitters Involved in Memory and Learning

The glutamate neurotransmitter system, especially the NMDA receptor, is very important in memory formation, and receptor antagonists and NMDA knockouts severely impair memory.[41,46] Adult mice lacking NMDA receptor-mediated synaptic currents and long-term potentiation in CA1 exhibit impaired storage and retrieval of spatial memory.[39] However, glutamate receptors, especially the NMDA receptor when over stimulated lead to neuronal injury or cell death. This is probably due to excessive Ca^{2+} influx through the receptor that eventually will start an apoptotic process.[47] Other neurotransmitters such as acetylcholine, GABA, serotonin (5HT) and dopamine, are also involved in memory function. 5HT1A and 5HT2C receptor knockout/mutant mice also show impaired spatial learning whereas 5HT1B knockout mice exhibit facilitation.[48] Reduction in both cholinergic and serotonergic functions causes severe memory impairment in young as well as in aged rats.[49] The dopamine system is also involved in learning and memory.[50] The GABA system is the major inhibitory system in the brain and $GABA_A$ receptor active substances, like GABA-steroids, benzodiazepine, barbiturates and alcohol can inhibit learning and memory in humans and animals.[51,52] In addition, $GABA_A$ receptor activation with propofol and THDOC (tetrahydrodeoxycorticosterone) can inhibit LTP induction.[43,44] There are also complex interactions between the 5HT, dopamine, acetylcholine and the GABA systems that are activated by neuroactive steroids and these interactions are involved in regulation of cognitive functions.[15]

5.2.2 $GABA_A$ Receptor System

The $GABA_A$ receptor is a transmembrane ion channel for chloride ions. It consists of five protein subunits. The $GABA_A$ receptor expresses different subunit compositions in different parts of the brain,[53] (Fig. 5.1). The subunit composition is related to different function of the specific part of the brain,[54] (Table 5.1). This anatomical diversity constitutes the very basis for the pathogenesis of different conditions. Steroids interact differently with the $GABA_A$ receptor depending on the subunit composition.[55] The subunit alpha5 is localized in high degree in the hippocampus,

Table 5.1 $GABA_A$ receptor subunit relation between disorders or symptoms to different brain areas and to different $GABA_A$ receptor subunit type ($GABA_A$-R).[54] The localization of the different subunits is shown in Fig. 5.1

	Brain region	$GABA_A$ –R type		
1. Memory and learning disruption	Hippocampus	α5		δ
2. Anxiety and mood disturbance	Hippocampus amygdale	α2 α4	β3	δ
3. Fatigue, sedation and exhaustion	Wide spread	α1		
4. Depression induced by stress		α1 α3	β1 β2	δ γ2
5. Eating disorders	Hypothalamus		β3	
6. Relapse in alcohol abuse		α6		
7. Balance and mobility disorders	Cerebellum	α6	β2	δ
8. Epilepsy and excitability disorders		α1	β3	

a key area for memory and learning. Alpha5 is shown to be important for learning and memory function (see Table 5.1 and Fig. 5.1.) since alpha5 knockout mice in comparison with wild-type mice show significantly better performance in a water maze model of spatial learning.[56] In addition, blockade of the $GABA_A$ receptor subunit alpha5 increased learning and memory.[57] $GABA_A$ receptor activation can inhibit LTP induction and NMDA receptors, which are involved in the regulation of hippocampal-dependent spatial memory.[41,46]

Several papers report changes in the $GABA_A$ receptor subunit composition and decreased GABA function after long-term exposure to $GABA_A$ receptor agonists.[55,58] It is a well-known fact that tolerance develops during long-term $GABA_A$ receptor exposure. Tolerance development is noted already after 90 min exposure to anesthetic dosages of allopregnanolone. Changes in the alpha4 subunit of the $GABA_A$ receptor in thalamus were related to the tolerance development.[59] GABA-steroid long-term treatment induces down regulation of the $GABA_A$ receptor function in mammalian cultured neurons both for neuroactive steroids and other $GABA_A$ receptor active drugs.[60,61] During pregnancy when allopregnanolone is high there is a decrease in $GABA_A$ receptor function and changes in subunit composition of the $GABA_A$ receptor. Inhibiting the synthesis of allopregnanolone blocked these changes.[62] Chronically high cortisol levels and GABA-steroids give irreversible cognitive damages.[2] A reduced sensitivity to benzodiazepines, alcohol and GABA-steroid is also seen in women with premenstrual dysphoric disorder (PMDD) during the luteal phase. This change in functional $GABA_A$ receptor sensitivity, as measured by reduced sedation and saccadic eye velocity response to GABA active compounds, also relates to symptom severity in PMDD patients.[63] This action of GABA-steroids is covered in another chapter of this book. In a rat model of PMDD, the GABA-steroid allopregnanolone up regulates the alpha4 subunit of the $GABA_A$ receptor in hippocampus parallel to the induction of anxiety and the anxiety induction was blocked if the animals were treated with alpha4 antisence.[64,65] We have further developed the model and noticed that the high risk-taking rats are the one that react with anxiety in the PMDD model.[66]

5.2.3 Stress

Chronic stress is known to impair memory.[67] In the persons with a chronically elevated adrenal steroid production the impaired memory is permanent and not possible to modulate.[2] Impaired memory is also reported in chronic stress syndromes like the "burn-out syndrome" and chronic stress. "Burn-out syndrome" is frequent in the patient history of patients with Alzheimer's disease.[3] During stress the production of cortisol and GABA-steroids increase in parallel.[18,22,68] In addition allo-THF, the 3alpha-hydroxy-5alpha metabolite of cortisol, enhance the effect of allopregnanolone on increased $GABA_A$ mediated chloride influx.[16] GABA-steroid production in the adrenals is regulated as cortisol and thus increased production occur during stress.[18] During chronic stress the long-term exposure of $GABA_A$ receptor activators results in similar changes as during prolonged exposure to benzodiasepines and alcohol, i.e., tolerance and dysfunction of the GABA-system. The situation during chronic stress is similar as during continuous medroxyprogesterone treatment in postmenopausal hormone therapy where the dementia frequency was doubled compared to placebo treatment.[1] The long-term and high activation of the GABA-system is an important factor in the etiology of stress induced memory impairment. Patients with AD have several changes in the adrenal regulation and response during stress compared to normal controls, indicating that the cortisol and GABA-steroid production are distorted in AD patients. Patients with Alzheimer's disease have similar cortisol and GABA-steroid response to adrenal stimulation as chronically stressed animals.[69,70] Patients with mild Alzheimer's disease have a high and non-suppressible production of cortisol and GABA-steroids.[31] Cortisol in chronic stress is related to the dementia development.[2] We know that the 3alpha-hydroxy-5alpha metabolites of cortisol interact with the $GABA_A$ receptor and enhance the effect of GABA-steroids on the $GABA_A$ receptor.[16]

5.2.4 Tolerance Development to Long-Term GABA-Steroid Exposure

After a long-term exposure to GABA agonists, down-regulation of the GABA-system is expected to occur with malfunction as a result.[71,72] A malfunctioning GABA-system is probably an important factor in the pathogenesis of stress-induced depression, "burn-out" syndrome and sex steroid-related depression.[73,74] The down-regulation occurs at different levels in a time-dependent manner: (1) desensitization; (2) receptor internalization; (3) receptor subunit degradation; and (4) changed expression of receptor mRNA.[71] Exposure to an agonist may earlier cause changes in receptor mRNA and thus a change in the $GABA_A$ receptor subunit composition.[65] In women with PMDD, reduced benzodiazepine, ethanol, and GABA-steroid sensitivity occurs during the luteal but not the follicular phase of the menstrual cycle.[75–77] In rodents GABA-steroids develop tolerance against several effects after repeated administration,[59] e.g., towards the GABA-steroid inhibition of learning even though some degree of inhibited learning remained.[78] It is well known that prolonged exposure to allopregnanolone changes the alpha4 subunit and the alpha1 subunit.[65] In fact, the $\alpha4\beta2\gamma2$ receptor is less sensitive to steroid modulation than $\alpha1\beta2\gamma2$ $GABA_A$ receptors.[55] Thus changes in subunit composition and/or down regulation of the whole $GABA_A$ receptor could be mechanisms for GABA-steroid induced disorders. We have developed animal models for tolerance studies and tolerance against allopregnanolone anesthesia is developed after 90 min related to changes in $GABA_A$ receptor subunit alpha4 in thalamus.[59]

5.2.5 *Sex and Stress Steroid Effect on Learning*

5.2.5.1 Estrogen

Estrogen has been shown to improve verbal memory and motor skills.[79,80] After ovarie-ectomy, women treated for 2 months with estrogen scored better on recall of a story than placebo-treated patients.[81] In women of fertile age, a decrease of ovarian function with gonadotropin releasing hormone agonists also induces verbal memory deficits, which are reversed by adding back estrogen.[82] Several epidemiological studies have indicated a protective role of estrogen on cognitive performance and risk of developing dementia.[83] This was the reason for starting the women's health initiative study to verify the epidemiological results in a clinical trial. In the study there was also a group of hysterectomized women treated with estrogen only. No significant effect was, however, seen on dementia frequency by estrogen-only treatment in hysterectomized women. The lack of effect by estrogen-only indicates that the dementia inducing effect by estrogen/progestagen treatment was caused by the progestagen.[1,6]

However, animal results show several biological effects in CNS by estradiol. In rats, dendrite spines, the primary loci of excitatory synapses, have been associated with neuronal plasticity.[84] Ovariectomy or gonadal steroid replacement affects spine density in the CA1 region.[85] Excitatory NMDA receptors increase in the CA1 region of the dorsal hippocampus after a 2-day estrogen supplementation.[86] It has been suggested that the estradiol-induced increase in dendritic spine density is mediated via stimulation of the glutamate system with activation of NMDA receptors[87] or via reduction of GABA neurotransmission in hippocampal neurons.[88]

5.2.5.2 Progesterone and Progestagens

Progesterone impairs spatial memory in Morris water maze and the progesterone metabolite and GABA-steroid allopregnanolone inhibits learning in rat studies.[7,89–91] Importantly, when progestagen is given continuously for a long time, as in hormone replacement therapy, the risk for dementia increases in postmenopausal women.[1]

In the study medroxyprogesterone acetate is combined with estrogen but, as no significant effect on dementia was noted by estrogen alone the results suggests that the dementia inducing effect was by the progestagen.[6] It is of interest that medroxyprogesterone acetate can induce anesthesia, probably via 3α-hydroxy-5α reduced metabolites that are active on the $GABA_A$ receptor.[4]

Cognitive disturbances are reported in pregnant women – including sleepiness, forgetfulness, reading difficulties, disorientation, poor concentration and poor coordination.[92] Complaints of memory impairment in some women often begin in the second trimester, when very high progesterone and allopregnanolone levels are present in the brain, and resolve soon after childbirth. Interestingly, decreased maternal brain size with slight ventricular increase has been shown late in pregnancy.[93] During the luteal phase of the menstrual cycle, when high circulating levels of progesterone and allopregnanolone are found, women with PMDD often show difficulty concentrating and impairment in working memory.[94] All hippocampal sub-regions are rich in $GABA_A$ receptors, and allopregnanolone can inhibit neural activity in the CA1 and the dentate gyrus areas of the hippocampus.[41,95]

We have investigated whether allopregnanolone influences amygdala-dependent emotional memory by modulating amygdala and/or hippocampal activity. A single progesterone dose was administered orally to healthy women in a double-blind, placebo-controlled, crossover design. Serum levels of progesterone and allopregnanolone were measured. The participants were asked to memorize and subsequently to recognize faces while their brain activity was measured with functional MRI (fMRI). The progesterone administration produced progesterone and allopregnanolone plasma concentrations similar to those during pregnancy, and the administration resulted in decreased recognition accuracy for faces. The imaging results show that the amygdala, hippocampus, and fusiform gyrus supported memory formation of faces. Importantly, the progesterone administration decreased the response to faces in the amygdala and fusiform gyrus during memory encoding. However, it did not modulate the differential activation between subsequently remembered and subsequently forgotten faces. A similar pattern of results was observed in the fusiform gyrus and prefrontal cortex during memory retrieval. These results suggest that progesterone impairs emotional memory by reducing amygdala activity during memory encoding. Furthermore, the results indicate that progesterone impairs memory by reducing the probability of engaging those neural mechanisms that support memory formation and retrieval without affecting the actual memory processing.[96] Preliminary findings from our own lab indicate that intravenous allopregnanolone administration, resulting in serum concentrations similar to third-trimester pregnancy, will impair episodic memory in healthy women, Kask in manuscript.[97]

5.2.5.3 Androgens

Testosterone is also metabolized to 3alpha-hydroxy steroids that are agonists on the $GABA_A$ receptor. Especially one metabolite is investigated namely androstanediol (2alpha-OH-androstane-17-ol). This steroid is weaker in its effect on the $GABA_A$

receptor than allopregnanolone and possesses only a part of allopregnanolones effect.[98] Androstanediol has in rats showed abuse potential effects and some of the abuse usage of anabolic steroids may be due to the rewarding effects of androstanediol or similar compounds.[99] Androgens can also be metabolized to estradiol in the brain and thereby may exert an estrogen like effect on memory and learning.[100] An enhancing effect on cognitive function has bee seen probably mediated via the hormonal androgen receptor in hippocampus.[101]

5.2.5.4 Effects on Memory and Learning of Other Types of $GABA_A$ Receptor Agonists

Permanent negative damage of memory is noted with long-term exposure by every type of $GABA_A$ receptor agonists, e.g., benzodiazepines, barbiturates and alcohol.[102] A meta-analysis for long-term benzodiazepine treatment show permanent memory damage noted after cessation of benzodiazepine treatment even after the data are corrected for confounding factors such as the disorders that were the reason for the treatment.[8] Alcohol is another $GABA_A$ receptor-enhancing substance and its long-term use has been shown to induce dementia.[10] The influence of alcohol on memory function remains after correction for confounders and end of drinking. In fact alcohol increases the production of allopregnanolone and GABA-steroids are suggested as mediators of the alcohol effect.[103] Barbiturates are also acting via the $GABA_A$ receptor and long-term usage is also known to induce permanent cognitive disturbances in animals.[9]

5.2.5.5 GABA-Steroids

As mentioned above GABA-steroids are $GABA_A$ receptor positive modulators of the GABA effect on the $GABA_A$ receptor.[13] Chemically they are 3alpha-hydroxy-5alpha/beta metabolites of progesterone (pregnanolone and allopregnanolone), testosterone (3alpha-5alpha-androstan-diol) and desoxycorticosterone (tetrahydro-desoxycorticosterone, THDOC). GABA-steroids enhance the GABA effect on chloride flux resulting in increased inhibition of neural activity.[13,104] GABA-steroids induce sedation and can be used as anesthetic drugs in humans.[105,106] GABA-steroids are antiepileptic[107,108] and in animal experiments they also possess an anxiolytic effect similar to benzodiazepines.[109] An anxiolytic effect of allopregnanolone has, however, never been shown in humans.

However, treatment with GABA-steroids, including allopregnanolone, impairs learning and memory.[7,90,91,110] A major decrease in spatial learning occurs after iv administration of allopregnanolone.[7] This effect is not caused by impairment of motor function, as swim speed is unaffected. An increase in swimming close to the pool wall was also found in most of the rats injected with allopregnanolone.[7] We found that the learning impairment occurred shortly after the injection but was not noted when the allopregnanolone concentration in hippocampus had decreased to half the impairing concentration. The result indicates that for the acute learning impairment the allopregnanolone concentration needed in hippocampus is high, while lower concentrations were below the threshold of acute learning impairment.[7] The concentrations needed for inducing acute learning impairment are probably much higher than the concentrations of medroxyprogesterone that induced dementia in women on postmenopausal hormone replacement therapy.[1] The sensitivity in the brain for $GABA_A$ receptor active compounds differs between individuals and therefore one can expect that only certain individuals will react with impaired memory and learning on low concentrations of GABA-steroids.[75–77] The hippocampus is a key brain area for learning and memory, especially spatial learning,[111] and from earlier studies we know that allopregnanolone can inhibit neural activity in the hippocampus.[40] Acetylcholine is known to enhance memory and the current symptomatic treatment for Alzheimer's disease is acetylcholine esterase inhibitors (AChEI) leading to increased levels of acetylcholine in the synapses. This facilitates the signal transmission, which leads to an improved cognition. The GABA-steroid allopregnanolone inhibits the acetylcholine release in hippocampus being one mechanism by which allopregnanolone can inhibit learning and memory.[45]

5.2.5.6 GABA-Steroid Antagonists

In contrast, steroids that act as $GABA_A$ receptor antagonists enhance learning and memory.[91,110] Thus pregnenolone sulfate and DHEAS improve memory in aging mice and prevent pharmacologically induced amnesia.[112] However, strong $GABA_A$ receptor antagonism is potentially dangerous, and might induce seizures.[113] Selective antagonism towards $GABA_A$ receptor agonistic neuroactive steroids would therefore be of great value. In vitro allopregnanolone-induced effects can be antagonized by 3beta-hydroxypregnane steroids.[114–116] Isoallopregnanolone is an isomer of allopregnanolone with a hydroxy group in the 3beta position and does not have any effect on the $GABA_A$ receptor on its own.[114] We showed for the first time that a 3beta-hydroxypregnane steroid (UC1011) can inhibit allopregnanolone-induced spatial learning impairment and GABA potentiation.[117] In vitro studies of UC1011 significantly decreased the allopregnanolone-induced increased chloride ion uptake seen in the presence of GABA in both cortex and hippocampus. Thus it is probable that the allopregnanolone-induced decreased learning is caused by increased $GABA_A$ activation, and that the UC1011 reduction of the allopregnanolone effect depends on decreased $GABA_A$ receptor activation. After 4 days of combined allopregnanolone and UC1011 treatment the rats with combined treatment had learned to find the platform equally rapid as the vehicle-treated rats that is 1 day later than the vehicle-treated control rats.[117] Animals that only received UC1011 learned the task just as well as control rats injected with vehicle, indicating that UC1011 had no effect on spatial learning by itself.[117] The group of rats that learned the task did not find the platform quicker because of better swim performance; there was no significant difference in swim speed between the groups. The in vitro antagonism of UC1011 on chloride ion uptake indicates that the antagonism of the allopregnanolone effect is at the $GABA_A$ receptor level.[117] This is supported by results with other 3beta-pregnan-steroids tested on recombinant $GABA_A$ receptors expressed in Xenopus oocytes,[118] as well as the isoallopregnanolone inhibition of allopregnanolone induced effects on the population spike in hippocampal slices[115] Importantly, the blockers have neither no influence on the effects by GABA itself nor any effect on the action of benzodiazepines or barbiturates, and consequently the function of the 3beta-pregnane compounds are specific.[114] In the hippocampus, the brain center for learning and memory, GABA-steroids inhibit neuronal activity in physiological concentrations and the inhibitory compounds act in molar range.[115] An injection of pregnenolone sulfate into hippocampus of rats initiates a release of acetylcholine whereas injection of the GABA-steroid allopregnanolone inhibits the acetylcholine release.[45]

5.3 Conclusion

Neuroactive steroids certainly influences memory and learning and are involved in the etiology of dementia. The detailed mechanism is not known but involvement of the $GABA_A$ receptor positive modulators including the GABA-steroids in the dementia etiology is supported both in clinical and experimental data. Compounds antagonizing the GABA-steroid effect may have a role in treatment strategy for cognitive disorders. The role of estradiol in the dementia etiology is less clear and not supported in clinical data but in experimental data and on certain memory functions estradiol seem to have a role.

Acknowledgements This work is supported by EU structural fund objective 1, Swedish research council Medicine (proj. 4X-11198), Västerbottens county, Umeå Municipal, Northern Sweden health region, Norrlands University Hospital, Umeå University foundations.

References

1. Shumaker SA, Legault C, Rapp SR, et al. Estrogen plus progestin and the incidence of dementia and mild cognitive impairment in postmenopausal women: the women's health initiative memory study: a randomized controlled trial. JAMA 2003; 289:2651–2662.
2. Lupien SJ, Fiocco A, Wan N, et al. Stress hormones and human memory function across the lifespan. Psychoneuroendocrinology 2005; 30:225–242.
3. Sandström A, Rhodin IN, Lundberg M, et al. Impaired cognitive performance in patients with chronic burnout syndrome. Biol Psychol 2005; 69:271–279.
4. Meyerson B. Relationship between the anesthetic and gestagenic action and estrus behaviour-inducing activity of different progestins. Endocrinology 1967; 81:369–374.
5. Henderson VW, Paganini Hill A, Emanuel CK, et al. Estrogen replacement therapy in older women. Comparisons between Alzheimer's disease cases and nondemented control subjects. Arch Neurol 1994; 51:896–900.

6. Shumaker SA, Legault C, Kuller L, et al. Conjugated equine estrogens and incidence of probable dementia and mild cognitive impairment in postmenopausal women: Women's Health Initiative Memory Study. JAMA 2004; 291:2947–2958.
7. Johansson IM, Birzniece V, Lindblad C, et al. Allopregnanolone inhibits learning in the Morris water maze. Brain Res 2002; 934:125–131.
8. Barker MJ, Greenwood KM, Jackson M, et al. Persistence of cognitive effects after withdrawal from long-term benzodiazepine use: a meta-analysis. Archives of Clinical Neuropsychology 2004; 19(3):437–454.
9. Mohammed AK, Wahlström G, Tiger G, et al. Impaired performance of rats in the Morris swim-maize test late in abstinence following long-term sodium barbital treatment. Drug Alcohol Depend 1987; 20(3):203–212.
10. Saunders PA, Copeland JR, Dewey ME, et al. Heavy drinking as a risk factor for depression and dementia in elderly men. Findings from the Liverpool longitudinal community study. Br J Psychiatry 1991; 159:213–216.
11. Maglóczki E, Pákáski M, Janka Z, et al. Risk factors of cognitive decline in residential care in Hungary. Int J Geriatr Psychiatry 2007; June 11 [epub ahead of print].
12. Solfrizzi V, D'Introno A, Colacicco AM, et al. Alcohol consumption, mild cognitive impairment, and progression to dementia. Neurology 2007; 68(21):1790–1799.
13. Majewska MD, Harrison NL, Schwartz RD, et al. Steroid hormone metabolites are barbiturate-like modulators of the GABA receptor. Science 1986; 232:1004–1007.
14. Wang MD, Landgren S, Backstrom T. The effects of allopregnanolone, pregnenolone sulphate and pregnenolone on the CA1 population spike of the rat hippocampus after 17 beta-oestradiol priming. Acta Physiol Scand 1997; 159(4):343–344.
15. Birzniece V, Bäckström T, Johansson IM, et al. Neuroactive steroid effects on cognitive functions with a focus on the serotonin and GABA systems. Brain Res Rev 2006; 51:212–239.
16. Strömberg J, Bäckström T, Lundgren P. Rapid non-genomic effect of glucocorticoid metabolites and neurosteroids on the gamma-aminobutyric acid-A receptor. Eur J Neurosci 2005; 21:2083–2088.
17. Celotti F, Melcangi RC, Martini L. The 5 alpha-reductase in the brain: molecular aspects and relation to brain function. Front Neuroendocrinol 1992; 13:163–215.
18. Purdy RH, Morrow AL, Moore Jr. PH, et al. Stress-induced elevations of gamma-aminobutyric acid type A receptor-active steroids in the rat brain. Proc Natl Acad Sci USA 1991; 88:4553–4557.
19. Bixo M, Andersson A, Winblad B, et al. Progesterone, 5alpha-pregnane-3,20-dione and 3alpha-hydroxy-5alpha-pregnane-20-one in specific regions of the human female brain in different endocrine states. Brain Res 1997; 764:173–178.
20. Wang M, Seippel L, Purdy RH, et al. Relationship between symptom severity and steroid variation in women with premenstrual syndrome: study on serum pregnenolone, pregnenolone sulfate, 5 alpha-pregnane-3,20-dione and 3 alpha-hydroxy-5 alpha-pregnan-20-one. J Clin Endocrinol Metab 1996; 81(3):1076–1082.
21. Luisi S, Petraglia F, Benedetto C, et al. Serum allopregnanolone levels in pregnant women: changes during pregnancy, at delivery, and in hypertensive patients. J Clin Endocrinol Metab 2000; 85:2429–2433.
22. Serra M, Pisu MG, Littera M, et al. Social isolation-induced decreases in both the abundance of neuroactive steroids and GABA(A) receptor function in rat brain. J Neurochem 2000; 75:732–740.
23. Rasmuson S, Andrew R, Nasman B, et al. Increased glucocorticoid production and altered cortisol metabolism in women with mild to moderate Alzheimer's disease. Biol Psychiatry 2001; 49(6):547–552.
24. Sutton J. Memory, In Ed. Zalta EN. The Stanford Encyclopedia of Philosophy (Summer 2004 Edition), http://plato.stanford.edu/archives/sum2004/entries/memory/
25. Morris R. Developments of a water-maze procedure for studying spatial learning in the rat. J Neurosci Meth 1984; 11:47–60.

26. D'Hooge R, De Deyn PP. Applications of the Morris water maze in the study of learning and memory. Brain Res Brain Res Rev 2001; 36:60–90.
27. American Psychiatric Association. Diagnostic and Statistical manual of mental disorders, 4th edn. Washington, DC, 1994.
28. Winblad B, Palmer K, Kivipelto M, et al. Mild cognitive impairment – beyond controversies, towards a consensus: report of the International Working Group on Mild Cognitive Impairment. J Intern Med 2004; 256(3):240–246.
29. Terry AV, Buccafusco JJ. The cholinergic hypothesis of age and Alzheimer's disease-related cognitive deficits: recent challenges and their implications for novel drug development. J Pharmacol Exp Ther 2003; 306(3):821–827.
30. Wenk GL. Neuropathologic changes in Alzheimer's disease: potential targets for treatment. J Clin Psychiatry 2006; 67 (Suppl 3):3–7.
31. Näsman B, Olsson T, Viitanen M, et al. A subtle disturbance in the feedback regulation of the hypothalamic-pituitary-adrenal axis in the early phase of Alzheimer's disease. Psychoneuroendocrinology 1995; 20(2):211–220.
32. Astur RS, Taylor LB, Mamelak AN, et al. Humans with hippocampus damage display severe spatial memory impairments in a virtual Morris water task. Behav Brain Res 2002; 132:7–84.
33. Farr SA, Flood JF, Morley JE. The effect of cholinergic, GABAergic, serotonergic, and glutamatergic receptor modulation on posttrial memory processing in the hippocampus. Neurobiol Learn Mem 2000; 73:150–167.
34. Teng E, Squire LR. Memory for places learned long ago is intact after hippocampal damage. Nature 1999; 400:675–677.
35. Maguire EA, Frackowiak RS, Frith CD. Recalling routes around london: activation of the right hippocampus in taxi drivers. J Neurosci 1997; 17:7103–7110.
36. Maguire EA, Gadian DG, Johnsrude IS, et al. Navigation-related structural change in the hippocampi of taxi drivers. Proc Natl Acad Sci USA 2000; 97:4398–4403.
37. Lynch MA. Long-term potentiation and memory. Physiol Rev 2004; 84:87–136.
38. Meiri N, Sun MK, Segal Z, et al. Memory and long-term potentiation (LTP) dissociated: normal spatial memory despite CA1 LTP elimination with Kv1.4 antisense. Proc Natl Acad Sci USA 1998; 95:15037–15042.
39. Tsien JZ, Huerta PT, Tonegawa S. The essential role of hippocampal CA1 NMDA receptor-dependent synaptic plasticity in spatial memory. Cell 1996; 87:1327–1338.
40. Nakazawa K, Quirk MC, Chitwood RA, et al. Requirement for hippocampal CA3 NMDA receptors in associative memory recall. Science 2002; 297:211–218.
41. Landgren S, Wang MD, Backstrom T, et al. Interaction between 3 alpha-hydroxy-5 alpha-pregnan-20-one and carbachol in the control of neuronal excitability in hippocampal slices of female rats in defined phases of the oestrus. Acta Physiol Scand 1998; 162:77–88.
42. O'Keefe J, Dostrovsky J. The hippocampus as a spatial map. Preliminary evidence from unit activity in the freely-moving rat. Brain Res 1971; 34:171–175.
43. Dubrovsky B, Tatarinov-Levin A, Harris J. Effects of the active neurosteroid allotetrahydrodeoxycorticosterone on long-term potentiation in the rat hippocampus: implications for depression. Prog Neuropsychopharmacol Biol Psychiatry 2004; 28:1029–1034.
44. Wei H, Xiong W, Yang S, et al. Propofol facilitates the development of long-term depression (LTD) and impairs the maintenance of long-term potentiation (LTP) in the CA1 region of the hippocampus of anesthetized rats. Neurosci Lett 2002; 324:181–184.
45. George O, Vallée M, Le Moal M, et al. Neurosteroids and cholinergic systems: implications for sleep and cognitive processes and potential role of age-related changes. Psychopharmacology (Berl) 2006; 186(3):402–413.
46. Riedel G, Platt B, Micheau J. Glutamate receptor function in learning and memory. Behav Brain Res 2003; 140:1–47.
47. Lipton SA. Pathologically-activated therapeutics for neuroprotection: mechanism of NMDA receptor block by memantine and S-nitrosylation. Curr Drug Targets 2007; 8(5):621–632.
48. Buhot MC, Martin S, Segu J. Role of serotonin in memory impairment. Ann Med 2000; 32:210–221.

49. Richter-Levin G, Segal M. Age-related cognitive deficits in rats are associated with a combined loss of cholinergic and serotonergic functions. Ann NY Acad Sci 1993; 695:254–257.
50. Devan BD, Goad EH, Petri HL. Dissociation of hippocampal and striatal contributions to spatial navigation in the water maze. Neurobiol Learn Mem 1996; 66:305–323.
51. Cain DP. Prior non-spatial pretraining eliminates sensorimotor disturbances and impairments in water maze learning caused by diazepam. Psychopharmacology (Berl) 1997; 130:313–319.
52. Holbrook AM, Crowther R, Lotter A, et al. Meta-analysis of benzodiazepine use in the treatment of insomnia. CMAJ 2000; 162:225–233.
53. Sieghart W, Sperk G. Subunit composition, distribution and function of GABA(A) receptor subtypes. Curr Top Med Chem 2002; 2:795–816.
54. Korpi ER, Debus F, Linden AM, et al. Does ethanol act preferentially via selected brain GABA(A) receptor subtypes? The current evidence is ambiguous. Alcohol 2007; 41(3):163–176.
55. Belelli D, Casula A, Ling A, et al. The influence of subunit composition on the interaction of neurosteroids with GABA(A) receptors. Neuropharmacology 2002; 43(4):651–661.
56. Collinson N, Kuenzi FM, Jarolimek W, et al. Enhanced learning and memory and altered GABAergic synaptic transmission in mice lacking the alpha 5 subunit of the $GABA_A$ receptor. J Neurosci 2002; 22:5572–5580.
57. Maubach K. 2003. GABA(A) receptor subtype selective cognition enhancers. J Med Chem 2003; 46(11):233–239.
58. Miller LG, Greenblatt DJ, Barnhill JG, et al. Chronic benzodiazepine administration. I. Tolerance is associated with benzodiazepine receptor downregulation and decreased gamma-aminobutyric acidA receptor function. J Pharmacol Exp Ther 1988; 246:170–176.
59. Birzniece V, Türkmen S, Lindblad C, et al. GABA(A) receptor changes in acute allopregnanolone tolerance. Eur J Pharmacol 2006; 535(1–3):125–134.
60. Yu R, Follesa P, Ticku MK. Down-regulation of the GABA receptor subunits mRNA levels in mammalian cultured cortical neurons following chronic neurosteroid treatment. Brain Res Mol Brain Res 1996; 41:163–168.
61. Yu R, Ticku MK. Effects of chronic pentobarbital treatment on the $GABA_A$ receptor complex in mammalian cortical neurons. J Pharmacol Exp Ther 1995; 275(3):1442–1446.
62. Concas A, Mostallino MC, Porcu P, et al. Role of brain allopregnanolone in the plasticity of gamma-aminobutyric acid type A receptor in rat brain during pregnancy and after delivery. Proc Natl Acad Sci USA 1998; 95:13284–13289.
63. Backstrom T, Andreen L, Birzniece V, et al. The role of hormones and hormonal treatments in premenstrual syndrome. CNS Drugs 2003; 17(5):325–342.
64. Gulinello M, Gong QH, Smith SS. Short-term exposure to a neuroactive steroid increases alpha4 GABA(A) receptor subunit levels in association with increased anxiety in the female rat. Brain Res 2001; 910(1–2):55–66.
65. Smith SS, Gong QH, Hsu FC, et al. GABA(A) receptor alpha4 subunit suppression prevents withdrawal properties of an endogenous steroid. Nature 1998; 392(6679):926–930.
66. Löfgren M, Johansson IM, Meyerson B, et al. Progesterone withdrawal effects in the open field test can be predicted by elevated plus maze performance. Horm Behav 2006; 50(2):208–215.
67. de Quervain DJ, Roozendaal B, McGaugh JL. Stress and glucocorticoids impair retrieval of long-term spatial memory. Nature 1998; 394:787–790.
68. Droogleever Fortuyn HA, van Broekhoven F, Span PN, et al. Effects of Ph.D. examination stress on allopregnanolone and cortisol plasma levels and peripheral benzodiazepine receptor density. Psychoneuroendocrinology 2004; 29(10):1341–1344.
69. Näsman B, Olsson T, Backstrom T, et al. Serum dehydroepiandrosterone sulfate in Alzheimer's disease and in multi-infarct dementia. Biol Psychiatry 1991; 30(7):684–690.
70. Näsman B, Olsson T, Fagerlund M, et al. Blunted adrenocorticotropin and increased adrenal steroid response to human corticotropin-releasing hormone in Alzheimer's disease. Biol Psychiatry 1996; 39(5):311–318.

71. Barnes EM Jr. Use-dependent regulation of $GABA_A$ receptors. Int Rev Neurobiol 1996; 39:53–76.
72. Yu R, Ticku MK. Chronic neurosteroid treatment decreases the efficacy of benzodiazepine ligands and neurosteroids at the gamma-aminobutyric acidA receptor complex in mammalian cortical neurons. J Pharmacol Exp Ther 1995; 275(2):784–789.
73. N-Wihlbäck AC, Sundström-Poromaa I, Bäckström T. Action by and sensitivity to neuroactive steroids in menstrual cycle related CNS disorders. Psychopharmacology (Berl) 2006; 186(3):388–401.
74. Drugan RC, Morrow AL, Weizman R, et al. Stress-induced behavioral depression in the rat is associated with a decrease in GABA receptor-mediated chloride ion flux and brain benzodiazepine receptor occupancy. Brain Res 1989; 487(1):45–51.
75. Sundstrom I, Ashbrook D, Bäckström T. Reduced benzodiazepine sensitivity in patients with premenstrual syndrome: a pilot study. Psychoneuroendocrinology 1997; 22:25–38.
76. Sundstrom I, Andersson A, Nyberg S, et al. Patients with premenstrual syndrome have a different sensitivity to a neuroactive steroid during the menstrual cycle compared to control subjects. Neuroendocrinology 1998; 67:126–138.
77. Nyberg S, Wahlstrom G, Backstrom T, et al. Altered sensitivity to alcohol in the late luteal phase among patients with premenstrual dysphoric disorder. Psychoneuroendocrinology 2004; 29(6):767–777.
78. Türkmen S, Löfgren M, Birzniece V, et al. Tolerance development to Morris water maze test impairments induced by acute allopregnanolone. Neuroscience 2006; 139(2):651–659.
79. Maki PM, Rich BJ, Rosenbaum RS. Implicit memory varies across the menstrual cycle: estrogen effects in young women. Neuropsychologia 2002; 40:518–529.
80. Maki PM, Zonderman AB, Resnick RM. Enhanced verbal memory in nondemented elderly women receiving hormone-replacement therapy. Am J Psychiatry 2001; 158:227–233.
81. Phillips SM, Sherwin BB. Effects of estrogen on memory function in surgically menopausal women. Psychoneuroendocrinology 1992; 17:485–495.
82. Sherwin BB, Tulandi T. "Add-back" estrogen reverses cognitive deficits induced by a gonadotropin-releasing hormone agonist in women with leiomyomata uteri. J Clin Endocrinol Metab 1996; 81:2545–2549.
83. Paganini-Hill A, Henderson VW. Estrogen replacement therapy and risk of Alzheimer disease. Arch Intern Med 1996; 156(19):2213–2217.
84. Harris KM, Kater SB. Dendritic spines: cellular specializations imparting both stability and flexibility to synaptic function. Annu Rev Neurosci 1994; 17:341–371.
85. Gould E, Woolley CS, Frankfurt M, et al. Gonadal steroids regulate dendritic spine density in hippocampal pyramidal cells in adulthood. J Neurosci 1990; 10:1286–1291.
86. Weiland NG. Estradiol selectively regulates agonist binding sites on the N-methyl-D-aspartate receptor complex in the CA1 region of the hippocampus. Endocrinology 1992; 131:662–668.
87. Woolley SC, McEwen BS. Estradiol regulates hippocampal dendritic spine density via an N-methyl-D-aspartate receptor-dependent mechanism. J Neurosci 1994; 14:7680–7687.
88. Murphy DD, Cole NB, Greenberger V, et al. Estradiol increases dendritic spine density by reducing GABA neurotransmission in hippocampal neurons. J Neurosci 1998; 18:2550–2559.
89. Chesler EJ. Acute administration of estrogen and progesterone impairs the acquisition of the spatial morris water maze in ovariectomized rats. Horm Behav 2000; 38(4):234–242.
90. Vallee M, Mayo W, Koob GF, et al. Neurosteroids in learning and memory processes. Int Rev Neurobiol 2001; 46:273–320.
91. Mayo W, Dellu F, Robel P, et al. Infusion of neurosteroids into the nucleus basalis magnocellularis affects cognitive processes in the rat. Brain Res 1993; 607:324–328.
92. Brett M, Baxendale M. Motherhood and memory: a review. Psychoneuroendocrinology 2001; 26:339–362.
93. Oatridge A, Holdcroft A, Saeed N, et al. Change in brain size during and after pregnancy: study in healthy women and women with preeclampsia. AJNR Am J Neuroradiol 2002; 23:19–26.

94. Man MS, MacMillan I, Scott J, et al. Mood, neuropsychological function and cognitions in premenstrual dysphoric disorder. Psychol Med 1999; 29:727–733.
95. Wisden W, Laurie DJ, Monyer H, et al. The distribution of 13 $GABA_A$ receptor subunit mRNAs in the rat brain. I. Telencephalon, diencephalon, mesencephalon. J Neurosci 1992; 12:1040–1062.
96. van Wingen GA, van Broekhoven F, Verkes RJ, et al. How progesterone impairs memory for biologically salient stimuli in healthy young women. *J. Neurosci.* 2007; 27(42):11416–23.
97. Kask K, Nilsson L-G, Bäckström T, et al. Allopregnanolone deteriorates episodic memory in healthy females. 2007; Manuscript.
98. Turner DM, Ransom RW, Yang JS, et al. Steroid anesthetics and naturally occurring analogs modulate the gamma-aminobutyric acid receptor complex at a site distinct from barbiturates. J Pharmacol Exp Ther 1989; 248:960–966.
99. Frye CA. Some rewarding effects of androgens may be mediated by actions of its 5alpha-reduced metabolite 3alpha-androstanediol. Pharmacol Biochem Behav 2007; 86(2):354–367.
100. Edinger KL, Frye CA. Androgens' effects to enhance learning may be mediated in part through actions at estrogen receptor-beta in the hippocampus. Neurobiol Learn Mem 2007; 87(1):78–85.
101. Edinger KL, Frye CA. Androgens' performance-enhancing effects in the inhibitory avoidance and water maze tasks may involve actions at intracellular androgen receptors in the dorsal hippocampus. Neurobiol Learn Mem 2007; 87(2):201–208.
102. Matthews DB, Morrow AL, Tokunaga S, et al. Acute ethanol administration and acute allopregnanolone administration impair spatial memory in the Morris water task. Alcohol Clin Exp Res 2002; 26:1747–1751.
103. Silvers JM, Tokunaga S, Berry RB, et al. Impairments in spatial learning and memory: ethanol, allopregnanolone, and the hippocampus. Brain Res Brain Res Rev 2003; 43(3):275–284.
104. Gee KW, Chang WC, Brinton RE, et al. GABA-dependent modulation of the Cl- ionophore by steroids in rat brain. Eur J Pharmacol 1987; 136:419–423.
105. Carl P, Hogskilde S, Nielsen JW, et al. Pregnanolone emulsion. A preliminary pharmacokinetic and pharmacodynamic study of a new intravenous anaesthetic agent. Anaesthesia 1990; 45(3):189–197.
106. Timby E, Balgård M, Nyberg S, et al. Pharmacokinetic and behavioral effects of allopregnanolone in healthy women. Psychopharmacology 2006; 186(3):414–424.
107. Backstrom T, Zetterlund B, Blom S, et al. Effects of intravenous progesterone infusions on the epileptic discharge frequency in women with partial epilepsy. Acta Neurol Scand 1984; 69(4):240–248.
108. Landgren S, Aasly J, Backstrom T, et al. The effect of progesterone and its metabolites on the interictal epileptiform discharge in the cat's cerebral cortex. Acta Physiol Scand 1987; 131(1):33–42.
109. Wieland S, Lan NC, Mirasedeghi S, et al. Anxiolytic activity of the progesterone metabolite 5 alpha-pregnan-3 alpha-o1-20-one. Brain Res 1991; 56(11):263–268.
110. Ladurelle NB, Eychenne D, Denton J, et al. Prolonged intracerebroventricular infusion of neurosteroids affects cognitive performances in the mouse. Brain Res 2000; 858:371–379.
111. Riedel G, Micheau J, Lam AG, et al. Reversible neural inactivation reveals hippocampal participation in several memory processes. Nat Neurosci 1999; 2:898–905.
112. Vallee M, Mayo W, Le Moal M. Role of pregnenolone, dehydroepiandrosterone and their sulfate esters on learning and memory in cognitive aging. Brain Res Brain Res Rev 2001; 37:301–312.
113. Kokate TG, Juhng KN, Kirkby RD, et al. Convulsant actions of the neurosteroid pregnenolone sulfate in mice. Brain Res 1999; 831:119–124.
114. Lundgren P, Stromberg J, Backstrom T Wang M. Allopregnanolone-stimulated GABA-mediated chloride ion flux is inhibited by 3beta-hydroxy-5alpha-pregnan-20-one (isoallopregnanolone). Brain Res 2003; 982:45–53.

115. Wang MD, Backstrom T, Landgren S. The inhibitory effects of allopregnanolone and pregnanolone on the population spike, evoked in the rat hippocampal CA1 stratum pyramidale in vitro, can be blocked selectively by epiallopregnanolone. Acta Physiol Scand 2000; 169:333–341.
116. Bäckström T, Wahlström G, Wahlström K, et al. Isoallopregnanolone; an antagonist to the anaesthetic effect of allopregnanolone in male rats. Eur J Pharmacol 2005; 512(1):15–21.
117. Turkmen S, Lundgren P, Birzniece V, et al. 3beta-20beta-dihydroxy-5alpha-pregnane (UC1011) antagonism of the GABA potentiation and the learning impairment induced in rats by allopregnanolone. Eur J Neurosci 2004; 20:1604–1612.
118. Wang M, He Y, Eisenman LN, et al. 3beta -hydroxypregnane steroids are pregnenolone sulfate-like GABA(A) receptor antagonists. J Neurosci 2002; 22:3366–3375.

Chapter 6
Estrogen, Cholinergic System and Cognition

Sonsoles de Lacalle[1], Bryan Hyler[1], and Thomas Borowski[2]

Abstract Disparities in the epidemiology of mental disorders in males and females provide indirect evidence for hormonal and other factors in disease etiology and course. For example, as hormonal levels fluctuate with the start of the menopause, women begin to experience menstrual cycle changes accompanied by a variety of distressing symptoms such as hot flashes, sleep problems, mood swings, anxiety, difficulty concentrating, disorientation, and memory lapses. These clinical signs suggest that hormonal changes impact both reproductive and non-reproductive areas of the brain.

This chapter focuses on the effects of estrogen on the basal forebrain cholinergic system. Considerable evidence indicates a close correlation between the integrity of the basal forebrain cholinergic system and cognitive and attentional behaviors, with estrogen playing a significant role since it produces an upregulation of cholinergic function, as well as neurite outgrowth and branching. Structural changes have been postulated as integral steps in cellular processes leading to information storage in the nervous system, and perhaps estrogen-induced neurite sprouting within cholinergic neurons could underlie the behavioral effects of estrogen treatment. Findings from our work and that of many others suggest that the specific influences of estrogen on the structure and function of the cholinergic system could explain its ability to maintain certain aspects of memory. These findings may shed light on why women are more susceptible to dementia after the menopause, and thus have important consequences for the quality of life of aging women.

Keywords Plasticity, estrogen, regeneration, memory, cholinergic system, basal forebrain

Abbreviations AD Alzheimer's disease; E2 estradiol; OVX ovariectomized; HACU high affinity choline uptake; ChAT choline acetyltransferase; NOVX non-ovariectomized; HDB nucleus of the horizontal limb of the diagonal band of broca;

[1]Department of Biomedical Sciences, Charles R. Drew University of Medicine and Sciences, Los Angeles, CA, USA

[2]Department of Psychology/Neuroscience, Pitzer College, Claremont, CA, USA

M.S. Ritsner and A. Weizman (eds.), *Neuroactive Steroids in Brain Function, Behavior and Neuropsychiatric Disorders*

P placebo; $p75^{NTR}$ p75 neurotrophin receptor; MS medial septum; ER estrogen receptor, α and β; VDB nucleus of the vertical limb of the diagonal band of broca; MAPK mitogen-activated protein kinase cascade

6.1 Introduction

Changing levels of hormones at puberty and during adult life determine morphological and neurochemical plasticity, not only in the parts of the nervous system that are organized by gonadal hormones during development, but also in other regions whose developmental dependence on hormones is not that clear. In fact, a growing body of evidence, accumulated over decades, supports a regulatory role for estrogen (estradiol, E2) on cognition/memory. In humans, this evidence springs from several sources, including sex differences in cognitive function, human menstrual cycle fluctuations in cognitive performance, randomized controlled trials and observational studies on the effects of E2 depletion and replacement in postmenopausal women, as well as imaging studies on E2 and cognition. While qualitative differences in cognitive skills between the sexes do not exist, quantitatively women tend to excel on tasks of verbal skills and memory, on perceptual speed and accuracy, and on fine motor skills.[1,2] Improved verbal working memory seems to be associated with periods of high E2 levels in the menstrual cycle.[3] In fact, the majority of studies that identified cycle stage, excluded anovulatory women and used gender-sensitive cognitive tests, found that women perform best on sexually dimorphic tests during the midluteal phase, further suggesting that E2 facilitates verbal memory and fine motor skills.[1,3–7]

Randomized control trials, while not completely consistent, nevertheless have shown that performance on 47% of memory measures was better in postmenopausal women who received E2 replacement therapy (see reviews[8,9]). Similarly, most observational and longitudinal studies also show that E2 users perform better on cognitive tests.[1] In fact, when considered in totality, 71% of the studies that examined the effect of E2 on cognitive functioning in humans found significant benefits on one or more neuropsychological tests of cognition.[1] For example, in postmenopausal women without dementia, E2 replacement boosts cognitive performance, measured with tests of immediate and delayed paragraph recall[10] and proper name recall.[11] Another study has shown that E2 use during the postmenopausal period was associated with higher scores on other neuropsychological measures of memory, abstract reasoning and language[12] in a multiethnic community-based cohort of non-demented older women. These results are consistent with several previous reports showing the benefits of E2 use on cognitive function in older women.[10,11,13–17] Although the results of Jacobs et al.[12] indicate that longer exposure to postmenopausal E2 therapy is associated with relatively higher scores on measures of cognition, women with very brief exposures also performed better that women with no history of E2 use.[18] It appears that administration of exogenous E2, even for a limited period of time, may delay or temporary halt neuronal changes associated with E2 deficiency. There is also evidence to suggest that there may be a critical period during the immediate postmenopausal years for protective effects of E2 on cognition,[19] whereas treatment too far past the menopausal transition may have less beneficial effect (see recent reviews[20–22]). In this regard, several studies have reported lack of correlation between E2 levels and cognition,[23,24] which highlights the physiological complexity of hormonal therapy.[25]

Data from experimental animals also indicate that E2 can affect performance on various measures of learning and memory. In ovariectomized (OVX) monkeys, E2 replacement significantly improved cognitive function,[26–28] particularly in visuospatial attention.[29] E2 improved the performance of rats on tasks that are dependent on spatial working memory, such as radial maze and water maze,[30–36] as well as in the operant alternation[31,37] and the active avoidance tasks.[38] This improvement was dependent on the maintenance of constant levels of E2 over a number of days or weeks, with either low, high or physiological levels reported effective.[39] Singh et al.[38] have described that active avoidance behavior in young OVX female rats is reduced by 40% relative to controls, while those animals receiving E2 not only demonstrated superior performance but also an accelerated rate of learning. In addition, it has been shown that E2 exposure prior to training was sufficient to improve working memory performance significantly,[30,31,40–42] suggesting that E2 may induce neuronal changes that persist beyond the period of exposure with functional consequences for behavior. For example, systemic or intrahippocampal administration of a water-soluble form of estradiol within 1 h after a place training session in the Morris water maze improved performance on a retention test 24 h later in both gonadally intact male and OVX female rats[43,44] through interaction with the cholinergic system. In contrast, E2 deprivation results in learning impairment and compromised cholinergic function.[38]

In most rodent studies, the effects of E2 on performance are modest.[39,45] This subtle effect of E2 on non-reproductive behaviors, particularly learning and memory, should not be surprising. Survival would be jeopardized if such functions fluctuated radically in response to the cyclic changes in steroid levels associated with reproductive functions. For this reason, the effects of E2 on learning and memory may be most evident, and crucial, when neural systems are compromised.

The neurotransmitter acetylcholine is directly involved in learning and memory.[46] While acknowledging its many behavioral targets,[47] in particular E2 could influence learning and memory performance through modulation of the basal forebrain cholinergic system,[48–50] which projects widely to the cerebral cortex and hippocampus.[51] Cholinergic neurotransmission could be enhanced by E2 at multiple levels, for example: (a) at the synthetic level, by increasing the synthesis or diminishing the degradation of acetylcholine; (b) at the re-uptake level, through influencing the transport and/or packaging of acetylcholine into vesicles; (c) through structural effects that would enhance synaptic connectivity; and (d) in aging and neurodegeneration, through facilitating the survival of otherwise degenerating cholinergic neurons, perhaps through action on growth factors and growth factor receptors. It has been shown that E2 treatment counteracts the amnesic effects of the cholinergic antagonist scopolamine in rats[31,37,52] and in monkeys.[27,29] In humans,

E2 pretreatment attenuates significantly the anticholinergic drug-induced impairment on attentional tasks.[49] It also prevents a decrease in both high-affinity choline uptake (HACU) and choline acetyltransferase (ChAT) activity in the hippocampus and cortex in OVX rats.[38,48] Moreover, higher levels of ChAT mRNA were demonstrated in cholinergic neurons of the basal forebrain following E2 administration.[53] Several other studies have shown the interaction between E2, cholinergic system and behavioral performance.[29,54–57] In this chapter we focus on the basal forebrain cholinergic system as the potential site of action and underlying mechanism for E2 beneficial influence on learning and memory through functional and structural effects. We set out to investigate whether E2 would have an impact on the cognitive performance of female rats, after a small lesion that mimics the degeneration of cholinergic neurons in Alzheimer's disease (AD), and whether changes in behavior could be correlated with structural plasticity of the cholinergic system.

6.2 Materials and Methods

OVX and intact (NOVX) female Fischer 344 rats (3 mo. old) were obtained from the National Institute of Aging animal breeding facility at Harlan. Bilateral ovariectomies were performed at Harlan 1 week before arrival. Rats were individually housed in standard laboratory cages and maintained on a 12 h-light/dark cycle (lights on at 0700 h). Subjects were fed approximately 12 g of standard laboratory chow daily to maintain adequate consumption of food rewards. Water was available ad libitum throughout the experiment. Animal care and procedures were conducted with approved animal care protocols and in accordance with the NIH Guide for the Care and Use of Laboratory Animals.

One week following arrival rats were deeply anesthetized and received a small unilateral lesion of the nucleus of the horizontal limb of the diagonal band of Broca (HDB) with 200 nl of 0.075 μg/μl 192 IgG-saporin (obtained from Chemicon), according to coordinates taken from the atlas of Paxinos and Watson[58]: −0.6 mm from Bregma; +2 mm from the midline; −8 mm from the surface of the dura. This site was chosen because it is the origin of a substantial cholinergic projection to the entorhinal cortex, and we have used it successfully in the past[59] to model the cholinergic fiber loss found in patients with AD. Four weeks after the lesion the animals were implanted subcutaneously with a pellet containing E2 or placebo (P), resulting in four experimental groups: OVX + E2, OVX + P, NOVX + E2, NOVX + P. The E2 pellets consisted of 0.25 mg of 17β-estradiol suspended in a slow release biodegradable matrix that was specially formulated to allow 2.7 μg/day release of 17β-estradiol over 60 days. Placebo pellets consisted of the biodegradable matrix without E2. Animals were trained and tested on a T-maze task commencing 60 days post-implant, using a black stainless steel T-maze apparatus.

Rats were first food deprived to 85% of their normal body weight and then exposed to five daily T-maze adaptation sessions once per day. Each adaptation session involved placing the animal in the T-maze baited with Kellogg's Fruitloops cereal pieces that were randomly distributed throughout the maze. Each daily adaptation session lasted approximately 10 min or until all the cereal pieces were consumed. Training began 24 h after the last T-maze adaptation session. One goal arm of the T-maze was blocked and the other arm was baited with two cereal pieces. Rats were trained to obtain the cereal pieces from the baited arm. Each training session consisted of six forced choices per day for 4 consecutive days. For each trial right and left arms were alternated in a random balanced fashion. Rats who did not acquire stable levels of T-maze performance were excluded from the experiment.

The delayed matching-to-position version of the T-maze task was used to examine the acquisition of spatial memory performance following E2 treatment. Each daily test session consisted of eight trial pairs. The first trial of each pair consisted of a forced choice in which one goal arm was blocked forcing the animal to enter the open arm to receive a cereal reward. The rat was then returned to the start box and after a 10 s delay was exposed to the second trial where both goal arms were open. Entering the arm that was previously visited resulted in food reward while entering the incorrect arm resulted in no reward and confinement to that arm for 10 s. A choice was defined as the animal placing four paws into a goal arm. Rats were run in squads of 4–6. After each trial pair the rat was returned to its cage for 5–10 min and the other animals were run in the maze. The order of the baited arms was alternated in a random balanced fashion for each test session such that the same arm would not be rewarded for more than three consecutive trial pairs. Subjects were tested in the T-maze task for 14 consecutive days. At the end of the testing, subjects were sacrificed with intracardial perfusion of paraformaldehyde under deep anesthesia, the brains were extracted and processed for histological analyses.

Forty micron-thick sections were stained immunocytochemically with a monoclonal antibody against the $p75^{NTR}$ receptor (clone 192; 1:7,500 from Oncogene Science), a membrane receptor that identifies basal forebrain cholinergic neurons. Sections were mounted on glass slides, dehydrated in a graded series of alcohol, cleared in xylene and coverslipped. Morphometric analysis was performed as described[60] and statistical comparisons (non-parametric, one-way analysis of variance (ANOVA) and t-test) were applied using StatView software, with $p < 0.05$.

For the behavioral analysis, choice accuracy (defined as the percentage of visits across the eight trial pairs to the arm previously baited) in OVX and NOVX female rats was collapsed over 2-day blocks and analyzed separately by a mixed ANOVA with Hormone Treatment as the between-group variable and Day as the within-group variable. Newman-Keuls multiple comparisons ($p < 0.05$) were conducted to evaluate effects of significant interactions.

6.3 Results

E2 treatment did not prevent cell loss induced with 192 IgG-saporin. To evaluate the effect of the immunotoxic lesions, HDB neurons expressing $p75^{NTR}$ were counted manually using the 40× objective of an Olympus microscope. In each animal, the

HDB region was identified in 4–5 sections, and carefully matched across brains. The boundaries for the HDB have been previously defined in our laboratory.[61] For each animal, the lesioned side was compared to the control side, using non-parametric statistics. In each of the four experimental groups, the lesioned side contained less cholinergic neurons than in the spared side, and this difference, although small (12–18%) was statistically significant (Table 6.1). To account for individual variability, we calculated the percent of cells in the lesioned side, compared to the contralateral (spared) side: lesioned × 100/control. The resulting number was used as an estimate of the effect of the lesion in each subject. If E2 had an effect on cell number, then compared to control (NOVX + P), we would see a larger loss in the OVX + P group than in the OVX + E2 group. However, a one-way ANOVA comparing the NOVX + P (as the intact control), OVX + P and OVX + E2 did not show statistically significant differences among the three groups ($F_{(2,14)} = 0.527$, $p > 0.05$), suggesting that the cholinergic neuronal loss was permanent, and E2 treatment had no recovery effect in this experimental paradigm.

By contrast, behavioral analysis provided strong evidence to support a role of E2 on cognition. In the NOVX group (Fig. 6.1A), ANOVA revealed a significant Hormone Treatment X Day interaction ($F_{(6,120)} = 3.27$, $p < 0.005$). Analysis of the effects involved in this interaction showed an impairment in choice accuracy in the NOVX + E2 rats relative to the NOVX + P group at day block 3 (days 5 and 6) and day block 5 to day block 7 (days 9–14). Choice accuracy in these animals was at or below chance for the duration of testing but was above chance (66%) by day 14. In contrast, choice accuracy in the NOVX + P group was 66% at day block 5 (days 9 and 10) and by day block 7 (days 13 and 14) was at 86%. In the OVX group, ANOVA revealed a significant effect of Day ($F_{(6,90)} = 45.15$, $p < 0.0001$) and Hormone Treatment ($F_{(1,15)} = 5.26$, $p < .04$). As shown in Fig. 6.1B, reinforced T-maze performance improved over the 14-day test period for both E2 and placebo treated OVX rats. OVX + E2 rats, however, made significantly more correct responses relative to the OVX + P animals over the 14-day test period. No significant Hormone Treatment X Day interaction was found ($F_{(6,90)} = 1.31$, ns), indicating that while E2 treatment enhanced T-maze performance, it did not affect the rate of acquisition.

Similar to the NOVX + E2 group, OVX + P rats also showed impairments in reinforced T-maze performance when compared to the NOVX + P control rats. Analysis of the learning curves of placebo treated OVX and NOVX rats revealed a significant Group X Day interaction ($F_{(6, 96)} = 2.90$, $p < .01$). Multiple comparisons

Table 6.1 Morphometric analysis of the effect of the 192 IgG-saporin lesion into the nucleus of the horizontal limb of the diagonal band of Broca

	Mean cell number ± SEM					
	Lesioned	Spared	% Change	Wilcoxon z-value	p	N
NOVX + P	284 ± 23	323 ± 27	−12	−2.023	0.0431	5
NOVX + E2	222 ± 17	261 ± 18	−15	−2.028	0.0425	5
OVX + P	244 ± 23	304 ± 39	−17	−1.991	0.0464	6
OVX + E2	331 ± 43	408 ± 60	−18	−2.201	0.0277	6

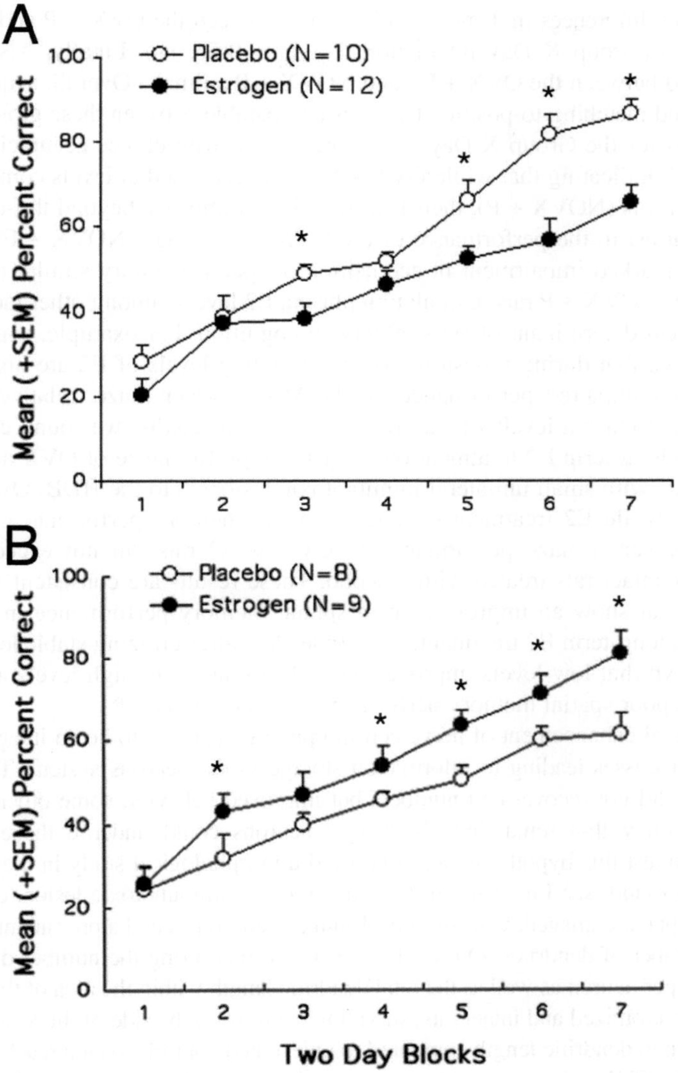

Fig. 6.1 Behavioral performance of non-ovariectomized (**A**) and ovariectomized (**B**) female rats in the T-maze, after a unilateral immunotoxic lesion of the nucleus of the horizontal limb of the diagonal band of Broca, using 192 IgG-saporin. Data reflects mean (±SEM) percent correct choices over 2 weeks period, collapsed into seven blocks (2 days per block). Statistical significance, provided in the text, is indicated with an asterisk

revealed impaired choice accuracy in the OVX + P rats at day block 5 that persisted for the remaining test days. Like the NOVX + E2, OVX + P animals were only able to attain choice accuracies at 61% at test block 7 (days 13 and 14). It should also be noted that T-maze performance in the NOVX + E2 rats was similar to that of the OVX + P group. A separate ANOVA with Group as the between-subjects variable showed no differences in T-maze performance between the OVX + P and NOVX + E2 groups (Group X Day interaction, $F_{(6, 108)} = 0.56$, ns). Finally, ANOVA was conducted between the OVX + E2 and NOVX + P animals. Overall, acquisition of the delayed matching-to-position task was comparable between these groups ($F_{(6, 102)} = 1.31$, ns for the Group X Day block interaction) with choice accuracies at 86% by day 14, indicating that while OVX + E2 rats performed at levels comparable to intact controls (NOVX + P), their learning did not improve beyond those levels.

In contrast to the performance of the OVX + E2 group, NOVX + E2 rats displayed a marked impairment in acquisition in spatial memory similar to the performance of OVX + P rats. Circulating plasma E2 levels, among other factors, may be a critical determinant of E2's effects on cognition. For example, some reports have shown that during proestrus, when circulating levels of E2 are high, female rats exhibit impaired performance on the Morris water maze, relative to estrus phase when plasma levels of E2 are low.[62,63] Unexpectedly, we found contrasting effects of long-term E2 treatment on the T-maze performance of OVX and NOVX female rats with small unilateral immunotoxic lesions into the HDB. Ovariectomy impaired, while E2 treatment enhanced, spatial memory performance in female rats. However, T-maze performance in OVX + E2 rats did not exceed that of gonadally intact rats treated with placebo. These results are consistent with other findings that show an improvement in spatial memory performance in a T-maze following long-term E2 treatment.[31,64–66] Indeed, studies utilizing stable levels of E2 have shown that low levels improve spatial learning while high levels are associated with poor spatial memory performance in some tasks.[67,68]

Structural enhancement of nerve cell morphology appears to be an integral step in cellular processes leading to information storage in the nervous system. Thus, if E2 treatment did not recover cell numbers but improved behavior, some other structural modification within remaining cholinergic neurons could underlie those cognitive effects. To test this hypothesis, we conducted a morphological study in the same animals, and found (see Fig. 1 in ref. 60), that after the immunotoxic lesion, cholinergic neurons appear damaged, with slightly shrunken cell bodies and a prominent reduction in the number of dendrites. Our analysis involved measuring the number of dendritic segments per neuron as well as the total dendritic length within the area of the HDB. In both ovariectomized and intact rats, surviving neurons on the side of the lesion showed a reduction in dendritic length compared to unlesioned controls, which reached a maximum in the OVX + P group, with an average 11% reduction in total length, compared to the unlesioned side of the NOVX + P group (see Fig. 2.A in ref. 60). Total dendritic length in lesioned neurons, compared to their own contralateral controls, showed an effect of the hormone. E2 treatment, whether from natural sources (the NOVX + P group) or from the pellet (the OVX + E2 group) was able to maintain dendritic arborization in lesioned neurons within levels not significantly different from control (see Fig. 2.B in ref. 60). By contrast, dendritic arborization remained significantly reduced in animals totally deprived of E2, which were also behaviorally impaired. These results are consistent with findings by Ferrini et al.[69] showing that in aged male mice E2 treatment increased neurite length in ChAT immunoreactive cells to levels similar to young animals, and also upregulated the expression of GAP-43 mRNA in aged mice.

6.4 Discussion

Taken together, results from these studies strongly suggest that, despite measurable cholinergic cell loss, E2 had a substantial effect on dendritic morphology of remaining basal forebrain cholinergic system, which we suggest may be related to the behavioral improvement found in the same animals. In particular, we hypothesize that E2 may play a role in the formation of new neurite branches, thus leading to the formation of new synapses.

An earlier report showed that 192 IgG-saporin-induced lesions of the medial septum (MS) and the diagonal band of Broca disrupted the ability of E2 to enhance acquisition of a delayed matching-to-position T-maze task.[65] By contrast, our findings demonstrate enhanced performance in this task, in OVX + E2 rats that received unilateral infusions of 192 IgG-saporin into the HDB. One possible explanation for this difference is the extent of the lesion: we selected a relatively low concentration of immunotoxin (.075 μg/μl) to induce partial (12–18%) destruction of cholinergic neurons specific to the HDB, while the other study involved an almost complete depletion of cholinergic neurons, from a larger area of the basal forebrain. Surprisingly, in our study the capacity of E2 to improve spatial learning in OVX rats was still present 2 weeks after E2 treatment had ended. Other reports have shown that E2 exposure prior to training was sufficient to improve spatial memory performance,[30,31,66] suggesting that E2 may induce neuronal changes that persist beyond the period of treatment, with functional consequences for behavior.

Our results also indicate that E2 treatment did not result in enhanced cholinergic neuron survival. Similarly, E2 did not prevent cell loss in basal forebrain cholinergic system following ibotenate or mechanical lesions,[70] but increased the number of ChAT positive cells in the nucleus basalis contralateral to the lesion, and in the MS ipsilateral to the lesion, 2 weeks after injury, congruent with an increase in ChAT protein within cholinergic cells that were previously undetected, as shown by others.[71,72] Another study[73] has shown that the number of ChAT-immunoreactive neurons in the area of the nucleus basalis correlated significantly with accuracy in an attentional task. It is tempting to speculate that this increased accuracy may be related to enhanced cholinergic innervation induced by E2, as described by others both in rats[74–76] and in monkeys.[77,78]

Because of their fundamental involvement in synaptic integration and synaptic plasticity, modifications in dendritic arbors profoundly shape the computational and storage capability of neurons, allowing a degree of flexibility that ensures the performance of the most appropriate computation under a given condition. From a behavioral perspective, this structural flexibility may be particularly useful to allow responses to fluctuating levels of hormones, even though it potentially renders the system more susceptible to errors, and thus to pathological conditions. Previous research has revealed that E2 modulates proteins specifically associated with axonal elongation,[79] enhance outgrowth of nerve processes in cultured neurons,[80,81] particularly in the basal forebrain,[82] promotes formation of dendritic spines and synapses[83] and enhances cortical cholinergic innervation following destruction of the nucleus basalis.[74] The ability of E2 to enhance cholinergic sprouting following cell loss may have significant clinical implications. E2 exposure is known to exert significant neuroprotective effects associated with cell loss that may depend on a number of factors including alterations in cell survival, axonal sprouting, regenerative processes, enhanced synaptic transmission and neurogenesis (see[84] for review). While the functional and morphological changes associated with cholinergic cell loss and E2 treatment are still not fully understood, our results show a marked deficit in spatial learning in OVX + P rats with partial cholinergic lesions of the HDB. While speculative, it is possible that the deficit in spatial learning in these animals may reflect an accelerated degeneration of surviving HDB cholinergic neurons as a result of E2 withdrawal brought on by ovariectomy. Furthermore, preserved spatial memory performance in the NOVX + P and in the OVX + E2 rats suggest that circulating levels of E2 may exert a neuroprotective effect on HDB neurons that survive the lesion, while maintaining their functional properties. It should be noted, however, that the neuroprotective effects of E2 might be dose-dependent.[85] High levels of E2 have been shown to be neurotoxic in some neuronal populations and to increase mortality following ischemic damage.[86] Commensurate with these reports we have also shown a significant impairment in spatial learning in NOVX + E2 animals, suggesting that circulating levels of E2 may interact dose-dependently with spared cholinergic neurons in our model.

The effect of E2 on the regulation of muscarinic receptors and ChAT activity was already described more than 25 years ago.[48,87,88] This finding was followed by a number of studies[65,71,89–94] showing that E2 can significantly enhance the expression and activity of ChAT, the synthetic enzyme responsible for the production of acetylcholine.[95] A number of studies in vitro have shown that E2 alters the cholinergic phenotype in basal forebrain by differentially regulating sodium-coupled HACU and ChAT activity.[96,97] In one study,[96] HACU was significantly increased by 37% in basal forebrain cholinergic neurons grown in the presence of a physiological dose of E2 (5 nM) from 4 to 10 days, whereas ChAT activity was not significantly changed in the presence of 5 or 50 nM E2 from 4 to 10 or 10–16 days in vitro. Newly synthesized acetylcholine was significantly increased 35% following 6 days of E2 treatment. These results in vitro do not match completely those described in vivo. Hormone replacement has significant short-term effects on cholinergic parameters within specific regions, but the effects vary according to the manner, regimen and duration of the treatment. For example, increases in ChAT activity, HACU or both have been detected in the hippocampus, frontal cortex, and olfactory bulbs depending on whether E2 was administered repeatedly, continuously or in combination with progesterone. Two weeks of continuous administration of E2 seemed to be the most effective regime to increase ChAT activity in olfactory bulbs and frontal cortex, and it increased HACU in the hippocampus and frontal cortex.[90] However, length of treatment may also be a key factor, and while studies examining the effects of short-term E2 treatment on ChAT expression in animal models are relatively consistent, there is conflicting evidence regarding the actions of long-term E2 replacement. Several studies have reported an increase in ChAT expression in OVX animals following 1, 4 or 8 weeks of E2 treatment,[78,98,99] while others found no change after 4 weeks or 6 months of E2.[90] The reason for this discrepancy is unknown, but it may be related to a decreased in E2 receptor (ER) expression following prolonged E2 exposure. Several studies have demonstrated significant decreases in ER mRNA and protein within different regions of the brain. Notably, Singh et al.[38] reported significantly greater ChAT activity in the frontal cortex of rats that received continuous E2 replacement for 25 weeks, compared to age-matched OVX controls, while Gibbs[90] found no change in ChAT activity in animals treated after 6 or 13 months of continuous E2 replacement. However, in a study by Gabor et al.[100] 6–7 weeks of E2 replacement significantly enhanced potassium-stimulated acetylcholine release.

In addition to dose and regime, timing of E2 administration may be crucial for its neuroprotective effects. Findings from an observation study[101] suggest that hormone therapy confers optimal cognitive benefits when initiated close in time to the menopausal transition (for a review see[102]). Our work also suggests that cholinergic neurons retain the ability to respond to E2 at least 6 weeks after OVX.

Regarding the molecular mechanisms responsible for E2 action on cholinergic neurons, the two known receptors, ERα and ERβ have been detected in the basal forebrain.[103] Colocalization studies by Shughrue et al.[104] demonstrate that ERα is the predominant form in basal forebrain cholinergic neurons, and provide an anatomical substrate for E2 action on this neurotransmitter system. Combined use of in vivo autoradiography and immunocytochemistry revealed that 22–41% of the ChAT neurons in the MS and in the nucleus of the vertical limb of the diagonal band of Broca (VDB) had nuclear uptake and retention of radiolabeled E2, while only 4% of the ChAT neurons in the nucleus basalis/substantia innominata region contained ER. Since ^{125}I-E2 has equal affinity for ERα and ERβ,[105] additional studies were conducted to characterize the nature of the E2-binding site seen in ChAT neurons, and showed that 23% of ChAT neurons in the HDB express ERα. Interestingly, many of the ChAT negative cells in the basal forebrain expressed ERβ. Other colocalization studies[106] have reported higher colocalization rates, but seem to reflect a methodological problem with the fixative causing unspecific immunoreactivity. Finally, the possibility of a direct relationship between ER and cholinergic function is reinforced by the genomic sequence of the human acetylcholinesterase and ChAT: they contain three potential ERE, present in a region upstream from the transcription start site, suggesting that E2 may regulate AChE and ChAT expression in humans.

E2 has also been shown to exert neuroprotective effects via non-classical mechanisms, involving activation of specific second messenger systems such as the mitogen-activated protein kinase pathway (MAPK).[82,96,97,107] Several studies have documented rapid membrane effects of E2 in cells that lack ER, resulting in neuroprotection or the activation of MAPK.[108–111] We have recently described[82] that MAPK mediates the E2-induced increase in neurite arborization in basal forebrain cholinergic neurons from female rats, grown in culture. Paralleling our results, an earlier study[96] had reported a MAPK-mediated increase (by 35%) in newly synthesized acetylcholine after 6 days treatment with E2.

The transport of choline into neuronal terminals through the HACU transporter is essential for cholinergic transmission,[112] as is the packaging of acetylcholine into vesicles via the vesicle acetylcholine transporter (VAChT). Electronmicroscopic studies[113] using immunocytochemical localization of VAChT and ERα have found that both are contained in presynaptic elements in the dorsal hippocampus. Approximately 3–17% of VAChT-immunoreactive profiles contain ERα, and VAChT presynaptic elements appear apposed to ERα-immunoreactive dendritic spines, presynaptic profiles and glia. The presence of ERα-immunoreactivity in cholinergic terminals suggests that E2 could rapidly affect the local release and/or reuptake of acetylcholine. Furthermore, the affiliation of cholinergic terminals with excitatory terminals near ERα-labeled dendritic spines or glial profiles suggests that alterations in acetylcholine release could indirectly affect E2-modulated structural plasticity.

E2 actions at the presynaptic level are also a source of modulatory influences on target regions, and affect patterns of synaptic connectivity more widely than would be expected from the distribution of E2 receptors. This is also the case in the basal forebrain cholinergic system, where E2 seems to elicit a sprouting response at the dendritic level in cholinergic neurons (which express E2 receptors), while also exerting a remote effect, that is, driving an increase in synaptic density in hippocampal neurons (which do not express E2 receptors) that are the target for cholinergic projections. Although the direct targets of E2 can be identified and localized based on the expression of the relevant receptors, it is thus important to acknowledge that the site of direct E2 action is not necessarily the same cell that displays morphological changes. E2-induced changes in hippocampal spine density have been correlated with changes in memory performance. The effect of E2 on cholinergic neurons at the level of presynaptic terminals may be responsible for those structural changes in hippocampal pyramidal cells, and a number of in vivo experiments suggest a role for the basal forebrain cholinergic system in the regulation of hippocampal dendritic spines. Evidence in this regard comes from work indicating that transsection of the fimbria/fornix, the fiber tract that contains the cholinergic axons from the basal forebrain to the hippocampus, blocks E2-induced spine formation,[114] and also from experiments showing that E2 delivered directly to the MS can increase spine density in the hippocampus,[92] as well as from electrophysiological data showing that, in animals with specific cholinergic lesions, E2 is much less effective in disinhibiting hippocampal pyramidal cells[57] than in the unlesioned controls.

6.5 Conclusion

Our findings and also those of others support a functional involvement of E2 in the interaction of the cholinergic system with learning and memory. We have described here that regulation of acetylcholine in the basal forebrain could underlie E2 effects on cognitive functions, through direct effects on cholinergic neurons, as well as through interactions with nerve growth factor and other neurotrophin receptors.[115] A recent behavioral study in rats supports this hypothesis[41] demonstrating that E2 deprivation results in an accelerated cognitive aging, whereas E2 replacement restored performance of working memory. These effects were blocked by scopolamine, suggesting a critical role of the cholinergic system in this process.

In addition, it is interesting to note that, during aging, this E2-related plastic mechanism may be activated. At least one study[116] has shown increased size of the Golgi apparatus in neurons from the human VDB in aging, with also increased ERα staining in the nucleus. This increase was even greater in AD, independently of age. The VDB sends cholinergic projections to the hippocampus, where there is marked decrease in ChAT and AChE activity in AD,[117,118] compared to controls. Pappas et al.[119] have shown that hippocampal ChAT correlates significantly with recent memory scores in AD. The presence of ERs in controls and the surprising increase in ERα in AD patients suggest that the VDB can also be a potential target for E2.

A sex difference in the onset of AD or the appearance of dementia is obvious only after the menopause, suggesting a pivotal role for E2 loss in the development of cognitive deficits. Further evidence comes from studies showing that in cerebrospinal fluid E2 levels are significantly lower in women with AD than in controls, and also that within the AD group, E2 levels were inversely correlated with β-amyloid 1–42 concentration.[120] While the exact mechanism by which E2 protects against neurodegeneration remains to be elucidated, evidence provided in this study suggests that the stimulation of neurotransmitter synthesis and release, and structural modifications in cholinergic neurons may play an important role in this process. It is well known that progressive decline in acetylcholine levels in cerebral cortex, hippocampus and striatum of AD patients are positively correlated with a decline in cognitive and memory functions.[119,121,122] Data from our present work and that of others suggest that E2 may act at multiple points to support cholinergic neurons and therefore their cognitive and behavioral functions.

Acknowledgements Work described in this paper was funded by USPHS grant AG19597. SL is also supported by USPHS grant P20 MD000545.

References

1. Sherwin BB. Estrogen and cognitive functioning in women. Endocr Rev 2003; 24:133–151.
2. Speck O, Ernst T, Braun J, et al. Gender differences in the functional organization of the brain for working memory. Neuroreport 2000; 11:2581–2585.
3. Rosenberg L, Park S. Verbal and spatial functions across the menstrual cycle in healthy young women. Psychoneuroendocrinology 2002; 27:835–841.
4. Phillips SM, Sherwin BB. Variations in memory function and sex steroid hormones across the menstrual cycle. Psychoneuroendocrinology 1992; 17:497–506.
5. Hampson E. Variations in sex-related cognitive abilities across the menstrual cycle. Brain Cogn 1990; 14:26–43.
6. Hampson E, Kimura D. Reciprocal effects of hormonal fluctuations on human motor and perceptual-spatial skills. Behav Neurosci 1988; 102:456–459.
7. Hampson E. Estrogen-related variations in human spatial and articulatory motor skills. Psychoneuroendocrinology 1990; 15:97–111.

8. Zec RF, Trivedi MA. The effects of estrogen replacement therapy on neuropsychological functioning in postmenopausal women with and without dementia: a critical and theoretical review. Neuropsychol Rev 2002; 12:65–109.
9. Rice MM, Graves AB, McCurry SM, et al. Estrogen replacement therapy and cognitive function in postmenopausal women without dementia. Am J Med 1997; 103:26S–35S.
10. Kampen DL, Sherwin BB. Estrogen use and verbal memory in healthy postmenopausal women. Obstet Gynecol 1994; 110:979–983.
11. Robinson D, Friedman L, Marcus R, et al. Estrogen replacement therapy and memory in older women. J Am Geriat Soc 1994; 42:919–922.
12. Jacobs DM, Tang MX, Stern Y, et al. Cognitive function in nondemented older women who took estrogen after menopause. Neurology 1998; 50:368–373.
13. Smith YR, Giordani B, Lajiness-O'Neill R, et al. Long-term estrogen replacement is associated with improved nonverbal memory and attentional measures in postmenopausal women. Fertil Steril 2001; 76:1101–1107.
14. Carlson LE, Sherwin BB. Steroid hormones, memory and mood in a healthy elderly population. Psychoneuroendocrinology 1998; 23:583–603.
15. Resnick SM, Metter EJ, Zonderman AB. Estrogen replacement therapy and longitudinal decline in visual memory – A possible protective effect? Neurology 1997; 49:1491–1497.
16. Yaffe K, Sawaya G, Lieberburg I, et al. Estrogen therapy in postmenopausal women: Effects on cognitive function and dementia. J Am Med Assoc 1998; 279:688–695.
17. Sherwin BB. Estrogenic effects on memory in women. Ann NY Acad Sci 1994; 743:213–230.
18. Krug R, Born J, Rasch B. A 3-day estrogen treatment improves prefrontal cortex-dependent cognitive function in postmenopausal women. Psychoneuroendocrinology 2006; 31:965–975.
19. Matthews K, Cauley J, Yaffe K, et al. Estrogen replacement therapy and cognitive decline in older community women. J Am Geriatr Soc 1999; 47:518–523.
20. Sherwin BB. Estrogen and cognitive aging in women. Neuroscience 2006; 138:1021–1026.
21. Henderson VW. Estrogen-containing hormone therapy and Alzheimer's disease risk: Understanding discrepant inferences from observational and experimental research. Neuroscience 2006; 138:1031–1039.
22. Pinkerton JV, Henderson VW. Estrogen and cognition, with a focus on Alzheimer's disease. Semin Reprod Med 2005; 23:172–179.
23. Almeida OP, Lautenschlager NT, Vasikaran S, et al. A 20-week randomized controlled trial of estradiol replacement therapy for women aged 70 years and older: effect on mood, cognition and quality of life. Neurobiol Aging 2006; 27:141–149.
24. Thal LJ, Thomas RG, Mulnard R, et al. Estrogen levels do not correlate with improvement in cognition. Arch Neurol 2003; 60:209–212.
25. Turgeon JL, McDonnell DP, Martin KA, et al. Hormone therapy: physiological complexity belies therapeutic simplicity. Science 2004; 304:1269–1273.
26. Rapp PR, Morrison JH, Roberts JA. Cyclic estrogen replacement improves cognitive function in aged ovariectomized rhesus monkeys. J Neurosci 2003; 23:5708–5714.
27. Tinkler GP, Voytko ML. Estrogen modulates cognitive and cholinergic processes in surgically menopausal monkeys. Prog Neuropsychopharmacol Biol Psychiatry 2005; 29:423–431.
28. Lacreuse A. Effects of ovarian hormones on cognitive function in nonhuman primates. Neuroscience 2006; 138:859–867.
29. Voytko ML. Estrogen and the cholinergic system modulate visuospatial attention in monkeys (macaca fascicularis). Behav Neurosci 2002; 116:187–197.
30. Daniel JM, Fader AJ, Spencer AL, et al. Estrogen enhances performance of female rats during acquisition of a radial arm maze. Horm Behav 1997; 32:217–225.
31. Fader AJ, Hendricson AW, Dohanich GP. Estrogen improves performance of reinforced T-maze alternation and prevents the amnestic effects of scopolamine administered systemically or intrahippocampally. Neurobiol Learn Mem 1998; 69:225–240.

32. Sandstrom NJ, Williams CL. Memory retention is modulated by acute estradiol and progesterone replacement. Behav Neurosci 2001; 115:384–393.
33. Luine VN, Richards ST, Wu VY, et al. Estradiol enhances learning and memory in a spatial memory task and effects levels of monoaminergic neurotransmitters. Horm Behav 1998; 34:149–162.
34. O'Neal MF, Means LW, Poole MC, et al. Estrogen affects performance of ovariectomized rats in a two-choice water-escape working memory task. Psychoneuroendocrinology 1996; 21:51–65.
35. Bimonte-Nelson HA, Francis KR, Umphlet CD, et al. Progesterone reverses the spatial memory enhancements initiated by tonic and cyclic oestrogen therapy in middle-aged ovariectomized female rats. Eur J Neurosci 2006; 24:229–242.
36. Rhodes ME, Frye CA. ERβ-selective SERMs produce mnemonic-enhancing effects in the inhibitory avoidance and water maze tasks. Neurobiol Learn Mem 2006; 85:183–191.
37. Dohanich GP, Fader AJ, Javorsky DJ. Estrogen and estrogen-progesterone treatments counteract the effect of scopolamine on reinforced T-maze alternation in female rats. Behav Neurosci 1994; 108:988–992.
38. Singh M, Meyer EM, Millard WJ, et al. Ovarian steroid deprivation results in a reversible learning impairment and compromised cholinergic function in female Sprague-Dawley rats. Brain Res 1994; 644:305–312.
39. Williams CL. Short-term but not long-term estradiol replacement improves radial-arm maze performance of young and aging rats. Soc Neurosci Abstr 1996; 22:1164.
40. Daniel JM, Dohanich GP. Acetylcholine mediates the estrogen-induced increase in NMDA receptor binding in CA1 of the hippocampus and the associated improvement in working memory. J Neurosci 2001; 21:6949–6956.
41. Markowska AL, Savonenko AV. Effectiveness of estrogen replacement in restoration of cognitive function after long-term estrogen withdrawal in aging rats. J Neurosci 2002; 22:10985–10995.
42. Daniel JM, Hulst JL, Berbling JL. Estradiol replacement enhances working memory in middle-aged rats when initiated immediately after ovariectomy but not after a long-term period of ovarian hormone deprivation. Endocrinology 2006; 147:607–614.
43. Packard MG, Kohlmaier JR, Alexander GM. Post-training intra-hippocampal estradiol injections enhance spatial memory in male rats: interaction with cholinergic systems. Behav Neurosci 1996; 110:626–632.
44. Packard MG, Teather LA. Posttraining estradiol injections enhance memory in ovariectomized rats: Cholinergic blockade and synergism. Neurobiol Learn Mem 1997; 68:172–188.
45. Luine VN, Rodriguez M. Effects of estradiol on radial arm maze performance of young and aged rats. Beha Neural Biol 1994; 62:230–236.
46. Hasselmo ME. The role of acetylcholine in learning and memory. Curr Opin Neurobiol 2006; 16:710–715.
47. Brann DW, Dhandapani K, Wakade C, et al. Neurotrophic and neuroprotective actions of estrogen: Basic mechanisms and clinical implications. Steroids 2007; 72:381–405.
48. Luine VN. Estradiol increases choline acetyltransferase activity in specific basal forebrain nuclei and projection areas of female rats. Exper Neurol 1985; 89:484–490.
49. Dumas J, Hancur-Bucci C, Naylor M, et al. Estrogen treatment effects on anticholinergic-induced cognitive dysfunction in normal postmenopausal women. Neuropsychopharmacology 2006; 31:2065–2078.
50. Gibbs RB. Estrogen and nerve growth factor-related systems in brain. In: Hormonal Restructuring of the Adult Brain. Annals of the New York Academy of Sciences, Vol. 743. November 14, 1994. Edited by New York Academy of Sciences, 1994:165–199.
51. Beninger RJ, Wirsching BA, Jhamandas K, et al. Animal studies of brain acetylcholine and memory. Arch Gerontol Geriatr Suppl 1989; 1:71–89.
52. Tanabe F, Miyasaka N, Kubota T, et al. Estrogen and progesterone improve scopolamine-induced impairment of spatial memory. J Med Dent Sci 2004; 51:89–98.

53. Gibbs RB, Wu DH, Hersh LB, et al. Effects of estrogen replacement on the relative levels of choline acetyltransferase, trkA, and nerve growth factor messenger RNAs in the basal forebrain and hippocampal formation of adult rats. Exper Neurol 1994; 129:70–80.
54. McGaughy J, Sarter M. Effects of ovariectomy, 192 IgG-saporin-induced cortical cholinergic deafferentation, and administration of estradiol on sustained attention performance in rats. Behav Neurosci 1999; 113:1216–1232.
55. Miller MM, Hyder SM, Assayag R, et al. Estrogen modulates spontaneous alternation and the cholinergic phenotype in the basal forebrain. Neuroscience 1999; 91:1143–1153.
56. Farr SA, Banks WA, Morley JE. Estradiol potentiates acetylcholine and glutamate-mediated post-trial memory processing in the hippocampus. Brain Res 2000; 864:263–269.
57. Rudick CN, Gibbs RB, Woolley CS. A role for the basal forebrain cholinergic system in estrogen-induced disinhibition of hippocampal pyramidal cells. J Neurosci 2003; 23:4479–4490.
58. Paxinos G, Watson C. The Rat Brain in Stereotaxic Coordinates. 1st ed. Sydney: Academic, 1982.
59. de Lacalle S, Kulkarni S, Wiley RG. Lesion induced transneuronal plasticity of the cholinergic innervation in the adult rat entorhinal cortex. Eur J Neurosci 1998; 10:1054–1062.
60. Saenz C, Dominguez R, de Lacalle S. Estrogen contributes to structural recovery after a lesion. Neurosci Lett 2006; 392:198–201.
61. Hartonian I, Mufson EJ, de Lacalle S. Long-term plastic changes in galanin innervation in the rat basal forebrain. Neuroscience 2002; 115:787–795.
62. Frye CA. Estrus-associated decrements in a water maze task are limited to acquisition. Physiol Behav 1995; 57:5–14.
63. Warren SG, Juraska JM. Spatial and non spatial learning across the rat estrous cycle. Behav Neurosci 1997; 111:259–265.
64. Gibbs RB. Estrogen replacement enhances acquisition of a spatial memory task and reduces deficits associated with hippocampal muscarinic receptor inhibition. Horm Behav 1999; 36:222–233.
65. Gibbs RB. Basal forebrain cholinergic neurons are necessary for estrogen to enhance acquisition of a delayed matching-to-position T-maze task. Horm Behav 2002; 42:245–257.
66. Heikkinen T, Puolivali J, Liu L, et al. Effects of ovariectomy and estrogen treatment on learning and hippocampal neurotransmitters in mice. Horm Behav 2002; 41:22–32.
67. Holmes MM, Wide JK, Galea LA. Low levels of estradiol facilitate, whereas high levels of estradiol impair working memory performance on the radial arm maze. Behav Neurosci 2002; 116:928–934.
68. Galea LA, Wide JK, Paine TA, et al. High levels of estradiol disrupt conditioned place preference learning, stimulus response learning and reference memory but have limited effects on working memory. Behav Brain Res 2001; 126:115–126.
69. Ferrini M, Bisagno V, Piroli G, et al. Effects of estrogens on choline-acetyltransferase immunoreactivity and GAP-43 mRNA in the forebrain of young and aging male rats. Cell Mol Neurobiol 2002; 22:289–301.
70. Aggarwal P, Gibbs RB. Estrogen replacement does not prevent the loss of choline acetyltransferase-positive cells in the basal forebrain following either neurochemical or mechanical lesions. Brain Res 2000; 882:75–85.
71. Gibbs RB. Effects of estrogen on basal forebrain cholinergic neurons vary as a function of dose and duration of treatment. Brain Res 1997; 757:10–16.
72. Gibbs RB, Pfaff DW. Effects of estrogen and fimbria/fornix transection on $p75^{NGFR}$ and ChAT expression in the medial septum and diagonal band of Broca. Exp Neurol 1992; 116:23–39.
73. McGaughy J, Dalley JW, Morrison CH, et al. Selective behavioral and neurochemical effects of cholinergic lesions produced by intrabasalis infusions of 192 IgG-saporin on attentional performance in a five-choice serial reaction time task. J Neurosci 2002; 22:1905–1913.
74. Horvath KM, Hartig W, Van der Veen R, et al. 17beta-Estradiol enhances cortical cholinergic innervation and preserves synaptic density following excitotoxic lesions to the rat nucleus basalis magnocellularis. Neuroscience 2002; 110:489–504.

75. Nakamura N, Fujita H, Kawata M. Effects of gonadectomy on immunoreactivity for choline acetyltransferase in the cortex, hippocampus, and basal forebrain of adult male rats. Neuroscience 2002; 109:473–485.
76. Gibbs RB, Nelson D, Anthony MS, et al. Effects of long term hormone replacement and of tibolone on choline acetyltransferase and acetylcholinesterase activities in the brains of ovariectomized, cynomologous monkeys. Neuroscience 2002; 113:907–914.
77. Tinkler GP, Tobin JR, Voytko ML. Effects of two years of estrogen loss or replacement on nucleus basalis cholinergic neurons and cholinergic fibers to the dorsolateral prefrontal and inferior parietal cortex of monkeys. J Comp Neurol 2004; 469:507–521.
78. Kritzer MF, Kohama SG. Ovarian hormones differentially influence immunoreactivity for dopamine β- hydroxylase, choline acetyltransferase, and serotonin in the dorsolateral prefrontal cortex of adult rhesus monkeys. J Comp Neurol 1999; 409:438–451.
79. Singer CA, Pang PA, Dobie DJ, et al. Estrogen increases GAP-43 (Neuromodulin) mRNA in the preoptic area of aged rats. Neurobiol aging 1996; 17:661–663.
80. Brinton RD, Tran J, Proffitt P, et al. 17 β-estradiol enhances the outgrowth and survival of neocortical neurons in culture. Neurochem Res 1997; 22:1339–1351.
81. Ferreira A, Caceres A. Estrogen-enhanced neurite growth: evidence for a selective induction of tau and stable microtubules. J Neurosci 1991; 11:392–400.
82. Dominguez R, Jalali C, de Lacalle S. Morphological effects of estrogen on cholinergic neurons in vitro involves activation of extracellular signal-regulated kinases. J Neurosci 2004; 24:982–990.
83. Segal M, Murphy D. Estradiol induces formation of dendritic spines in hippocampal neurons: functional correlates. Horm Behav 2001; 40:156–159.
84. Garcia-Segura LM, Azcoitia I, DonCarlos LL. Neuroprotection by estradiol. Prog Neurobiol 2001; 63:29–60.
85. Desjardins GC, Beaudet A, Meaney MJ, et al. Estrogen-induced hypothalamic β-endorphin neuron loss: a possible model for hypothalamic aging. Exp Gerontol 1995; 30:253–267.
86. Sudo S, Wen TC, Desaki J, et al. β-Estradiol protects hippocampal CA1 neurons against transient forebrain ischemia in gerbil. Neurosci Res 1997; 29:345–354.
87. Luine VN, Khylchevskaya RI, McEwen BS. Effect of gonadal steroids on activities of monoamine oxidase and choline acetylase in rat brain. Brain Res 1975; 86:293–306.
88. McEwen BS, Biegon A, Davis PG, et al. Steroid hormones: humoral signals which alter brain cell properties and functions. Recent Prog Horm Res 1982; 38:41–92.
89. Gibbs RB, Aggarwal P. Estrogen and basal forebrain cholinergic neurons: Implications for brain aging and Alzheimer's disease- related cognitive decline. Horm Behav 1998; 34:98–111.
90. Gibbs RB. Effects of gonadal hormone replacement on measures of basal forebrain cholinergic function. Neuroscience 2000; 101:931–938.
91. Gibbs RB. Effects of ageing and long-term hormone replacement on cholinergic neurones in the medial septum and nucleus basalis magnocellularis of ovariectomized rats. J Neuroendocrinol 2003; 15:477–485.
92. Lam TT, Leranth C. Role of the medial septum diagonal band of Broca cholinergic neurons in oestrogen-induced spine synapse formation on hippocampal CA1 pyramidal cells of female rats. Eur J Neurosci 2003; 17:1997–2005.
93. Bora SH, Liu Z, Kecojevic A, et al. Direct, complex effects of estrogens on basal forebrain cholinergic neurons. Exp Neurol 2005; 194:506–522.
94. Grandholm A-CE, Ford KA, Hyde LA, et al. Estrogen restores cognition and cholinergic phenotype in an animal model of Down syndrome. Physiol Behav 2002; 77:371–385.
95. McMillan PJ, Singer CA, Dorsa DM. The effects of ovariectomy and estrogen replacement on trkA and choline acetyltransferase mRNA expression in the basal forebrain of the adult female Sprague-Dawley rat. J Neurosci 1996; 16:1860–1865.
96. Pongrac JL, Gibbs RB, DeFranco DB. Estrogen-mediated regulation of cholinergic expression in basal forebrain neurons requires extracellular-signal-regulated kinase activity. Neuroscience 2004; 124:809–816.

97. Szego EM, Barabas K, Balog J, et al. Estrogen induces estrogen receptor alpha-dependent cAMP response element-binding protein phosphorylation via mitogen activated protein kinase pathway in basal forebrain cholinergic neurons in vivo. J Neurosci 2006; 26:4104–4110.
98. Pan Y, Anthony M, Clarkson TB. Effect of estradiol and soy phytoestrogens on choline acetyltransferase and nerve growth factor mRNAs in the frontal cortex and hippocampus of female rats. Proc Soc Exp Biol Med 1999; 221:118–125.
99. Watters JJ, Dorsa DM. Transcriptional effects of estrogen on neuronal neurotensin gene expression involve cAMP/protein kinase A-dependent signaling mechanisms. J Neurosci 1998; 18:6672–6680.
100. Gabor R, Nagle R, Johnson DA, et al. Estrogen enhances potassium-stimulated acetylcholine release in the rat hippocampus. Brain Res 2003; 962:244–247.
101. Resnick SM, Henderson VW. Hormone therapy and risk of Alzheimer disease: a critical time. JAMA 2002; 288:2170–2172.
102. Maki PM. Hormone therapy and cognitive function: Is there a critical period for benefit? Neuroscience 2006; 138:1027–1030.
103. Shughrue PJ, Lane MV, Merchenthaler I. Comparative distribution of estrogen receptor-alpha and -β mRNA in the rat central nervous system. J Comp Neurol 1997; 388:507–525.
104. Shughrue PJ, Scrimo PJ, Merchenthaler I. Estrogen binding and estrogen receptor characterization (ERα and ERβ) in the cholinergic neurons of the rat basal forebrain. Neuroscience 2000; 96:41–49.
105. Shughrue PJ, Lane MV, Merchenthaler I. Biologically active estrogen receptor-β: evidence from in vivo autoradiographic studies with estrogen receptor α-knockout mice. Endocrinology 1999; 140:2613–2620.
106. Miettinen RA, Kalesnykas G, Koivisto EH. Estimation of the total number of cholinergic neurons containing estrogen receptor-α in the rat basal forebrain. J Histochem Cytochem 2002; 50:891–902.
107. McMillan PJ, LeMaster AM, Dorsa DM. Tamoxifen enhances choline acetyltransferase mRNA expression in rat basal forebrain cholinergic neurons. Mol Brain Res 2002; 103:140–145.
108. Singer CA, Figueroa-Masot XA, Batchelor RH, et al. The mitogen-activated protein kinase pathway mediates estrogen neuroprotection after glutamate toxicity in primary cortical neurons. J Neurosci 1999; 19:2455–2463.
109. Singh M, Setalo G, Jr., Guan X, et al. Estrogen-induced activation of mitogen-activated protein kinase in cerebral cortical explants: convergence of estrogen and neurotrophin signaling pathways. J Neurosci 1999; 19:1179–1188.
110. Singh M, Setalo G, Jr., Guan X, et al. Estrogen-induced activation of the mitogen-activated protein kinase cascade in the cerebral cortex of estrogen receptor-a knock-out mice. J Neurosci 2000; 20:1694–1700.
111. Pedata F, Giovannelli L, Pepeu G. GM1 ganglioside facilitates the recovery of high-affinity choline uptake in the cerebral cortex of rats with a lesion of the nucleus basalis magnocellularis. J Neurosci Res 1984; 12:421–427.
112. Sarter M, Parikh V. Choline transporters, cholinergic transmission and cognition. Nat Rev Neurosci 2005; 6:48–56.
113. Towart LA, Alves SE, Znamensky V, et al. Subcellular relationships between cholinergic terminals and estrogen receptor-alpha in the dorsal hippocampus. J Comp Neurol 2003; 463:390–401.
114. Leranth C, Shanabrough M, Horvath TL. Hormonal regulation of hippocampal spine synapse density involves subcortical mediation. Neuroscience 2000; 101:349–356.
115. Mufson EJ, Cai WJ, Jaffar S, et al. Estrogen receptor immunoreactivity within subregions of the rat forebrain: neuronal distribution and association with perikarya containing choline acetyltransferase. Brain Res 1999; 849:253–274.
116. Ishunina TA, Swaab DF. Increased neuronal metabolic activity and estrogen receptors in the vertical limb of the diagonal band of Broca in Alzheimer's disease: relation to sex and aging. Exp Neurol 2003; 183:159–172.

117. Henke H, Lang W. Cholinergic enzymes in neocortex, hippocampus and basal forebrain of non-neurological and senile dementia of Alzheimer-type patients. Brain Res 1983; 267:281–291.
118. Allen SJ, MacGowan SH, Tyler S, et al. Reduced cholinergic function in normal and Alzheimer's disease brain is associated with apolipoprotein E4 genotype. Neurosci Lett 1997; 239:33–36.
119. Pappas BA, Bayley PJ, Bui BK, et al. Choline acetyltransferase activity and cognitive domain scores of Alzheimer's patients. Neurobiol Aging 2000; 21:11–17.
120. Schonknecht P, Pantel J, Klinga K, et al. Reduced cerebrospinal fluid estradiol levels are associated with increased β-amyloid levels in female patients with Alzheimer's disease. Neurosci Lett 2001; 307:122–124.
121. Auld DS, Kornecook TJ, Bastianetto S, et al. Alzheimer's disease and the basal forebrain cholinergic system: relations to β-amyloid peptides, cognition, and treatment strategies. Prog Neurobiol 2002; 68:209–245.
122. Wilcock GK, Esiri MM, Bowen DM, et al. Alzheimer's disease: correlation of cortical choline acetyltransferase activity with the severity of dementia and histological abnormalities. J Neurol Sci 1982; 57:407–417.

Chapter 7
Local Production of Estrogen and its Rapid Modulatory Action on Synaptic Plasticity

Suguru Kawato[1,2,3,4], Yasushi Hojo[1,2,3], Hideo Mukai[1,2,3], Gen Murakami[1,2], Mari Ogiue-Ikeda[1,4], Hirotaka Ishii[1,2], and Tetsuya Kimoto[1,2]

Abstract It has long been a common sense that sex hormones are synthesized in the gonads, and reach the brain via the blood circulation. In contrast with this view, the authors demonstrate that estrogen and androgen are also synthesized locally in the hippocampus of adult animals, from cholesterol through dehydroepiandrosterone in hippocampal neurons. These neurosteroids are synthesized by cytochrome P450s and hydroxysteroid dehydrogenases and 5α-reductase. The expression levels of enzymes are as low as 1/200–1/50,000 of those in endocrine organs, preventing quantitative investigations. Localization of P450(17α) and P450arom is observed in synapses of principal glutamatergic neurons, in addition to endoplasmic reticulum, suggesting synaptocrine machanisms.

Because several nanomolar estrogen and androgen are observed in the hippocampus, they are expected to have physiological functions. Estrogen modulates memory-related synaptic plasticity not only slowly, but also rapidly in the hippocampus. Molecular mechanisms of rapid action via membrane receptors have not been well elucidated in comparison with those of delayed action via genomic processes. We here describe rapid modulation of representative synaptic plasticity, i.e., long-term depression (LTD), long-term potentiation (LTP) and spinogenesis, by 17β-estradiol, selective estrogen agonists.

We demonstrate that 1–10 nM estradiol induced rapid enhancement of LTD within 1 h in CA1, CA3 and dentate gyrus (DG). On the other hand, the modulation of LTP by estradiol is not statistically significant. The total density of spines is increased in CA1 pyramidal neurons, within 2 h after application of estradiol. The total density of thorns (postsynaptic spine-like structure) is, however, decreased by estradiol in CA3

[1] Department of Biophysics and Life Sciences, Graduate School of Arts and Sciences, The University of Tokyo, Komaba, Meguro, Tokyo 153–8902, Japan

[2] Core Research for Evolutional Science and Technology Project of Japan Science and Technology Agency, The University of Tokyo

[3] Bioinformatics Project of Japan Science and Technology Agency, The University of Tokyo

[4] Project of Special Coordinate Funds for Promoting Science and Technology, The University of Tokyo

M.S. Ritsner and A. Weizman (eds.), *Neuroactive Steroids in Brain Function, Behavior and Neuropsychiatric Disorders*

pyramidal neurons. Both the increase of spines in CA1 and the decrease of thorns in CA3 are driven by Erk MAP kinase. Only agonist of estrogen receptor ERalpha induces the same enhancement/suppression effect as estradiol on both LTD and spinogenesis in CA1 and CA3. ERbeta agonist induces completely different results.

Estrogen receptor ERalpha localizes in spines and presynapses of principal glutamatergic neurons in CA1, CA3 and DG. The same ERalpha is also located in nuclei. Identification of ERalpha is successfully performed using purified RC-19 antibody. Attention must be paid to the fact that non-purified ERalpha antisera often react significantly with unknown proteins, resulting in wrong staining different from real ERalpha distribution. Identification of kinases/phosphatases in downstream of ERalpha as well as other synaptic estrogen receptors is essential to advance the field.

Keywords Estradiol, estrogen, androgen, neurosteroid, synaptic plasticity, estrogen receptor, hippocampus, LTP, LTD

Abbreviations ACSF artificial cerebrospinal fluid; AMPA α-amino-3-hydroxy-5-methylisoxazole-4- propionic acid; CHO Chinese hamster ovary; CNQX 6-cyano-7-nitroquinoxaline-2,3-dione; DHEA dehydroepiandrosterone; DPN diarylpropionitrile; GPR30 G-protein-coupled receptor 30; HSD hydroxysteroid dehydrogenase; LC-MS/MS liquid chromatography-tandem mass spectrometry; LTD long-term depression; LTP long-term potentiation; MALDI-TOF matrix-assisted laser desorption ionization -time-of-flight; MCF-7 human breast cancer cell line; NMDA *N*–methyl-D-aspartate; PBR peripheral benzodiazepine receptor; PSD postsynaptic density; PREG pregnenolone; PPT (propyl-pyrazole-triyl)tris-phenol; RIA radioimmunoassay; StAR steroidogenic acute regulatory protein

7.1 Hippocampal Synthesis of Estrogen and Androgen

7.1.1 Introduction

In recent years, increasing evidence has accumulated to support the local endogenous synthesis of estrogens, androgens and nonsteroidal sex hormones in the mammalian brain, in areas such as the hippocampus.[1–5] Over many decades it has been believed that steroid hormones are synthesized only in the body, though peptide hormones or amines are observed to be synthesized not only in the body, but also in the brain. In the 1980s, Baulieu and co-workers have proposed a neurosteroid hypothesis, suggesting that pregnenolone, progesterone and DHEA may be endogenously synthesized in the brain due to the finding that pregnenolone (PREG) and dehydroepiandrosterone (DHEA) has been found in the mammalian brain at concentrations greater than that in plasma.[6,7] Because the concentration of PREG and DHEA does not decrease after adrenalectomy and castration, many experiments have been performed with the aim of demonstrating the de novo synthesis of DHEA within the brain.[7,8]

Direct demonstration of steroidogenesis in the mammalian brain had, however, long been not successful over decades, due to the extremely low levels of steroidogenic proteins in the brain.[9] Therefore, sex steroids had not been considered to be brain-derived steroids, and rather thought to reach the brain exclusively via blood circulation after crossing the blood brain barrier.[10] This belief had been supported by many reports suggesting the absence of cytochrome P450(17α) (DHEA synthase) in adult mammalian brain[11,12] and also by the observation of the complete disappearance of testosterone in the brain within 1 day after castration.[10]

Extensive investigations have been performed to examine neuromodulatory actions of gonadal sex hormones in the brain such as the hippocampus, because the hippocampus is attractive as a center for learning and memory.[13–18] Many scientists had, however, not seriously considered that memory formation process might favor the local hippocampal steroidogenesis rather than gonadal hormones which reach the brain via circulation. A main reason for this is due to extremely low level of expressions for P450s and hydroxysteroid dehydrogenases (HSD) in the hippocampus, much lower than 1/100–1/1,000 of their levels in the endocrine organs.

7.1.2 mRNA Expression of Steroidogenic Enzymes

Highly sensitive molecular biology investigations are necessary for determination of the presence of steroidogenic enzymes, because of the very low level of expression of the mRNAs in the cerebrum and cerebellum.[9] It is essential to improve the sensitivity of measurements by nearly 1,000-fold for immunostaining, Western blot, RT-PCR as well as steroid metabolism assay.

Taken collectively from many studies, the relative level of mRNA expressed in the hippocampus has been suggested to be lowest for cytochrome P450scc and 3β-hydroxysteroid dehydrogenase (3β-HSD), and highest for steroidogenic acute regulatory protein (StAR) and 5α-reductase, with those of P450(17α) and P450arom expressed at an intermediate level (see Table 7.1).

The expression level of cytochrome P450scc (CYP11A1) mRNA is extremely low, preventing many scientists to believe the physiological significance of neurosteroid synthesis. P450scc is expressed in the brain is reported to be only 10^{-4}–10^{-5} of that in the adrenal gland.[12,19] Our analysis showed approximately 1/50,000 for the P450scc level in the adult hippocampus as compared with that in the adrenal gland (Table 7.1).[20] As a result, the presence of P450scc mRNA could be demonstrated only by the RT-PCR method. The ribonuclease (RNase) protection assay for P450scc in the hippocampus, which was performed using a ^{32}P-labelled rat P450scc antisense riboprobe, however, yielded no detectable specific hybridization signals.[21] On the other hand, StAR is most abundant, therefore, not only the PCR-amplification but also the RNase protection assay demonstrated the presence of StAR transcripts with an expression level of approximately 1/200 of the level in the adrenal gland.[21,22]

The mRNAs for cytochrome P450(17α) (CYP17A) had not been detected in adult rat brain by either RNase protection assays or RT-PCR.[12] The expression of the mRNA for P450(17α) had been reported by many laboratories as only transient,

Table 7.1 Comparison of relative mRNA expression level for steroidogenic enzymes in the adult rat (3 months)

	Hippo[a]	Hypo	Adrenal/testis/ovary/liver
P450scc	1[b]	3	50000 (Ad)
P450(17α)	1	3	300 (Te)
P450arom	1	3	300 (Ov)
17β-HSD (type 1)	1	3	200 (Ov)
17β-HSD (type 3)	1	5	300 (Te)
3β-HSD (type 1)	1	3	5000 (Ov)
5α-reductase (type 1)	1	2	5 (Li)
5α-reductase (type 2)	1	1/3	200 (Pro)
ERα	1	5	15 (Ov)
ERβ	1	4	80 (Ov)

[a] Hippocampus (Hippo), hypothalamus (Hypo), adrenal gland (Ad), testis (Te), ovary (Ov), liver (Li), and prostate (Pro) are compared

[b] The level in the hippocampus is normalized to be 1. The level of mRNA expression is approximate value obtained from semiquantitative RT-PCR analyses

occurring during rat embryonic and neonatal development.[23–25] We overcame this difficulty by carefully choosing the sequence of primer pairs which have high specificity by minimizing Gibbs free energy upon recombination of a 3 -primer with cDNA, using computer calculation.[4] In the hippocampal tissues from adult male rats aged 3 months, we observed that P450(17α) transcripts expressed approximately 1/300,[4] when compared with those expressed in the testis.

The role of P450arom (CYP19) (estrogen synthase) in the hippocampus had also not been easy to elucidate. Many studies had reported the absence of P450arom in the adult rat and mouse hippocampus. The significant expression of mRNA for P450arom in the pyramidal and granule neurons of the adult rat hippocampus, however, has recently been demonstrated using in situ hybridization.[26] The level of the mRNA expression in the adult mouse hippocampus is approximately half of that in neonatal stages.[27] We observed the P450arom transcripts expressed approximately 1/300,[4] as compared with those expressed in the ovary by using carefully designed primer pairs for RT-PCR.

The presence of mRNAs for 17β-hydroxysteroid dehydrogenase (17β-HSD) type 1 and 3 is demonstrated in the human and rat hippocampus.[28] We investigated the expression level of mRNA transcripts for 17β-HSD (types 1–4) using carefully designed primer pairs in the hippocampus from adult male rats. The mRNA level of 17β-HSD transcripts observed was approximately 1/200, relative to the level in the ovary for 17β-HSD (type 1), and 1/300, relative to the level in the testis for 17β-HSD (type 3), respectively.[4]

The localization in neurons of several steroidogenic proteins is demonstrated by means of in situ hybridization. For example, mRNAs for both StAR and 3β-hydroxysteroid dehydrogenase (3β-HSD) mRNA (10^{-2} for StAR and 10^{-3} for 3β-HSD of the levels in the adrenal gland) have been observed to be localized along the pyramidal cell layer in CA1-CA3 and the granule cell layer in the dentate gyrus (DG) of rats[21] and mice.[22]

Glial cells have been considered to play an important role in steroidogenesis, as many reports have indicated the presence of mRNA for P450scc, P450(17α), 3β-HSD, and 17β-HSD in cultures of astrocytes and oligodendrocytes from embryonic and neonatal brains.[6,23,24,29] Although similar levels of P450(17α) mRNA are expressed in both astrocytes and neurons in primary cell cultures from the brain of neonatal rats, a much lower metabolic activity is observed in neurons than astrocytes for the conversion of PREG to DHEA.[23,24] The very low metabolic activity of neurons might be caused by cytosine arabinoside applied to cultures in order to suppress the proliferation of glial cells. These investigations are available on primary glial cell cultures which are easily prepared from embryonic and neonatal brains. However, information regarding the synthesis system of neurosteroids in "adult" rat brain is not directly available from these cell culture studies.

7.1.3 Neuronal Localization of Steroidogenic Proteins

The role of neurons in steroid synthesis in mammalian brains had long been difficult to determine, although some reports suggested the expression of several steroidogenic enzymes in rat brain.[30,31] The belief of the absence of P450(17α) had been partly due to the fact that many attempts to demonstrate the immunohistochemical reactivity in the rat brain had been unsuccessful for almost two decades.[11] We overcame many difficulties of nonspecific immunostaining by using affinity column-purified antibodies[32,33] (instead of using non-purified antisera) in order to avoid cross-reaction with IgG with unknown proteins having similar antigen sequences, as well as using fresh frozen slices of hippocampus (instead of using paraffin sections) from adult male rats. A significant localization of cytochromes P450scc, P450(17α) and P450arom was observed in pyramidal neurons in CA1-CA3, as well as in granule cells in DG, by means of the immunohistochemical staining of hippocampal slices from adult rats (12 weeks) (Fig. 7.1).[1–4] The co-localization of immunoreactivity against P450s and NeuN (marker of neuron) confirmed the presence of P450s in these neurons.[1,2,4] StAR was co-localized with P450s.[1,22] These results imply that pyramidal neurons and granule cells are equipped with complete steroidogenic systems which catalyze the conversion of cholesterol to PREG, DHEA, testosterone and estradiol.

An immunoelectron microscopic analysis using postembedding immunogold method is very useful to determine the intraneuronal localization of P450(17α) in the hippocampal neurons of adult male rats. Surprisingly, we observed that both P450(17α) and P450arom were localized not only in the endoplasmic reticulum but also in the presynaptic region as well as the postsynaptic region of pyramidal neurons in the CA1 and CA3 regions and of granule neurons in DG (Fig. 7.2). These results suggest "synaptocrine" mechanisms of synthesis of estrogens and androgens, in addition to classical microsomal synthesis of sex steroids.

The existence of these steroidogenic proteins was confirmed by Western immunoblot analyses.[1,2,4] The molecular weights obtained for P450scc, P450(17α) and P450arom were identical to those obtained from peripheral steroidogenic organs.

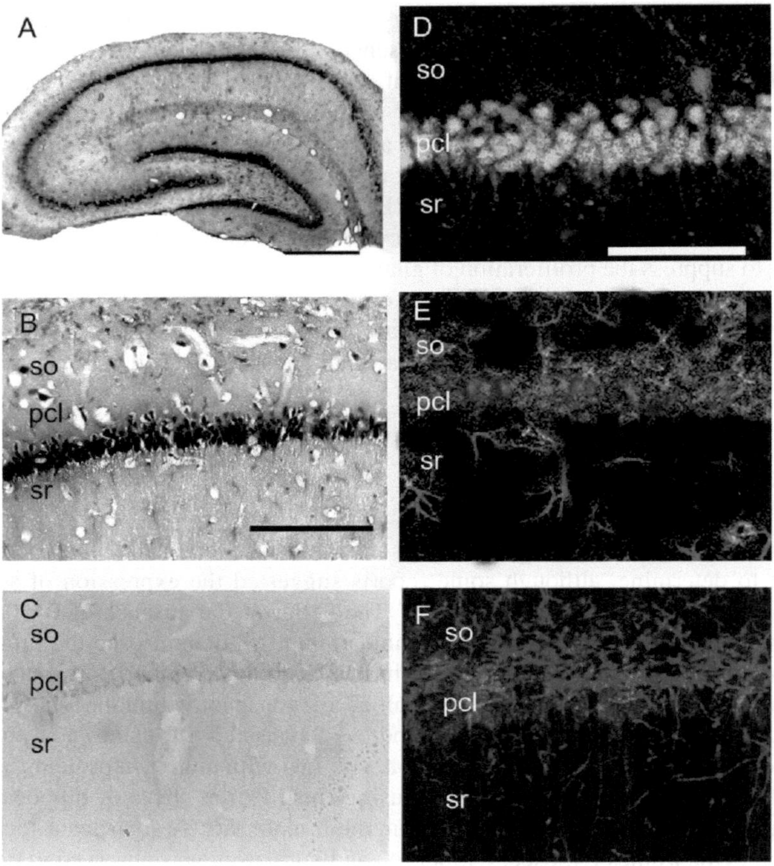

Fig. 7.1 Immunohistochemical staining of cytochrome P45017α in the hippocampal slices of adult male rat. (**A**) the coronal section of the whole hippocampus. (**B**) the CA1. (**C**) the CA1 anti-P45017α IgG was presaturated with P45017α protein. (**D**) fluorescence dual staining of P45017α (green) and NeuN (red). (**E**) fluorescence dual staining of P45017α (green) and GFAP (red). (**F**) fluorescence dual staining of P45017α (green) and myelin basic protein (red). Superimposed regions of green and red fluorescence are represented by yellow. so, stratum oriens; pcl, pyramidal cell layer; sr, stratum radiatum. Scale bar 800 μm for A, and 120 μm for B–F (From Hojo et al.[4])

The relative levels of these P450s in the hippocampus were approximately 1/500 (P450scc) and 1/300 (P450(17α) and P450arom) of that in the testis (P450scc and P450(17α)) and the ovary (P450arom), respectively.

Astroglial cells and oligodendroglial cells was not stained significantly by antibodies of P450scc, P450(17α) and P450arom in adult and developmental hippocampus.[1,2,4] The activity of neurosteroidogenesis in glial cells is probably much lower than that of neurons. In earlier study, the white matter including glial cells had been stained with anti-P450scc antisera, throughout the adult rat brain.[34] This white matter staining of P450scc antisera is likely to be an artifact which is due to the nonspecific adsorption of the non-purified bovine antisera in rat hippocampus.

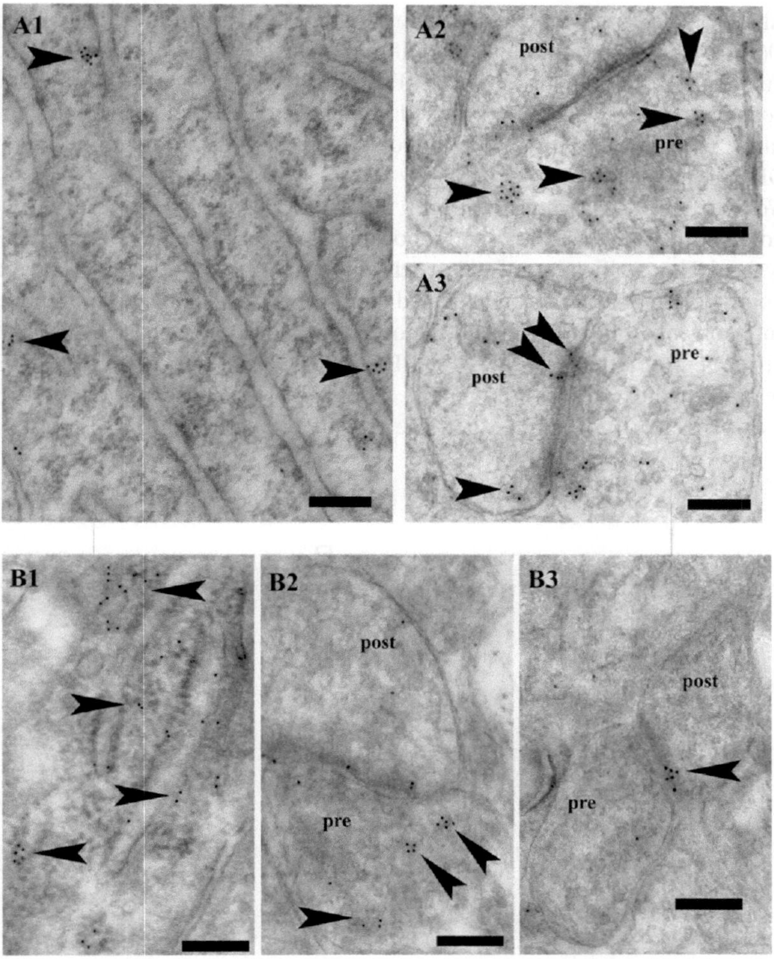

Fig. 7.2 Immunoelectron microscopic analysis of the distribution of P45017α (A1–A3) and P450arom (B1–B3) within axospinous synapses, in the strata radiatum of the hippocampal CA1 region at the central region of the rostrocaudal level. Gold particles (indicated with arrow heads) were observed to be localized in the endoplasmic reticulum (A1 and B1), the presynaptic region (A2 and B2), and the postsynaptic region (A3 and B3) of pyramidal neurons. In the axon terminal (A2 and B2), gold particles were associated with small synaptic vesicles (A2 and B2). In dendritic spines, gold particles were found within the head of the spine (A3 and B3). Pre, presynaptic region; Post, postsynaptic region; Scale bar 200 nm (From Hojo et al.[4])

7.1.4 Synthesis of Estrogen and Androgen

A direct demonstration of the neuronal synthesis of DHEA and testosterone in adult mammals is for the first time reported by our group.[1,2,4] It had been assumed that testosterone are supplied to the male brain such as hypothalamus, via the blood circulation, where they are converted to estradiol by P450arom.[6,10] The absence of P450(17α) activity in the brain of adult mammals had been reported in a number of studies.[10,11,35,36] Incubations of [^{3}H]-PREG with brain slices, homogenates and microsomes, primary cultures of mixed glial cells, or astrocytes and neurons from rat and mouse embryos, had failed to produce a radioactive metabolite ^{3}H-DHEA.[10]

We succeeded in demonstration of the synthesis of DHEA, testosterone and estradiol in the adult (12 weeks) hippocampal slices by means of careful HPLC analysis.[2,4] The purification of neurosteroids from very fatty brain tissues requires the combination of several sophisticated methods, which included purification with organic solvent, solid column chromatography, and normal phase HPLC.[1,4,37] The significant conversion from [^{3}H]-PREG to ^{3}H-DHEA, from ^{3}H-DHEA to ^{3}H-androstenediol, ^{3}H-testosterone and ^{3}H-estradiol was observed after incubation with the slices for 5 h (Fig. 7.3).[4] The rate of production for ^{3}H-estradiol from

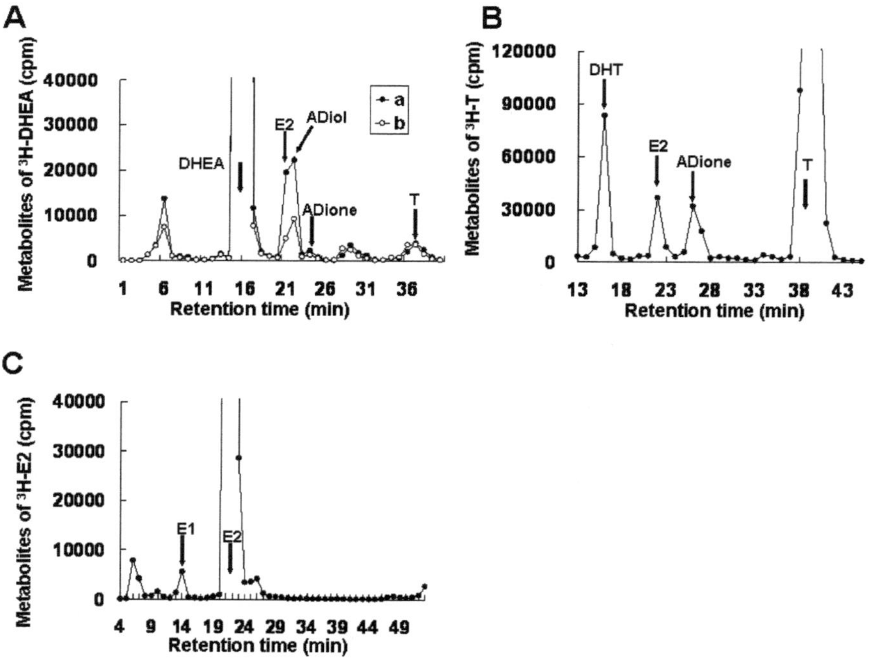

Fig. 7.3 Synthesis of estradiol and testosterone in adult hippocampal slices after incubation of slices for 5 h at 30 °C. (**A**) HPLC analysis shows the profile of ^{3}H-DHEA metabolites in the absence (line a) or in the presence (line b) of fadrozole (inhibitor of P450arom). In fadrozole-treated samples, E2 production was suppressed but androstenediol was still present at nearly the same position. (**B**) ^{3}H-testosterone metabolites. (**C**) ^{3}H-estradiol metabolites. DHT (dihydrotestosterone), E2 (estradiol), T (testosterone), ADiol (androstenedione), E1 (estrone), and U (unknown metabolites). The vertical axis indicates ^{3}H radioactivity (cpm) (Modified from Hojo et al.[4])

^{3}H-testosterone was very slow, and the production rate of ^{3}H-dihydrotestosterone from ^{3}H-testosterone was much more rapid than that of estradiol. These activities were abolished by the application of specific inhibitors of cytochrome P450s. Suprisingly, ^{3}H-estradiol was extremely stable and not significantly converted to other steroid metabolites such as estrone. On the other hand, dihydrotestosterone was rapidly converted to 3α, 5α-androstanediol.

We determined the concentration of DHEA and 17β-estradiol as well as PREG in acute hippocampal slices from adult male rats by means of RIA, after careful purification of steroids with normal phase HPLC. The basal concentrations of PREG, DHEA and estradiol, in the male rat hippocampus were approximately 18, 0.3 and 0.6 nM, which were 6–10 times greater than those typical of plasma.[1,2,4] However, more recently we very much improved the determination of steroids by using liquid chromatography-tandem-mass spectrometry (LC/MS/MS) which has a high specificity and accuracy. A much higher value of 8 nM was obtained for estradiol.[38] The concentration of testosterone in the hippocampus (17 nM) was only slightly higher than that of circulating testosterone (15 nM) in male rats. To demonstrate the rapid net production of neurosteroids upon synaptic stimulation, the NMDA-induced production of PREG and estradiol was investigated in hippocampal slices.[1,2,4] Upon stimulation with NMDA for 30 min, the hippocampal level of PREG and estradiol increased to approximately twofold of the basal levels. This implies that the NMDA-induced Ca^{2+} influx drives net production of PREG and estradiol. Estradiol synthesis is also demonstrated in cultured hippocampal slices from neonatal rats in the absence and presence of letrozole, an inhibitor of P450arom. After 4 days treatment with letrozole, the amount of estradiol released into the medium was significantly decreased.[5]

Interestingly, PREG sulfate and DHEA sulfate have been reported to be absent in the rat brain as measured by direct mass spectrometric analysis, although cholesterol sulfate is present.[39–41] In many previous publications, PREG sulfate or DHEA sulfate had been determined indirectly, i.e., measuring PREG or DHEA after solvolysis of water soluble fractions which may contain some lipoidal derivatives of PREG different from sulfated steroids.[1,6,7,39,41] Because numerous publications have reported that sulfated steroids are important participants in neuromodulation, these results merit careful consideration.[10,42,43]

Is the local concentration of brain neurosteroids sufficiently high to allow action as local mediators? Based on RIA determination, the concentration of estradiol detected in the hippocampus was only about 0.6 nM (basal) and 1.3 nM after the NMDA-stimulation, respectively.[4] However, with very accurate LC-MS/MS analysis, basal level of estradiol was determined to be roughly 8 nM.[44] This level is sufficient to allow estradiol to act as local mediators that modulate synaptic plasticity.[13,14,16,45,46]

Functional differences for estradiol produced from circulating T and estradiol produced from hippocampus-derived T may be differences in the time-dependence of their levels. Brain is filled with circulating testosterone (for male), or estradiol (for female) whose level slowly changes depending on the circadian rhythm, while the endogenous synthesis of estradiol (for both male and female) is a transient event depending on neural activity.[4]

7.2 Rapid Modulation of Synaptic Plasticity by Estrogen

7.2.1 Introduction

Extensive investigations have been performed to examine neuromodulatory actions of gonadal sex hormones in the brain such as the hippocampus, because the hippocampus is attractive as a center of learning and memory. Investigations have been often focused on their role of slow modulation of spinogenesis (in a timescale of days).[13–18] Importance of the rapid effect of estrogen is also suggested from many observations on modulation of electrophysiological properties of the hippocampus.[13,14,47,48] Many scientists had, however, not seriously considered that these rapid modulations might favor the local hippocampal steroidogenesis rather than gonadal hormones which reach the brain traveling for long distances. A main reason for this is due to weak activity of sex steroid production of the hippocampus. However, the low levels of synthesized steroids are sufficient for the local usage within small neurons (i.e., intracrine system). This intracrine system contrasts with the endocrine organs in which high expression levels of steroidogenic enzymes are necessary in order to supply steroids to many other organs via the blood circulation. For brain-derived sex hormones, the rapid modulation (within 1–2h) of synaptic plasticity and cognitive functions are essential functions.

7.2.2 Long-Term Depression (LTD) and Long-Term Potentiation (LTP)

17β-Estradiol may rapidly modulate several different types of synaptic plasticity of neurons. One is synaptic transmission such as LTD or LTP, and another is spinogenesis. LTD and LTP probe the characteristics of preformed synapses, whereas spinogenesis analyzes not only spine-synapses (spines forming synapses) but also free spines (spines without forming synapses). Estradiol-induced modulation of LTD or LTP occurs only in preexistent synapses, because newly generated spines by estradiol-treatments do not form new synapses within 2h, as judged from no increase in the baseline magnitude of EPSP signal during 2h of estradiol perfusion.[49]

Evidence is emerging that estradiol exerts a rapid influence (0.5–1 h) on synaptic transmission of hippocampal slices from adult rats (3 months), as demonstrated by electrophysiology.[14,16,45,46,48] In case of the occasionally observed enhancement of LTP by 1–10nM estradiol in CA1 pyramidal neurons, a baseline increase by 20–30% has always been observed upon the onset of 10nM estradiol perfusion in the initial slope of the excitatory postsynaptic potential (EPSP), which has been attendant upon a further increase to approx. 160% upon high-frequency tetanic stimulation of Schaffer collaterals of hippocampus from adult rat (3 months) (Fig. 7.4).[13,14,50] However, without this 20–30% baseline increase in EPSP slope (before the tetanic stimulation), the enhancement of LTP by estradiol is not apparent, with regard

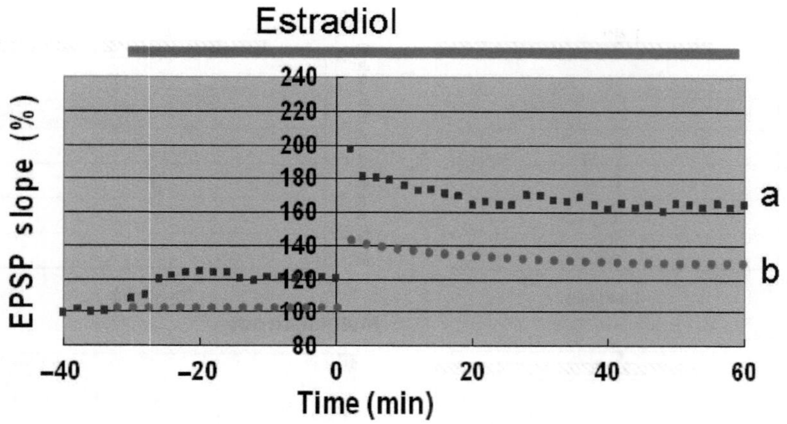

Fig. 7.4 Rapid modulation of LTP by 17β-estradiol in CA1 of hippocampal slices from adult male rats. In some slices such as curve a (blue square, n = 5) (observed for less than five slices out of 100 slices), pre-perfusion of 10 nM estradiol for 30 min at 30 °C increased the baseline slope of the excitatory postsynaptic potential (EPSP) to approximately 120–130%. Upon tetanic stimulation (100 Hz, 1 s, at t = 0) of the Schaffer collaterals, EPSP slope was significantly increased to the final level of 164.4 ± 12.6% (LTP-induction). In many slices such as curve b (red circle, n = 6), when an increase in the baseline EPSP slope was not occurred by perfusion of 10 nM estradiol (observed for nearly 95 slices out of 100 slices), EPSP slope was increased to 132.1 ± 7.8% upon tetanic stimulation. The difference between curves a and b is primarily due to approximately 20–30% increase in the baseline EPSP slope of curve a, upon estradiol perfusion. Red bar above the graph indicates period of time during which estradiol was administered. Acute slices from adult male (3 months) Wistar rats were investigated with a conventional electrophysiological setup (Modified from Kawato[50])

to the pure tetanic stimulation-induced LTP. In other words, the magnitude of pure tetanic stimulation-induced LTP (approximately 130%) is nearly the same between the presence and the absence of 10 nM estradiol (Fig. 7.4).[46,50] It should be noted that in 3–4 weeks puberty rats, 10 nM estradiol even suppressed LTP-induction down to the same level as that for adult rats.[16,46] Estradiol effects on LTP seem to be significantly dependent on the age of rats.

In memory processing, not only LTP (memory forming mechanism) but also LTD is essential. Mutant mice, which show enhanced LTP and suppressed LTD, have shown impaired learning of Morris water maze.[51] This suggests that LTD may be required to "correct" wrong memories formed by initial LTP processes, which store not only correct information but also wrong information.

We found that LTD was very sensitive to 17β-estradiol treatments in hippocampal slices from adult male rats. We demonstrated, for the first time, a significant rapid enhancement of LTD by 1–10 nM estradiol perfusion in CA1, CA3 and DG (Fig. 7.5).[49] Recordings were performed using 64 planar multielectrodes particularly arranged to stimulate the Schaffer collaterals in the stratum radiatum of CA1, the recurrent collateral fibers in the stratum radiatum of CA3, and the medial perforant pathways in the molecular layer of DG. LTD was induced pharmacologically by the transient application (3 min) of NMDA. This LTD was induced by the activation of phosphatase

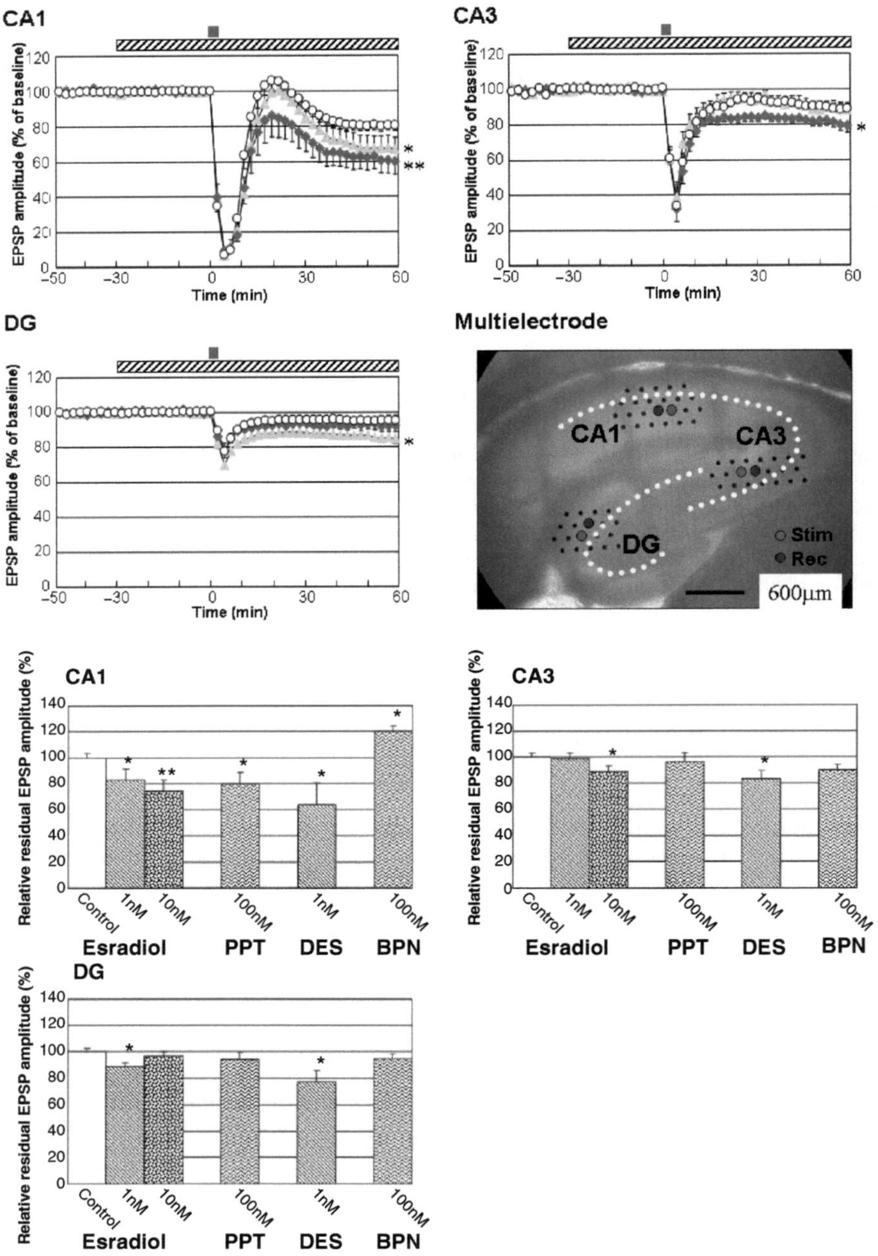

Fig. 7.5 Rapid modulation of LTD by 17β-estradiol in hippocampal slices from adult male rats. (Upper CA1, CA3 and DG) Time dependence of maximal EPSP amplitude in CA1 (CA1), CA3 (CA3) and DG (DG). Estradiol concentration was 0 nM (open circle), 1 nM (yellow closed triangle) and 10 nM (red closed diamond), respectively. (Multielectrode) Custom-made 64 multielectrode probe (MED64, Panasonic, Japan) with the hippocampal slice. Stimulation (red circle) and

(continued)

due to a moderate Ca^{2+} influx through NMDA receptors.[52] LTD is effectively induced by the transient application of NMDA for adult hippocampus, whereas low frequency stimulation cannot induce LTD in adult slices. Low frequency stimulation can induce LTD in slices from animals younger than 2 weeks. The plateau EPSP amplitude at 60 min after NMDA application was 80.4% (CA1), 88.8% (CA3) and 95.1% (DG), respectively. A 30 min preperfusion of 10 nM estradiol significantly enhanced LTD resulting in the residual EPSP amplitude of 59.7% (CA1), 79.1% (CA3) and 92.2% (DG) (Fig. 7.5).[49] Investigations using specific estrogen agonists indicated that the contribution of ERα (but not ERβ) was essential to these estradiol effects. Propyl-pyrazole-trinyl-phenol (PPT, ERα agonist)[53] at 100 nM exhibited a significant LTD enhancement in CA1, while diarylpropionitrile (DPN, ERβ agonist)[53] did induce a suppression of LTD in CA1, implying that the contribution of ERβ was opposite to that of ERα in the estradiol effect on LTD. Taken collectively, estradiol-bound ERα may activate phosphatase at the moderate Ca^{2+} concentration of around 0.7–1 μM induced upon 30 μM NMDA application,[54] and facilitated dephosphorylation of AMPA receptors may induce enhancement of LTD. On the other hand, estradiol-bound ERα is not functional in LTP modulation at the transiently high Ca^{2+} concentration of around 5–12 μM under tetanic stimulation,[47,54–56] because phosphorylation of AMPA receptors by CaM kinase II is a dominant process at the high Ca^{2+} concentration.

7.2.3 Spinogenesis

Modulation of spinogenesis is another essential action of estrogen in memory processes, involving production of new spines that creates sites for new neuronal contacts. We demonstrated that dendritic spines were rapidly modulated by estradiol application, using single spine analysis of Lucifer-Yellow-injected neurons in hippocampal slices from adult male rats (3 months).[47,57–59] Following a 2 h treatment with estradiol in the stratum radiatum of CA1 region, the treated dendrites have significantly more spines at 1 nM estradiol (1.31 spines/μm) than dendrites at 0 nM

Fig. 7.5 (continued) recording (blue circle) electrodes are indicated. Here, 100% EPSP amplitude refers to the EPSP value at t = –40 min prior to NMDA stimulation, irrespective of the test condition. LTD was induced by 30 μM NMDA perfusion at time t = 0 to 3 min (closed bar above the graph). Hatched bar above the graph indicates period of time during which estradiol was administered. The LTD enhancement was reproducibly observed in more than 90 slices out of 100 slices. (Lower CA1, CA3 and DG) Comparison of modulation effect of LTD by 17β-estradiol and agonists in the CA1 (CA1), CA3 (CA3) and DG (DG) of hippocampal slices. Vertical axis is relative EPSP amplitude at t = 60 min, where EPSP amplitude at t = 60 min of the control slice without drug application is taken as 100%. From left to right, 17β-estradiol (estradiol), PPT, DES and DPN at indicated concentrations. Note that coperfusion of 1 μM ICI with 10 nM 17β-estradiol did not suppress the enhancing effect of LTD by estradiol (data not shown). The significance of the estradiol effect was confirmed at 60 min via statistical analysis using ANOVAs (*, $p < 0.05$; **, $p < 0.01$) (Modified from Mukai et al.[49])

estradiol (0.85 spines/μm) (Fig. 7.6A).[49,60] PPT induced a significant enhancement of the spine density to 1.20 spines/μm. However, DPN increased the spine density only slightly (0.95 spines/μm). Blocking of ERα by ICI 182,780 completely suppressed the enhancing effect of estradiol on the spine density. Blocking of phosphorylation of Erk MAP kinase by PD98059 completely prevented the estradiol-induced spinogenesis.[58] Taken together, the enhancement of the spine density is probably induced by activation of Erk MAP kinase by estradiol-bound ERα at the basal low Ca^{2+} concentration of around 0.1–0.2 μM in resting neurons.[61] When the Ca^{2+} concentration in spines was further decreased by blocking NMDA receptors with MK-801, the enhancing effect by estradiol was completely suppressed. Function of estradiol-bound ERα therefore needs the basal level of Ca^{2+} concentration of around 0.1–0.2 μM. The morphological changes in CA1 spines occurred by 2 h estradiol-treatments. In control slices (0 nM estradiol), the relative population of spines was approximately 24% for mushroom spine, 62% for thin spine, 1% for filopodium, and 13% for stubby spine. Upon 1 nM estradiol treatment, the density of thin spine was selectively increased, from 0.57 to 0.97 spines/μm, while the density of mushroom and stubby was not significantly altered (Fig. 7.6A).

The spine density is not always increased but in some cases decreased by the estradiol-treatment. The estradiol-induced spinogenesis is highly region specific and heterogeneous. In fact, in CA3 pyramidal neurons, the total density of thorns of thorny excrescences (spine-like postsynaptic structures in the stratum lucidum of CA3, having contacts with mossy fiber terminals originated from granule cells) decreased dramatically to approx. 70% upon a 2 h treatment of 1 nM estradiol (Fig. 7.6B).[59] PPT significantly decreased the density of thorns from 2.19 to 1.66 thorns/μm, but DPN did not significantly change the density of thorns (Fig. 7.6B). Blocking of Erk MAP kinase by PD98059 completely prevented the estradiol-induced decrease of thorns. Taken together, in the stratum lucidum of CA3, the decrease of the thorn density is probably induced by activation of Erk MAP kinase by estradiol-bound ERα at the basal Ca^{2+} concentration of around 0.1–0.2 μM. When the Ca^{2+} concentration was decreased to below the basal level by blocking AMPA receptors with CNQX, or by changing the outer medium to Ca^2 free ACSF, the supression effect of estradiol was completely inhibited (Fig. 7.6B). These results suggest that the decrease of thorns requires the basal Ca^{2+} concentration which is kept by spontaneous postsynaptic Ca^{2+} fluctuation via voltage activated calcium channels depending upon AMPA receptor-mediated spontaneous voltage fluctuations, because the spontaneous Ca^{2+} influx within thorny excrescences occurs mainly via voltage activated calcium channels.[62–66] Note that blocking of NMDA receptors by MK-801 did not prevent the estradiol-induced decrease of thorns. This may be due to much smaller contribution of NMDA receptors to the spontaneous Ca^{2+} influx within thorns than that of voltage activated calcium channels. Function of estradiol-bound ERα therefore needs the basal level of Ca^{2+} concentration around 0.1–0.2 μM.

We always use acute hippocampal slices in order to examine the direct effect of estradiol on glutamatergic neurons within slices. Results from in vivo investigations using whole rat may reflect not only direct but also indirect effects of estradiol on glutamatergic neurons via cholinergic or serotonergic neurons, projecting to the hippocampus.[67,68]

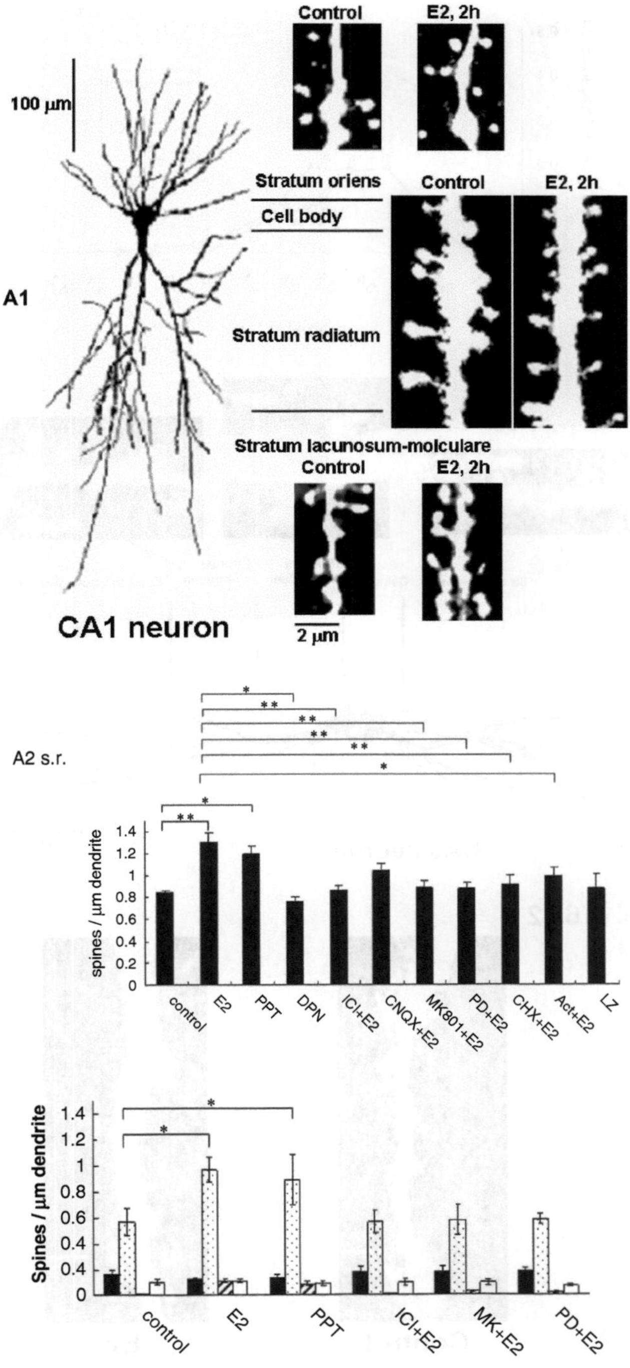

Fig. 7.6 (continued)

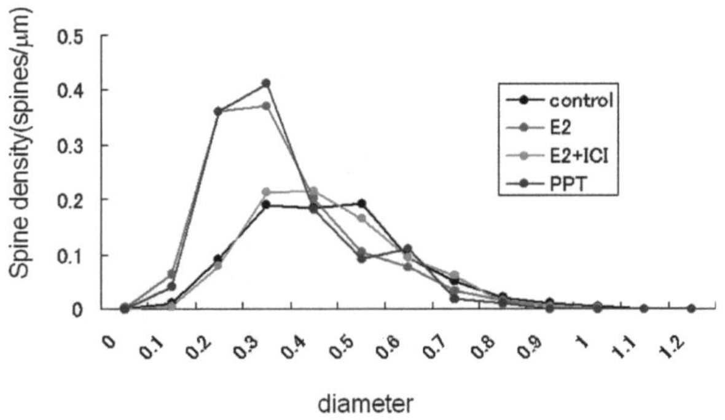

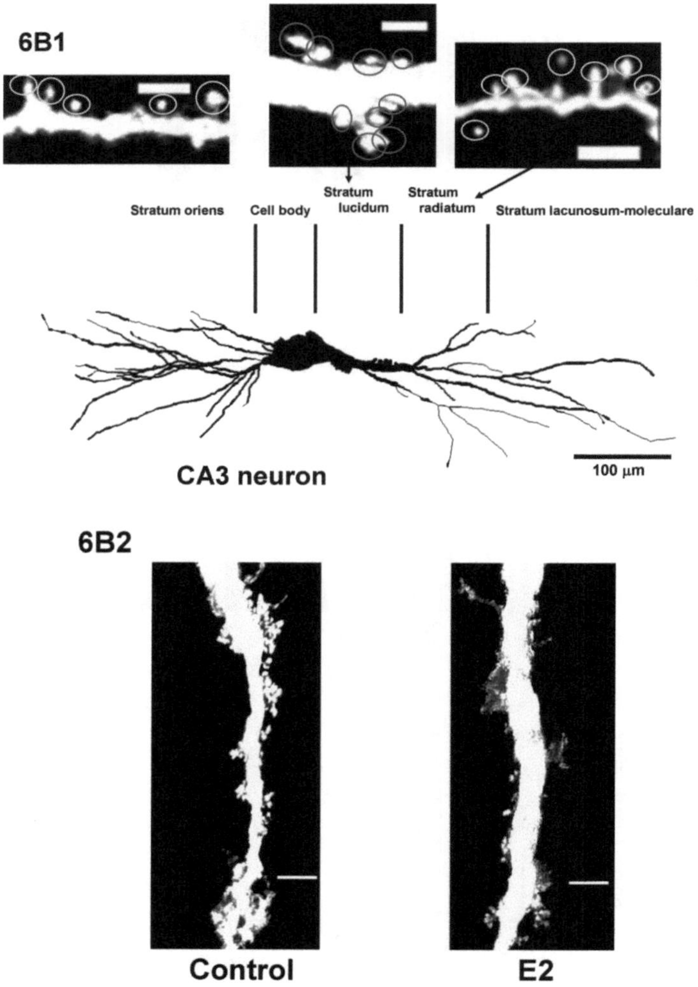

Fig. 7.6 (continued)

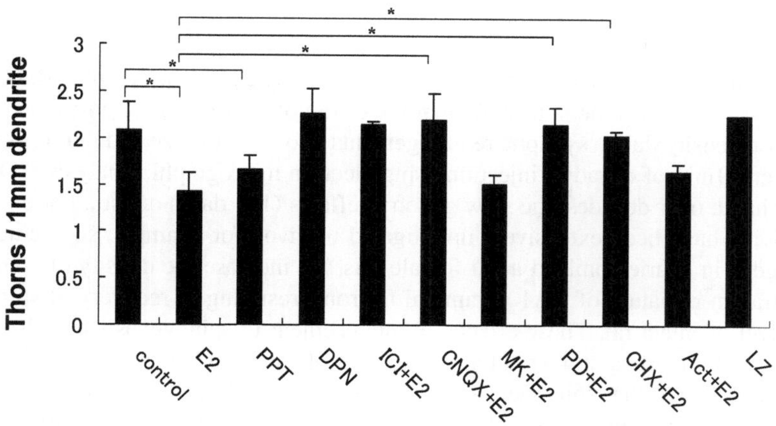

Fig. 7.6 (continued) Changes in the density and morphology of spines in CA1 (A1–A4) or thorns in CA3 (B1–B3) upon treatments of 17β-estradiol (E2) and drugs in hippocampal slices from adult male rats. Spines/thorns were analyzed along the dendrites of pyramidal neurons. (A1) Confocal micrographs showing spines along the secondary dendrites of hippocampal CA1 pyramidal neurons. (Left) A whole image of Lucifer Yellow-injected CA1 neuron. Vertical bar 100 μm. (Right) Maximal intensity projections onto XY plane from z-series confocal micrographs, showing spines along the dendrites. From top to bottom, spines along the basal dendrite in stratum oriens (Control and E2), spines along the apical dendrite in stratum radiatum (Control and E2), and spines along the apical dendrite in stratum lacunosum-moleculare (Control and E2), horizontal bar 2 μm. Slices were treated in artificial cerebrospinal fluid (ACSF) for 2 h without drugs (Control) or with 1 nM E2 (E2). ACSF contains 2 mM Ca^{2+} and 2 mM Mg^{2+}. (A2) Effect of drug treatments on the total spine density of CA1 neurons in the stratum radiatum (sr). Vertical axis is the average number of spines per 1 μm of dendrite. A 2 h treatment in ACSF without drugs (Control), with 1 nM E2 (E2), with 100 nM PPT (PPT), with 100 nM DPN (DPN), with 1 nM E2 and 1 μM ICI 182,780 (ICI + E2), with 1 nM E2 and 20 μM CNQX (CNQX + E2), with 1 nM E2 and 50 μM MK-801 (MK + E2), with 1 nM E2 and 50 μM PD98059 (PD + E2), with 1 nM E2 and 20 μM CHX (CHX + E2), with 1 nM E2 and 4 μM Actinomycin D (Act + E2), with 1 nM E2 and 100 nM letrozole (LZ + E2). (A3) Density of four subtypes of spines in the stratum radiatum. A 2 h treatment in ACSF without drugs (Control group), with 1 nM E2 (E2 group), with 100 nM PPT (PPT group), with 1 nM E2 and 1 μM ICI (ICI + E2 group), with 1 nM E2 and 50 μM MK-801 (MK + E2 group), with 1 nM E2 and 50 μM PD98059 (PD + E2 group). In each group, from left to right, mushroom (black column), thin (dotted column), filopodium (hatched column), and stubby (open column). (A4) Histogram of spine head diameters after a 2 h treatment in ACSF without drugs (control, black), with 1 nM E2 (red), with 1 nM E2 and ICI (green), and with 100 nM PPT (blue) (B1) Maximal intensity projections onto XY plane from z-series confocal micrographs, showing thorns and spines along the primary and secondary dendrites of hippocampal CA3 pyramidal neurons. (Upper left) spines along the basal dendrite in stratum oriens, bar 2 μm. (Upper middle) thorns on thorny excrescences along the primary/secondary dendrite in stratum lucidum, bar 2 μm. (Upper right) spines along the apical dendrite in stratum radiatum, bar 2 μm. Thorny excrescences have bulbous-shaped huge heads named thorns (red circles) which are different from spines with separated distribution (yellow circles). (Lower image) a whole image of Lucifer Yellow-injected CA3 neuron. Horizontal bar 100 μm. (B2) Maximal intensity projections onto XY plane from z-series confocal micrographs, showing thorns in the stratum lucidum without drug-treatments (control) and thorns after estradiol-treatments (E2). Bar 5 μm. (B3) Effect of drug treatments on the average number of thorns per 1 μm dendritic segment. A 2 h treatment in ACSF without estradiol (control), with 1 nM estradiol (E2), with 100 nM PPT (PPT), with 100 nM DPN (DPN), with 1 nM E2 and 1 μM ICI 182,780 (ICI + E2), with 1 nM estradiol and 20 μM CNQX (E2 + CNQX), with 1 nM estradiol and 50 μM MK-801 (E2 + MK), with 1 nM estradiol and 20 μM PD98059 (E2 + PD), with 1 nM E2 and 20 μM CHX (CHX + E2), with 1 nM E2 and 4 μM Actinomycin D (Act + E2), and with 1 nM E2 and 100 nM letrozole (LZ + E2). Statistical significance (*, $p < 0.05$) (Modified from Mukai et al.[49], Murakami et al.[58] and Tsurugizawa et al.[59]]

The rapid effect of estrogens has also been observed in vivo. Leranth, MacLusky and co-workers have demonstrated that the estradiol (60 µg/kg) increases the spine-synapse density due to synaptic rearrangements in ovariectomized adult rats as rapid as after 30 min of estradiol injection using electron micrographic analysis.[69] On the other hand, over decades, the slow genomic effects (1–4 days) of estradiol on spine plasticity, have been extensively investigated in vivo. For example, supplement of estrogens in ovariectomized adult female rats,[69–72] increase the density of spines in the stratum radiatum of CA1 pyramidal neurons, resulting in recovery of spines to the level of intact rat. These effects of enhancement in spinogenesis has also been observed as rapid as at 4.5 h after sc injection of estrogen.[69] In vitro investigations have also shown that spine density in CA1 increases following several days' treatment of cultured hippocampal slices with exogenous estradiol.[15] The contribution of endogenous estradiol has been reported by Rune and co-workers who demonstrated that the suppression of endogenous estradiol synthesis by letrozole treatments for 4 days significantly decreased the density of spines, spine-synapses, spinophilin (spine marker protein) and synaptophysin (presynaptic marker) in the stratum radiatum of the CA1 region in cultured slices.[5] No increase in the density of spines, spine-synapses and spinophilin expression was seen after exogenous application of 100 nM estradiol to the medium of slice cultures that had not been treated with letrozole. Application of 100 nM estradiol, however, induced rescue effect which restored the synaptophysin expression that had been once decreased by letrozole.

7.2.4 Synaptic Membrane Receptors

What is the receptor of 17β-estradiol that mediates rapid actions (0.5–2 h) on synaptic plasticity in the hippocampus? Putative synaptic membrane estrogen receptors remain poorly defined. Many attempts have been made to identify membrane estrogen receptors. At the present stage, the most probable candidates for membrane estrogen receptors may be ERα, ERβ and GPR30.

Why are classical nuclear type receptors ERα and ERβ candidates? Because ICI do not suppress estradiol–induced rapid modulation of electrophysiological properties such as LTD, LTP, and kinate-induced currents, classical estrogen receptors are suggested to be not involved in these modulations (Fig. 7.5).[45] However, these results do not eliminate the possibility that ERα and ERβ could drive these synaptic transmissions. ICI has been indicated to display its effect by inhibiting dimerization of ERα and ERβ. If dimerization processes are not involved in rapid modulation of electrophysiological phenomena, then ICI cannot block these phenomena. On the other hand, rapid enhancement of spinogenesis via ERα was significantly blocked by ICI (Fig. 7.6B),[49] therefore dimerization processes occur for synaptic ERα in spinogenesis.

After several years of careful investigations, we successfully identified the membrane estrogen receptor ERα localized in the spines of hippocampal pyramidal and granule neurons by means of immunoelectron microscopic analysis as well as

Western blot analysis using affinity-column-purified anti-ERα antibody RC-19.[49] A post-embedding immunogold electron microscopic analysis demonstrated the synaptic localization of ERα in the glutamatergic neurons in CA1, CA3 and DG (Fig. 7.7). ERα was also localized in the nuclei. Western blot analysis demonstrated that ERα (67 kDa) and Erk MAP kinase were tightly associated with postsynaptic density fractions (PSD). Because the estradiol-induced modulation of LTD and spine density appeared so rapidly in the time range of 1–2 h, the synaptic ERα observed at PSD and postsynaptic compartments probably plays an essential role in driving rapid processes. It should be noted that specific binding of purified RC-19 antibody to real ERα (67 kDa) in the hippocampus was verified using MALDI-TOF mass-spectrometric analysis of RC-19 reacted proteins as well as the absence of reactivity of RC-19 with ERαKO mice hippocampus (Fig. 7.8 Upper A–F).[49] These

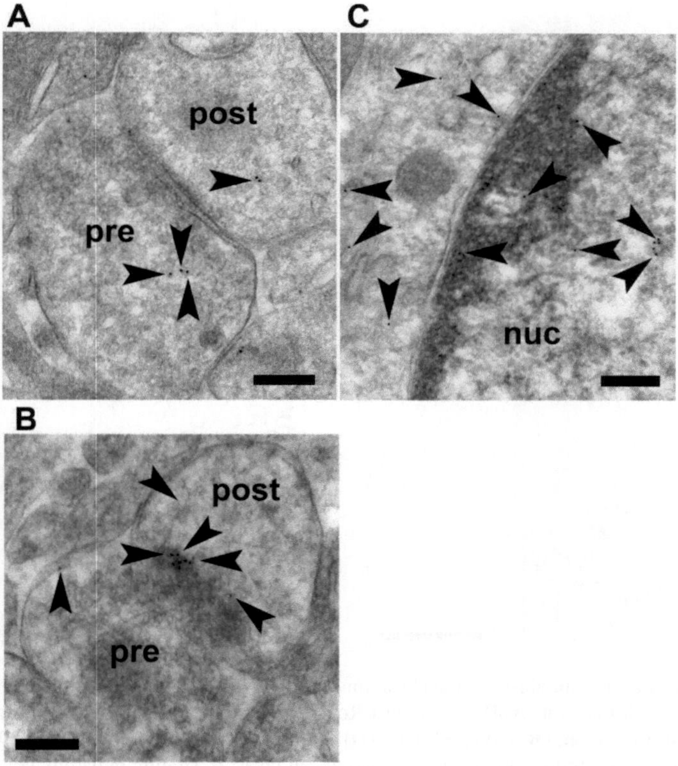

Fig. 7.7 Immunoelectron microscopic analysis of the distribution of ERα within axospinous synapses, in the stratum radiatum of the hippocampal slices from adult male rat. (**A**) Gold particles (arrowheads) were localized in the pre- and postsynaptic regions in CA1. (**B**) In dendritic spines, gold particles were associated with PSD regions in CA1. (**C**) Gold particles were also localized in the nuclei in CA1. Pre, presynaptic region; Post, postsynaptic region; Scale bar: 200 nm (Modified from Mukai et al.[49])

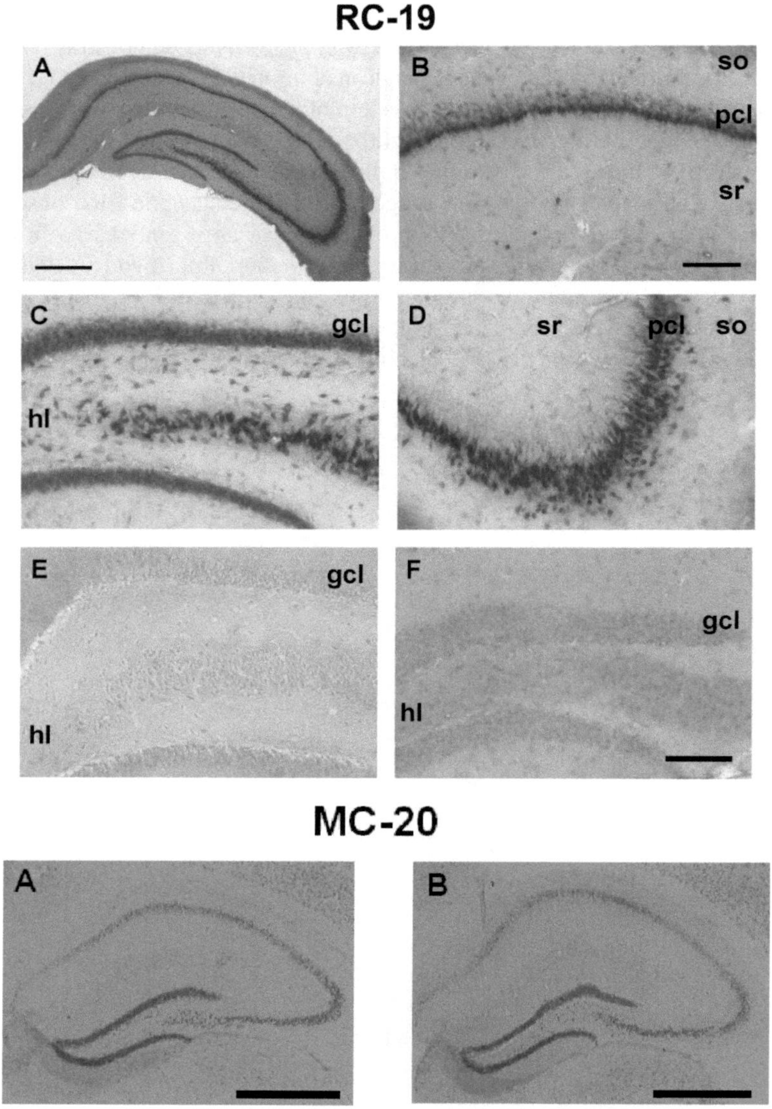

Fig. 7.8 (Upper) Immunohistochemical staining of ERα with RC-19 antibody in the hippocampal slices from adult male rat (**A–D**) and adult ERαKO mouse (**F**). (**A**) coronal section of the whole hippocampal formation; (**B**) CA1; (**C**) DG; (**D**) CA3; (**E**) pre adsorption with RC–19; (**F**) DG of ERαKO mouse. so, stratum oriens; pcl, pyramidal cell layer; sr, stratum radiatum; gcl, granule cell layer; hl, hilus. Scale bar, 500 μm for A, and 200 μm for B–E. (Lower) Immunohistochemical staining with MC-20 antiserum in the hippocampal slices from adult ERαKO (**A**) and wild male mice (**B**). (**A**) and (**B**) the coronal section of the whole hippocampal formation. Scale bar, 50 μm (Modified from Mukai et al.[49])

analyses are essential in the hippocampus, because we found that non-purified MC-20 antisera, frequently used in previous investigations, often reacted with 62 kDa unknown proteins in the brain and did not significantly react with real ERα (67 kDa).[49] AS409, another frequently used antisera did mainly react with unknown proteins different from real ERα.[49] Non-purified antisera may largely react with proteins having amino acid sequences similar to the real antigen in the hippocampus in which extremely low level of ERα is expressed as compared with that in the ovary (Fig. 7.8 Lower A and B). Suprisingly, ERα antisera are often examined for their reactivity only in endocrine organs such as the ovary in which ERα is highly expressed. Therefore, staining of interneurons and no staining of primary neurons with non-purified antisera (such as MC-20 or AS409) probably do not show real ERα distribution in the hippocampus. Antisera should be purified before application to the hippocampus.

ERα knock-out mice may be useful to investigate the participation of ERα in modulation of synaptic plasticity. However, so far no data are available for ERα knock-out mice. Electrophysiological investigations are performed by using knock-down mice (not knock out mice) by Moss and coworkers.[45,73] They have reported no essential contribution of ERα to estradiol-induced rapid enhancement of the kainate currents of CA1 neurons. They reach this conclusion due to the observation of very small difference in estradiol effect on the kainate currents between wild-type and ERα-Neo knock-down mice which have been constructed by the method of Neomycin insertion into exon 1 (the previously named exon 2).[74] It should be noted that in Neomycin-insertion ERα-Neo knock-down mice, N-terminal-modified ERα (61 kDa) is expressed.[74–76] Because the N-terminal-modified ERα is demonstrated to be still active on estradiol binding and drives genomic processes,[74–76] the participation of ERα in electrophysiological properties of the CA1 cannot be excluded from their investigations. Therefore, it is necessary to investigate real ERα knock-out mice which are, for example, deleted in the whole exon 2 of the mouse ERα gene.[77] Note that nomenclature of ERα exon changes recently, and the current exon 1 and exon 2[75,76] correspond to the previous exon 2 and exon 3, respectively.[77]

ERβ has been reported to associate with membranes in genetically expressed CHO cells and MCF-7 cells.[78,79] Several investigations of immunostaining of ERβ have suggested extranuclear expression of ERβ including dendritic appearance in the hippocampal principal neurons.[80] ERβ is, however, not yet identified as synaptic membrane receptors. Subcellular immunostaining patterns of these reports might reflect relatively minor expression of ERβ and major expression of unknown proteins, due to multiple reactivity of non-purified ERβ antisera to several unknown proteins in Western blot analysis of hippocampal tissues. The purity of commercially available ERβ antisera may be worse than that of ERα antisera as judged from Western blot analysis.

Recently transmembrane G-protein coupled estrogen receptor GPR30 has been identified in the plasma membrane of SKBR3 breast cancer cells that lack ERα and ERβ[81] as well as in endoplasmic reticulum membranes of COS7 after genetic expression of GPR30 fused with Green Fluorescent Protein.[82] Because expression of GPR30 has also been observed in the hippocampal neurons,[83] further investigations may reveal its contribution to rapid estradiol modulation of synaptic plasticity.

7.3 Synaptocrinology

Based on experimental observations, we illustrate in Fig. 7.9, a hypothetical model for the synaptic synthesis of brain steroid and the modulation of the synaptic plasticity of neurons by brain steroid. Brain steroid synthesis proceeds in the following manner. First, glutamate release from the presynapse induces a Ca^{2+} influx through the NMDA receptors. The Ca^{2+} influx drives StAR or peripheral benzodiazepine receptor (PBR)[84] to transport cholesterol into the mitochondria, where P450scc converts cholesterol to pregnenolone. the conversion of "pregnenolone → DHEA → androstenediol → testosterone → estradiol, or testosterone → dihydrotestosterone → androstanediol" is performed at postsynaptic density as well as endoplasmic reticulum by P450(17α), 3β-HSD, 17β-HSD,

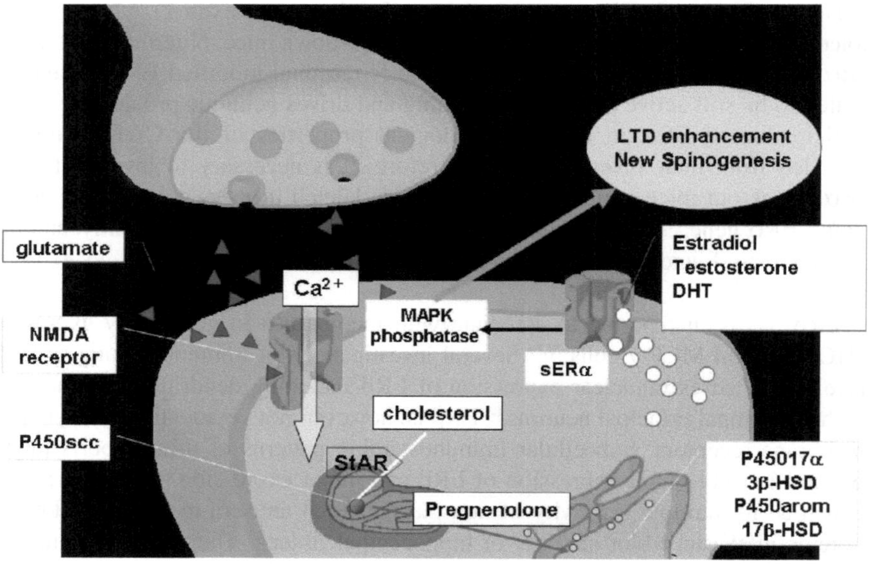

Fig. 7.9 Schematic illustration for the synaptic synthesis of sex hormones, and the modulation of the synaptic plasticity of neurons by estradiol. AMPA type of glutamate receptors are omitted for clarity. StAR, PBR and P450scc are present in the mitochondria. P450(17α), 3β-HSD, 17β-HSD and P450arom are localized in the membranes in the synaptic compartment. The site of rapid action for estradiol is synaptic ERα. Synaptic ERβ might also function. The site of delayed action for estradiol is ERα present in cytoplasm and nuclei

P450arom, 5α-reductase and 3α-HSD. Produced estradiol binds to synaptic ERα and drives signaling pathway including kinases (such as Erk MAP kinase) or phosphatases, finally resulting in modulation of AMPA receptors or NMDA receptors. Note that brain steroids are synthesized also in endoplasmic reticulum and mitochondria in neurons. Genomic pathway via nuclear ERα receptors is also function in delayed estradiol effects such as neuroprotection, spinogenesis, keeping homeostasis, etc.

References

1. Kimoto T, Tsurugizawa T, Ohta Y, et al. Neurosteroid synthesis by cytochrome p450-containing systems localized in the rat brain hippocampal neurons: N-methyl-D-aspartate and calcium-dependent synthesis. Endocrinology 2001; 142:3578–3589.
2. Kawato S, Hojo Y, Kimoto T. Histological and metabolism analysis of P450 expression in the brain. Methods Enzymol 2002; 357:241–249.
3. Kawato S, Yamada M, Kimoto T. Brain neurosteroids are 4th generation neuromessengers in the brain: cell biophysical analysis of steroid signal transduction. Adv Biophys 2003; 37:1–48.
4. Hojo Y, Hattori TA, Enami T, et al. Adult male rat hippocampus synthesizes estradiol from pregnenolone by cytochromes P45017alpha and P450 aromatase localized in neurons. Proc Natl Acad Sci USA 2004; 101:865–870.
5. Kretz O, Fester L, Wehrenberg U, et al. Hippocampal synapses depend on hippocampal estrogen synthesis. J Neurosci 2004; 24:5913–5921.
6. Baulieu EE. Neurosteroids: of the nervous system, by the nervous system, for the nervous system. Recent Prog Horm Res 1997; 52:1–32.
7. Corpechot C, Robel P, Axelson M, et al. Characterization and measurement of dehydroepiandrosterone sulfate in rat brain. Proc Natl Acad Sci USA 1981; 78:4704–4707.
8. Robel P, Bourreau E, Corpechot C, et al. Neuro-steroids: 3 beta-hydroxy-delta 5-derivatives in rat and monkey brain. J Steroid Biochem 1987; 27:649–655.
9. Warner M, Gustafsson JA. Cytochrome P450 in the brain: neuroendocrine functions. Front Neuroendocrinol 1995; 16:224–236.
10. Baulieu EE, Robel P. Dehydroepiandrosterone (DHEA) and dehydroepiandrosterone sulfate (DHEAS) as neuroactive neurosteroids. Proc Natl Acad Sci USA 1998; 95:4089–4091.
11. Le Goascogne C, Sananes N, Gouezou M, et al. Immunoreactive cytochrome P-450(17 alpha) in rat and guinea-pig gonads, adrenal glands and brain. J Reprod Fertil 1991; 93:609–622.
12. Mellon SH, Deschepper CF. Neurosteroid biosynthesis: genes for adrenal steroidogenic enzymes are expressed in the brain. Brain Res 1993; 629:283–292.
13. Bi R, Broutman G, Foy MR, et al. The tyrosine kinase and mitogen-activated protein kinase pathways mediate multiple effects of estrogen in hippocampus. Proc Natl Acad Sci USA 2000; 97:3602–3607.
14. Foy MR, Xu J, Xie X, et al. 17beta-estradiol enhances NMDA receptor-mediated EPSPs and long-term potentiation. J Neurophysiol 1999; 81:925–929.
15. Pozzo-Miller LD, Inoue T, Murphy DD. Estradiol increases spine density and NMDA-dependent Ca2+ transients in spines of CA1 pyramidal neurons from hippocampal slices. J Neurophysiol 1999; 81:1404–1411.
16. Shibuya K, Takata N, Hojo Y, et al. Hippocampal cytochrome P450s synthesize brain neurosteroids which are paracrine neuromodulators of synaptic signal transduction. Biochim Biophys Acta 2003; 1619:301–316.
17. Woolley CS. Estrogen-mediated structural and functional synaptic plasticity in the female rat hippocampus. Horm Behav 1998; 34:140–148.

18. Woolley CS, McEwen BS. Estradiol regulates hippocampal dendritic spine density via an N-methyl-D-aspartate receptor-dependent mechanism. J Neurosci 1994; 14:7680–7687.
19. Sanne JL, Krueger KE. Expression of cytochrome P450 side-chain cleavage enzyme and 3 beta-hydroxysteroid dehydrogenase in the rat central nervous system: a study by polymerase chain reaction and *in situ* hybridization. J Neurochem 1995; 65:528–536.
20. Murakami G, Tanabe N, Ishii HT, et al. Role of cytochrome p450 in synaptocrinology: endogenous estrogen synthesis in the brain hippocampus. Drug Metab Rev 2006; 38:353–369.
21. Furukawa A, Miyatake A, Ohnishi T, Ichikawa Y. Steroidogenic acute regulatory protein (StAR) transcripts constitutively expressed in the adult rat central nervous system: colocalization of StAR, cytochrome P-450SCC (CYP XIA1), and 3beta-hydroxysteroid dehydrogenase in the rat brain. J Neurochem 1998; 71:2231–2238.
22. King SL, Marks MJ, Grady SR, et al. Conditional expression in corticothalamic efferents reveals a developmental role for nicotinic acetylcholine receptors in modulation of passive avoidance behavior. J Neurosci 2003; 23:3837–3843.
23. Zwain IH, Yen SS. Neurosteroidogenesis in astrocytes, oligodendrocytes, and neurons of cerebral cortex of rat brain. Endocrinology 1999; 140:3843–3852.
24. Zwain IH, Yen SS. Dehydroepiandrosterone: biosynthesis and metabolism in the brain. Endocrinology 1999; 140:880–887.
25. Compagnone NA, Bulfone A, Rubenstein JL, Mellon SH. Steroidogenic enzyme P450c17 is expressed in the embryonic central nervous system. Endocrinology 1995; 136:5212–5223.
26. Wehrenberg U, Prange-Kiel J, Rune GM. Steroidogenic factor-1 expression in marmoset and rat hippocampus: co-localization with StAR and aromatase. J Neurochem 2001; 76:1879–1886.
27. Ivanova T, Beyer C. Ontogenetic expression and sex differences of aromatase and estrogen receptor-alpha/beta mRNA in the mouse hippocampus. Cell Tissue Res 2000; 300:231–237.
28. Beyenburg S, Watzka M, Blumcke I, et al. Expression of mRNAs encoding for 17beta-hydroxisteroid dehydrogenase isozymes 1, 2, 3 and 4 in epileptic human hippocampus. Epilepsy Res 2000; 41:83–91.
29. Jung-Testas I, Hu ZY, Baulieu EE, Robel P. Neurosteroids: biosynthesis of pregnenolone and progesterone in primary cultures of rat glial cells. Endocrinology 1989; 125:2083–2091.
30. Koenig HL, Schumacher M, Ferzaz B, et al. Progesterone synthesis and myelin formation by Schwann cells. Science 1995; 268:1500–1503.
31. Tsutsui K, Ukena K, Usui M, et al. Novel brain function: biosynthesis and actions of neurosteroids in neurons. Neurosci Res 2000; 36:261–273.
32. Jakab RL, Horvath TL, Leranth C, et al. Aromatase immunoreactivity in the rat brain: gonadectomy-sensitive hypothalamic neurons and an unresponsive "limbic ring" of the lateral septum-bed nucleus-amygdala complex. J Steroid Biochem Mol Biol 1993; 44:481–498.
33. Shinzawa K, Ishibashi S, Murakoshi M, et al. Relationship between zonal distribution of microsomal cytochrome P-450s (P-450(17)alpha,lyase and P-450C21) and steroidogenic activities in guinea-pig adrenal cortex. J Endocrinol 1988; 119:191–200.
34. Le Goascogne C, Robel P, Gouezou M, et al. Neurosteroids: cytochrome P-450scc in rat brain. Science 1987; 237:1212–1215.
35. Mensah-Nyagan AG, Do-Rego JL, Beaujean D, et al. Neurosteroids: expression of steroidogenic enzymes and regulation of steroid biosynthesis in the central nervous system. Pharmacol Rev 1999; 51:63–81.
36. Kibaly C, Patte-Mensah C, Mensah-Nyagan AG. Molecular and neurochemical evidence for the biosynthesis of dehydroepiandrosterone in the adult rat spinal cord. J Neurochem 2005; 93:1220–1230.
37. Wang MD, Wahlstrom G, Backstrom T. The regional brain distribution of the neurosteroids pregnenolone and pregnenolone sulfate following intravenous infusion. J Steroid Biochem Mol Biol 1997; 62:299–306.
38. Kawato S, Ogiue-Ikeda M, Tanabe N, et al. Rapid modlulation of long-term depression and spinogenesis by endocrine disrupters in adult rat hippocampus. In 4th International Meeting Steroids and Nerbous System. Torino, 2007.

39. Liu S, Sjovall J, Griffiths WJ. Neurosteroids in rat brain: extraction, isolation, and analysis by nanoscale liquid chromatography-electrospray mass spectrometry. Anal Chem 2003; 75:5835–5846.
40. Higashi T, Sugitani H, Yagi T, Shimada K. Studies on neurosteroids XVI. Levels of pregnenolone sulfate in rat brains determined by enzyme-linked immunosorbent assay not requiring solvolysis. Biol Pharm Bull 2003; 26:709–711.
41. Liere P, Akwa Y, Weill-Engerer S, et al. Validation of an analytical procedure to measure trace amounts of neurosteroids in brain tissue by gas chromatography-mass spectrometry. J Chromatogr B Biomed Sci Appl 2000; 739:301–312.
42. Vallee M, Mayo W, Darnaudery M, et al. Neurosteroids: deficient cognitive performance in aged rats depends on low pregnenolone sulfate levels in the hippocampus. Proc Natl Acad Sci USA 1997; 94:14865–14870.
43. Wu FS, Gibbs TT, Farb DH. Pregnenolone sulfate: a positive allosteric modulator at the N-methyl-D-aspartate receptor. Mol Pharmacol 1991; 40:333–336.
44. Hojo Y, Nakajima K, Nakanishi H, et al.: Synthesis brain steroids and localization of P450s in the hippocampal neurons. In 20th IUBMB International Congress of Biochemistry and Molecular Biology, 2006:2PA-311.
45. Gu Q, Moss RL. 17 beta-Estradiol potentiates kainate-induced currents via activation of the cAMP cascade. J Neurosci 1996; 16:3620–3629.
46. Ito K, Skinkle KL, Hicks TP. Age-dependent, steroid-specific effects of oestrogen on long-term potentiation in rat hippocampal slices. J Physiol 1999; 515 (Pt 1):209–220.
47. Mukai H, Tsurugizawa T, Ogiue-Ikeda M, et al. Local neurosteroid production in the hippocampus: influence on synaptic plasticity of memory. Neuroendocrinology 2006; 84:255–263.
48. Teyler TJ, Vardaris RM, Lewis D, Rawitch AB. Gonadal steroids: effects on excitability of hippocampal pyramidal cells. Science 1980; 209:1017–1018.
49. Mukai H, Tsurugizawa T, Murakami G, et al. Rapid modulation of long-term depression and spinogenesis via synaptic estrogen receptors in hippocampal principal neurons. J Neurochem 2007; 100:950–967.
50. Kawato S. Endocrine disrupters as disrupters of brain function: a neurosteroid viewpoint. Environ Sci 2004; 11:1–14.
51. Migaud M, Charlesworth P, Dempster M, et al. Enhanced long-term potentiation and impaired learning in mice with mutant postsynaptic density-95 protein. Nature 1998; 396:433–439.
52. Lee HK, Kameyama K, Huganir RL, Bear MF. NMDA induces long-term synaptic depression and dephosphorylation of the GluR1 subunit of AMPA receptors in hippocampus. Neuron 1998; 21:1151–1162.
53. Harrington WR, Sheng S, Barnett DH, et al. Activities of estrogen receptor alpha- and beta-selective ligands at diverse estrogen responsive gene sites mediating transactivation or transrepression. Mol Cell Endocrinol 2003; 206:13–22.
54. Yang SN, Tang YG, Zucker RS. Selective induction of LTP and LTD by postsynaptic [Ca2+]i elevation. J Neurophysiol 1999; 81:781–787.
55. Lisman J. A mechanism for the Hebb and the anti-Hebb processes underlying learning and memory. Proc Natl Acad Sci USA 1989; 86:9574–9578.
56. Ogiue-Ikeda M, Tanabe N, Mukai H, et al. Rapid Modulation of Synaptic Plasticity by Estrogens as well as Endocrine Disrupters in Hippocampal Neurons. Brain Res Rev 2007 (in press).
57. Komatsuzaki Y, Murakami G, Tsurugizawa T, et al. Rapid spinogenesis of pyramidal neurons induced by activation of glucocorticoid receptors in adult male rat hippocampus. Biochem Biophys Res Commun 2005; 335:1002–1007.
58. Murakami G, Tsurugizawa T, Hatanaka Y, et al. Comparison between basal and apical dendritic spines in estrogen-induced rapid spinogenesis of CA1 principal neurons in the adult hippocampus. Biochem Biophys Res Commun 2006; 351:553–558.
59. Tsurugizawa T, Mukai H, Tanabe N, et al. Estrogen induces rapid decrease in dendritic thorns of CA3 pyramidal neurons in adult male rat hippocampus. Biochem Biophys Res Commun 2005; 337:1345–1352.

60. Mukai H, Tsurugizawa, T, Murakami G, Kominami S, Ishii T, Ogiue-Ikeda M, Takata N, Tanabe N, Furukawa A, Hojo Y, Morrison JH, Janssen WGM, Rose JA, Chambon P, Kato S, Izumi S, Yamazaki T, Kimoto T, Kawato S. Rapid Modulation of Long-term Depression and Spinogenesis Depending on Synaptic Estrogen Receptors in Principal Neurons of Hippocampus. J. Neurochem. 2007; 100:950–967.
61. Ishii H, Shibuya K, Ohta Y, et al. Enhancement of nitric oxide production by association of nitric oxide synthase with N-methyl-D-aspartate receptors via postsynaptic density 95 in genetically engineered Chinese hamster ovary cells: real-time fluorescence imaging using nitric oxide sensitive dye. J Neurochem 2006; 96:1531–1539.
62. Monaghan DT, Holets VR, Toy DW, Cotman CW. Anatomical distributions of four pharmacologically distinct 3H-L-glutamate binding sites. Nature 1983; 306:176–179.
63. Baude A, Nusser Z, Molnar E, et al. High-resolution immunogold localization of AMPA type glutamate receptor subunits at synaptic and non-synaptic sites in rat hippocampus. Neuroscience 1995; 69:1031–1055.
64. Fritschy JM, Weinmann O, Wenzel A, Benke D. Synapse-specific localization of NMDA and GABA(A) receptor subunits revealed by antigen-retrieval immunohistochemistry. J Comp Neurol 1998; 390:194–210.
65. Reid CA, Fabian-Fine R, Fine A. Postsynaptic calcium transients evoked by activation of individual hippocampal mossy fiber synapses. J Neurosci 2001; 21:2206–2214.
66. Reid CA. The role of dendritic spines: comparing the complex with the simple. Eur J Pharmacol 2002; 447:173–176.
67. Leranth C, Petnehazy O, MacLusky NJ. Gonadal hormones affect spine synaptic density in the CA1 hippocampal subfield of male rats. J Neurosci 2003; 23:1588–1592.
68. Leranth C, Shanabrough M, Horvath TL. Hormonal regulation of hippocampal spine synapse density involves subcortical mediation. Neuroscience 2000; 101:349–356.
69. MacLusky NJ, Luine VN, Hajszan T, Leranth C. The 17alpha and 17beta isomers of estradiol both induce rapid spine synapse formation in the CA1 hippocampal subfield of ovariectomized female rats. Endocrinology 2005; 146:287–293.
70. Gould E, Woolley CS, Frankfurt M, McEwen BS. Gonadal steroids regulate dendritic spine density in hippocampal pyramidal cells in adulthood. J Neurosci 1990; 10:1286–1291.
71. Woolley CS, Gould E, Frankfurt M, McEwen BS. Naturally occurring fluctuation in dendritic spine density on adult hippocampal pyramidal neurons. J Neurosci 1990; 10:4035–4039.
72. Woolley CS, McEwen BS. Estradiol mediates fluctuation in hippocampal synapse density during the estrous cycle in the adult rat. J Neurosci 1992; 12:2549–2554.
73. Gu Q, Korach KS, Moss RL. Rapid action of 17beta-estradiol on kainate-induced currents in hippocampal neurons lacking intracellular estrogen receptors. Endocrinology 1999; 140:660–666.
74. Couse JF, Curtis SW, Washburn TF, et al. Analysis of transcription and estrogen insensitivity in the female mouse after targeted disruption of the estrogen receptor gene. Mol Endocrinol 1995; 9:1441–1454.
75. Kos M, Denger S, Reid G, et al. Down but not out? A novel protein isoform of the estrogen receptor alpha is expressed in the estrogen receptor alpha knockout mouse. J Mol Endocrinol 2002; 29:281–286.
76. Pendaries C, Darblade B, Rochaix P, et al. The AF-1 activation-function of ERalpha may be dispensable to mediate the effect of estradiol on endothelial NO production in mice. Proc Natl Acad Sci USA 2002; 99:2205–2210.
77. Dupont S, Krust A, Gansmuller A, et al. Effect of single and compound knockouts of estrogen receptors alpha (ERalpha) and beta (ERbeta) on mouse reproductive phenotypes. Development 2000; 127:4277–4291.
78. Pedram A, Razandi M, Levin ER. Nature of functional estrogen receptors at the plasma membrane. Mol Endocrinol 2006; 20:1996–2009.
79. Razandi M, Pedram A, Greene GL, Levin ER. Cell membrane and nuclear estrogen receptors (ERs) originate from a single transcript: studies of ERalpha and ERbeta expressed in Chinese hamster ovary cells. Mol Endocrinol 1999; 13:307–319.

80. Milner TA, Ayoola K, Drake CT, et al. Ultrastructural localization of estrogen receptor beta immunoreactivity in the rat hippocampal formation. J Comp Neurol 2005; 491:81–95.
81. Thomas P, Pang Y, Filardo EJ, Dong J. Identity of an estrogen membrane receptor coupled to a G protein in human breast cancer cells. Endocrinology 2005; 146:624–632.
82. Revankar CM, Cimino DF, Sklar LA, et al. A transmembrane intracellular estrogen receptor mediates rapid cell signaling. Science 2005; 307:1625–1630.
83. Brailoiu E, Dun SL, Brailoiu GC, et al. Distribution and characterization of estrogen receptor G protein-coupled receptor 30 in the rat central nervous system. J Endocrinol 2007; 193:311–321.
84. Papadopoulos V. Peripheral-type benzodiazepine/diazepam binding inhibitor receptor: biological role in steroidogenic cell function. Endocr Rev 1993; 14:222–240.

Chapter 8
Effects of Estradiol and DHEA on Morphological Synaptic Plasticity

Tibor Hajszan[1,2], Zsofia Hoyk[1], Luis Miguel Garcia-Segura[3], and Arpad Parducz[1]

Abstract Experimental data demonstrate that the nervous system is widely influenced by sex hormones, and the brain is continuously shaped by the environment via the changing hormone milieu throughout the entire life. The effects of gonadal hormones extend beyond regulating gonadotropin secretion; they are able to alter the structure of the adult nervous system by changing neuron and synapse numbers, as well as dendritic and synaptic morphology. These structural modifications are believed to serve as a morphological basis for varying behavior and cellular activity. In this review, we discuss the hormonally induced synaptic remodeling in different hypothalamic nuclei and in the hippocampus, focusing on specificity of action, time course, and functional consequences of synaptic alterations. It has been shown that the effect of estradiol is highly specific. In the arcuate nucleus, spine synapses and GABAergic axo-somatic terminals are affected, while in case of the hippocampus, estrogen induces the formation of spine synapses. Morphological and electrophysiological data show that synaptic remodeling could be very rapid. The neurosteroid dehydroepiandrosterone can also induce synaptic plasticity, reproducing the robust synaptogenic effect of estrogen. The underlying mechanisms, however, show considerable gender differences. Nevertheless, the many favorable responses to dehydroepiandrosterone by the brain and the periphery make this neurosteroid a promising substitute for estrogen in hormone therapy.

Keyword Hypothalamus, hippocampus, gonadal steroids, neurosteroids, synaptogenesis

Abbreviations ARC arcuate nucleus of the hypothalamus; DHEA dehydroepiandrosterone; ER estrogen receptor; IR immunoreactive; NCAM neural cell adhesion molecule; PSA polysialic acid; VMN ventromedial nucleus of the hypothalamus

[1] Institute of Biophysics, Biological Research Center, Hungarian Academy of Sciences, Szeged, Hungary

[2] Department of Obstetrics, Gynecology and Reproductive Sciences, Yale University School of Medicine, New Haven, CT, USA

[3] Instituto Cajal, C.S.I.C., Madrid, Spain

M.S. Ritsner and A. Weizman (eds.), *Neuroactive Steroids in Brain Function, Behavior and Neuropsychiatric Disorders*

8.1 Introduction

Synaptic plasticity is the remarkable ability of the nervous system to modify the number, morphology, and activity of synapses, thereby modulating the response of a neuron to given inputs. At the beginning of the 20th century, it has already been suggested that the changes are of structural character, i.e., modification in neuronal connectivity occurs as a consequence of the mental activity.[1] Later, this idea has been contested, and more emphasis was given to the so-called behavioral plasticity, which presumed a function-dependent modulation of the strength of existing synapses without any structural change in the neural circuits. Recent experimental data clearly indicate that there is indeed a synaptic turnover as an adaptation to the changing external and internal environment. The magnitude and distribution of these processes is more expressed in the immature animal, although the adult brain also retains a remarkable capacity for undergoing morphological and functional modifications. It has become clear that several factors are able to influence synaptic remodeling; and among these, the gonadal steroids represent a group of endogenous compounds with unparalleled synaptogenic potential. In this review, we will first discuss the role of sex steroids in the shaping of brain circuitry during development ("organizational effects"), while hormone-induced structural remodeling in the adult brain ("activational effects") will be considered in greater detail afterwards. Finally, we will summarize findings from our laboratories on the synaptogenic actions of dehydroepiandrosterone (DHEA), which parallel the effects of estrogens and androgens.

8.2 Organizational Effects

The presence of sexually dimorphic structures in the vertebrate brain suggests the involvement of gonadal steroids in the development of the nervous system. The early reports on these differences[2–4] have already discussed the possible role of sexual hormones and the effect was demonstrated by neonatal castration and hormone substitution. Describing the gross sexual dimorphism in the vertebrate brain, Nottebohm and Arnold[5] have demonstrated that three vocal control areas of songbirds are significantly larger in males than in females. The possible organizational effect of hormones has also been shown in another sexually dimorphic structure: in the spinal nucleus of the bulbocavernosus of rats, where the number of neurons are higher in males. Prenatal treatment with testosterone propionate was able to masculinize the number of neurons in the spinal nucleus of the bulbocavernosus of females.[6]

The structural sexual dimorphism may manifest itself at different other organizational levels such as dendritic structure,[7] neuronal membrane organization[8] and neuronal connectivity[9–13] and opiate receptor numbers.[14]

Sexual dimorphism was found in the synaptic organization of the neuropil in the ventrolateral subdivision of the ventromedial nucleus (VMN). Pozzo Miller and

Aoki[15] have described that the numerical densities of spine and shaft synapses in the adult male are higher than in the female. A dimorphic pattern in the numerical density of spine synapses occurs as early as day 5, and is present throughout postnatal life. Neonatal treatment of female rats with testosterone increased the numerical density of axodendritic synapses, inducing a pattern similar to the adult male. On the other hand, administration of tamoxifen to newborn male rats significantly reduced the numerical density of spine synapses to levels comparable to normal female rats.

These experimental data clearly indicate that gonadal steroids have well-defined organizational effects on different parts of the central nervous system.[16] The effects are permanent and are acting only during development, mainly in the fetal-neonatal period when estrogens and aromatizable androgens modulate the neuronal development and the formation of neuronal circuits. As a result, sexually dimorphic structures develop in several areas of the central nervous system.

8.3 Activational Effects

According to recent studies, the effect of gonadal steroids is not limited to the early developmental stages. Neuro-glial plastic changes and synaptic remodeling have been observed in the adult nervous system, and experiments on ovariectomized and hormone treated animals indicate that estrogen plays a key role in the plastic processes. These effects are transitory and can be reversed if the hormonal milieu is altered, i.e., plastic changes affecting brain functions may occur even in the adult brain in a very short timescale. It has been demonstrated that estradiol is involved in determining dendritic and axon length, as well as their branching patterns in specific areas of the central nervous system. The phenomenon is well documented, but we have only limited knowledge of the mechanisms involved in gonadal steroid-induced neuroplasticity and especially, little information has been available about the time course of the changes.

It has been hypothesized that gonadal steroids may alter behavior in adulthood by changing the activity of neural circuits. This is the reason why early studies have been focused on hormonally induced neurochemical changes, such as synthesis of neurotransmitters and neuropeptides, or changes in receptor numbers.[17,18] However, experimental data indicate that these effects include well-defined morphological changes, as well. According to DeVoogd and Nottebohm,[19] dendritic growth can be induced by gonadal hormones in ovariectomized adult female canaries. In a group of androgen-sensitive motoneurons that mediate male copulatory functions, decreases in androgen levels after castration of adult rats produced dramatic structural changes, decreasing both the dendritic length and soma size of these motoneurons. These alterations were reversed by androgen replacement. Kurz et al.[20] have provided evidence that physiological androgen level changes in adulthood are associated with significant structural and functional modifications of these neurons. De Vries et al.[21,22] have given an estimation concerning the timescale of changes by showing that gonadectomy of adult male rats results in a decrease in vasopressin fiber density over a period of 15 weeks in those areas where the fibers probably are derived from the bed nucleus of the stria terminalis. The original fiber density could be restored by testosterone replacement within 5 weeks.

8.4 Hormonally Induced Synaptic Changes in Hypothalamic Nuclei

Considering the fact that gonadal hormones have been found to induce plastic changes, it was logical to expect that these changes occur mainly in brain areas involved in the neuronal control of reproductive or neuroendocrine events. This is the reason why the early research, looking for signs of structural plasticity, has concentrated on the hypothalamus. The reports indicate that lactation-induced changes in the magnocellular oxytocinergic neurons and their astrocytes are accompanied by changes in the synaptic input onto these neurons. This kind of synaptic plasticity has been characterized as a difference in the incidence of shared synapses, with an increase in frequency already at parturition and a decrease after weaning.[23–25] A more detailed study[26] has revealed that the synapses are not equally affected; the density of GABA immunoreactive terminals, mainly those impinging on somas, was significantly greater in lactating rats. The increase was more obvious in the mid and posterior regions of the supraoptic nucleus.

Experiments using confocal microscopy[27] demonstrate that estradiol benzoate treatment (2 days) of ovariectomized rats increases dendritic spine density by 48% in the ventrolateral VMN but had no effect in the dorsal part of the nucleus. The primary dendrites of VMN neurons were differentially affected by estrogen. Estrogen treatment increased spine density twofold on the short primary dendrites but did not affect spine density on long primary dendrites.

Synaptic plasticity in the anteroventral periventricular nucleus of cycling female rats has been studied by Langub et al.[28] The authors have found a 39% increase in axosomatic synapses upon anteroventral periventricular nucleus neurons between proestrus and estrus day. In the subsequent 24 h to metestrus, the number of synapses decreased by 22%, while ovariectomy resulted in more axosomatic synapses in the nucleus. They have also described that estrogen receptor-immunoreactive (ER-IR) neurons received significantly more synaptic input at proestrus and estrus than did non-ER-IR neurons. They have concluded that the ER-positive and ER-negative neurons are differentially innervated.

The majority of data concerning synaptic remodeling in developing and adult neuroendocrine brain refers to the arcuate nucleus (ARC). Arai and Matsumoto[29] have provided evidence on the effect of neonatal estrogen treatment on synapse formation during postnatal development of female rats. In isolated ARC, 3 weeks estradiol benzoate treatment reversed deafferentation-induced loss of axo-dendritic synapses.[30] These pioneer experiments clearly indicate that neural plasticity in this nucleus is significantly enhanced by the hormone.[12] Olmos et al.[31] have reported that the nucleus exhibits a phasic remodeling, which can be linked to hormonal variations during the estrous cycle. The number of axo-somatic synapses on arcuate neurons decreases between the morning and afternoon of proestrus day, remains low on estrus morning, and then rises again to the metestrus/diestrus day level. The changes in synapses are not accompanied by the appearance of degenerative terminals; therefore, the reduction in number of synaptic contacts could involve some sort of a displacement of synapses from the soma of the neurons rather than a degenerative loss. The circulating level of estradiol peaks on the proestrus morning and this suggests a possible triggering role for the hormone. Other experiments have provided direct evidences for this hypothesis,[32,33] although these early studies have been focusing on synaptic changes observed on larger timescale.

We have performed a more detailed analysis of estradiol-induced synaptic remodeling and revealed that the effect is specific, because not all types of synapses are involved.[34] Quantitative postembedding immunocytochemical analysis of the ARC showed that the majority of the axo-somatic synaptic terminals on arcuate neurons of ovariectomized rats are GABA-immunoreactive. The administration of a single dose of 17β-estradiol resulted in a significant decrease in the number of these GABA-IR synapses. In contrast, estradiol administration had no significant effect on the number of non-immunoreactive axo-somatic synapses. These data clearly show that physiological levels of estradiol induce a remodeling of GABAergic inhibitory inputs on arcuate neurons. A more detailed analysis has shown that estradiol treatment does not induce changes in the number of axo-dendritic arcuate synapses, but there is a significant increase in the volume density of excitatory spine synapses.[35]

The arcuate area of the rat hypothalamus has previously been shown to have increased neuronal activity in vivo corresponding to the estradiol-induced luteinizing hormone rise in the proestrus afternoon.[36] This raises the important question of whether the structural synaptic remodeling induced by the hormone is correlated with functional changes in neuronal activity. The most plausible hypothesis is that the sum of a decreased number of axo-somatic inhibitory synaptic inputs and an increased number of excitatory synapses on dendritic spines in the arcuate nucleus will result in an increased neuronal excitability.[35] Indeed, electrophysiological recordings showed that estradiol increases firing frequency in a subpopulation of arcuate neurons. This finding suggests that the morphological synaptic changes induced by estrogen are linked to functional modifications in neuronal activity.

The cellular and molecular background of the hormonally induced synaptic remodeling is a complex phenomenon, because it involves not only the neuronal, but also astrocytic elements. Several authors have demonstrated that there is a negative correlation between the level of glial ensheathing and the number of synaptic inputs to hypothalamic neurons in a variety of species, suggesting that this may be related to a common mechanism underlying one aspect of synaptic plasticity.[33,37,38] During remodeling, the synaptic terminals remain in the proximity of neuronal somas, although glial processes are interposed between the pre- and postsynaptic membranes preventing the formation of synaptic contacts. This observation suggests the involvement of membrane–membrane interactions, which has been demonstrated by Hoyk et al.,[39] providing strong evidence that the embryonic form of NCAM is a necessary prerequisite for estrogen-induced phasic remodeling of synapses in the adult female arcuate nucleus, the process requiring the presence of PSA-NCAM on cell surfaces.

According to freeze-fracture studies,[40] the structure of the arcuate neuronal membrane exhibits reversible changes in adult females following the fluctuation in plasma estradiol levels during the ovarian cycle. The density of the intramembrane particles in ovariectomized animals decreased following 17β-estradiol injection and this was associated with an increased number of exo-endocytotic images (coated pits). As these organelles are thought to play a role in trafficking molecules to and from the membrane, it can be concluded that estradiol induces an increased turnover of membrane components.[41]

Although the functional importance of the hormonally induced plastic changes is generally accepted, experimental data on the time-course of structural synaptic remodeling are sparse. Earlier studies have focused on the phenomenon itself and the effects were reported following long-term hormonal treatment. The fact that synaptic changes are linked to hormonal variations during the ovarian cycle indicates that morphological alterations may take place in hours rather than days, but so far, no direct experimental evidence has been published supporting this hypothesis. Electrophysiological recordings, however, clearly show that such synaptic remodeling can be induced in less than 1 h. Kis et al.[42] have recorded a significant increase in the spontaneous activity of ARC neurons 25 min after the ip administration of 17β-estradiol. The authors used an in situ preparation, i.e., all synaptic connections of the cells remained intact and they were able to follow the development of the hormonal effect on the same units. The onset of the estradiol action around 25 min makes it unlikely that the changes are the consequence of a direct membrane effect and the authors have proposed that the enhancement of firing may be the result of reduced inhibitory synaptic input to arcuate neurons.

According to recent studies,[43,44] not only 17β-estradiol, but also the neurosteroid DHEA is able to induce changes in hippocampal spine synapses. The estradiol precursor, DHEA has a well-defined effect on remodeling of axo-somatic synapses in the ARC, as well. Similar to estrogen, DHEA treatment also resulted in a significant decrease in the number of GABAergic terminals, while the non-GABAergic population was not affected. When the animals were pretreated with the selective nonsteroid aromatase inhibitor, letrozole, this remodeling did not occur, suggesting that the DHEA effect may be mediated by aromatization. Considering the molecular mechanisms of hormone-induced synaptoplastic changes, the local synthesis of estradiol seems to be important. This may ensure a well-timed and sufficiently high local concentration of the hormone, which is essential for the quick and precise regulation of the process.

8.5 Estrogen-Induced Synaptic Remodeling in the Hippocampus

Studies on the effects of estrogens on cognitive performance in animals provided the first clue that gonadal steroids might also affect higher brain functions in adulthood. Estrogen levels are positively associated with learning and memory performance in rats,[45] as well as in primates and humans.[46–48] While the mechanisms underlying these responses have remained unknown, circumstantial evidence suggests the involvement of the hippocampal circuitry.[49] Gould et al.[50] have demonstrated that estrogen treatment of ovariectomized rats increases the density of spines on the dendrites of CA1 pyramidal neurons. This effect was potentiated within 5 h by progesterone administration, suggesting that similar changes could happen dynamically during the course of the normal female reproductive cycle.[51] Cyclical changes in dendritic morphology reflected profound and rapid (<24 h) alterations in the density of spine synapses, primarily under the control of circulating estrogen levels.[52] Subsequent work has confirmed and extended these initial observations in primates and laboratory rodents.[53–55] Strikingly, in rats, the positive effects of estrogen on CA1 spine synapse density are profoundly sexually differentiated. Thus, administration of estrogen to ovariectomized rats resulted in a nearly twofold increase in the density of CA1 spine synapses. In males, however, not only the gonadectomy-induced "base" spine synapse density level is generally lower than in females, but also the synaptoplastic effect of estrogen treatment is completely lacking.[56] The basis for this sex difference remains uncertain for now.

Experimental data[50,51] strongly indicate that induction of spine synapses is a dynamic process, occurring within hours, in the timeframe of physiological changes in steroid hormone levels during the female reproductive cycle. However, the majority of hormone replacement studies have administered estrogen over periods of days, rather than hours; so it has remained uncertain whether the observed changes in synapse density might represent a part of the mechanisms involved in altered cognitive function, or a delayed consequence of hormone exposure. Work by Sandstrom and Williams[57] has demonstrated that the enhancement of a working memory version of the Morris water maze spatial memory task is within the timeframe of previously reported estrogen-induced increases in spine density, suggesting that these plastic changes might well contribute to the behavioral responses. However, the question of what might happen at shorter time intervals has remained unanswered. This problem has been thrown into sharp focus by the observations of Luine et al.,[58] who have found that estrogen treatment enhances performance on tests of object placement and object recognition memory, within less than 4 h. Given the above-described evidence for rapid estradiol-induced synaptic remodeling, it seemed possible that the mechanisms underlying these behavioral responses might include effects on synaptic plasticity.

Studies in ovariectomized rats injected with estradiol have revealed that changes in hippocampal synaptic plasticity do indeed occur with remarkable rapidity. Surprisingly, dramatic effects on synapse numbers were observed within 30 min of estradiol administration.[59] Since the steroid was injected subcutaneously in oil, resulting in a slow release of the hormone into the circulation, the initial synaptic responses to estradiol may occur almost immediately after estrogen enters the brain.[59] Even more surprisingly, the increase in synapse density at 30 min was significantly *greater* than at 4.5 h after estradiol injection. This again is consistent with the view that synapses respond dynamically to changing estradiol levels, rather than over a prolonged period, which might be expected if the synaptic effects were secondary to estrogen-induced changes in gene expression. Dose-response analysis revealed that relatively high doses of estradiol are required to induce the rapid synaptoplastic effect. Thus, higher doses of 45–60μg/kg were only effective, which produced supraphysiological circulating levels of estradiol.[59] The fact that such doses are required to detect any significant synaptic modifications is important because it provides an additional link between increases in synapse number and the behavioral effects of the steroid. Earlier studies have demonstrated that the object placement test is critically dependent on the integrity of the hippocampal circuitry.[49] Interestingly, Luine et al.[58] have found that significant enhancement of object placement performance in ovariectomized rats is only observed at higher doses of 17β-estradiol (30–60μg/kg), mirroring the requirement for relatively high doses of estrogen to elicit rapid effects on CA1 spine synapses.

It is yet to be known how these rapid structural synaptic responses in CA1 are triggered by estrogen. The rapidity of estrogen-induced synapse formation seems to preclude genomic effects.[60] A more likely scenario is that extranuclear response cascades[61,62] triggered by estradiol may rapidly alter the interactions between pre- and postsynaptic elements in CA1. In this regard, the pharmacology of the responses is consistent with the hypothesis that membrane-associated estrogen receptors may be one of the initial targets of estradiol. Thus, the rapid effects of 17β-estradiol were mimicked by its 17α isomer[59] that has a much lower affinity for the nuclear estrogen receptors, ERα and ERβ.[63] As Wade et al.[64] have shown, association of ERα or ERβ with the plasma membrane results in their enhanced sensitivity to 17α-estradiol. Various studies have demonstrated the presence of ERα-IR in axon terminals within CA1[65,66] and revealed the existence of ERβ in synaptosomes and synaptic membranes of the mouse hippocampus.[67] A reasonable hypothesis to explain the rapid actions of estrogen on CA1 spine synapse density might therefore be that high doses of estradiol activate ERs located in target synapses, within CA1 itself, thereby triggering signaling mechanisms that result in fast synaptic proliferation.

8.6 Hippocampal Synaptogenic Effects of DHEA

In primates, one of the most abundant circulating androgens, DHEA, is a product of adrenal steroidogenesis. In both men and women, adrenal production of DHEA increases before puberty, plateaus in adulthood, but then declines between the third an seventh decades of life.[68] Several studies have demonstrated that DHEA has effects on cognitive function in human beings as well as laboratory animals.[69] Thus, DHEA-replacement therapy may have beneficial cognitive effects in aging, particularly in cases of low endogenous DHEA secretion.[69,70] Because hippocampal spine synapse formation has been implicated in memory[71] as a potential cellular mechanism to explain the behavioral effects of sex steroids, a key remaining question has been whether DHEA is capable of inducing hippocampal spine synapse growth. In addition, DHEA is a very weak androgen, but it could be converted to more powerful androgens and estrogens, raising the possibility of sex differences in the mechanisms of the synaptogenic effect of DHEA.[72]

We have reported that DHEA reproduces the effects of both estrogens and androgens on CA1 spine synapse density in both female and male gonadectomized rats.[43,44] This hippocampal synaptogenic action of DHEA may contribute to the beneficial cognitive effects of this hormone. Regarding potential mechanisms, however, the picture is more complicated. The potent antiandrogen, flutamide, did not prevent hippocampal spine synapse growth in response to DHEA administration in either sex (Figs. 8.1 and 8.2). This is not so surprising in light of the weak androgenicity of DHEA, but clearly indicates that DHEA either interacts with other receptor mechanisms or acts via one of its metabolites with androgenic or estrogenic bioactivity.[68]

We have reported earlier that the hippocampal synaptogenic response to androgens is smaller in females than it is in males,[56,73] suggesting that the female may be less sensitive to the effects of androgens. However, the female is clearly more sensitive than the male to the effects of locally synthesized estrogens. In line with this, the aromatase enzyme inhibitor, letrozole, almost completely blocked the effects of testosterone on CA1 spine synapse density in females,[73] and completely blocked the effects of DHEA in ovariectomized rats, suggesting that the response was mediated via intracerebral conversion of DHEA to estradiol.[43] In males, by contrast, aromatization does not appear to play a significant role: the synaptic effects of DHEA in males were completely unaffected by letrozole administration (Fig. 8.2). This is consistent with our previous observation that estradiol has no significant effect on CA1 spine synapse density in ORCH rats.[56] Thus, although DHEA induces quantitatively similar numbers of CA1 spine synapses in males and females, different mechanisms mediate the responses. In females, DHEA-induced increase in CA1 spine synapse density is mediated almost completely via intracerebral estrogen

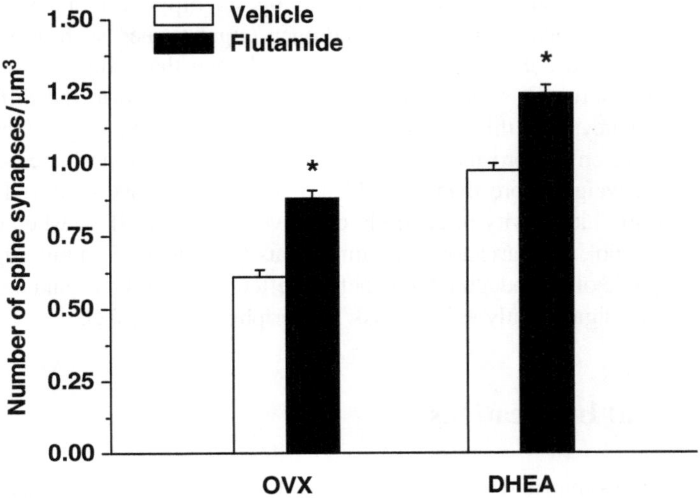

Fig. 8.1 Effects of DHEA and flutamide administration on CA1 spine synapse density in ovariectomized (OVX) female rats. Columns labeled with asterisk are significantly different compared to the vehicle-treated groups (Scheffe test, $p > 0.05$)

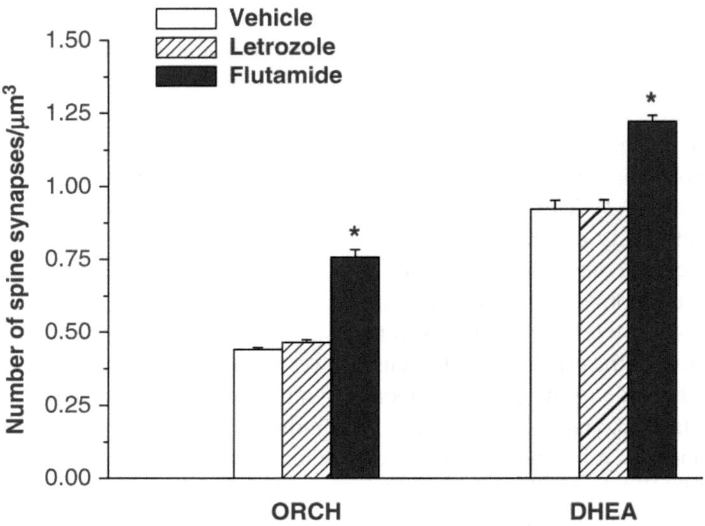

Fig. 8.2 Effects of DHEA, letrozole, and flutamide administration on CA1 spine synapse density in orchidectomized (ORCH) male rats. Columns labeled with asterisk are significantly different compared to the vehicle-treated groups (Scheffe test, $p > 0.05$)

biosynthesis. In males, on the other hand, neither androgen receptors nor aromatization is involved, and the exact mechanisms via which DHEA exerts its hippocampal synaptogenic activity remain to be elucidated.

Despite the uncertainties in the underlying mechanisms, DHEA possesses unmistakable values as a potential agent for human hormone therapy. First, it is beneficial for cognition,[69] which could be explained by its hippocampal synaptogenic activity[43,44]; and second, in peripheral androgen-target tissues such as the ventral prostate, DHEA is a considerably less potent androgen than either testosterone or 5α-dihydrotestosterone.[74,75] Our data for prostate weight confirm the difference in androgenic potency: with this relatively short-term treatment paradigm, there was no effect of DHEA on the prostate, whereas 5α-dihydrotestosterone increased ventral prostate organ weight more than twofold.[44] Moreover, we have observed similarly reduced uterotrophic activity of centrally effective doses of DHEA, when compared to that of estradiol, in ovariectomized female rats.[43] All these findings suggest that DHEA is capable of reproducing the beneficial effects of androgens and estrogens in the brain with a significantly reduced risk for peripheral side effects.

8.7 Clinical Implications

That CA1 spine synapse density is upregulated by androgens as well as estrogens has implications for both the physiological role of androgens in regulating hippocampal function, as well as for the potential contribution of these hormones to the maintenance of normal hippocampal structure, in aging and disease. Here, we will focus on the possible implications of the data for androgen replacement therapy.

Two specific properties of androgens and especially DHEA make them particularly attractive as potential therapeutic modalities, for use when endogenous gonadal steroid secretion falls below normal. The first property, which has been extensively commented on by others,[68] is that the ability of these steroids to act as precursors for local metabolism raises the possibility of "targeted" androgen treatment, designed to achieve a selective biological response. For example, aromatizable androgens with little intrinsic androgenic bioactivity, such as DHEA, could theoretically be used to deliver estrogen precisely where it is needed in the brain, through local metabolism, without exposing other tissues in the body to either high circulating estrogen levels, or potentially unacceptable anabolic effects. Since responses to DHEA are observed naturally, in both sexes, there is no theoretical impediment to the use of DHEA for treatment of hormonal deficiency syndromes associated with aging in men, as well as women.[76,77]

Second, the unusual specificity of the hippocampal synaptoplastic response to androgens raises the possibility of distinguishing pharmacologically between the effects of androgens on the hippocampus, as opposed to the masculinizing effects of these steroids elsewhere in the body. Obviously, a great deal remains to be done to determine whether such distinctions may be feasible, let alone therapeutically useful. Nevertheless, it seems to be clear that induction of CA1 spine synapse growth by DHEA can be achieved, without systemic hormone effects.[43,44] Thus, combinations of either 5α-dihydrotestosterone and flutamide or DHEA and flutamide result in very high CA1 dendritic spine synapse densities. However, these same treatments have no significant effect on ventral prostate weight.[44] It will be of considerable interest, in future studies, to determine whether the high CA1 spine synapse density observed after combination DHEA/flutamide administration results in a concomitant enhancement of hippocampal function, in terms of improvements in working memory and/or effects on mood state.[78]

Acknowledgments This work was supported by Hungarian National Office for Research and Technology RET 08/2004, ETT 476/2006, E-438/2006 (A.P.), a National Institute of Mental Health Grant MH074021 (T.H.), and a 2007 NARSAD Young Investigator Award (T.H.).

References

1. Ramon y Cajal S. Histologie du Systéme Nerveux de l'Homme et des Vertébrés. Paris: Maloine, 1911.
2. Pfaff DW. Morphological changes in the brains of adult male rats after neonatal castration. J Endocrinol 1966; 36:415–416.
3. Dorner G, Staudt J. Structural changes in the preoptic anterior hypothalamic area of the male rat, following neonatal castration and androgen substitution. Neuroendocrinology 1968; 3:136–140.
4. Raisman G, Field PM. Sexual dimorphism in the preoptic area of the rat. Science 1971; 173:731–733.

5. Nottebohm F, Arnold AP. Sexual dimorphism in vocal control areas of the songbird brain. Science 1976; 194:211–213.
6. Breedlove SM, Arnold AP. Hormonal control of a developing neuromuscular system. II. Sensitive periods for the androgen-induced masculinization of the rat spinal nucleus of the bulbocavernosus. J Neurosci 1983; 3:424–432.
7. Greenough WT, Carter CS, Steerman C, et al. Sex differences in dentritic patterns in hamster preoptic area. Brain Res 1977; 126:63–72.
8. Garcia-Segura LM, Baetens D, Naftolin F. Sex differences and maturational changes in arcuate nucleus neuronal plasma membrane organization. Brain Res 1985; 351:146–149.
9. Raisman G, Field PM. Sexual dimorphism in the neuropil of the preoptic area of the rat and its dependence on neonatal androgen. Brain Res 1973; 54:1–29.
10. Matsumoto A, Arai Y. Sexual dimorphism in "wiring pattern" in the hypothalamic arcuate nucleus and its modification by neonatal hormonal environment. Brain Res 1980; 190:238–242.
11. Matsumoto A, Arai Y. Effect of androgen on sexual differentiation of synaptic organization in the hypothalamic arcuate nucleus: an ontogenetic study. Neuroendocrinology 1981; 33:166–169.
12. Matsumoto A, Arai Y. Neuronal plasticity in the deafferented hypothalamic arcuate nucleus of adult female rats and its enhancement by treatment with estrogen. J Comp Neurol 1981; 197:197–205.
13. Perez J, Naftolin F, Garcia Segura LM. Sexual differentiation of synaptic connectivity and neuronal plasma membrane in the arcuate nucleus of the rat hypothalamus. Brain Res 1990; 527:116–122.
14. Hammer RP, Jr. The sexually dimorphic region of the preoptic area in rats contains denser opiate receptor binding sites in females. Brain Res 1984; 308:172–176.
15. Pozzo Miller LD, Aoki A. Stereological analysis of the hypothalamic ventromedial nucleus. II. Hormone-induced changes in the synaptogenic pattern. Brain Res Dev Brain Res 1991; 61:189–196.
16. Arnold AP, Breedlove SM. Organizational and activational effects of sex steroids on brain and behavior: a reanalysis. Horm Behav 1985; 19:469–498.
17. Cardinali DP, Vacas MI. Mechanisms underlying hormone effects on pineal function: a model for the study of integrative neuroendocrine processes. J Endocrinol Invest 1978; 1:89–96.
18. Romano GJ, Krust A, Pfaff DW. Expression and estrogen regulation of progesterone receptor mRNA in neurons of the mediobasal hypothalamus: an in situ hybridization study. Mol Endocrinol 1989; 3:1295–1300.
19. DeVoogd TJ, Nottebohm F. Sex differences in dendritic morphology of a song control nucleus in the canary: a quantitative Golgi study. J Comp Neurol 1981; 196:309–316.
20. Kurz EM, Sengelaub DR, Arnold AP. Androgens regulate the dendritic length of mammalian motoneurons in adulthood. Science 1986; 232:395–398.
21. de Vries GJ, Buijs RM, Sluiter AA. Gonadal hormone actions on the morphology of the vasopressinergic innervation of the adult rat brain. Brain Res 1984; 298:141–145.
22. Commins D, Yahr P. Adult testosterone levels influence the morphology of a sexually dimorphic area in the Mongolian gerbil brain. J Comp Neurol 1984; 224:132–140.
23. Theodosis DT, Poulain DA, Vincent JD. Possible morphological bases for synchronisation of neuronal firing in the rat supraoptic nucleus during lactation. Neuroscience 1981; 6:919–929.
24. Hatton GI, Tweedle CD. Magnocellular neuropeptidergic neurons in hypothalamus: increases in membrane apposition and number of specialized synapses from pregnancy to lactation. Brain Res Bull 1982; 8:197–204.
25. Theodosis DT, Poulain DA. Evidence for structural plasticity in the supraoptic nucleus of the rat hypothalamus in relation to gestation and lactation. Neuroscience 1984; 11:183–193.
26. Gies U, Theodosis DT. Synaptic plasticity in the rat supraoptic nucleus during lactation involves GABA innervation and oxytocin neurons: a quantitative immunocytochemical analysis. J Neurosci 1994; 14:2861–2869.

27. Calizo LH, Flanagan-Cato LM. Estrogen selectively regulates spine density within the dendritic arbor of rat ventromedial hypothalamic neurons. J Neurosci 2000; 20:1589–1596.
28. Langub MC, Jr., Maley BE, Watson RE, Jr. Estrous cycle-associated axosomatic synaptic plasticity upon estrogen receptive neurons in the rat preoptic area. Brain Res 1994; 641:303–310.
29. Arai Y, Matsumoto A. Synapse formation of the hypothalamic arcuate nucleus during postnatal development in the female rat and its modification by neonatal estrogen treatment. Psychoneuroendocrinology 1978; 3:31–45.
30. Matsumoto A, Arai Y. Synaptogenic effect of estrogen on the hypothalamic arcuate nucleus of the adult female rat. Cell Tissue Res 1979; 198:427–433.
31. Olmos G, Naftolin F, Perez J, et al. Synaptic remodeling in the rat arcuate nucleus during the estrous cycle. Neuroscience 1989; 32:663–667.
32. Garcia-Segura LM, Baetens D, Naftolin F. Synaptic remodelling in arcuate nucleus after injection of estradiol valerate in adult female rats. Brain Res 1986; 366:131–136.
33. Naftolin F, Garcia-Segura LM, Keefe D, et al. Estrogen effects on the synaptology and neural membranes of the rat hypothalamic arcuate nucleus. Biol Reprod 1990; 42:21–28.
34. Parducz A, Perez J, Garcia-Segura LM. Estradiol induces plasticity of gabaergic synapses in the hypothalamus. Neuroscience 1993; 53:395–401.
35. Parducz A, Hoyk Z, Kis Z, et al. Hormonal enhancement of neuronal firing is linked to structural remodelling of excitatory and inhibitory synapses. Eur J Neurosci 2002; 16:665–670.
36. Yeoman RR, Jenkins AJ. Arcuate area of the female rat maintained in vitro exhibits increased afternoon electrical activity. Neuroendocrinology 1989; 49:144–149.
37. Witkin JW, Ferin M, Popilskis SJ, et al. Effects of gonadal steroids on the ultrastructure of GnRH neurons in the rhesus monkey: synaptic input and glial apposition. Endocrinology 1991; 129:1083–1092.
38. Garcia-Segura LM, Chowen JA, Parducz A, et al. Gonadal hormones as promoters of structural synaptic plasticity: cellular mechanisms. Prog Neurobiol 1994; 44:279–307.
39. Hoyk Z, Parducz A, Theodosis DT. The highly sialylated isoform of the neural cell adhesion molecule is required for estradiol-induced morphological synaptic plasticity in the adult arcuate nucleus. Eur J Neurosci 2001; 13:649–656.
40. Garcia-Segura LM, Hernandez P, Olmos G, et al. Neuronal membrane remodelling during the oestrus cycle: a freeze-fracture study in the arcuate nucleus of the rat hypothalamus. J Neurocytol 1988; 17:377–383.
41. Parducz A, Szilagyi T, Hoyk S, et al. Neuroplastic changes in the hypothalamic arcuate nucleus: the estradiol effect is accompanied by increased exoendocytotic activity of neuronal membranes. Cell Mol Neurobiol 1996; 16:259–269.
42. Kis Z, Horvath S, Hoyk Z, et al. Estrogen effects on arcuate neurons in rat. An in situ electrophysiological study. Neuroreport 1999; 10:3649–3652.
43. Hajszan T, MacLusky NJ, Leranth C. Dehydroepiandrosterone increases hippocampal spine synapse density in ovariectomized female rats. Endocrinology 2004; 145:1042–1045.
44. MacLusky NJ, Hajszan T, Leranth C. Effects of dehydroepiandrosterone and flutamide on hippocampal CA1 spine synapse density in male and female rats: implications for the role of androgens in maintenance of hippocampal structure. Endocrinology 2004; 145:4154–4161.
45. Dohanich GP. Gonadal steroids, learning and memory. In: Pfaff DW, Arnold AP, Etgen AM, et al. (eds). Hormones, brain and behavior. San Diego, CA: Academic, 2002:265–327.
46. Yaffe K, Lui LY, Grady D, et al. Cognitive decline in women in relation to non-protein-bound oestradiol concentrations. Lancet 2000; 356:708–712.
47. Yaffe K, Lui LY, Zmuda J, et al. Sex hormones and cognitive function in older men. J Am Geriatr Soc 2002; 50:707–712.
48. McGaugh JL, Roozendaal B. Role of adrenal stress hormones in forming lasting memories in the brain. Curr Opin Neurobiol 2002; 12:205–210.
49. Mumby DG, Gaskin S, Glenn MJ, et al. Hippocampal damage and exploratory preferences in rats: memory for objects, places, and contexts. Learn Mem 2002; 9:49–57.
50. Gould E, Woolley CS, Frankfurt M, et al. Gonadal steroids regulate dendritic spine density in hippocampal pyramidal cells in adulthood. J Neurosci 1990; 10:1286–1291.

51. Woolley CS, Gould E, Frankfurt M, et al. Naturally occurring fluctuation in dendritic spine density on adult hippocampal pyramidal neurons. J Neurosci 1990; 10:4035–4039.
52. Woolley CS, McEwen BS. Estradiol mediates fluctuation in hippocampal synapse density during the estrous cycle in the adult rat. J Neurosci 1992; 12:2549–2554.
53. Leranth C, Shanabrough M, Horvath TL. Hormonal regulation of hippocampal spine synapse density involves subcortical mediation. Neuroscience 2000; 101:349–356.
54. Leranth C, Shanabrough M, Redmond DE, Jr. Gonadal hormones are responsible for maintaining the integrity of spine synapses in the CA1 hippocampal subfield of female nonhuman primates. J Comp Neurol 2002; 447:34–42.
55. Hao J, Janssen WG, Tang Y, et al. Estrogen increases the number of spinophilin-immunoreactive spines in the hippocampus of young and aged female rhesus monkeys. J Comp Neurol 2003; 465:540–550.
56. Leranth C, Petnehazy O, MacLusky NJ. Gonadal hormones affect spine synaptic density in the CA1 hippocampal subfield of male rats. J Neurosci 2003; 23:1588–1592.
57. Sandstrom NJ, Williams CL. Spatial memory retention is enhanced by acute and continuous estradiol replacement. Horm Behav 2004; 45:128–135.
58. Luine VN, Jacome LF, MacLusky NJ. Rapid enhancement of visual and place memory by estrogens in rats. Endocrinology 2003; 144:2836–2844.
59. MacLusky NJ, Luine VN, Hajszan T, et al. The 17alpha and 17beta isomers of estradiol both induce rapid spine synapse formation in the CA1 hippocampal subfield of ovariectomized female rats. Endocrinology 2005; 146:287–293.
60. McEwen BS, Coirini H, Schumacher M. Steroid effects on neuronal activity: when is the genome involved? Ciba Found Symp 1990; 153:3–12.
61. Bi R, Foy MR, Vouimba RM, et al. Cyclic changes in estradiol regulate synaptic plasticity through the MAP kinase pathway. Proc Natl Acad Sci USA 2001; 98:13391–13395.
62. Znamensky V, Akama KT, McEwen BS, et al. Estrogen levels regulate the subcellular distribution of phosphorylated Akt in hippocampal CA1 dendrites. J Neurosci 2003; 23:2340–2347.
63. Kuiper GG, Carlsson B, Grandien K, et al. Comparison of the ligand binding specificity and transcript tissue distribution of estrogen receptors alpha and beta. Endocrinology 1997; 138:863–870.
64. Wade CB, Robinson S, Shapiro RA, et al. Estrogen receptor (ER)alpha and ERbeta exhibit unique pharmacologic properties when coupled to activation of the mitogen-activated protein kinase pathway. Endocrinology 2001; 142:2336–2342.
65. Milner TA, McEwen BS, Hayashi S, et al. Ultrastructural evidence that hippocampal alpha estrogen receptors are located at extranuclear sites. J Comp Neurol 2001; 429:355–371.
66. Towart LA, Alves SE, Znamensky V, et al. Subcellular relationships between cholinergic terminals and estrogen receptor-alpha in the dorsal hippocampus. J Comp Neurol 2003; 463:390–401.
67. Nishio M, Kuroki Y, Watanabe Y. Subcellular localization of estrogen receptor beta in mouse hippocampus. Neurosci Lett 2004; 355:109–112.
68. Labrie F, Luu-The V, Labrie C, et al. Endocrine and intracrine sources of androgens in women: inhibition of breast cancer and other roles of androgens and their precursor dehydroepiandrosterone. Endocr Rev 2003; 24:152–182.
69. Vallee M, Mayo W, Le Moal M. Role of pregnenolone, dehydroepiandrosterone and their sulfate esters on learning and memory in cognitive aging. Brain Res Brain Res Rev 2001; 37:301–312.
70. Huppert FA, Van Niekerk JK. Dehydroepiandrosterone (DHEA) supplementation for cognitive function. Cochrane Database Syst Rev 2001; CD000304.
71. Kasai H, Matsuzaki M, Noguchi J, et al. Structure-stability-function relationships of dendritic spines. Trends Neurosci 2003; 26:360–368.
72. Beck SG, Handa RJ. Dehydroepiandrosterone (DHEA): a misunderstood adrenal hormone and spine-tingling neurosteroid? Endocrinology 2004; 145:1039–1041.

73. Leranth C, Hajszan T, MacLusky NJ. Androgens increase spine synapse density in the CA1 hippocampal subfield of ovariectomized female rats. J Neurosci 2004; 24:495–499.
74. Labrie C, Simard J, Zhao HF, et al. Stimulation of androgen-dependent gene expression by the adrenal precursors dehydroepiandrosterone and androstenedione in the rat ventral prostate. Endocrinology 1989; 124:2745–2754.
75. Wright AS, Thomas LN, Douglas RC, et al. Relative potency of testosterone and dihydrotestosterone in preventing atrophy and apoptosis in the prostate of the castrated rat. J Clin Invest 1996; 98:2558–2563.
76. Almeida OP. Sex playing with the mind. Effects of oestrogen and testosterone on mood and cognition. Arq Neuropsiquiatr 1999; 57:701–706.
77. Delhez M, Hansenne M, Legros JJ. Testosterone and depression in men aged over 50 years. Andropause and psychopathology: minimal systemic work-up. Ann Endocrinol (Paris) 2003; 64:162–169.
78. Hajszan T, MacLusky NJ, Leranth C. Short-term treatment with the antidepressant fluoxetine triggers pyramidal dendritic spine synapse formation in rat hippocampus. Eur J Neurosci 2005; 21:1299–1303.

Chapter 9
Pregnane Steroids and Short-Term Neural Plasticity

Yuri B. Saalmann[1] **and Mike B. Calford**[2]

Abstract Gamma-aminobutyric acid (GABA) is the major inhibitory transmitter in the brain, and its fast effects are mediated by the GABA-A receptor. It is well known, from pharmacological manipulations, that many exogenous agents alter the efficacy of GABA-A receptors. For example, benzodiazepines increase the effect of GABA and some β-carbolines reduce the effect of GABA at these receptors. Increasing the strength of neuronal inhibition can prevent seizures, reduce anxiety and be neuroprotective. There are also endogenous mechanisms that increase efficacy. For example, more GABA-A receptors can be synthesized and inserted into synapses, but this requires up to 1 h or more. On a shorter timescale, GABAergic inhibition can be potentiated by steroids, e.g., allopregnanolone, synthesized *de novo* in neural tissue or derived from peripheral endocrine organs. The widespread distribution of these neuroactive steroids across the brain suggests an extensive role in short-term neural plasticity.

Keywords Allopregnanolone, allotetrahydrodeoxycorticosterone, GABA, gain control, homeostasis

Abbreviations Allopregnanolone 3α-hydroxy-5α-pregnan-20-one; alloTHDOC 3α,21-dihydroxy-5α-pregnan-20-one; CNS central nervous system; DBI diazepam binding inhibitor; 5α-DHDOC 5α-dihydrodeoxycorticosterone; 5α-DHP 5α-dihydroprogesterone; DNLL dorsal nucleus of the lateral lemniscus; GABA γ-aminobutyric acid; 3αHSD 3α-hydroxysteroid dehydrogenase; 5-HT 5-hydroxytryptamine; ICC inferior colliculus; IPSC inhibitory post-synaptic current; KA kainic acid; MBR mitochondrial benzodiazepine receptor; P450scc cytochrome P450 side chain cleavage enzyme; SNR substantia nigra pars reticulata; StAR steroid acute regulatory protein; Wm white matter

[1]Department of Psychology, Royal Holloway, University of London, Egham, Surrey, UK

[2]School of Biomedical Sciences and Hunter Medical Research Institute, University of Newcastle, Callaghan, New South Wales, Australia

M.S. Ritsner and A. Weizman (eds.), *Neuroactive Steroids in Brain Function, Behavior and Neuropsychiatric Disorders*

9.1 Introduction

Changes in the efficacy of synaptic transmission between neurons is thought to underlie learning and memory formation,[1,2] a number of disease states and the counteractions for homeostasis,[3] and aging processes.[4] These changes, also known as plasticity, take various forms, occurring over different timescales. Known mechanisms include changes in the amount of transmitter released (which can increase in less than 1 s and last minutes[5]), to the synthesis of proteins such as receptors, taking tens of minutes or more and having long-term effects.[6] In this chapter, we will focus on a mechanism for the plasticity of inhibitory inputs, lasting minutes to tens of minutes. Critical during this period are the endogenous pregnane steroids 3α-hydroxy-5α-pregnan-20-one (also known as 3α,5α-tetrahydroprogesterone or allopregnanolone; Fig. 9.1) and 3α,21-dihydroxy-5α-pregnan-20-one (also known as 3α,5α-TetraHydroDeOxyCorticosterone or alloTHDOC). These are potent modulators of inhibitory neurotransmission.

9.2 Pregnane Steroid Synthesis and Metabolism

Pregnane steroids are derived from cholesterol (Fig. 9.2). The putative rate-limiting step in their synthesis is the conversion of cholesterol to pregnenolone by the cytochrome P450 side chain cleavage enzyme[7] (P450scc). This is dependent upon the steroid acute regulatory protein (StAR), in co-operation with the diazepam binding inhibitor (DBI) and the mitochondrial benzodiazepine receptor (MBR), facilitating cholesterol transport to P450scc in the inner mitochondrial membrane.[8,9] The aforementioned neurosteroid synthetic machinery, and the enzymes downstream in the

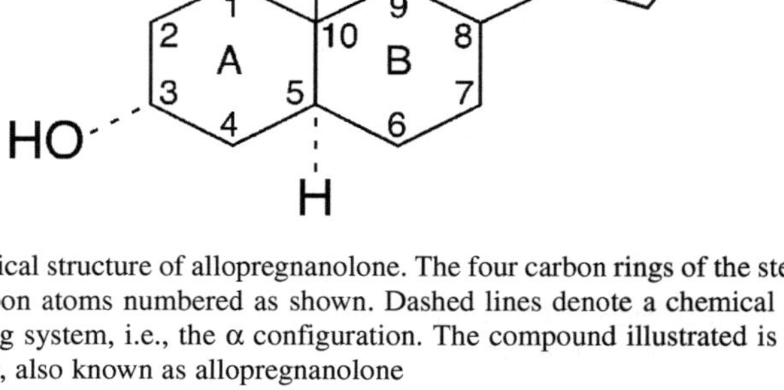

Fig. 9.1 Chemical structure of allopregnanolone. The four carbon rings of the steroid are labelled **A–D**, with carbon atoms numbered as shown. Dashed lines denote a chemical group below the plane of the ring system, i.e., the α configuration. The compound illustrated is 3α-hydroxy-5α-pregnan-20-one, also known as allopregnanolone

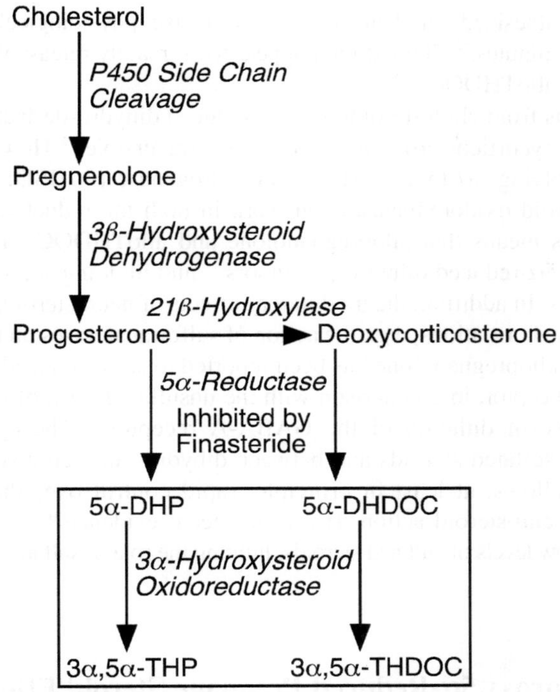

Fig. 9.2 The pathway for the synthesis of pregnane steroids that enhance GABAergic inhibition. The 5α-reduced pregnane steroids are boxed, and enzymes are in italics. 3α-hydroxysteroid oxidoreductase is also known as 3α-hydroxysteroid dehydrogenase. Abbreviations: DHP, dihydroprogesterone; THP, tetrahydroprogesterone; DHDOC, dihydrodeoxycorticosterone; THDOC, tetrahydrodeoxycorticosterone (From[45]. With permission)

pathway leading to 3α-hydroxy,5α-reduced neurosteroid production, are present in the central nervous system (StAR and P450scc[10]; DBI[11]; MBR[12]; P450scc and 3β-hydroxysteroid dehydrogenase[13]; 21β-hydroxylase[14,15]; 5α-reductase and 3α-hydroxysteroid dehydrogenase[16]); and these enzymes are viable as physiologically relevant levels of neurosteroids persist within the central nervous system after removal of peripheral sources (for 5α-reduced neurosteroids, see[17,18]).

Progesterone and deoxycorticosterone are converted by 5α-reductase into 5α-dihydroprogesterone (5α-DHP) and 5α-dihydrodeoxycorticosterone (5α-DHDOC) respectively. These dihydro derivatives regulate gene expression through progesterone receptors.[19] A possible influence of 5α-DHP and 5α-DHDOC over GABA-A receptors should also be noted, but this has not been consistently demonstrated.[19–21] 5α-DHP and 5α-DHDOC can be further reduced to allopregnanolone and alloTHDOC respectively. These tetrahydro derivatives are amongst the most potent known ligands of GABA-A receptors.[20,22] Allopregnanolone and alloTHDOC are efficacious at concentrations above a few nanomoles per litre, with levels in the brain of male and female rats inducible to between 10 and 100 nmol/l.[21] These 3α-hydroxy,5α-reduced steroids are synthesized on demand,[23] and effective physiological levels can be reached within minutes.[24] There do not appear to be readily releasable pools of allopreganolone or alloTHDOC.[25,26]

The reactions from cholesterol to the 5α-reduced dihydro-derivatives of progesterone and deoxycorticosterone are effectively irreversible.[27] However, the enzymatic step involving 3α-hydroxysteroid dehydrogenase (3αHSD; also known as 3α-hydroxysteroid oxidoreductase) can work in both the reductive and oxidative directions. This means that allopregnanolone and alloTHDOC can be converted back into their 5α-reduced dihydro-precursors,[19] and in doing so, attenuate GABA agonistic effects. In addition, the modulatory effects of neurosteroids can be altered by chemical transformations, e.g., the action of sulfotransferases on the steroid. The sulfate ester of allopregnanolone has been reported to show reduced potentiation of the GABA-A receptor, in comparison with the unsulfated form of this steroid,[28] or display negative modulation of the GABA-A receptor.[29] The cycling between sulfated and unsulfated steroids and between dihydro- and tetrahydro-5α-reduced neurosteroids allows, at least in principle, rapid control over the duration and magnitude of neurosteroid action. However, recent evidence[30,31] suggests that the brain contains low levels of sulfated steroids, limiting the role of sulfation as a regulatory mechanism.

9.3 3α-Hydroxy,5α-Reduced Pregnane Steroid Effects at the GABA-A Receptor

Fast inhibitory synaptic transmission in the CNS is largely mediated by chloride conductances through receptor-mediated chloride channel complexes. At higher levels the GABA-A receptor dominates, while at brainstem levels the glycine receptor is the major fast inhibitory receptor-iontophore. At the GABA-A channel, typically, a fast rising current is observed followed by a decay that is often best modelled biphasically (with a fast and slow component; e.g.[32]). 3α-hydroxy,5α-reduced neurosteroids have been shown to prolong the decay of these inhibitory post-synaptic currents (IPSCs), both evoked and spontaneous, recorded from hippocampal,[33] cerebellar[34] and hypothalamic[35] neurons. According to a study by Zhu and Vicini[36] in which high concentrations of GABA were applied in pulses to patched cerebellar granule cells, mimicking vesicular release, the second phase of the inhibitory current decay with a slow time constant, thought to reflect movements in and out of desensitised states of GABA-A receptors,[37] is affected by alloTHDOC.[36] Specifically, the 3α-hydroxy,5α-reduced neurosteroid increased late channel openings and the probability of an open channel state.

When low, non-saturating levels of GABA are maintained, which may be relevant to situations where a tonic GABA concentration influences neuronal excitability (such as at extra-synaptic GABA-A receptors), progesterone metabolites can increase mean channel open time, due to an increased probability of longer-duration channel openings and an increased frequency of single channel openings.[38] In addition, at concentrations (micromolar) much higher than that required for potentiating GABA-evoked currents, a direct effect of progesterone metabolites upon GABA-A receptor-mediated chloride flux has been reported.[39] For the sake of completeness, it should be mentioned that 3α-hydroxy,5α-reduced neurosteroids have been shown to modulate nicotinic acetylcholine[40] and 5-HT_3 receptors.[41] However, as the concentrations required in these instances were in the high micromolar range and unlikely to be of physiological relevance, one concludes that the 3α-hydroxy,5α-reduced neurosteroids are highly selective for the GABA-A receptor.

The GABA-A receptor is formed from a combination of five of the following subunits: α_{1-6}, β_{1-3}, γ_{1-3}, δ, ε, π, θ and ρ_{1-3}. The subunit composition of the GABA-A receptor determines its pharmacological profile, including sensitivity to neurosteroids.[22] The most common brain receptor type, consisting of $\alpha_1\beta_2\gamma_2$ subunits, is sensitive to neurosteroid action, with 5α-reduced neurosteroid modulation of GABAergic function in the nanomolar range.[22] However, GABA-A receptor subunit make-up varies across the brain, suggesting regional neurosteroid sensitivity differences.

9.4 Plasticity Mediated by Endogenous 3α-Hydroxy, 5α-Reduced Pregnane Steroids

The levels of endogenous 3α-hydroxy,5α-reduced pregnane steroids have been shown to change during stress[24] and depression,[42] over the oestrus cycle,[43] and following ethanol ingestion.[44] Although these changes may be unrelated, with different roles for the 3α-hydroxy, 5α-reduced steroids under the different conditions, there is evidence for a common underlying determinant, i.e., a general role for 3α-hydroxy, 5α-reduced pregnane steroids as regulators of neuronal inhibition. This section will focus on a series of our experiments where pregnane steroid synthesis was upregulated to counteract reduced neuronal inhibition in an experimental model.

We looked for a means to separate GABAergic neuronal populations, in order to be able to selectively manipulate each subpopulation. This involved finding a brain area that receives inhibitory input from both local interneurons and at least one source of long-range projection neurons. A source of long-range projections could then be deactivated, leaving the local inhibitory influences intact. The auditory midbrain structure, the central nucleus of the inferior colliculus (ICC), satisfies these criteria and so was adopted as the anatomical substrate for our experiments. The ICC receives a major GABAergic input from the contralateral dorsal nucleus of the lateral lemniscus (DNLL), which conveys an ipsilateral-ear dominated signal. There are additional sources of GABAergic inhibition, which are dominated by contralateral-ear stimulation, including a minor projection from the ipsilateral DNLL and a large interneuron population in ICC[46–48] (Fig. 9.3). We recorded single neuron activity and evoked potentials, a neuronal population measure, in the ICC before and after permanently deactivating the contralateral DNLL with kainic acid. This allowed us to measure change in efficacy of the remaining GABAergic inputs to ICC neurons.

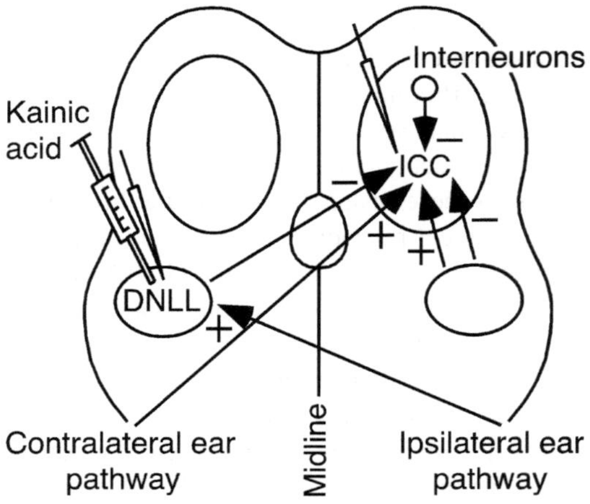

Fig. 9.3 The experimental paradigm used to show inhibitory synaptic plasticity. Neural recordings were made from the central nucleus of the inferior colliculus (ICC) before and after kainic-acid-induced deactivation of the contralateral dorsal nucleus of the lateral lemniscus (DNLL). Contra- and ipsilateral designations refer to the ICC from which we recorded. The major inhibitory input to the ICC, derived from the contralateral DNLL, is activated by sounds at the ipsilateral ear. We tested whether the remaining sources of inhibition in the ICC, predominantly activated by sounds at the contralateral ear, were upregulated to counter the effect of DNLL deactivation. Brainstem and pontine auditory nuclei are not shown, but the effective binaural differentiation of their output is represented by arrows indicating ipsilateral-ear- and contralateral-ear-dominated inputs to the midbrain (From[45]. With permission)

The results of this study[45] can be illustrated by presenting changes in responses of single neurons over the time-course of the manipulation. Figure 9.4A shows the change in excitability of a single ICC neuron over the experimental time-course. The activity of the ICC neuron increased immediately after deactivation of the inhibitory DNLL input (10 min after the kainic acid injection). However, this initial increase was quickly followed by a reduction in neuronal excitability: the activity decreased in minutes and remained near the baseline level for the duration of the experiment. Hence the experimentally induced increase in neuronal excitability was counteracted. Important in this interpretation is the fact that the removed inhibition had been activated by ipsilateral-ear stimulation, but the delayed reduction in response was demonstrated for contralateral-ear stimulation; there was no recovery of ipsilateral-ear-stimulated inhibition. Figure 9.4B presents an example of the response from a single neuron recorded from one of a group of rats that were pretreated with finasteride, an inhibitor of 5α-reductase. As before, DNLL deactivation caused an immediate increase in the activity of the ICC neuron. However, no counteracting reduction in excitability occurred when the synthesis of allopregnanolone and alloTHDOC was blocked. The neuron behaved abnormally (>10 min after the kainic acid injection), as shown by persistently elevated responses to sound at the contralateral ear, excitation to ipsilateral-ear stimulation (which usually causes inhibition), and high spontaneous activity. Thus, we conclude that 5α-reduced pregnane steroids mediated the counteracting plasticity in animals that

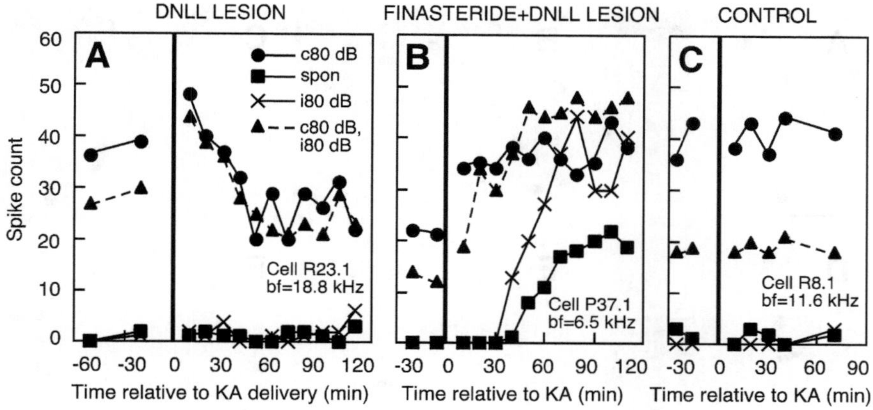

Fig. 9.4 Homeostasis through inhibitory synaptic plasticity mediated by 5α-reduced pregnane steroids. (**A–C**) Three ICC neurons are featured: one from a DNLL-lesioned rat (**A**), one from a DNLL-lesioned rat pretreated with finasteride (**B**), and one from a control rat where kainic acid (KA) was injected outside the ICC (**C**). Neuronal responses are plotted against time relative to KA delivery. (**A**) Prior to the KA injection, the ICC neuron was inhibited by sounds at the ipsilateral ear (solid cf. dashed line). Initially after the KA-induced DNLL lesion, the ipsilateral-ear-driven inhibition disappeared, and spike counts increased. However, from 20 to 50 min post-KA, spike counts decreased, and thereafter remained just below the baseline level. (**B**) In contrast, the initial effect of KA was not followed by a decrease in ICC excitability in finasteride-treated animals. Although inhibition was reduced in the ICC after lesioning the contralateral DNLL, the raised spike counts persisted at all times after the KA injection. At 30 min post-KA and beyond, sounds at the ipsilateral ear evoked abnormal excitatory neuronal responses and spontaneous activity manifested. (**C**) When KA was injected outside the DNLL, spike counts in the ICC changed little over the duration of the experiment. This control showed that the changes in the ICC occurred due to the deactivation of the inhibitory input from the contralateral DNLL and not due to a non-specific effect of KA (From[45]. With permission)

had not received finasteride. These steroid effects were also apparent on ensembles of neurons, i.e., evoked potentials. It should be noted that finasteride caused little or no change in the baseline response properties of neurons, as has been previously reported.[44] This may be because the basal turnover of 5α-reduced pregnane steroids is longer than a few hours (i.e., the duration of the aforementioned experiments).

We next used an anti-allopregnanolone/alloTHDOC antibody to perform immunohistochemistry and immunopositive-cell counts[45] (Fig. 9.5). We compared four treatment groups: (1) rats that received no treatment, (2) finasteride-treated rats, (3) rats euthanased 30 min after DNLL deactivation, and (4) finasteride-treated rats euthanased 30 min after DNLL deactivation. The number of steroid-positive cells in the ICC was increased 30 min after DNLL deactivation (i.e., treatment group (3)), in comparison with the other treatment groups, which were not different from each other. In contrast, the number of steroid-positive cells did not change in other brain areas, e.g., another midbrain area, the substantia nigra pars reticulata, is featured in Fig. 9.5D–F. These experiments show that the synthesis of allopregnanolone and/or alloTHDOC can be locally upregulated in order to counter increased neuronal excitability.

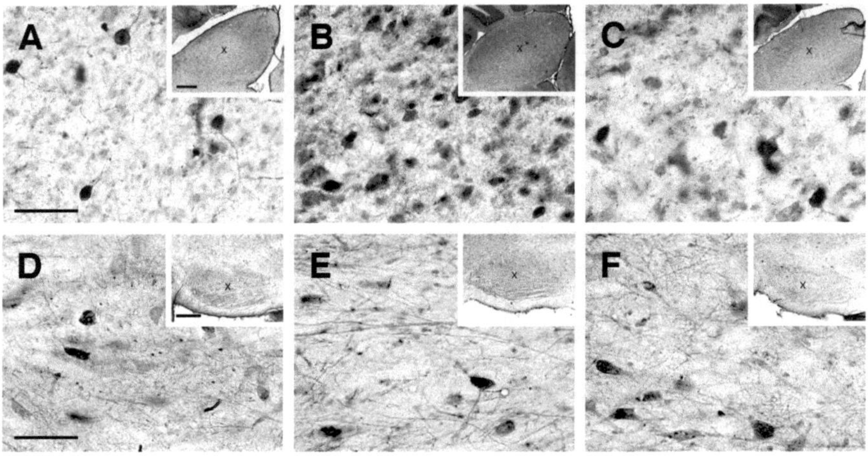

Fig. 9.5 Local increase in 3α-hydroxy,5α-reduced pregnane steroids in response to reduced GABAergic inhibition. Matching areas of the ICC (**A–C**) and SNR (**D–F**) from one normal rat, one DNLL-lesioned rat, and one DNLL-lesioned rat pretreated with finasteride (400×; scale bar = 50 μm). Darkly labeled cells are immunopositive for allopregnanolone/alloTHDOC. Insets show the whole inferior colliculus and substantia nigra in sagittal sections, with the area shown in **A–F** marked with an "X" (25×; scale bar = 500 μm). The number of immunopositive cells increased in the ICC 30 min after the kainic acid injection into the contralateral DNLL (**B**) compared with the baseline (**A**). This increase was blocked by finasteride (**C**). (**D–F**) The number of immunopositive cells in the SNR did not differ across treatment groups (From[45]. With permission)

9.5 Cellular Distribution of 3α-Hydroxy,5α-Reduced Pregnane Steroids

Until recently, glia had been considered the predominant source of 3α-hydroxy, 5α-reduced pregnane steroids in the brain. However, recent studies have shown that neurons are not only sources of 3α-hydroxy,5α-reduced pregnane steroids, but that glial contributions may be small. First, de novo synthesis of allopregnanolone and alloTHDOC is possible in neurons, as they contain the synthetic enzymes P450scc,[51,52] 3β-hydroxysteroid dehydrogenase,[51] P450 2D (which can act as a 21-hydroxylase),[53] 5α-reductase[49,54] and 3αHSD.[49,54] Second, allopregnanolone and alloTHDOC have themselves been shown present in neurons[45,50] (Figs. 9.6 and 9.7). Third, the critical synthetic enzymes 5α-reductase and 3αHSD,[49] and allopregnanolone and alloTHDOC[45,50] (Fig. 9.8), were not detected in glia in recent studies. Recent data derived from our immunohistochemical study localising 3α-hydroxy, 5α-reduced pregnane steroids (allopregnanolone and alloTHDOC cannot be separated with this method)[45,50] and that of Agis-Balboa and colleagues localising mRNA for 5α-reductase and 3αHSD (in situ hybridization)[49] show a striking similarity in the distribution of critical synthetic enzymes and their end products allopregnanolone and alloTHDOC.

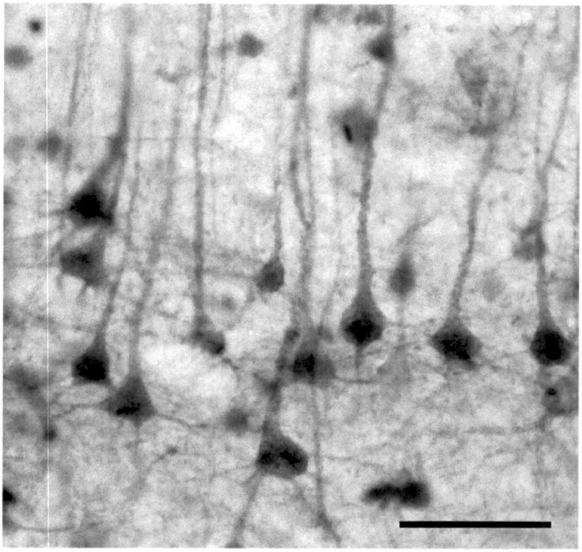

Fig. 9.6 Pyramidal cells in the cerebral cortex. Cell bodies and thick dendrites were labeled by the anti-allopregnanolone/alloTHDOC antibody. Sagittal section, 1.5 mm lateral; scale bar = 50 μm (From[50])

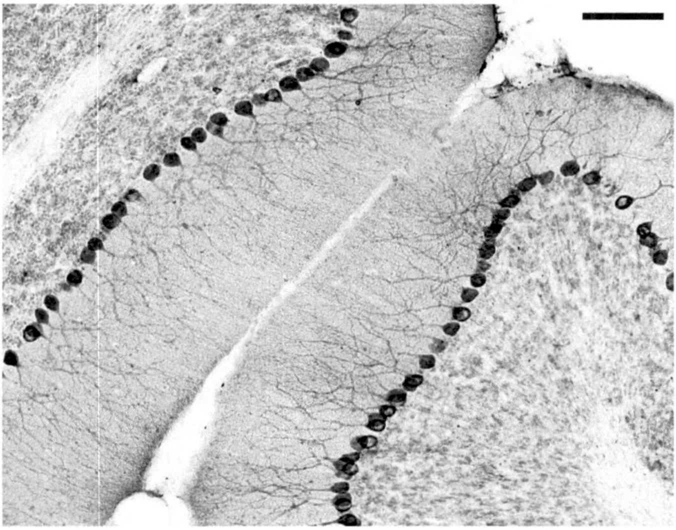

Fig. 9.7 Purkinje cells of the cerebellum. Cerebellar labeling was not homogenous, with labeling concentrated in the soma and principal dendrites of Purkinje cells. Sagittal section; scale bar = 50 μm (From[50])

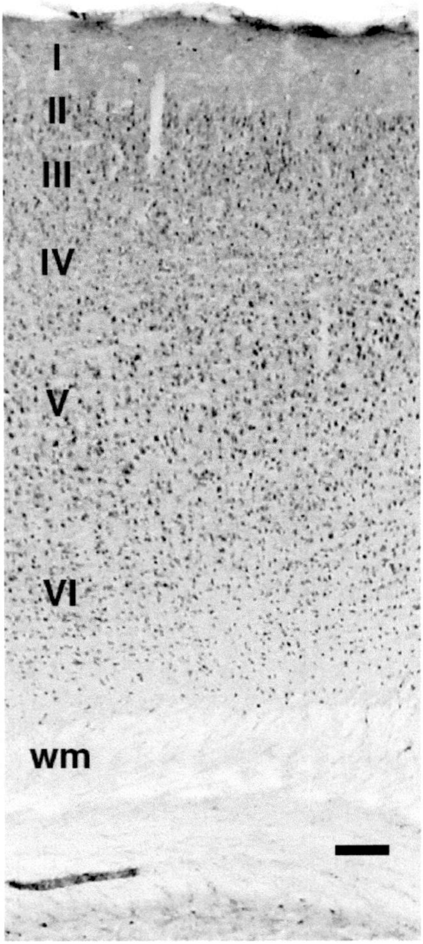

Fig. 9.8 S1Tr region of the cerebral cortex. Layer I and white matter (wm) showed little labeling. Coronal section, bregma −2.6 mm; scale bar = 100 μm (From[50])

There is selectivity in the types of neurons and subcellular compartments that contain allopregnanolone and alloTHDOC.[45,50] These steroids are present in excitatory neurons, e.g., pyramidal cells (Fig. 9.6), and inhibitory projection neurons, e.g., cerebellar Purkinje cells (Fig. 9.7). In contrast, allopregnanolone and alloTHDOC have not been demonstrated in local inhibitory neurons. This means that the steroids are present in the cells that they most likely modulate. Within steroid-containing cells, the steroids localise to the cell body and thick dendrites (Fig. 9.6). Inhibitory input to the cell body and thick dendrites has the potential to greatly influence action potential generation. Thus, allopregnanolone and alloTHDOC are well positioned to modulate cellular excitability.

Allopregnanolone and alloTHDOC are widespread across the brain, present in autonomic, limbic, motor and sensory pathways. This distribution has been shown by radioimmunoassay,[55,56] gas chromatography-mass fragmentography,[17,57–59] and immunohistochemistry.[45,50,60] However, the concentration of allopregnanolone and alloTHDOC varies between brain regions. The olfactory bulb, striatum and cerebral cortex have high levels of these steroids, while the brainstem reticular formation has low levels.[17,50,55–60] There are also variations within regions, notably allopregnanolone and alloTHDOC are confined to pyramidal cells in the hippocampus and Purkinje cells in the cerebellum[50] (Fig. 9.7).

9.6 Conclusion

3α-Hydroxy,5α-reduced pregnane steroids have long been known to potentiate GABAergic inhibition[20] and to change in concentration under various circumstances.[24,42–44] For example, pregnane steroid concentrations in the brain increase during stressful conditions,[24] and this increase in steroids likely serves an anxiolytic function. However, there is now evidence that allopregnanolone and alloTHDOC play a general role in regulating neuronal excitability.[45] This includes homeostatic actions, particularly under conditions of high neural excitability that could lead to neurotoxicity or seizures. The steroids are able to act within minutes to balance excitatory and inhibitory inputs in complex neural circuits. There are also recent studies showing that allopregnanolone and alloTHDOC, although widely distributed across the brain, are found mainly in projection neurons.[45,49,50] Thus, these steroids are present at their likely sites of action.

References

1. Horn G. Pathways of the past: the imprint of memory. Nat Rev Neurosci 2004; 5:108–120.
2. Lynch MA. Long-term potentiation and memory. Physiol Rev 2004; 84:87–136.
3. Turrigiano GG, Nelson SB. Homeostatic plasticity in the developing nervous system. Nat Rev Neurosci 2004; 5:97–107.
4. Hedden T, Gabrieli JD. Insights into the ageing mind: a view from cognitive neuroscience. Nat Rev Neurosci 2004; 5:87–96.
5. Zucker RS, Regehr WG. Short-term synaptic plasticity. Annu Rev Physiol 2002; 64:355–405.
6. Klann E, Dever TE. Biochemical mechanisms for translational regulation in synaptic plasticity. Nat Rev Neurosci 2004; 5:931–942.
7. Warner M, Gustafsson JA. Cytochrome P450 in the brain: neuroendocrine functions. Front Neuroendocrinol 1995; 16:224–236.
8. Bose H, Lingappa VR, Miller WL. Rapid regulation of steroidogenesis by mitochondrial protein import. Nature 2002; 417:87–91.
9. West LA, Horvat RD, Roess DA, Barisas BG, Juengel JL, Niswender GD. Steroidogenic acute regulatory protein and peripheral-type benzodiazepine receptor associate at the mitochondrial membrane. Endocrinology 2001; 142:502–505.

10. King SR, Manna PR, Ishii T, Syapin PJ, Ginsberg SD, Wilson K, Walsh LP, Parker KL, Stocco DM, Smith RG, Lamb DJ. An essential component in steroid synthesis, the steroidogenic acute regulatory protein, is expressed in discrete regions of the brain. J Neurosci 2002; 22:10613–10620.
11. Alho H, Costa E, Ferrero P, Fujimoto M, Cosenza-Murphy D, Guidotti A. Diazepam-binding inhibitor: a neuropeptide located in selected neuronal populations of rat brain. Science 1985; 229:179–182.
12. Anholt RR, Murphy KM, Mack GE, Snyder SH. Peripheral-type benzodiazepine receptors in the central nervous system: localization to olfactory nerves. J Neurosci 1984; 4:593–603.
13. Sanne JL, Krueger KE. Expression of cytochrome P450 side-chain cleavage enzyme and 3β-hydroxysteroid dehydrogenase in the rat central nervous system: a study by polymerase chain reaction. J Neurochem 1995; 65:528–536.
14. Iwahashi K, Kawai Y, Suwaki H, Hosokawa K, Ichikawa Y. A localization study of the cytochrome P-450(21)-linked monooxygenase system in adult rat brain. J Steroid Biochem Mol Biol 1993; 44:163–169.
15. Kishimoto W, Hiroi T, Shiraishi M, Osada M, Imaoka S, Kominami S, Igarashi T, Funae Y. Cytochrome P450 2D catalyze steroid 21-hydroxylation in the brain. Endocrinology 2004; 145:699–705.
16. Li X, Bertics PJ, Karavolas HJ. Regional distribution of cytosolic and particulate 5alpha-dihydroprogesterone 3alpha-hydroxysteroid oxidoreductases in female rat brain. J Steroid Biochem Mol Biol 1997; 60:311–318.
17. Cheney DL, Uzunov D, Costa E, Guidotti A. Gas chromatographic-mass fragmentographic quantitation of 3 alpha-hydroxy-5 alpha-pregnan-20-one (allopregnanolone) and its precursors in blood and brain of adrenalectomized and castrated rats. J Neurosci 1995; 15:4641–4650.
18. Baulieu EE. Neurosteroids: a novel function of the brain. Psychoneuroendocrinology 1998; 23:963–987.
19. Rupprecht R, Reul JM, Trapp T, van Steensel B, Wetzel C, Damm K, Zieglgansberger W, Holsboer F. Progesterone receptor-mediated effects of neuroactive steroids. Neuron 1993; 11:523–530.
20. Majewska MD, Harrison NL, Schwartz RD, Barker JL, Paul SM. Steroid hormone metabolites are barbiturate-like modulators of the GABA receptor. Science 1986; 232:1004–1007.
21. Paul SM, Purdy RH. Neuroactive steroids. FASEB J 1992; 6:2311–2322.
22. Lambert JJ, Belelli D, Harney SC, Peters JA, Frenguelli BG. Modulation of native and recombinant GABA(A) receptors by endogenous and synthetic neuroactive steroids. Brain Res Brain Res Rev 2001; 37:68–80.
23. Reddy DS, Rogawski MA. Stress-induced deoxycorticosterone-derived neurosteroids modulate GABA(A) receptor function and seizure susceptibility. J Neurosci 2002; 22:3795–3805.
24. Purdy RH, Morrow AL, Moore PH Jr, Paul SM. Stress-induced elevations of gamma-aminobutyric acid type A receptor-active steroids in the rat brain. Proc Natl Acad Sci USA 1991; 88:4553–4557.
25. Corpechot C, Young J, Calvel M, Wehrey C, Veltz JN, Touyer G, Mouren M, Prasad VV, Banner C, Sjovall J, Baulieu EE, Robel P. Neurosteroids: 3 alpha-hydroxy-5 alpha-pregnan-20-one and its precursors in the brain, plasma, and steroidogenic glands of male and female rats. Endocrinology 1993; 133:1003–1009.
26. Korneyev A, Guidotti A, Costa E. Regional and interspecies differences in brain progesterone metabolism. J Neurochem 1993; 61:2041–2047.
27. Celotti F, Melcangi RC, Martini L. The 5 alpha-reductase in the brain: molecular aspects and relation to brain function. Front Neuroendocrinol 1992; 13:163–215.
28. El-Etr M, Akwa Y, Robel P, Baulieu EE. Opposing effects of different steroid sulfates on GABAA receptor-mediated chloride uptake. Brain Res 1998; 790:334–338.
29. Park-Chung M, Malayev A, Purdy RH, Gibbs TT, Farb DH. Sulfated and unsulfated steroids modulate gamma-aminobutyric acidA receptor function through distinct sites. Brain Res 1999; 830:72–87.

30. Mitamura K, Yatera M, Shimada K. Quantitative determination of pregnenolone 3-sulfate in rat brains using liquid chromatography/electrospray ionization-mass spectrometry. Anal Sci 1999; 15:951–955.
31. Liere P, Pianos A, Eychenne B, Cambourg A, Liu S, Griffiths W, Schumacher M, Sjovall J, Baulieu EE. Novel lipoidal derivatives of pregnenolone and dehydroepiandrosterone and absence of their sulfated counterparts in rodent brain. J Lipid Res 2004; 45:2287–2302.
32. Edwards FA, Konnerth A, Sakmann B. Quantal analysis of inhibitory synaptic transmission in the dentate gyrus of rat hippocampal slices: a patch-clamp study. J Physiol 1990; 430:213–249.
33. Spigelman I, Li Z, Liang J, Cagetti E, Samzadeh S, Mihalek RM, Homanics GE, Olsen RW. Reduced inhibition and sensitivity to neurosteroids in hippocampus of mice lacking the GABA(A) receptor delta subunit. J Neurophysiol 2003; 90:903–910.
34. Cooper EJ, Johnston GA, Edwards FA. Effects of a naturally occurring neurosteroid on GABAA IPSCs during development in rat hippocampal or cerebellar slices. J Physiol 1999; 521:437–449.
35. Brussaard AB, Kits KS, Baker RE, Willems WP, Leyting-Vermeulen JW, Voorn P, Smit AB, Bicknell RJ, Herbison AE. Plasticity in fast synaptic inhibition of adult oxytocin neurons caused by switch in GABA(A) receptor subunit expression. Neuron 1997; 19:1103–1114.
36. Zhu WJ, Vicini S. Neurosteroid prolongs GABAA channel deactivation by altering kinetics of desensitized states. J Neurosci 1997; 17:4022–4031.
37. Jones MV, Westbrook GL. The impact of receptor desensitization on fast synaptic transmission. Trends Neurosci 1996; 19:96–101.
38. Twyman RE, Macdonald RL. Neurosteroid regulation of GABAA receptor single-channel kinetic properties of mouse spinal cord neurons in culture. J Physiol 1992; 456:215–245.
39. Callachan H, Cottrell GA, Hather NY, Lambert JJ, Nooney JM, Peters JA. Modulation of the GABAA receptor by progesterone metabolites. Proc R Soc Lond B Biol Sci 1987; 231:359–369.
40. Bullock AE, Clark AL, Grady SR, Robinson SF, Slobe BS, Marks MJ, Collins AC. Neurosteroids modulate nicotinic receptor function in mouse striatal and thalamic synaptosomes. J Neurochem 1997; 68:2412–2423.
41. Wetzel CH, Hermann B, Behl C, Pestel E, Rammes G, Zieglgansberger W, Holsboer F, Rupprecht R. Functional antagonism of gonadal steroids at the 5-hydroxytryptamine type 3 receptor. Mol Endocrinol 1998; 12:1441–1451.
42. Uzunova V, Sheline Y, Davis JM, Rasmusson A, Uzunov DP, Costa E, Guidotti A. Increase in the cerebrospinal fluid content of neurosteroids in patients with unipolar major depression who are receiving fluoxetine or fluvoxamine. Proc Natl Acad Sci USA 1998; 95:3239–3244.
43. Palumbo MA, Salvestroni C, Gallo R, Guo AL, Genazzani AD, Artini PG, Petraglia F, Genazzani AR. Allopregnanolone concentration in hippocampus of prepubertal rats and female rats throughout estrous cycle. J Endocrinol Invest 1995; 18:853–856.
44. VanDoren MJ, Matthews DB, Janis GC, Grobin AC, Devaud LL, Morrow AL. Neuroactive steroid 3alpha-hydroxy-5alpha-pregnan-20-one modulates electrophysiological and behavioral actions of ethanol. J Neurosci 2000; 20:1982–1989.
45. Saalmann YB, Morgan IG, Calford MB. Neurosteroids involved in regulating inhibition in the inferior colliculus. J Neurophysiol 2006; 96:3064–3073.
46. Roberts RC, Ribak CE. An electron microscopic study of GABAergic neurons and terminals in the central nucleus of the inferior colliculus of the rat. J Neurocytol 1987; 16:333–345.
47. Gonzalez-Hernandez T, Mantolan-Sarmiento B, Gonzalez-Gonzalez B, Perez-Gonzalez H. Sources of GABAergic input to the inferior colliculus of the rat. J Comp Neurol 1996; 372:309–326.
48. Zhang DX, Li L, Kelly JB, Wu SH. GABAergic projections from the lateral lemniscus to the inferior colliculus of the rat. Hear Res 1998; 117:1–12.
49. Agis-Balboa RC, Pinna G, Zhubi A, Maloku E, Veldic M, Costa E, Guidotti A. Characterization of brain neurons that express enzymes mediating neurosteroid biosynthesis. PNAS 2006; 103:14602–14607.

50. Saalmann YB, Kirkcaldie MT, Waldron S, Calford MB. Cellular distribution of the GABAA receptor-modulating 3alpha-hydroxy, 5alpha-reduced pregnane steroids in the adult rat brain. J Neuroendocrinol 2007; 19:272–284.
51. Tsutsui K, Sakamoto H, Ukena K. Biosynthesis and action of neurosteroids in the cerebellar Purkinje neuron. J Steroid Biochem Mol Biol 2003; 85:311–321.
52. Patte-Mensah C, Kappes V, Freund-Mercier MJ, Tsutsui K, Mensah-Nyagan AG. Cellular distribution and bioactivity of the key steroidogenic enzyme, cytochrome P450side chain cleavage, in sensory neural pathways. J Neurochem 2003; 86:1233–1246.
53. Miksys S, Rao Y, Hoffmann E, Mash DC, Tyndale RF. Regional and cellular expression of CYP2D6 in human brain: higher levels in alcoholics. J Neurochem 2002; 82:1376–1387.
54. Patte-Mensah C, Penning TM, Mensah-Nyagan AG. Anatomical and cellular localization of neuroactive 5 alpha/3 alpha-reduced steroid-synthesizing enzymes in the spinal cord. J Comp Neurol 2004; 477:286–299.
55. Bernardi F, Salvestroni C, Casarosa E, Nappi RE, Lanzone A, Luisi S, Purdy RH, Petraglia F, Genazzani AR. Aging is associated with changes in allopregnanolone concentrations in brain, endocrine glands and serum in male rats. Eur J Endocrinol 1998; 138:316–321.
56. Bixo M, Andersson A, Winblad B, Purdy RH, Backstrom T. Progesterone, 5alpha-pregnane-3,20-dione and 3alpha-hydroxy-5alpha-pregnane-20-one in specific regions of the human female brain in different endocrine states. Brain Res 1997; 764:173–178.
57. Uzunov DP, Cooper TB, Costa E, Guidotti A. Fluoxetine-elicited changes in brain neurosteroid content measured by negative ion mass fragmentography. Proc Natl Acad Sci USA 1996; 93:12599–12604.
58. Uzunova V, Ceci M, Kohler C, Uzunov DP, Wrynn AS. Region-specific dysregulation of allopregnanolone brain content in the olfactory bulbectomized rat model of depression. Brain Res 2003; 976:1–8.
59. Weill-Engerer S, David JP, Sazdovitch V, Liere P, Eychenne B, Pianos A, Schumacher M, Delacourte A, Baulieu EE, Akwa Y. Neurosteroid quantification in human brain regions: comparison between Alzheimer's and nondemented patients. J Clin Endocrinol Metab 2002; 87:5138–5143.
60. Frye CA, Vongher JM. 3alpha,5alpha-THP in the midbrain ventral tegmental area of rats and hamsters is increased in exogenous hormonal states associated with estrous cyclicity and sexual receptivity. J Endocrinol Invest 1999; 22:455–464.

Chapter 10
Steroidogenesis and Neuroplasticity in the Songbird Brain

Colin J. Saldanha[1] **and Barney A. Schlinger**[2]

Abstract The vertebrate brain is a source and target of steroid hormones. Songbirds have long provided an array of structural and behavioral endpoints towards understanding how steroid molecules affect neuroanatomy and neurophysiology. More recently, our conceptualization of the brain has expanded to include the CNS as a potent source of these very steroids. Here we review recent findings about the expression of steroidogenic enzymes in the songbird brain with particular emphasis upon the role of neurosteroidogenesis on the plasticity of brain circuits. We include examples of natural neuroplasticity in laboratory and field studies. Additionally, we discuss the role of neurosteroidogenesis on the outcomes of pathological brain damage. These two areas of research have provided novel and fundamentally restructured hypotheses as to the role of neurosteroidogenesis in brain function. Notable among recent findings from such studies, are the consideration of alternate substrates for steroidogenic enzymes, an expansion of the suite of enzymes found at sites of neuronal recruitment and the expression of these enzymes in additional cell types and ultrastructural compartments. The powerful link among ethology, physiology, anatomy and cell biology is exemplified in these vertebrates and renders the songbird an enduring model for the study of the role of *de novo* steroid synthesis on the plasticity of brain structure and function.

Keywords Testosterone, estrogen, neurogenesis, stroke, synapse

Abbreviations mMAN medial portion of the magnocellular nucleus of the anterior nidopallium; lMAN lateral portions of the magnocellular nucleus of the anterior nidopallium; HVC high vocal center; RA robust nucleus of the arcopallium; T testosterone; E_2 estradiol; BrdU 5-bromo-2-deoxyuridine; P_{450}CYP19a cyctochrome

[1] Department of Biological Sciences and Program in Cognitive Science Lehigh University, Bethlehem, PA

[2] Department of Physiological Science and Laboratory for Neuroendocrinology University of California, Los Angeles, USA

M.S. Ritsner and A. Weizman (eds.), *Neuroactive Steroids in Brain Function, Behavior and Neuropsychiatric Disorders*

aromatase; DHEA dehydroepiandrosterone; 3β–HSD 3β-hydroxysteroid dehydrogenase/isomerase; AE androstenedione; DHT dihydrotestosterone; AR androgen-receptors; ER estrogen-receptors; P_{450}CYP17 cytochrome 17α-hydroxylase; StAR steroidogenic acute regulatory protein; P_{450} CYP11A1 side-chain cleavage enzyme; GABA gamma-aminobutyric acid

The landmark studies of Berthold[1] demonstrated the profound influence of gonadal secretions on secondary sexual characteristics and behavior in birds. Since then, the vertebrate brain and peripheral organs have been well documented as targets for circulating gonadal steroids or sex-steroids. In fact, the term "endocrinology" was coined to describe chemical messengers that were capable of potently affecting distant target tissues, via passage through the vasculature. The pivotal role of sex-steroid synthesis and action in the regulation of behavior, continue as fields of intense investigation.

This chapter will focus on one of many fundamental clarifications in our understanding of hormone–behavior relationships. Specifically, we will focus on the many roles of the brain in steroid metabolism and steroid synthesis and how these actions potentiate or complement neural actions of peripheral hormones. Notably, this clarification, has been considerably honed and elaborated using the songbird as a model for steroid synthesis and action. We discuss these findings in the context of natural and pathological neural plasticity, two areas where the considerable influence of brain steroid synthesis (neurosteroidogenesis) is obvious. We begin by describing in broad strokes natural changes in the cytoarchitecture of the passerine brain. We then describe work done in our laboratories towards understanding neural changes associated with mechanical brain damage. Lastly, we provide an overview of recent work on the role of neurosteroidogenesis in these two forms of neuroplasticity.

10.1 Natural Plasticity of the Songbird Brain

Songbirds are likely the most recently evolved of all birds and include some 4,000 species worldwide. They are recognized for the complex, often beautiful songs they produce, and by the fact that they learn these songs from their fathers or from tutors.[2] Songs are produced by airflow across a muscular and membranous syrinx (unlike our larynx). The syringeal muscles are in turn controlled by a rather remarkable set of neural circuits. This song circuit, an interconnected group of nuclei in the brains of all songbirds is well studied as a target for newly mitotic neurons during adulthood.[3,4] These largely telencephalic nuclei form a sensory or acquisition pathway comprised of Area X in the parolfactory lobe, the medial and lateral portions of the magnocellular nucleus of the anterior nidopallium (medial and lateral MAN), and the high vocal center (HVC). HVC is also part of a motor pathway containing the robust nucleus of the arcopallium (RA) that projects to motor nuclei in the hindbrain.[3] These two major pathways sub-serve the acquisition and production of song respectively and contain additional projections among most of the nuclei described above. Several of these nuclei exhibit profound sexual dimorphisms in keeping with dimorphisms in song learning and singing.[3,4] Importantly for our consideration here, several song nuclei can demonstrate dramatic seasonal cycles in volume, cell number, and cell size.[5]

The songbird brain demonstrates annual cycles in neuronal death, neurogenesis, site-specific neuron recruitment, neuron size, dendritic aborization, and synapto genesis.[5,6] These changes have been exhaustively described in nuclei of the song circuit and, to a lesser extent, in the songbird hippocampus.[7,8] Several photoperiodic passerines increase the volume of the high vocal center HVC, RA, Area X and the lMAN in the summer relative to the winter. This volume increase is due to a combination of increased cell number, cell size, cell spacing, dendritic arborization, and/or synaptic density.[5]

Some songbirds also possess impressive spatial memory capabilities with their hippocampus largely responsible for memory acquisition.[9,10] Moreover, the capacity to form certain kinds of spatial memories can change across seasons and there is evidence that the neural substrate underlying these memories changes seasonally as well. For example, in adult songbirds that cache food in autumn and winter, the plasticity of the hippocampus reflects the seasonal increases and decreases in memory function required for accurate food caching and retrieval.[9] In a wild North American songbird, the black-capped chickadee, mitotic neurons are known to migrate into the adult hippocampus,[11] thereby increasing hippocampal volume via changes in neuron number.[7,8,12] Interestingly, in contrast to what is observed in the song system, new neurons invade the passerine hippocampus during the autumn and winter. This suggests the possibility that different parts of the adult songbird brain recruit new neurons in a season- and behavior-specific manner. In a recent study, we compared cell density and volume of the hippocampus among adult black-capped chickadees sacrificed in February, June, or October. These combined data on the song system and hippocampus point to the exaggerated capability of the adult passerine brain to reorganize behaviorally relevant circuitry during adulthood. The strongly seasonal nature of cell birth, migration, and death in the songbird brain suggests the involvement of seasonally regulated reproductive hormones. The marked sexual dimorphisms in song circuit structure and function may suggest the involvement of steroids in particular. As described below, several structural and behavioral endpoints of song circuit plasticity are potently affected by steroid synthesis within the brain itself.

The role of neurosteroidogenesis on passerine neuroplasticity is perhaps exemplified by estrogenic influences of neurons in the adult song circuit. Circulating testosterone (T) is important in this influence, however, many effects of T appear mediated via its conversion to estradiol (E_2) at neural loci. Administration of T or its metabolite E_2 increases the size and number of neurons in HVC and Area X in a number of species including dark-eyed juncos (*Junco hyemalis*) and song sparrows (*Zonotrichia leucophrys gambeli*). Either steroid also increases the volume of RA, but this influence appears to involve the size and spacing of cells.[13–16] The

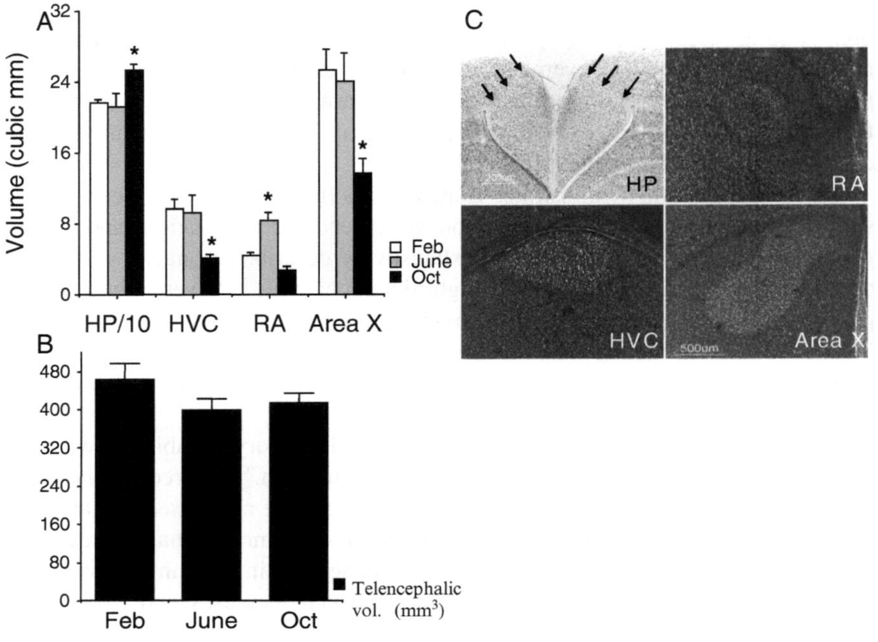

Fig. 10.1 Histograms showing the volume of the (**A**) hippocampus and song nuclei (HVC, RA, Area X), and (**B**) telenephalon in male black-capped chickadees during different times of the year. Photographs of the brain areas measures appear in (**C**). Arrows depict the Lateral borders of the hippocampus

arborization of neurons within RA is profoundly affected by E_2 in the adult canary, with the number of dendrites and synaptic profiles increasing with steroid administration.[17] In support of this idea, administration of the aromatase inhibitor fadrozole itself decreases the vernal increase of HVC volume in song sparrows.[16] Given that the brain is an abundant source of aromatization in the passerine brain[18,19] it is likely that this aromatization occurs within the brain.

10.2 Pathological Brain Injury and Songbird Neuroplasticity

In recent years our laboratories and collaborators have studied songbirds as models for brain injury, injury-induced degeneration and recovery from brain damage. Because they breed readily in captivity one songbird species, the zebra finch, has been the focus of much of this work. Perhaps the high levels of naturally occurring neuroplasticity described above make the songbird an excellent model for these studies. Indeed, studies on the effects and modulators of mechanical brain damage in the zebra finch have focused on questions that involve secondary degeneration[20,21] and injury-induced cell proliferation.[22]

Mechanical or anoxic/ischemic injury to the vertebrate brain results in an orchestrated neural response initiated by the pyknotic death of cells at the site of primary damage. Perhaps not surprisingly, the compromise of cell membranes in these dying cells results in an increase in extracellular concentrations of various factors.[23] In mammals, this primary pyknosis (and related increases in extracellular signals) is followed by a wave of apoptotic secondary degeneration.[23–30] The amplitude of this wave at various discrete time-points is affected by numerous factors including cations, cytokines, adhesion molecules and steroid hormones.[31,32] Finally, cell proliferation and recruitment to the site of primary damage may be considered one of the terminal steps in the orchestrated response of the vertebrate brain to injury. Neuronal recruitment to sites of brain damage are likely mediated by the local expression of adhesion molecules and trophic factors.[28,30] It is likely that this neuronal recruitment is key in the brain repair. Thus, the data suggest that the mediation or pyknosis, apoptosis, cell proliferation, and/or migration are all likely candidates for mediation by steroid hormones.

In birds and mammals primary damage to the neuropil is associated with an upregulation of aromatase transcription and translation in reactive glia around the site of injury.[20,21,33–35] Several lines of evidence suggest that the upregulation of aromatase in reactive glia is clearly functional. First, following a mechanical injury to the avian hippocampus there is a striking spatial relationship between reactive astrocytes in which aromatase in upregulated and newly born cells (labeled with the mitotic marker BrdU) around the lesion site.[36] Moreover, treatment with an aromatase inhibitor substantially reduces the number of BrdU-labled cells within this field of reactive glia.[22] Furthermore, inhibition of aromatase around the site of damage decreases apoptotic secondary degeneration.[20] The mechanism behind this effect is apparently mediated by estrogen production since co-administration of 17β-estradiol with fadrozole results in less degeneration than fadrozole alone.[21]

Injury-induced aromatase expression is not limited to reactive glia, but also includes radial glia.[36] The functional role of aromatase expression in radial glia remains to be studied more extensively. However, appositions between aromatase-positive radial glia and mitotic cells around the site of injury suggest a role for neurosteroidogenesis in general and estrogen synthesis in particular in the migration of cells following brain injury. Indeed, aromatization may play a key role in the proliferation of cells along the ventricles in response to brain trauma. As discussed above, Lee et al., [22] recently documented decreases in injury-induced proliferation following administration of the aromatase-inhibitor fadrozole, an effect that is mitigated by simultaneous estradiol administration.

The aforementioned areas of study involve natural and pathological fluctuations in various indices of structural reorganization within the passerine brain. Notable among these is the recruitment of new neurons to specific sites within the song circuit, the hippocampus and sites of brain injury. Most of the work in our laboratories has focused on aromatase (CYP19a) and its role in estrogen provision. Localized estrogen synthesis is known to influence several aspects of neuroplasticity in the songbird including the vernal increases in the volume of song nuclei[16] and injury

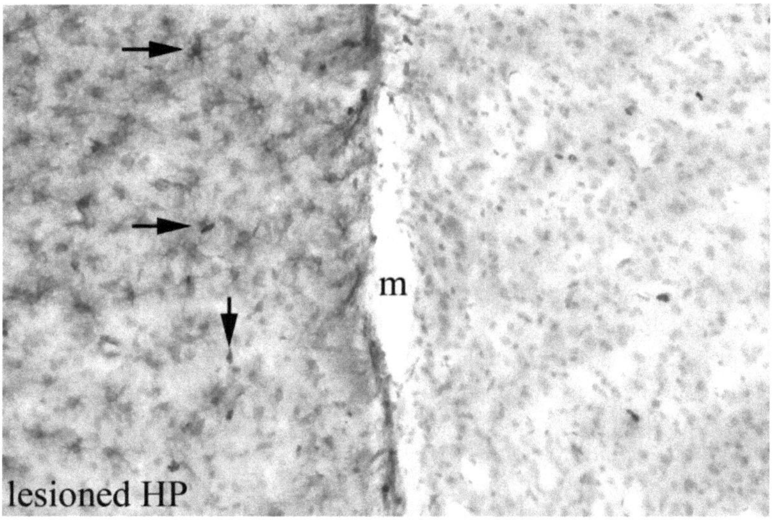

Fig. 10.2 Injury to the adult zebra finch hippocampus results in increased numbers of BrdU-labeled cells and aromatase-expressing astrocytes ipsilateral and proximal to the lesion. Coronal section depicting aromatase immunoreactivity (brown; horizontal arrows) and BrdU Immunoreactivity (black; vertical arrow) in an adult zebra finch hippocampus 48 h after a penetrating injury, 24 h after BrdU administration. m = midline

induced cell proliferation.[22] Questions remain, however, as to the specific cellular mechanisms that support this steroidal mediation of neuroplasticity.

Two intriguing questions arising from all the studies described above are: given that most of the described effects on passerine neuroplasticity arise from estrogenic effects (1) what is the substrate utilized by aromatase during times of the year when circulating androgens are basal and (2) why do birds with ample circulating E need to aromatize androgens within the brain. Here we describe recent findings in our laboratories that address possible (alternative) substrates for aromatization in the songbird and the presence of steroidogenic enzymes in novel ultrastructural compartments of neurons.

10.3 3β-Hydroxysteroid Dehydrogenase, DHEA and Neuroplasticity

Steroid control of avian neuroplasticity may involve non-gonadal steroids. One possibility includes the hormone DHEA likely secreted by the avian adrenals or synthesized within the songbird brain itself. DHEA and its sulfated ester (DHEA-S) are synthesized and secreted in large amounts by the human adrenals[38,39] but are present in low levels of many other vertebrate species. Although these hormones circulate in abundance in humans, we are just beginning to appreciate their roles as endocrine signals. DHEA and DHEA (S) can have direct actions on the mammalian brain, but we know that they also function as prohormones, available for conversion into more active androgens or estrogens by enzyme catalyzed reactions. Labrie and colleagues have coined the term "intracrinology" to account for the synthesis and actions of the active downstream products of DHEA in target tissues.[40]

Unlike many other non-primate species, songbirds have relatively high levels of DHEA in blood. In captive reproductively active zebra finches, DHEA circulates in both stressed and unstressed males and females at around 1 ng/ml, similar to or even greater than T.[41] In the case of wild song sparrows, DHEA is present in the blood of males year round with levels during the non-breeding season substantially higher than either T or E_2.[41] Although not a songbird per se, the spotted antbird, a suboscine passerine bird (closely related to the oscine species that are the focus of this manuscript) also has detectable circulating DHEA. Like the song sparrow, DHEA is present in blood of virtually all wild and captive non-breeding antbirds, at levels greater than either T or E_2.[42]

DHEA also has documented effects on brain and behavior in songbirds. In a series of studies using the wild North American song sparrow, Kiran Soma and his colleagues showed that the treatment of non-breeding males with DHEA stimulated both singing behavior and significant growth of the song nucleus HVC.[43] As stated above, DHEA circulates at higher levels than T in non-breeders so DHEA may be the natural hormonal stimulates of territorial behavior and song in this species during the non-breeding season. During molt, when birds are noticeably non-aggressive, less DHEA is present in blood than at all other times of year.[41] Presumably, during the non-breeding season, only tissues that could further activate DHEA would be sensitive to its stimulatory effects whereas reproductive tissues would remain unstimulated. There are additional costs to males when T is elevated in blood,[44] and these might be avoided by secretion of the prohormone DHEA. Interestingly, estrogens also seem to stimulate territorial song in non-breeding song sparrows[45,46] and can stimulate growth of nucleus HVC.[16] We have evidence that the brain has the requisite enzymatic machinery to utilize DHEA as a substrate for the formation of active estrogens (and androgens) in the songbird brain (see below), including in the brain of male song sparrows (unpublished data).

As discussed above, DHEA might have direct neural actions or might serve as the substrate for the formation of more active steroidal molecules. The enzyme immediately responsible for the activational transformation of DHEA is 3β-hydroxysteroid dehydrogenase/isomerase (3B-HSD).[47] This enzyme converts DHEA into androstenedione (AE) a weak androgen that is readily converted into the more potent testosterone (T) by a common isoform of the enzyme 17β-HSD. In many tissues, most notably in the brain, T is metabolized into 5α-dihydrotestosterone (DHT) by the enzyme 5α-reductase or into E_2 by the enzyme aromatase. DHT and E_2 bind to androgen-receptors (AR) and estrogen-receptors (ER), respectively. Aromatase and 5α-reductase are widespread and abundant enzymes in the songbird brain (see above), so T derived from DHEA can potentially stimulate both androgen- or estrogen-dependent actions on the brain. We have considerable evidence that 3β-HSD, the enzyme that initiates this cascade starting with DHEA is expressed and active in the songbird brain.

We first measured 3B-HSD in dissociated primary cell cultures of the zebra finch brain.[48] Cells grown in vitro for 7–30 days are composed of both neurons and glia and exhibit considerable expression and/or activity of aromatase and 5a-reductase (and also 5B-reductase an androgen inactivation pathway.[49,50] If similar cultures were incubated with ^{3}H-DHEA we could extract 3H-AE from the media and cells as well as the expected downstream products of 3H-5a-androstanedione and 3H-estrone. We employed several steps to confirm each of these reactions. As a final confirmation of the presence of 3β-HSD, we also incubated these neuronal cell cultures with 3H-pregnenolone and confirmed the presence of ^{3}H-progesterone.[48]

While useful, cell cultures might not accurately reflect the in vivo condition. We have obtained additional evidence that 3β-HSD is expressed and active in the intact songbird brain. First, we have identified activity in homogenates of the zebra finch brain in both adult[43] and in developing birds.[51] As described above, homogenates were incubated with ^{3}H-DHEA but we added the NAD+ cofactor which is necessary for 3β-HSD reaction.[47] We then used several procedures to confirm 3β-HSD activity as well as subsequent enzymatic transformations by 5α-and 5β-reductase and aromatase. 3β-HSD activity is evident in all brain regions in zebra finches at all ages and in both sexes. Interestingly in adults, activity was generally higher in the brains of females than in males and 3β-HSD in females could be rapidly down-regulated by stress.[43] In many brains regions, ^{3}H-DHEA was the substrate for the ultimate formation of ^{3}H-estrogens. Because of the well-described effects of estrogens on development and plasticity of the songbird brain, it is this coordinated activity of 3β-HSD and aromatase that turn DHEA into estrogen that may have particular significance in songbird neuroendocrine signaling.

To ensure that experiments using homogenates accurately measure steroidogenic enzyme activity, we add cofactors required by these reactions. If these cofactors were, however, limited in brain, then these reactions may not proceed in vivo at rates measured in homogenates. Homogenates also artificially mix enzymes that might ordinarily be spatially separated. Recently, we have developed the capacity to measure enzyme activity in live slices of the songbird brain.[52] Under these conditions, slices are incubated with ^{3}H-DHEA but no cofactors are added. Again we have observed that ^{3}H-estrogens and ^{3}H-5α-DHT were produced. We conclude that 3β-HSD, aromatase and 5a-reductase are functionally colocalized in the songbird brain and that the biochemical conditions are appropriate in vivo to allow the prohormone DHEA to be the substrate for the formation of active androgens and estrogens. This capacity to convert DHEA into active sex steroids might also give circulating DHEA a role in neural protection or repair where E_2 and glial aromatization are known to be important.

10.4 Neurosteroid Synthesis in the Songbird Brain

DHEA circulates in blood of songbirds, so we assume that, like humans, it is produced in the avian adrenals. The synthesis of androgens, including DHEA, requires expression of the enzyme 17α-hydroxylase (P450CYP17). Although we have detected small amounts of this enzyme in the zebra finch adrenals,[53] it is certainly a minor adrenal enzyme. Unlike humans, birds synthesize and secrete corticosterone as their primary adrenal stress hormone and this hormone does not require CYP17.[54] Cortisol, the primary human glucocorticoid, does require CYP17, so it is an abundant enzyme in our adrenals.[55] It is possible that the songbird brain, in addition to the adrenals, synthesize some functional DHEA. We have several lines of evidence that all the enzymes and transporters required to make active sex steroids from cholesterol are expressed in the songbird brain. Sterodogensis involves the cholesterol transport by StAR (steroidogenic acute regulatory protein) into specialized mitochindria that express the side-chain cleavage enzyme (P_{450} CYP11A1). This enzyme converts cholesterol into pregnenolone. StAR and CYP11A1 cDNA fragments can be amplified using rt-PCR from mRNA extracted from various regions of the developing and adult zebra finch brain.[51,56] Transcripts for both these enzymes are detected on Northern blots and by in situ hybridization analyses. We have further cloned cDNAs for these enzymes and confirm that they encode functional, full-length transcripts similar in sequence to those of all other species examined. Mapping studies of the adult and juvenile zebra finch brain indicate that more often than not, these enzymes show a spatial brain organization.[51,56] These results suggest that the proteins that initiate steroidogenesis are present together in the songbird brain. We are pursuing additional studies identify the functional proteins for StAR and CYP11A1.

As mentioned previously, 3B-HSD converts pregnenolone into progesterone and we know this enzyme is active in the songbird brain. Additional studies confirm that, like StAR and CYP11A1, the mRNA for 3β-HSD is widely expressed in the developing and adult songbird brain.[51,56] These studies also show that CYP17, the enzyme that converts pregnenolone into DHEA, is expressed in the zebra finch brain.[56,57] As for the other enzymes and transporters, we have evidence from PCR, Northern blot and in situ hybridization analyses for neural CYP 17 expression. In a few brain regions, all of these steroidogenic factors show spatial overlap. In the first 5 days after zebra finches hatch, all of the factors are expressed in an overlapping pattern near the germinal zone of the lateral ventricles[56] where neurogenesis is most active.[58,59] In adults, different enzymes show different patterns of overlap with some regions, such as nucleus HVC expressing many factors (StAR, CYP11A1, 3β-HSD and CYP17) with other regions expressing other factors (including aromatase). Altogether, these results establish the songbird brain as a neurosteroidogenic tissue, capable of using cholesterol as a substrate and synthesizing a wide variety of steroids including progestins, androgens and estrogens. Just how songbirds naturally synthesize various steroids and then functionally utilize steroids from the gonads, adrenals and brain remains of matter of intense investigation by our laboratories.

The best evidence for a neurosteroid function for songbirds is in the growth of the masculine neural song system in the zebra finch. In this species, adult males sing and possess a song system greatly larger and more complex than do adult females who do not sing.[3] Exogenous E_2 can stimulate the growth of a masculine circuit when given to young females, but there is no clear evidence that males make more active sex steroid peripherally in young males.[60] One possibility is that the brain of males synthesizes more E_2 than do females and this neuroestrogen that masculinizes the neural song system. Slices of the developing zebra finch brain appear to synthesize E_2 de novo, and E_2 seems to induce some at least some masculine growth in vitro.[61] As mentioned previously, it may also be that neurosteroids stimulate behavior and brain growth in non-breeding birds when gonadal steroids would inappropriately stimulate reproductive behaviors. Similarly, sex steroid influence some hippocampal-based learning and memory processes in non-reproductive songbirds,[62] and these events may be stimulated by neurosteroids.

10.5 Neurosteroid Modulation of Neurotransmitter Systems

There is considerable evidence for direct steroid modulation of neurostransmitter systems,[63] and these may be especially important actions of neurosteroids or steroids produced in brain from peripheral DHEA. We have focused on the aromatase reaction, its capacity to convert circulating T into E_2, its coordinate function with 3β-HSD to synthesize estrogens from DHEA and its position as the terminal reaction of the steroidogenic cascade that might begin in brain with cholesterol and lead to the formation of estrogens. We have much evidence that aromatase exists in synaptic terminals in the avian brain[64,65] and this localization may be particularly widespread and abundant in the brain of songbirds.[66,67] In other species, estrogens can directly influence neurotransmitter systems or influence neuronal architecture that might increase or decrease neural activation.[68] Estrogenic modulation of neurotransmission may be particularly important in songbirds where estrogens are widely produced in synapses.

Birds have additional steroid transformations in brain that point to alternate mechanisms for steroidal modulation of neurostransmission. Two enzymes, 5α- and 5β-reductase are expressed widely, and often quite abundantly, throughout the avian brain including the brains of songbirds.[37,69–71] Initially, these enzymes were thought to activate and inactivate T respectively, by converting T into either the active androgen 5α-dihydrotestosterone (DHT)[72] or the presumed inactive 5β-DHT.[73] While there is excellent evidence that both of these transformations of T do occur and are biologically important, these enzyme are also expressed in many parts of the avian brain that lack androgen-receptors. In these brain regions these enzymes likely have little functional action on T. Both of these enzymes can also utilize progesterone as a substrate, in addition to T. If acted upon by either 5α- or 5β-reductase and 3α-hydroxylase, progesterone is converted into isoforms of the compound allopregnanolone that strongly potentiate GABA-actions on the $GABA_A$-receptor to hyperpolarize target neurons.[74]

We have examined the electrophysiological and behavioral effects of both isoforms of allopregnanolone in the zebra finch. As has been shown in other species, these isoforms do indeed potentiate GABA-induced hyperpolarization of hippocampal and telencephalic neurons, even at low nM concentrations.[75] Moreover, exogenous treatment of adult finches with both isoforms of allopregnanolone induce a rapid state of anesthesia in adult zebra finches with the 5β-3α isoform more potent than the 5α-3α isoform. For example, after 100 μg injections, birds injected with the 5β-isomer began to lose balance on average after 178 s, whereas birds receiving the 5α-3α isomer lost balance after 748 s (Schlinger, unpublished). Interestingly, after injections with 300, 400, 500 and 600 μg of T birds showed no significant loss of balance but birds did lose balance after injections with the same doses of Progesterone. It is likely that this progesterone was converted into allopregnanolone in vivo to exert its anesthetic actions. Additional studies are required to determine the full extent to which these metabolites naturally regulate neurotransmission in avian and non-avian models.

10.6 Conclusions and Future Directions

Songbirds have emerged as extraordinary models to visualize several forms of adult neuroplasticity, most of which have clear functional endpoints. As a consequence, studies of songbird neurobiology and behavior give us added hope that there is a inherent capacity of the vertebrate brain, including the brain of humans, to undergo important transformations in adults that might be utilized naturally, or might be captured and activated clinically, to offset damage the brain might experience from degenerative disease, or from percussive or chemical-induced injury. Studies of songbirds also reveal a striking capacity for steroid metabolism, steroid synthesis and steroid action and these neurochemical adaptations seem related to the observed neuroplasticity. Continued study of songbirds stands to shed even more light on the highly conserved and highly adaptive connection between sex steroids and plasticity of the vertebrate brain.

The songbird has long served as an excellent model towards understanding how the structure and function of the vertebrate brain is molded by steroid hormones. Early work on this model focused upon the actions of gonadal steroids on forebrain structure during development. Most of these studies were aimed at elucidating the mechanisms responsible for sexually dimorphic brain anatomy and physiology.[3] Subsequently, the steroidal modulation of appetitive reproductive behaviors was intensely studied in the songbird,[5] but the source of the steroids in question remained peripheral.

More recent studies showing the abundant and widespread expression of numerous steroidogenic enzymes in the passerine brain have raised important questions as to how locally synthesized steroids may affect the developing and adult brain. The first among these enzymes was aromatase[18,69] and have more recently been followed with StAR, CYP11A1, 3B-HSD and CYP17.[56] The functional role(s) played by neurosteroidogenesis in the songbird span natural cycles in cell proliferation and death and/or alterations in cell recruitment and turnover following brain injury.

The idea that the songbird brain possesses all the requisite enzymes to synthesize the entirety of known gonadal sex steroids is remarkable. To date, this finding[56] has been convincingly demonstrated in the developing zebra finch. This expression may exert a profound influence on the development of the songbird brain. We are beginning to ask similar questions of the adult songbird brain. Given the reliable and robust upregulation of aromatase in glial cells following brain injury,[20,22,33] one may consider the possibility that other neurosteroidogenic enzymes may be expressed under similar conditions. These experiments will undoubtedly continue to hone and clarify our understanding of neurosteroidogenesis in particular and neuroendocrinology in general.

10.7 From Ethology to Clinical Medicine: A Conservative Speculation

Much of the molecular, neuroanatomical, and physiological studies conducted on songbird brain, rely heavily on elegant behavioral observations of native species in their natural habitats followed by rigorous experimental comparative psychology. A notable characteristic in this evolution of understanding remains the resilience of fundamental concepts that are transferable from wild species, through feral populations, and even to domesticated and laboratory-reared animals. It is even reasonable to consider the possibility that scientific concepts that emerge from research on the songbird brain may inform human clinical medicine. As described above, the songbird brain is an enduring research model for neuroplasticity including naturally occurring oscillations in neuron birth, survival, and death. The mechanisms that underlie these dramatic oscillations may be critical in refining our approaches to human clinical medicine. Specifically, much effort is currently focused upon the mitigation of primary and secondary brain degeneration associated with age and pathologies like Alzheimer's disease, Stroke, and Parkinson's disease. Accumulating evidence strongly implicates neurosteroidogenesis in several aspects of natural and pathological neuroplasticity of the songbird brain. It may be possible to continue to reveal important effectors of human neuro-degeneration and perhaps even – regeneration by studying the neuroanatomical, cell-specific, and subcellular localization of steroidogenic enzymes in the passerine brain.

Bibliography

1. Berthold AA. Transplantation der Hoden. Arch Anat Physiol Wissensch Med 1849; 42–46.
2. Marler, P. Three models of song learning:evidence from behavior. J. Neurobiol 1997; 33:501–516.
3. Nottebohm F, Arnold AP. Sexual dimorphism in vocal control areas of the songbird brain. Science 1976; 194:211–213.
4. Brenowitz EA. Comparative approaches to the avian song system. J Neurobiol 1997; 33:517–531.
5. Goodson JL, Saldanha CJ, Hahn TP, Soma KK. Recent advances in behavioral neuroendocrinology: insights from studies on birds. Horm Behav 2005; 48:461–473.

6. Nottebohm F. The road we travelled: discovery, choreography, and significance of brain replaceable neurons. Ann NY Acad Sci 2004; 1016:628–658.
7. Smulders TV, Sasson AD, DeVoogd TJ. Seasonal variation in hippocampal volume in a food-storing bird, the black-capped chickadee. J Neurobiol 1995; 27:15–25.
8. Smulders TV, Shiflett MW, Sperling AJ, DeVoogd TJ. Seasonal changes in neuron numbers in the hippocampal formation of a food-hoarding bird: the black-capped chickadee. J Neurobiol 2000; 44:414–422.
9. Sherry DF, Vaccarino AL, Buckenham K, Herz RS. The hippocampal complex of food-storing birds. Brain Behav Evol 1989; 34:308–317.
10. Patel SN, Clayton NS, Krebs JR. Hippocampal tissue transplants reverse lesion-induced spatial memory deficits in zebra finches (Taeniopygia guttata). J Neurosci 1997; 17:3861–3869.
11. Barnea A, Nottebohm F. Recruitment and replacement of hippocampal neurons in young and adult chickadees: an addition to the theory of hippocampal learning. Proc Natl Acad Sci USA 1996; 93:714–718.
12. Hoshooley JS, Sherry DF. Neuron production, neuron number, and structure size are seasonally stable in the hippocampus of the food-storing black-capped chickadee (Poecile atricapillus). Behav Neurosci 2004; 118:345–355.
13. Gulledge CC, Deviche P. Autoradiographic localization of opioid receptors in vocal control regions of a male passerine bird (Junco hyemalis). J Comp Neurol 1995; 356:408–417.
14. Tramontin AD, Brenowitz EA. Seasonal plasticity in the adult brain. Trends Neurosci 2000; 23:251–258.
15. Tramontin AD, Wingfield JC, Brenowitz EA. Androgens and estrogens induce seasonal-like growth of song nuclei in the adult songbird brain. J Neurobiol 2003; 57:130–14016.
16. Soma KK, Tramontin AD, Featherstone J, Brenowitz EA. Estrogen contributes to seasonal plasticity of the adult avian song control system. J Neurobiol 2004; 58:413–422.
17. DeVoogd T, Nottebohm F. Gonadal hormones induce dendritic growth in the adult avian brain. Science 1981; 214:202–204.
18. Schlinger BA, Arnold AP. Brain is the major site of estrogen synthesis in a male songbird. Proc Natl Acad Sci USA 1991; 88:4191–4194.
19. Saldanha CJ, Popper P, Micevych PE, Schlinger BA. The passerine hippocampus is a site of high aromatase: inter- and intraspecies comparisons. Horm Behav 1998; 34:85–97.
20. Wynne RD, Saldanha CJ. Glial aromatization decreases neural injury in the zebra finch (*Taeniopygia guttata*): influence on apoptosis. Journal of Neuroendocrinology 2004; 16(8): 676–683.
21. Saldanha CJ, Rohmann KN, Coomaralingam L, Wynne RD. Estrogen provision by reactive glia decreases apoptosis in the zebra finch (Taeniopygia guttata). J Neurobiol 2005; 64:192–201.
22. Lee DW, Fernando G, Peterson RS, Allen TA, Schlinger BA. Estrogen mediation of injury-induced cell birth in neuroproliferative regions of the adult zebra finch brain. Dev. Neurobiol. 2007; 67(8):1107–1117.
23. Arvin B, Neville LF, Barone FC, Feuerstein GZ. The role of inflammation and cytokines in brain injury. Neurosci Biobehav Rev 1996; 20:445–452.
24. Jorgensen MB. The role of signal transduction in the delayed necrosis of the hippocampal CA1 pyramidal cells following transient ischemia. Acta Neurol Scand Suppl 1993; 143:1–20.
25. Sato M, Chang E, Igarashi T, Noble LJ. Neuronal injury and loss after traumatic brain injury: time course and regional variability. Brain Res 2001; 917:45–54.
26. Benkovic SA, O'Callaghan JP, Miller DB. Sensitive indicators of injury reveal hippocampal damage in C57BL/6J mice treated with kainic acid in the absence of tonic-clonic seizures. Brain Res 2004; 1024:59–76.
27. Garcia-Ovejero D, Azcoitia I, Doncarlos LL, Melcangi RC, Garcia-Segura LM. Glia-neuron crosstalk in the neuroprotective mechanisms of sex steroid hormones. Brain Res Brain Res Rev 2005; 48:273–286.
28. Pekny M, Nilsson M. Astrocyte activation and reactive gliosis. Glia 2005; 50:427–434.

29. Wise PM, Dubal DB, Rau SW, Brown CM, Suzuki S. Are estrogens protective or risk factors in brain injury and neurodegeneration? Reevaluation after the women's health initiative. Endocr Rev 2005; 26:308–312.
30. Benkovic SA, O'Callaghan JP, Miller DB. Regional neuropathology following kainic acid intoxication in adult and aged C57BL/6J mice. Brain Res 2006; 1070:215–231.
31. Liberto CM, Albrecht PJ, Herx LM, Yong VW, Levison SW. Pro-regenerative properties of cytokine-activated astrocytes. J Neurochem 2004; 89:1092–1100.
32. Mrak RE, Griffin WS. Glia and their cytokines in progression of neurodegeneration. Neurobiol Aging 2005; 26:349–354.
33. Peterson RS, Saldanha CJ, Schlinger BA. Rapid upregulation of aromatase mRNA and protein following neural injury in the zebra finch (Taeniopygia guttata). J Neuroendocrinol 2001; 13:317–323.
34. Azcoitia I, Sierra A, Veiga S, Garcia-Segura LM. Aromatase expression by reactive astroglia is neuroprotective. Ann NY Acad Sci 2003; 1007:298–305.
35. Garcia-Segura LM, Veiga S, Sierra A, Melcangi RC, Azcoitia I. Aromatase: a neuroprotective enzyme. Prog Neurobiol 2003; 71:31–41.
36. Peterson RS, Lee DW, Fernando G, Schlinger BA. Radial glia express aromatase in the injured zebra finch brain. J Comp Neurol 2004; 475:261–269.
37. Soma KK, Schlinger BA, Wingfield JC, Saldanha CJ. Brain aromatase, 5 alpha-reductase, and 5 beta-reductase change seasonally in wild male song sparrows: Relationship to aggressive and sexual behavior. J Neurobiol 2003; 56:209–221.
38. Migeon CJ, Keller AR, Lawrence B, Shepart TH. Dehydroepiandrosterone and androsterone in human plasma: effects of age and sex: day-to-day and diurnal variations. J Clin Endocr Metab 1957; 17:1051–1062.
39. Odell WD, Parker LN. Control of adrenal androgen production. Endocr Res 1984; 10:617–630.
40. Labrie F, Bélanger A, Simard J, Van L-T, Labrie C. DHEA and peripheral androgen and estrogen formation: intracinology. Ann NY Acad Sci 1995; 774:16–28.
41. Soma KK, Wingfield JC. Dehydroepiandrosterone in songbird plasma: seasonal regulation and relationship to territorial aggression. Gen Comp Endocrinol 2001; 123:144–155.
42. Hau M, Stoddard ST, Soma KK. Territorial aggression and hormones during the non-breeding season in a tropical bird. Horm Behav 2004; 45:40–49.
43. Soma KK, Wissman AM, Brenowitz EA, Wingfield JC. Dehydroepiandrosterone (DHEA) increases territorial song and the size of an associated brain region in a male songbird. Horm Behav 2002; 41:203–212.
44. Alonso-Alvarez C, Bertrand S, Sorci G. Energetic reserves, leptin and testosterone: a refinement of the immunocompetence handicap hypothesis. Biol. Lett. 2007; 3(3):271–274.
45. Soma KK, Tramontin AD, Wingfield JC. Oestrogen regulates male aggression in the non-breeding season. Proc R Soc Lond B Biol Sci 2000; 267:1089–1096.
46. Soma KK, Sullivan KA, Tramontin AD, Saldanha CJ, Schlinger BA, Wingfield JC. Acute and chronic effects of an aromatase inhibitor on territorial aggression in breeding and nonbreeding male song sparrows. J Comp Phys A 2000; 186:759–769.
47. Payne AH, Hales DB. Overview of steroidogenic enzymes in the pathway from cholesterol to active steroid hormones. Endocr Revs 2004; 25:947–970.
48. Vanson A, Arnold AP, Schlinger BA. 3B-Hydroxysteroid dehydrogenase/Isomerase and Aromatase Activity in Primary Cultures of Developing Zebra Finch Telencephalon: Dehydroepiandrosterone as Substrate for Synthesis of Androstenedione and Estrogens. Gen Comp Endocrinol 1996; 102:342–350.
49. Schlinger BA, Amur-Umarjee S, Shen P, Campagnoni AT, Arnold AP. Neuronal and non-neuronal aromatase in primary cultures of developing zebra finch telencephalon. J Neurosci 1994; 14:7541–7552.
50. Wade J, Schlinger BA Arnold AP. Aromatase and 5-reductase activity in cultures of developing zebra finch brain: an Investigation of sex and regional differences. J Neurobiol 1995; 27:240–251.

51. London S, Monks DA, Wade J, Schlinger BA. Widespread capacity for steroid synthesis in the avian brain and song system. Endocrinology 2006; 147:5975–5987.
52. Tam H, Schlinger BA. Activities of 3B-HSD and aromatase in slices of the adult and developing zebra finch brain. Gen Comp Endocrinol 2007; 150:26–33.
53. Schlinger BA, Lane NI, Grisham W, Thompson L. Androgen synthesis in a songbird: a study of Cyp17 (17 alpha-hydroxylase/C17,20-lyase) activity in the zebra finch. Gen Comp Endocrinol 1999; 113:46–58.
54. Wingfield, JC Silverin B. Ecophysiological studies of hormone-behavior elations in birds. In: Pfaff DW, Arnold AP, Etgen AM, Fahrbach SE, Rubin RT (eds). Hormones, brain and behavior, Vol. 2.. San Diego, CA: Academic, 2002, pp. 587–647..
55. Miller WL, Achus RJ, Geller D. The regulation of 17,20 lyase activity. Steroids 1997; 62(1):133–142.
56. London S, Schlinger BA. Steroidogenic enzymes along the ventricular proliferative zone in the developing songbird brain. J Comp Neurol 2007; 502:507–521.
57. London SE, Boulter J, Schlinger BA. Cloning of the zebra finch androgen synthetic enzyme CYP17: a study of its neural expression throughout posthatch development. J Comp Neurol 2003; 467:496–508.
58. DeWulf V, Bottjer SW. Age and sex differences in mitotic activity within the zebra finch telencephalon. J Neurosci 2002; 22:4080–4094.
59. DeWulf V, Bottjer SW. Neurogenesis within the juvenile zebra finch telencephalic ventricular zone: a map of proliferative activity. J Comp Neurol 2005; 481:70–83.
60. Schlinger BA. Sexual differentiation of avian brain and behavior: current views on gonadal hormone-dependent and independent mechanisms. Ann Rev Physiol 1998; 60:407–429.
61. Holloway CC, Clayton DE. Estrogen synthesis in the male brain triggers development of the avian song control pathway in vitro. Nat Neurosci. 2001; 4:170–175
62. Oberlander JG, Schlinger BA, Clayton NS, Saldanha CJ. Neural aromatization accelerates the acquisition of spatial memory via an influence on the songbird hippocampus. Horm Behav 2004; 45:250–258.
63. Baulieu EE. Neurosteroids: of the nervous system, by the nervous system, for the nervous system. Rec Prog Horm Res 1997; 52:1–32.
64. Schlinger BA, Callard GV. Localization of aromatase in synaptosomal and microsomal subfractions of quail (Coturnix coturnix japonica) brain. Neuroendocrinol 1989; 49:434–441.
65. Naftolin F, Horvath TL, Jakab RL, Leranth C, Harada N, Balthazart J. Aromatase immunoreactivity in axon terminals of the vertebrate brain: an immunocytochemical study on quail, rat, monkey and human tissues. Neuroendocrinol 1996; 63:149–155.
66. Peterson RS, Yarram L, Schlinger BA, Saldanha CJ. Aromatase is presynaptic and sexually-Dimorphic in the adult zebra finch brain. Proc Roy Soc Lond B 2005; 272:2089–2096.
67. Rohman KN, Schlinger BA, Saldanha CJ. The subcellular compartmentalization of aromatase is sexually dimorphic in the adult zebra finch brain. Dev. Neurobiol 2006; 67:1–9.
68. Gould E, Woolley CS, Frankfurt M, McEwen BS. Gonadal steroids regulate dendritic spine density in hippocampal pyramidal cells in adulthood. J Neurosci 1990; 10:1286–1291.
69. Vockel A, Pröve E, Balthazart J. Sex- and age-related differences in the activity of testosterone-metabolizing enzymes in microdissected nuclei of the zebra finch brain. Brain Res 1990; 511:291–302.
70. Schlinger BA, Amur Umarjee S, Campagnoni AT, Arnold AP. 5 beta-reductase and other androgen-metabolizing enzymes in primary cultures of developing zebra finch telencephalon. J Neuroendocrinol 1995; 7:187–192.
71. Soma KK, Bindra RK, Gee J, Wingfield JC, Schlinger BA. Androgen-metabolizing enzymes show region-specific changes across the breeding season in the brain of a wild songbird. J Neurobiol 1999; 41:176–188.
72. Martini L. The 5alpha-reduction of testosterone in the neuroendocrine structures. Biochemical and biophysical implications. Endocr Revs 1982; 3:1–25.
73. Hutchison JB, Steimer T. Brain 5beta-reductase: a correlate of behavioral sensitivity to androgen. Science 1981; 213:244–246.

74. Majewska MD. Neurosteroids: endogenous bimodal modulators of the GABAA receptor. Mechanism of action and physiological significance. Prog Neurobio 1992; 38:379–395.
75. Carlisle HJ, Hales TG, Schlinger BA. Characterization of neuronal zebra finch GABA(A) receptors: steroid effects. J Comp Physiol A 1998; 182:531–538.

Chapter 11
Dehydroepiandrosterone, as Endogenous Inhibitor of Neuronal Cell Apoptosis: Potential Therapeutic Implications in Neurodegenerative Diseases

Ioannis Charalampopoulos[1], Christos Tsatsanis[2], Andrew N. Margioris[2], Elias Castanas[3], and Achille Gravanis[1]

Abstract Dehydroepiandrosterone (DHEA) is synthesized in the adrenals and the brain. Additionally, DHEA is produced at high concentrations in the human embryo enhancing neuronal development. Its production rate and levels in serum, brain and adrenals decrease gradually with advancing age. This decline was associated to age-related neuronal dysfunction and degeneration, suggesting a neuroprotective effect of endogenous DHEA against noxious agents. This hypothesis is substantiated by experimental findings showing that DHEA protect neural crest derived cells against serum deprivation-induced apoptosis with EC_{50} 1.8 nM. This potent antiapoptotic effect of DHEA is mediated by G-protein coupled specific membrane binding sites, the subsequent activation of prosurvival kinases Src and PKC and transcription factors CREB and NF-κB, upstream effectors of the antiapoptotic Bcl-2 proteins. These findings suggest that DHEA may act as an endogenous neuroprotective factor, during development and in adulthood. The decline of DHEA levels during ageing may leave the brain unprotected against neurotoxic challenges. The DHEA specific membrane binding sites conferring neuroprotection offer a new target for developing synthetic DHEA analogs with antiapoptotic and neuroprotective properties, deprived of endocrine toxicity.

Keywords DHEA, apoptosis, neurons, neuroprotection

Abbreviations DHEA dehydroepiandrosterone; DEX dexamethasone; DHT dehydrotestosterone; PTX pertussis toxin

[1] Departments of Pharmacology, School of Medicine, University of Crete, Heraklion

[2] Departments of Clinical Chemistry, School of Medicine, University of Crete, Heraklion

[3] Departments of Experimental Endocrinology, School of Medicine, University of Crete, Heraklion

M.S. Ritsner and A. Weizman (eds.), *Neuroactive Steroids in Brain Function, Behavior and Neuropsychiatric Disorders*

11.1 Introduction

Dehydroepiandrosterone (DHEA) and its sulfate ester DHEAS are mainly produced in the zona reticularis of human adrenal cortex. Adrenal secretion of DHEA(S) increases during adrenarche. Maximal values of circulating DHEA(S) are reached between the ages of 20 and 30 years, thereafter, their levels decrease markedly[1,2] and at 70 years of age, they are at approximately 20% of their peak values. Furthermore, stressful conditions, such as major depression, chronic psychological stress or chronic inflammatory diseases result to decreased levels of adrenal DHEA(S).[3,4] The decline of DHEA and DHEAS levels during ageing was associated to neuronal dysfunction and degeneration,[5–7] most probably because these steroids protect CNS neurons against noxious agents.[8–11] Indeed, DHEA protects rat hippocampal neurons against NMDA-induced excitotoxicity.[11] The decline of brain concentrations of these neurosteroids was also associated to age-related neurodegenerative conditions. Indeed, recent studies investigating the physiopathological significance of neurosteroids in Alzheimer's disease (AD), have shown a significant decline of neurosteroid concentrations in individual brain regions of AD patients compared to aged nondemented controls.[12,13] DHEAS was significantly lower in the striatum, cerebellum, and the hypothalamus. Additionally, a significant negative correlation was found between the levels of phosphorylated tau proteins and DHEAS in the hypothalamus.[12] These studies suggest a possible endogenous neuroprotective role of these neurosteroids in AD. It is also of interest that DHEA is able to potentiate locomotor activity of hemiparkinsonian monkeys, improving symptomatic treatment of the moderately and severely impaired MPTP animals.[14,15]

11.2 Dehydroepiandrosterone Protects Neural Crest-Derived Cells Against Apoptosis, via the Anti-Apoptotic Bcl-2 Proteins

We have recently shown that neurosteroids, such as DHEA and DHEAS protect against apoptosis the neural crest derived PC12 cells. Indeed, DHEA protect, in a time- and dose-dependent fashion, PC12 cells from serum deprivation induced apoptosis, with EC_{50} at 1.8 nM.[16] Structure–activity relation (SAR) analysis of the anti-apoptotic effects of DHEA structurally related androstenes showed that conformations 3α-OH, 3-keto, Δ^4 (double bond at C4–C5) are inactive. Thus, the Δ^4–3-keto steroids including testosterone, progesterone, corticosterone, and 4-androsten-3β-ol-17-one, 5-androsten-3α-ol-17-one, and 5-androsten-3,17-dione do not have any anti-apoptotic activity in serum-starved cells. Hydroxylation at C7 (7α-hydroxy-DHEA, 7β-hydroxy-DHEA) or at C17 (hermaphrodiol) resulted in loss of anti-apoptotic activity.

The protective effect of DHEA use the major prosurvival pathway in sympathetic neurons, the anti-apoptotic Bcl-2 proteins. Indeed, DHEA induce the expression of the anti-apoptotic Bcl-2 and Bcl-xL proteins.[16] The role of these proteins appears to be crucial since inhibition of their production by antisense oligonucleotides (directed towards the translation initiation site of the Bcl-2 transcript) resulted in an almost complete abolition of the protective effect of DHEA. DHEA may protect neural-crest derived cells, by closely controlling the expression of anti-apoptotic Bcl-2 proteins, both at transcriptional and post-translational levels, activating the prosurvival transcription factors CREB and NF-κB, as well as the PKCαβ kinase (Fig. 11.1). Indeed, the promoter regions of anti-apoptotic Bcl-2 and Bcl-xL genes contain the cAMP-response element (CRE) and the NF-κB sensitive motif and transcription factors CREB and NF-κB[17,18] have been identified as positive regulators of Bcl-2 and Bcl-xL gene expression, been involved in neuroprotective and survival mechanisms of central and peripheral neurons. DHEA rapidly within 30 min induces the activation and translocation of NF-κB into the nucleus of serum-deprived PC12 cells and affect the phosphorylation/activation of CREB protein. Phosphorylation of Bcl-2 at serine 70 is required for its anti-apoptotic function.[19] It is now well documented in various biological systems that phosphorylation of

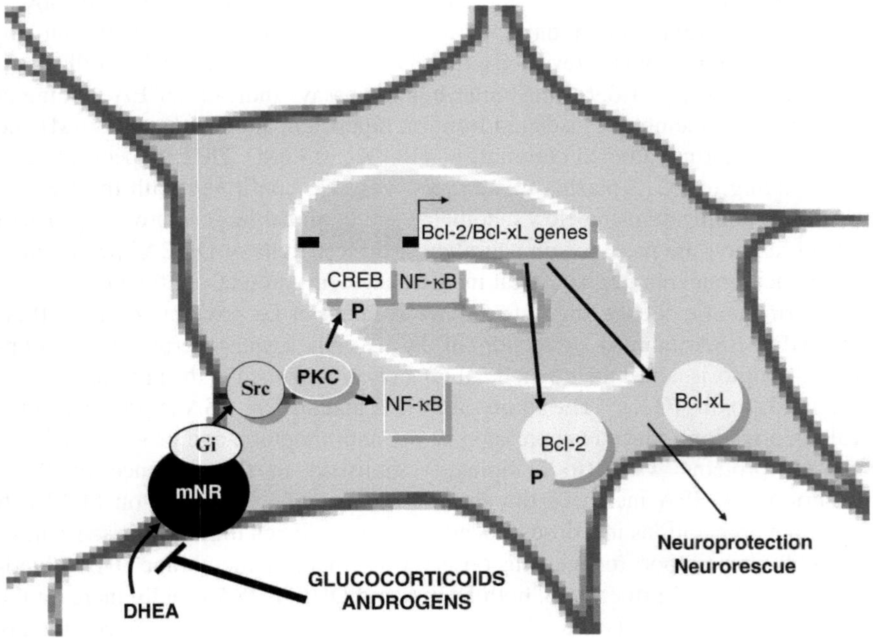

Fig. 11.1 Mechanism of the neuroprotective actions of DHEA. Dehydroepiandrosterone protects neural-crest derived cells against apoptosis, by tightly controlling the expression of anti-apoptotic Bcl-2 proteins, both at transcriptional and post-translational level. They bind on Gi protein associated membrane binding sites (mNR), downstream activating the prosurvival Src-PKC kinases, which then activate the prosurvival transcription factors CREB and NF-κB, and stimulate the expression of anti-apoptotic Bcl-2 proteins. Glucocorticoids and androgens act as endogenous antagonists of the DHEA actions, via competition on the mNR sites. These specific membrane binding sites offer a new target for developing antiapoptotic, neuroprotective agents, synthetic analogs of DHEA, deprived of their endocrine properties, thus with limited endocrine toxicity

Bcl-2 is afforded by αβ forms of PKC. DHEA induces also post-translational activation of antiapoptotic Bcl-2 proteins, by increasing within 10 min the levels of phosphorylated PKCαβ.[16]

11.3 The Neuroprotective Effect of DHEA is Mediated via G-Protein-Coupled Membrane-Binding Sites

The rapid onset of these actions supports the hypothesis that DHEA may utilize a membrane receptor system, although an intracellular receptor cannot be excluded. Membrane impermeable DHEA conjugated to bovine serum albumin (BSA), a molecule with no intracellular penetrance abilities, can protect PC12 cells against serum deprivation-induced apoptosis with an apparent IC_{50} of 1.5 nM, in a manner similar to that of unconjugated DHEA (1.8 nM), strongly suggesting the involvement of specific membrane binding sites.[20] Furthermore, DHEA-BSA effectively mimicked DHEA actions on anti-apoptotic Bcl-2 proteins, by preventing its down-regulation by serum-deprivation. Saturation binding assays of [^{3}H]DHEA on isolated PC12 cell membranes revealed a rapidly saturable (30 min) binding of DHEA, with an apparent K_D of 0.9 nM. Similar binding assays indicate DHEA binding at high affinity on membranes isolated from rat hippocampal cells (K_D: 61.9 nM) and from human normal adrenal chromaffin cells (K_D: 0.1 nM). DHEA specific membrane binding to PC12 plasma membranes was also confirmed with the DHEA-BSA-FITC conjugate using flow cytometry and confocal laser microscopy. These findings suggest the presence on neural crest-derived cells of DHEA-specific high affinity membrane binding sites that mediate the neuroprotective effect of DHEA.

Experimental evidence suggests the involvement of Gi proteins in the DHEA- and DHEA-BSA-induced protection of PC12 cells against serum deprivation-induced apoptosis, since their beneficial effect was abolished in the presence of 10^{-6} M pertussis toxin (PTX).[20] The ability of DHEA and DHEA-BSA to protect against serum deprivation-induced suppression of the anti-apoptotic and pro-survival Bcl-2/Bcl-xL proteins was also completely abolished in the presence of PTX. Furthermore, DHEA increased the specific binding of [^{35}S]GTPγS on PC12 cell membrane preparations in a dose-dependent manner. Taken together, these findings provide further support for previous observations that link membrane DHEA binding to Gi proteins. Furthermore, both DHEA and DHEA-BSA rapidly increase the phosphorylation of Src (within 5 min of exposure), an effect that was completely reversed by the Gi inhibitor PTX.[20] It should be noted here that activation of the Src-PKC pathway induces NF-κB activity and PC12 cell survival.[21] These observations considered together, suggest that DHEA may exert its protective effects, activating G protein associated membrane binding sites, and the subsequent activation of prosurvival Src-PKC kinases, leading finally to mobilization of transcription factor NF-κB and the production of anti-apoptotic Bcl-2 proteins (Fig. 11.1).

Several structurally related steroids including the synthetic estrogen diethylstilbestrol, progestin ORG2058, and the pregnane allopregnanolone are unable to compete with [^{3}H]DHEA for membrane binding at concentrations ranging from 1 pM to 1 μM. On the other hand, the DHEA sulfate ester (DHEAS) is an efficient competitor with an IC_{50} of 1.3 nM, i.e., similar to that of DHEA (1.5 nM) and to the K_D of DHEA binding (0.9 nM). Interestingly, glucocorticoids and androgens possess a 10- to 15-fold lower affinity for the DHEA membrane binding sites. However, these steroids completely lack protective anti-apoptotic effects on PC12 cells.[20] These observations support the hypothesis that androgens and glucocorticoids may act as endogenous antagonists of DHEA. Indeed, both DEX or DHT completely reverses the protective effects of DHEA, as well as its stimulatory effect on the anti-apoptotic Bcl-2 proteins and pro-survival Src activation.[20] It is thus possible that glucocorticoids and androgens act as antagonists of DHEA by binding to common membrane binding sites.

Glucocorticoids and testosterone can exert neurotoxic effects in a range of tissues.[22–24] Indeed, chronic exposure to glucocorticoids has been associated with decreased cognitive performance, attenuation of synaptic efficacy and neuronal atrophy. Elevation of glucocorticoids during ageing is also associated with cognitive impairment and hippocampal atrophy. Furthermore, earlier animal experiments have shown that overexposure to glucocorticoids during prolonged periods of stress is detrimental to central nervous system neurons, especially in aged animals, affecting mainly the hippocampus. Cumulative exposure to corticosteroids or to chronic stress, with its ensuing increase in corticosteroid levels, caused degenerative loss of pyramidal neurons in the hippocampus, and reduced cell numbers in CA1 and CA3 and subsequent deficits in memory function and cognition in rats. Meanwhile, the greater tendency to nigrostriatal dopaminergic neurotoxicity and neurodegeneration in Parkinson's disease observed in males[25] may be partially attributed to the neurotoxic effect of testosterone. In recent studies, testosterone has been shown to possess a deleterious effect on ischemic stroke in a focal ischemia model, whereas acute testosterone depletion exerts a neuroprotective effect, suggesting that testosterone could also contribute to gender differences in the outcome of stroke.[26] These findings considered together suggest that part of neurotoxic effects of glucocorticoids and testosterone could be attributed to their antagonistic effect on the neuroprotective effect of endogenous DHEA. The decline of brain DHEA levels during ageing and in Alzheimer's disease[1–4] might exacerbate this phenomenon, rendering neurons more vulnerable to glucocorticoid and androgen toxicity. Furthermore, glucocorticoid neurotoxicity becomes more pronounced in aged subjects since cortisol levels in the CSF increase in the course of normal ageing, as well as in relatively early stages of Alzheimer's disease.[27,28]

11.4 DHEA in Neurodegenerative Processes: Potential Therapeutic Implications of DHEA Analogs

Neurodegeneration, the progressive loss of nerve cells, occurs in aging and in neurodegenerative disorders, such as Alzheimer's, Parkinson's, Amyotrophic lateral Sclerosis (ALS), Multiple Sclerosis and Huntington's disease, and in stroke, head and spinal trauma.[29,30] Primarily, these diseases are characterized by chronic and progressive loss of neurons in discrete areas of the brain or the peripheral nerves, causing dementia, loss of memory, loss of sensory or motor capability, decreased overall quality of life and well-being, disability, and eventually, premature death. For most neurodegenerative diseases, currently there is little or no treatment; at best, treatments are symptomatic in nature and do not prevent or slow the progression of disease.[31] A crucial approach of controlling neurodegenerative diseases is thus to slow or halt progression, and ultimately, to prevent the onset of the disease process. Pharmacological manipulations for neurorescue, neurorepair, or neuroprotection may also offer new therapeutic interventions.

Neuronal cell death by apoptosis is the "end-point" of many human neurological disorders, including Alzheimer's, Parkinson's and Huntington's diseases, stroke/ trauma, multiple and amyotrophic lateral sclerosis.[29,30] Indeed, apoptotic death of hippocampal and cortical neurons is responsible for the symptoms of Alzheimer's disease; death of midbrain neurons that use the neurotransmitter dopamine underlies Parkinson's disease; Huntington's disease involves the death of neurons in the striatum, which control body movements; and death of lower motor neurons manifests as amyotrophic lateral sclerosis. Additionally, brain ischemia and trauma induce necrosis of a small brain area, which then propagates neuronal cell loss by apoptosis to a larger brain area, due to the neurotoxic material released by the necrotic cells. Apoptotic neuronal cell loss is also observed in the ageing brain, as a physiological process.[29] With the identification of mechanisms that prevent neuronal apoptosis come new approaches for preventing and treating neurodegenerative disorders.

Neurons do not divide. Therefore, considerable pressure was placed on the nervous system during evolution to develop mechanisms that guard against neuronal death. The marked symptoms of neurodegenerative disorders emphasize the importance of mechanisms that promote neuron survival and plasticity. There are several prominent anti-apoptotic signalling pathways. Neurotrophic factors have been identified that can protect neurons against apoptosis by activating receptors linked through kinase cascades to production of cell-survival-promoting proteins.[29,30] For example, brain-derived neurotrophic factor (BDNF), nerve growth factor (NGF) and basic fibroblast growth factor (bFGF) can prevent death of cultured neurons, in part by stimulating production of anti-apoptotic, prosurvival proteins, such the Bcl-2 family members. Cytokines such as tumour necrosis factor-a (TNFa), ciliary neurotrophic factor (CNTF) and leukaemia inhibitory factor (LIF) can prevent neuronal death in experimental models of natural neuronal death and neurodegenerative disorders. Several antiapoptotic neurotrophic factors and cytokines use survival signaling pathways involving the PI3K/Akt kinases, downstream activating prosurvival transcription factors NF-κB, and CREB.

We have recently shown that DHEA protect neural crest derived cells against apoptosis at low, nanomolar concentrations (1 nM), activating within minutes the pro-survival factors NF-κB and CREB, two upstream effectors of anti-apoptotic Bcl-2 proteins.[16] Furthermore, DHEA at nanomolar concentrations stimulate acutely (10 min) the secretion of catecholamines, via induction of the depolymerization and disassembly of submembrane actin cytoskeleton.[32] The neuroprotective effects of DHEA are mediated by specific G-protein associated membrane binding sites, independent of NMDA or $GABA_A$ receptors.[20] These specific membrane binding sites offer a new target for developing antiapoptotic, neuroprotective agents, synthetic analogs of DHEA, deprived of their endocrine properties, thus with limited endocrine toxicity. Our group is now testing a large chemical library of synthetic DHEA analogs for their anti-apoptotic, neuroprotective effects using in vitro neuronal cell cultures, challenged with neurotoxic agents and growth factor deprivation. A number of highly effective compounds are synthesized, which protect neuronal cells against neurotrophin deprivation induced apoptosis, with EC_{50} in the range of 0.1–1.0 nM. These compounds bind with high affinity to recently described by our group membrane DHEA binding sites (K_D at nM concentration), and mimic endogenous neurosteroids in inducing prosurvival anti-apoptotic Bcl-2 proteins. Their prosurvival neuroprotective actions are structure-specific, confined mainly to conformation substitution 3α-OH-Δ.[5] They lack the classical androgenic or estrogenic effects, since they are not subjective to metabolism by the enzymes 3β- and 17β-hydroxysteroid dehydrogenases. Our findings suggest that these chemical entities might prove lead molecules for the synthesis of novel neuroprotective agents.

11.5 Conclusion

Our findings suggest that DHEA may act as an endogenous neuroprotective factor, during development and in adulthood. The decline of DHEA levels during ageing may leave the brain unprotected against neurotoxic challenges. The DHEA specific membrane binding sites conferring neuroprotection offer a new target for developing synthetic DHEA analogs with antiapoptotic and neuroprotective properties, deprived of endocrine toxicity.

References

1. Orentreich N, Brind JL, Rizer RL, et al. Age changes and sex differences in serum dehydroepiandrosterone sulfate concentrations throughout adulthood. J Clin Endocrinol Metab 1984; 59:551–555.
2. Be'langer A, Candas B, Dupont A, et al. Changes in serum concentrations of conjugated and unconjugated steroids in 40- to 80-year old men. J Clin Endocrinol Metab 1994; 79:1086–1090.
3. Wolkowitz OM, Epel, ES, Reus VI. Stress hormone-related psychopathology: pathophysiological and treatment implications. World J Biol Psychiatry 2001; 2:115–143.
4. Straub RH, Lehle K, Herfarth H, Weber, et al. Dehydroepiandrosterone in relation to other adrenal hormones during an acute inflammatory stressful disease state compared with chronic inflammatory disease: role of interleukin-6 and tumour necrosis factor. Eur J Endocrinol 2002; 146:365–374.
5. Sapolsky RM. Stress, the aging brain and the mechanism of neuron death. Cambridge, MA: MIT, 1992.

6. Seals DR, Esler M. Human ageing and the sympathoadrenal system. J Physiol 2000; 528:407–417.
7. Esler M, Lambert G, Kaye D, et al. Influence of ageing on the sympathetic nervous system and adrenal medulla at rest and during stress. Biogerontology 2002; 3:45–49.
8. Bastianetto S, Ramassamy C, Poirier, et al. Dehydroepiandrosterone (DHEA) protects hippocampal cells from oxidative stress-induced damage. Brain Res Mol Brain Res 1999; 66:35–41.
9. Cardounel A, Regelson W, Kalimi M. Dehydroepiandrosterone protects hippocampal neurons against neurotoxin-induced cell death: mechanism of action. Proc Soc Exp Biol Med 1999; 22:145–149.
10. Lapchak PA, Chapman DF, Nunez SY, et al. Dehydroepiandrosterone sulfate is neuroprotective in a reversible spinal cord ischemia model: possible involvement of GABA(A) receptors. Stroke 2000; 31:1953–1956.
11. Kimonides VG, Khatibi NH, Svendsen CN, et al. Dehydroepiandrosterone (DHEA) and DHEA-sulfate (DHEAS) protect hippocampal neurons against excitatory amino acid-induced neurotoxicity. Proc Natl Acad Sci USA 1998; 95:1852–1857.
12. Weill-Engerer S, David JP, Sazdovitch V, et al. Neurosteroid quantification in human brain regions: comparison between Alzheimer's and nondemented patients. J Clin Endocrinol Metab 2002; 87:5138–5143.
13. Schumacher M, Weill-Engerer S, Liere P, et al. Steroid hormones and neurosteroids in normal and pathological aging of the nervous system. Prog Neurobiol 2003; 71:3–29.
14. Belanger N, Gregoire L, Bedard P, et al. Estradiol and dehydroepiandrosterone potentiate levodopa-induced locomotor activity in 1-methyl-4-phenyl-1,2,3,6-tetrahydropyridine monkeys. Endocrine 2003; 1:97–101.
15. Belanger N, Gregoire L, Bedard PJ, et al. DHEA improves symptomatic treatment of moderately and severely impaired MPTP monkeys. Neurobiol Aging 2006; 11:1684–1693.
16. Charalampopoulos I, Tsatsanis C, Dermitzaki E, et al. Dehydroepiandrosterone and allopregnanolone protect sympathoadrenal cells against apoptosis, via Bcl-2 antiapoptotic proteins. Proc Natl Acad Sci USA 2004; 101:8209–8214.
17. Riccio A, Ahn S, Davenport CM, et al. Mediation by a CREB family transcription factor of NGF-dependent survival of sympathetic neurons. Science 1999; 286:2358–2361.
18. Tamatani M, Che YH, Matsuzaki H, et al. Tumor necrosis factor induces Bcl-2 and Bcl-x expression through NFkappaB activation in primary hippocampal neurons. J Biol Chem 1999; 274:8531–8538.
19. Ruvolo P, Deng X, Carr B, et al. A functional role for mitochondrial protein kinase Calpha in Bcl2 phosphorylation and suppression of apoptosis. J Biol Chem 1998; 273:25436–25442.
20. Charalampopoulos I, Alexaki VI, Lazaridis I, et al. G protein-associated, specific membrane binding sites mediate the neuroprotective effect of Dehydroepiandrosterone. FASEB J 2006; 20:577–579.
21. Wooten M, Seibenhener ML, Neidich K, et al. Mapping of atypical protein kinase C within the nerve growth factor signaling cascade: relationship to differentiation and survival of PC12 cells. Mol Cell Biol 2000; 20:4494–4504.
22. Swaab DF, Bao AM, Lucassen PJ. The stress system in the human brain in depression and neurodegeneration. Ageing Res Rev 2005; 2:141–194.
23. Sapolsky RM. Glucocorticoid toxicity in the hippocampus: reversal by supplementation with brain fuels. J Neurosci 1986; 6:2240–2244.
24. Sapolsky RM, Krey LC, McEwen BC. The neuroendocrinology of stress and aging: the glucocorticoid cascade hypothesis. Endocrine Revs 1986; 7:284–301.
25. Marder K, Tang MX, Mejia H, et al. Risk of Parkinson's disease among first-degree relatives: a community-based study. Neurology 1996; 47:155–160.
26. Yang SH, Perez E, Cutright J, et al. Testosterone increases neurotoxicity of glutamate in vitro and ischemia-reperfusion injury in an animal model. J Appl Physiol 2002; 1:195–201.
27. Swaab DF, Raadsheer FC, Endert E, et al. Increased cortisol levels in aging and Alzheimer's disease in postmortem cerebrospinal fluid. J Neuroendocrinol 1994; 6:681–687.

28. Umegaki H, Ikari H, Nakahata H, et al. Plasma cortisol levels in elderly female subjects with Alzheimer's disease: a cross-sectional and longitudinal study. Brain Res 2000; 881:241–243.
29. Mattson M. Apoptosis in neurodegenerative disorders. Nature Rev Mol Cell Biol 2000; 1:120–129.
30. Kratnic S, Mechawar N, Reix S, et al. Molecular basis of programmed cell death involved in neurodegeneration. Trends Neurosci 2006; 28:670–676.
31. Forman MS, Trojanowski J, Lee V. Neurodegenerative diseases: a decade of discoveries paves the way for therapeutic breakthroughs. Nature Med 2004; 10:1055–1063.
32. Charalampopoulos I, Dermitzaki E, Vardouli L, et al. Dehydroepiandrosterone and allopregnanolone directly stimulate catecholamine production via induction of tyrosine hydroxylase and secretion by affecting actin polymerization. Endocrinology 2005; 146:3309–3318.

Chapter 12
DHEA and DHEA-S, and their Functions in the Brain and Adrenal Medulla

Alexander W. Krug, Christian G. Ziegler, and Stefan R. Bornstein

Abstract Produced by the adrenal glands and the brain, neuroactive steroids dehydroepiandrosterone and its sulfate ester – DHEA and DHEA-S – are the most abundant hormones in the human body. DHEA and DHEA-S are known to exert major physiological and pathological effects on cognitive functions and memory. Interestingly, DHEA and DHEA-S levels gradually decline during aging; this decrease has been associated with neuronal degeneration processes and dysfunction. Moreover, DHEA has recently been identified as a regulator in neuronal stem cell proliferation. In animal models, DHEA has shown protective properties against a variety of diseases including obesity, diabetes, immune disorders, atherosclerosis and cancer. There is some evidence of the beneficial effects of DHEA in patients with adrenal insufficiency.

However, the exact effects of DHEA and DHEA-S on brain and adrenomedullary function and the signaling pathways mediating these effects are not fully understood. Recent studies have provided evidence that DHEA can act via G-protein-associated, membrane-bound receptors, and that DHEA acts as a gamma-aminobutyric acid type A ($GABA_A$) antagonist. Moreover, these steroids are able to potentiate glutamate action in neuronal cells. Additional studies are needed to develop a more complete picture of DHEA and DHEA-S functions in human physiology and pathology.

Keywords Dehydroepiadrosterone, neuroactive steroids, brain and adrenomedullary function

Abbreviations ACTH adrenocorticotropic hormone; Allo allopregnanolone; AR androgen receptor; DHEA dehydroepiandrosterone; DHEA-S dehydroepiandrosterone sulfate ester; EGF epidermal growth factor; ER α/β estrogen receptors α/β; $GABA_A$ gamma-aminobutyric acid type A; HST sulfotransferase; KA kainic acid; LIF leukemia inhibitory factor; MAPK ERK1/2 mitogen activated protein kinases ERK1/2; MPTP 1-methyl-4-phenyl-1,2,3,6-tetrahydropyridine; NMDA *N*-methyl-D-aspartate

Carl Gustav Carus University Hospital, Department of Medicine III, 01307 Dresden, Germany

M.S. Ritsner and A. Weizman (eds.), *Neuroactive Steroids in Brain Function, Behavior and Neuropsychiatric Disorders*

NGF nerve growth factor; PREG pregnenolone; PS pregnenolone sulfate; PROG progesterone; P450c17 17α-hydroxylase/c17, 20 lyase; SULT2A1 DHEA-S sulfatase 1; SNAP-25 synaptosome-associated protein of 25 kD; STS sulfohydrolase; VAMP-2/synaptobrevin vesicle-associated membrane protein-2; VSMC vascular smooth muscle cells; 17β-HSD 17β-hydroxysteroid dehydrogenase; 3β-HSD 3β-hydroxysteroid dehydrogenase-isomerase

12.1 General Aspects on DHEA and DHEA-S

DHEA and its sulfate ester, DHEA-S, are produced in the inner layer of the adrenal cortex, the *zona reticularis*. This zone is tightly intermingled with adrenomedullary chromaffin cells, enabling strong paracrine interactions. These interactions may play an important role during early adrenogenesis, but also in the adult adrenal gland and in states of cancerogenesis in this tissue. Furthermore, as the most abundant hormone in the human body, DHEA-S has been shown to have neuroprotective and beneficial effects in various disease states, and also can stimulate neuronal stem cell proliferation.[1–3] In this regard, DHEA has been shown to be antiapoptotic on sympathoadrenal medulla cells via Bcl-2 proteins. Furthermore, the pro-survival effect of DHEA-S and allopregnanolone (Allo) was independent of *N*-methyl-D-aspartate, GABAA, sigma1, or estrogen receptor.[4]

The metabolism of DHEA-S to DHEA in vivo occurs within specific anatomical compartments via the actions of the enzyme DHEA-S sulfatase, SULT2A1. This enzymatic activity facilitates the conversion of hydrophilic DHEA-S to the hydrophobic species DHEA, which can then be further metabolized to other steroid hormones. DHEA has a shorter half-life of 1–3 h, while the half-life of DHEA-S is 10–20 h.[5] DHEA-S is secreted in very large amounts (3.5–20 mg daily), and mainly by the adrenal cortex.[6] The daily production rate of DHEA is 6–8 mg, whereby 50% is secreted by the *zona reticularis* in both genders. Furthermore, DHEA and DHEA-S secretions are mainly modulated by adrenocorticotropic hormone (ACTH).[7] DHEA and DHEA-S are widely distributed throughout the body. The amount of DHEA-S in the brain exceeds those in the adrenals, spleen, kidneys, testes, liver and plasma, based on data from male rats.[8]

12.2 DHEA Function on Cells in the Adrenal Medulla

DHEA-S is still the least understood hormone produced by the adrenal gland; levels of circulating DHEA-S follow a trend during development and aging that is distinct from that of other adrenal steroids or gonad-derived estrogens and androgens.[9] Thus, levels of DHEA and DHEA-S are very high in fetal adrenal, and DHEA is a major product of the adrenal primordium during this stage in life. Directly after birth, DHEA levels decrease and remain low for the first 5 years of life. After that, the levels of this hormone rise again during adrenarche at 5–6 years of age, reaching a peak in early adulthood before declining again.[9,10] Adrenache itself reflects the development of a new histological layer of the adrenal cortex, the *zona reticularis*.[11] The mechanisms that regulate DHEA-S formation for life and the reasons for the gender differences are still unclear. While circulating concentrations of DHEA and DHEA-S rise progressively, cortisol and ACTH concentrations do not change significantly, indicating that adrenarche is more than just global activation of the pituitary–adrenal axis.[12]

Throughout human life, the normal range of DHEA is 7–31 nM, whereas DHEA-S ranges from 1.3 to 6.8 μM.[13] Thus, the amount of DHEA-S is 250–300 times as high as for DHEA; it has long been argued that DHEA-S simply serves as a reservoir for DHEA without any other effects, and only DHEA is considered to be biologically active. More recent studies have shown that DHEA-S is a marker of adrenal function. Additionally, DHEA-S is directly associated with increased body mass index (BMI) and somatic development, while the production of DHEA-S is associated with the development of the *zona reticularis*. During life, changes in the expression of the DHEA converting enzyme SULT2A1 may be responsible for the differences in DHEA blood concentrations, which might also be important in the development and function of the adrenal medulla.[14]

12.3 Preliminary Own Studies of DHEA and DHEA-S Effects on Cells

12.3.1 Progenitor Cells of the Medulla

In one of our own current ongoing studies, we have demonstrated an effect of DHEA and DHEA-S on the in vitro propagation of bovine chromaffin progenitor cells. We have demonstrated that a small population of chromaffin cells survives in suspension, generating floating spherical colonies referred to as chromaffin spheres (unpublished data). Chromaffin spheres are highly enriched in undifferentiated cells, as demonstrated by the ability of single cells isolated from chromaffin spheres to generate new colonies and different kind of cells when cultured under varying conditions.

Culturing bovine chromaffin spheres should test whether local environments are crucial for inducing differentiation of sympathoadrenal (SA) progenitor cells into sympathetic neurons and chromaffin cells. These data further demonstrated that in addition to glucocorticoids, DHEA and DHEA-S might play a role in chromaffin cell development.

12.3.2 Primary Juvenile and Adult Cells in the Adrenal Medulla

In a further study focused on juvenile and adult primary bovine chromaffin cells, we demonstrated an age-dependent sensitivity of these cells to growth factors leukemia inhibitory factor (LIF) and epidermal growth factor (EGF). Here, DHEA and DHEA-S were able to modulate the proliferation induced by these growth factors, which are known to be involved in adrenal biology, in a differential manner. DHEA reduced the proliferation in both populations of cells, whereas DHEA-S exclusively increased the proliferation provoked by EGF in adult-cell cultures. These effects were not mediated by androgen or estrogen receptors.

These data support the view that adrenomedullary cells are controlled by complex interactions between several factors released by the adrenal cortex such as growth factors and steroid hormones including DHEA and DHEA-S. These data strongly support the assumption that the aging process is associated with declines in the levels of hormones and trophic factors; the loss of adrenomedullary function with advancing age as well as tumor formation could be due to an imbalance in the equilibrium of the paracrine pro/anti-proliferative factors.[14] Concordantly, studies by Charalampopoulos et al.[15] have demonstrated that neuroactive steroids such as DHEA, DHEA-S as well as Allo exert a direct tonic effect on adrenal catecholamine synthesis and secretion. These data associate the adrenomedullary malfunction observed in old age with neuroactive steroids.

12.3.3 Chromaffin Cancer Cells (PC12 Cells)

We were interested in elucidating the possible effects of DHEA and DHEA-S on adrenomedullary tumor cells. In this next step, we used a well-established model, rat chromaffin pheochromcytoma PC12 tumor cells. In this study, we demonstrated that nerve growth factor (NGF)-promoted cell survival in serum deprived PC12 cells as well as neuronal differentiation of the cells was decreased by DHEA in a dose-dependent manner. Thus, DHEA reduced neurite outgrowth of NGF-stimulated cells and decreased expression of neuronal marker proteins such as SNAP-25 and VAMP-2. Concordantly, this reduction in neuronal cell parameters was associated with a shift of the cells to a more endocrine phenotype as shown by elevated catecholamine release of DHEA-treated cells in the presence, but not in the absence of NGF. DHEA also enhanced the expression of the neuroendocrine marker protein chromogranin A. On a molecular level, we demonstrated the involvement of ERK1/2 MAPK. This signaling pathway is known to be involved in proliferation and differentiation processes, and we also demonstrated that NGF-induced ERK1/2 phosphorylation was counterbalanced by DHEA, an effect that was even more pronounced by the DHEA-sulfate ester, DHEA-S.

Taken together, these data on medullary chromaffin cancer cells indicate a decisive role of DHEA and DHEA-S in the neuroendocrine differentiation of chromaffin cells, an effect previously thought to be mediated only by glucocorticoids.[16,17] The importance of this observation seen in our in vitro model fits well with our previous in vivo studies.

In summary, actual studies by our group found evidence for an important impact of DHEA and DHEA-S on all developmental steps of the adrenal medulla, starting from chromaffin progenitor cells to primary juvenile and adult chromaffin cells up to chromaffin tumor cells.

12.3.4 DHEA, DHEA-S Act as Neurosteroids in the Central Nervous System and Brain

Neurosteroids are steroid hormones synthesized in the central and peripheral nervous system either de novo from cholesterol or by in situ metabolism of blood-borne precursors; they accumulate in the nervous system independently of classical steroidogenic gland secretion rates.[18] Thus, compounds such as DHEA, androstendione, pregnenolone (PREG), their sulfate esters and lipoid conjugated forms, as well as reduced metabolites of progesterone (PROG) and deoxycorticosterone have been found at high levels in the brain long after gonadectomy and adrenalectomy.[19,20] The brain also contains steroid-metabolizing enzymes such as sulfotransferases and sulfohydrolases that convert classic steroid hormones to a variety of neuroactive compounds.

The first step in the biosynthesis of neurosteroids is the conversion of cholesterol to PREG. This reaction is catalyzed by the cytochrome P450 cholesterol side-chain cleavage (P450SCC) in three successive chemical reactions – 20α-hydroxylation, 22-hydroxylation and scission of the C20–C22 carbon bond in cholesterol. The products of this reaction are PREG and isocaproic acid. PREG can be converted to DHEA via cytochrome P450c17. Both PREG and DHEA are 3β-hydroxy-Δ5-steroids present in neural tissue in the free forms along with their sulfate ester forms.[21–23] 17β-hydroxysteroid dehydrogenase (17β-HSD) and 3β-hydroxysteroid dehydrogenase-isomerase (3β-HSD) mediate conversion of DHEA into androgens. Aromatase converts testosterone to estradiol, while 5α-reductase converts testosterone to dihydrotestosterone.[21,24]

The neurotrophic effects of DHEA-S were first reported by Roberts et al.[25] When cultures of dissociated brain cells of mouse embryos were cultured with either DHEA or DHEA-S, prominent increases were found in the numbers of neurofilament-positive neurons and glial fibrillary acid protein-positive astrocytes, with extensions of the processes of both types of cells. A study by Compagnone and Mellon[26] has made further inroads on the neurotrophic function of DHEA-S, demonstrating that the neurological effects in rodents involve neuronal excitability, plasticity and neuroprotection. DHEA and DHEA-S have been found in brains from many species, and recent studies have shown that enzymes crucial for their synthesis, especially P450c17 (17α-hydroxylase/c17, 20 lyase), are expressed in a developmentally regulated, region-specific fashion in the developing rodent brain. One region of embryonic P450c17 expression, the neocortical subplate, has been suggested to play a role in guiding cortical projections to their appropriate targets. In particular, DHEA-S plays an important role in extended brain maturation among humans, and also affects mood and acts at neuron receptor sites. Furthermore, it increases activity of the amygdala and hippocampus, and promotes synaptogenesis within the cortex. As a neurosteroid in rodents, DHEA-S is produced and used in small amounts by neurons in the brain. The discovery that neuroactive steroids exert rapid, direct effects upon the function of both excitatory and inhibitory neurotransmitter receptors has raised the possibility that endogenous neurosteroids may also play a regulatory role in synaptic transmission by modulating the balance between excitatory and inhibitory neurotransmission. The sites to which neuroactive steroids bind may also serve as targets for the discovery of therapeutic neuromodulators.[27]

As neurosteroids, DHEA and DHEA-S show also various neuroprotective properties. Thus, DHEA can lead to protection of mouse striatal DA neurons against 1-methyl-4-phenyl-1,2,3,6-tetrahydropyridine (MPTP) and protection of rat hippocampal hilar neurons against kainic acid (KA)-induced death.[28] DHEA-S involves the protection of hippocampal neurons against glutamate-induced neurotoxicity, partial neuroprotection against 1-methyl-4-phenylpridinium, colchicine, glutamate- and NMDA-induced neurotoxicity, and protection of cultured rat cerebellar granule cells against oxygen and glucose deprivation. Studies have also documented neuroprotection in an animal model of reversible spinal cord ischemia.[29] In addition, DHEA-S has been shown to promote the survival of newly formed neurons and to prevent corticosterone-induced suppression.[3] Furthermore, due to its oxygen-free radical scavenger properties, DHEA-S clearly seems to have neuronal protective activity.[30] Some investigators suggest that the neuroprotective activity of DHEA against excitotoxicity-induced neuronal death is mediated by its conversion to testosterone, and further on, to estradiol by aromatase.[28,31–33] It has also been hypothesized that 7α-hydroxylation of DHEA mediates its neuroprotective activity.[34,35] Furthermore, it seems that the sulfohydrolase (STS) and sulfotransferase (HST) activity are key regulators of DHEA-S effects in the developing brain.[36]

In summary, this data suggest that DHEA-S concentrations can be particularly high in the brain; DHEA-S and related steroids can be synthesized de novo in the CNS and in brain glial cells from cholesterol or from steroid precursors through mechanisms partly independent of peripheral steroidogenic endocrine glands, gonads or adrenal glands. Furthermore, these compounds show neuroprotective activities on various nerve cells, and may also potentially exert clinically relevant effects in the CNS in general. Thus, DHEA and DHEA-S may be active in protecting the brain from insult due to neurodegeneration and/or ischemia. Additionally, these locally produced compounds may work in concert to promote, guide, and refine axonal growth and synaptic connections in the developing neocortex. DHEA is mainly involved in axonal growth and functional networks, while at least some of these effects seem to be mediated by NMDA receptors.[36]

12.3.5 DHEA Has Many Beneficial Effects and Low Levels of DHEA Are Associated with Aging, Disease States, and Have a Negative Impact on the Nerve System and Brain

DHEA blood levels in humans reach a maximum at an age of round 30 years, and then steadily decline. Furthermore, there are clear gender differences in DHEA-S concentrations. Abnormally low levels of serum DHEA have been documented in patients with Alzheimer's disease compared to healthy controls of the same age.[37]

Clinically, DHEA-S supplementation is rare, while beneficial effects for patients suffering from adrenal insufficiency, Addison disease, schizophrenia and dementia-related diseases, more generally, are well-documented. Three studies have shown a significant association between low DHEA-S levels and the presence of depressive syndromes.[38–40] Beneficial effects of DHEA have been documented in patients with major depression,[41] Addison's disease[42] and HIV.[43] Furthermore, a multitude of beneficial effects include the prevention of cancer, heart disease, diabetes and obesity.[44] Furthermore, DHEA may be used at supraphysiological concentrations in the treatment of breast cancer,[45] and may also inhibit the proliferation of myeloma cells in patients with myeloma.[46] Additionally, DHEA-S stimulates neuronal outgrowth and development, while improving glial survival, learning and memory. DHEA supplementation in rodents has shown many beneficial effects such as protection against obesity, diabetes, cancer, cardiovascular and autoimmune diseases.[47]

The beneficial effects of DHEA supplementation during stress and trauma could occur via direct effects on immune cells or indirectly via the suppression of cortisol effects or cortisol production.[48] Since DHEA inhibits the expression of molecules involved in the inflammatory process, it might be used as an alternative treatment for chronic inflammatory diseases such as arthrosclerosis.[49]

In summary, DHEA and DHEA-S have beneficial effects against various disease states. DHEA acts on T and B lymphocytes differentially in the immune system,[50] shows anti-obesity effects and effects in the cardiovascular system.[7] Furthermore, as neurosteroids, DHEA/DHEA-S may benefit patients suffering from depression, lead to improvements in specific symptoms of sadness, guilt, discouragement, tiredness and have a positive effect on memory and sleep.

12.3.6 Potential Molecular Path of Action (Receptors, Signaling Cascades, Non-Genomic Pathways)

Mechanisms by which neurosteroids affect neuronal and brain functions include genomic actions mediated by nuclear steroid receptors as well as non-genomic actions mediated by neurotransmitter receptors.[51] Furthermore, the genomic and non-genomic effects of steroids within the central nervous system provide the molecular basis for a broad spectrum of steroid action on neuronal function and plasticity.[52]

Generally, neurosteroids such as DHEA-S show neural synthesis from cholesterol. DHEA-S then acts at the site of various synaptic receptors, including $GABA_A$, sigma-1 and *N*-methyl-D-aspartate (NMDA) receptors. Furthermore, DHEA-S is a negative allosteric antagonist of the $GABA_A$ receptor. Inhibition of $GABA_A$ receptor function by DHEA-S is not voltage-dependent, and the sites of action of sulfated neurosteroids on $GABA_A$ receptors remains unclear. Other studies have shown that DHEA can act directly at the androgen receptor (AR) and the estrogen α/β receptors,[53,54] and also affects glutamate release. The neurosteroid DHEA-S therefore enhances spontaneous glutamate release in rat prelimbic cortex by activating dopamine D1 and sigma-1 receptors.[55] These studies have further shown that sigma-1 receptor blockers partly inhibit the effect in the prelimbic cortex but completely abolish it in the hippocampus. Additionally, a NMDA receptor antagonist has no impact on the effects of DHEA-S; on the contrary, D1 receptor antagonists completely block the effect in the prelimbic cortex. The reduction in fearfulness and anxiety observed is based on DHEA activity at the AR. Since the hippocampus is rich in estrogen receptors, DHEA-S might act there via ERβ to promote hippocampal activity, improve memory, increase retention and sharpen neuronal connections.[11] DHEA is known to be a potential signaling molecule for neocortical organization during development.[26] Other studies have shown that nM concentrations of pregnenolone sulfate inhibits, while DHEA-S enhances NMDA induced H-norepinephrine release from preloaded hippocampal slices. The effects of both steroids are sensitive to pertussis toxin and *o*-receptor antagonists.[56]

The average concentration of DHEA-S is around 35 nM in post-mortem human brain, but the highest level was measured in the pituitary at 29–71 ng/g. Whether sulfated neurosteroids ever attain the nanomolar to micromolar concentrations required for some modulatory effects upon receptors remains unclear. This would be more likely if pregnenolone sulfate (PS) or DHEA-S is produced locally in response to stimulation. The presence of steroid sulfotransferases in neurons is well-documented, but the physiological role of sulfated neurosteroids would be easier to understand if specific antagonists were available.

Many direct actions of DHEA have been documented in other cell systems, although a specific receptor has not been successfully cloned. DHEA, for example, inhibits intracellular calcium release in β-cells via a plasma membrane-dependent mechanism.[57] High-affinity binding sites were found on T lymphocytes, bovine endothelial cells and vascular smooth muscle cells. The DHEA-specific receptor in human vascular smooth muscle cells (VSMC) mediates signaling via ERK1, suggesting an impact in the pathogenesis of atherosclerosis.[58] On endothelial cells, DHEA activates eNOS potentially via a G-protein-coupled plasma membrane receptor. Additionally, binding to G-protein sites confers neuroprotection.[59]

Taken together, these studies provide evidence of mechanisms by which DHEA and DHEA-S exert biological actions, show that they have specific functions other than as sex steroid precursors, and mediate their effects via non-classic steroid hormone receptors.

12.4 Beneficial Effects of Neurosteroids

Although the exact role of how sulfated neurosteroids exert their physiological role is still unclear, investigating the modulatory effects of these neuroactive steroids has identified novel potential targets for pharmacological intervention. In particular, pregnenolone sulfate (PS) and DHEA-S enhance memory retention in mice[60–62] while blocking learning deficits induced by NMDA antagonists or scopolamine,[63–66] suggesting that sulfated neurosteroids or analogous may be useful in treatment of cognitive impairment. Weaver et al. have illustrated the potential for utilizing pregnenolone hemisuccinate and related neuroactive steroids for the treatment of ischemic brain injury, chronic pain, and seizures,[67,68] while Sadri-Vakili et al.[69] illustrated its potential for modulating cocaine-induced behavioral activation and possibly cocaine addiction.

12.5 Summary and Conclusion

In this communication, we have provided evidence from studies by others as well as our own studies that DHEA and DHEA-S can fulfill important functions in the adrenal medulla. Especially during adrenogenesis, we have demonstrated a significant influence on chromaffin progenitor cells, juvenile and adult chromaffin cells as well as chromaffin tumor cells.

In addition, acting as neurosteroids, activities of these compounds include maturation of brain regions, growth of neurons, myelinization and synaptogenesis in the central nervous system.

Many beneficial effects have been documented for DHEA and DHEA-S, especially on the immune system, the cardiovascular system as well as the brain and CNS. Nevertheless, one should always be aware that most studies on DHEA and DHEA-S were conducted on rodents although their adrenals are not capable of DHEA synthesis.[70] Thus, circulating levels of these androgens are several orders of magnitude lower than in humans, and no age-dependent decline has been reported.[71]

References

1. Suzuki M, Wright LS, Marwah P, Lardy HA, Svendsen CN. Mitotic and neurogenic effects of dehydroepiandrosterone (DHEA) on human neural stem cell cultures derived from the fetal cortex. Proc Natl Acad Sci USA 2004; 101:3202–3207.
2. Kimonides VG, Khatibi NH, Svendsen CN, Sofroniew MV, Herbert J. Dehydroepiandrosterone (DHEA) and DHEA-sulfate (DHEA-S) protect hippocampal neurons against excitatory amino acid-induced neurotoxicity. Proc Natl Acad Sci USA 1998; 95:1852–1857.
3. Karishma KK, Herbert J. Dehydroepiandrosterone (DHEA) stimulates neurogenesis in the hippocampus of the rat, promotes survival of newly formed neurons and prevents corticosterone-induced suppression. Eur J Neurosci 2002; 16:445–453.
4. Charalampopoulos I, Tsatsanis C, Dermitzaki E, Alexaki VI, Castanas E, Margioris AN, Gravanis A. Dehydroepiandrosterone and allopregnanolone protect sympathoadrenal medulla cells against apoptosis via antiapoptotic Bcl-2 proteins. Proc Natl Acad Sci USA 2004; 101:8209–8214.
5. Rosenfeld RS, Rosenberg BJ, Hellman L. Direct analysis of dehydroisoandrosterone in plasma. Steroids 1975; 25:799–805.
6. Burger HG. Androgen production in women. Fertil Steril 2002; 77(Suppl 4):S3–S5.
7. Kroboth PD, Salek FS, Pittenger AL, Fabian TJ, Frye RF. DHEA and DHEA-S: a review. J Clin Pharmacol 1999; 39:327–348.
8. Corpechot C, Baulieu EE, Robel P. Testosterone, dihydrotestosterone and androstanediols in plasma, testes and prostates of rats during development. Acta Endocrinol (Copenh) 1981; 96:127–135.

9. Rainey WE, Parker CR, Jr., Rehman K, Carr BR. The adrenal genetic puzzle: how do the fetal and adult pieces differ? Endocr Res 2002; 28:611–622.
10. Rainey WE, Carr BR, Sasano H, Suzuki T, Mason JI. Dissecting human adrenal androgen production. Trends Endocrinol Metab 2002; 13:234–239.
11. Campbell B. Adrenarche and the evolution of human life history. Am J Hum Biol 2006; 18:569–589.
12. Auchus RJ, Rainey WE. Adrenarche – physiology, biochemistry and human disease. Clin Endocrinol (Oxf) 2004; 60:288–296.
13. Chen F, Knecht K, Birzin E, Fisher J, Wilkinson H, Mojena M, Moreno CT, Schmidt A, Harada S, Freedman LP, Reszka AA. Direct agonist/antagonist functions of dehydroepiandrosterone. Endocrinology 2005; 146:4568–4576.
14. Sicard F, Ehrhart-Bornstein M, Corbeil D, Sperber S, Krug AW, Ziegler CG, Rettori V, McCann SM, Bornstein SR. Age-dependent regulation of chromaffin cell proliferation by growth factors, dehydroepiandrosterone (DHEA), and DHEA sulfate. Proc Natl Acad Sci USA 2007; 104:2007–2012.
15. Charalampopoulos I, Dermitzaki E, Vardouli L, Tsatsanis C, Stournaras C, Margioris AN, Gravanis A. Dehydroepiandrosterone sulfate and allopregnanolone directly stimulate catecholamine production via induction of tyrosine hydroxylase and secretion by affecting actin polymerization. Endocrinology 2005; 146(8):3309–3318 (epub 2005 April 28).
16. Huber K, Combs S, Ernsberger U, Kalcheim C, Unsicker K. Generation of neuroendocrine chromaffin cells from sympathoadrenal progenitors: beyond the glucocorticoid hypothesis. Ann NY Acad Sci 2002; 971:554–559.
17. Finotto S, Krieglstein K, Schober A, Deimling F, Lindner K, Bruhl B, Beier K, Metz J, Garcia-Arraras JE, Roig-Lopez JL, Monaghan P, Schmid W, Cole TJ, Kellendonk C, Tronche F, Schutz G, Unsicker K. Analysis of mice carrying targeted mutations of the glucocorticoid receptor gene argues against an essential role of glucocorticoid signalling for generating adrenal chromaffin cells. Development 1999; 126:2935–2944.
18. Baulieu EE. Neurosteroids: of the nervous system, by the nervous system, for the nervous system. Recent Prog Horm Res 1997; 52:1–32.
19. Corpechot C, Robel P, Axelson M, Sjovall J, Baulieu EE. Characterization and measurement of dehydroepiandrosterone sulfate in rat brain. Proc Natl Acad Sci USA 1981; 78:4704–4707.
20. Cheney DL, Uzunov D, Costa E, Guidotti A. Gas chromatographic-mass fragmentographic quantitation of 3 alpha-hydroxy-5 alpha-pregnan-20-one (allopregnanolone) and its precursors in blood and brain of adrenalectomized and castrated rats. J Neurosci 1995; 15:4641–4650.
21. Mellon SH, Griffin LD, Compagnone NA. Biosynthesis and action of neurosteroids. Brain Res Brain Res Rev 2001; 37:3–12.
22. Plassart-Schiess E, Baulieu EE. Neurosteroids: recent findings. Brain Res Brain Res Rev 2001; 37:133–140.
23. Baulieu EE, Robel P, Schumacher M. Neurosteroids: beginning of the story. Int Rev Neurobiol 2001; 46:1–32.
24. Mellon SH, Vaudry H. Biosynthesis of neurosteroids and regulation of their synthesis. Int Rev Neurobiol 2001; 46:33–78.
25. Roberts E, Bologa L, Flood JF, Smith GE. Effects of dehydroepiandrosterone and its sulfate on brain tissue in culture and on memory in mice. Brain Res 1987; 406:357–362.
26. Compagnone NA, Mellon SH. Dehydroepiandrosterone: a potential signalling molecule for neocortical organization during development. Proc Natl Acad Sci USA 1998; 95:4678–4683.
27. Gibbs TT, Russek SJ, Farb DH. Sulfated steroids as endogenous neuromodulators. Pharmacol Biochem Behav 2006; 84:555–567.
28. Veiga S, Garcia-Segura LM, Azcoitia I. Neuroprotection by the steroids pregnenolone and dehydroepiandrosterone is mediated by the enzyme aromatase. J Neurobiol 2003; 56:398–406.

29. Lapchak PA, Chapman DF, Nunez SY, Zivin JA. Dehydroepiandrosterone sulfate is neuroprotective in a reversible spinal cord ischemia model: possible involvement of GABA(A) receptors. Stroke 2000; 31:1953–1956; discussion 1957.
30. Khalil A, Fortin JP, LeHoux JG, Fulop T. Age-related decrease of dehydroepiandrosterone concentrations in low density lipoproteins and its role in the susceptibility of low density lipoproteins to lipid peroxidation. J Lipid Res 2000; 41:1552–1561.
31. Cyr M, Calon F, Morissette M, Grandbois M, Di Paolo T, Callier S. Drugs with estrogen-like potency and brain activity: potential therapeutic application for the CNS. Curr Pharm Des 2000; 6:1287–1312.
32. Rao ML, Kolsch H. Effects of estrogen on brain development and neuroprotection – implications for negative symptoms in schizophrenia. Psychoneuroendocrinology 2003; 28(Suppl 2):83–96.
33. Azcoitia I, Sierra A, Veiga S, Honda S, Harada N, Garcia-Segura LM. Brain aromatase is neuroprotective. J Neurobiol 2001; 47:318–329.
34. Jellinck PH, Lee SJ, McEwen BS. Metabolism of dehydroepiandrosterone by rat hippocampal cells in culture: possible role of aromatization and 7-hydroxylation in neuroprotection. J Steroid Biochem Mol Biol 2001; 78:313–317.
35. Morfin R, Starka L. Neurosteroid 7-hydroxylation products in the brain. Int Rev Neurobiol 2001; 46:79–95.
36. Compagnone NA, Mellon SH. Neurosteroids: biosynthesis and function of these novel neuromodulators. Front Neuroendocrinol 2000; 21:1–56.
37. Rigaud AS, Pellerin J. Neuropsychic effects of dehydroepiandrosterone. Ann Med Interne (Paris) 2001; 152(Suppl 3):IS43–49.
38. Barrett-Connor E, von Muhlen D, Laughlin GA, Kripke A. Endogenous levels of dehydroepiandrosterone sulfate, but not other sex hormones, are associated with depressed mood in older women: the Rancho Bernardo study. J Am Geriatr Soc 1999; 47:685–691.
39. Berr C, Lafont S, Debuire B, Dartigues JF, Baulieu EE. Relationships of dehydroepiandrosterone sulfate in the elderly with functional, psychological, and mental status, and short-term mortality: a French community-based study. Proc Natl Acad Sci USA 1996; 93:13410–13415.
40. Wolkowitz OM, Reus VI, Keebler A, Nelson N, Friedland M, Brizendine L, Roberts E. Double-blind treatment of major depression with dehydroepiandrosterone. Am J Psychiatry 1999; 156:646–649.
41. Yaffe K, Ettinger B, Pressman A, Seeley D, Whooley M, Schaefer C, Cummings S. Neuropsychiatric function and dehydroepiandrosterone sulfate in elderly women: a prospective study. Biol Psychiatry 1998; 43:694–700.
42. Arlt W, Callies F, van Vlijmen JC, Koehler I, Reincke M, Bidlingmaier M, Huebler D, Oettel M, Ernst M, Schulte HM, Allolio B. Dehydroepiandrosterone replacement in women with adrenal insufficiency. N Engl J Med 1999; 341:1013–1020.
43. Rabkin JG, Ferrando SJ, Wagner GJ, Rabkin R. DHEA treatment for HIV + patients: effects on mood, androgenic and anabolic parameters. Psychoneuroendocrinology 2000; 25:53–68.
44. Arlt W. Dehydroepiandrosterone and ageing. Best Pract Res Clin Endocrinol Metab 2004; 18:363–380.
45. Gayosso V, Montano LF, Lopez-Marure R. DHEA-induced antiproliferative effect in MCF-7 cells is androgen- and estrogen receptor-independent. Cancer J 2006; 12:160–165.
46. Liu S, Ishikawa H, Li FJ, Ma Z, Otsuyama K, Asaoku H, Abroun S, Zheng X, Tsuyama N, Obata M, Kawano MM. Dehydroepiandrosterone can inhibit the proliferation of myeloma cells and the interleukin-6 production of bone marrow mononuclear cells from patients with myeloma. Cancer Res 2005; 65:2269–2276.
47. Eich DM, Nestler JE, Johnson DE, Dworkin GH, Ko D, Wechsler AS, Hess ML. Inhibition of accelerated coronary atherosclerosis with dehydroepiandrosterone in the heterotopic rabbit model of cardiac transplantation. Circulation 1993; 87:261–269.
48. Butcher SK, Lord JM. Stress responses and innate immunity: aging as a contributory factor. Aging Cell 2004; 3:151–160.

49. Gutierrez G, Mendoza C, Zapata E, Montiel A, Reyes E, Montano LF, Lopez-Marure R. Dehydroepiandrosterone inhibits the TNF-alpha-induced inflammatory response in human umbilical vein endothelial cells. Atherosclerosis 2007; 190:90–99.
50. Sakakura Y, Nakagawa Y, Ohzeki T. Differential effect of DHEA on mitogen-induced proliferation of T and B lymphocytes. J Steroid Biochem Mol Biol 2006; 99:115–120.
51. Majewska MD, Harrison NL, Schwartz RD, Barker JL, Paul SM. Steroid hormone metabolites are barbiturate-like modulators of the GABA receptor. Science 1986; 232:1004–1007.
52. Monnet FP, Maurice T. The sigma1 protein as a target for the non-genomic effects of neuro(active)steroids: molecular, physiological, and behavioral aspects. J Pharmacol Sci 2006; 100:93–118.
53. Iruthayanathan M, Zhou YH, Childs GV. Dehydroepiandrosterone restoration of growth hormone gene expression in aging female rats, in vivo and in vitro: evidence for actions via estrogen receptors. Endocrinology 2005; 146:5176–5187.
54. Labrie F, Belanger A, Luu-The V, Labrie C, Simard J, Cusan L, Gomez JL, Candas B. DHEA and the intracrine formation of androgens and estrogens in peripheral target tissues: its role during aging. Steroids 1998; 63:322–328.
55. Dong LY, Cheng ZX, Fu YM, Wang ZM, Zhu YH, Sun JL, Dong Y, Zheng P. Neurosteroid dehydroepiandrosterone sulfate enhances spontaneous glutamate release in rat prelimbic cortex through activation of dopamine D1 and sigma-1 receptor. Neuropharmacology 2007; 52:966–974.
56. Monnet FP, Mahe V, Robel P, Baulieu EE. Neurosteroids, via sigma receptors, modulate the [3H]norepinephrine release evoked by *N*-methyl-D-aspartate in the rat hippocampus. Proc Natl Acad Sci USA 1995; 92:3774–3778.
57. Liu D, Ren M, Bing X, Stotts C, Deorah S, Love-Homan L, Dillon JS. Dehydroepiandrosterone inhibits intracellular calcium release in beta-cells by a plasma membrane-dependent mechanism. Steroids 2006; 71:691–699.
58. Williams MR, Ling S, Dawood T, Hashimura K, Dai A, Li H, Liu JP, Funder JW, Sudhir K, Komesaroff PA. Dehydroepiandrosterone inhibits human vascular smooth muscle cell proliferation independent of ARs and ERs. J Clin Endocrinol Metab 2002; 87:176–181.
59. Charalampopoulos I, Alexaki VI, Lazaridis I, Dermitzaki E, Avlonitis N, Tsatsanis C, Calogeropoulou T, Margioris AN, Castanas E, Gravanis A. G protein-associated, specific membrane binding sites mediate the neuroprotective effect of dehydroepiandrosterone. FASEB J 2006; 20:577–579.
60. Flood JF, Morley JE, Roberts E. Memory-enhancing effects in male mice of pregnenolone and steroids metabolically derived from it. Proc Natl Acad Sci USA 1992; 89:1567–1571.
61. Flood JF, Morley JE, Roberts E. Pregnenolone sulfate enhances post-training memory processes when injected in very low doses into limbic system structures: the amygdala is by far the most sensitive. Proc Natl Acad Sci USA 1995; 92:10806–10810.
62. Markowski M, Ungeheuer M, Bitran D, Locurto C. Memory-enhancing effects of DHEA-S in aged mice on a win-shift water escape task. Physiol Behav 2001; 72:521–525.
63. Urani C, Doldi M, Crippa S, Camatini M. Human-derived cell lines to study xenobiotic metabolism. Chemosphere 1998; 37:2785–2795.
64. Mathis C, Vogel E, Cagniard B, Criscuolo F, Ungerer A. The neurosteroid pregnenolone sulfate blocks deficits induced by a competitive NMDA antagonist in active avoidance and lever-press learning tasks in mice. Neuropharmacology 1996; 35:1057–1064.
65. Meziane H, Mathis C, Paul SM, Ungerer A. The neurosteroid pregnenolone sulfate reduces learning deficits induced by scopolamine and has promnestic effects in mice performing an appetitive learning task. Psychopharmacology (Berl) 1996; 126:323–330.
66. Maurice T, Junien JL, Privat A. Dehydroepiandrosterone sulfate attenuates dizocilpine-induced learning impairment in mice via sigma 1-receptors. Behav Brain Res 1997; 83:159–164.
67. Weaver CE, Jr., Marek P, Park-Chung M, Tam SW, Farb DH. Neuroprotective activity of a new class of steroidal inhibitors of the *N*-methyl-D-aspartate receptor. Proc Natl Acad Sci USA 1997; 94:10450–10454.

68. Weaver CE, Land MB, Purdy RH, Richards KG, Gibbs TT, Farb DH. Geometry and charge determine pharmacological effects of steroids on *N*-methyl-D-aspartate receptor-induced Ca(2+) accumulation and cell death. J Pharmacol Exp Ther 2000; 293:747–754.
69. Sadri-Vakili G, Johnson DW, Janis GC, Gibbs TT, Pierce RC, Farb DH. Inhibition of NMDA-induced striatal dopamine release and behavioral activation by the neuroactive steroid 3alpha-hydroxy-5beta-pregnan-20-one hemisuccinate. J Neurochem 2003; 86:92–101.
70. Miller WL. Androgen biosynthesis from cholesterol to DHEA. Mol Cell Endocrinol 2002; 198:7–14.
71. Arlt W, Hewison M. Hormones and immune function: implications of aging. Aging Cell 2004; 3:209–216.

Chapter 13
Neurosteroids in the Aging Brain

Rael D. Strous

Abstract In addition to the important function neurosteroids play in neurodevelopment and as endogenous neuroprotectors, they play a pivotal role in the pathophysiology and management of memory and aging conditions. The relevance of neurosteroids to aging conditions is multifactorial and encompass changes of a variety of neurosteroid levels including DHEA, DHEA-S, pregnenolone, allopregnanolone, estradiaol and progesterone. Age-related decline in neurosteroid levels may induce a range of age-related neuronal dysfunction and degeneration due to loss of the neurosteroid protective and neurogenerative effect. Decreasing levels of neurosteroids with age may also leave the vulnerable brain unprotected against neurotoxic challenges and particularly predisposed to various age-related illnesses. These may include cardiovascular and metabolic diseases, immune disorders, malignancies, and neurological dysfunction. Since a variety of neurosteroids decrease in levels during the course of aging, it has frequently been suggested that replacement of these neurosteroids would either reduce or prevent age-related illness processes. While several studies have investigated administration of other neurosteroids, supplemental DHEA in particular has been hypothesized and observed by some to reverse many features of the aging process including memory, mood, energy levels and overall quality of life. However, the administration of DHEA to aging patients remains very controversial since many well-controlled studies remain inconclusive. A great need remains for development of improved evidence-based pharmaceutical compounds as technology improves and management of aging conditions advances with the possibility of safer more efficacious compounds based on neurosteroid activity.

Keywords Aging, neurosteroids, DHEA, neuroprotection

Beer Yaakov Mental Health Center, Sackler Faculty of Medicine, Tel Aviv University, Israel

M.S. Ritsner and A. Weizman (eds.), *Neuroactive Steroids in Brain Function, Behavior and Neuropsychiatric Disorders*

13.1 Changing in Neurosteroid Levels with Age

In addition to the important function neurosteroids play in neurodevelopment and as endogenous neuroprotectors, they play a pivotal role in the pathophysiology and management of memory and aging conditions. The relevance of neurosteroids to aging conditions is multifactorial. This refers to several areas which will be discussed.

The neurosteroids of particular importance in aging conditions, namely dehydroepiandrosterone (DHEA), its sulfate ester DHEAS, and allopregnanolone, are produced in the adrenals and the brain. With age, their rate of production decreases gradually. Thus levels in serum and adrenals decline as a consequence and it is assumed that a similar effect occurs in levels in the brain. However, while this is a most likely effect, there is little evidence confirming changes of steroid levels in the aging human brain. This would be important to substantiate since changes in blood levels of steroids with age do not necessarily reflect changes in their brain levels.[1] DHEA and DHEA-S levels in particular decrease (some have shown even up to fourfold) between its peak in the late twenties to those aged 85 and above. In men, this decrease appears to continue even after the age of 85 and it has been shown that DHEA-S attains an asymptotic low at advanced ages (over the age of 85). Similar to levels at younger ages, DHEAS values are higher in men than in women in the elderly. It remains unclear however whether the age-associated decrease of neurosteroids such as DHEA/S is related to organ-specific failure at the adrenal level or the gonads, or rather related to changes in feedback or regulatory mechanisms.[2]

Since DHEA and DHEA-S exhibit well-described antiglucocorticoid effects and direct activity on membrane-bound neurotransmitters (e.g., gamma-aminobutyric acid-A receptors), decrease of these neurosteroids with age would be expected to result in important effects related to age-associated variations in brain function.[3] Subsequently, evidence exists that age-related declines in several neurosteroid levels induces a range of age-related neuronal dysfunction and degeneration. Most likely this process transpires due to their CNS protective effect against noxious agents being attenuated.[4] This age-related decline has lead in particular to the suggestion that supplementation with neurosteroids, DHEA more specifically, may serve as an anti-aging compound.[5]

While DHEA has been the neurosteroid most studied in relation to the aging process, pregenenolone is also affected by age. Significantly lower levels in the hippocampus and other brain areas of single adult and aged (22- to 24-month-old) male Sprague-Dawley rats have been observed in aged rats. However, values have been noted to be widely scattered and, in about half of rats, similar ranges were observed to those of young rats.[6]

Similarly, studies have shown that men may demonstrate an age-related decrease in allopregnanolone. Interestingly however, serum levels in women do not change with age and correlate with P levels during the menstrual cycle.[7] Thus at least in a rat model, the aging process results in a significant age-related allopregnanolone decrease in brain cortex and pituitary, with an age-related increase in testes, adrenals and serum.[8] Others have noted serum levels of estradiaol to decrease markedly after menopause in females in contrast to levels of DHEA and DHEAS, which decrease gradually with age. In contrast these changes with age are accompanied by no significant change of serum cortisol levels. CSF concentrations of E2 have also been noted to decrease with age while CSF cortisol levels appear to increase with age.[9]

Serum levels of progesterone do not change to any major extent in aging men.[10] In contrast, levels of progesterone metabolite 3α, 5α-TH progesterone significantly decrease in males over the age of 40. This is in turn in contrast to females whose levels practically remain the same between the ages of 20 and 70.[7]

13.2 Changes in Neuroprotective Effect with Age

The decrease in neurosteroid levels with age has great significance for a number of processes in which neurosteroids are involved. Arguably the most important of these with relevance to aging is that of the loss of the neurosteroid protective and neurogenerative effect. Physiological functions attributed to specific neurosteroids include regulation of myelinization, neuroprotection, and growth of axons and dendrites.[11] This is important since the normally aging nervous system maintains some capacity for regeneration and it has been proposed that age-dependent changes in the nervous system and cognitive dysfunctions can be reversed to some extent by the administration of steroids. The aging nervous system also remains sensitive to the neuroprotective effects of steroids.[1] Thus with the diminishing levels in aging, this effect is slowly lost often with evident results. Neurosteroids involved in this process include most prominently DHEA, DHEA-S, allopregnanolone, pregnenolone and progesterone.

It has been proposed that diminished levels of DHEA and allopregnanolone are associated with age-related neuronal dysfunction and degeneration since these neurosteroids protect central nervous system (CNS) neurons against noxious agents. More specifically, DHEA(S) is known to protect hippocampal neurons against NMDA-induced excitotoxicity in a rat model, whereas allopregnanolone mediates NMDA-induced excitotoxicity in human neurons.[4] In addition, DHEA, DHEAS, and allopregnanolone protect chromaffin cells and the sympathoadrenal PC12 cells (an established model for the study of neuronal cell apoptosis and survival) against serum deprivation-induced apoptosis.[4] Thus neurosteroids appear to function as endogenous neuroprotective factors and their physiological decreased levels with aging lead to the brain being unprotected against neurotoxic challenges.[4] DHEA in particular has been shown to exhibit neurotrophic and neuroprotective activity.[1]

Although only a relatively few studies have investigated the biological significance of progesterone (PROG) synthesis in the brain, it is believed that autocrine/paracrine functions of locally synthesized PROG play an important role in the maintenance of neurons and in the development of myelin sheaths.[12] In animal mouse models, the neuroprotective effects of PROG have been noted in model of spinal cord motoneuron degeneration, where PROG administration diminished neuropathological changes in spinal motor-neurons and had beneficial effects on muscle strength and animal survival.[12] It is thought that PROG exhibits neuroprotective effects by regulating expression of specific genes in neurons and glial cells, which may become hormone-sensitive after injury or with age. PROG also promotes myelination in the brain. Since PROG is also known to decrease in the aging individual,[13] the neuroprotective and promyelinating effects of this neurosteroid may provide important therapeutic possibilities for the prevention and treatment of neurodegenerative diseases, for accelerating regenerative processes and for preserving cognitive functions during aging.[12]

Further neurosteroids with neuroprotective effects include pregnenolone-sulphate (PREG-S). There is evidence that PREG-S administered centrally significantly increase neurogenesis. It is hypothesized that this effect is associated with the negative modulator properties of this steroid at the GABA(A) receptor level. Furthermore it is thought that these neurosteroids serve a vital role for in the modulation of cerebral plasticity, mainly on hippocampal neurogenesis.[14] In addition, in aging rats, hippocampal PREG-S has been observed to be critical with respect to memory performance.[15] It is thought that pregnenolone exerts its neuroprotective effect either indirectly by its conversion to progesterone or alternatively by direct action on spinal cord neurons.[1]

Estrogens have also demonstrated neuroprotective effects including neurogenesis,[16] neuronal survival,[17] and anti-excitotoxicity.[18] Thus diminished levels in the post menopausal woman would place such a subpopulation of individuals at particular risk for these phenomena theoretically reversible by means of estrogen supplementation.

13.3 Specific Changes in Neurosteroid Levels Relating to the Aging Process

Various illness states and known syndromes of medical disease in the aging brain are known to be associated with changes in neurosteroid levels. Some have suggested that decreasing levels of neurosteroids with age may leave the vulnerable brain unprotected against neurotoxic challenges and particularly predispose it to various age-related illnesses. In animal models, for example, studies indicate that the neurosteroids PREG-S, DHEA and DHEA-S display memory-enhancing properties (e.g.[6,19]). Evidence in humans remains unclear, however, many neurosteroid physiological variables are implicated in memory and normal and disease aging processes including the well-known decrease of DHEA with age. Evidence also exists demonstrating low plasma levels of DHEA and DHEA-S in Alzheimer's disease and cardiovascular dementia in some but not all studies (e.g.[2]). Indeed, it has been proposed that DHEA deficiency may expedite the development of some diseases that are common in the elderly.[20]

Several bodies of research have demonstrated an inverse relationship between DHEA/S levels in men in particular and age-related illnesses, including cardiovascular and metabolic diseases, immune disorders, malignancies, and neurological dysfunction. This has lead to significant focus on the potential role of DHEA administration in ameliorating age-associated illnesses.[21]

The metabolism of DHEA has been investigated in the aging brain comparing Alzheimer patients and non-demented controls. Findings have indicated that DHEA is converted to Delta5-androstene-3beta,17beta-diol and to 7alpha-OH-dehydroepiandrosterone in frontal cortex, hippocampus, amygdala, cerebellum and striatum of both Alzheimer's patients and controls. Interestingly, formation of these metabolites within distinct brain regions inversely correlated with the density of beta-amyloid deposits.[1] Various neurosteroids measured in the serum have also been correlated with features of the Alzheimer's brain including cortical beta-amyloid peptides and phosphorylated tau proteins. Others have suggested that the biosynthesis of 7alpha-hydroxy-DHEA and/or Delta5-androstene-3beta,17beta-diol is likely to regulate DHEA cerebral concentrations and may contribute to the control of DHEA activity in the aging brain including in Alzheimer's disease.[22]

In general, a trend toward diminished neurosteroid levels has been noted in Alzheimer's disease patients' brain regions compared with controls. Thus, PREGS and DHEAS appear to be significantly lower in the striatum and cerebellum, and DHEAS is also significantly reduced in the hypothalamus. In addition, a significant negative correlation has been observed between levels of cortical beta-amyloid peptides and PREGS in the striatum and cerebellum and between levels of phosphorylated tau proteins and DHEAS in the hypothalamus. High levels of key proteins implicated in the formation of plaques and neurofibrillary tangles are also correlated with decreased brain levels of PREGS and DHEAS. This latter finding implies a potential neuroprotective role of these neurosteroids in Alzheimer's disease.[23]

However, while observations suggesting a relationship between neurosteroids and the aging brain and various illnesses in particular are interesting, many findings are contradictory and further research is demanded in order to determine definitively any relationship.

13.4 Treatment Effects Related to Neurosteroids in the Aging Process

Since a variety of neurosteroids decrease in levels during the course of aging, it has frequently been suggested that replacement of these decreased levels of neurosteroids would either retard or stop the decline of these neurosteroids and thus either reduce or prevent age-related illness processes. This would be important since changes in the aging nervous system are subtle and it is conceivable that many of these changes may be reversible with resultant improvement of cognitive function by means of pharmacological intervention such as neurosteroid administration.[1]

Supplemental DHEA in particular has been hypothesized to reverse many features of the aging process including memory, mood, energy levels and overall quality of life. Thus beneficial effects of DHEA have been suggested as an anti-aging steroid with regard to its stimulation of the immune system and its anti-diabetes, anti-atherosclerosis, anti-dementia (neurosteroid), anti-obesity and anti-osteoporosis effects.[24] DHEA and its principal metabolite DHEA-S serve other functions not necessarily directly related to illness effects. For example, DHEA-S (however not DHEA) is able, at pharmacological concentrations, to stimulate melatonin production.[25]

While some have noted that DHEA in appropriate replacement doses appears to have remedial effects with respect to its ability to induce an anabolic growth factor, increase muscle strength and lean body mass, activate immune function, and enhance quality of life in aging men and women,[26,27] the administration of DHEA to aging patients remains very controversial.[28] This is since several randomized, placebo-controlled clinical trials of healthy individuals aged 60 years and over treated with physiological doses of DHEA (50–100 mg/day) have yielded very few positive results. Furthermore, many studies of this nature remain of short duration (maximum 1 year) with a low number of study participants involved.[29] Thus, in various metanalyzes of the subject even those with age-related decline in DHEA secretion show little benefit from DHEA administration.[5] Furthermore, while observations suggest that DHEA replacement seems to be well tolerated with an absence of significant side effects, studies of DHEA administration offer little support for an improvement in memory or other aspects of cognitive function following DHEA treatment in normal older people.[30] A further recent Cochrane meta-analysis concluded that the little evidence there is from controlled trials does not support a beneficial effect of DHEA supplementation on cognitive function of non-demented middle-aged or elderly people.[31] However, these results do not exclude the possibility that DHEA treatment may one day prove beneficial in specific subgroups of elderly subjects.[21] At least in certain subpopulations such as both early and late postmenopausal women, chronic DHEA administration is capable of modifying circulating levels of androgens and progestins by modulating the age-related changes in adrenal function.[32]

With respect to progesterone, Schumacher et al.,[12] has proposed that its neuroprotective and promyelinating properties may lead to therapeutic options for the prevention and treatment of neurodegenerative diseases, for accelerating regenerative processes and for preserving cognitive functions during aging. This would be theoretically be possible since the aging nervous system maintains some capacity for regeneration and age-dependent changes in the nervous system and cognitive dysfunctions may be reversed by such neurosteroid administration.[1]

While controversial based on questionable efficacy and side effects, several studies have shown beneficial effects of estrogen replacement therapy in age associated decrease in cognitive function, memory decline and management of Alzheimer's disease symptomatology.[33–35] Since testosterone levels also decline with age, androgen replacement therapy also remains a hotly debated subject in aging men. It appears that at least in hypogonadal men, androgen replacement therapy has a particular role and at least one report has described beneficial effects of testosterone in the improvement of mood as well as verbal and spatial memory in aging men.[36]

13.5 Conclusion

As in a variety of normal and pathological conditions in humans, neurosteroids have great relevance to the period of aging in a range of areas.[37] There is a great need for development of improved evidence-based pharmaceutical compounds particularly as technology improves and management advances of aging conditions with the possibility of safer and more efficacious compounds based on neurosteroid activity. These may include compounds that increase synthesis of endogenous neurosteroids of benefit to the aging system such as mitochondrial benzodiazepine receptors ligands and steroid sulfatase inhibitors.[1] Further research in the rapidly developing area of neurosteroids in the aging is mandated in order to improve quality of life and longetivity in this critically important and increasingly enlarging aged population.

References

1. Schumacher M, Weill-Engerer S, Liere P, et al. Steroid hormones and neurosteroids in normal and pathological aging of the nervous system. Prog Neurobiol 2003; 71:3–29.
2. Birkenhäger-Gillesse EG, Derksen J, Lagaay AM. Dehydroepiandrosterone sulphate (DHEAS) in the oldest old, aged 85 and over. Ann NY Acad Sci 1994; 719:543–552.
3. Guazzo EP, Kirkpatrick PJ, Goodyer IM, et al. Cortisol, dehydroepiandrosterone (DHEA), and DHEA sulfate in the cerebrospinal fluid of man: relation to blood levels and the effects of age. J Clin Endocrinol Metab 1996; 81:3951–3960.
4. Charalampopoulos I, Alexaki VI, Tsatsanis C, et al. Neurosteroids as endogenous inhibitors of neuronal cell apoptosis in aging. Ann NY Acad Sci 2006; 1088:139–152.
5. Allolio B, Arlt W. DHEA treatment: myth or reality? Trends Endocrinol Metab 2002; 13:288–294.
6. Vallee M, Mayo W, Darnaudery M, et al. Neurosteroids: deficient cognitive performance in aged rats depends on low pregnenolone sulfate levels in the hippocampus. Proc Natl Acad Sci USA 1997; 94:14865–14870.
7. Genazzani AR, Petraglia F, Bernardi F, et al. Circulating levels of allopregnanolone in humans: gender, age, and endocrine influences. J Clin Endocrinol Metab 1998; 83:2099–20103.
8. Bernardi F, Salvestroni C, Casarosa E, Nappi RE, et al. Aging is associated with changes in allopregnanolone concentrations in brain, endocrine glands and serum in male rats. Eur J Endocrinol 1998; 138:316–321.
9. Murakami K, Nakagawa T, Shozu M, et al. Changes with aging of steroidal levels in the cerebrospinal fluid of women. Maturitas 1999; 33:71–80.
10. Belanger A, Candas B, Dupont A, et al. Changes in serum concentrations of conjugated and unconjugated steroids in 40- to 80-year-old men. J Clin Endocrinol Metab 1994; 79:1086–1090.
11. Compagnone NA, Mellon SH. Neurosteroids: biosynthesis and function of these novel neuromodulators. Front Neuroendocrinol 2000; 21:1–56.
12. Schumacher M, Guennoun R, Robert F, et al. Local synthesis and dual actions of progesterone in the nervous system: neuroprotection and myelination. Growth Horm IGF Res 2004; 14(Suppl A):S18–S33.
13. Schumacher M, Guennoun R, Ghoumari A, et al. Novel perspectives for progesterone in hormone replacement therapy, with special reference to the nervous system. Endocr Rev 2007; 28:387–439.

14. Mayo W, George O, Darbra S, et al. Individual differences in cognitive aging: implication of pregnenolone sulfate. Prog Neurobiol 2003; 71:43–48.
15. Baulieu EE, Robel P, Schumacher M. Neurosteroids: beginning of the story. Int Rev Neurobiol 2001; 46:1–32.
16. Gould E, Tanapat P, Rydel T, et al. Regulation of hippocampal neurogenesis in adulthood. Biol Psychiatry 2000; 48:715–720.
17. Wise PM, Dubal DB. Estradiol protects against ischemic brain injury in middle-aged rats. Biol Reprod 2000; 63:982–985.
18. Goodman Y, Bruce AJ, Cheng B, et al. Estrogens attenuate and corticosterone exacerbates excitotoxicity, oxidative injury, and amyloid beta-peptide toxicity in hippocampal neurons. J Neurochem 1996; 66:1836–1844.
19. Vallee M, Mayo W, Koob GF, et al. Neurosteroids in learning and memory processes. Int Rev Neurobiol 2001; 46:273–320.
20. Watson RR, Huls A, Araghinikuam M, et al. Dehydroepiandrosterone and diseases of aging. Drugs Aging 1996; 9:274–291.
21. Perrini S, Laviola L, Natalicchio A, et al. Associated hormonal declines in aging: DHEAS. J Endocrinol Invest 2005; 28(3 Suppl):85–93.
22. Weill-Engerer S, David JP, Sazdovitch V, et al. In vitro metabolism of dehydroepiandrosterone (DHEA) to 7alpha-hydroxy-DHEA and Delta5-androstene-3beta,17beta-diol in specific regions of the aging brain from Alzheimer's and non-demented patients. Brain Res 2003; 969:117–125.
23. Weill-Engerer S, David JP, Sazdovitch V, et al. Neurosteroid quantification in human brain regions: comparison between Alzheimer's and nondemented patients. J Clin Endocrinol Metab 2002; 87:5138–5143.
24. Nawata H, Yanase T, Goto K, et al. Adrenopause Horm Res 2004; 62(Suppl 3):110–114.
25. Djeridane Y, Touitou Y. Ex vivo studies on the acute and chronic effects of DHEA and DHEA-sulfate on melatonin synthesis in young- and old-rat pineal glands. Steroids 2004; 69:343–349.
26. Yen SS, Morales AJ, Khorram O. Replacement of DHEA in aging men and women. Potential remedial effects. Ann NY Acad Sci 1995; 774:128–142.
27. Nawata H, Yanase T, Goto K, et al. Mechanism of action of anti-aging DHEA-S and the replacement of DHEA-S. Mech Ageing Dev 2002; 123:1101–1106.
28. Genazzani AD, Lanzoni C, Genazzani AR. Might DHEA be considered a beneficial replacement therapy in the elderly? Drugs Aging 2007; 24:173–185.
29. Legrain S, Girard L. Pharmacology and therapeutic effects of dehydroepiandrosterone in older subjects. Drugs Aging 2003; 20:949–967.
30. Huppert FA, Van Niekerk JK. Dehydroepiandrosterone (DHEA) supplementation for cognitive function. Cochrane Database Syst Rev 2001; (2):CD000304.
31. Grimley Evans J, Malouf R, Huppert F, et al. Dehydroepiandrosterone (DHEA) supplementation for cognitive function in healthy elderly people. Cochrane Database Syst Rev 2006; (4): CD006221.
32. Genazzani AR, Pluchino N, Begliuomini S, et al. Long-term low-dose oral administration of dehydroepiandrosterone modulates adrenal response to adrenocorticotropic hormone in early and late postmenopausal women. Gynecol Endocrinol 2006; 22:627–635.
33. Brenner DE, Kukull WA, Stergachis A, et al. Postmenopausal estrogen replacement therapy and the risk of Alzheimer's disease: a population-based case-control study. Am J Epidemiol 1994; 140:262–267.
34. Jacobs DM, Tang MX, Stern Y, et al. Cognitive function in nondemented older women who took estrogen after menopause. Neurology 1998; 50:368–373.
35. Resnick SM, Maki PM, Golski S, et al. Effects of estrogen replacement therapy on PET cerebral blood flow and neuropsychological performance. Horm Behav 1998; 34:171–182.
36. Sternbach H. Age-associated testosterone decline in men: clinical issues for psychiatry. Am J Psychiatry 1998; 155:1310–1318.
37. Strous RD, Maayan R, Weizman A. The relevance of neurosteroids to clinical psychiatry: from the laboratory to the bedside. Eur Neuropsychopharmacol 2006; 16:155–169.

Part II
Neuroactive Steroids and Neuropsychiatric Disorders

Chapter 14
Dehydroepiandrosterone and Pregnenolone Alterations in Schizophrenia

Michael S. Ritsner[1], Anatoly Gibel[1], Yael Ratner[1], and Abraham Weizman[2]

Abstract Neurosteroids such as pregnenolone (PREG), dehydroepiandrosterone (DHEA) and their sulfates (PREGS and DHEAS) are reported to have amodulatory effect on neuronal excitability and synaptic plasticity. They also have many other functions associated with response to stress, mood regulation and cognitive performance. Furthermore, these neurosteroids have been linked to, and their levels are altered in neuropsychiatric disorders. This chapter pays special attention to alterations of these neurosteroids in schizophrenia and in particular to their association with the clinical and neurocognitive manifestations. The authors also provide substantial information on the metabolism, mode of action of PREG and DHEA, and hope that this review will contribute to a deeper understanding of the role of these neurosteroids in schizophrenia. The chapter concludes with a list of topics for suggested future research directions.

Keywords Pregnenolone, pregnenolone sulfate, dehydroepiandrosterone, dehydroepiandrosterone sulfate, metabolism, mode of action, schizophrenia, symptoms, anxiety, mood and cognitive impairment

Abbreviations ACTH adrenocorticotropic hormone; AD Alzheimer's disease; AMPA alpha-amino-3-hydroxy-5-methyl-4-isoxazolepropionic acid; CNS central nervous system; DA dopamine; DHEA dehydroepiandrosterone; DHEAS dehydroepiandrosterone sulfate; DHEA(S) both DHEA and DHEAS; GABA gamma-aminobutyric acid; $GABA_A$ gamma-aminobutyric acid receptor type A; HPA hypothalamic–pituitary–adrenal axis; NMDA *N*-methyl-D-aspartate; P450SCC cytochrome P450 cholesterol side-chain cleavage; PANSS positive and negative

[1] Department of Psychiatry, The Rappaport Faculty of Medicine, Technion – Israel Institute of Technology, and Acute Department, Haifa, and Sha'ar Menashe Mental Health Center, Hadera, Israel

[2] Sackler Faculty of Medicine, Tel Aviv University, Laboratory of Biological Psychiatry, Felsenstein Medical Research Center, Research Unit, Geha Mental Health Center, Israel

M.S. Ritsner and A. Weizman (eds.), *Neuroactive Steroids in Brain Function, Behavior and Neuropsychiatric Disorders*

symptom scale; PREG pregnenolone; PREGS pregnenolone sulfate; PREG(S) both PREG and PREGS; PREG/DHEA both PREG(S) and DHEA(S); PROG progesterone; SANS scale for the assessment of negative symptoms

14.1 Introduction

Pregnenolone (PREG), dehydroepiandrosterone (DHEA), and their sulfates (PREGS, DHEAS) (together abbreviated as PREG(S) and DHEA(S)) are neurosteroids. They display multiple effects on the central nervous system (CNS) and may act as potential signaling molecules for neocortical organization during neuronal development.[1,2] In particular, neurosteroids can interact with various neurotransmitter systems to promote neuronal remodeling; they regulate growth of neurons, enhance myelinization and synaptogenesis in the CNS, affect synaptic functioning, and show neuroprotective properties.[3–5] Furthermore, these neurosteroids have been found in the mammalian brain at considerably higher concentrations than typically detected in serum or plasma.[6,7] There is evidence that neurosteroids may be involved in the vulnerability to develop neuropsychiatric disorders such as dementia, mood disorders, substance abuse and others (see for review[8–10]). As described below, clinical studies suggested that PREG(S) and DHEA(S) may be involved in the pathophysiology of schizophrenia, and in some of its manifestations.

14.2 Biosynthesis and Metabolism

Current knowledge concerning PREG and DHEA metabolism, the enzymes mediating these reactions, and their localization has been summarized in recent publications.[11–14] Briefly, trophic hormones activate a chain of reactions that lead to the hydrolysis of cholesterol esters into free cholesterol, and the transport of cholesterol into mitochondria where it is converted to pregnenolone (PREG) by cytochrome P450 side-chain cleavage (P450 scc; now referred to as CYP11A1). This is the first step in steroidogenesis.[15] The CYP11A1 localized in the dmitochonria, and is controlled by adrenocorticotropic, follicle stimulating, and luteinizing hormones. Pregnenolone undergoes further steroid metabolism in one of three possible ways (Fig. 14.1).

To start with, PREG can be converted to progesterone (PROG). The critical enzyme step is twofold using a 3β-hydroxysteroid dehydrogenase and a delta 4–5 isomerase. The latter transfers the double bond on the A ring from C5 to C4. Progesterone is the entry into the delta-4-pathway resulting in production of 17-OH-PROG, and androstenedione (ANDRO).[16] Cortisol, corticosterone, aldosterone, and allopregnanolone are also derived from progesterone or its derivatives. Alternatively, PREG can be converted to 17-OH-PREG by the enzyme 17α-hydroxylase (CYP17A1). The following step is conversion of 17-OH-PREG to DHEA (3β-hydroxy-5-androsten-17-one) by the microsomal cytochrome P450c17 enzyme

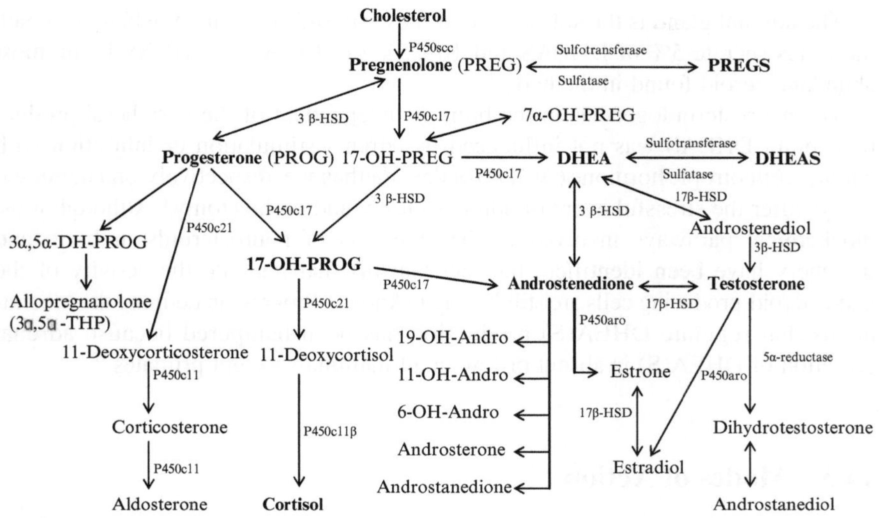

Fig. 14.1 Biosynthesis of pregnenolone, and metabolic pathway that demonstrates the synthesis of PREG, DHEA, and other steroids

Abbreviations: Andro = androstenedione; PREG = pregnenolone; PREGS = pregnenolone sulfate; 17-OH-PREG = 17-dihydropregnenolone; 7α-OH-PREG = 7α-dihydropregnenolone; DHEA = dehydroepiandrosterone; DHEAS = dehydroepiandrosterone sulfate; PROG = progesterone; 17-OH-PROG = 17-dihydroprogesterone; THP = tetrahydroprogesterone (allopregnanolone); 3α,5α-THP = 3α,5α-tetrahydroprogesterone; 3α,5α-DH-PROG = 3α,5α-dihydroprogesterone; P450scc = P450 side-chain cleavage enzyme (now referred to as CYP11A1); P450c17 = 17α-hydroxylase/17,20-desmolase (CYP17A1); P450c21 = 21α-hydroxylase; P450c11 = 11α-hydroxylase; P450c11β = 17β-hydroxylase; HSD = hydroxysteroid dehydrogenase; 17β-HSD = 17β-hydroxysteroid dehydrogenase; 3β-HSD = 3β-hydroxysteroid dehydrogenase; P450aro = P450 aromatase; 5α-reductase = 5α-hydroxysteroid reductase

(17α-hydroxylase/17,20-desmolase) in the brain and in the adrenals.[17,18] Lastly, hydroxysteroid sulfotransferases convert PREG to PREGS, and DHEA to DHEAS, whereas steroid sulfatases convert PREGS to PREG, and DHEAS to DHEA.[19,20]

DHEA serves as a precursor of androstenedione, testosterone, as well as of approximately 50% of androgens in adult men, 75% of active estrogens in premenopausal women, and 100% of active estrogens after menopause.[21] Androstenedione is synthesized by 3β-hydroxysteroid dehydrogenase/D5–D4-isomerase from DHEA, and, as mentioned above, from progesterone DHEAvia 17-OH-progesterone.[16,17] The conversion of androstenedione to testosterone is catalyzed by 17β-hydroxysteroid dehydrogenase.[22,23] Because DHEA concentrations do not decrease after adrenalectomy and castration, many experiments were performed with the intent of demonstrating the de novo synthesis of DHEA within the brain. DHEA biosynthesis has been found in cultured glial cells and neurons from the neonatal rat brain. In particular, the major producers of DHEA were found to be astrocytes, whereas oligodendrocytes are predominantly the producer of PREG.[18]

The adrenal gland is the sole source of these steroids in women, whereas in men the testes secrete 5% of DHEAS and 10–25% of DHEA.[24,25] DHEAS is the most abundant steroid found in the body.[26]

The neurosteroidogenesis in the brain is independent of the peripheral production; brain DHEAS was not influenced by adrenal stimulation or inhibition with adrenocorticotropic hormone (ACTH) or dexamethasone, respectively, and increased 2 days after the stressful event of adrenalectomy and orchiectomy.[6] Although most biochemical pathways involved in the synthesis of neurosteroids in brain and periphery have been identified, the mechanisms that regulate the activity of the neurosteroid-producing cells are still largely unknown. Progress in defining the mechanisms that regulate DHEA(S) production has been hampered because adrenal secretion of DHEA(S) is absent or low in all mammals except primates.[20]

14.3 Modes of Action

The nongenomic actions of neurosteroids in the nervous system have been described by several reviews.[11,14,26–28] Overall, the mechanisms that have been proposed for the effects of neurosteroids may be classified as either indirect by metabolism to biologically active metabolites, which then interact with their cognate receptors,[29] or direct activty including alteration of plasma membrane biophysical properties,[30,31] as well as intracellular enzyme activity.[32] Table 14.1 summarizes possible mechanisms of these neurosteroids' neuropsychiatric effects, such as modulation of neurotransmitter receptors, anti-stress effects, neuroprotective properties, cognitive-enhancing effects, androgenic and estrogenic activities, and neuropsychopharmacological effects.

14.3.1 Modulation of Neurotransmitter Receptors

The evidence supporting a receptor-dependent basis for the direct physiological effects of DHEA(S) was recently reviewed by Widstrom and Dillon.[33] The data supporting an intracellular receptor for DHEA(S) are relatively weak and do not allow us to determine whether DHEA(S) directly, or a metabolite of DHEA(S), acts as a direct receptor ligand. Although data strongly support the possible presence of a plasma membrane receptor for DHEA, such putative receptor is yet to be identified, isolated and characterized. However, neuroactive steroids have been shown to modulate neuronal excitability via their interaction with ion channel-coupled $GABA_A$ and NMDA, as well as the G-protein-coupled sigma 1 (σ_1) receptors.

$GABA_A$ receptors. PREG displays mixed modulatory effects (positive and negative) on the γ-aminobutyric acid type A ($GABA_A$) receptor,[34,35] whereas the PREGS and DHEAS are negative noncompetitive modulators with $GABA_A$-antagonistic properties[36–39] inhibiting the GABA-induced chloride ion transport.[40,41] On the other

Table 14.1 Summary of the main mechanisms that have been proposed for the effects of PREG(S) and DHEA(S)

Effects	Modes of action and properties		Reference
Modulation of neurotransmitter receptors	$GABA_A$ receptors	PREGS and DHEAS display antagonistic properties	[36,39]
		PREG displays mixed modulatory effect (positive and negative)	[34,35]
		DHEAS is a negative noncompetitive modulator, inhibiting the GABA-induced chloride ion transport currents in a dose-dependent manner	[40,41]
	NMDA receptors	PREGS, and DHEA (S) have been shown also as positive allosteric modulators	[44,48]
		PREGS enhances NMDA-receptor-independent long-term potentiation at hippocampal CA1 synapses	[228]
	Sigma-1 receptors	PREG and DHEA display neuromodulatory effects, they act as an agonist sigma-1 receptors	[45,49,50, 229,230]
	Neuropsycho-pharmacological effects	PREG and PREGS have antidepressant effects	[51–53]
		PREG and PREGS have anxiolytic effects	
		PREG has analgesic, anticonvulsant, sedative, hypnotic and anesthetic properties	
Anti-stress effects	HPA axis	DHEA(S) mediate of the HPA axis adaptation to stress	[57–59]
		Injection of DHEAS decreases the stress reactivity in male rats exposed to repeated stress	[61]
		DHEA is an ACTH-regulated steroid that possesses anti-glucocorticoid properties	[57,62–64]
Neuroprotective effects	Positive neurons	DHEA(S) increase the numbers of neurofilament-positive neurons and glial fibrillary acid protein-positive astrocytes	[73]
	Neuronal survival	Supplement of DHEA(S) increase neuronal survival and differentiation in mouse brain cells in cultures	[74]
	DHEA(S) protect neurons against	CuSO4-induced oxidative damage	[2,231,232]
		Hyperglycemia-induced toxicity	
		Neurotoxic insults inflicted by the excitatory amino acid glutamate	

(continued)

Table 14.1 (continued)

Effects	Modes of action and properties		Reference
		Neurotoxic effects of cortisol on hippocampal cells	[233]
		Glutamate and amyloid β-protein toxicity	[234]
		Glucocorticoid toxicity	[63]
		Oxidative stress-induced damage	[235]
		Anoxia in rodent models	[217]
		Apoptosis via antiapoptotic Bcl-2 proteins	[236]
	Neurodegeneration	DHEA(S) exhibit reduction of neurodegeneration	[34,237]
	Organizing the neocortex	DHEA(S) developmentally regulated synthesis in vivo may play crucial and different roles in organizing the neocortex	[77]
	Mitochondrial membranes	DHEA protects mitochondrial membranes from anoxia-reoxygenation oxidative stress	[238]
	Neurogenesis	DHEA regulates neurogenesis in the hippocampus	[64]
	Neural stem cells	DHEA is involved in the maintenance and division of human neural stem cells	[239]
	Apoptosis	DHEA(S) inhibits apoptosis in human peripheral blood lymphocytes	[240]
	Glutamate and amyloid beta protein	PREG has neuroprotective effects against both glutamate and amyloid beta protein neuropathology and glutamate neurotoxicity	[78]
Cognitive-enhancing effects	Memory	PREG and DHEA enhance post-training memory processes in mice	[83,84]
		PREG and PREGS enhance memory in rodents	[90]
		Memory performance is correlated with PREGS levels in the hippocampus of 24-month-old rats	[94]
		Memory-enhancing effects of PREGS and DHEAS in preclinical studies have been attributed to their NMDA-agonistic properties	[91,92]
		Steroid sulfatase inhibitors may enhance cognitive function following chronic treatment by increasing the concentration of excitatory neurosteroids such as DHEAS in the brain	[97]
		Steroid sulfatase inhibition can potentiate the memory enhancing properties of DHEAS	[96,98]
Androgenic and estrogenic activities		DHEA has a predominance of androgenic over estrogenic activity	[104]
	The androgen receptor	DHEA(S), and its major metabolites act genomically through the androgen receptor	[105]

hand, it has been shown that GABA, by activating the $GABA_A$ receptor complex, inhibits the activity of neurosteroidogenic enzymes in the brain.[42] Recent animal studies have shown that olanzapine and clozapine administration are associated with decreases in the $GABA_A$-negative modulators of DHEA(S) levels, respectively.[32,43]

NMDA receptors. The *N*-methyl-D-aspartate (NMDA) receptors play a critical role in synaptogenesis, synaptic plasticity, and excitotoxicity. PREGS, DHEA, and DHEAS have been shown to be positive allosteric modulators of NMDA receptors.[44,45] Furthermore, treatment with DHEAS increases the number of NMDA receptors[4] and potentiates NMDA-evoked noradrenaline release via σ_1 receptors.[45] PREGS inhibits stimulus-evoked excitatory postsynaptic currents via presynaptic inhibition of protein kinase A in rat prelimbic cortical neurons.[46] Horak et al.[47] reported that responses mediated by recombinant rat NMDA receptors, expressed in human embryonic kidney HEK293 cells, are differentially affected by PREG. PREGS enhances glutamatergic synaptic transmission by facilitating presynaptic calcium currents.[48]

Sigma 1 (σ_1) receptors. PREG displays neuromodulatory effects on σ_1 receptor[49]; it acts through a G-protein-coupled σ_1-like receptor to enhance short-term facilitation in adult hippocampal neurons.[50] DHEA may also modulate NMDA-evoked norepinephrine release via σ_1 receptors.[45]

14.3.2 Neuropsychopharmacological Properties

Studies in animals have suggested that PREG and PREGS have antidepressant[51] and anxiolytic effects.[52] As has already been mentioned, the modulation of neurotransmitter receptors, in particular the $GABA_A$, the NMDA and the σ_1 receptors, is involved in the psychopharmacological effects of neurosteroids and explains their anxiolytic, antidepressive and sedative effects. More specifically, the naturally occurring PREG can enhance $GABA_A$ receptor efficiency in a direct manner, and consequently as anxiolytic, muscle relaxant, analgesic, anticonvulsant, sedative, hypnotic and anaesthetic properties.[53]

14.3.3 Anti-stress Effects

Neurosteroids influence the neurochemical responses to both acute[54] and chronic[55,56] stressors. For instance, DHEA(S) exhibits anti-stress properties,[57,58] in particular, as a mediator of the HPA axis adaptation to stress.[59] Specifically, DHEA increases following adrenocorticotropic hormone (ACTH) administration[60] in humans, and therefore persistent DHEA elevations may reflect a prolonged upregulation of the HPA axis activity. In male rats exposed to repeated stress, the decrease in stress reactivity produced by subcutaneous injection of DHEAS was observed 1–6 days postinjection and involved central regulatory mechanisms.[61]

In addition, DHEA is an ACTH-regulated steroid that possesses anti-glucocorticoid properties,[57,62–64] supporting a role for DHEA in HPA axis regulation. The secretion of both cortisol and DHEA(S) is regulated by ACTH, but there may be dissociation, for example during chronic stress or medical illness.[65,66] Hechter and associates[67] postulated that DHEA maintains cortisol homeostasis by serving as an antiglucocorticoid in humans, and, therefore, DHEA might prevent or slow the development of the diverse pathological processes potentiated by prolonged increased cortisol secretion. In the presence of normal cortisol levels, low DHEA(S) levels could result in "functional hypercortisolaemia". Therefore, the antagonistic action of DHEA on brain cortisol,[68,69] and the cortisol-lowering effect of DHEA administration[70] may provide a relevant physiological basis for the examination of the ratio of cortisol to DHEA(S) molar ratios. Indeed, as there is a wide interindividual variability in plasma DHEA(S) levels, cortisol to DHEA(S) molar ratios are more informative than DHEA(S) values alone.[67,71] Intracellular mechanism of neurosteroid interaction with glucocorticoids has not been elucidated. Basta-Kaim et al.[72] evaluated effects of some neurosteroids on functional activity of the glucocorticoid receptor (GR) in vitro. Obtained data indicated that PREG derivatives might attenuate some glucocorticoid effects via inhibition of GR-mediated gene transcription.

14.3.4 Neuroprotective Properties

DHEA(S) neurotrophic effects were first reported by Roberts et al.[73] When cultures of dissociated brain cells of mouse embryos were cultured with either DHEA or DHEAS, prominent increases were found in the numbers of neurofilament-positive neurons and glial fibrillary acid protein-positive astrocytes, with extensions of the processes of both types of cells. Also, supplement of DHEA(S) greatly increased neuronal survival and differentiation and reduced astroglial proliferation rates in mouse brain cells in cultures.[74] Treating adult male rats with subcutaneous pellets of DHEA increased the number of newly formed cells in the dentate gyrus of the hippocampus, and also antagonized the suppression of corticosterone. In other words, DHEA regulates neurogenesis in the hippocampus and modulates the inhibitory effect of increased glucocorticoids on both the formation of new neurons and on their survival.[64] Further evidence that the development and regeneration of the nervous system are regulated by PREG(S) and DHEA(S) was established by Schumacher et al.[75,76] Compagnone and Mellon[77] examined the roles of DHEA(S) on the growth of neocortical neurites and determined the mechanisms by which they mediate their effects. While DHEA promoted axonal growth, DHEAS promoted dendritic growth and cell clustering. They suggest that DHEA(S) developmentally regulated synthesis in vivo may play crucial and different roles in organizing the neocortex.

It has been reported that PREG has neuroprotective effects against both glutamate and amyloid beta protein neuropathology and glutamate neurotoxicity,[78] while

PREGS and pregnenolone hemisuccinate (PHS) augment NMDA neurotoxicity in rat-cultured neurons. The effects of neuroactive steroids on AMPA cytotoxicity display features distinct from the effects on NMDA cytotoxicity, in particular, PREGS attenuates AMPA cytotoxicity in rat cortical neurons.[79]

Many studies have shown that DHEA(S) protect certain neuronal populations against various agent induced toxicities, anoxia, damage, oxidative stress, and apoptosis (Table 14.1). Neuroprotective effects of DHEA may be in part mediated by their conversion to testosterone and by the consecutive conversion of testosterone to estradiol by the enzyme aromatase. This enzyme is induced in reactive astrocytes after different forms of neurodegenerative lesions, and the resultant local production of estradiol in the brain has been shown to be neuroprotective. The participation of aromatase in the neuroprotective effect of these neurosteroids has been assessed in a study that suggests that estradiol formation by aromatase mediates neuroprotective effects of pregnenolone and DHEA against excitotoxic-induced neuronal death in the hippocampus.[80] The neuroprotective effect of DHEA appears to be mediated by G-protein-coupled, specific membrane binding sites.[81] Campbell[82] suggests three ways in which DHEAS may play a role in human brain maturation: (1) increasing activity of the amgydala; (2) increasing activity of the hippocampus; and (3) promoting synaptogenesis within the cortex.

14.3.5 Cognitive-Enhancing Effects

Flood et al.[83,84] indicated enhancement of post-training memory processes after injecting low doses of PREG or DHEA into limbic system structures of the mouse brain. Robel et al.[85] have found a positive correlation between the performances of 2-year-old rats in the two-trial recognition task and the concentrations of PREGS in the hippocampus, namely animals which performed best had the highest steroid levels. Additional experiments with animals show that systemic or intra-cerebral administration of PREG and PREGS enhance memory in rodents by increasing the animal's natural performance, or by antagonizing pharmacologically induced amnesia.[86–90] In particular, memory-enhancing effects of PREGS and DHEAS in preclinical studies have been attributed to their NMDA-agonistic properties.[91,92]

Furthermore, working memory-enhancing effects of DHEAS were demonstrated in aged mice on a win-shift water escape task.[93] Vallee et al.[94] conclude from the data in animals and humans that neurosteroids impact cognitive functioning in aging subjects; it seems that the PREG/DHEA display memory-enhancing properties in aged rodents.

The mechanisms by which neurosteroids influence cognitive functioning were studied in various species.

- Rhodes et al.[95,96] demonstrated that the DHEAS could enhance the release of acetylcholine from neurons in the hippocampus of anesthetized rats. These results also suggest that DHEAS, rather than DHEA, enhances brain cholinergic function and that steroid sulfatase inhibition may become an important tool for enhancing neuronal functions, such as memory, mediated by excitatory neurosteroids. Furthermore, chronic treatment with steroid sulfatase inhibitors (DU-14) may enhance cognitive function by increasing the concentration of excitatory neurosteroids, such as DHEAS, in the brain.[97] Such elevation in brain DHEAS enhances memory capability.[98]

- Extensive evidence indicates that glucocorticoid hormones influence cognitive performance, mainly activating glucocorticoid-sensitive pathways that enhance the consolidation of long-term memory.[99] Glucocorticoid effects on memory consolidation involve noradrenergic activation of the basolateral amygdala. In addition, it has been suggested that an increased glucocorticoid secretion either in response to stressful stimuli or via exogenous administration impairs the hippocampal dependent declarative memory, whereas amygdala-dependent emotional learning appeared to be enhanced by glucocorticoid stimulation.[100] The antiglucocorticoid actions of DHEA are particularly interesting with regard to its putative memory-enhancing properties.
- Pregnenolone sulfate has been shown to reverse age-related deficits in spatial memory performance and to have protective effects on memory in different models of amnesia. PREGS stimulates the release of gonadotropin-releasing hormone from GT1–GT7 hypothalamic neurons, through NMDA receptors.[101]
- The specific interaction of DHEA(S) with the hippocampal glutamatergic neurotransmission, provided a promising mechanism to mediate the modulation of cognitive processes. Systemic injections of DHEAS significantly affected the electrophysiological indices of hippocampal plasticity by enhancing the primed burst but not the long-term potentiation in the hippocampus.[102]
- It has been hypothesized that DHEA has neuroprotective effects, which are relevant to cognitive performance.[103]

14.3.6 Androgenic and Estrogenic Activities

DHEA itself has a three- to tenfold predominance of androgenic over estrogenic activity, determined by using the reversal of the effects of gonadectomy in male (ventral prostate weight), and female rats (uterine weight).[104] A recent study demonstrated that in addition to its well-known effects on cell surface receptors, DHEA and its major metabolites, including DHEAS, androstenedione, and androstenediol, acts genomically through the androgen receptor.[105] These results provide the first systematic comparison of: (a) the androgenic activity of DHEA and its sulfated and hydroxylated metabolites, (b) the relative androgenicity of DHEA itself vs the established androgens androstenedione, and androstenediol, and (c) the ability of DHEA and its major metabolites to promote the androgen receptor intracellular trafficking. These findings contribute to our understanding of the possible intracellular processes mediating the genomic effects of DHEA through the androgen receptor.

14.4 Age and Gender Differences

A remarkable feature of PREG(S) and DHEA(S) levels in both men and women is that they decrease with age, as indicated by several cross-sectional studies.[106,107] DHEA(S) concentrations rise during puberty, peak in the early twenties in men and in the thirties in women, and decline from the third decade onward. In elderly populations they are reduced to 20–30% of peak levels in young adulthood.[108–110] The developmental and age-related changes in DHEA(S) are not paralleled by any other steroid hormone, suggesting that the mechanisms regulating DHEA(S) formation are unique.[20] The decline of DHEAS concentrations with aging has led to the suggestion that DHEAS could play a role in aging and may be implicated in longevity.

Women tend to have lower DHEAS concentrations and higher DHEA/DHEAS ratios than men.[111] The difference in gender-specific DHEAS concentrations exists during each decade from 20 to 69 years.[112] It has been suggested that higher DHEA/DHEAS molar ratios in women result from greater in vivo metabolism of DHEAS to DHEA, a possible consequence of the female hormone environment.[113] Glei et al.[114] used data from a large, nationally representative sample of older Taiwanese (aged 54 and older in 2000) to investigate sex differences in the relationship between DHEAS and various health outcomes. Results reveal that higher levels of DHEAS are associated with fewer mobility limitations (especially for women), better cognitive function (among women but not among men), and better self-rated health (significant only for men but of similar magnitude for women). These findings are in contrast to previous studies conducted in the USA and Europe that generally find stronger associations for men than for women.

14.5 Clinical Importance of DHEA and Pregnenolone

In 1954, Migeon and Plager[115] extracted DHEA from human plasma. DHEA(S) concentrations in the human brain were found to be much higher than in the peripheral circulation but also exceeded their very low cerebrospinal fluid levels, ranging from about 1% to 5% of the corresponding plasma concentrations. DHEAS plasma concentrations are 100 or more times higher than DHEA and approximately 5–10 times higher than cortisol plasma concentrations.[116] Yildiz et al.[117] found that genetic factors account for between 40% and 50% of the overall variation in DHEAS levels in women with polycystic ovary syndrome.

The involvement of PREG(S) and DHEA(S) in health and disease is not fully established. A growing body of studies has demonstrated the significant relationship between the various somatic and psychiatric conditions and alterations in blood concentrations of neurosteroids. Reduction in circulating DHEA and DHEAS has been associated with a number of age-related conditions including coronary artery disease,[118,119] memory impairment,[120] and type 2 diabetes.[121] Significant lower DHEA or DHEAS levels were observed in the group of patients with fatigue and progressive multiple sclerosis,[122] as well as, among patients with untreated polymyalgia rheumatica/giant cell arteritis.[123] Serum DHEAS is increased and DHEA-response to corticotrophin-releasing hormone is decreased in the hyperthyroid state compared with the euthyroid state.[124]

There is evidence that DHEA(S) is involved in an organism's response to acute psychological stress. For example, DHEA secretion is stimulated by a first time parachute jump in healthy volunteers.[125] The plasma DHEAS level is increased in healthy humans by acute stress as well.[126] Increased morning plasma levels of DHEA and DHEAS were reported in post-traumatic stress disorder (PTSD) patients,[60,127] and in anorexic and bulimic patients.[128] In both anorexic and bulimic patients, plasma levels of neuroactive steroids did not correlate with depressive symptoms, aggressiveness, or clinical measures of nutritional status, such as body weight and body mass index.[128] Yehuda et al.[129] examined plasma DHEA(S) in 40 male veterans with and without PTSD, and determined their relationships to PTSD symptom severity and symptom improvement. The PTSD group showed significantly higher plasma DHEA and non-significantly higher DHEAS levels, controlling for age. Regression analyses demonstrated that DHEA and DHEAS levels could be predicted by symptom improvement and coping.

Morning DHEA levels were significantly higher in the women with borderline personality disorder than in controls[130] (Monteleone et al.[131] found that, before dexamethasone administration, both cortisol and DHEA plasma concentrations were significantly increased in anorexic patients as compared to controls. Moreover, while cortisol levels after dexamethasone administration were suppressed in anorexic patients to values similar to normal controls, DHEA concentrations, although significantly decreased, remained higher than in normal controls. These data support the view that in anorexic patients, the increased production of DHEA may be linked to mechanisms other than the enhanced CRF/ACTH drive. DHEA levels are raised in chronic fatigue syndrome (CFS) without depression and correlate with the degree of self-reported disability.[132]

Clinical studies demonstrated low levels of PREG in individuals with major depression,[133] while depressed women suffering from anxiety-depressive disorder treated with fluoxetine showed significantly higher values of plasma PREG levels compared to the control group.[134] Reported findings demonstrate that eight male patients with generalized anxiety disorder have significantly lower levels of PREGS than control subjects.[135] Similarly, PREGS levels were significantly lower in 12 non-medicated male patients with generalized social phobia than in 12 matched healthy male controls.[136] Serum concentrations of PREGS were significantly lower in 24 patients with hypothyroidism than in age- and sex-matched 43 healthy controls. By contrast, in 22 patients with hyperthyroidism, serum PREGS concentrations were significantly higher.[137] Premenopausal women treated for alcohol addiction exhibited elevated serum PREGS and PREGS/PREG ratio.[138]

Depressed patients showed significantly higher values of plasma DHEA and DHEAS than controls.[139–142] Takebayashi et al.[139] investigated plasma DHEAS and cortisol levels in 12 patients with unipolar depression and 11 matched normal controls. The depressed patients showed significantly higher values of plasma DHEAS than the controls. After 4 weeks of treatment with antidepressants (mainly clomipramine), the high plasma DHEAS levels recovered.

14.6 Cortisol to DHEA(S) Molar Ratios

As has already been mentioned, since there is a wide interindividual variability in both cortisol and neurosteroid plasma levels, cortisol to DHEA(S) molar ratios are more informative than DHEA(S) values alone.[67]

Cortisol and the DHEA(S) are involved in response to stress and anxiety. Indeed, a significant shift from cortisol to DHEA occurred during stress exposure in healthy humans.[125] Among a non-clinical sample of 46 men (aged 62–76), elevated cortisol to DHEA ratio was found associated with high anxiety ($r=0.46$; $p=0.004$).[120] The PTSD patients showed a significantly lower cortisol to DHEA molar ratio, controlling for age. Regression analyses demonstrated that the cortisol to DHEA molar ratio was predicted by severity of childhood trauma and current symptom severity.[129] Evidence from preclinical and clinical studies suggests that the low cortisol to DHEAS molar ratio has a buffering effect on stress and anxiety, pointing to the possible existence of a relationship between fear-potentiated startle, cortisol and DHEAS activity. Fear-potentiated startle was larger in individuals with high cortisol to DHEAS molar ratio compared to those with a low ratio. Multiple regression analyses revealed that fear-potentiated startle was positively associated with cortisol and negatively associated with DHEAS. There was no significant correlation between DHEAS and cortisol levels.[143] Cognitive-behavioral stress management buffers decrease in DHEAS and increase in the cortisol to DHEAS ratio and reduce mood disturbance and perceived stress among HIV-seropositive men.[144] Adolescent girls with conduct disorder, particularly aggressive subtype, have lower cortisol to DHEA ratios than girls with non-aggressive conduct disorder.[145]

Some studies in depressed patients found significantly higher cortisol to DHEA(S) ratios compared to controls,[146,147] while other studies failed to find abnormal ratios.[148,149] A high salivary cortisol to DHEA ratio at presentation in conjunction with disappointing life events has been shown to predict the persistence of major depression.[68] Furthermore, Michael et al.[142] present data in which the salivary cortisol to DHEA ratio clearly differentiates depressed patients from controls – 82.5% of depressed subjects have ratios above the 85th percentile of the control group. Therefore decreased levels of DHEA and a consequent elevation of the cortisol to DHEA ratio appear to reflect an additional state abnormality in adult depression.

Depressed patients showed significantly higher cortisol to DHEA ratios than controls.[142,150,151] Goodyer et al.[152] reported that in the community, adolescents at high risk for persistent major depression may be distinguished from those with sporadic forms by the 0800 h salivary cortisol to DHEA ratio prior to onset of illness. The value of this ratio was higher in elderly subjects with major depression, than in age-matched healthy controls.[153] The molar cortisol to DHEA ratio from saliva samples was significantly higher in drug-free depressed patients than in the healthy comparison subjects. This ratio correlated positively with length of recent depressive episode. Gallagher, and Young[154] suggest that elevated cortisol to DHEA ratios may be a state marker of depressive illness and may contribute to the associated deficits in learning and memory. Concentrations of serum cortisol, as well as the cortisol to DHEA molar ratio are significantly increased in depressed patients with comorbid borderline personality disorder, as compared with the healthy group. Elevated concentrations of serum cortisol, and the cortisol to DHEA molar ratios might indicate a state marker in these patients and might contribute to long-term metabolic alterations that have also been associated with major depression.[155]

Dementia patients showed a trend for higher cortisol to DHEAS molar ratios than controls.[156] The cortisol to DHEA molar ratio is dramatically altered in HIV-infected men, particularly when displaying syndromes of malnutrition and lipodystrophy, and this ratio remains elevated whatever the antiretroviral treatment, including highly active antiretroviral therapy.[157]

14.7 Schizophrenia

14.7.1 Case–Control Comparisons

Comparison of the values of blood DHEA and DHEAS levels of schizophrenia patients with healthy controls were found to differ between studies, ranging from normal,[158–162] to low,[163,164] and to high[165–169] levels.

Tourney and Erb[170] examined temporal variations in DHEA, cortisol, testosterone, and androstenedione, in chronically ill, unmedicated schizophrenia patients and in healthy control subjects at four time points during the day. DHEA was significantly below normal in the morning for the schizophrenia group, but within the range of the controls at subsequent measurements during the day. There were no significant differences between the two subject groups with regard to cortisol, testosterone, and androstenedione. Erb et al.[171] were able to identify, with 100% accuracy, ten schizophrenia patients from within a group of 20 subjects, 10 of whom were control subjects, by measuring plasma DHEA diurnal rhythms and analyzing the data using a linear function equation. Identification with such high level of accuracy is in contrast to most studies of altered biological functioning in schizophrenia, which have generally yielded inconsistent results. However, Oades and Schepker[165] reported elevated plasma DHEAS levels in medicated young men with psychosis, but not in young women, as compared to healthy control subjects.

Table 14.2 summarises recent comparative studies of the blood DHEA(S) concentrations. As can be seen, serum DHEA and DHEAS concentrations range from 15.7 to 90.9 nmol/l, and from 4,928 to 12,777 nmol/l, respectively, among schizophrenia patients, as well as, from 24.0 to 68.8 nmol/l, and from 5,375 to 13,477 nmol/l among healthy subjects, respectively. Meta-analysis of differences in mean

Table 14.2 Serum DHEA and DHEAS concentrations among schizophrenia patients and control subjects

Study (sample)	Schizophrenia	Patients (controls)	DHEA (nmol/l)		DHEAS (nmol/l)		Reference
			Patients	Controls	Patients	Controls	
S_1	Chronic[a]	40 (15)[1]	61.1 (28.3)	68.8 (24.1)	12,777 (12,949)	13,447 (6,104)	[161]
S_2	First episode[b]	37 (27)[1]	55.4 (26.0)*	40.4 (30.8)	10,721 (5,155)*	5,357 (3,670)	[166]
S_3	Chronic	23 (23)[1]	90.9 (61.4)*	24.0 (17.9)	–	–	[167]
S_4	Chronic[c]	21 (14)[2]	41.6 (23.3)	42.7 (21.6)	5,817 (3,424)*	9,333 (4,715)	[168]
			50.1 (21.4)	45.6 (22.8)	4,928 (2,600)	9,671 (3,385)	
			63.7 (33.0)	46.6 (28.7)	7,187 (5,048)	9,333 (4,470)	
			53.6 (25.2)	39.1 (17.1)	7,643 (4,724)	9,594 (4,787)	
S_5	Chronic	55 (0)	56.0 (27.1)	–	4,775 (3,391)	–	[183]
S_{6a}	Chronic[d]	43 (20)[3]	49.6 (50.8)	39.9 (20.7)	7,132 (5,034)*	8,157 (4,819)	[172]
S_{6b}			41.3 (40.9)	43.1 (21.8)	5,654 (3,921)	8,441 (3,982)	
S_{6c}			57.6 (58.2)	41.0 (21.8)	6,843 (5,212)	8,344 (3,983)	
Combined sample (S_1, S_5, S_{6a})	Chronic[e]	136 (35)	52.5 (36.5)	51.5 (26.3)	7,915 (8,400)**	1,0174 (6,049)	[161,168,172]

Significance: *$p < 0.05$; **$p = 2.6$, $p = 0.010$

Mean values and (SD) are shown:

[1] Single measure; [2] Observations were at four time points (upon entry into the study, and after 2, 4 and 8 weeks); [3] Observations were at three time points (upon entry into the study, and after 2 and 4 weeks)

Age at examination (mean ± SD, years): [a] Patients = 38.0 ± 11.1, controls = 35.1 ± 7.6; [b] patients = 27.3 ± 6.1, controls = 28.1 ± 6.8; [c] patients = 37.3 ± 9.8, controls = 36.0 ± 9.1; [d] patients = 34.1 ± 9.2, controls = 37.2 ± 8.7; [e] patients = 36.2 ± 10.0, controls = 36.3 ± 8.2

concentrations of serum DHEA(S) between schizophrenia patients and control subjects show significant non-zero effect ($p < 0.001$), and significant heterogeneity of data ($p < 0.001$; Table 14.3).Ritsner and associates[161] compared serum DHEA(S), and the cortisol levels of 40 medicated schizophrenia inpatients with 15 healthy control subjects (study S_1). Levels of these hormones did not significantly differ between patients and healthy controls, however, the cortisol to DHEA(S) molar ratios were higher in schizophrenia patients than in the healthy comparison subjects. Likewise, Goyal et al.[160] and Yilmaz et al.[162] did not find significant differences in serum DHEAS levels between schizophrenia and healthy control subjects.

Table 14.3 Meta-analysis of mean difference in serum DHEA and DHEAS levels between schizophrenia patients and control subjects (a random effects model)

Samples	Mean difference	Standard error	95% confidence interval Lower	Upper	Weight[a] (percent)
Serum DHEA levels					
S_1	−7.7	8.3	−24.3	8.9	18.3
S_2	15.0	7.1	0.8	29.2	19.0
S_3	66.9	13.3	40.0	93.8	14.8
S_{6a}	9.7	11.8	−14.0	33.4	15.8
S_{6b}	−1.8	9.8	−21.3	17.7	17.3
S_{6c}	16.6	13.5	−10.4	43.6	14.7
Average	15.0	9.2	−3.1	33.1	
Non-directional zero-effect test: Chi-square=32.7, df=6, $p < 0.001$[b]					
Effect-equality (heterogeneity) test: Cochran's Q=24.9, df=5, $p < 0.001$[c]					
Serum DHEAS levels					
S_1	−670.0	3494.6	−7679.3	6339.3	12.3
S_2	5364.0	1162.1	3041.1	7686.9	22.2
S_{6a}	−1025.0	1344.6	−3713.8	1663.8	21.4
S_{6b}	−2787.0	1066.4	−4919.4	−654.6	22.5
S_{6c}	−1501.0	1316.1	−4132.7	1130.7	21.6
Average	−64.4	1732.6	−3460.2	3331.3	
Non-directional zero-effect test: Chi-square=30.0, df=5, $p < 0.001$[b]					
Effect-equality (heterogeneity) test: Cochran's Q=30.1, df=4, $p < 0.001$[c]					

[a]The last column shows the relative weight (in percent) used to calculate the weighted average. These values indicate the influence of each study on the weighted average

[b]The χ^2 value tests the null hypothesis that all effects are zero vs the alternative that all studies had the same, non-zero effect

[c]This tests the null hypothesis that all effects are equal (homogeneous) versus the alternative that at least one effect differed from the others (heterogeneous). This test facilitated the choice between the use of a fixed effect (homogeneous) model and a random effects (heterogeneous) model

Strous et al.[166] investigated DHEA(S) and cortisol blood levels in 37 first-episode schizophrenia subjects and 27 normal age- and sex-matched controls (study S_2). Baseline DHEA and DHEAS levels were significantly higher in schizophrenia patients. The authors concluded that first-episode-of-schizophrenia patients might develop a neurosteroid response to the first onset of psychosis. However, this assumption has not been supported by a number of recent studies, which found elevated DHEA levels in chronic schizophrenia patients as well.[167–169]

When serum concentrations of cortisol and DHEA(S) were examined repeatedly during a 4-week period, ANCOVA indicated that patients had normal DHEA ($p=0.24$), decreased DHEAS ($p=0.011$) concentrations, and elevated cortisol levels ($p < 0.001$) and cortisol to DHEA(S) molar ratios ($p < 0.001$) as compared to healthy controls[172]; study S_6 (Fig. 14.2). No significant "group × time" interactions were found. As expected, age associated significantly negatively with serum DHEAS levels. However, when small samples of medicated males with chronic paranoid schizophrenia and matched healthy controls were followed up for a 2-month period,

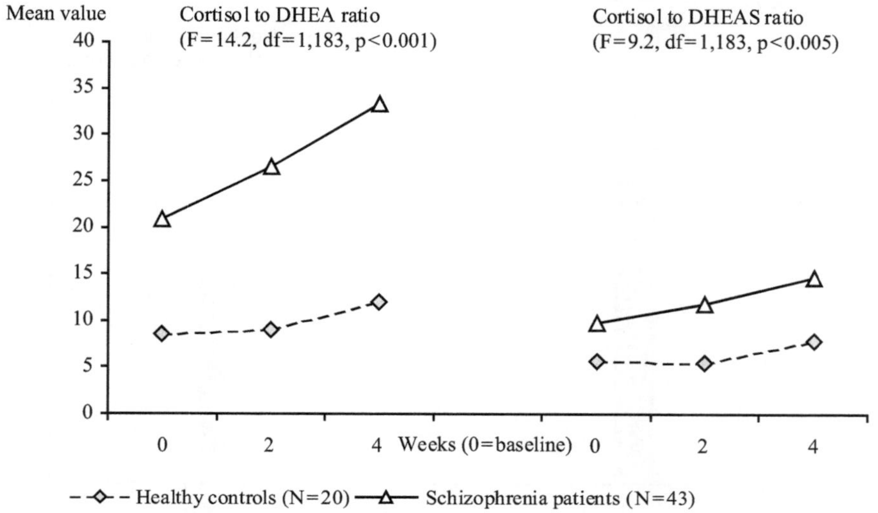

Fig. 14.2 Comparison of cortisol to DHEA and cortisol to DHEAS molar ratios between schizophrenia patients and healthy controls

mean serum concentrations of DHEA were increased (p=0.025) and those of DHEAS decreased ($p < 0.001$) in the patients, compared to the control subjects (study S_4).[168] Again, concentrations of these neurosteroids did not significantly change during the 2-month duration of the study either among schizophrenia patients, or healthy controls. Thus, between-group hormonal differences remained stable throughout the study. Elevated serum DHEA levels in this small sample of patients is supported by di Michele et al.[167] who found strongly elevated plasma DHEA in a small group of schizophrenia outpatients compared with a small sample of healthy control subjects (study S_3).

Recently Gallagher et al.[169] found elevated plasma DHEA levels in schizophrenia patients compared with healthy controls and with bipolar patients (each group consisting of 20 subjects). It should be noted that steroid levels in this study were measured from plasma samples collected at 30 min intervals between 1:00 p.m. and 4:00 p.m.

14.7.2 Combined Sample

This sample includes 136 schizophrenia patients and 35 healthy controls from studies S_1, S_5, and S_6. Comparisons of these subgroups revealed the following findings.

- Schizophrenia patients had normal serum DHEA levels (z=0.5, p=0.62), but that of DHEAS was decreased (z=2.6, p=0.01) compared to healthy controls in

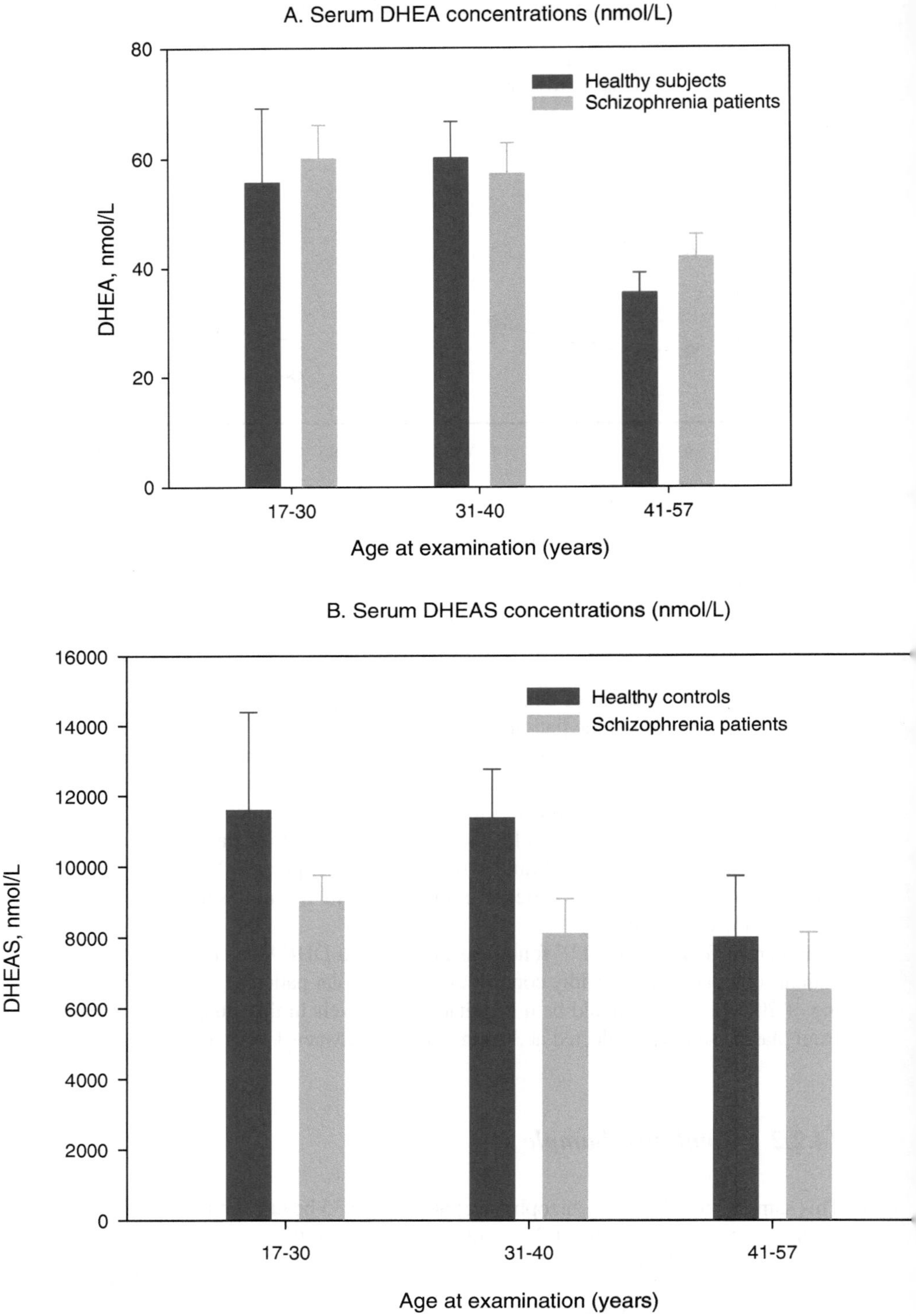

Fig. 14.3 (continued)

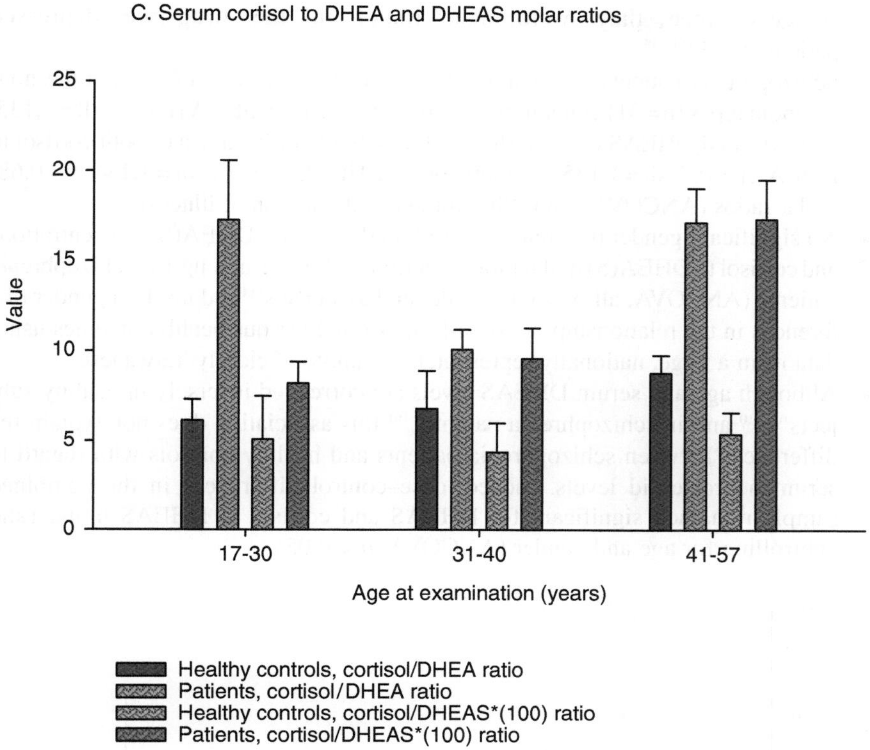

Fig. 14.3 Circulating DHEA(S) and cortisol to DHEA(S) molar ratios of 136 schizophrenia patients compared to 35 healthy subjects. (**A**) Serum DHEA concentrations (nmol/l). (**B**) Serum DHEAS concentrations (nmol/l). (**C**) Serum cortisol to DHEA and cortisol to DHEAS molar ratios

the same age subgroups (Fig. 14.3a and b). Considering that blood DHEAS concentrations are 100–200 times higher than DHEA concentrations, these findings suggest that schizophrenia patients are characterized by deficit peripheral DHEA(S), and elevated cortisol to DHEA(S) molar ratios.

- Both cortisol to DHEA (15.3 ± 16.3) and DHEAS (12.1 ± 12.7) molar ratios were significantly elevated among schizophrenia patients as compared to healthy controls in the same age subgroups (7.4 ± 6.7, and 4.8 ± 5.8, respectively, both $p < 0.001$, Fig. 14.3c). Increased cortisol levels in schizophrenia patients were reported by a number of studies,[173–175] but were contradicted by other reports.[176,177] Elevation of cortisol to DHEA(S) ratios may be due to imbalance between cortisol and these neurosteroids ("functional hypercortisolaemia"). Elevated cortisol to DHEA(S) molar ratios do not seem to be specific to schizophrenia subjects since they were also found in the drug-free depressed patients.[142,150,151,155]

- Schizophrenia patients with paranoid type (n = 105) did not differ from the non-paranoid types (n = 31) either in serum concentrations of DHEA (F = 0.3, df = 1,135, $p = 0.61$), and DHEAS (F = 0.4, df = 1,134, $p = 0.50$), or in values of both cortisol to DHEA (F = 0.7, df = 1,135, $p = 0.40$) or to DHEAS (F = 0.3, df = 1,134, $p = 0.68$) molar ratios (ANCOVA controlling for age and duration of illness).
- No significant gender differences were found in serum DHEA(S) concentrations and cortisol to DHEA(S) molar ratios, controlling for age, among 136 schizophrenia patients (ANCOVA, all $ps > 0.05$). Glei and associates[114] did not find gender differences in the relationship between DHEAS and various health outcomes using data from a large, nationally representative sample of elderly Taiwanese.
- Although age and serum DHEAS levels are correlated inversely in healthy subjects[108,110] and in schizophrenia patients,[161] this association does not explain the differences between schizophrenia patients and healthy controls with regard to serum neurosteroid levels. Indeed, case–control differences in the combined sample remained significant for DHEAS and cortisol to DHEAS molar ratio controlling for age and gender (ANCOVA, $p < 0.05$).

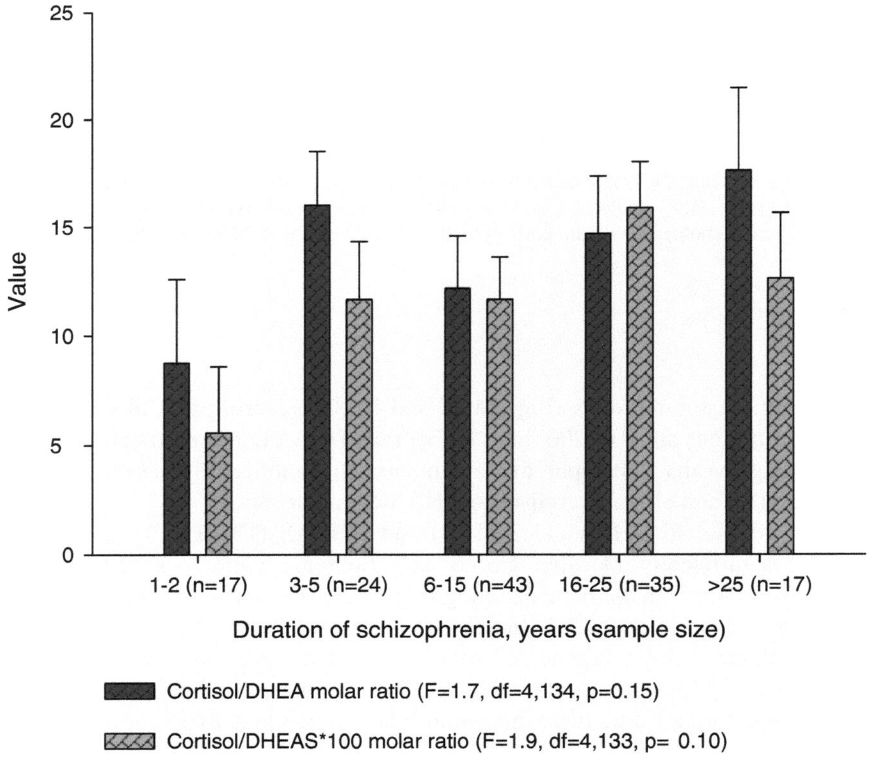

Fig. 14.4 Serum cortisol to DHEA(S) molar ratios by duration of schizophrenia

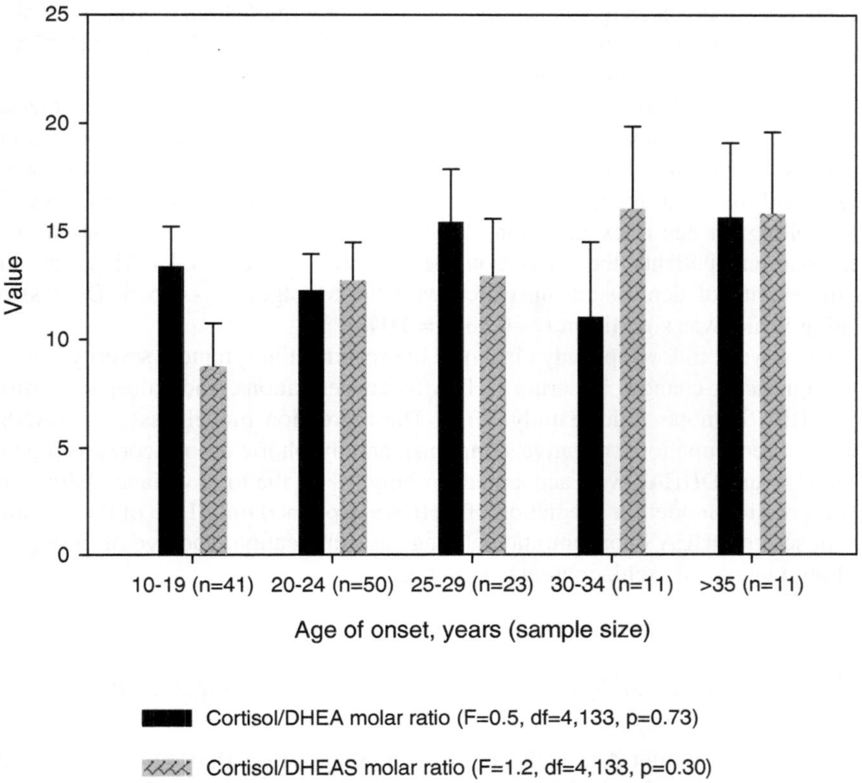

Fig. 14.5 Serum cortisol to DHEA(S) molar ratios by age of onset of schizophrenia

- Alterations in DHEA(S) and molar ratios did not show any association with duration of schizophrenia and age of onset (Figs. 14.4 and 14.5).

It is important to examine the relationship of DHEA(S) alterations in schizophrenia with some illness manifestations such as age at onset, smoking status, substance abuse, comorbidity, severity of symptoms, anxiety, distress, depressive mood, cognitive impairment, and with responsiveness to antipsychotic treatment and others factors.

14.8 Symptom Severity

Although several studies have demonstrated negative correlations between DHEA(S) concentrations and severity of depression among patients with major depression,[142] and in abstinent alcoholics,[178] some of the literature indicated the controversial role of DHEA(S) in the ongoing symptomatology of schizophrenia. Harris et al.[159] found that, when controlling for age, higher DHEA levels and/or higher DHEA to cortisol ratios were significantly correlated with lower symptom ratings on the Brief Psychiatric Rating Scale in 17 medicated chronic schizophrenia or schizoaffective inpatients. However, these findings were not being replicated using larger samples of chronic shizophrenia patients, and both PANSS dimensions (study S_1),[161] and the Scale for the Assessment of Negative Symptoms (SANS) scores.[179] Controlling for age at examination, duration of illness and age of illness onset of schizophrenia patients, the correlation coefficient of cortisol to DHEAS molar ratio with severity of depression, measured with the Montgomery-Åsberg Depression Rating Scale, was significant ($r=0.33$, $p=0.048$).[161]

The 4-week follow-up study displayed improvement in symptom severity, but not any significant changes in serum DHEA(S) concentrations, and values of cortisol to DHEA(S) molar ratios (study S_6).[172] The regression model based on PANSS autistic preoccupations, negative symptoms, and dysphoric mood scores as predictors of serum DHEA levels accounted for only 9% of the total variance. Similarly, the regression model for prediction of DHEAS explained only 15% of the variability in serum DHEAS concentrations by age at examination, and age of onset, and did not include any schizophrenia symptoms.

14.8.1 Severity of Symptoms Versus Self-reported Anxiety

Hsiao[180] provides pilot data indicating that morning serum DHEAS concentrations were positively correlated with anxiety Severity, measured with the Hospital Anxiety and Depression Scale after controlling for age, gender and body-mass index in 23 medication-free outpatients experiencing a major depressive episode. Schizophrenia patients, when compared to controls, displayed higher scores for state and trait anxiety, anger expression index, and emotional and somatic distress, which positively correlated with the PANSS activation factor and dysphoric mood ratings (r ranged from 0.31 to 0.69, $p<0.05$–<0.001) (study S_1).[161] When age at examination, duration of illness and age of illness onset were partialled from the correlation matrix, correlation coefficients of cortisol to DHEAS molar ratio were significant, with severity of state and trait anxiety ($r=0.43$, $p=0.008$ and $r=0.40$, $p=0.014$, respectively), trait anger ($r=0.41$, $p=0.012$), angry temperament ($r=0.46$, $p=0.004$), anger expression index ($r=0.36$, $p=0.033$), and hostility ($r=0.42$, $p=0.010$).

Therefore, we assumed that variability in serum cortisol to DHEA(S) molar ratios may be associated with trait-anxiety and age-related variables rather than with severity of schizophrenia symptoms. This assumption was tested by the prediction of serum DHEA(S) levels and cortisol to DHEA(S) molar ratios from severity of symptoms and anxiety in three analyses (study S_6).[172] The regression model, which accounted for 21% of the total variance in serum cortisol to DHEA molar ratio, indicates two negative indicators: trait-anxiety (partial $R^2=14\%$), and negative symptoms (partial $R^2=6.2\%$). The next model explains 15% of the variability in serum cortisol to DHEAS molar ratio, it includes trait-anxiety (partial $R^2=7.3\%$), and negative symptoms (partial $R^2=3.4\%$) with age at examination and age of illness onset (partial $R^2=12.7\%$). Thus, it seems to confirm our assumption, since serum cortisol to DHEA(S) molar ratios are associated with trait-anxiety (partial $R^2=8–14\%$) at least, twice as much as with negative symptoms (partial $R^2=3–6\%$). Furthermore, age and age of onset account for 12.7% of variability of cortisol to DHEAS ratio.

Emotional distress and anxiety scores show a significantly confounding effect on the association between serum DHEA levels and severity of symptomatology[172] that should be taken into consideration when discussing the discrepancy between DHEA findings as it appears in the literature.

Thus, while somewhat controversial, there is no well-documented evidence demonstrating the association between the severity of psychotic symptoms, which is considered to be an important factor for schizophrenia, and the fluctuations in serum DHEA(S) levels and cortisol to DHEA(S) molar ratios in schizophrenia. Additional support for these conclusions comes from DHEA augmentation studies. In the first study that included 15 schizophrenia subjects receiving 100 mg/day DHEA and 12 patients receiving placebo, DHEA appeared to be effective in the management of negative, depressive and anxiety symptoms of schizophrenia inpatients.[181] The authors demonstrated a significant correlation of increase in DHEA(S) plasma levels with improvement in negative symptoms but no correlation with improvement in depressive and anxiety symptoms. However, the sample size was small, medication was not standardized, study duration was only 6 weeks and the clinical parameters investigated remained limited. When 20 patients with chronic schizophrenia, stabilized on olanzapine, received DHEA (150 mg/day) for a period of 12 weeks compared to placebo augmentation (n=20), improvement of negative symptoms as measured by SANS scale, but without the PANSS dimensions, was observed.[182] Correlation analysis between change of blood DHEA(S) levels and change of clinical scores (SANS) showed no statistical significance. These findings, however, do not replicate previous findings by the same group[181] showing improvement in depression and anxiety, and correlations between change in DHEA levels and SANS ratings. Lastly, a randomized double-blind, placebo-controlled crossover augmenting treatment with DHEA (200 mg/day), for 6 weeks did not improve severity of symptoms.[183] In light of these inconsistent findings it seems that DHEA augmentation strategy in the management of schizophrenia symptomatology appeared to be premature and insufficiently substantiated (see Chapter 17).

14.9 Pregnenolone

PREG through 17-OH-PREG can be converted to DHEA (Fig. 14.1). Low levels of serum PREG were reported previously in patients with dementia,[106] generalized anxiety disorder,[135] and social phobia.[136] It was hypothesized that alterations in neurosteroids in schizophrenia may also include PREG. Therefore, serum PREG and DHEA(S) levels in 15 chronic medicated schizophrenia patients were compared with 12 healthy subjects at four time points: at the start of the study, after 2, 4 and 8 weeks.[184] During the study period, improvement on the state–anxiety scale scores was observed, while severity of symptoms did not change significantly. The main results of this study were:

- According to Hotelling's test schizophrenia patients had significantly decreased PREG (1.26 ± 0.7 nmol/l) compared to healthy subjects across four time points (1.64 ± 1.0 nmol/l, $t=4.6$, $p=0.033$).
- In the three-way ANCOVA model, when controlling for age and sex, serum PREG levels remain significantly and stably decreased over the four time points ($F=4.6$, $df=1,104$, $p=0.035$). No significant "group × time" interactions or main gender effect were found.
- The levels of DHEA and the DHEA to PREG molar ratio are higher among patients compared to controls; within- and between-groups hormonal differences remained stable throughout the study.
- Serum PREG concentrations appear to be associated with trait-anxiety scores independent of schizophrenia symptoms.

The molecular mechanisms underlying the regulation of PREG conversion to DHEA in schizophrenia are still unclear. There could be an increase in P450-17α-hydroxylase activity (DHEA to PREG molar ratio) converting PREG to DHEA. Thus, altered serum PREG and DHEA levels appear to be associated with anxiety and emotional distress scores, but not substantiated with schizophrenia symptoms. PREG and DHEA to PREG molar ratio may represent a trait-like marker of impaired hormonal response to stress in schizophrenia. Although the role of PREG and DHEA in non-specific response to distress and anxiety or in the pharmacotherapy of schizophrenia is as yet unclear, this report supports the assumption that PREG and DHEA concentrations are not related to specific diagnoses, but to more general psychological states, like anxiety and distress, that can occur in various disorders. Limitations of the study include the relatively small sample size of patients and use of polypharmacotherapy and benzodiazepines, and a relatively small number of healthy control subjects. Since schizophrenia is a chronic disorder with various different clinical signs and symptoms and a fluctuating course, a long-term follow-up is needed to substantiate our findings. A further longitudinal large-scale case-control comparison of PREG and DHEA levels in treated and an untreated, as well as in washed-out schizophrenia patients, is warranted.

14.10 Androstenedione, Testosterone, and Progesterone

Another source for the discrepancies of DHEA findings in schizophrenia may be connected to the alterations in DHEA's metabolites. Therefore, serum DHEA's metabolites (DHEAS, androstenedione, testosterone), as well as, progesterone, and prolactin concentrations together with severity of symptoms, emotional distress, and antipsychotic treatment were examined in a small sample of medicated chronic paranoid schizophrenia patients and healthy controls.[168] The main results of this study are:

- Schizophrenia patients had significantly higher serum concentration of androstenedione, but the concentration of progesterone and testosterone showed normal levels compared with healthy controls.
- Patients were also characterized by elevated androstenedione to DHEAS molar ratio, and lowered testosterone to androstenedione molar ratios compared with healthy controls.
- Elevated androstenedione to DHEAS ratios among schizophrenia patients were independent of emotional distress, state- and trait-anxiety. In contrast, case–control differences in serum DHEA and androstenedione levels disappeared when these psychometric properties were held constant.
- Canonical correlation analysis demonstrates multivariate association of the alterations in DHEA(S) and metabolites with several symptoms. In the first pair of canonical variates dysphoric mood, positive and activation symptom scores were positively associated with serum DHEA(S) and androstenedione levels together, as well as with androstenedione to DHEA and testosterone to androstenedione molar ratios.

Limitations of the study include the relatively small sample size and the relatively small number of control subjects, as well as lack of a parallel group of untreated patients.

14.11 Cognitive Impairment

It is now widely accepted that cognitive impairment, particularly in the areas of attention, memory, and executive functions, represents an essential feature of schizophrenia.[185,186] Such impairment appears to be present at early stages of the illness,[187,188] and appears to be relatively stable over time,[189,190] but questions remain about the factors underlying cognitive impairment. Current theories link efficacy in cognition to unique effects on glutamatergic neurotransmission in the prefrontal cortex, cortical dopaminergic and cholinergic function and improved patterns of connectivity in the brain during cognitive task performance.[191] Lower levels of estrogen are associated with reduced performance in cognitive function, especially verbal performance and executive functioning.[192] Intensive studies suggest that PREG/DHEA could be involved in learning and memory processes (see Table 14.1).

Human studies, however, have reported contradictory results. Cognitive dysfunctions have been associated with low DHEAS levels, high DHEAS levels, or high DHEA levels; while in other studies, no relationship was found.[94,193]

- The data offer no support at present for an improvement in memory or other aspects of cognitive function following DHEA treatment in normal older people.[194–196]

- The decline in endogenous DHEAS concentration is independent of cognitive status and cognitive decline in healthy aging men.[197] Fonda et al.[198] based on population-based cohort data (n = 981) concluded that the direct effects of hormones on cognition are not significant when salient factors are considered.
- A marked decrease in the DHEAS plasma levels have been observed in Alzheimer's disease (AD) patients.[199–201] However, no correlation was found between the plasma DHEAS, age, and mini-mental test score in Alzheimer's disease.[200]
- PREG and/or PREGS levels are low in dementia.[106] de Bruin et al.[156] have studied the relationship between DHEAS, cortisol, and cognitive function in a population of dementia patients (n = 29), age-matched controls (n = 46), and younger subjects (n = 11). There was no difference between DHEAS levels of dementia subjects and age-matched controls.
- In schizophrenia cognitive dysfunction has been associated with low DHEAS levels,[202] high DHEAS levels[203] or high DHEA levels,[202] or no association[94] was found. Harris et al.[159] reported that higher DHEA levels were significantly correlated with better performance on attention and working memory ($r = 0.55$, $p < 0.05$), and DHEAS levels were positively correlated with semantic memory ($r = 0.51$, $p < 0.05$) in a small sample of medicated chronic schizophrenia or schizoaffective inpatients. Comparison between studies is complicated by different methodologies and measures of cognition.

14.11.1 Correlation Analysis

The following tools were used: the computerized batteries the Cambridge Automated Neuropsychological Test Battery (CANTAB),[204] the Wisconsin Card Sorting Test (WCST),[205] and the baseline data about 55 schizophrenia patients who participated in the previously reported DHEA augmentation trial.[183] Overall 12 neuropsychological tests were grouped into five cognitive domains: movement skills, attention, memory, learning, and executive function: Motor Screening (MOT), Big/Little Circle (BLC), Reaction Time (RTI), Matching to Sample Visual Search (MTS), Delayed Matching to Sample (DMS), Pattern Recognition Memory (PRM), Spatial Recognition Memory (SRM), Spatial Span (SSP), Rapid visual information processing (RVP), Spatial working memory (SWM), Intra/Extra Dimensional Set Shift (IED), and Stockings of Cambridge (SOC).

Performance of CANTAB tasks was presented as the standard z-score that compared each patient with the mean performance of healthy subjects matched by age and sex. The CANTAB program was used to calculate z-scores on the basis of the extensive normative database included in CANTAB. A negative value indicates poorer than average performance. Table 14.4 depicts correlation coefficients of the CANTAB tasks and neurosteroids. As can be seen, movement skills associated negatively with serum DHEA concentrations ($r = 0.46$–0.48, $p < 0.01$), and positively with cortisol to DHEA ($r = 0.44$–0.56, $p < 0.01$–0.001), and androstenedione to DHEA ($r = 0.37$–0.40, $p < 0.01$) molar ratios. At the same time, better performance

Table 14.4 Spearman correlation coefficients between the CANTAB tasks (z-scores) and serum neurosteroids concentrations of schizophrenia patients (baseline assessment)

CANTAB tasks and indices	DHEA (nmol/l)	DHEAS (nmol/l)	Androstenedione (nmol/l)	Cortisol to DHEA ratio	Cortisol to DHEAS*100 ratio	Androstenedione to DHEA ratio
Global cognitive index	–0.12	0.08	–0.01	0.24	0.03	0.19
Movement skills						
Skills index	–0.46***	–0.14	–0.25	0.56***	0.20	0.37**
Big/little circle[a]	–0.24	–0.06	–0.14	0.44**	0.19	0.18
Reaction time[b]	–0.48**	–0.21	–0.24	0.46***	0.20	0.40**
Attention and memory						
Attention/memory index	0.02	0.19	0.09	–0.01	–0.09	0.10
Matching to sample visual search[c]	–0.14	0.03	0.07	0.10	–0.01	0.28*
Frontal/executive tasks						
Executive function index	0.28*	0.34*	0.20	–0.16	–0.21	–0.12
Intra/extradimensional set shift[d]	0.14	0.20	0.29*	–0.22	–0.22	0.05
Intra/extradimensional set shift[e]	0.28*	0.15	0.33*	–0.23	–0.05	–0.09
Stockings of Cambridge[f]	0.40**	0.18	0.24	–0.23	–0.08	–0.21

CANTAB's task metrics: [a]percent correct; [b]five choice reaction time is the speed with which the subject releases the press pad in response to a stimulus in any one of five locations (milliseconds); [c]total number correct; [d]stages completed (There are nine stages to be completed in this task in the clinical setup. Subjects completing all stages are deemed to have "passed the test".); [e]total errors (This is a measure of the subject's efficiency in attempting the test.); [f]subsequent thinking time (This measure reflects the subject's speed of movement after the initial move has been made for 2, 3, 4 and 5 move problems.)
Significance: *$P<0.05$, **$P<0.01$, ***$P<0.001$

Table 14.5 Spearman correlation coefficients between the computerized Wisconsin Card Sorting Test measures (T-scores) and serum neurosteroids concentrations among schizophrenia patients (baseline assessment)

Circulating neurosteroids	Total errors	Perseverative responses	Perseverative errors	Non-perseverative errors
DHEA (nmol/)	0.30*	0.30*	0.28*	0.33*
DHEAS (nmol/l)	0.04	0.07	0.06	0.09
Progesterone (nmol/l)	0.26	0.22	0.21	0.18
17-HO-Progesterone (nmol/)	0.48***	0.42**	0.41**	0.43**
Androstenedione (nmol/l)	0.46***	0.45***	0.42**	0.40**
Cortisol (nmol/l)	–0.22	–0.22	–0.20	–0.22
Cortisol to DHEA ratio	–0.38**	–0.38**	–0.36*	–0.38**
Cortisol to DHEAS ratio	–0.12	–0.13	–0.11	–0.17
DHEAS to DHEA ratio	–0.19	–0.15	–0.16	–0.18
Prolactin (nmol/l)	–0.15	–0.21	–0.18	–0.08

Significance: $*p < 0.05$; $**p < 0.01$; $***p < 0.001$

on executive tasks was associated with higher serum levels of DHEA(S) and androstenedione ($r=0.28$–0.40, $p<0.05$).

This association was substantiated with the WCST (Table 14.5). In addition to DHEA and androstenedion, significant correlation coefficients were found between the WCST measures (T-scores) and serum concentrations of 17-OH-progesterone, and cortisol to DHEA molar ratio.

14.11.2 Hypofrontality

Hypofrontality refers to impaired or diminished prefrontal cortex function and this phenotypic expression of a reproducible neurocognitive abnormality has been noted in several domains including neuropsychological and electrophysiological testing, functional brain imaging, molecular biology and neuropathological specimens.[206] While the hypofrontality of schizophrenia is thought to be a function of a dysfunctional neurodevelopmental process, others have focused more on the possibility of it reflecting a neurodegenerative process. Whatever the case, it appears that hypofrontality sets in motion a cascade of aberrant neurocircuitry connectivity, most notably abnormal corticolimbic circuitry and abnormal corticomesotemporal lobe circuitry.[207,208]

Although the "hypofrontality hypothesis" has been supported by many neuroimaging studies that have shown association between reduced left frontal regional cerebral blood flow (rCBF) and neuropsychological impairment on the WCST,[209,210] the results are still controversial, mainly due to methodological aspects and the heterogeneity of the patient samples. To measure WCST specific and reproducible reduced cerebral activations in schizophrenic patients, Riehemann et al.[211] revised the study design and patient recruitment, respectively. The findings are in line with previous PET and SPECT studies and confirm the "hypofrontality" hypothesis, as well as, that hypofrontality is not caused by neuroleptic medication. Moreno-Iniguez et al.[212] studied the regional cortical blood flow (rCBF), both at rest and during the WCST performance (activation) by comparing 21 schizophrenia patients with 18 healthy subjects. The data support the hypothesis of hypofrontality at rest and during activation, which implies the incapacity of schizophrenic patients to increase the frontal CBF while performing the WCST. Recent studies confirm the common hypothesis of schizophrenic hypofrontality at rest and during WCST performance.[212,213] It is still unknown whether other biological mechanisms influence prefrontal functioning, in particular at the neuroendocrine level. One such factor of potential influence is that of the neurosteroid system, and DHEA(S) in particular.

The objective of our next analysis was to determine associations of hypofrontality (as expressed by impaired the WCST function), with blood neurosteroids, in a group of medicated schizophrenia patients. This was carried out in an attempt to clarify whether hypofrontality may be considered an expression of neurocognitive dysfunction. Participants with chronic schizophrenia or schizoaffective disorder were drawn from the patient population of two Mental Health Centers (see details[183]). Deficit prefrontal functions were assessed using tasks from the computerized WCST. Raw WCST scores were transformed to demographically corrected *T* scores with cuff-of-point for impaired (≤44) and normal (>44) performance.[205] Since a higher number of perseverative errors have been suggested to implicate a deficit in task switching and inhibitory functions in schizophrenia,[214] we used perseverative error *T* scores in order to separate patients into two subgroups with impaired (hypofrontality) and normal executive performance. Thus, according to WCST perseverative errors *T* scores patients were separated into two subgroups with impaired (hypofrontality, n = 18) and normal (n = 33) executive performance.

Table 14.6 Serum neurosteroid levels of schizophrenia patients with impaired and normal performance on the Wisconsin Card Sorting Test (perseverative errors, T-score: <44 = impaired, >44 = normal range)

	Impaired performance (n = 18)		Normal performance (n = 33)		Significance[a]	
Serum neurosteroid levels	Mean	SD	Mean	SD	t	p
DHEA (nmol/l)	42.9	23.9	51.4	26.4	1.2	0.28
DHEAS (nmol/l)	4,835	3,516	5,042	3,506	0.2	0.84
Progesterone (nmol/l)	9.0	3.5	14.6	13.6	2.2	0.032
17-OH-Progesterone (nmol/l)	8.0	4.2	13.8	8.0	3.3	0.002
Androstenedione (nmol/l)	7.3	3.4	9.3	2.3	2.4	0.019
Cortisol (nmol/l)	666.0	319.2	526.2	201.1	1.7	0.11
Cortisol to DHEA molar ratio	18.6	10.6	13.0	8.2	2.2	0.027
Cortisol to DHEAS*100 molar ratio	20.8	15.6	17.1	15.4	0.8	0.42
DHEAS to DHEA molar ratio	136.0	108.1	123.2	127.1	0.4	0.71
Prolactin (nmol/l)	40.8	34.0	30.3	27.0	1.1	0.27

[a]ANCOVA model tests the main effect of the executive function (impaired vs normal performance) by gender effects and controlling for age, PANSS, anxiety (state and trait) and side effect (ESRS) scores

Patients with impaired executive performance were older (41.2 ± 8.2 years), than patients with normal performance (33.7 ± 8.7 years; $t=3.0$, $p=0.004$). However, no significant differences were found between these groups in gender ($\chi^2=1.5$, df = 1, $p=0.22$), length of education ($t=0.2$, $p=0.81$), paranoid vs non-paranoid type of schizophrenia ($\chi^2=0.4$, df = 1, $p=0.53$), illness duration ($t=1.8$, $p=0.085$), number of admissions ($t=1.7$, $p=0.082$), and type of antipsychotic agents ($\chi^2=4.6$, df = 1, $p=0.43$). The daily mean dose of antipsychotics (CPZ equivalent) was 648.1 (SD = 469.6) mg/day, and was comparable in both patient subgroups.

Table 14.6 compares serum basal levels of hormones between patients with WCST hypofrontality and normal executive performance. As can be seen, patients who had hypofrontality features were characterised by lower serum levels of progesterone ($p=0.032$), 17-OH-progesterone ($p=0.002$), and androstenedione ($p=0.019$), as well as, higher value of cortisol to DHEA molar ratio ($p=0.027$) compared with those without hypofrontality. No significant between-group differences were detected in serum levels of DHEA(S), cortisol, and prolactin.

14.12 Antipsychotic Agents

Since most schizophrenia patients were medicated, the possibility still exists that the antipsychotic treatment could affect DHEA(S) circulatory levels in some schizophrenia patients. A previous study in medicated and non-medicated schizophrenia patients has shown a dysregulation of the HPA axis, as assessed with dexamethasone suppression test/corticotropin releasing hormone stimulation test. Treated patients showed a lesser magnitude of the HPA axis dysfunction compared to the untreated patients.[215] Recent animal studies have shown that olanzapine and clozapine administration are associated with an increase in the brain GABA-positive neurosteroid allopregnanolone and decreases in the GABA-negative modulators DHEA(S) levels, respectively.[39,43] Marx et al.[216] investigated potential PREG alterations in rat hippocampus and cerebral cortex following clozapine, olanzapine, and other second generation agents as a candidate of mechanism contributing to antipsychotic efficacy. They reported that clozapine markedly elevates PREG in rat hippocampus, cerebral cortex, and serum; hippocampal levels were strongly correlated with serum levels ($r=0.987$). It is possible that olanzapine- and fluoxetine-induced PREG elevations may contribute to the antidepressant actions of these agents.[217]

Our team repeatedly examined the possible effect of the antipsychotic treatment on DHEA(S) circulatory levels in schizophrenia patients. For instance, patients who received first-, second-generation and both types of antipsychotic agents had no significant differences in the cortisol to DHEA, and cortisol to DHEAS molar ratios. Futhermore, no significant correlations were found between these molar ratios with daily doses of antipsychotic agents.[161] Likewise, DHEA to PREG molar ratio, serum concentrations of PREG, and DHEA did not reach a significant level between schizophrenia patients who received first-, second-generation and both types of antipsychotic agents.[168] Patients who received first-, second-generation antipsychotic agents had comparable serum DHEA(S) and their metabolites' levels.[172] However, it should be noted that antipsychotic treatment and elevated prolactin might stimulate the adrenal cortex to secrete DHEA. Indeed, plasma levels of prolactin and DHEA were found elevated during sulpiride treatment in four out of five healthy men compared with those of the controls.[218] Thus, the possible contribution of antipsychotic treatment to alterations in DHEA metabolism in our schizophrenia patients could not be assessed in the present study, since we did not have a parallel group of untreated patients. Raised prolactin levels are a common consequence of antipsychotic treatment. We found strong evidence for a significant association between prolactin and alterations in DHEA metabolism. Specifically, findings indicate (a) elevated serum prolactin levels among patients compared with controls; (b) significant association of prolactin levels with serum DHEA(S) concentrations, and androstenedione to DHEA(S) molar ratios, but not with androstenedione, progesterone and testosterone concentrations; and (c) alterations in DHEAS and androstenedione concentrations, DHEAS to DHEA and androstenedione to DHEAS molar ratios in schizophrenia are independent, while elevated serum DHEA concentrations and reduced testosterone to androstenedione ratio are related to hyperprolactinemia. However, the direct role of prolactin in the regulation of DHEA(S) is unclear, and the alterations in both prolactin and neurosteroids may be related to independent effects of antipsychotic treatment.

14.12.1 Adverse Side Effects

Only one study reported that blood DHEA levels were inversely correlated with extrapyramidal symptomatology in 17 medicated chronic schizophrenia or schizoaffective inpatients, specifically parkinsonian symptoms.[159] However, this association has not being replicated, and a randomized double-blind, placebo-controlled, crossover 6-week study does not support DHEA's value as an effective adjunct in the treatment of extrapyramidal side effects in schizophrenia (study S_5).[183] In addition, alterations in DHEA(S), and androstenedione in schizophrenia subjects were not related to extrapyramidal side effects (study S_4).[168]

14.12.2 Response to Antipsychotic Treatment

The next study examined the effect of antipsychotics on serum hormones in schizophrenia patients admitted for exacerbation of psychosis, and compared between those who responded to antipsychotic treatment and those who did not respond.[219] The patients were treated with stable doses of antipsychotic agents up to 2 weeks prior to entering the study and for the 4-week duration of the study, after which they were classified as either responders or non-responders to treatment. This study has demonstrated that:

- Responders had significantly higher serum cortisol levels and cortisol to DHEA(S) molar ratios compared with non-responders.

- These differences remained significant across three examinations controlling for gender, age, severity of symptoms and emotional distress, benzodiazepines, type or dosage of antipsychotic agents, and background variables.
- Using hierarchical logistic regression analysis cortisol to DHEA(S) molar ratios were found to be significant predictors of responsivity to antipsychotic treatment. This model accounted for 42% of the variance and correctly classified responders and non-responders – 87.8% of the 43 patients. Values of serum DHEA(S) and gender demonstrated no significant contribution that accounted for prediction of responsiveness to antipsychotic treatment in schizophrenia patients.
- No significant canonical correlations were observed between changes from baseline through end-of-study in hormonal values and PANSS factors and emotional distress scores among both responders and non-responders.

These findings suggest that an imbalance in serum DHEA(S) and cortisol may be related to responsiveness to antipsychotic treatment. It is notable that the group of responders showed improvement in symptom severity after 4 weeks of treatment, though cortisol to DHEA(S) molar ratios and serum cortisol concentrations did not significantly change. Therefore, we may assume that despite improvement, these patients either still remained in the exacerbation stage of illness, or that these hormonal alterations represent a trait-like marker. It should also be noted, that responders had significantly higher basal levels of cortisol, DHEAS and cortisol to DHEA molar ratio values compared with non-responders. Further studies are needed to test these assumptions. The complex relationship between exacerbated chronic schizophrenia, stress, circulatory DHEA(S), HPA activity and response to antipsychotic treatment merits further investigation. Limitations of the study include the predominance of men among the schizophrenia patients. Another potential confounder relates to the lack of assessment of diurnal rhythmicity of hormones. Monitoring over a longer period of time is required. Future studies should include examination of cortisol to DHEA(S) ratios during the administration of DHEA. Understanding the complex interactions of DHEA(S) with cortisol, and other stress-responsive systems may help explain the variability in the time-course and treatment responsivity observed in schizophrenia.

14.13 Conclusions

While somewhat controversial, there is considerable evidence that PREG(S) and DHEA(S) may be altered in schizophrenia.

1. Compared to healthy controls, schizophrenia patients are characterized by relative deficit of blood PREG, and DHEA(S) concentrations, elevated androstenedione levels, and the cortisol to DHEA(S) molar ratios.
2. Variability in serum neurosteroids, and cortisol to DHEA(S) molar ratios are attributed to trait-anxiety and age; they show a weak association with severity of schizophrenia symptoms.

3. Better performance on executive tasks was associated with the higher serum levels of DHEA(S) and androstenedione in schizophrenia.
4. Schizophrenia patients who had hypofrontality features were characterized by lower serum levels of progesterone, 17-OH-progesterone, and androstenedione, as well as by higher value of cortisol to DHEA molar ratio compared with those without hypofrontality.
5. Blood level of DHEA(S) seems associated with responsiveness to antipsychotic treatment: responders to antipsychotic agents had significantly higher serum cortisol levels and cortisol to DHEA(S) molar ratios compared with non-responders.
6. Blood concentrations of prolactin are associated with those of DHEA(S), and androstenedione to DHEA(S) molar ratios, but not with levels of androstenedione, progesterone and testosterone. The direct role of prolactin in the regulation of DHEA(S) is unclear.
7. Demographic, and background variables, daily dose and type of antipsychotic compounds did not explain differences in blood levels of these neurosteroids between schizophrenia patients and healthy subjects.

In conclusion, we suggest that alterations in DHEA(S) and cortisol to DHEA(S) molar ratios in schizophrenia are associated with impaired stress-response and may lead to dysregulated neurotransmission, resulting in chronic and progressive deterioration in emotional, cognitive, and psychosocial functions of patients hypothesized.[161, 168,172] Indeed, several lines of evidence indicate that schizophrenia may be associated with dysregulated HPA axis,[220] and abnormal hormonal responses to stress.[103,221] There are findings that DHEA(S) may play an important role in stress response[58,61] as mediators of the HPA axis adaptation to extreme stress and the psychiatric symptoms, for instance those associated with post-traumatic stress disorder.[59,60,127] However, the precise significance of these neurosteroid alterations in schizophrenia is currently unclear, and a number of questions remain.

- Could these alterations be the result of enhanced biosynthesis and/or decreased metabolism to downstream neuroactive steroids, reflecting changes in the regulation of biosynthetic pathways? The molecular mechanisms underlying the alterations in PREG/DHEA metabolism in schizophrenia are still unclear (see Fig. 14.1). There could be abnormalities in the DHEA pathways themselves. Decreased DHEA sulfotransferase activity and/or increased sulfatase activity in schizophrenia patients may lead to reduced conversion of DHEA to DHEAS and to the observed elevated DHEA and reduced DHEAS concentrations in the serum. In addition, high activity of the 3β-hydroxysteroid dehydrogenase (androstenedione to DHEA ratio) and deficiency in the 17β-hydroxysteroid dehydrogenase (testosterone to androstenedione ratio) enzymes may enhance the conversion of DHEA into androstenedione.
- Could these changes potentially alter the regulation of excitatory and/or inhibitory neurotransmission? It has become increasingly clear that neuromodulatory effects of DHEA(S) on $GABA_A$, NMDA and σ_1 receptors in the brain,[4,27,34,38] and the antiglucocorticoid properties of DHEA[222] may contribute to dysregulation of

behavioral functions such as an emotional response to stress, anxiety, aggressive behavior, learning and memory[223,224] (see Table 14.1).

- It is still possible that the PREG/DHEA findings are not related to a diagnostic specificity, and that the elevated cortisol to DHEA(S) molar ratios in schizophrenia patients may be associated with impaired stress-response, and may lead to dysregulated neurotransmission, resulting in chronic and progressive deterioration in cognitive, emotional and psychosocial functions. Indeed, clinical investigations produced evidence of the involvement of neuroactive steroids in conditions such as fatigue during pregnancy, premenstrual syndrome, epilepsy, dementia, depressive and anxiety disorders. The antiglucocorticoid properties of DHEA may also contribute to an upregulation of HPA axis responses and mitigate possible deleterious effects of high cortisol levels on the brain in schizophrenia subjects. In addition, Marx et al.[225] investigated neuroactive steroid levels in post-mortem brain tissue from subjects with schizophrenia, bipolar disorder, nonpsychotic depression, and control subjects to determine if neuroactive steroids are altered in these disorders. They found that PREG and DHEA are elevated in subjects with schizophrenia and bipolar disorder in both posterior cingulate and parietal cortex. Furthermore, the concentrations of PREG and PREGS were of the same order of magnitude, and about tenfold larger than in plasma.
- Consistent with the stress-vulnerability model we may also assume that elevated cortisol to DHEA(S) molar ratios are a trait-like condition rather than a state-dependent phenomenon in schizophrenia, and that alterations in serum DHEA(S) and cortisol levels among schizophrenia patients may be related to additional unknown factors. Thus, elevated serum cortisol to DHEA(S) molar ratios were attributed to trait-anxiety and age rather than to clinical symptoms. The findings may indicate persistent dysfunction of the HPA axis that is independent of the patients' clinical state.
- What is the meaning of these hormonal differences between responders and non-responders to antipsychotic treatment in schizophrenia? We suggest that imbalances in serum DHEA(S) and cortisol may be related to pathophysiological processes in schizophrenia, particularly, to responsiveness to antipsychotic treatment. Interestingly, the group of responders showed improvement in symptom severity after 4 weeks of treatment, though cortisol to DHEA(S) molar ratios and serum cortisol concentrations did not significantly change. Therefore, we may assume that despite improvement, these patients either still remained in the exacerbation stage of illness, or that these hormonal alterations might represent a trait-like marker. It should be noted, that responders had significantly higher basal levels of cortisol, DHEAS and cortisol to DHEA(S) ratio values compared with non-responders. Further studies are needed to test these assumptions.

The current understanding of the origin of schizophrenia is mainly based on the multifactorial-threshold model of genetic liability and the neurodevelopmental model.[226] Accumulating evidence suggests that schizophrenia may involve a "two-hit" process.[227]

1. In this two-hit process, genetic load, adverse embryonic events, and perinatal events may be considered a neurodevelopmental first hit that leads to vulnerability to schizophrenia.
2. Hormonal events, such as altered neurosteroid biosynthesis during development could act as a second hit, facilitating excitotoxicity or oxygen radical formation in combination with environmental factors.

We speculate that PREG/DHEA abnormalities may increase the sensitivity of "primed" schizophrenia patients (i.e., patients with a first-hit neurodevelopmental insult) to a second "hit" (perhaps an unusual biochemical or psychological stressor) that results in clinically evident schizophrenia. Alternatively, these variable hormonal changes may simply reflect the subtle architectonic disturbances of the developing brain in schizophrenia (i.e., faulty patterns of lamination) and the neurodevelopmental disturbance of synaptic connectivity. Thus, the hormonal alterations may be nonspecific biomarkers reflecting disturbances of cortical transmission that impact the hypothalamus and, subsequently, the pituitary–adrenal circuit.

Following are a number of issues that require further studies.

- The complex relationship between exacerbated chronic schizophrenia, stress, circulatory DHEA(S), HPA activity and response to antipsychotic treatment.
- Long-term monitoring of the relationship between the hormonal alterations and the psychotic symptomatology, depressive-anxiety manifestations, aggressive-suicidal behavior, negative symptoms and cognitive deficits in first-onset and chronic schizophrenia patients.
- Future studies should monitor the hormonal alterations in drug-free patients, as well as include examination of cortisol to DHEA(S) ratios during the administration of DHEA. Further research is necessary to clarify the relationship of neurosteroids with cognitive and quality of life impairments.
- Well-designed brain imaging studies are needed to explore the relationship between blood PREG/DHEA(S) levels and neuroimaging findings including the amygdala, insula, ventral striatum, and ventral regions of the anterior cingulate gyrus and prefrontal cortex.

In summary, experimental and clinical observations support the speculation that neurobiological alterations in PREG/DHEA neurosteroids and their derivates are related to the pathophysiology of schizophrenia. Based on the accumulated evidence from animal and human studies, it is also possible to conclude that PREG/DHEA play a significant role in the expressions of stress response, anxiety, and cognitive deficit in schizophrenia. Finally, these insights underscore the need for developing novel treatment strategies such as neuroprotective strategies using neurosteroids and other compounds, which should help overcome the limitations of current antipsychotic drugs and improve the cognitive deficits and negative symptoms, as well as functioning and quality of life outcomes of people affected by schizophrenia. Neuroprotective effects of these agents as add-on therapies merit further investigation in schizophrenia and related disorders.

Acknowledgments The authors wish to express their deepest gratitude for the scientific contributions of their colleagues Dr. Rael Strous, Ms. Rachel Maayan and to Dr. Edward Ram for doing the hormone assays, as indicated in the referenced papers. The authors would also like to thank Micky Gerchak for editorial services. This work was partly supported by generous grants from the Stanley Foundation.

References

1. Baulieu EE, Robel P. Dehydroepiandrosterone and dehydroepiandrosterone sulfate as neuroactive neurosteroids. J Endocrinol 1996; 150(Suppl):221–239.
2. Mao X, Barger SW. Neuroprotection by dehydroepiandrosterone-sulfate: role of an NFkappaB-like factor. Neuroreport 1998; 9:759–763.
3. Friess E, Schiffelholz T, Steckler T, Steiger A. Dehydroepiandrosterone – a neurosteroid. Eur J Clin Invest 2000; 30(Suppl 3):46–50.
4. Wen S, Dong K, et al. Treatment with dehydroepiandrosterone sulfate increases NMDA receptors in hippocampus and cortex. Eur J Pharmacol 2001; 430:373–374.
5. Johnson M, Bebb R, Sirrs S. Uses of DHEA in aging and other disease states. Ageing Res Rev 2002; 1:29–41.
6. Corpechot C, Robel P, Axelson M, Sjovall J, Baulieu EE. Characterization and measurement of dehydroepiandrosterone sulfate in rat brain. Proc Natl Acad Sci USA 1981; 78:4704–4707.
7. Baulieu EE. Neurosteroids: of the nervous system, by the nervous system, for the nervous system. Recent Prog Horm Res 1997; 52:1–32.
8. Epperson CN, Wisner KL, Yamamoto B. Gonadal steroids in the treatment of mood disorders. Psychosom Med 1999; 61:676–697.
9. Sundstrom Poromaa I, Smith S, Gulinello M. GABA receptors, progesterone and premenstrual dysphoric disorder. Arch Women Ment Health 2003; 6:23–41.
10. Eser D, Schule C, Baghai TC, et al. Neuroactive steroids in depression and anxiety disorders: clinical studies. Neuroendocrinology 2006; 84:244–254.
11. Compagnone NA, Mellon SH. Neurosteroids: biosynthesis and function of these novel neuromodulators. Front Neuroendocrinol 2000; 21:1–56.
12. Stoffel-Wagner B. Neurosteroid metabolism in the human brain. Eur J Endocrinol 2001; 145:669–679.
13. Stoffel-Wagner B. Neurosteroid biosynthesis in the human brain and its clinical implications. Ann NY Acad Sci 2003; 1007:64–78.
14. Labrie F, Luu-The V, Belanger A, et al. Is dehydroepiandrosterone a hormone? J Endocrinol 2005; 187:169–196.
15. Le Goascogne C, Jo DH, Corpéchot C. Neurosteroids: biosynthesis, metabolism and function of pregnenolone and dehydroepiandrosterone in the brain. J Steroid Biochem 1991; 40:71–81.
16. Rupprecht R. Neuroactive steroids: mechanisms of action and neuropsychopharmacological properties. Psychoneuroendocrinology 2003; 28:139–168.
17. Baulieu EE. Neurosteroids: a novel function of the brain. Psychoneuroendocrinology 1998; 23:963–987.
18. Zwain IH, Yen SS. Dehydroepiandrosterone: biosynthesis and metabolism in the brain. Endocrinology 1999; 140:880–887.
19. Falany CN, Comer KA, Dooley TP, Glatt H. Human dehydroepiandrosterone sulfotransferase. Purification, molecular cloning, and characterization. Ann NY Acad Sci 1995; 774:59–72.
20. Rainey WE, Carr BR, Sasano H, Suzuki T, Mason JI. Dissecting human adrenal androgen production. Trends Endocrinol Metab 2002; 13:234–239.

21. Regelson W, Kalimi M. Dehydroepiandrosterone (DHEA) – the multifunctional steroid. II. Effects on the CNS, cell proliferation, metabolic and vascular, clinical and other effects. Mechanism of action? Ann NY Acad Sci 1994; 719:564–575.
22. King DS, Sharp RL, Vukovich MD, et al. Effect of oral androstenedione on serum testosterone and adaptations to resistance training in young men: a randomized controlled trial. JAMA 1999; 281:2020–2028.
23. Broeder CE, Quindry J, Brittingham K, et al. The Andro Project: physiological and hormonal influences of androstenedione supplementation in men 35 to 65 years old participating in a high-intensity resistance training program. Arch Intern Med 2000; 160:3093–3104.
24. Bird CE, Masters V, Clark AF. Dehydroepiandrosterone sulfate: kinetics of metabolism in normal young men and women. Clin Invest Med 1984; 7:119–122.
25. Longcope C. Dehydroepiandrosterone metabolism. J Endocrinol 1996(Suppl); 150:125–127.
26. Kroboth PD, Salek FS, Pittenger AL, et al. DHEA and DHEA-S: a review. J Clin Pharmacol 1999; 39:327–348.
27. Rupprecht R. The neuropsychopharmacological potential of neuroactive steroids. J Psychiatr Res 1997; 31:297–314.
28. Mellon SH, Griffin LD, Compagnone NA. Biosynthesis and action of neurosteroids. Brain Res Rev 2001; 37:3–12.
29. Labrie F, Belanger A, Van LT, et al. DHEA and the intracrine formation of androgens and estrogens in peripheral target tissues: its role during aging. Steroids 1998; 63:322–328.
30. Morissette M, Dicko A, Pezolet M, et al. Effect of dehydroepiandrosterone and its sulfate and fatty acid ester derivatives on rat brain membranes. Steroids 1999; 64:796–803.
31. Peng W, Hoidal JR, Farrukh IS. Role of a novel KCa opener in regulating K channels of hypoxic human pulmonary vascular cells. Am J Respir Cell Mol Biol 1999; 20:737–745.
32. Gordon G, Mackow MC, Levy HR. On the mechanism of interaction of steroids with human glucose 6-phosphate dehydrogenase. Arch Biochem Biophys 1995; 318:25–29.
33. Widstrom RL, Dillon JS. Adrenal androgens in human physiology. Bruce RC (editor in chief); William ER, Bruce RC (guest editors) Semin Reprod Med 2004; 22:289–298.
34. Majewska MD. Neurosteroids: endogenous bimodal modulators of the GABA-A receptor. Mechanism of action and physiological significance. Prog Neurobiol 1992; 38:379–395.
35. Bitran D, Shiekh M, McLeod M. Anxiolytic effect of progesterone is mediated by the neurosteroid allopregnanolone at brain $GABA_A$ receptors. J Neuroendocrinol 1995; 7:171–177.
36. Majewska MD, Demigören S, Spivak CE, London ED. The neurosteroid dehydroepiandrosterone sulfate is an allosteric antagonist of the $GABA_A$ receptor. Brain Res 1990; 526:143–146.
37. Demirgoren S, Majewska MD, Spivak CE, London ED. Receptor binding and electrophysiological effects of dehydroepiandrosterone sulfate, an antagonist of the GABAA receptor. Neuroscience 1991; 45:127–135.
38. Debonnel G, Bergeron R, de Montigny C. Potentiation by dehydroepiandrosterone of the neuronal response to N-methyl-D-aspartate in the CA3 region of the rat dorsal hippocampus: an effect mediated via sigma receptors. J Endocrinol 1996(Suppl) 150:33–42.
39. Marx CE, Jarskog LF, Lauder JM, et al. Neurosteroid modulation of embryonic neuronal survival in vitro following anoxia. Brain Res 2000; 871:104–112.
40. Paul SM, Purdy RH. Neuroactive steroids. FASEB J 1992; 6:2311–2322.
41. Lambert JJ, Belelli D, Hill-Venning C, Peters JA. Neurosteroids and $GABA_A$ receptor function. Trends Pharmacol Sci 1995; 16:295–303.
42. Do-Rego JL, Mensah-Nyagan GA, Beaujean D, Vaudry D, et al. gamma-Aminobutyric acid, acting through gamma-aminobutyric acid type A receptors, inhibits the biosynthesis of neurosteroids in the frog hypothalamus. Proc Natl Acad Sci USA 2000; 97:13925–13930.
43. Nechmad A, Maayan R, Ramadan E, et al. Clozapine decreases rat brain dehydroepiandrosterone and dehydroepiandrosterone sulfate levels. Eur Neuropsychopharmacol 2003; 13:29–31.
44. Fahey JM, Lindquist DG, Pritchard GA, Miller LG. Pregnenolone sulfate potentiation of NMDA-mediated increases in intracellular calcium in cultured chick cortical neurons. Brain Res 1995; 669:183–188.

45. Monnet FP, Mahé V, Robel P, Baulieu EE. Neurosteroids, via sigma receptors, modulate the [^{3}H]norepinephrine release evoked by N-methyl-D-aspartate in the rat hippocampus. Proc Natl Acad Sci USA 1995; 92:3774–3778.
46. Sun JL, Dong YL, Fu YM, et al. Neurosteroid pregnenolone sulfate inhibits stimulus-evoked EPSC via presynaptic inhibition of protein kinase A in rat prelimbic cortical neurons. Neuropharmacology 2005; 49:389–399.
47. Horak M, Vlcek K, Chodounska H, Vyklicky L, Jr. Subtype-dependence of N-methyl-D-aspartate receptor modulation by pregnenolone sulfate. Neuroscience 2006; 137:93–102.
48. Hige T, Fujiyoshi Y, Takahashi T. Neurosteroid pregnenolone sulfate enhances glutamatergic synaptic transmission by facilitating presynaptic calcium currents at the calyx of Held of immature rats. Eur J Neurosci 2006; 24:1955–1966.
49. Bergeron R, de Montigny C, et al. Potentiation of neuronal NMDA response induced by dehydroepiandrosterone and its suppression by progesterone: effects mediated via sigma receptors. J Neurosci 1996; 16:1193–1202.
50. Schiess AR, Partridge LD. Pregnenolone sulfate acts through a G-protein-coupled sigma1-like receptor to enhance short term facilitation in adult hippocampal neurons. Eur J Pharmacol 2005; 518:22–29.
51. Reddy DS, Kaur G, Kulkarni SK. Sigma (σ_1) receptor mediated antidepressant-like effects of neurosteroids in the Porsolt forced swim test. Neuroreport 1998; 9:3069–3073.
52. Melchior CL, Ritzmann RF. Dehydroepiandrosterone is an anxiolytic in mice on the plus maze. Pharmacol Biochem Behav 1994; 47:437–444.
53. Belelli D, Lambert JJ. Neurosteroids: endogenous regulators of the GABA(A) receptor. Nat Rev Neurosci 2005; 6:565–575.
54. Morrow AL, Devaud LL, Purdy RH, Paul SM. Neuractive steroid modulators of the stress response. In: Chrousos GP, McCarty R, Pacák K, Cizza G, Sternberg E, Gold PW, Kvetnansky R (eds). Stress: basic mechanisms and clinical implications. New York: The New York Academy of Sciences, 1995, 257–272.
55. Dong E, Matsumoto K, Uzunova V, et al. Brain 5alpha-dihydroprogesterone and allopregnanolone synthesis in a mouse model of protracted social isolation. Proc Natl Acad Sci USA 2001; 98:2849–2854.
56. Wolkowitz OM, Epel ES, et al. Stress hormone-related psychopathology: pathophysiological and treatment implications. World J Biol Psychiatry 2001; 2:115–143.
57. Hu Y, Cardounel A, et al. Anti-stress effects of dehydroepiandrosterone: protection of rats against repeated immobilization stress-induced weight loss, glucocorticoid receptor production, and lipid peroxidation. Biochem Pharmacol 2000; 59:753–762.
58. Boudarene M, Legros JJ, et al. Study of the stress response: role of anxiety, cortisol and DHEAs. Encephale 2002; 28:139–146.
59. Rasmusson AM, Vythilingam M, Morgan CA 3rd. The neuroendocrinology of posttraumatic stress disorder: new directions. CNS Spectr 2003; 8:651–566, 665–667.
60. Rasmusson AM, Vasek J, Lipschitz DS, et al. An increased capacity for adrenal DHEA release is associated with decreased avoidance and negative mood symptoms in women with PTSD. Neuropsychopharmacology 2004; 29:1546–1557.
61. Obut TA, Ovsyukova MV, Cherkasova OP. Prolonged decrease in stress reactivity caused by dehydroepiandrosterone sulfate. Bull Exp Biol Med 2006; 141:571–573.
62. Kalimi M, Shafagoj Y, Loria R. Anti-glucocorticoid effects of dehydroepiandrosterone (DHEA). Mol Cell Biochem 1994; 131:99–104.
63. Kimonides VG, Spillantini MG, et al. Dehydroepiandrosterone antagonizes the neurotoxic effects of corticosterone and translocation of stress-activated protein kinase 3 in hippocampal primary cultures. Neuroscience 1999; 89:429–436.
64. Karishma KK, Herbert J. Dehydroepiandrosterone (DHEA) stimulates neurogenesis in the hippocampus of the rat, promotes survival of newly formed neurons and prevents corticosterone-induced suppression. Eur J Neurosci 2002; 16:445–453.
65. Parker LN, Levin ER, Lifrak ET. Evidence for adrenocortical adaptation to severe illness. J Clin Endocrinol Metab 1985; 60:947–952.

66. Parker L, Eugene J, Farber D, et al. Dissociation of adrenal androgen and cortisol levels in acute stress. Horm Metab Res 1985; 17:209–212.
67. Hechter O, Grossman A, Chatterton RT. Relationship of dehydroepiandrosterone and cortisol in disease. Med Hypotheses 1997; 49:85–91.
68. Goodyer IM, Herbert J, Altham PM. Adrenal steroid secretion and major depression in 8- to 16-year-olds. III. Influence of cortisol/DHEA ratio at presentation on subsequent rates of disappointing life events and persistent major depression. Psychol Med 1998; 28:265–273.
69. Goodyer IM, Park RJ, Netherton CM, Herbert J. Possible role of cortisol and dehydroepiandrosterone in human development and psychopathology. Br J Psychiatry 2001; 179:243–249.
70. Kroboth PD, Amico JA, Stone RA, et al. Influence of DHEA administration on 24-hour cortisol concentrations. J Clin Psychopharmacol 2003; 23:96–99.
71. Van Broekhoven F, Verkes RJ. Neurosteroids in depression: a review. Psychopharmacology (Berl) 2003; 165:97–110.
72. Basta-Kaim A, Budziszewska B, Jaworska-Feil L, et al. Effects of neurosteroids on glucocorticoid receptor-mediated gene transcription in LMCAT cells – A possible interaction with psychotropic drugs. Eur Neuropsychopharmacol 2007; 17:37–45.
73. Roberts E, Bologa L, Flood JF, Smith GE. Effects of dehydroepiandrosterone and its sulfate on brain tissue in culture and on memory in mice. Brain Res 1987; 406:357–362.
74. Bologa L, Sharma J, et al. Dehydroepiandrosterone and its sulfated derivative reduce neuronal death and enhance astrocytic differentiation in brain cell cultures. J Neurosci Res 1987; 17:225–234.
75. Schumacher M, Robel P, Baulieu EE. Development and regeneration of the nervous system: a role for neurosteroids. Dev Neurosci 1996; 18:6–21.
76. Schumacher M, Guennoun R, Robel P, Baulieu EE. Neurosteroids in the Hippocampus. Neuronal Plasticity and Memory 1997; 2:65–78.
77. Compagnone NA, Mellon SH. Dehydroepiandrosterone: a potential signalling molecule for neocortical organization during development. Proc Natl Acad Sci USA 1998; 95:4678–4683.
78. Gursoy E, Cardounel A, et al. Pregnenolone protects mouse hippocampal (HT-22) cells against glutamate and amyloid beta protein toxicity. Neurochem Res 2001; 26:15–21.
79. Shirakawa H, Katsuki H, Kume T, et al. Pregnenolone sulphate attenuates AMPA cytotoxicity on rat cortical neurons. Eur J Neurosci 2005; 21:2329–2335.
80. Veiga S, Garcia-Segura LM, et al. Neuroprotection by the steroids pregnenolone and dehydroepiandrosterone is mediated by the enzyme aromatase. J Neurobiol 2003; 56:398–406.
81. Charalampopoulos I, Alexaki VI, Tsatsanis C, et al. Neurosteroids as endogenous inhibitors of neuronal cell apoptosis in aging. Ann NY Acad Sci 2006; 1088:139–152.
82. Campbell B. Adrenarche and the evolution of human life history. Am J Hum Biol 2006; 18:569–589.
83. Flood JF, Morley JE, Roberts E. Memory-enhancing effects in male mice of pregnenolone and of steroids metabolically derived from it. Proc Natl Acad Sci USA 1992; 89:1567–1571.
84. Flood J., et al. Pregnenolone sulfate enhances post-training memory processes when injected in very low doses into limbic system structures. Proc Nat Acad Sci USA 1995; 92: 10806–108010.
85. Robel P, Young J, Corpechot C, et al. Biosynthesis and assay of neurosteroids in rats and mice: functional correlates. J Steroid Biochem Mol Biol 1995; 53:355–360.
86. Mayo W, Dellu F, Robel P, et al. Infusion of neurosteroids into the nucleus basalis magnocellularis affects cognitive processes in the rat. Brain Res 1993; 607:324–328.
87. Isaacson RL, Varner JA, Baars J-M, de Wied D. The effects of pregnenolone sulfate and ethylestrenol on retention of a passive avoidance task. Brain Res 1995; 689:79–84.
88. Mathis C, Paul SM, Crawley JN. The neurosteroid pregnenolone sulfate blocks NMDA antagonist-induced deficits in a passive avoidance memory task. Psychopharmacology (Berl) 1994; 116:201–206.
89. Cheney DL, Uzunov D, Guidotti A. Pregnenolone sulfate antagonizes dizocilpine amnesia: role for allopregnanolone. Neuroreport 1995; 6:1697–1700.

90. Melchior CL, Ritzmann RF. Neurosteroids block the memory-impairing effects of ethanol in mice. Pharmacol Biochem Behav 1996; 53:51–56.
91. Bowlby MR. Pregnenolone sulfate potentiation of N-methyl-d-aspartate receptor channels in hippocampal neurons. Mol Pharmacol 1993; 43:813–819.
92. Mathis C, Vogel E, Cagniard B, et al. The neurosteroid pregnenolone sulfate blocks deficits induced by a competitive NMDA antagonist in active avoidance and lever-press learning tasks in mice. Neuropharmacology 1996; 35:1057–1064.
93. Markowski M, Ungeheuer M, Bitran D, Locurto C. Memory-enhancing effects of DHEAS in aged mice on a win-shift water escape task. Physiol Behav 2001; 72:521–525.
94. Vallee M, Mayo W, Le Moal M. Role of pregnenolone, dehydroepiandrosterone and their sulfate esters on learning and memory in cognitive aging. Brain Res Brain Res Rev 2001; 37:301–312.
95. Rhodes ME, Li PK, et al. Enhancement of hippocampal acetylcholine release by the neurosteroid dehydroepiandrosterone sulfate: an in vivo microdialysis study. Brain Res 1996; 733:284–286.
96. Rhodes ME, Li PK, et al. Enhanced plasma DHEAS, brain acetylcholine and memory mediated by steroid sulfatase inhibition. Brain Res 1997; 773:28–32.
97. Johnson DA, Rhodes ME, et al. Chronic steroid sulfatase inhibition by (p-O-sulfamoyl)-N-tetradecanoyl tyramine increases dehydroepiandrosterone sulfate in whole brain. Life Sci 1997; 61:355–359.
98. Li PK, Rhodes ME, et al. Memory enhancement mediated by the steroid sulfatase inhibitor (p-O-sulfamoyl)-N-tetradecanoyl tyramine. Life Sci 1997; 60:45–51.
99. Lupien SJ, Fiocco A, Wan N, et al. Stress hormones and human memory function across the lifespan. Psychoneuroendocrinology 2005; 30:225–242.
100. Roozendaal B, McGaugh JL. Amygdaloid nuclei lesions differentially affect glucocorticoid-induced memory enhancement in an inhibitory avoidance task. Neurobiol Learn Mem 1996; 65:1–8.
101. El-Etr M, Akwa Y, Baulieu EE, Schumacher M. The neuroactive steroid pregnenolone sulfate stimulates the release of gonadotropin-releasing hormone from GT1–7 hypothalamic neurons, through N-methyl-D-aspartate receptors. Endocrinology 2006; 147:2737–2743.
102. Diamond DM, Branch BJ, et al. The neurosteroid dehydroepiandrosterone sulfate (DHEAS) enhances hippocampal primed burst, but not long-term, potentiation. Neurosci Lett 1996; 202:204–208.
103. Wolf OT, Kirschbaum C. Actions of dehydroepiandrosterone and its sulfate in the central nervous system: effects on cognition and emotion in animals and humans. Brain Res Rev 1999; 30:264–288.
104. Labrie C, Flamand M, Belanger A, Labrie F. High bioavailability of dehydroepiandrosterone administered percutaneously in the rat. J Endocrinol 1996; 150(Suppl):107–118.
105. Mo Q, Lu SF, et al. Dehydroepiandrosterone and its metabolites: differential effects on androgen receptor trafficking and transcriptional activity. J Steroid Biochem Mol Biol 2006; 99:50–58.
106. Roberts E, Fitten LJ. Serum steroid levels in two old men with Alzheimer's disease (AD) before and after oral administration of dehydroepiandrosterone (DHEA). Pregnenolone synthesis may be rate limiting in aging. In: Kalimi M, Regelson W (eds). The biological role of dehydroepiandrosterone (DHEA). Berlin: de Gruyter, 1990, pp. 43–63.
107. Morley JE, Kaiser F, et al. Potentially predictive and manipulable blood serum correlates of aging in the healthy human male: progressive decreases in bioavailable testosterone, dehydroepiandrosterone sulfate, and the ratio of insulin-like growth factor 1 to growth hormone. Proc Natl Acad Sci USA 1997; 94:7537–7542.
108. Orentreich N, Brind JL, Vogelman JH, et al. Long-term longitudinal measurements of plasma dehydro-epiandrosterone sulfate in normal men. J Clin Endocrinol Metab 2002; 75:1002–1004.
109. Labrie F, Belanger A, Cusan L, et al. Marked decline in serum concentrations of adrenal C19 sex steroid precursors and conjugated androgen metabolites during aging. J Clin Endocrinol Metab 1997; 82:2396–2402.

110. Nafziger AN, Bowlin SJ, et al. Longitudinal changes in dehydroepiandrosterone concentrations in men and women. J Lab Clin Med 1998; 131:316–323.
111. Carlström K, Brody S, Lunell N-O, et al. Dehydroepiandrosterone sulphate and dehydroepiandrosterone in serum: differences related to age and sex. Maturitas 1988; 10:297–306.
112. Orentreich N, Brind JL, Rizer RL, et al. Age changes and sex differences in serum dehydroepiandrosterone sulfate concentrations throughout adulthood. J Clin Endocrinol Metab 1984; 59:551–555.
113. Zumoff B, Bradlow HL. Sex difference in the metabolism of dehydroisoandrosterone sulfate. J Clin Endocrinol Metab 1980; 51:334–336.
114. Glei DA, Goldman N, Weinstein M, Liu IW. Dehydroepiandrosterone sulfate (DHEAS) and health: does the relationship differ by sex? Exp Gerontol 2004; 39:321–331.
115. Migeon CJ, Plager JE. Identification and isolation of dehydroisoandrosterone from peripheral human plasma. J Biol Chem 1954; 209:767–772.
116. Guazzo EP, Kirkpatrick PJ, et al. Cortisol, dehydroepiandrosterone (DHEA), and DHEA sulfate in the cerebrospinal fluid of man: relation to blood levels and the effects of age. J Clin Endocrinol Metab 1996; 81:3951–3960.
117. Yildiz BO, Goodarzi MO, et al. Heritability of dehydroepiandrosterone sulfate in women with polycystic ovary syndrome and their sisters. Fertil Steril 2006; 86:1688–1693.
118. Feldman HA, Johannes CB, et al. Low dehydroepiandrosterone sulfate and heart disease in middle-aged men: cross-sectional results from the Massachusetts Male Aging Study. Ann Epidemiol 1998; 8:217–228.
119. Feldman HA, Johannes CB, et al. Low dehydroepiandrosterone and ischemic heart disease in middle-aged men: prospective results from the Massachusetts male aging study. Am J Epidemiol 2001; 153:79–89.
120. van Niekerk JK, Huppert FA, et al. Salivary cortisol and DHEA: association with measures of cognition and well-being in normal older men, and effects of three months of DHEA supplementation. Psychoneuroendocrinology 2001; 26:591–612.
121. Barrett-Connor E, Ferrara A. Dehydroepiandrosterone, dehydroepiandrosterone sulfate, obesity, waist-hip ratio, and noninsulin-dependent diabetes in postmenopausal women: the Rancho Bernardo Study. J Clin Endocrinol Metab 1996; 81:59–64.
122. Tellez N, Comabella M, et al. Fatigue in progressive multiple sclerosis is associated with low levels of dehydroepiandrosterone. Mult Scler 2006; 12:487–494.
123. Narvaez J, Bernad B, et al. Low serum levels of DHEAS in untreated polymyalgia rheumatica/giant cell arteritis. J Rheumatol 2006; 33:1293–1298.
124. Yamakita N, Murai T, Kokubo Y, et al. Dehydroepiandrosterone sulphate is increased and dehydroepiandrosterone-response to corticotrophin-releasing hormone is decreased in the hyperthyroid state compared with the euthyroid state. Clin Endocrinol (Oxf) 2001; 55:797–803.
125. Oberbeck R, Benschop RJ, Jacobs R, et al. Endocrine mechanisms of stress-induced DHEA-secretion. J Endocrinol Invest 1998; 21:148–153.
126. Morgan CA 3rd, Southwick S, Hazlett G, et al. Relationships among plasma dehydroepiandrosterone sulfate and cortisol levels, symptoms of dissociation, and objective performance in humans exposed to acute stress. Arch Gen Psychiatry 2004; 61:819–825.
127. Spivak B, Maayan R, Kotler M, et al. Elevated circulatory level of GABA(A) – antagonistic neurosteroids in patients with combat-related post-traumatic stress disorder. Psychol Med 2000; 30:1227–1231.
128. Monteleone P, Luisi M, Colurcio B, et al. Plasma levels of neuroactive steroids are increased in untreated women with anorexia nervosa or bulimia nervosa. Psychosom Med 2001; 63:62–68.
129. Yehuda R, Brand SR, Golier JA, Yang RK. Clinical correlates of DHEA associated with post-traumatic stress disorder. Acta Psychiatr Scand 2006; 114:187–193.
130. Jogems-Kosterman BJ, de Knijff DW, Kusters R, van Hoof JJ. Basal cortisol and DHEA levels in women with borderline personality disorder. J Psychiatr Res 2006 [epub ahead of print].
131. Monteleone P, Luisi M, Martiadis V, et al. Impaired reduction of enhanced levels of dehydroepiandrosterone by oral dexamethasone in anorexia nervosa. Psychoneuroendocrinology 2006; 31:537–542.

132. Cleare AJ, O'Keane V, et al. Levels of DHEA and DHEAS and responses to CRH stimulation and hydrocortisone treatment in chronic fatigue syndrome. Psychoneuroendocrinology 2004; 29:724–732.
133. George MS, Guidotti A, Rubinow D, et al. CSF neuroactive steroids in affective disorders: pregnenolone, progesterone, and DBI. Biol Psychiatry 1994; 35:775–780.
134. Bicikova M, Tallova J, Hill M, et al. Serum concentrations of some neuroactive steroids in women suffering from mixed anxiety-depressive disorder. Neurochem Res 2000; 25:1623–1627.
135. Semeniuk T, Jhangri GS, Le Melledo JM. Neuroactive steroid levels in patients with generalized anxiety disorder. J Neuropsychiatry Clin Neurosci 2001; 13:396–398.
136. Heydari B, Le Melledo JM. Low pregnenolone sulphate plasma concentrations in patients with generalized social phobia. Psychol Med 2002; 32:929–933.
137. Tagawa N, Tamanaka J, et al. Serum dehydroepiandrosterone, dehydroepiandrosterone sulfate, and pregnenolone sulfate concentrations in patients with hyperthyroidism and hypothyroidism. Clin Chem 2000; 46:523–528.
138. Hill M, Popov P, et al. Altered profiles of serum neuroactive steroids in premenopausal women treated for alcohol addiction. Steroids 2005; 70:515–524.
139. Takebayashi M, Kagaya A, Uchitomi Y, et al. Plasma dehydroepiandrosterone sulfate in unipolar major depression. Short communication. J Neural Transm 1998; 105:537–542.
140. Heuser I, Deuschle M, Luppa P, et al. Increased diurnal plasma concentrations of dehydroepiandrosterone in depressed patients. J Clin Endocrinol Metab 1998; 83:3130–3133.
141. Maayan R, Yagorowski Y, Grupper D, et al. Basal plasma dehydroepiandrosterone sulfate level: a possible predictor for response to electroconvulsive therapy in depressed psychotic inpatients. Biol Psychiatry 2000; 48:693–701.
142. Michael A, Jenaway A, Paykel ES, Herbert J. Altered salivary dehydroepiandrosterone levels in major depression in adults. Biol Psychiatry 2000; 48:989–995.
143. Grillon C, Pine DS, Baas JM, et al. Cortisol and DHEA-S are associated with startle potentiation during aversive conditioning in humans. Psychopharmacology (Berl.) 2006; 186:434–441.
144. Cruess DG, Antoni MH, Kumar M, et al. Cognitive-behavioral stress management buffers decreases in dehydroepiandrosterone sulfate (DHEA-S) and increases in the cortisol/DHEA-S ratio and reduces mood disturbance and perceived stress among HIV-seropositive men. Psychoneuroendocrinology 1999; 24:537–549.
145. Pajer K, Tabbah R, Gardner W, et al. Adrenal androgen and gonadal hormone levels in adolescent girls with conduct disorder. Psychoneuroendocrinology 2006; 31:1245–1256.
146. Osran H, Reist C, Chen CC, et al. Adrenal androgens and cortisol in major depression. Am J Psychiatry 1993; 150:806–809.
147. Scott LV, Salahuddin F, Cooney J, et al. Differences in adrenal steroid profile in chronic fatigue syndrome, in depression and in health. J Affect Disord 1999; 54:129–137.
148. Reus VI, Wolkowitz OM, Frederick S. Antiglucocorticoid treatments in psychiatry. Psychoneuroendocrinology 1997; 22(Suppl 1):121–124.
149. Fabian TJ, Dew MA, Pollock BG, et al. Endogenous concentrations of DHEA and DHEA-S decrease with remission of depression in older adults. Biol Psychiatry 2001; 50:767–774.
150. Goodyer IM, Herbert J, et al. Adrenal secretion during major depression in 8- to 16-year-olds. I. Altered diurnal rhythms in salivary cortisol and dehydroepiandrosterone (DHEA) at presentation. Psychol Med 1996; 26:245–256.
151. Young AH, Gallagher P, Porter RJ. Elevation of the cortisol-dehydroepiandrosterone ratio in drug-free depressed patients. Am J Psychiatry 2002; 159:1237–1239.
152. Goodyer IM, Herbert J, Tamplin A. Psychoendocrine antecedents of persistent first-episode major depression in adolescents: a community-based longitudinal enquiry. Psychol Med 2003; 33:601–610.
153. Ferrari E, Mirani M, et al. Cognitive and affective disorders in the elderly: a neuroendocrine study. Arch Gerontol Geriatr 2004 (Suppl 9):171–182.

154. Gallagher P, Young A. Cortisol/DHEA ratios in depression. Neuropsychopharmacology 2002; 26:410.
155. Kahl KG, Bens S, Ziegler K, et al. Cortisol, the cortisol-dehydroepiandrosterone ratio, and pro-inflammatory cytokines in patients with current major depressive disorder comorbid with borderline personality disorder. Biol Psychiatry 2006; 59:667–671.
156. de Bruin VM, Vieira MC, Rocha MN, Viana GS. Cortisol and dehydroepiandosterone sulfate plasma levels and their relationship to aging, cognitive function, and dementia. Brain Cogn 2002; 50:316–323.
157. Christeff N, Nunez EA, Gougeon ML. Changes in cortisol/DHEA ratio in HIV-infected men are related to immunological and metabolic perturbations leading to malnutrition and lipodystrophy. Ann NY Acad Sci 2000; 917:962–970.
158. Brophy MH, Rush AJ, Crowley G. Cortisol, estradiol, and androgens in acutely ill paranoid schizophrenics. Biol Psychiatry 1983; 18:583–590.
159. Harris DS, Wolkowitz OM, Reus VI. Movement disorder, memory, psychiatric symptoms and serum DHEA levels in schizophrenic and schizoaffective patients.World J Biol Psychiatry 2001; 2:99–102.
160. Goyal RO, Sagar R, Ammini AC, et al. Negative correlation between negative symptoms of schizophrenia and testosterone levels. Ann NY Acad Sci 2004; 1032:291–294.
161. Ritsner M, Maayan R, Gibel A, et al. Elevation of the cortisol/dehydroepiandrosterone ratio in schizophrenia patients. Eur Neuropsychopharmacol 2004; 14:267–273.
162. Yilmaz N, Herken H, Cicek HK, et al. Increased levels of nitric oxide, cortisol and adrenomedullin in patients with chronic schizophrenia. Med Princ Pract 2007; 16:137–141.
163. Tourney G, Hatfield L. Plasma androgens in male schizophrenics. Arch Gen Psychiatry 1972; 27:753–755.
164. Oertel GW, Benes P, Schirazi M, et al. Interaction between dehydroepiandrosterone, cyclic adenosine-3',5'-monophosphate and glucose-6-phosphate-dehydrogenase in normal and diseased subjects. Experientia 1974; 30:872–873.
165. Oades RD, Schepker R. Serum gonadal steroid hormones in young schizophrenic patients. Psychoneuroendocrinology 1994; 19:373–385.
166. Strous RD, Maayan R, Lapidus R, et al. Increased circulatory dehydroepiandrosterone and dehydroepiandrosterone-sulphate in first-episode schizophrenia: relationship to gender, aggression and symptomatology. Schizophr Res 2004; 71:427–434.
167. di Michele F, Caltagirone C, Bonaviri G, et al. Plasma dehydroepiandrosterone levels are strongly increased in schizophrenia. J Psychiatr Res 2005; 39:267–273.
168. Ritsner M, Gibel A, Ram E, et al. Alterations in DHEA metabolism in schizophrenia: two-month case-control study. Eur Neuropsychopharmacol 2006; 16:137–146.
169. Gallagher P, Watson S, Smith MS, et al. Plasma cortisol-dehydroepiandrosterone (DHEA) ratios in schizophrenia and bipolar disorder. Schizophr Res 2007; 90:258–265.
170. Tourney G, Erb JL. Temporal variations in androgens and stress hormones in control and schizophrenic subjects. Biol Psychiatry 1979; 14:395–404.
171. Erb JL, Kadane JB, Tourney G, et al. Discrimination between schizophrenic and control subjects by means of plasma dehydroepiandrosterone measurements. J Clin Endocrinol Metab 1981; 52:181–186.
172. Ritsner M, Gibel A, Maayan R, et al. State and trait related predictors of serum cortisol to DHEA(S) molar ratios and hormone concentrations in schizophrenia patients. Eur Neuropsychopharmacol 2007; 17:257–264.
173. Ceskova E, Drybcak P, Hrobar P, et al. The changes of biological markers and treatment efficacy in schizophrenia. Prog Neuropsychopharmacol Biol Psychiatry 2001; 25:323–335.
174. Ceskova E, Kasparek T, Zourkova A, Prikryl R. Dexamethasone suppression test in first-episode schizophrenia. Neuro Endocrinol Lett 2006; 27:433–437.
175. Zhang XY, Zhou DF, Cao LY, et al. Cortisol and cytokines in chronic and treatment-resistant patients with schizophrenia: association with psychopathology and response to antipsychotics. Neuropsychopharmacology 2005; 30:1532–1538.

176. Kaneda Y, Fujii A, Ohmori T. The hypothalamic-pituitary-adrenal axis in chronic schizophrenic patients long-term treated with neuroleptics. Prog Neuropsychopharmacol Biol Psychiatry 2002; 26:935–938.
177. Taherianfard M, Shariaty M. Evaluation of serum steroid hormones in schizophrenic patients. Indian J Med Sci 2004; 58:3–9.
178. Heinz A, Weingartner H, George D, et al. Severity of depression in abstinent alcoholics is associated with monoamine metabolites and dehydroepiandrosterone-sulfate concentrations. Psychiatry Res 1999; 89:97–106.
179. Shirayama Y, Hashimoto K, Suzuki Y, Higuchi T. Correlation of plasma neurosteroid levels to the severity of negative symptoms in male patients with schizophrenia. Schizophr Res 2002; 58:69–74.
180. Hsiao CC. Positive correlation between anxiety severity and plasma levels of dehydroepiandrosterone sulfate in medication-free patients experiencing a major episode of depression. Psychiatry Clin Neurosci 2006; 60:746–750.
181. Strous RD, Maayan R, Lapidus R, et al. Dehydroepiandrosterone augmentation in the management of negative, depressive, and anxiety symptoms in schizophrenia. Arch Gen Psychiatry 2003; 60:133–141.
182. Strous RD, Stryjer R, Maayan R, et al. Analysis of clinical symptomatology, extrapyramidal symptoms and neurocognitive dysfunction following dehydroepiandrosterone (DHEA) administration in olanzapine treated schizophrenia patients: a randomized, double-blind placebo controlled trial. Psychoneuroendocrinology 2007; 32:96–105.
183. Ritsner MS, Gibel A, Ratner Y, et al. Improvement of sustained attention and visual and movement skills, but not clinical symptoms, after dehydroepiandrosterone augmentation in schizophrenia: a randomized, double-blind, placebo-controlled, crossover trial. J Clin Psychopharmacol 2006; 26:495–499.
184. Ritsner M, Maayan R, Gibel A, Weizman A. Differences in blood pregnenolone and dehydroepiandrosterone levels between schizophrenia patients and healthy subjects. Eur Neuropsychopharmacol 2007; 17:358–365.
185. Censits DM, Ragland JD, Gur RC, Gur RE. Neuropsychological evidence supporting a neurodevelopmental model of schizophrenia: a longitudinal study. Schizophr Res 1997; 24:289–298.
186. Heinrichs RW. The primacy of cognition in schizophrenia. Am Psychol 2005; 60:229–242.
187. Hutton SB, Puri BK, Duncan LJ, et al. Executive function in first-episode schizophrenia. Psychol Med 1998; 28:463–473.
188. Brewer WJ, Pantelis C, Anderson V, et al. Stability of olfactory identification deficits in neuroleptic-naive patients with first-episode psychosis. Am J Psychiatry 2001; 158:107–115.
189. Rund BR. A review of longitudinal studies of cognitive functions in schizophrenia patients. Schizophr Bull 1998; 24:425–435.
190. Hoff AL, Svetina C, Shields G, et al. Ten year longitudinal study of neuropsychological functioning subsequent to a first episode of schizophrenia. Schizophr Res 2005; 78:27–34.
191. Lewis DA, Moghaddam B.Cognitive dysfunction in schizophrenia: convergence of gamma-aminobutyric acid and glutamate alterations. Arch Neurol 2006; 63:1372–1376.
192. Ko YH, Joe SH, Cho W, et al. Estrogen, cognitive function and negative symptoms in female schizophrenia. Neuropsychobiology 2006; 53:169–175.
193. Kalmijn S, Launer LJ, Stolk RP, et al. A prospective study on cortisol, dehydroepiandrosterone sulfate, and cognitive function in the elderly. J Clin Endocrinol Metab 1998; 83: 3487–3492.
194. Wolf OT, Naumann E, Hellhammer DH, Kirschbaum C. Effects of dehydroepiandrosterone (DHEA) replacement in elderly men on event related potentials (ERPs), memory and well-being. J Gerontol A Biol Sci Med Sci 1998; 53:385–390.
195. Wolf OT, Kudielka BM, Hellhammer DH, et al. Opposing effects of DHEA replacement in elderly subjects on declarative memory and attention after exposure to a laboratory stressor. Psychoneuroendocrinology 1998; 23:617–629.

196. Huppert FA, van Niekerk JK. Dehydroepiandrosterone (DHEA) supplementation for cognitive function (Cochrane Review). Oxford: The Cochrane Library (Issue 2), Update Software 2000.
197. Moffat SD, Zonderman AB, Harman SM, et al. The relationship between longitudinal declines in dehydroepiandrosterone sulfate concentrations and cognitive performance in older men. Arch Intern Med 2000; 160:2193–2198.
198. Fonda SJ, Bertrand R, O'Donnell A, et al. Age, hormones, and cognitive functioning among middle-aged and elderly men: cross-sectional evidence from the Massachusetts Male Aging Study. J Gerontol A Biol Sci Med Sci 2005; 60:385–390.
199. Murialdo G, Barreca A, Nobili F, et al. Relationships between cortisol, dehydroepiandrosterone sulphate and insulin-like growth factor-I system in dementia. J Endocrinol Invest 2001; 24:139–146.
200. Armanini D, Vecchio F, Basso A, et al. Alzheimer's disease: pathophysiological implications of measurement of plasma cortisol, plasma dehydroepiandrosterone sulfate, and lymphocytic corticosteroid receptors. Endocrine 2003; 22:113–118.
201. Genedani S, Rasio G, Cortelli P, et al. Studies on homocysteine and dehydroepiandrosterone sulphate plasma levels in Alzheimer's disease patients and in Parkinson's disease patients. Neurotox Res 2004; 6:327–332.
202. Yanase T, Fukahori M, Taniguchi S,et al. Serum dehydroepiandrosterone (DHEA) and DHEA-sulfate (DHEA-S) in Alzheimer's disease and in cerebrovascular dementia. Endocr J 1996; 43:119–123.
203. Morrison MF, Redei E, TenHave T, et al. Dehydroepiandrosterone sulfate and psychiatric measures in a frail, elderly residential care population. Biol Psychiatry 2000; 47:144–1450.
204. CANTAB, Cambridge Automated Neuropsychological Test Battery. http://www.cantab.com/cantab/site/home.acds).
205. Heaton RK, Wisconsin Card Sorting Test Manual. Revised and expanded. Odessa, FL: Psychological Assessment Resources, 1993.
206. Jacquy J, Wilmotte J, et al. Cerebral blood flow patterns studied by rheoencephalography in schizophrenia. Neuropsychobiology 1976; 2:94–103.
207. Weinberger DR, Berman KF. Prefrontal function in schizophrenia: confounds and controversies. Philos Trans R Soc Lond B Biol Sci 1996; 351:1495–1503.
208. Hill K, Mann L, et al. Hypofrontality in schizophrenia: a meta-analysis of functional imaging studies. Acta Psychiatr Scand 2004; 110:243–256.
209. Paulman RG, Devous MD, Sr, et al. Hypofrontality and cognitive impairment in schizophrenia: dynamic single-photon tomography and neuropsychological assessment of schizophrenic brain function. Biol Psychiatry 1990; 27:377–399.
210. Davidson LL, Heinrichs RW. Quantification of frontal and temporal lobe brain-imaging findings in schizophrenia: a meta-analysis. Psychiatry Res 2003; 122:69–87.
211. Riehemann S, Volz HP, et al. Hypofrontality in neuroleptic-naive schizophrenic patients during the Wisconsin Card Sorting Test – a fMRI study. Eur Arch Psychiatry Clin Neurosci 2001; 251:66–71.
212. Moreno-Iniguez M, Ortuno F, et al. Regional cerebral blood flow SPECT study, at rest and during Wisconsin Card Sorting Test (WCST) performance, in schizophrenia naive patients or treated with atypical neuroleptics. Actas Esp Psiquiatr 2005; 33:343–351.
213. Ortuno F, Moreno-Iniguez M, et al. Cortical blood flow during rest and Wisconsin Card Sorting Test performance in schizophrenia. Wien Med Wochenschr 2006; 156:179–184.
214. Hazlett EA, Buchsbaum MS, et al. Hypofrontality in unmedicated schizophrenia patients studied with PET during performance of a serial verbal learning task. Schizophr Res 2000; 43:33–46.
215. Lammers CH, Garcia-Borreguero D, Schmider J, et al. Combined dexamethasone/corticotropin-releasing hormone test in patients with schizophrenia and in normal controls. II. Biol Psychiatry 1995; 38:803–807.
216. Marx CE, Shampine LJ, et al. Clozapine markedly elevates pregnenolone in rat hippocampus, cerebral cortex, and serum: candidate mechanism for superior efficacy? Pharmacol Biochem Behav 2006; 84:598–608.

217. Marx CE, Shampine LJ, et al. Olanzapine and fluoxetine administration and coadministration increase rat hippocampal pregnenolone, allopregnanolone and peripheral deoxycorticosterone: implications for therapeutic actions. Pharmacol Biochem Behav 2006; 84:609–617.
218. Oseko F, Morikawa K, Nakano A, et al. Effect of chronic hyperprolactinemia induced by sulpiride on plasma dehydroepiandrosterone (DHA) in normal men. Andrologia 1986; 18:523–528.
219. Ritsner M, Gibel A, Maayan R, et al. Cortisol/dehydroepiandrosterone ratio and responses to antipsychotic treatment in schizophrenia. Neuropsychopharmacology 2005; 30:1913–1922.
220. Jakovljevic M, Muck-Seler D, Pivac N, Crncevic Z. Platelet 5-HT and plasma cortisol concentrations after dexamethasone suppression test in patients with different time course of schizophrenia. Neuropsychobiology 1998; 37:142–145.
221. Walker EF, Diforio D. Schizophrenia: a neural diathesis-stress model. Psychology Review 1997; 104:667–685.
222. Bradlow HL, Murphy J, Byrne JJ. Immunological properties of dehydroepiandrosterone, its conjugates, and metabolites. Ann NY Acad Sci 1999; 876:91–101.
223. Rupprecht R, Holsboer F. Neuroactive steroids: mechanisms of action and neuropsychopharmacological perspectives. Trends Neurosci 1999; 22:410–416.
224. Le Melledo JM, Baker GB. Neuroactive steroids and anxiety disorders. J Psychiatry Neurosci 2002; 27:161–165.
225. Marx CE, Stevens RD, et al. Neuroactive steroids are altered in schizophrenia and bipolar disorder: relevance to pathophysiology and therapeutics. Neuropsychopharmacology 2006; 31:1249–1263.
226. Sawa A, Snyder SH. Schizophrenia: diverse approaches to a complex disease. Science 2002; 296:692–695.
227. Anderson JE, O'Donnell BF, McCarley RW, Shenton ME. Progressive changes in schizophrenia: do they exist and what do they mean? Restor Neurol Neurosci 1998; 12:175–184.
228. Sabeti J, Nelson TE, Purdy RH, Gruol DL.Steroid pregnenolone sulfate enhances NMDA-receptor-independent long-term potentiation at hippocampal CA1 synapses: role for L-type calcium channels and sigma-receptors. Hippocampus 2007 [epub ahead of print].
229. Maurice T, Phan VL, Urani A, et al. Neuroactive neurosteroids as endogenous effectors for the sigma1 (sigma1) receptor: pharmacological evidence and therapeutic opportunities. Jpn J Pharmacol 1999; 81:125–155.
230. Maurice T, Urani A, Phan VL, Romieu P. The interaction between neuroactive steroids and the sigma1 receptor function: behavioral consequences and therapeutic opportunities. Brain Res Brain Res Rev 2001; 37:116–132.
231. Boccuzzi G, Aragno M, Seccia M, et al. Protective effect of dehydroepiandrosterone against copper-induced lipid peroxidation in the rat. Free Radic Biol Med 1997; 22:1289–1294.
232. Aragno M, Brignardello E, Tamagno E, et al. Dehydroepiandrosterone administration prevents the oxidative damage induced by acute hyperglycemia in rats.J Endocrinol 1997; 155:233–240.
233. Kimonides VG, Khatibi NH, et al. Dehydroepiandrosterone (DHEA) and DHEA-sulfate (DHEAS) protect hippocampal neurons against excitatory amino acid-induced neurotoxicity. Proc Natl Acad Sci USA 1998; 95:1852–1857.
234. Cardounel A, Regelson W, et al. Dehydroepiandrosterone protects hippocampal neurons against neurotoxin-induced cell death: mechanism of action. Proc Soc Exp Biol Med 1999; 222:145–149.
235. Bastianetto S, Ramassamy C, et al. Dehydroepiandrosterone (DHEA) protects hippocampal cells from oxidative stress-induced damage. Brain Res Mol Brain Res 1999; 66:35–41.
236. Charalampopoulos I, Tsatsanis C, et al. Dehydroepiandrosterone and allopregnanolone protect sympathoadrenal medulla cells against apoptosis via antiapoptotic Bcl-2 proteins. Proc Natl Acad Sci USA 2004; 101:8209–8214.
237. Majewska MD. Neuronal actions of dehydroepiandrosterone. Possible roles in brain development, aging, memory, and affect. Ann NY Acad Sci 1995; 774:111–120.

238. Morin C, Zini R, Simon N, Tillement JP. Dehydroepiandrosterone and alpha-estradiol limit the functional alterations of rat brain mitochondria submitted to different experimental stresses. Neuroscience 2002; 115:415–424.
239. Suzuki M, Wright LS, Marwah P, et al. Mitotic and neurogenic effects of dehydroepiandrosterone (DHEA) on human neural stem cell cultures derived from the fetal cortex. Proc Natl Acad Sci USA. 2004; 101:3202–3207.
240. Takahashi H, Nakajima A, et al. Dehydroepiandrosterone (DHEA) and its sulfate (DHEAS) inhibit the apoptosis in human peripheral blood lymphocytes. J Steroid Biochem Mol Biol 2004; 88:261–264.
241. Kurata K, Takebayashi M, et al. Beta-estradiol, dehydroepiandrosterone, and dehydroepiandrosterone sulfate protect against N-methyl-D-aspartate-induced neurotoxicity in rat hippocampal neurons by different mechanisms. J Pharmacol Exp Ther 2004; 311:237–245.

Chapter 15
Neurosteroids in Cortical Development and the Etiology of Schizophrenia

Samantha S. Gizerian

Abstract Schizophrenia is thought to be a neurodevelopmental disorder, and both genetic and non-genetic risk factors associated with the disease are related to neural development and function. However, no one mechanism has yet been proposed that explains the entire pathology of the disorder. One pathway that is clearly involved, however, is signaling via GABA. Post-mortem investigations reveal deficits in GABA transmission and altered localization of some GABAergic interneurons in the brains of schizophrenic patients. Furthermore, GABAergic signaling during early cortical development influences neuronal migration and later enhances synapse formation. Modulators of GABA signaling, including the neuroactive steroids, are therefore critical for normal neurodevelopment. Neurosteroids, including allopregnanolone, are potent, endogenous, *stress-responsive* modulators of $GABA_A$ receptor function. Cortical neurosteroid levels are dynamically regulated across embryonic and early postnatal development, and may differentially affect $GABA_A$ receptor function across distinct ontological periods. Furthermore, neuroactive steroid levels are increased in response to gestational stress or infection, both of which are risk factors associated with the development of a schizophrenia-vulnerable phenotype. The unique situation of neurosteroids at the junction of stress and GABA transmission makes them an important mechanism linking stress and neurodevelopmental vulnerability, as in the etiology of schizophrenia.

Keywords Allopregnanolone, neurosteroids, schizophrenia, stress, neurodevelopment

Abbreviations AMPA alpha-amino-3-hydroxy-5-methyl-4-isoxazolepropionic acid; BrdU bromodeoxyuridine; Ca calcium; Cl chlorine; CNS central nervous system; CRF corticotropin-releasing factor; DA dopamine; E embryonic day; EAAT3 excitatory amino acid transporter type 3; GABA gamma-aminobutyric acid; GAD glutamic acid decarboxylase; HPA hypothalamic–pituitary–adrenal; MD medial dorsal nucleus of the thalamus; Mg magnesium; MPFC medial prefrontal cortex; MRI magnetic resonance

Department of Biomedical Sciences, Charles R. Drew University of Medicine and Science

M.S. Ritsner and A. Weizman (eds.), *Neuroactive Steroids in Brain Function, Behavior and Neuropsychiatric Disorders*

imaging; NADPH nicotinamide adenine dinucleotide phosphate; NMDA *n*-methyl D-aspartate; P postnatal day; PCP phencyclidine; PFC prefrontal cortex; PPI prepulse inhibition of the acoustic startle reflex; PV-IR parvalbumin immunoreactivity; TH tyrosine hydroxylase; VTA ventral tegmental area; WM white matter.

15.1 Introduction

Since the initial scientific description of the disorder then termed *Dementia Praecox*, physicians and scientists have regarded schizophrenia as the prototype of severe mental illness. For over a century, however, this disorder has resisted scientific discernment despite its compelling pathophysiology. Schizophrenia has defied etiological theories of a single causal agent, genetic contributor, or pathogenic mechanism; no single theory yet proposed can account for all aspects of the disorder. It is widely believed that schizophrenia is a genetically mediated neurodevelopmental disorder[1]; however, investigators are unable to explain how and why the disorder remains asymptomatic for up to three decades before expressing itself via a chronic, recurrent, and deteriorating course.

Non-specific stress is the most consistent factor in development of a schizophrenia-vulnerable phenotype. The most commonly associated factor is prolonged labor,[2,3] supporting the notion that generalized "stress" is a factor in producing a schizophrenia-vulnerable phenotype.[4] The temporal lag between exposure to early risk factors and the clinical expression of symptoms may be a consequence of plasticity in the CNS or may arise from normal post-pubertal developmental processes that unmask pre-established neurodevelopmental abnormalities. The linkage between the disorder and the frontal cortex may also contribute to this quiescent period. Frontal cortical maturation is unique in that it is longer and later than most neuronal development periods, and therefore has potential to be more, and uniquely vulnerable. Furthermore, the abstract functions processed by the frontal cortex and commonly disrupted in schizophrenia are manifested relatively late in development, as compared to perceptual and other cognitive functions. In humans, cortical development extends over years, with the final stages of pruning and maturation occurring in the second decade of life. Therefore, an ideal candidate mechanism in the pathogenesis of schizophrenia would be *non-specifically sensitive to stress, exhibit considerable functional plasticity, and participate in the late stages of neural development.* The unique characteristics of neurosteroid modulation of GABAergic neurotransmissions make it an ideal candidate mechanism in the pathogenesis of schizophrenia.

15.2 The Prefrontal Cortex Is Uniquely Susceptible to Neurosteroid Insult

The prefrontal cortex (PFC) in mammals is defined as the area of cortex that receives direct and indirect projections from the medial dorsal (MD) nucleus of the thalamus.[5,6] The PFC is an agranular area of cortex that lacks cortical layer IV, and receives the majority of thalamic inputs in layers III and V–VI instead. The timeline and process of PFC development is distinct from other cortical areas: it develops more slowly than other areas of cortex. While most cortical areas display an apparent adult pattern of lamination and connectivity by P4 in the rodent, the prefrontal cortex is still undergoing lamination and differentiation after P7.[7] Full adult-type connectivity in the PFC is not complete until around postnatal day 30 (P30), when the dopaminergic afferents from the ventral tegmental area (VTA) arrive and form synapses in the region.[7] Similarly, the frontal cortex is the last cortical area to develop fully in the human brain, and continues to mature throughout childhood and young adulthood.

Migratory cues for excitatory neurons in the developing PFC may also contribute to its unique susceptibility to neurosteroid effects. Unlike granular areas of cortex, where thalamic afferents arrive at the cortical plate and follow their target cells into their terminal layers before initiating synaptogenesis, thalamic projections to prefrontal cortex invade the developing cortex ahead of their target cells and may serve to direct their migration towards their destination layer before initiating synaptogenesis. The majority of these thalamic axons originate in the medial dorsal nucleus of the thalamus and make topographically organized connections in the PFC.[6]

These topographically arranged projections between the MD and PFC form a monosynaptic feedback loop which is regulated by modulatory input from GABAergic interneurons in layers III and V–VI as well as dopaminergic neurons in layer III of the PFC[8] (Fig. 15.1). Axons from the medial dorsal nucleus travel

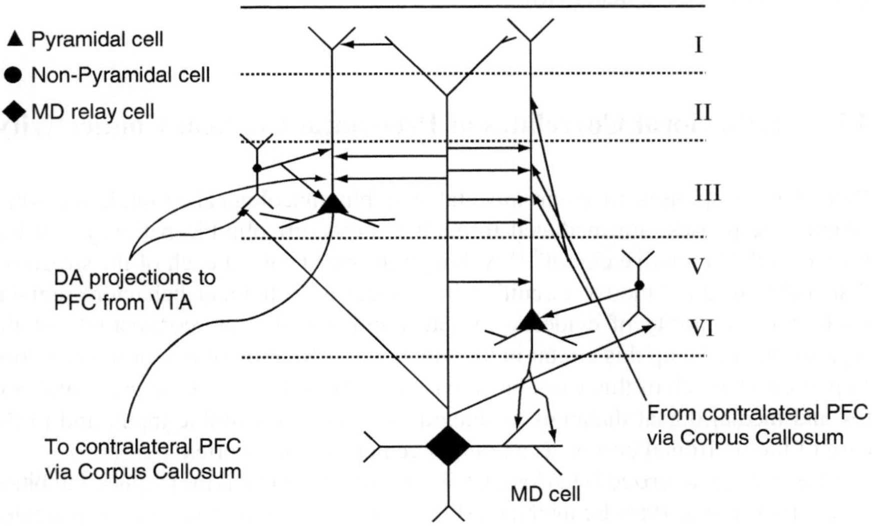

Fig. 15.1 Connectivity in the Prefrontal Cortex. Schematic depicts the connectivity pattern of the normal prefrontal cortex. Medial dorsal relay neurons project axons to layers I, III, and V/VI of prefrontal cortex, where they synapse on the dendrites and soma of pyramidal cells in layers III and V/VI. The majority of medial dorsal afferents synapse in layer III, where the medial dorsal nucleus is the exclusive source of thalamic afferents. These connections are modulated by input from GABAergic interneurons in layers III and V/VI. Pyramidal cells in layer III project to contralateral cortex via the corpus callosum while pyramidal cells in the deep layers send afferents back to the medial dorsal nucleus. MD – medial dorsal nucleus of the thalamus, PFC – prefrontal cortex

forward to the developing prefrontal cortex where they initially invade the lower half of the cortical plate at around embryonic day 18 (E18). These axons remain in the cortical plate until their target neurons emerge from the subplate layer and lead these targets out into the developing cortical laminae. Thalamic afferents form synapses with their cortical targets in layer III of PFC between E4 and E7, and the reciprocal afferents from the pyramidal cells of layer VI of PFC make their thalamic connections in the MD between P4 and P10, significantly later than other corticothalamic connectivity, which displays an adult-type pattern by P4.[6] Thalamic axons may influence the migration of their target neurons, setting up a unique vulnerability to developmental perturbations in the early postnatal period in this region. Whether this pattern of migration is unique to the rodent or holds true for human development has not been clearly demonstrated.

GABAergic neurons of the prefrontal cortex are born in the medial ganglionic eminence and migrate tangentially through the developing cortex to their eventual target location. GABA and its receptors have been detected in the embryonic cortex as early as E12.[9] During the period of cortical neuron migration, autocrine GABA signaling via the $GABA_A$ receptor stimulates neurite outgrowth.[10] $GABA_A$ signaling has been shown to be excitatory in immature neurons, and may strengthen developing synapses by removing the Mg^{2+} block from NMDA receptors before AMPA receptors become functional.[11,12] Thus, signal transduction via cortical GABA neurons may play an important role in the final steps of cortical migration, differentiation, and connectivity.

15.3 Behavioral Correlates of Prefrontal Cortical Connectivity

Behavioral responses to environmental and pharmacological stimuli, including adaptive responses, are mediated through a mesocorticolimbic pathway that has been called "the motive circuit".[13] Although the exact role of each of the structures that make up this "motive circuit" with respect to individual behaviors remains unclear, the majority of evidence to date implicates the interconnectivity of the regions that make up this circuit in the functional regulation of response behaviors. In particular, much of this work has focused on the influence of the prefrontal cortex and medial dorsal thalamus as integration centers for limbic input, and in the case of the prefrontal cortex, as an output center to motor cortex.

One well-characterized behavior related to the motive circuit is the prepulse inhibition of the startle reflex. Prepulse inhibition (PPI) occurs when a weak stimulus is presented just before a strong stimulus of the same modality. Attending to the weak stimulus is thought to abridge or "gate" the response to the weak stimulus; hence this phenomenon is also known as sensory gating and can occur in any sensory modality. Aberrant sensory gating, as determined by measuring PPI in the auditory or touch modalities, is associated with focal lesions in prefrontal cortex and medial dorsal thalamus. Disruptions in PPI can also be caused by excess dopaminergic transmission (such as after amphetamine or apomorphine administration)[14] or a decrease in NMDA function (as after PCP exposure)[15] in the prefrontal cortex. Increased GABA transmission in the medial dorsal nucleus has also been shown to block PPI.[16] Furthermore, aberrant PPI is commonly found among neuropsychiatric patients and is a feature of many animal models of psychosis.[17–20] Hippocampal damage has also been linked to reduced PPI in rodents.[21] Although no direct evidence points to an exact mechanism, it is thought that the hippocampus influences PPI indirectly via its association with the nucleus accumbens. PPI may also be affected by late developmental events, such as puberty. Lipska and coworkers[21] showed that PPI was not significantly different in neonatal ventral hippocampal lesioned animals and controls prepubertally (P35), but PPI was reduced significantly in lesioned animals as compared to controls in adulthood (P56).

Locomotor activity is another adaptive response behavior that is regulated by the motive circuit. In response to increased dopamine (DA) transmission in the prefrontal cortex (such as after amphetamine or cocaine administration), total locomotor activity including ambulations and stereotypies increases. Repeated exposure to enhanced DA transmission leads to even greater increases in locomotor activity, a phenomenon known as behavioral sensitization.[22] Sensitization is thought to reflect alterations in DA signaling in the motive circuit; specifically, increased DA signaling between the VTA and nucleus accumbens and decreased DA input to the PFC from the VTA. The result of this change in DA signaling is a disinhibition of the motive circuit, resulting in augmented behavior.[13] Enhanced locomotor response to dopamine may also occur without repeated administration of DA agonists. Increased activity may be present as a result of a single DA challenge after pharmacological or surgical manipulation of the motive circuit, indicating that the manipulation has altered the DA signaling pathways of the motive circuit. Because DA innervation of the prefrontal cortex occurs late in development (around P30),[7] dopamine-modulated behavioral measures may show changes after puberty that were not present prepubertally. Similar to their results in PPI experiments, Lipska and coworkers[23] showed that while lesioned and control animals had similar responses to a single dose of amphetamine at P35, lesioned animals were hyper responsive to amphetamine compared to controls at P56.

15.4 Neurosteroids Affect Cortical Connectivity

Neurosteroids are synthesized de novo in brain from cholesterol or other precursors. Endogenous neurosteroid levels are regulated via activation of the HPA axis and increase in response to stress. The neurosteroid allopregnanolone (3α-hydroxy-5α-pregnan-20-one) is a potent endogenous modulator of $GABA_A$ receptor function, and has less activity at classical nuclear steroid receptors. Allopregnanolone is more potent than benzodiazepines and barbiturates at the $GABA_A$ receptor,[24] acting to facilitate and/or maintain the channel opening; potentiating the chloride flux through the channel. Levels of allopregnanolone in brain correlate with circulating progesterone levels, which are high in pregnancy (>20 ng/g). Therefore, the fetus is exposed to high levels of allopregnanolone in utero and this exposure may influence the function of $GABA_A$ receptors during prenatal development.[25]

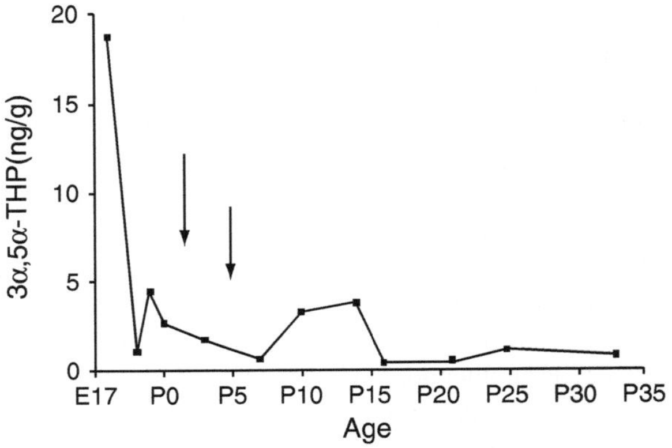

Fig. 15.2 Allopregnanolone (3α, 5α-THP) levels fluctuate in the perinatal and early postnatal period. Arrows indicate endogenous allopregnanolone levels at the time of exogenous neonatal allopregnanolone administration on P2 or P5 (Adapted from Grobin and Morrow[113])

Allopregnanolone levels change across development, dropping before birth and remaining low until around P14[26] (Fig. 15.2). Much of the cortex, therefore, develops at a time when allopregnanolone levels are relatively high. Because endogenous neurosteroid levels change across development, allopregnanolone may play a role in GABA signaling actions in normal development. Altered allopregnanolone levels may similarly affect GABA signaling in abnormal CNS development. The prefrontal cortex may be preferentially vulnerable to perturbations in allopregnanolone levels due to the relatively low amount of endogenous allopregnanolone present during its postnatal maturation.

GABA signaling plays a role in directing neural migration via its chemoattractive effects and influence on neurite outgrowth, and may influence synapse building before excitatory AMPA receptors are present in a functional form.[10–12] As a potent, endogenous modulator of $GABA_A$ signaling that is clearly present in the developing cortex; allopregnanolone almost certainly plays a key role in these GABA-mediated developmental processes. Although the potential role of neurosteroids in development has yet to be explored fully, allopregnanolone's influence on $GABA_A$ transmission may buffer fluctuations in GABA tone as neurons mature or contribute to the development of synapses by increasing the $GABA_A$-mediated depolarization of immature neurons. Moreover, the presence of allopregnanolone in the developing CNS supports cell survival and the establishment of connectivity.[27]

15.5 Neurosteroid Levels and Stress

GABAergic neurosteroids are components of the complex stress-response system. Allopregnanolone is a metabolite of stress-responsive steroid hormones produced in the adrenal glands. As a potent, positive modulator of $GABA_A$

receptor function with anxiolytic activity, allopregnanolone has the potential to be a part of a negative feedback loop or act as an endogenous coping mechanism. Neurosteroid levels in brain and plasma are markedly increased in response to stressful stimuli including swim stress, ethanol and caffeine administration, and social isolation.[24,28–31] Following exposure to stress, neurosteroid levels rapidly increase in brain to pharmacologically relevant concentrations.[24] Accordingly, increased brain levels of allopregnanolone feed back on peripheral measures of stress such as cortisol and CRF production.[32] Moreover, GABAergic neurosteroids can regulate CNS stress responses specific to frontal cortical areas such as dopamine metabolism.[33,34] In adult animals, GABAergic neurosteroid administration may also be protective against the detrimental effects of chronic stress exposure.[35] Thus, cortical GABAergic neurosteroid levels are responsive to stress and may participate in a centrally mediated feedback loop located in the frontal cortex.

15.6 Neurosteroids Link Stress and Neurodevelopmental Vulnerability

Prenatal stress is frequently considered to be correlated with vulnerability to developing schizophrenia and other neurodevelopmental disorders. Stresses associated with the development of schizophrenia can be gestational, including maternal infection, malnutrition, emotional upheaval, and prolonged labor[36–38]; or environmental, including urban birth[39–40] and birth in the late winter or early spring.[41] However, a definitive connection between mechanisms of the stress response and neurodevelopmental deficits seen in schizophrenic patients has not been found. Allopregnanolone's effects on $GABA_A$ signaling represent a candidate mechanism whereby stress could alter neurodevelopment. Fetal levels of allopregnanolone and neurosteroidogenic enzymes increase in response to stress activation of the HPA axis[42] and infection, and basal levels of allopregnanolone have been shown to remain higher than normal throughout life in animals exposed to prenatal infection,[25] which could result in permanent effects on cortical $GABA_A$ signaling.

Changes in allopregnanolone levels could therefore lead to altered development resulting from changes in $GABA_A$-mediated processes. Disrupted or altered neuronal migration is considered to be a risk factor for autism and mental retardation as well as schizophrenia.[43–45] Cortical hypofunction resulting from decreased synaptic connectivity has been implicated in sleep disorders, vulnerability to Alzheimer's disease, Tourette's syndrome, and schizophrenia.[46–49] While the role of neurosteroids in the etiology of these disorders, if any, is unclear, the possibility remains that neurosteroid-mediated effects on GABA signaling in development could underlie a variety of mechanisms that strongly impact neurodevelopmental outcomes.

15.7 Anatomical and Functional Deficits Associated with Schizophrenia

Many of the functional deficits in patients with schizophrenia are attributable to structural abnormalities in the frontal cortex. Patients with schizophrenia consistently perform poorly on certain cognitive tasks, such as those requiring spatial memory or working memory, that are subserved by circuitry involving the dorsal frontal cortex.[50] Imaging studies have demonstrated extensively that patients fail to show normal activation of the PFC when they attempt to perform such tasks.[51] Functional MRI studies using both visual and auditory tasks have reported dorsolateral prefrontal activation during the processing of visual stimuli, and activation in superior temporal regions during auditory target detection. Many postmortem studies have reported a 5–10% reduction in cortical thickness, with a corresponding increase in cell density, but no change in total neuron number, in the dorsal PFC of patients with schizophrenia.[52] Decreased levels of the presynaptic protein synaptophysin have also been observed post-mortem in the PFC of patients with schizophrenia.[53] However, post-mortem studies have found no evidence of gliosis, consistent with the hypothesized role of neurodevelopmental factors in the pathogenesis and pathophysiology of schizophrenia.[54]

One of the most consistent findings in post-mortem schizophrenic brain has been a deficit in selected cortical GABA neurons. Decreased thickness of the cortical layers and the loss of neuropil[55] throughout the cortex are hallmarks of the schizophrenic brain. Loss of GABAergic functional markers (GAD65, NADPH diaphorase) has been observed specifically in frontal cortical regions of post-mortem brains of schizophrenics.[56] Post-mortem studies have also shown an increase in the number of $GABA_A$ receptors in brains of schizophrenics,[57] and imaging studies using benzodiazepine receptor antagonists have revealed selective deficits in schizophrenic patients.[58] Additionally, data from unmedicated and first-episode patients suggest that these deficits may be independent of drug treatment.[59] Moreover, neuropsychiatric drugs including the atypical antipsychotics Clozapine and Olanzapine and the anti-depressant Fluoxetine have been shown to increase neurosteroid levels in cortex, which may contribute to their therapeutic effects.[60,61] Thus, GABAergic neurotransmission within certain cortical regions is likely to be compromised in the brain of schizophrenic patients prior to treatment and endogenous effectors of this system, such as the neurosteroids, are plausible candidate molecules in the regulation of neurodevelopment leading to increased risk for schizophrenia.

The medial dorsal nucleus of the thalamus has also been implicated in developmental models of psychosis.[62,63] Anatomical changes in the MD have been associated with changes in behavior and cognition in human patients. MD-specific lesions after stroke have also been shown to result in decreases in episodic memory and acquisition of new material.[64,65] Subtle changes in the morphology of the MD are commonly found to be present in schizophrenia. Several groups have reported 20–40% decreases in the number of MD neurons in the post-mortem brains of schizophrenic patients.[66–69] Lewis and co-workers[53, 69,70] also demonstrated alterations in the number and localization of both parvalbumin-positive neurons and varicosities arising from the medial dorsal thalamus in the prefrontal cortex of schizophrenic patients in a post-mortem examination.

15.8 Anatomical Correlates of Neonatal Allopregnanolone Exposure

In order to study the effects of increased allopregnanolone exposure on the developing brain, we exposed neonatal rats to a high dose of allopregnanolone via a single injection at a specific timepoint during the first postnatal week.[71] The dose of allopregnanolone that was chosen approximates the level of allopregnanolone in circulation after an extremely stressful event, such as prolonged maternal separation.[72] However, by directly administering allopregnanolone, rather than inducing stress to raise allopregnanolone levels via activation of the HPA axis, experiments based on this model focus specifically on the influence of changing allopregnanolone levels in CNS development. This precludes potential confounds associated with other downstream effects of the stress response.

Initial experiments[26] with this model indicated that allopregnanolone exposure on P1 or P5, but not on P2, alters the final localization of parvalbumin-positive interneurons in the PFC. The number of parvalbumin-positive GABAergic interneurons neurons in layer III of the prefrontal cortex was decreased, while the number of parvalbumin-positive cells was increased in layers V and VI. Parvalbumin-immunoreactive neuropil was also decreased in layer III of the prefrontal cortex (AC Grobin, unpublished observation), after neurosteroid insult on P1 or P5, indicating a possible decrease in the number of glutamatergic parvalbumin-expressing thalamic afferents terminating there. Altered localization and decreased neuropil were only found in the parvalbumin-expressing interneuron population. Calbindin- and Calretinin-positive interneuron populations were not affected by increased allopregnanolone exposure. The altered localization and decreased neuropil were found in conjunction with decreased [^{3}H]MK-801 binding in PFC, but no change in $GABA_A$ receptor-mediated Cl^- uptake, consistent with altered cortical connectivity.

These experiments constituted the first notion that allopregnanolone might be responsible for mediating stress-related effects in development, and could perhaps play a role in neural maldevelopment as well. Further investigations therefore focused on the PFC and MD and the connectivity between these structures, as well as the functional outcome of disruptions in these structures. Moreover, as initial findings demonstrated clear phenotypic differences in outcome between animals exposed to increased levels of allopregnanolone on P2 and P5, subsequent experiments focused on these two timepoints because of the relative difference in native allopregnanolone levels at each of these developmental times (see Fig. 15.2).

The adult medial dorsal thalamus is altered after a single neonatal neurosteroid administration late in the first postnatal week of life.[71] The total number of neurons in the medial dorsal thalamus is decreased after neonatal neurosteroid administration. However, the number of parvalbumin-immunopositive cells is increased and the total number of EAAT3-immunopositive (glutamatergic) cells is unchanged. Thus, the relative proportion of parvalbumin-immunopositive cells in medial dorsal thalamus is increased in animals exposed to neurosteroid at P5 and not P2. The temporal and cell-type specificity of altered neonatal neurosteroid levels suggests that neurosteroid levels may play a role in the development of the medial dorsal thalamus.

A significant decrease in neuron number in medial dorsal thalamus was seen only in P5 treated animals. This is similar to decreases in medial dorsal thalamus neuron number previously observed after neonatal prefrontal cortical lesions at similar time points.[73,74] Furthermore, decreased thalamic cell number may result in fewer axonal projections to thalamic receptive fields such as the prefrontal cortex. Hence, providing a plausible explanation for previously observed decreases in prefrontal cortical parvalbumin-immunoreactivity[26] and consistent with the notion that abnormal neurosteroid levels may lead to a disruption in thalamocortical communication.

Prefrontal cortex and the medial dorsal nucleus are connected by a monosynaptic, glutamatergic loop comprised of thalamic inputs to superficial dendrites of pyramidal cells that provide reciprocal projections to medial dorsal nucleus from their deep-layer cell bodies.[8] Therefore, a decrease in the patency of these connections could arise either from thalamic axons that fail to reach their cortical targets or from deep layer pyramidal axons unable to make their connections in the medial dorsal nucleus.

The lack of effect of neurosteroid administration on the entire glutamatergic neuron population, as measured by EAAT3 immunoreactivity, is consistent with the functional sparing found after neonatal prefrontal lesion at the same time point.[75] The apparent stability of the glutamatergic population of the medial dorsal nucleus after neurosteroid administration may also be significant because previous reports[73–77] have been unable to demonstrate altered connectivity supporting spared function, in spite of complete severance of direct communication between these two structures. Thus, retained function in the prefrontal/medial dorsal circuit after communication between the regions is disrupted may reflect redundant, alternative, or compensatory mechanisms utilizing this neuron population.

Because no significant change in the EAAT3-immunopositive population was observed, it is tempting to speculate that the loss of cells seen here represents a loss in the non-glutamatergic neuron population. However, there is a dearth of non-glutamatergic neurons in the medial dorsal nucleus, and the exact proportion of this population is unknown but presumed to be very small (JL Price, personal communication (2003)). Therefore these cells may only be responsible for a small fraction of the neurons that are lost. Population decreases are likely represented by glutamatergic relay neurons that do not express the EAAT3 marker, which displays a high degree of variability in expression across the midbrain.[78,79]

The number of parvalbumin-positive cells was increased in the medial dorsal thalamus after neonatal allopregnanolone administration on P5.[71] An increase in parvalbumin-positive cells could result from a change in the proportion of cells expressing parvalbumin, an increase in the level of parvalbumin expression in the cell, or a change in the type of cell expressing parvalbumin in the medial dorsal nucleus. Given the observed decrease in total cell number in the medial dorsal thalamus, the observed change in parvalbumin-positive cells likely indicates an increase in the proportion of medial dorsal thalamic neurons expressing parvalbumin. Because the onset of parvalbumin expression occurs up to 3 weeks after neurosteroid administration,[80] parvalbumin expression in the MD could be regulated by through an intermediate mechanism. Further experiments are needed to determine if allopregnanolone can directly or indirectly regulate the transcription or translation of parvalbumin in neurons.

The temporal specificity of neurosteroid-induced alterations in medial dorsal thalamic morphology is consistent with the notion that developmental stage is important for sensitivity to neurosteroid exposure. Neurosteroid exposure on P5 but not on P2 increased the number of parvalbumin-positive cells in medial dorsal thalamus. On P2 neither thalamocortical nor corticothalamic connections between medial dorsal thalamus and prefrontal cortex are present. By P5, however, thalamic afferents are beginning to find their topographical layer III cortical targets and the reciprocal projection from layer VI pyramidal cell is developing.[6] Because the inhibitory neurons of the prefrontal cortex may influence the growth of thalamic afferents during development,[10] alterations in interneuron localization, as described above, may impact the ability of thalamic afferents to form and maintain synapses on cortical pyramidal cells. The decrease in parvalbumin-positive neuropil of prefrontal cortex could represent a decrease in thalamic axons. The decreased number of medial dorsal thalamic neurons observed after P5 neurosteroid exposure is consistent with such a decrease in thalamocortical projections. Moreover, the lack of change observed after neurosteroid exposure on P2 suggests that developmental milestones such as the initiation of thalamocortical connectivity are important for neurosteroid-mediated effects.

The increase in parvalbumin-immunopositive cells may reflect an attempt by the neurons of the medial dorsal nucleus to increase the number of functional synapses present on prefrontal cortical neurons. It is plausible to speculate that a compensatory mechanism local to the medial dorsal nucleus may respond to a decrease in functional thalamocortical synapses in an attempt to increase the number of these synapses in cortex, and that the number of parvalbumin-positive cells at maturity may also be increased as a result of this response. Figure 15.3 shows how the decrease in medial dorsal thalamic total neuron number and the increase in parvalbumin-positive cells could be integrated with neurosteroid-induced alterations in prefrontal cortex to reflect a decrease in thalamocortical connectivity.

Bromodeoxyuridine (BrdU) birthdating on E17 in rodents revealed that allopregnanolone exposure during the first postnatal week shifts the localization of cortical neurons at maturity without altering other measures of cortical morphology (Fig. 15.4) (SS Gizerian, JA Lieberman, and AC Grobin, previously unpublished data). In these experiments, animals were exposed to a single dose of BrdU (50 mg/kg) or vehicle at E17, and exogenously exposed to allopregnanolone or vehicle during the first postnatal week as described previously.[71] At P21, the mean individual distance of BrdU-labeled cells from the medial pial surface was found to be

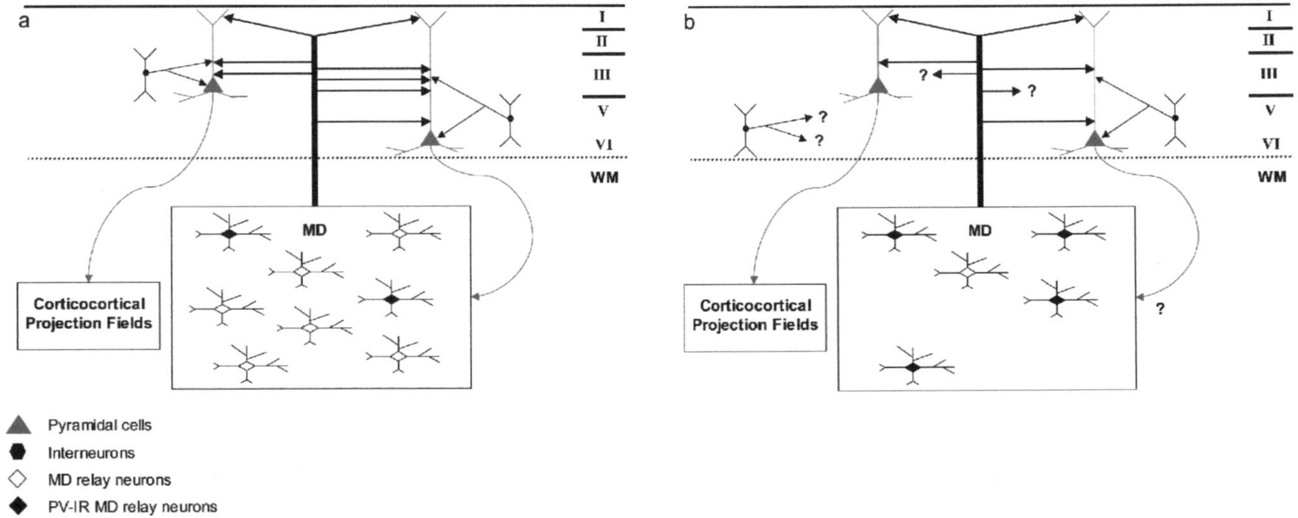

Fig. 15.3 Connectivity between the medial dorsal nucleus of the thalamus and prefrontal cortex is altered by neonatal neurosteroid administration. (**a**) Schematic summarizes the pattern of connectivity between the normal prefrontal cortex and medial dorsal nucleus. Note the proportion of PV-IR cells in the MD. See Fig. 15.1 for more details. (**b**) Schematic summarizes anatomical findings in the prefrontal cortex and medial dorsal nucleus after exogenous neonatal allopregnanolone exposure. PV-IR is increased in the medial dorsal nucleus. The MD relay neuron population is decreased, which may result in a decrease in thalamic afferents making synapses in prefrontal cortex, particularly in layer III, and may also be responsible for a portion of the reduced parvalbumin-positive neuropil. The previously described[71] shift in GABAergic interneurons from layer III to layer V/VI may lead to alterations in the connectivity of those neurons. Axons from the prefrontal cortex may be less likely to find their thalamic targets. Question marks indicate that further experiments are needed to determine if allopregnanolone alters the connectivity of this circuit. Roman numerals indicate cortical layers in prefrontal cortex. (Adapted from Gizerian et al.[71]) PV-IR – parvalbumin immunoreactivity, WM – white matter, MD – medial dorsal nucleus of the thalamus

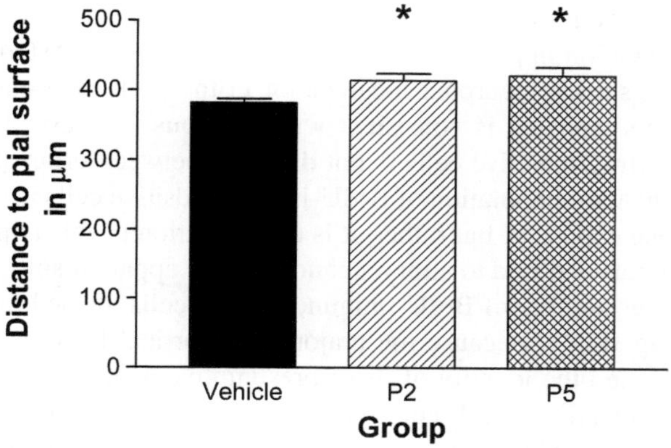

Fig. 15.4 Neonatal neurosteroid administration shifts the migration of BrdU-labeled cells toward the deep layers of the PFC. Timed pregnant dams received a single injection of BrdU (50 mg/kg, ip) on E17. Pups received a single injection of allopregnanolone (10 mg/kg) or vehicle on P2 or P5, and were allowed to survive to P21. Fixed 100 μm sections through the PFC were processed for BrdU immunoreactivity. Individual distances of cells from the pial surface were increased significantly in treated animals compared to vehicle (one-way ANOVA with Tukey's post hoc test, $p = 0.0013$, $F = 6.743$, $^{*}p < 0.05$)

significantly shifted away from the pial surface in the prefrontal cortex of both P2- and P5-exposed animals as compared to controls. This finding reflected that a greater number of BrdU-immunopositive cells were located *farther* from the pial surface in each experimental group as compared to controls. The width of the characteristic stained band of cells was not different between groups and no differences were observed between groups in stereological estimates of the number of BrdU-positive cells observed, the volume of the medial frontal cortex, or the BrdU-labeled cell density of the medial frontal cortex. The width of the medial frontal cortex, as measured between the pial surface and the layer VI/white matter boundary was also unchanged between groups.

BrdU-immunopositive cells are presumed in these experiments to be neurons by P21 because subsequent cell divisions would render BrdU immunoreactivity nearly undetectable in other cell types, including glia, as described by Bredy and coworkers.[81] Measures of distance from the pial surface were chosen for these experiments rather than assigning laminar identities because distance is a less subjective measure.

Because the localization of parvalbumin-positive neurons was shifted in the prefrontal cortex,[26] and Lewis et al.[69] demonstrated that parvalbumin-expressing interneurons have also been shown to be mislocalized toward deeper cortical layers in the cortex of schizophrenics, we investigated the effect of allopregnanolone exposure on parvalbumin-expressing cells; specifically, the distance of cells found to be immunopositive for both BrdU and parvalbumin from the pial surface was examined in adjacent sections. No differences were found between groups in the distance of the double stained cells to the pial surface or the number of double stained cells present. However, as these cells were still dividing on E17, it is possible that these cells do not represent the majority of the parvalbumin-expressing interneuron population in prefrontal cortex,[82] warranting further study.

The apparent shift in neuronal localization in the prefrontal cortex following allopregnanolone exposure is consistent with previously observed phenotypes.[26] The number of BrdU-positive cells is not different between groups; therefore the survival of the general population of BrdU-immunopositive cells is not affected by allopregnanolone exposure, but rather, it is the migration of this population that is affected. In animals exposed to allopregnanolone, the apparent shift in localization is due to a greater number of BrdU-immunopositive cells located farther from the pial surface (Fig. 15.4). Because the majority of cortical neurons are excitatory pyramidal cells, and the identity of these presumptive neurons is not known, this apparent mislocalization could represent a decrease in radial migration of excitatory neurons and/or alterations in the postnatal inward radial migration of interneurons in cortex.[83] The width of the typical BrdU-labeled band of cells in cortex was unchanged between groups, indicating that allopregnanolone exposure affected the majority of labeled cells, rather than specific subpopulations, which would result in a wider band of labeled cells as unaffected cells moved farther toward the pial surface.

Less consistent with previous results,[26] however, is the finding that localization of cortical neurons is altered by allopregnanolone exposure at either P2 or P5. However, previous experiments only investigated the localization of interneurons, and the population of cells expressing BrdU in these studies is likely made up of both interneurons and pyramidal neurons, indicating that neonatal neurosteroid exposure may affect both radial and tangential migration in cortex. The possibility that allopregnanolone exposure alters radial migration as well as tangential migration is intriguing. Autocrine and paracrine GABA signaling plays an important role in directing migration of cortical neurons from the sub-ventricular zone to the cortical plate via $GABA_A$ and $GABA_B$ receptors.[84,85] Allopregnanolone increases signaling through $GABA_A$ receptors, and may affect signaling through $GABA_B$ receptors via common downstream effectors (including internal Ca^{2+} levels), resulting in decreased neuritogenesis and chemoattraction to sources of GABA.[84, 86,87]

Signaling via $GABA_A$ receptors is important for migration and synapse formation in the developing brain. $GABA_A$ receptors are expressed as early as E12.[9] Immature neurons have an increased E_{Cl}, therefore $GABA_A$ signaling is excitatory during the prenatal and early postnatal period.[12] Autocrine GABA signaling via the $GABA_A$ receptor is thought to stimulate neurite outgrowth.[10] Furthermore, because AMPA receptors are not fully functional until the second postnatal week, it has been speculated that $GABA_A$-mediated signaling removes the Mg^{2+} block at NMDA receptors, driving glutamatergic synapse formation.[11] An increase in $GABA_A$ transmission in response to increased levels of allopregnanolone could result in altered localization of migrating neurons as a result of changes in neurite outgrowth and/or alteration in the number and distribution of synapses in areas where synapse formation is taking place at the time of neurosteroid insult.

15.9 Behavioral Correlates of Neonatal Allopregnanolone Exposure

The mesocorticolimbic system is a circuit of cortical and subcortical structures that links limbic processing centers to pyramidal and extrapyramidal motor systems to direct adaptive behavioral responses to environmental or pharmacological stimuli.[13] The function of the mesocorticolimbic system is regulated by the balance of excitatory and inhibitory signaling in the PFC, which mediates the release of dopamine (DA) from the VTA.[13] In the normal adult brain, these functions include sensorimotor gating (including prepulse inhibition of the acoustic startle reflex) and adaptive locomotor behavior.[22,88] Changes in mesocorticolimbic-related behaviors are associated with decreased inhibition in the PFC. Increased glutamatergic output from the PFC leads to increased DA transmission from the VTA,[89–91] resulting in reduced sensorimotor gating and increased locomotor response to amphetamine exposure, respectively. Decreased inhibition in the prefrontal cortex may result from reductions in DA and/or GABA signaling.[92,93] In particular, Japha and Koch[94] showed that infusion of picrotoxin into the medial PFC reduced sensorimotor gating in a site-specific manner, indicating that GABA signaling in the MPFC is important in the regulation of this phenomenon. Schizophrenia is also associated with decreased prefrontal inhibition and changes in mesocorticolimbic-related behaviors,[95] including reduced PPI and hyper responsiveness to amphetamine exposure.[18]

The acoustic startle response is an effective cross-species measure of sensorimotor gating. When a small prepulse precedes the startle stimulus, the response of the animal is decreased. This phenomenon is known as prepulse inhibition (PPI) and frequently shown to be disrupted in schizophrenics and their close relatives.[17–20] Animals exposed to allopregnanolone during the first postnatal week, as described, demonstrate disrupted prepulse inhibition differentially across development in a temporally specific manner.[114] PPI was measured at four different timepoints: before puberty (P20), puberty (P40), young adulthood (P60), and maturity (P80). Animals exposed to allopregnanolone on P2 showed disrupted PPI both early in development and after puberty, while P5-exposed animals displayed reduced PPI only in adulthood.

Comparison of startle amplitude values between types of trials (no stimulus, non-prepulse stimulus, and prepulse-stimulus) for each animal indicated that all subjects had relatively normal hearing. Because these animals were subjected to repeated measures of PPI (20 days apart) between P20 and P60, care was taken to ensure that habituation to the testing did not interfere with analysis. Habituation to startle is defined as a reduction in startle after repeated presentation of the stimulus.[96] Mean startle amplitude was not decreased in any trial after P20 except trials including clozapine pretreatment.

The atypical antipsychotic clozapine has previously been shown to ameliorate reductions in PPI in schizophrenic patients and in animal models of psychosis.[97] Clozapine is thought to exert its actions on PPI via its antagonism of D2 and D4 dopamine receptors. Interestingly, however, the actions of clozapine on PPI are more robust than more specific antagonists of either D2 or D4 receptors.[97] Pretreatment with clozapine on P87 enhanced prepulse inhibition in the P2-exposed group only, at every prepulse intensity.

Changes in locomotor response to amphetamine after neonatal drug exposure have been observed as early as P8,[98] indicating that adult levels of dopamine in the prefrontal cortex are not critical for the normal pattern of locomotor response to amphetamine in the pre-adolescent period. In order to determine the pattern of locomotor responsiveness to amphetamine, animals were tested at P20 (prepuberty), P40 (puberty), P60 (young adulthood), and P80 (maturity). Baseline activity was measured for 30 min prior to amphetamine administration. Baseline locomotor activity recorded prior to amphetamine administration was not changed in either P2- or P5-exposed animals as compared to controls at any age tested. Interestingly, locomotor activity measures display a pattern of disruption similar to that seen in PPI testing in similarly treated animals. That is, animals exposed to exogenous allopregnanolone on P2 were hyper responsive to amphetamine exposure early in development and in adulthood, while those exposed to allopregnanolone on P5 were only hyper responsive in adulthood.

These data demonstrate clearly that adult behavior is altered after a single neonatal dose of allopregnanolone. The pattern of onset of behavioral disruption indicates that neonatal neurosteroid exposure alters the mature function of the mesocorticolimbic circuitry. Furthermore, the ability of clozapine to ameliorate changes in PPI supports the notion that neonatal allopregnanolone exposure is relevant to the mechanisms underlying the development of schizophrenia.

Reduced PPI and increased locomotor response to amphetamine were present as early as P20 in animals exposed to exogenous allopregnanolone on P2, consistent with decreased inhibition in the PFC. Previous studies have demonstrated that DA-mediated inhibition in the PFC is associated with both reduced PPI and hyper responsiveness to amphetamine exposure[13,99]; however, the role of GABA-mediated inhibition in these behaviors remains unclear. Our data[114] indicate that GABA-mediated inhibition is important for both PPI and locomotor response to amphetamine, at least at juvenile timepoints when DA signaling is extremely low due to the immaturity of axons that will form the dopaminergic connections in the PFC. While these axons are present in the PFC as early as P1[7]; the actual DA content of these axons is extremely low until about P30.

As in humans, mature DA signaling in rodent PFC begins around the time of puberty. Prior to this time, cortical inhibition is largely due to GABA signaling in the PFC, but the relative importance of GABA and DA signaling in the regulation of these behaviors at all developmental timepoints is not clear. Japha and Koch[94] showed that an infusion of the $GABA_A$ receptor-specific antagonist picrotoxin into the adult medial PFC was sufficient to disrupt PPI in a region-specific manner. These authors speculated that a reduction in GABA-mediated inhibition from the medial PFC enhances the DA output of the VTA and nucleus accumbens, impairing the PPI response. Because DA-mediated inhibition is not present at adult levels in the PFC at P20, our observations of reduced PPI and increased locomotor response to amphetamine at P20 are consistent with this notion. The absence of behavioral deficits at P20 in the P5-exposed group could similarly indicate that GABA-mediated inhibition in the prefrontal cortex is not altered by neonatal neurosteroid exposure on P5.

At P80, both groups display reduced PPI and locomotor hyper responsivity to amphetamine as compared to controls, although the observed changes are more robust in the P2-exposed animals. This could be due to a more severe functional impairment in the P2 group. The increased severity of behavioral changes in the P2 group compared to the P5-exposed group is highly significant because previous studies have not observed anatomical changes in either the cortex or thalamus of these animals.[26,71] Functional changes could therefore be due to differences in local synapse function or signal transduction within the PFC. Decreased function may also result from a relatively subtle anatomical change that is difficult to observe using standard imaging techniques.

Interestingly, neither experimental group was observed to demonstrate behavioral changes at P40 or P60 in either paradigm. This could manifest from a transient change in the function of the circuit during this intermediate period of development that compensated for the previously observed altered function in the P2 animals, and alternatively, could mask decreased inhibition in the P5 group (Fig. 15.5). One factor that could account for this pattern of behavior is the expression of DA receptors in the prefrontal cortex. DA neurons from the VTA are present in PFC as early as P1, but DA signaling does not begin until around the time of puberty.[7] Expression

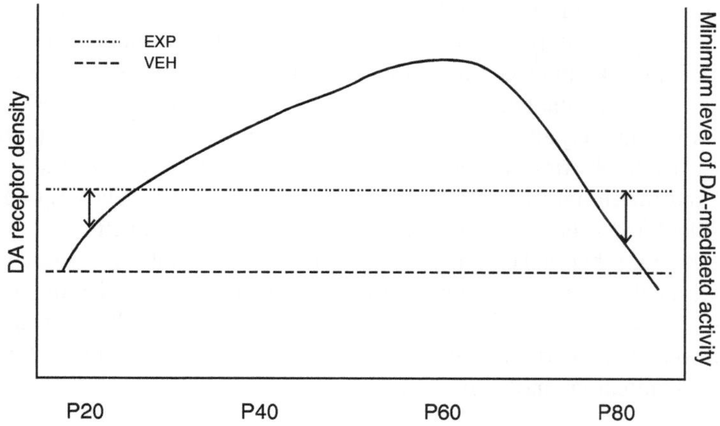

Fig. 15.5 Developmental alterations in dopamine receptor density may compensate for allopregnanolone-associated deficits in inhibitory signaling in the PFC. Andersen et al.[100] demonstrated that the dopamine (DA) receptor density increases and then decreases across the peripubertal period in rodents (curve). Horizontal lines approximate the minimum amount of DA-mediated inhibition in the PFC that is required for normal behavioral response to stimulus in vehicle-exposed (bottom line) and experimental (allopregnanolone-exposed, top line) animals. Where the curve falls below the minimum level, deficits in behavioral response are measured, as indicated by the arrows. (Adapted from Grobin et al.[115]). EXP indicates experimental groups injected with allopregnanolone on P2 or P5; VEH indicates vehicle-injected littermates

of DA receptors (D1 and D2) also increases across this period, with a spike in receptor expression between P30 and P80.[100] If this increase in DA receptor expression is associated with an increase in DA-mediated inhibition, the transient nature of the increase could account for the lack of effect at P40 and P60.

This is also consistent with the idea that perinatal stress alters the trajectory of development in the PFC and prevents the normal sequence of postnatal events from being completed. Early life stress decreases the stress response in adult animals, including plasticity to insult and neurogenesis.[101] Animals with a decreased ability to respond to stress also have decreased dopaminergic tone in areas relevant to control of PPI and locomotor response to amphetamine,[102] including decreased inhibition in the prefrontal cortex.[103] Because allopregnanolone levels increase in response to stress,[24] exogenous increases in allopregnanolone may mimic stress-related outcomes. Moreover, allopregnanolone's putative role in the development and maintenance of neurons and neuronal circuitry[27] is consistent with changes in cortical allopregnanolone levels resulting in decreased cortical plasticity and neurogenesis, and therefore altered allopregnanolone exposure during neurodevelopment may contribute to diminished cortical function. Bakshi and Geyer[104] demonstrated that stressful rearing conditions (post-weaning social isolation) induced deficits in PPI in rats at post-pubertal ages (P49) and hyperlocomotion as early as P35. Reduced PPI and increased locomotor response to amphetamine are associated with disinhibition in the prefrontal cortex, and could result from a stress- or allopregnanolone exposure-induced decrease in dopaminergic transmission in the PFC.

The difference in the anatomical and behavioral phenotypes between animals receiving neurosteroid on P2 and animals receiving neurosteroid on P5 indicates that the competency of the developing brain to respond to insult changes across the first postnatal week; thus, the timing of a perinatal stress insult is important for both anatomical and functional outcome. Unlike other areas of cortex, the PFC laminates and forms its mature pattern of connections after birth. On P2, the deep layers of the cortex are laminating in PFC, and by P5, the middle layers of the cortex are forming and the initial synaptic connections between the deep layers of PFC and other cortical and subcortical areas are being made.[6] By contrast, the other structures of the brain that make up the mesocorticolimbic circuitry are already present in approximately their mature form at this time, including the TH-immunopositive afferents from the VTA.[6,105–107] The relative immaturity of the PFC during the first postnatal week may render it more vulnerable than other areas to insults that disrupt migration, lamination, and/or synapse formation.

15.10 Significance

The neurosteroid allopregnanolone plays a complex role in CNS development. Experimental evidence to date is consistent with allopregnanolone's putative role as a link between stress and schizophrenia-vulnerable neurodevelopment. The dynamic response of the CNS to increased allopregnanolone exposure at distinct timepoints in early postnatal development is also consistent with this role. The distinct phenotypes observed at maturity after allopregnanolone exposure on P2 or P5 may represent shifting regional vulnerability to insult or reflect changes in intrinsic ability to respond to neurosteroid administration. Native regulation of allopregnanolone levels may also play a role in the differential response to neonatal allopregnanolone insult. Allopregnanolone levels on P2 are very low; decreasing from P0 levels, and reaching their lowest point at P5–P6, before increasing again from P10 to P15. It is possible that the normal pattern of expression of neurosteroids in cortex is required for proper ontogeny, including mechanisms of cortical migration. For example, because allopregnanolone levels decrease across the first postnatal week, the P5 cortex may be more vulnerable to exogenous allopregnanolone exposure, which would approximate a shift in the temporal pattern of allopregnanolone exposure. Furthermore, allopregnanolone withdrawal has been shown to alter the subunit composition of $GABA_A$ receptors[108,109]; therefore, increased allopregnanolone exposure during the withdrawal period (the first postnatal week) may differentially affect the ability of cortical neurons to respond to $GABA_A$ mediated developmental signals depending on the temporal pattern of subunit expression in local $GABA_A$ receptors.

Allopregnanolone sensitivity also increases across the early postnatal period, and is associated with changes in the subunit composition of the maturing $GABA_A$ receptors.[110] Changes in the subunit composition of $GABA_A$ receptors on migrating neurons or the migratory substrate may also impact the response of the migrating cells to allopregnanolone. For example, increases in the expression of the $\alpha1$ subunit and concomitant decreases in the expression of the $\alpha2$ subunit of the $GABA_A$ receptor are associated with increased sensitivity to allopregnanolone in the postnatal animal.[110] Recent work from Shen et al.[111] also indicates that allopregnanolone may even potentiate neuronal excitability depending on the subunit composition of the $GABA_A$ receptors present ($\alpha4\beta2\delta$ receptors), which may contribute to differential response to allopregnanolone exposure across development as the expression patterns of the $\alpha4\beta2\delta$ receptors wax and wane.

The ontological state of the cortex may also be important in the development of distinct phenotypes after allopregnanolone exposure at different times. The prefrontal cortex develops in a manner unlike most other areas of cortex, both in the timing and progress of its ontogeny. Because endogenous allopregnanolone levels are low at the time of PFC development, migration and lamination of this area of cortex may be preferentially vulnerable to changes in allopregnanolone levels. As ontogeny progress, the region of the developing cortex that is vulnerable to the effects of allopregnanolone shifts from deeper structures to more superficial ones.

Allopregnanolone's effects on $GABA_A$ signaling represent a candidate mechanism whereby stress could alter neurodevelopment. Allopregnanolone levels increase in response to stress activation of the HPA axis and infection, and basal levels of allopregnanolone have been shown to remain higher than normal throughout life in animals exposed to prenatal infection[25] (AC Grobin and JH Gilmore, personal communication (2005)), which could result in permanent effects on $GABA_A$ signaling. Furthermore, exogenously increasing allopregnanolone levels to those observed after an acute stressor during a critical period in development results in thalamocortical deficits that are consistent both anatomically and functionally with findings in schizophrenic patients.

Relatively little is known about how alterations in the various processes of neuronal migration affect function later in life, and the ability to study these processes in humans is limited, therefore the relevance of migration and synapse formation to neurodevelopmental disease is unclear, but the data are somewhat compelling. Deficits in migration and synapse formation lead to changes in both anatomy and function in animal models, and have been linked to a number of neurodevelopmental disorders in humans, including schizophrenia. Several clinical and post-mortem observations in schizophrenic patients, such as enlarged ventricles, heterotopic neuron placement, and mislocalized interneurons are all putatively related to decreased neuronal migration. However, a neurodevelopmental mechanism resulting in decreased migration that is consistent with proposed mechanisms of etiology has not yet been elucidated.

Allopregnanolone appears to affect migration in a development-specific manner. Increased exposure to allopregnanolone results in the mislocalization of PFC neurons, consistent with a role for allopregnanolone in the regulation of neuronal migration in the normal development of the cortex. Moreover, another potential downstream effect of altered migration is a decrease in functional thalamocortical synapses, or decreased inhibitory regulation of these synapses by mislocalized cortical interneurons; potentially leading to withdrawal of trophic support in the MD[112] and loss of cells in the MD nucleus, another common finding in schizophrenic patients.[66,67] These data suggest that increased neonatal allopregnanolone exposure may be a viable model for the alterations in neuronal migration, mislocalization of cortical neurons, and thalamocortical deficiencies, similar to those seen in schizophrenic patients.

Acknowledgements The author gratefully acknowledges personal communications from AC Grobin, JH Gilmore, and JL Price, and technical assistance in the preparation of this manuscript from JL Ervin.

References

1. Harrison PJ, Weinberger DR. Schizophrenia genes, gene expression, and neuropathology: on the matter of their convergence. Mol. Psychiatry 2005; 10(1):40–68.
2. Kinney DK, Jacobsen B, Jansson L, Faber B, Tramer SJ, Suozzo M. Winter birth and biological family history in adopted schizophrenics. Schizophr Res 2000; 44:95–103.
3. McNeil TF, Cantor-Graae E, Weinberger DR. Relationship of obstetric complications and differences in size of brain structures in monozygotic twin pairs discordant for schizophrenia. Am J Psychiatry 2000; 157:203–212.
4. Benes FM. The role of stress and dopamine-GABA interactions in the vulnerability for schizophrenia. J Psychiatr Res 1997; 31:257–275.
5. Rose JE, Woolsey CN. The Orbitofrontal Cortex and its connections with the Medial Dorsal nucleus in rabbit, sheep, and cat. Res Publ Ass Nerv Ment Dis 1948; 27:210–232.

6. Van Eden CG. Development of connections between the mediodorsal nucleus of the thalamus and the prefrontal cortex in the rat. J Comp Neurol 1986; 244:349–359.
7. Van Eden CG, Uylings H. Postnatal cytoarchetectonic development of the prefrontal cortex in the rat. J Comp Neurol 1985; 241:253–267.
8. Kuroda M, Yokofujita J, Murakami K. An ultrastructural study of the neural circuit between the prefrontal cortex and the mediodorsal nucleus of the thalamus. Prog Neurobiol 1998; 54:417–458.
9. Del Rio JA, Soriano E, Ferrer I. Development of GABA-immunoreactivity in the neocortex of the mouse. J Comp Neurol 1992; 326:501–526.
10. Maric D, Liu QY, Maric I, Chaudry S, Chang YH, Smith SV, Sieghart W, Fritschy JM, Barker JL. GABA expression dominates neuronal lineage progression in the embryonic rat neocortex and facilitates neurite outgrowth via GABA(A) autoreceptor/Cl-channels. J Neurosci 2001; 21:2343–2360.
11. Owens DF, Liu X, Kriegstein AR. Changing properties of GABA (A) receptor-mediated signaling during early neocortical development. J Neurophysiol 1999; 82:570–583.
12. Owens DF, Kriegstein AR. Is there more to GABA than synaptic inhibition? Nat Rev Neurosci 2002; 3:715–727.
13. Pierce RC, Kalivas PW. A circuitry model of the expression of behavioral sensitization to amphetamine-like psychostimulants. Brain Res Brain Res Rev 1997; 25:192–216.
14. Swerdlow NR, Lipska BK, Weinberger DR, Braff DL, Jaskiw GE, Geyer MA. Increased sensitivity to the sensorimotor gating-disruptive effects of apomorphine after lesions of medial prefrontal cortex or ventral hippocampus in adult rats. Psychopharmacology (Berl) 1995; 122:27–34.
15. Schwabe K, Koch M. Role of the medial prefrontal cortex in N-methyl-D-aspartate receptor antagonist induced sensorimotor gating deficit in rats. Neurosci Lett 2004; 355:5–8.
16. Kodsi MH, Swerdlow NR. Regulation of prepulse inhibition by ventral pallidal projections. Brain Res Bull 1997; 43:219–228.
17. Braff D, Stone C, Callaway E, Geyer M, Glick I, Bali L. Prestimulus effects on human startle reflex in normals and schizophrenics. Psychophysiology 1978; 15:339–343.
18. Braff DL, Geyer MA, Swerdlow NR. Human studies of prepulse inhibition of startle: normal subjects, patient groups, and pharmacological studies. Psychopharmacology (Berl) 2001; 156:234–258.
19. Cadenhead KS, Geyer MA, Braff DL. Impaired startle prepulse inhibition and habituation in patients with schizotypal personality disorder. Am J Psychiatry 1993; 150:1862–1867.
20. Cadenhead KS, Serper Y, Braff DL. Transient versus sustained visual channels in the visual backward masking deficits of schizophrenia patients. Biol Psychiatry 1998; 43:132–138.
21. Lipska BK, Swerdlow NR, Geyer MA, Jaskiw GE, Braff DL, Weinberger DR. Neonatal excitotoxic hippocampal damage in rats causes post-pubertal changes in prepulse inhibition of startle and its disruption by apomorphine. Psychopharmacology (Berl) 1995; 122:35–43.
22. Wolf ME. The role of excitatory amino acids in behavioral sensitization to psychomotor stimulants. Prog Neurobiol 1998; 54:679–720.
23. Lipska BK, Jaskiw GE, Weinberger DR. Postpubertal emergence of hyperresponsiveness to stress and to amphetamine after neonatal excitotoxic hippocampal damage: a potential animal model of schizophrenia. Neuropsychopharmacology 1993; 9:67–75.
24. Purdy RH, Morrow AL, Moore PH, Jr., Paul SM. Stress-induced elevations of gammaaminobutyric acid type A receptor-active steroids in the rat brain. Proc Natl Acad Sci USA 1991; 88:4553–4557.
25. Billiards SS, Walker DW, Canny BJ, Hirst JJ. Endotoxin increases sleep and brain allopregnanolone concentrations in newborn lambs. Pediatr Res 2002; 52:892–899.
26. Grobin AC, Heenan EJ, Lieberman JA, Morrow AL. Perinatal neurosteroid levels influence GABAergic interneuron localization in adult rat prefrontal cortex. J Neurosci 2003; 23:1832–1839.
27. Griffin LD, Gong W, Verot L, Mellon SH. Niemann-Pick type C disease involves disrupted neurosteroidogenesis and responds to allopregnanolone. Nat Med 2004; 10:704–711.

28. Barbaccia ML, Roscetti G, Trabucchi M, Cuccheddu T, Concas A, Biggio G. Neurosteroids in the brain of handling-habituated and naive rats: effect of CO_2 inhalation. Eur J Pharmacol 1994; 261:317–320.
29. Barbaccia ML, Affricano D, Trabucchi M, Purdy RH, Colombo G, Agabio R, Gessa GL. Ethanol markedly increases "GABAergic" neurosteroids in alcohol-preferring rats. Eur J Pharmacol 1999; 384:R1–R2.
30. VanDoren MJ, Matthews DB, Janis GC, Grobin AC, Devaud LL, Morrow AL. Neuroactive steroid 3α-hydroxy-5α-pregnan-20-one modulates electrophysiological and behavioral actions of ethanol. J Neurosci 2000; 20:1982–1989.
31. Kehoe P, Mallinson K, McCormick CM, Frye CA. Central allopregnanolone is increased in rat pups in response to repeated, short episodes of neonatal isolation. Dev Brain Res 2000; 124:133–136.
32. Owens MJ, Ritchie JC, Nemeroff CB. 5α-Pregnane-3α,21-diol-20-one (THDOC) attenuates mild stress-induced increases in plasma corticosterone via a non-glucocorticoid mechanism: comparison with alprazolam. Brain Res 1992; 573:353–355.
33. Grobin AC, Roth RH, Deutch AY. Regulation of the prefrontal cortical dopamine system by the neuroactive steroid 3α, 21-dihydroxy-5α-pregnane-20-one. Brain Res 1992; 578:351–356.
34. Motzo C, Porceddu ML, Maira G, Flore G, Concas A, Dazzi L, Biggio G. Inhibition of basal and stress-induced dopamine release in the cerebral cortex and nucleus accumbens of freely moving rats by the neurosteroid allopregnanolone. J Psychopharmacol 1996; 10:266–272.
35. Patchev VK, Montkowski A, Rouskova D, Koranyi L, Holsboer F, Almeida OFX. Neonatal treatment of rats with the neuroactive steroid tetrahydrodeoxycorticosterone (THDOC) abolishes the behavioral and neuroendocrine consequences of adverse early life events. J Clin Invest 1997; 99:962–966.
36. Brixey SN, Gallagher BJ, III, McFalls JA, Jr., Parmelee LF. Gestational and neonatal factors in the etiology of schizophrenia. J Clin Psychol 1993; 49:447–456.
37. Gilmore JH, Jarskog LF. Exposure to infection and brain development: cytokines in the pathogenesis of schizophrenia. Schizophr Res 1997; 24:365–367.
38. Marcelis M, Van Os J, Sham P, Jones P, Gilvarry C, Cannon M, McKenzie K, Murray R. Obstetric complications and familial morbid risk of psychiatric disorders. Am J Med Genet 1998; 81:29–36.
39. Pedersen CB, Mortensen PB. Evidence of a dose-response relationship between urbanicity during upbringing and schizophrenia risk. Arch Gen Psychiatry 2001; 58:1039–1046.
40. Pedersen CB, Mortensen PB. Family history, place and season of birth as risk factors for schizophrenia in Denmark: a replication and reanalysis. Br J Psychiatry 2001; 179:46–52.
41. Battle YL, Martin BC, Dorfman JH, Miller LS. Seasonality and infectious disease in schizophrenia: the birth hypothesis revisited. J Psychiatr Res 1999; 33:501–509.
42. Hirst JJ, Yawno T, Nguyen P, Walker DW. Stress in pregnancy activates neurosteroid production in the fetal brain. Neuroendocrinology 2006; 84(4):264–74.
43. Arnold SE. Neurodevelopmental abnormalities in schizophrenia: insights from neuropathology. Dev Psychopathol 1999; 11:439–456.
44. Keller F, Persico AM. The neurobiological context of autism. Mol Neurobiol 2003; 28:1–22.
45. Sherr EH. The ARX story (epilepsy, mental retardation, autism, and cerebral malformations): one gene leads to many phenotypes. Curr Opin Pediatr 2003; 15:567–571.
46. Nofzinger EA. What can neuroimaging tell us about sleep disorders? Sleep Med 2004; 5(Suppl 1):S16–S22.
47. Ricceri L, Minghetti L, Moles A, Popoli P, Confaloni A, De Simone R, Piscopo P, Scattoni ML, di Luca M, Calamandrei G. Cognitive and neurological deficits induced by early and prolonged basal forebrain cholinergic hypofunction in rats. Exp Neurol 2004; 189(1):162–172.
48. Segawa M. Neurophysiology of Tourette's syndrome: pathophysiological considerations. Brain Dev 2003; 25(Suppl 1):S62–S69.

49. Weinberger DR, Berman KF. Prefrontal function in schizophrenia: confounds and controversies. Phil Trans R Soc Lond B Biol Sci 1996; 351(1346):1495–1503.
50. McCarley RW, Wible CG, Frumin M, Hirayasu Y, Levitt JJ, Fischer IA, Shenton ME. MRI anatomy of schizophrenia. Biol Psychiatry 1999; 45:1099–1119.
51. Goldman-Rakic PS, Selemon LD. Functional and anatomical aspects of prefrontal pathology in schizophrenia. Schizophr Bull 1997; 23:437–458.
52. Selemon LD, Goldman-Rakic PS. The reduced neuropil hypothesis: a circuit based model of schizophrenia. Biol Psychiatry 1999; 45:17–25.
53. Lewis DA. GABAergic local circuit neurons and prefrontal cortical dysfunction in schizophrenia. Brain Res Brain Res Rev 2000; 31:270–276.
54. Harrison PJ. The neuropathology of schizophrenia. A critical review of the data and their interpretation. Brain 1999; 122(Pt 4):593–624.
55. Lewis DA. Development of the prefrontal cortex during adolescence: insights into vulnerable neural circuits in schizophrenia. Neuropsychopharm 1997; 16(6):385–398.
56. Akbarian S, Bunney WE, Jr., Potkin SG, Wigal SB, Hagman JO, Sandman CA, Jones EG. Altered distribution of nicotinamide-adenine dinucleotide phosphate-diaphorase cells in frontal lobe of schizophrenics implies disturbances of cortical development. Arch Gen Psychiatry 1993; 50:169–177.
57. Benes FM, Vincent SL, Alsterberg G, Bird ED, SanGiovanni JP. Increased GABAA receptor binding in superficial layers of cingulate cortex in schizophrenics. J Neurosci 1993; 12:924–929.
58. Ball S, Busatto GF, David AS, Jones SH, Hemsley DR, Pilowsky LS, Costa DC, Ell PJ, Kerwin RW. Cognitive functioning and $GABA_A$/benzodiazepine receptor binding in schizophrenia: a ^{123}I-Iomazenil SPET study. Biol Psychiatry 1998; 43:107–117.
59. Kegeles LS, Humaran TJ, Mann JJ. In vivo neurochemistry of the brain in schizophrenia as revealed by magnetic resonance spectroscopy. Biol Psychiatry 1998; 44:382–398.
60. Marx CE, Shampine LJ, Duncan GE, VanDoren MJ, Grobin AC, Massing MW, Madison RD, Bradford DW, Butterfield MI, Lieberman JA, Morrow AL. Clozapine markedly elevates pregnenolone in rat hippocampus, cerebral cortex, and serum: candidate mechanism for superior efficacy? Pharmacol Biochem Behav 2006; 84(4):598–608.
61. Marx CE, Shampine LJ, Khisti RT, Trost WT, Bradford DW, Grobin AC, Massing MW, Madison RD, Butterfield MI, Lieberman JA, Morrow AL. Olanzapine and fluoxetine administration and coadministration increase rat hippocampal pregnenolone, allopregnanolone and peripheral deoxycorticosterone: implications for therapeutic actions. Pharmacol Biochem Behav 2006 Aug; 84(4):609–617.
62. Lillrank SM, Lipska BK, Weinberger DR. Neurodevelopmental animal models of schizophrenia. Clin Neurosci 1995; 3:98–104.
63. Aggleton JP, Saunders RC. The relationships between temporal lobe and diencephalic structures implicated in anterograde amnesia. Memory 1997; 5:49–71.
64. Aggleton JP, Brown MW. Episodic memory, amnesia, and the hippocampal-anterior thalamic axis. Behav Brain Sci 1999; 22:425–444.
65. Byne W, Buchsbaum MS, Mattiace LA, Hazlett EA, Kemether E, Elhakem SL, Purohit DP, Haroutunian V, Jones L. Postmortem assessment of thalamic nuclear volumes in subjects with schizophrenia. Am J Psychiatry 2002; 159:59–65.
66. Pakkenberg B. The volume of the mediodorsal thalamic nucleus in treated and untreated schizophrenics. Schizophr Res 1992; 7:95–100.
67. Pakkenberg B. Leucotomized schizophrenics lose neurons in the mediodorsal thalamic nucleus. Neuropathol Appl Neurobiol 1993; 19:373–380.
68. Popken GJ, Bunney WE, Jr., Potkin SG, Jones EG. Subnucleus-specific loss of neurons in medial thalamus of schizophrenics. Proc Natl Acad Sci USA 2000; 97:9276–9280.
69. Lewis DA, Cruz DA, Melchitzky DS, Pierri JN. Lamina-specific deficits in parvalbumin-immunoreactive varicosities in the prefrontal cortex of subjects with schizophrenia: evidence for fewer projections from the thalamus. Am J Psychiatry 2001; 158:1411–1422.

70. Woo TU, Miller JL, Lewis DA. Schizophrenia and the parvalbumin-containing class of cortical local circuit neurons. Am J Psychiatry 1997; 154:1013–1015.
71. Gizerian SS, Morrow AL, Lieberman JA, Grobin AC. Neonatal neurosteroid administration alters parvalbumin expression and neuron number in medial dorsal thalamus of adult rats. Brain Res 2004; 1012:66–74.
72. Zimmerberg B, McDonald BC. Prenatal alcohol exposure influences the effects of neuroactive steroids on separation-induced ultrasonic vocalizations in rat pups. Pharmacol Biochem Behav 1996; 55:541–547.
73. Kolb B. Recovery from early cortical damage in rats. I. Differential behavioral and anatomical effects of frontal lesions at different ages of neural maturation. Behav Brain Res 1987; 25:205–220.
74. Van Eden CG, Rinkens A, Uylings HB. Retrograde degeneration of thalamic neurons in the mediodorsal nucleus after neonatal and adult aspiration lesions of the medial prefrontal cortex in the rat. Implications for mechanisms of functional recovery. Eur J Neurosci 1998; 10:1581–1589.
75. Kolb B, Nonneman AJ. Sparing of function in rats with early prefrontal cortex lesions. Brain Res 1978; 151:135–148.
76. Kolb B, Elliott W. Recovery from early cortical damage in rats. II. Effects of experience on anatomy and behavior following frontal lesions at 1 or 5 days of age. Behav Brain Res 1987; 26:47–56.
77. Van Eden CG, van Hest A, van Haaren F, Uylings HB. Effects of neonatal mediodorsal thalamic lesions on structure and function of the rat prefrontal cortex. Brain Res Dev Brain Res 1994; 80:26–34.
78. Torp R, Hoover F, Danbolt NC, Storm-Mathisen J, Ottersen OP. Differential distribution of the glutamate transporters GLT1 and rEAAC1 in rat cerebral cortex and thalamus: an in situ hybridization analysis. Anat Embryol (Berl) 1997; 195:317–326.
79. Berger UV, Hediger MA. Comparative analysis of glutamate transporter expression in rat brain using differential double in situ hybridization. Anat Embryol (Berl) 1998; 198:13–30.
80. Alcantara S, Ferrer I, Soriano E. Postnatal development of parvalbumin and calbindin D28K immunoreactivities in the cerebral cortex of the rat. Anat Embryol (Berl) 1993; 188:63–73.
81. Bredy TW, Grant RJ, Champagne DL, Meaney MJ. Maternal care influences neuronal survival in the hippocampus of the rat. Eur J Neurosci 2003; 18:2903–2909.
82. Lavdas AA, Grigoriou M, Pachnis V, Parnavelas JG. The medial ganglionic eminence gives rise to a population of early neurons in the developing cerebral cortex. J Neurosci 1999; 19:7881–7888.
83. Hevner RF, Daza RA, Englund C, Kohtz J, Fink A. Postnatal shifts of interneuron position in the neocortex of normal and reeler mice: evidence for inward radial migration. Neuroscience 2004; 124:605–618.
84. Behar TN, Li YX, Tran HT, Ma W, Dunlap V, Scott C, Barker JL. GABA stimulates chemotaxis and chemokinesis of embryonic cortical neurons via calcium-dependent mechanisms. J Neurosci 1996; 16:1808–1818.
85. Behar TN, Smith SV, Kennedy RT, McKenzie JM, Maric I, Barker JL. GABA(B) receptors mediate motility signals for migrating embryonic cortical cells. Cereb Cortex 2001; 11:744–753.
86. Behar TN, Schaffner AE, Scott CA, O'Connell C, Barker JL. Differential response of cortical plate and ventricular zone cells to GABA as a migration stimulus. J Neurosci 1998; 18:6378–6387.
87. Behar TN, Schaffner AE, Scott CA, Greene CL, Barker JL. GABA receptor antagonists modulate postmitotic cell migration in slice cultures of embryonic rat cortex. Cereb Cortex 2000; 10:899–909.
88. Swerdlow NR, Geyer MA, Braff DL. Neural circuit regulation of prepulse inhibition of startle in the rat: current knowledge and future challenges. Psychopharmacology (Berl) 2001; 156:194–215.

89. Ellenbroek BA, Budde S, Cools AR. Prepulse inhibition and latent inhibition: the role of dopamine in the medial prefrontal cortex. Neuroscience 1996; 75:535–542.
90. Gruen RJ, Wenberg K, Selim M, Friedhoff AJ, Bradberry CW. Novelty-associated locomotion: correlation with cortical and sub-cortical GABAA receptor binding. Eur J Pharmacol 1996; 309:115–120.
91. Varshavskaya VM, Ivanova ON, Yakimovskii AF. Motor behavior in rats after separate and combined administration of GABAergic agents into the neostriatum. Neurosci Behav Physiol 2004; 34:293–298.
92. Swerdlow NR, Pitcher L, Noh HR, Shoemaker JM. Startle gating in rats is disrupted by chemical inactivation but not D2 stimulation of the dorsomedial thalamus. Brain Res 2002; 953:246–254.
93. An JJ, Bae MH, Cho SR, Lee SH, Choi SH, Lee BH, Shin HS, Kim YN, Park KW, Borrelli E, Baik JH. Altered GABAergic neurotransmission in mice lacking dopamine D2 receptors. Mol Cell Neurosci 2004; 25:732–741.
94. Japha K, Koch M. Picrotoxin in the medial prefrontal cortex impairs sensorimotor gating in rats: reversal by haloperidol. Psychopharmacology (Berl) 1999; 144:347–354.
95. Benes FM, Berretta S. GABAergic interneurons: implications for understanding schizophrenia and bipolar disorder. Neuropsychopharmacology 2001; 25:1–27.
96. Groves PM, Thompson RF. Habituation: a dual-process theory. Psychol Rev 1970; 77:419–450.
97. Swerdlow NR, Varty GB, Geyer MA. Discrepant findings of clozapine effects on prepulse inhibition of startle: is it the route or the rat? Neuropsychopharmacology 1998; 18:50–56.
98. Collins SL, Montano R, Izenwasser S. Nicotine treatment produces persistent increases in amphetamine-stimulated locomotor activity in periadolescent male but not female or adult male rats. Brain Res Dev Brain Res 2004; 153:175–187.
99. Zavitsanou K, Cranney J, Richardson R. Dopamine antagonists in the orbital prefrontal cortex reduce prepulse inhibition of the acoustic startle reflex in the rat. Pharmacol Biochem Behav 1999; 63:55–61.
100. Andersen SL, Thompson AT, Rutstein M, Hostetter JC, Teicher MH. Dopamine receptor pruning in prefrontal cortex during the periadolescent period in rats. Synapse 2000; 37:167–169.
101. Mirescu C, Peters JD, Gould E. Early life experience alters response of adult neurogenesis to stress. Nat Neurosci 2004; 7:841–846.
102. van den Buuse M, Morris M, Chavez C, Martin S, Wang J. Effect of adrenalectomy and corticosterone replacement on prepulse inhibition and locomotor activity in mice. Br J Pharmacol 2004; 142:543–550.
103. Lindley SE, Bengoechea TG, Schatzberg AF, Wong DL. Glucocorticoid effects on mesotelencephalic dopamine neurotransmission. Neuropsychopharmacology 1999; 21:399–407.
104. Bakshi VP, Geyer MA. Ontogeny of isolation rearing-induced deficits in sensorimotor gating in rats. Physiol Behav 1999; 67:385–392.
105. Wise SP, Jones EG. Development studies of thalamocortical and commissural connections in the rat somatic sensory cortex. J Comp Neurol 1978; 178:187–208.
106. Park M, Kitahama K, Geffard M, Maeda T. Postnatal development of the dopaminergic neurons in the rat mesencephalon. Brain Dev 2000; 22(Suppl 1):S38–S44.
107. Antonopoulos J, Dori I, Dinopoulos A, Chiotelli M, Parnavelas JG. Postnatal development of the dopaminergic system of the striatum in the rat. Neuroscience 2002; 110:245–256.
108. Smith SS, Gong QH, Li X, Moran MH, Bitran D, Frye CA, Hsu FC. Withdrawal from 3alpha-OH-5alpha-pregnan-20-One using a pseudopregnancy model alters the kinetics of hippocampal GABAA-gated current and increases the GABAA receptor alpha4 subunit in association with increased anxiety. J Neurosci 1998; 18:5275–5284.
109. Smith SS. Withdrawal properties of a neuroactive steroid: implications for GABA(A) receptor gene regulation in the brain and anxiety behavior. Steroids 2002; 67:519–528.
110. Mtchedlishvili Z, Sun CS, Harrison MB, Kapur J. Increased neurosteroid sensitivity of hippocampal GABAA receptors during postnatal development. Neuroscience 2003; 118:655–666.

111. Shen H, Gong QH, Aoki C, Yuan M, Ruderman Y, Dattilo M, Williams K, Smith SS. Reversal of neurosteroid effects at alpha4beta2delta GABAA receptors triggers anxiety at puberty. Nat Neurosci 2007; 10(4):469–477.
112. Asavaritikrai P, Lotto B, Anderson G, Price DJ. Regulation of cell survival in the developing thalamus: an in-vitro analysis. Exp Neurol 2003; 181(1):39–46.
113. Grobin AC, Morrow AL. 3Alpha-hydroxy-5alpha-pregnan-20-one levels and GABA (A) receptor-mediated 36Cl (–) flux across development in rat cerebral cortex. Brain Res Dev Brain Res 2001; 131:31–39.
114. Gizerian SS, Moy SS, Lieberman JA, Grobin AC. Neonatal neurosteroid administration results in development-specific alterations in prepulse inhibition and locomotor activity. Psychopharmacology (Berl) 2006; 186(3):334–342.
115. Grobin AC, Gizerian S, Lieberman JA, Morrow AL. Perinatal allopregnanolone influences prefrontal cortex structure, connectivity and behavior in adult rats. Neuroscience 2006; 138(3):809–819.

Chapter 16
Neurosteroid Perturbation and Neuropsychiatric Symptoms in Schizophrenia: From the Mechanisms to the Treatment

Flavia di Michele[1], Carlo Caltagirone[2], and Gianfranco Spalletta[2]

Abstract Schizophrenia is a psychiatric disorder with a complex pathophysiology, involving many biochemical abnormalities (Dopaminergic, GABAergic, Glutamatergic) and neurodevelopmental alterations resulting in a disconnectivity in the brain. Since neurosteroids modulate neurotransmitter systems as well as growth factor expression implicated in schizophrenia, recent research has focused on examining the role that neurosteroids play in the illness phenomenology.

Beginning with clinical evidence of variation in symptom severity over the menstrual cycle of schizophrenics, it has been shown that Progesterone and its reduced metabolite tetrahydroprogesterone (THP) affect dopamine-mediated behaviour, suggesting their antipsychotic-like profile. Indeed, THP, the most potent endogenous positive modulator of $GABA_A$ receptors, is able to regulate the Dopaminergic release, via GABA. Interestingly, low cerebral levels of THP were found in a post-mortem study of schizophrenics. Moreover, a positive correlation between THP levels and aggressiveness/hostility, and a negative correlation between negative symptoms and testosterone levels have been evidenced in schizophrenics. Also, low estradiol levels in schizophrenic women have been related with a major vulnerability to psychosis.

Finally, progesterone and dihydroprogesterone, acting on the genome, may modulate growth factor expression (i.e., BDNF), and regulate the expression of D5 receptors, suggesting their likely neuroprotective and neuroplastic role.

This chapter summarizes the current understanding of these neurosteroids in schizophrenia, their potential antipsychotic role, and discusses new avenues for this area of schizophrenia research.

Keywords Progesterone, DHP, THP, estradiol, schizophrenia, psychosis

Abbreviation GABA γ-aminobutyric acid; DA dopamine; DHP dihydroprogesterone; BDNF brain-derived neurotrophic factor; THP tetrahydroprogesterone;

---

[1]Mental Health Service & Department of Neuroscience, Tor Vergata University, Rome, Italy

[2]IRCCS Santa Lucia Foundation, Department of Clinical and Behavioural Neurology, Rome; Department of Neuroscience, Tor Vergata University, Rome, Italy

M.S. Ritsner and A. Weizman (eds.), *Neuroactive Steroids in Brain Function, Behavior and Neuropsychiatric Disorders*

NMDA *N*-methyl-D-aspartate; 5α-R1 5α-reductase type 1; 3α-HSOR-2 3α-HydroxySteroid-OxidoReductase type 2; NADP nicotinamide adenine dinucleotide phosphatase; PFC pre-frontal cortex; DHEA dehydroepiandrosterone; DHEAS dihydroepiandrosterone sulphate; THDOC tetrahydrodesossicorticosterone

16.1 Introduction

Epidemiological studies suggest that the onset of psychotic symptoms may be related to changes in the secretion of gonadal hormones. These hormones have not only been implicated in the pathophysiology of schizophrenia but also in sex differences at onset (earlier in adolescent boys than in adolescent girls, and during post-menopausal period in women), prognosis (women seem to fare better than men) and treatment.[1–4] A fascinating hypothesis[5] concerning the development of schizophrenia during puberty suggests that this disorder may be related to a disturbance of the balance between inhibitory systems in the anterior basal forebrain, via γ-aminobutyric acid (GABA), serotonin, dopamine (DA), and excitatory systems in response to the flood of gonadal hormones to the brain throughout the reproductive period.

During the reproductive period, the severity of clinical symptoms in women have been shown to vary across the menstrual cycle.[1] In particular, significant improvement in psychotic symptoms has been observed during the luteal phase, compared with other days of the menstrual cycle.[6] Moreover, there is a difference between pre- and post-menopausal women with an increased vulnerability for the onset of schizophrenic episodes after menopause.[2] Women appear more vulnerable to psychotic illness even in the period after birth (postpartum psychosis). Therefore, Rupprecht et al.[7] state that a sudden drop of steroid concentrations (particularly progesterone) may contribute to the development of such disorders and a steroid replacement might be of therapeutic value.

16.2 Neuronal Pathophysiology

Progesterone and its reduced metabolite dihydroprogesterone (DHP) by binding intracellular steroid hormone receptors, regulate gene expression providing the molecular basis for a broad spectrum of neuronal function and plasticity, such as neurodevelopment and neuroprotection. For instance, these steroids can promote the synthesis of growth factors (i.e., brain-derived neurotrophic factor – BDNF-) or other specific proteins responsible for myelin formation.[8–16] Interestingly, the second progesterone reduced metabolite in the enzymatic chain, named tetrahydroprogesterone (THP, or allopregnanolone), unlike its precursors, exerts a non-genomic action, functioning as a neuromodulator of the synaptic transmission by binding the ionotropic membrane receptors (*N*-methyl-D-aspartate –NMDA-, $GABA_A$).[17] This steroid may, nevertheless, affect transcription via progesterone receptors through the generation of the metabolite DHP (after intracellular oxidation) that is ligand for intracellular steroid receptors.[8] Therefore, the metabolic pathway regulating the production of the mentioned neuroactive reduced metabolites of progesterone seems to play a key role in the homeostasis of these steroids, which is necessary for their correct functioning.

The 5α-Reductase type 1 (5α-R1) and the 3α-HydroxySteroid-OxidoReductase type 2 (3α-HSOR-2) represent the key enzymes in the production of DHP (neurotrophic agent) and THP (neuromodulator).[18–20] Specifically, progesterone is converted to DHP through an irreversible reaction mediated by 5α-R1. A second reversible reaction, mediated by 3α-HSOR2 with the cofactors NADP(H), NAD(P), allows the conversion of DHP to THP and viceversa.[21,22] Thus, these neuroactive steroids regulate neuronal function through their concurrent influence on neuronal excitability and gene expression.[7]

Interestingly, fluctuations of neurosteroid concentrations may also influence the plasticity of receptors which are regulated by neurosteroids themselves. For instance, it has been shown that THP, which is the most potent endogenous positive allosteric modulator at the $GABA_A$ receptors, may itself modulate the expression of the gene encoding for the α_4 subunit of the $GABA_A$ receptor (reduced levels of THP enhance transcription of the gene). Moreover, there is evidence of a heterogeneity of THP sensitivity to different $GABA_A$ receptor isoforms. Therefore, a fine feedback system between the neuromodulator THP and $GABA_A$ receptors must be considered another mechanism (in addition to the metabolic pathway) by which the neurosteroid function is regulated.[23–26]

It is interesting to note that gender differences (male–female) exist in the expression of the enzymes that regulate the neurosteroids biosynthesis. For instance, Mitev et al.[27] state that the enzyme 3α-HSOR presents a different expression in males and in females during puberty, but not during adulthood. This data is very suggestive considering the similar age-related sex differential expressivity in schizophrenia, where in males the age at onset during puberty is earlier than in females.[3,4]

16.3 Steroids May Influence Psychotic Symptom Expression

The study of the progesterone reduced metabolites (DHP and THP) in schizophrenia has only recently begun. The evidence that they act as inhibitory $GABA_A$ and excitatory NMDA receptors as well as at the genomic level, demonstrating neuroprotective and neurotrophic effects, render them candidate modulators of the pathophysiology of schizophrenia and relevant to the treatment of this disorder. So far, to our knowledge, only three clinical studies[28–30] have been carried out in this field.

Spalletta et al.[29] evaluated the association between aggressive behaviour and THP plasma levels in chronic schizophrenic subjects, considering that aggressiveness has been, to some extent, associated with schizophrenia. Little is known about THP properties on aggressive behaviour, although the potent sedative and anxiolytic properties of THP, shared with benzodiazepines, are well noted.[17,31,32] It has been often suggested that agents acting on the $GABA_A$ receptor complex may be biologic modulators of aggression. It has, indeed, been observed[33–35] that low quantities of benzodiazepines or alcohol may heighten aggression in humans as well as in animals, whereas higher doses decrease aggression. Likewise, administration of low quantities of THP in mice increases aggressive behaviour in a dose-dependent way, whereas higher doses of THP or co-administration of THP and alcohol may reverse this effect.[36,37] This suggests a complex key role for THP in the action of GABAergic agonists on postsynaptic $GABA_A$ receptor function, that would result in dose-dependent biphasic changes in aggressive behaviour.

Surprisingly, we found a strong positive correlation between aggressive and hostile behaviour and THP plasma concentrations in our sample of schizophrenic patients treated with antipsychotics only. On the other hand, since high levels of THP or a co-administration with another GABA agonist may reverse the effect on aggression, it is conceivable to find a diminution of aggression in patients with schizophrenia treated with added benzodiazepines.[29]

Preclinical studies evidenced that progesterone administration dose-dependently decreased locomotor activity in male Wistar rats, the same way as antipsychotic medications did.[38] In contrast to haloperidol, progesterone neither produced catalepsy nor antagonized amphetamine-induced stereotypy, indicating that progesterone would not produce extrapyramidal side effects and therefore it might have a side effect profile similar to that of atypical antipsychotics. However, both progesterone and haloperidol effectively restored the disruption of the prepulse inhibition of the acoustic startle response that was evoked by apomorphine. This behavioural profile of progesterone is compatible with the sedative properties of its metabolite THP, enhancing the function of GABA. There is also the possibility that progesterone itself, as well as THP, share some properties with atypical antipsychotics, which may be relevant for the development and treatment of psychotic disturbances, e.g., postpartum psychosis.[38] Accordingly, progesterone has been shown to modulate the D5 receptors expression in hypothalamic ANP neurons probably through an indirect action, namely potentiating the estrogen-mediated effects which increase the D5 receptors expression. Therefore, since a reduction of D1 receptor family (which includes D5) in some cerebral areas has been implicated in schizophrenia, it is conceivable that progesterone and its reduced metabolites (DHP and THP), together with estrogens, exert a protective role against the incidence of psycosis.[39] Furthermore, regarding estrogens, Kurata et al.[40] demonstrated that acute treatment with beta-estradiol dose dependently inhibited NMDA-induced intracellular Ca(2+) increases, which strongly correlated with its neuroprotective effect via L-type voltage-gated calcium channels.

It is interesting to note that low estradiol levels in schizophrenic women have been related to a major vulnerability to psychosis.[41,42] Moreover, Ko et al.[43] showed that lower levels of estrogen in schizophrenic women of reproductive age were associated with more severe negative symptomatology as well as with reduced performance in cognitive function, especially verbal performance and executive functioning. Similar results have been found in male patients with chronic schizophrenia regarding testosterone levels. Namely, a negative correlation between negative symptoms and testosterone levels have been evidenced.[44]

16.4 Potential Therapeutic Properties of Steroids

It has been suggested that gonadal steroids may be useful in the treatment of schizophrenia. However, although beneficial effects of progesterone in monotherapy or in association with estrogens, have been demonstrated in some case reports of psychotic disorders (postpartum, premenstrual), a review of the literature[45] does not confirm the efficacy of these steroids in psychoses. Therefore, it is not clear if gonadal steroids may be useful in association with neuroleptics, to reduce neuroleptic doses, especially in women susceptible to side effects of neuroleptics. Indeed, Kulkarni et al.[46] in an open label study showed that schizophrenic women improved more rapidly when receiving 17β-estradiol as an adjunct to neuroleptic therapy when compared with neuroleptic treatment alone, while a more recent study[47] has not confirmed the hypothesis that a combined estradiol/antipsychotic therapy is superior to an antipsychotic monotherapy for relapse prevention.

On the other hand, pre-clinical studies have shown encouraging results. For instance, Gizerian et al.[48] evidenced that THP administration in neonatal rats modify the physiological development of the prefrontal cortex (PFC) and of the medio-dorsal thalamus, indicating that this steroid is important for an appropriate development of the thalamo-cortical system and that it may play a role in neurodevelopmental pathologies, such as schizophrenia. A possible mechanism to explain this function is that THP, which is the most potent endogenous positive modulator of the $GABA_A$ receptors, may regulate the neurotrophin expression, such as BDNF, via GABA, since there is evidence that an increment of the GABAergic signal is responsible for an increase of BDNF levels in the cortex and hippocampus.[49–51] Moreover, THP may modulate the neurotrophin expression, such as BDNF, also through a genomic action after oxidation into the compound DHP.[8]

Thus, the well-noted neuroprotective effect of progesterone[9,11,13,15] might be attributed to its reduced metabolites, DHP and THP, that would therefore exert an important role not just during the neurodevelopment but also in the neurodegenerative diseases.[52] Interestingly, a neurotrophin alteration (which warrants survival, plasticity, differentiation and neuronal connectivity, as well as synaptogenesis) has been implicated in the pathogenesis of schizophrenia.[53] Particularly, reductions of both BDNF levels and of neurotrophin receptor mRNAs in the PFC of patients with schizophrenia have been recently reported.[54,55] Moreover, Akaneya et al.[56] demonstrated that the DAergic neurons are specifically protected by BDNF because of a selective acid vulnerability.

Besides, many preclinical studies have evidenced the importance of the GABA function in controlling some aspects of the neurodevelopment. GABA and its receptors are already present in the cortex at the first stages of embryogenesis.[57] During the period of neuronal migration, the GABA signal mediated by $GABA_A$ receptors stimulates the neuritis growth.[58] It may also be involved in the synapsis development through the removal of the magnesium blockade at the NMDA receptors.[59]

In schizophrenia, alterations of PFC, such as a reduction of arborizations, dendritic spines and dimensions of pyramidal cells,[60] responsible for a functional disconnectivity, as well as a reduction of GABAergic chandelier neurons[61] have been reported. Moreover, the cortical GABAergic interneurons produce reelin, a glycoprotein that during the neurodevelopment is responsible for a correct neuronal positioning after migration.[62] The finding of a decrease of reelin expression in PFC and hippocampus in schizophrenic subjects has been suggested by Impagnatiello et al.[63] as a putative vulnerability factor of schizophrenia. Moreover, a decrease in glutamic acid decarboxylase 67 expression, the enzyme responsible for the GABA synthesis, has been described by Guidotti et al.[64] in schizophrenia.

In addition to its action via GABA and on the genome (after conversion to DHP), THP may exert a neuroprotective effect also by attenuating the excitotoxicity and NMDA-mediated apoptosis[16] because it is an antagonist at the NMDA receptor.[17,32]

Therefore, since an anatomic degeneration of inhibitory GABAergic interneurons, together with the existence of disinhibited glutamatergic neurotransmission and the subsequent cascade of excitotoxic events resulting from NMDA receptor hypofunction, has been postulated in schizophrenia,[65] with THP being the most important endogenous neuromodulator at the $GABA_A$ receptors as well as a NMDA antagonist, it is clear that it may play some role in the pathophysiology of schizophrenia and may represent a novel target for therapeutic compounds.

Interestingly, two atypical antipsychotic agents, olanzapine and clozapine, but not haloperidol and risperidone, may increase the concentrations of THP and progesterone in rat brain.[66–68] In addition, clozapine, but not haloperidol, is able to reduce DHEA and DHEAS (GABA-antagonists) concentrations in rat brain.[69] Therefore, it is conceivable that the highly sedative and the antidepressant effect of clozapine and olanzapine may also be GABA-mediated, and not only histamine(H1)-mediated. Moreover, it is possible that the antipsychotic effect of these drugs is partially mediated by the neurosteroid modifications induced by them. An essential role of THP in the olanzapine antipsychotic action has been hypothesized by Ugale et al.[70] using two experimental paradigms in roolents (apomorphine induced climbing and conditioned avoiding response). These results are in line with the hypothesis of Benes and Berretta[71] of a reduction of the GABAergic neurotransmission as a physiopathological basis for schizophrenia. Therefore, a THP increment, and a reduction of the GABA-antagonists DHEA e DHEAS, pharmacologically induced, could increase the GABAergic tone in the cortex, resulting in an improvement of psychotic symptoms.

On the other hand, a clinical study carried out on schizophrenic patients has not confirmed the above data. No significant changes in circulating concentrations of THP and THDOC (Tetrahydrodesossicorticosterone) were observed in the course of clozapine administration in spite of the patients' good clinical response to the drug. Thus it could be deduced that the therapeutic efficacy of clozapine is not linked to changes in the baseline concentrations of THP and THDOC.[28]

In spite of this, a recent post-mortem study by Marx et al.[30] has demonstrated increased DHEA levels in the parietal cortex and in the posterior cingolate of schizophrenics and bipolar patients compared with normal subjects, while THP levels resulted lower in the parietal cortex of schizophrenics compared to normal controls. Thus, the authors highlighted the possible physiopathological role of these neurosteroids in schizophrenia and their therapeutic potential.

Although the mechanism by which the GABA-agonists neurosteroids exert a neuroleptic-like action is not clear, it has been demonstrated that THP is able to modulate the central DAergic neurotransmission through its direct action on $GABA_A$ receptors. Indeed, Czlonkowska et al.[72] stated that intracerebroventricular injections of midazolam and THP significantly decreases the concentration of DA metabolites as well as the DA turnover in the mouse striatum. Moreover, an increase of THP brain levels would suppress the DA release (basal and stress-induced) in rat cerebral cortex and in nucleus accumbens.[73,74] Similar findings have also been reported by Calabresi et al.[75] for benzodiazepine agonists that can suppress DAergic activity in the striatum of 6-hydroxydopamine-lesioned rats, an animal model of Parkinson's disease, via $GABA_A$ receptors.

Furthermore, Khisti et al.[76,77] stated that THP would promote haloperidol-like behavioural changes. In particular, it would influence DA-mediated behaviours, such as catalepsy, when administered in high doses, while low ip dose of THP (1.5 mg/kg) did not induce catalepsy.

On the other hand, Dazzi et al.[78] stated that the depletion of cortical THP would potentiate DA release. These results have been confirmed by Jaworska-Feil et al.[79] who have also evidenced the opposite effect exerted by the GABA-antagonist steroids.

Overall, the results mentioned above support the hypothesis of a physiopathological role for these neurosteroids in psychosis pathogenesis and, consequently, a potential role for the treatment.

16.5 Conclusions

There are many gaps in our knowledge of the mechanisms underlying schizophrenia pathogenesis. All of the models proposed so far to describe the pathophysiology of schizophrenia have missing links, and new models are necessary to develop new treatment approaches. Starting from the fact that the onset of psychosis may be related to changes of gonadal hormone secretion here we described a new avenue that involves progesterone and its neuroactive reduced metabolites DHP in regulating gene expression and, particularly, THP as a neuromodulator of the NMDA and $GABA_A$ synaptic transmission. These neuroactive steroids may also influence receptor plasticity. Thus, there is evidence that part of the heterogeneity of schizophrenia pathogenesis and symptom expression, such as aggressive behaviour, negative symptoms, cognition, and motor activity, may be explained by neuroactive steroid perturbation. This indicates that the potential beneficial therapeutic effects of steroids, at least in association with antipsychotic drugs, should be further investigated in schizophrenia and other forms of non-schizophrenia psychosis.

References

1. Hallonquist MV, Seeman M, Lang NA, et al. Variation in symptom severity over the menstrual cycle of schizophrenics. Biol Psychiatry 1993; 33:207–209.
2. Hafner H, Riecher-Rossler A, Wan der Heiden K, et al. Generating and testing a causal explanation of the gender difference in age at first onset of schizophrenia. Psychol Med 1993; 23:925–940.
3. Jablesky A. Epidemilogy of schizophrenia: the global burden of disease and disability. Eur Arch Psychiatry Clin Neurosci 2000; 250:274–285.
4. Spauwen J, Krabbendam L, Lieb R, et al. Sex differences in psychosis: normal or pathological? Schizophr Res 2003; 62:45–49.
5. Stevens JR. Schizophrenia: reproductive hormones and the brain. Am J Psychiatry 2002; 159:713–719.
6. Bergemann N, Parzer P, Runnebaum B, et al. Estrogen, menstrual cycle phases, and psychopathology in women suffering from schizophrenia. Psychol Med 2007; 37:1427–1436.
7. Rupprecht R, di Michele F, Hermann B, et al. Neuroactive steroids: molecular mechanisms of action and implications for neuropsychopharmacology. Brain Res Rev 2001; 37:59–67.
8. Rupprecht R, Reul JM, Trapp T, et al. Progesterone receptor-mediated effects of neuroactive steroids. Neuron 1993; 11:523–530.
9. Garcia-Estrada J, Del Rio JA, Luquin S, et al. Gonadal hormones down-regulate reactive gliosis and astrocyte proliferation after a penetrating brain injury. Brain Res 1993; 628:271–278.
10. Koenig HL, Schumacher M, Ferzas B, et al. Progesterone synthesis and myelin formation by Schwann cells. Science 1995; 268:1500–1563.
11. Gibbs RB. Levels of trkA and BDNF mRNA, but not NGF mRNA, fluctuate across the estrous cycle and increase in response to acute hormone replacement. Brain Res 1998; 787:259–268.
12. Vongher JM, Frye CA. Progesterone in conjunction with estradiol has neuroprotective effects in an animal model of neurodegeneration. Pharmacol Biochem Behav 1999; 64:777–785.
13. Schumacher M, Akwa Y, Guennoun R, et al. Steroid synthesis and metabolism in the nervous system: trophic and protective effects. J Neurocytol 2000; 29:307–326.
14. di Michele F, Lekieffre D, Pasini A, et al. Increased neurosteroids synthesis after brain and spinal cord injury. Neurosci Lett 2000; 284:65–68.
15. Callier S, Morissette M, Grandbois M, et al. Neuroprotective properties of 17beta-estradiol, progesterone, and raloxifene in MPTP C57Bl/6 mice. Synapse 2001; 41:131–138.
16. Lockhart EM, Warner DS, Pearlstein RD, et al. Allopregnanolone attenuates *N*-methyl-D-aspartate-induced excitotoxicity and apoptosis in the human NT2 cell line in culture. Neurosci Lett 2002; 328:33–36.
17. Majewska MD. Neurosteroids: endogenous bimodal modulators of the GABAA receptor. Mechanism of action and physiological significance. Prog Neurobiol 1992; 38:379–395.
18. Krause JE, Karavolas HJ. Subcellular localization of hypothalamic progesterone metabolizing enzymes and evidence for distinct NADH- and NADPH-linked 3 alpha-hydroxysteroid oxidoreductase activities. J steroid Biochem 1980; 13:271–280.
19. Karavolas HJ, Hodges D. Neuroendocrine metabolism of progesterone and related progestin steroids and neuronal activity. Ciba Foundation Symposium 1990; 153:22–55.
20. Celotti F, Melcangi RC, Martini L. The 5 alpha reductase in the brain: molecular aspects and relation to brain function. Front Neuroendocrine 1992; 13:163–215.
21. Penning TM, Burczynski ME, Jez JM, et al. Human 3alpha-hydroxysteroid dehydrogenase isoforms (AKR1C1-AKR1C4) of the aldo-keto reductase superfamily: functional plasticity and tissue distribution reveals roles in the inactivation and formation of male and female sex hormones. Biochem J 2000; 351:67–77.
22. Stoffel-Wagner B. Neurosteroid metabolism in the human brain. Eur J Endocrinol 2001; 145:669–679.

23. Maitra R, Reynolds JN. Modulation of GABA(A) receptor function by neuroactive steroids: evidence for heterogeneity of steroid sensitivity of recombinant GABA(A) receptor isoforms. Can J Physiol Pharmacol 1998; 76:909–920.
24. Smith SS, Gong QH, Hsu FC, et al. GABA(A) receptor α_4 subunit suppression prevents withdrawal properties of an endogenous steroid. Nature1998; 392:926–930.
25. Grobin AC, Morrow AL. 3Alpha-hydroxy-5alpha-pregnan-20-one exposure reduces GABA(A) receptor α_4 subunit mRNA levels. Eur J Pharmacol 2000; 409:R1–R2.
26. Biggio G, Follesa P, Sanna E, et al. GABAA-receptor plasticity during long-term exposure to a withdrawal from progesterone. Int Rev Neurobiol 2001; 46:207–241.
27. Mitev YA, Darwish M, Wolf SS, et al. Gender differences in the regulation of 3-alpha-hydroxysteroid dehydrogenase in rat brain and sensitivity to neurosteroid-mediated stress protection. Neurosci 2003; 120:541–549.
28. Monteleone P, Fabrazzo M, Serra M, et al. Long-term treatment with clozapine does not affect morning circulating levels of allopregnanolone and THDOC in patients with schizophrenia: a preliminary study. J Clin Psychopharmacol 2004; 24:437–440.
29. Spalletta G, Romeo E, Bonaviri G, et al. Preliminary evidence for association between aggression, hostility and 5alpha-3alpha-tetrahydroprogesterone levels in schizophrenia. J Psychiatry Neurosci 2005; 30:49–52.
30. Marx CE, Stevens RD, Shampine LJ, et al. Neuroactive steroids are altered in schizophrenia and bipolar disorder: relevance to pathophysiology and therapeutics. Neuropsychopharm 2006; 31:1249–1263.
31. Paul SM, Purdy RH. Neuroactive steroids. FASEB J 1992; 6:2311–2322.
32. Baulieu EE. Neurosteroids: a novel function of the brain. Psychoneuroendocrinology 1998; 23:963–987.
33. Rodgers RJ, Waters AJ. Benzodiazepines and their antagonists: a pharmacoethological analysis with particular reference to effects on aggression. Neurosci Behavioural Rev 1985; 9:21–35.
34. Miczek KA, Weerts EM, DeBold JF. Alcohol, benzodiazepine- GABA(A) receptor complex and aggression: ethological analysis of individual differences in rodents and primates. J Stud Alcohol Suppl 1993; 11:170–179.
35. Bond AJ, Silveira JC. The combination of alprazolam and alcohol on behavioral aggression. J Stud Alcohol Suppl 1993; 11:30–39.
36. Van Erp AM, Miczek KA. Increased aggression after ethanol self-administration in male resident rats. Psychopharmacol 1997; 327:97–101.
37. Fish EW, Faccidomo S, DeBold JF, et al. Alcohol, allopregnanolone and aggression in mice. Psychopharmacol 2001; 153:473–483.
38. Rupprecht R, Koch M, Montowski A, et al. Assessment of neuroleptic-like properties of progesterone. Psychopharm 1999; 143:29–38.
39. Lee D, Wang L, dong P, Tran T, et al. Progesterone modulation of D5 receptor expression in hypothalamic ANP neurons, the role of estrogen. Mol Psychiatry 2001; 6:112–117.
40. Kurata K, Yamawaki S, Morinobu S, et al. Beta-estradiol, dehydroepiandrosterone, and dehydroepiandrosterone sulfate protect against *N*-methyl-D-aspartate-induced neurotoxicity in rat hippocampal neurons by different mechanisms. J Pharmacol Exp Ther 2004; 311:237–245.
41. Oades RD, Schepker R. Serum gonadal steroid hormones in young schizophrenic patients. Psychoneuroendocrinology 1994; 19:373–385.
42. Bergemann N, Mundt C, Parzer P, et al. Plasma concentrations of estradiol in women suffering from schizophrenia treated with conventional versus atypical antipsychotics. Schizophr Res 2005; 73:357–366.
43. Ko YH, Joe SH, Jung SW, et al. Estrogen, cognitive function and negative symptoms in female schizophrenia. Neuropsychobiology 2006; 53:169–175.
44. Ko YH, Jung SW, Joe SH, et al. Association between serum testosterone levels and the severity of negative symptoms in male patients with chronic schizophrenia. Psychoneuroendocrinology 2007; 32:385–391.
45. Granger AC, Underwood MR. Review of the role of progesterone in the management of postnatal mood disorders. J Psychosom Obstet Gynaecol 2001; 22:49–55.

46. Kulkarni J, de Castella A, Smith D, et al. A clinical trial of the effects of estrogen in acutely psychotic women. Schizophrenia Res 1996; 20:247–252.
47. Bergemann N, Mundt C, Parzer P, et al. Estrogen as an adjuvant therapy to antipsychotics does not prevent relapse in women suffering from schizophrenia: results of a placebo-controlled double-blind study. Schizophr Res 2005; 74:125–134.
48. Gizerian SS, Morrow AL, Lieberman JA, et al. Neonatal neurosteroid administration alters parvalbumin expression and neuron number in medial dorsal thalamus of adults rats. Brain Res 2004; 1012:66–74.
49. Berninger B, Marty S, Zafra BM, et al. GABAergic stimulation switches from enhancing to repressing BDNF expression in rat hippocampal neurons during maturation in vitro. Development 1995; 121:2327–2335.
50. Marty S, Berninger B, Carroll P, et al. GABAergic stimulation regulates the phenotype of hippocampal interneurons through the regulation of BDNF. Neuron 1996; 16:565–570.
51. Owens DF, Kriegstein AR. Is there more to GABA than synaptic inhibition? Nat Rev Neurosci 2002; 3:715–727.
52. di Michele F, Longone P, Romeo E, et al. Decreased plasma and CSF content of Neuroactive steroids in Parkinson's disease. Neurol Sci 2003; 24:172–173.
53. Spalletta G, Tomaiuolo F, Marino V, et al. Schizophrenia as a brain misconnection syndrome: a white matter voxel-based morphometry study. Schizophr Res 2003; 64:15–23.
54. Weickert CS, Hyde TM, Lipska BK, et al. Reduced brain-derived neurotrophic factor in prefrontal cortex of patients with schizophrenia. Mol Psychiatry 2003; 8:592–610.
55. Weickert CS, Ligons DL, Romanczyk T, et al. Reductions in neurotrophin receptor mRNAs in the prefrontal cortex of patients with schizophrenia. Mol Psychiatry 2005; 10:637–650.
56. Akaneya Y, Takahashi M, Hatanaka H. Selective acid vulnerability of dopaminergic neurons and its recovery by brain-derived neurotrophic factor. Brain Res 1995; 704:175–183.
57. DelRio JA, Soriano E, Ferrer I. Development of GABA immunoreactivity in the neocortex of the mouse. J Com Neurol 1992; 326:501–526.
58. Maric D, Liu QY, Maric I, et al. GABA expression dominates neuronal lineage progression in the embryonic rat neocortex and facilitates neurite outgrowth via GABA(A) autoreceptor/Cl-channels. J Neurosci 2001; 21:2343–2360.
59. Owens DF, Liu X, Kriegstein AR. Changing properties of GABA(A) receptor-mediated signaling during early neocortical development. J Neurophysiol 1999; 82:570–583.
60. Glantz LA Lewis DA. Decreased dendritic spine density on prefrontal cortical pyramidal neurons in schizophrenia. Arch Gen Psychiatry 2000; 57:65–73.
61. Pierri JN, Chaudry AS, Woo TU, et al. Alterations in chandelier neuron axon terminals in the prefrontal cortex of schizophrenic subjects. Am J Psychiatry 1999; 156:1709–1719.
62. Curran T, D'arcangelo G. Role of reelin in the control of brain development. Bran Res Brain Res Rev 1998; 26:285–294.
63. Impagnatiello F, Guidotti AR, Pesold C. A decrease of reelin expression as a putative vulnerability factor of schizophrenia. Proc Natl Acad Sci USA 1998; 95:15718–15723.
64. Guidotti A, Aiuta J, Davis JM, et al. Decrease in reelin and glutamic acid decarboxylase 67 expression in schizophrenia and bipolar disorder. Arch Gen Psychiatry 2000; 57:1061–1069.
65. Deutsch SI, Rosse RB, Schwartz BL, et al. A revised excitotoxic hypothesis of schizophrenia: therapeutic implications. Clin Neuropharmacol 2001; 24:43–49.
66. Marx CE, Duncan GE, Gilmore JH, et al. Olanzapine increases allopregnanolone in the rat cerebral cortex. Biol Psychiatry 2000; 47:1000–1004.
67. Marx CE, Van Doren MJ, Duncan GE, et al. Olanzapine and clozapine increase the GABAergic neuroactive steroid allopregnanolone in rodens. Neuropsychopharm 2003; 28:1–13.
68. Barbaccia ML, Affricano D, Purdy RH, et al. Clozapine, but not haloperidol, increases brain concentrations of neuroactive steroids in the rat. Neuropsychopharmacol 2001; 25:489–497.
69. Nechmad A, Maayan R, Ramadan E, et al. Clozapine decreases rat brain dehydroepiandrosterone and dehydroepiandrosterone sulfate levels. Eur Neuropsychopharmacol 2003; 13:29–31.

70. Ugale RR, Hirani K, Morelli M, et al. Role of neuroactive steroid allopregnanolone in antipsychotic-like action of olanzapine in rodents. Neuropsychopharm 2004; 29:1597–1609.
71. Benes FM, Berretta S. GABAergic interneurons: implication for understanding schizophrenia and bipolar disorder. Neuropsychopharm 2001; 25:1–27.
72. Czlonkowska AI, Krzascik P, Sienkiewicz-Jarosz H, et al. A Tolerance to the anticonvulsant activity of midazolam and allopregnanolone in a model of picrotoxin seizures. Eur J Pharmacol 2001; 425:121–127.
73. Motzo C, Porceddu ML, Maira G, et al. Inhibition of basal stress-induced dopamine release in the cerebral cortex ans nucleus accumbens of freely moving rats by the neurosterid allopregnanolone. J Psychopharm 1996; 10:266–272.
74. Laconi M, Cabrera RJ. Effect of centrally injected allopregnanolone on sexual receptivity, luteinizing hormone release, hypothalamic dopamine turnover, and release in female rats. Endocrine 2002; 17:77–83.
75. Calabresi P, Centonze D, Gubellini P, et al. Synaptic transmission in the striatum: from plasticity to neurodegeneration. Progress Neurobiol 2000; 61:231–265.
76. Khisti RT, Mandhane SN, Chopde CT. The neurosteroid allopregnanolone induces catalepsy in mice. Neurosci Lett 1998; 251:85–88.
77. Khisti RT, Despande LS, Chopde CT. The neurosteroid allopregnanolone affects dopamine-mediated behavior in rodents. Psychopharmacol 2002; 161:120–128.
78. Dazzi L, Serra M, Vacca G, et al. Depletion of cortical allopregnanolone potentiates stress-induced increase in cortical dopamine output. Brain Res 2002; 932:135–139.
79. Jaworska-Feil L, Budziszewska B, Leskiewicz M, et al. Opposite effects of inhibitory and excitatory neurosteroids on [3H]dopamine release from rat nucleus accumbens. Pol J Pharmacol 1998; 50:449–452.

Chapter 17
Dehydroepiandrosterone Administration in Treating Medical and Neuropsychiatric Disorders

High Hopes, Disappointing Results, and Topics for Future Research

Michael S. Ritsner

Dedicated to Shmuel and Basia Ritsner, my wonderful parents and friends

Abstract Studies have shown that dehydroepiandrosterone (DHEA) influences multiple systems and disease processes in animals and humans. Many researchers around the world have explored the therapeutic role of DHEA. This chapter provides the main facts on biological effects of DHEA administration followed by review of recently reported findings from DHEA clinical trials in schizophrenia. Alterations in DHEA metabolism in schizophrenia are not well understood. Despite higher hopes after preliminary randomized and placebo-controlled cross-sectional trials, a crossover study failed to find any significant effect of DHEA administration on both positive and negative symptoms, on side effects of antipsychotic agents, or on quality of life measures. However, while still preliminary, there is evidence regarding improvement in some neurocognitive functions due to DHEA administration. This chapter also examines the influence of DHEA administration on blood concentrations of neuroactive steroids, symptomatology, neurocognitive functions, side effects and quality of life measures in schizophrenia. A special section of the chapter addresses the association between serum DHEA(S) concentrations on performance of neurocognitive task in schizophrenia patients during the DHEA trial. Deficit in the current knowledge in this area are identified, and suggestions for future research are provided.

Keywords Dehydroepiandrosterone, dehydroepiandrosterone sulfate, schizophrenia, treatment, cognition, side effects, depression, anxiety, medical disorders

Department of Psychiatry, the Rappaport Faculty of Medicine, Technion – Israel Institute of Technology (Haifa), and Shaar Menashe Mental Health Center (Hadera), Israel

M.S. Ritsner and A. Weizman (eds.), *Neuroactive Steroids in Brain Function, Behavior and Neuropsychiatric Disorders*

Abbreviations AIMS abnormal involuntary movement scale; ATPase adenosine triphosphatase; BMD bone mineral density; BMI body mass index; CGI-S clinical global impression scale-severity; CGI-I clinical global impression-improvement; CDSS Calgary depression scale for schizophrenia; DHEA dehydroepiandrosterone; DHEAS dehydroepiandrosterone sulfate; DHEA(S) both DHEA and DHEAS; EPS medication-induced extrapyramidal symptoms; ESRS extrapyramidal symptom rating scale; FGAs first-generation antipsychotic agents; HDRS Hamilton depression rating scale; LHA life history of aggression; NMDA *N*-methyl-D-aspartate; PANSS positive and negative symptom scale; PMW postmenopausal women; SANS scale for the assessment of negative symptoms; SAS Simpson–Angus extrapyramidal symptom scale; SGAs second-generation antipsychotic agents; Q-LES-Q quality of life enjoyment and satisfaction questionnaire; QLS quality of life scale for rating the schizophrenic deficit syndrome; CANTAB Cambridge automated neuropsychological test battery; BLC big/little circle; RTI reaction time; MTS matching to sample visual search; DMS delayed matching to sample; PRM pattern recognition memory; SRM spatial recognition memory; SSP spatial span; PAL paired associates learning; RVP rapid visual information processing; SWM spatial working memory; IED intra/extra dimensional set shift; SOC stockings of Cambridge; VMS visual/movement skills index; AM attention/memory index; EE executive functions index; GCI global cognitive index

17.1 Introduction

Dehydroepiandrosterone (DHEA) and its sulphate form (DHEAS) [both DHEA(S)] are the most abundant circulating neurosteroids. Although the physiological roles of DHEA(S) are still not fully understood, they have many functions associated with neuronal excitability and synaptic plasticity, stress, mood and neurocognitive performance. DHEA is increasingly available commercially as a supplement aimed at slowing aging processes, and improving various medical and neuropsychiatric conditions. However, there is scant evidence to support the use of DHEA for this purpose.

Two treatment models have been applied in the DHEA administration studies:

- *A replacement model* with physiological (25–50 mg) or near-physiological (100 mg) daily doses of DHEA[1–5].
- *A pharmacological model* of treatment using more than 100 mg/day DHEA. Few reports are available regarding the use of higher DHEA doses (up to 1,600 mg/day),[6] and supraphysiologic doses (1,600 mg–2.25 g).[7]

It should be noted that the perception of DHEA differs around the world. In the USA it is considered only a mere dietary supplement and is sold “over the counter”, while in many European countries it is considered a “true hormone” that has yet to be approved for use as a hormonal treatment by the European health authorities.[8]

This chapter reviews briefly the main findings of DHEA administration in animal studies and in treating medical and neuropsychiatric disorders followed by a special focus on treating schizophrenia.

17.2 Effects of DHEA Administration in Animals

Although the exact mechanisms of action and clinical roles of DHEA and DHEAS in animal studies remain unclear, administration of DHEA and DHEAS has shown various behavioral effects (Fig. 17.1).

Evidence suggests that:

- Subcutaneous injections of DHEAS attenuate *stress reactivity* in male rats exposed to repeated stress,[9] supporting the hypothesis that neurosteroids could be involved in the termination of the stress response.
- Administration of DHEA has *antidepressive effects* in the mouse forced swim test model of depression.[10,11]
- DHEA has been demonstrated to exert a *neuroprotective action* enhancing maze performance and memory in mice.[12] Chronic DHEA administration has a modulatory action on the learning of the adult male rats.[13] DHEAS is also neuroprotective when administered either before or after injury in a focal cortical cold lesion model.[14]
- Treatment with DHEA *reduces oxidative stress* in synaptosomes isolated from the brain of rats.[15] In healthy animals, DHEA exerts an antioxidant effect in the colon, by reducing the tissue susceptibility to oxidation of both lipids and proteins.[16]
- Treatment with DHEA stimulates significantly the oxidative energy metabolism in brain mitochondria of rats thus emphasizing the role of DHEA in *brain development in postnatal life*.[17,18] Recently, Patel and co-authors[19] examined the effects of treatment with DHEA (0.2 or 1.0 mg/kg body weight for 7 days) on the oxidative energy metabolism in the brain mitochondria of old and young adult rats. DHEA treatment stimulated ATPase activity in a dose-dependent manner in young adult rats whereas in the old rats the effect on ATPase activity was marginal.
- Administration of DHEA to animals *reduces visceral fat accumulation*, and improves insulin resistance in experimental models of diet-induced obesity.[20,21]
- Administration of DHEA *ameliorates murine antigen-induced arthritis*.[22] DHEA improves murine osteoblast growth and bone tissue morphometry via the mitogen-activated protein kinase signaling pathway independent of either the androgen receptor or the estrogen receptor.[23]
- DHEA has a possible *anticancer effect*.[24–26]
- DHEA *decreases cocaine self-administration* and reinstatement of cocaine-seeking behavior in rats.[27]

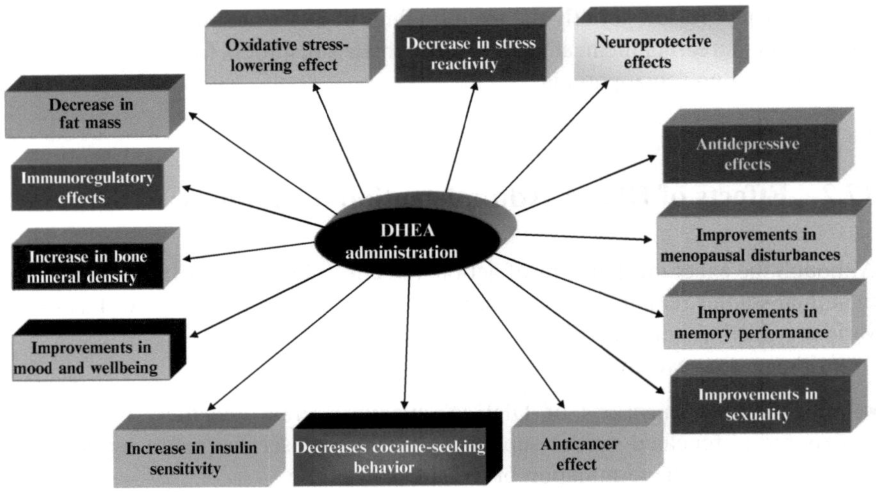

Fig. 17.1 Biological effects of DHEA administration in preclinical and clinical studies

Considering the fact that experimental data clearly demonstrate multiple effects of DHEA(S) on different parts of the brain and body, it is logical to expect that DHEA administration may be useful for humans. This is the reason why several DHEA trials were conducted on samples of healthy elderly subjects and on subjects with various disorders.

17.3 DHEA Administration in Humans

Clinical trials with the use of oral DHEA have been extensively studied in healthy elderly, postmenopausal women, in various somatic and neuropsychiatric disorders (see for review[8,28–33]).

17.3.1 Healthy Elderly

During the normal life cycle, levels of DHEA(S) progressively decline, starting with production of large quantities in the fetus, stopping at birth, resuming during ages 5–7 years, and increasing throughout puberty to maximal production in the 1920s (this age-associated decrease has been termed "adrenopause"). Since many age-related disturbances have been reported to begin with the decline of DHEA(S) levels, this has prompted speculation about a relationship between relative "DHEA deficiency" in older age and diseases of aging as well as the possibility of deriving benefits from administration of DHEA.[34]

DHEA administration in older people was based on the assumption that restoring the circulating levels of DHEA to those found in young people may have anti-ageing effects and improve well-being, and cognitive and sexual functions. However, reported results have been controversial, with negative findings for mood and cognitive function in trials of 2 weeks duration,[35–37] and a pronounced benefit to physical and psychological well-being after a 3-month trial.[2,38] Serum DHEAS concentrations were restored to those found in young adults within 2 weeks of DHEA replacement, and were sustained throughout the 6 months of the study. In elderly men administered 100 mg DHEA for 3 months Flynn et al.[39] did not observe any effect on body composition, serum prostate-specific antigen or urological function.

Van Niekerk et al.[40] conducted a clinical trial in a sample of 46 healthy men aged 62–76, who received either 50 mg DHEA daily for the first 13 weeks, followed by placebo *for the last 13 weeks*, or vice versa, in a randomise double-blind cross-over trial design. No significant effects of DHEA were observed on any of the trial outcomes, providing no support for benefits of DHEA supplementation for cognition or well-being in normal older men in the shorter-term. No increase in well-being or sexuality was noted *after 4 months* of DHEA supplementation (50 mg/day) in healthy 50- to 70-year-old men with low, aging-related endogenous serum DHEAS.[41] Other studies in elderly subjects,[42,43] suggested that DHEA supplementation could have the potential to improve impaired well-being, mood and sexuality, but that it is unlikely to enhance an already normal state.

Kahn and Halloran[44] reported that a 6-month oral DHEA treatment (90 mg/day) did not affect bone turnover in 26 middle-aged and elderly men as compared to a placebo group. Later, Villareal, and Holloszy[45] used a randomized, double-blind, placebo-controlled design on 56 elderly persons aged 65–78 years, who were randomly assigned to receive 50 mg/day of DHEA or placebo for 6 months. The authors concluded that DHEA replacement could play a role in prevention and treatment of the metabolic syndrome associated with abdominal obesity.

In one study 6 months of DHEA supplementation had no effect on bone mineral density,[2,3] while slight increases in bone mineral density (BMD) were observed in other studies.[46,47] Baulieu et al.[46] published the results of a randomized placebo-controlled 12-month study of 280 men and women aged 60–79 years, during which test subjects were given 50 mg DHEA replacement. BMD improved at the femoral neck and radius, and there was a fall in the levels of serum collagen telopeptide. Likewise, only women reported an increase in libido, sexual function and satisfaction. Oral DHEA did not affect bone turnover in 43 healthy middle-aged to elderly men (56–80 years old) when used for *a 6-month period* in doses targeted to restore circulating levels of the steroid to that seen in young adults.[44]

A 12-month DHEA replacement therapy of 70 women and 70 men, aged 60–88 years, with low serum DHEAS levels improved hip BMD in older adults and spine BMD in older women.[48]

Genazzani et al.[49] evaluated the endocrine, neuroendocrine and clinical consequences of a long-term (*12 months*), low-dose of DHEA replacement therapy (25 mg/day) in a group of aging men who presented the clinical characteristics of partial androgen deficiency. The Kupperman score indicated a progressive improvement in mood, fatigue and joint pain. This study demonstrates that DHEA is able to cause significant changes in the hormonal profile and clinical symptoms and can counteract the age-related decline of endocrine and neuroendocrine functions.

Reiter et al.[50] found significant improvement in erectile function and other aspects of sexuality during DHEA treatment in 40- to 60-year-old men with erectile dysfunction and low endogenous DHEAS levels. Thus, one may conclude that, while DHEA may improve impaired well-being, mood and sexuality it does not further enhance already normal or near-normal performance.

Most of the studies using DHEA in healthy, age-advanced volunteers focused on potential metabolic effects on symptoms, usually associated with aging such as hyperlipidemia, decreased insulin sensitivity, increased fat mass, reduced muscle mass, and decreased BMD. In one study that used a daily dose of 100 mg DHEA, a slight but significant high-density lipoproteins reduction in men was also observed.[39] However, several double-blind trials did not find significant changes in body composition.[38,39,46]

Overall, at present, the data does not support improvement in memory or other aspects of cognitive function following DHEA treatment in normal older people.[29]

17.3.2 Postmenopausal Women

In recent years, a great deal of research has focused on the effects of DHEA administration on postmenopausal women (PMW). Menopause is the event in a woman's life that induces a dramatic change in the steroid milieu. DHEA administration has been reported to restore both the androgenic and estrogenic environment. For instance, Stomati et al.[51] investigated the effects of a 6-month DHEA administration (50 mg/day) in early PMW (50–55 years, with normal weight [BMI = 20–24] and overweight [BMI = 26–30]) and late PMW (60–65 years, with normal weight and overweight)]. The endometrial thickness did not show significant modifications in any of the groups of PMW during this trial.

Finckh et al.[52] examined the effect of DHEA on amelioration of fibromyalgia in a double-blind crossover study. A total of 52 PMW were randomized to DHEA treatment (50 mg/day) or placebo *for a 3-month period*, with a 1-month washout period in between. After 3 months of treatment with DHEA, median DHEAS blood levels had tripled, but there was no improvement in well-being, pain, fatigue, cognitive dysfunction, functional impairment, depression, or anxiety, or in objective measurements made by physicians. Androgenic side effects (greasy skin, acne, and increased growth of body hair) were more common during the DHEA treatment period ($p = 0.02$). Thus, DHEA does not improve quality of life, pain, fatigue, cognitive function, mood, or functional impairment in fibromyalgia.

Administration of low doses of DHEA (25 mg) positively modulates several endocrine parameters in early and late PMW, by inducing the increase of the androgenic, estrogenic, and progestogenic milieu and reducing the climateric symptoms, similar to estroprogestin replacement therapy.[53,54] Genazzani and associates[53,54] suggest that DHEA supplementation is a more effective replacement therapy than a simple "dietary supplement".

Therefore, current evidence from randomized, controlled trials in healthy elderly persons including several cohorts of postmenopausal women does not justify the use of DHEA.[55] DHEA administration can delay and correct age-related disorders only to a limited degree.[56] In view of growing popularity of DHEA therapy, and the potential for long-term neuroprotective effects of DHEA(S), it is suggested that further high quality trials should be conducted in which the duration of DHEA treatment is longer than 1 year, and the number of participants is large enough to provide statistically adequate results.[57,58]

17.3.3 Medical Conditions

Clinical trials have investigated the potential efficacy of DHEA therapy in myotonic dystrophy,[59] systemic lupus erythematosus,[60,61] osteoporosis,[62] advanced HIV infection,[63] hypopituitary patients,[64] and other conditions[65] with inconsistent results.[30,33,42] DHEA daily dosages in all these trials ranged from 50 to 500 mg/day for 3–16 weeks with no serious side effects.

DHEA administration, however, has been significantly beneficial in patients with adrenal insufficiency (Addison's disease) representing the pathophysiological model of complete DHEA deficiency. In particular, DHEA administration in women with adrenal insufficiency significantly improved their overall well-being and mood following 4 months of treatment.[66,67] Arlt et al.[68] went on to investigate the effects of DHEA replacement (50 mg/day) in 24 women with primary and secondary adrenal insufficiency employing a double-blind, placebo-controlled, randomized crossover design. Treatment with DHEA raised low serum concentrations of DHEA(S), androstenedione, and testosterone from the initially low levels into the normal range. Libe et al.[69] reported findings from a 4-month, randomized, placebo-controlled DHEA therapy (50 mg/day) study in 20 patients with hypoadrenalism. After 4 months, serum DHEAS levels were raised both in men and in women treated with DHEA. No changes in glucose metabolism and insulin sensitivity were observed. A significant reduction in body fat mass percentage after DHEA administration was observed. DHEA did not cause any relevant variation of subjective health scales and sexuality in both either gender. This study confirms that DHEA may be beneficial for female patients with hypoadrenalism, primarily in restoring androgen levels.

Thus, DHEA therapy has been studied in treatment of several medical conditions with variable but inconclusive evidence of efficacy, with the exception of *adrenal insufficiency*. DHEA shows some promise in observational clinical studies and laboratory experiments, but large-scale human studies are still needed in order find the answers to key questions.

17.3.4 Neuropsychiatric Disorders

The early clinical investigations produced evidence for involvement of neuroactive steroids in conditions such as fatigue during pregnancy, premenstrual syndrome, postpartum depression, depressive and anxiety disorders, epilepsy, and dementia (see for review[28,70–72]).

DHEA has potential therapeutic applications in the treatment of depression due to anti-GABAergic and NMDA-stimulating effects[73–75] that have been confirmed in several studies. A 6-week open-label, and double-blind, randomized, placebo-controlled oral administration of DHEA (90mg/day) found that DHEA might be useful for improving psychological well-being, and for reducing symptomatology measured with the Hamilton Depression Rating Scale (HDRS) in people with major depression.[76,77] More recently, Schmidt et al.[78] evaluated the efficacy of a 6-week long *monotherapy* treatment with DHEA (90 mg/day for 3 weeks and 450 mg/day for 3 weeks) or 6 weeks of placebo for midlife-onset depression ($n=46$) in a double-blind, randomized, placebo-controlled, crossover trial. Significant improvement was observed during DHEA treatment on the HDRS and the Center for Epidemiologic Studies Depression Scale ratings compared with both baseline ($p=0.01$) and placebo treatment ($p=0.01$). A 50% or greater reduction in baseline HDRS scores was observed in 23 subjects after DHEA treatment and in 13 subjects after placebo treatments. Six weeks of DHEA treatment was also associated with significant improvements in Derogatis Interview for Sexual Functioning scores as compared to baseline and placebo conditions. The authors concluded that DHEA might be an effective treatment for midlife-onset major and minor depression.

Hsiao[79] examined pre- and post-treatment changes in plasma DHEA levels and pre- and post-treatment differences in depression scores (HDRS) following successful antidepressant therapy for major depression with venlafaxine. Plasma DHEA levels decreased from baseline to remission was significant ($p=0.017$). After controlling age and gender, a significant correlation ($p=0.044$) was found between the pre- and post-treatment difference in HDRS scores and the pre- and post-treatment difference in DHEA concentrations. This study provides the first clinical evidence identifying that differences in pre- and post-treatment plasma DHEA levels were significantly and positively correlated with differences in pre- and post-treatment depression scores, following successful therapy with venlafaxine for major depression in remitters (non-remitters were not examined).

Bloch and co-authors[80] evaluated the efficacy of DHEA in the treatment of midlife-onset dysthymia in a double-blind, randomized crossover treatment study which included 3 weeks of 90mg DHEA, 3 weeks of 450mg DHEA, and 6 weeks of placebo. Fifteen patients completed the study. A robust effect of DHEA on mood was observed as compared with placebo. Sixty percent of the patients on DHEA improved at the end of the 6-week treatment period compared with 20% on placebo. The symptoms that improved most significantly were anhedonia, loss of energy, lack of motivation, emotional "numbness," sadness, inability to cope, and worry. DHEA showed no specific effects on cognitive function or sleep disturbance, although a type II error could not be ruled out. This pilot study suggests that DHEA is an effective treatment for midlife-onset dysthymia.

Oral DHEA administration (100–400 mg/day) appears to be a useful treatment for non-major depression among men and women with HIV/AIDS that is superior to placebo in reducing depressive symptoms.[81] On the basis of clinicians' ratings, the true drug effect in the study group was about 30% (DHEA response rate minus placebo response rate). The doses of DHEA used and the DHEAS serum levels achieved in this study clearly reflect a pharmacological rather than replacement model of treatment.

Several mechanisms of action of DHEA have been proposed in order to explain mood-enhancing effects. DHEA is a precursor to testosterone and estradiol, each of which independently has been associated with mood effects. In addition, DHEA is a neurosteroid that modulates neuronal excitability by means of specific interactions with neurotransmitter receptors that are known to modulate mood.[82] DHEA therapy may be effective in treating patients with adrenal insufficiency and depression, however, clinical trials investigating its efficacy in other conditions have not provided consistent findings.

17.4 DHEA Administration in Schizophrenia

Schizophrenia is characterized by a wide array of positive, negative, and mood-related psychopathological symptoms (depressive, anxiety), impaired cognitive functions, psychological (emotional and somatic) distress, poor quality of life, and decline in general functioning. Although antipsychotic agents are an indispensable component of the treatment, the development of more effective treatments is an important and desirable goal. A promising direction is the use of neurosteroids. The role of alterations in DHEA metabolism in schizophrenia is not clear and given to much controversy. Chapter 14 reviewed studies investigating blood DHEA levels in patients with schizophrenia. Despite the uncertainties in the underlying mechanisms, DHEA administration has been hypothesized to enhance symptoms, side effects, and cognitive and quality of life impairments in schizophrenia patients. This assumption was examined in several well-controlled studies.[83–86]

17.4.1 Cross-Sectional DHEA Trials

In the first clinical trial 27 adult schizophrenia patients were drawn from an inpatient population, and were randomly assigned to receive either DHEA daily ($n = 15$) or a placebo ($n = 12$) in addition to a constant dosage of antipsychotic agents *for the 6-week trial period.*[83] The starting dose (first 2 weeks) was 25 mg/day; this was followed by 50 mg/day for the next 2 weeks, and then 100 mg/day for the remainder of the trial. The authors noted a decrease in *anxiety, depression and negative symptoms* among the DHEA-treated patients and conclude that DHEA supplementation may be useful in alleviating these symptoms in schizophrenia patients. Interestingly, following DHEA augmentation circulating DHEA and DHEAS levels were correlated with improvement in negative symptoms, but not with improvement in depressive and anxiety symptoms the study are presented in Table 1[84], and that significant change of scores in DHEA patients was present on the SANS, SAS, and BAS scales. However, this table does not present significance for group comparisons.

In a second study the authors investigated the effect of DHEA administration during a period of only *7 days* on medication-induced extrapyramidal symptoms (EPS) among inpatients with schizophrenia or schizoaffective disorder.[85] Patients were randomized in double-blind fashion to receive either 100 mg DHEA or placebo in addition to a constant dosage of antipsychotic medications. Analysis was performed on 30 patients (15 patients in each arm). The authors reported that DHEA showed *a favorable effect on parkinsonism* with a significant time effect ($p<0.0001$) as well as a significant group by time interaction ($p<0.05$). No change was noted in akathisia. Change of DHEA blood levels was negatively associated with both changes of parkinsonism and total EPS ratings ($p<0.05$). This study demonstrates a significant effect of DHEA administration on medication-induced extrapyramidal symptoms, with improvement observed particularly in parkinsonian symptoms.

A third study by the same group included 40 patients with chronic schizophrenia stabilized on olanzapine were randomized in double-blind fashion to receive either DHEA (150 mg/day) or placebo augmentation for a period of *12 weeks*.[84] Sixteen patients receiving DHEA and 15 patients receiving placebo completed the study. The analysis *of negative symptoms* measured with the Scale for the Assessment of Negative Symptoms (SANS) revealed a significant treatment by time interaction ($F=4, 28$, $df=6{,}144$, $p<0.001$), *but not between DHEA and placebo arms.* At the same time, DHEA augmentation was not superior to placebo in improving the scores of the Positive and Negative Symptom Scale (PANSS), measures of side effects [the Simpson–Angus Extrapyramidal Symptom Scale (SAS), Barnes Akathisia Scale (BAS), and Abnormal Involuntary Movements Scale (AIMS)], in cognitive performance (Mindstreams battery), and aggressive behavior (the Life History of Aggression scale).

Several methodological concerns should be mentioned regarding this trial.

1. Scales data were analyzed by using 2×7 repeated measures ANOVA models with main factor of treatment (DHEA vs placebo) and a repeated measurements factor of time (weeks 0–12). Unfortunately, findings concerning significance of the main treatment factor (DHEA vs placebo) were not reported. For instance, the authors wrote that data on clinical scales scores at baseline and at the end of the study are presented in Table 17.1, and that significant change of scores in DHEA patients was present on the SANS, SAS, and BAS scales. However, Table 17.1 does not present significance for group comparisons.
2. Furthermore, the ANOVA for repeated measures and correlation coefficients are presented without the Bonferroni correction for multiple testing.
3. In the discussion section the authors noted that the efficacy of DHEA augmentation in the management of negative symptoms of schizophrenia was observed on

the SANS scale and not the negative subscale of the PANSS. However, both the negative symptom subscale of the PANSS (PANSS-NS) and the SANS are highly correlated.[87,88] Welham et al.[87] have reported that the PANSS-NS variates explain 58% of the SANS variance, while the SANS variates explained 85% of the PANSS-NS variance. In order to avoid the risk of misclassifying negative symptoms, and to obtain more meaningful conclusions many researches used ratings of both PANSS and SANS scales.[89–92]

4. In this study effect size as the partial eta squared (η^2) was calculated based on the ANCOVA models (computed with SPSS statistical software). The η^2 means the proportion of total variability attributable to a factor (the group) by time interaction. The effect size in this study was very small (<0.20), in particular, the η^2 was only 0.172, which means that "the group by time interaction on SANS factor" accounted for 17.2% of the overall (effect + error) variance. Likewise, the η^2 on the SAS, and the BAS scores were 0.063, and 0.093, respectively. *This means that 3 months of DHEA administration failed to detect any efficacy in treating symptoms of schizophrenia, and side effects (SAS, BAS).*
5. Finally, since blood DHEA(S) levels among DHEA treated patients compared to the placebo group failed to reach significance ($p > 0.05$), the relationship of DHEA therapy with reduction of negative symptom ratings remains doubtful.

Thus, we conclude that these cross-sectional DHEA trials[83–85] did not replicate one another in terms of the depressive and anxiety symptoms, and in the medication-induced adverse side effects. They did not show a consistent and unequivocal significant favorable effect of DHEA administration on negative symptoms compared to placebo.

17.4.2 A Crossover DHEA Trial

In order to resolve some of the concerns that have risen in the cross-sectional trials, a randomized, double-blind, placebo-controlled *crossover* study was conducted in two mental health centers.[86] During this trial patients received either DHEA (200 mg/day) or placebo in identical capsules *for 6 weeks* following which they were switched to either placebo or DHEA for a further 6 weeks. Patients continued to receive their regular treatment with daily doses of antipsychotic medication kept constant for at least 2 weeks prior to entering the study and throughout the study period.

Figure 17.2 presents a flow diagram of the study population. A total of 76 patients underwent screening for this study. Sixty-two patients were randomized; seven patients withdrew their consent following randomization, but prior to having received the first dose of medication for the study. Thus 55 patients completed the trial (41M, 14F). Their average age was 36.4 years (SD = 9.6, range = 20–53 years), and 35 of them met the DSM-IV criteria for paranoid subtype, 6 for disorganized, 9 for undifferentiated, 4 for schizoaffective and 1 for catatonic subtypes. Thus,

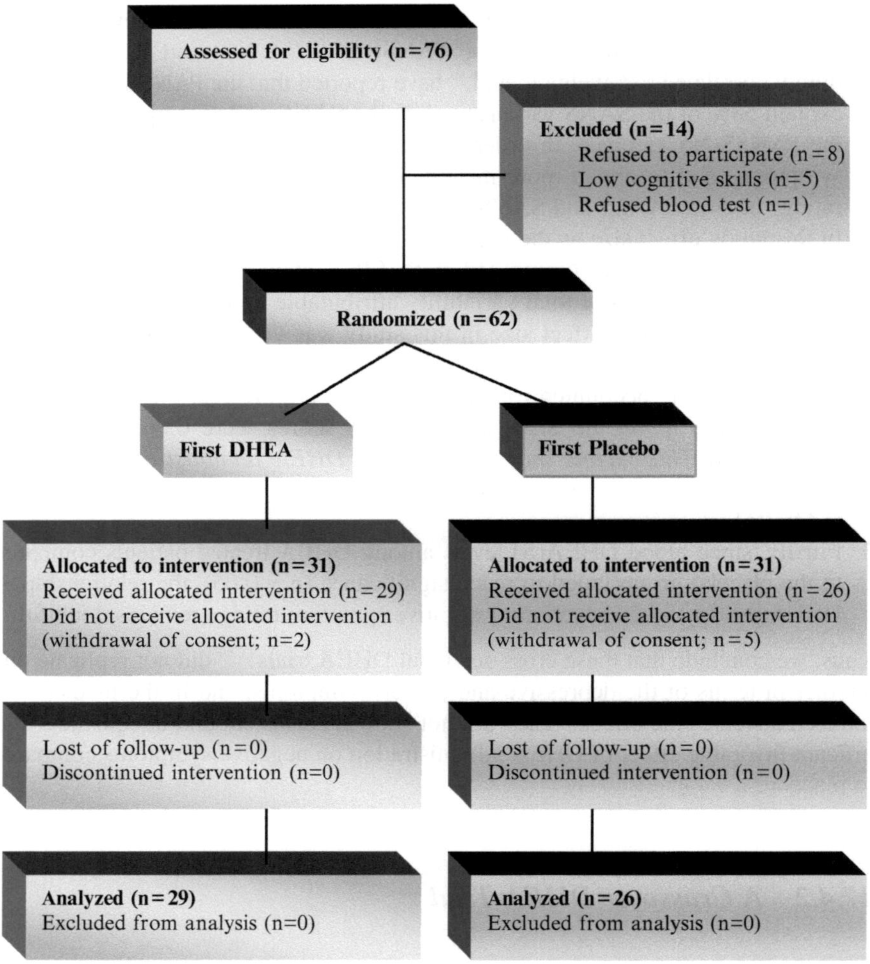

Fig. 17.2 Flow chart of the study recruiting and randomization procedures

55 schizophrenia patients were randomly allocated to two treatment groups, one receiving DHEA for 6 weeks, the other placebo, and then crossing over to the opposite option.

At baseline, there were no significant differences with respect to age, gender, marital status, body mass index (BMI), smoking history, age of illness onset, number of hospital admissions, length of current admission, illness duration, PANSS scores, circulating DHEA(S) levels or antipsychotic medication dose between those who were randomly selected to receive DHEA first and those who were selected to receive placebo first (Table 17.1; all $p>0.05$).

The *crossover analysis* revealed no statistically significant treatment effect of DHEA on severity of illness symptoms (PANSS), side effects (AIMS and ESRS),

Table 17.1 Baseline characteristics of schizophrenia patients ("crossover trial")

Variables	Value	DHEA first ($n = 29$)		Placebo first ($n = 26$)		Significance
		N	Percent	N	Percent	
Gender	Male	22	75.9	19	73.1	$\chi^2 = 0.06$, df = 1, $p = 0.81$
	Female	7	24.1	7	26.9	
Marital status	Never married	17	58.6	19	73.1	$\chi^2 = 1.3$, df = 2, $p = 0.52$
	Married	4	13.8	2	7.7	
	Others	8	27.6	5	19.2	
DSM-IV schizophrenia	295.1, 295.2	4	13.8	3	11.5	$\chi^2 = 4.9$, df = 3, $p = 0.76$
	295.3	19	65.5	16	61.5	
	295.7	2	6.9	2	7.7	
	295.9	4	13.8	5	19.2	
		Mean	SD	Mean	SD	
Age at examination	Years	34.3	9.2	38.8	9.6	$t = 1.8$; $p = 0.085$
Body mass index (BMI)		26.1	5.4	26.7	6.0	$t = 0.4$; $p = 0.72$
Smoking history	Cigarettes/day	26.3	12.2	24.1	11.4	$t = 0.6$; $p = 0.57$
Age of onset	Years	21.6	7.3	22.0	6.2	$t = 0.2$; $p = 0.87$
Number admissions		8.0	7.6	8.2	6.6	$z = 0.3$; $p = 0.82$
Length of stay	Months	39.3	48.6	70.8	74.0	$z = 1.7$; $p = 0.090$
Illness duration	Years	12.2	7.6	16.8	9.9	$t = 1.9$; $p = 0.056$
Baseline: PANSS, total	Score	91.6	17.1	95.6	17.2	$t = 0.9$; $p = 0.38$
DHEA	nmol/l	48.1	29.9	48.4	28.9	$t = 0.04$; $p = 0.97$
DHEAS	nmol/l	5,590	4,042	4,032	2,281	$t = 1.7$; $p = 0.089$
Antipsychotic agents, daily doses[a]	CPZ, mg/day	560	378	728	573	$t = 1.3$; $p = 0.21$

PANSS – Positive and Negative Symptom Scale[127]; DHEA and DHEAS – dehydroepiandrosterone and its sulfate ester; CPZ – chlorpromazine equivalent (mg/day)

[a] Of the total number of patients treated with first-generation agents (FGAs), 6 received haloperidol (mean dose = 14.6 mg/day, SD = 13.6), 5 perphenazine (mean dose = 35.2 mg/day, SD = 9.1), 6 zuclopenthixol (M = 31.3 mg/day, SD = 22.7), 9 clothiapine (mean dose = 97.8 mg/day, SD = 85.1), 2 levomepromazine (N = 2; mean dose = 62.5 mg/day, SD = 53.0), 1 sulpiride (400 mg/day), 4 haloperidol decanoate (mean dose = 8.6 mg/day, SD = 5.6), 10 zuclopenthixol decanoate (mean dose = 21.4 mg/day, SD = 10.4), 2 flupenthixol decanoate (mean dose = 3.5 mg/day, SD = 0.7), 6 fluphenazine decanoate (mean dose = 2.5 mg/day, SD = 0.5), and 2 plenfluridol (mean dose = 4.5 mg/day, SD = 2.1) (total percentages exceed 100% because 19 patients received more than 1 FGAs, and 11 patients received second-generation agents, SGAs). Of the patients treated with SGAs, 11 received risperidone (M = 4.3 mg/day, SD = 1.6), 6 olanzapine (M = 18.3 mg/day, SD = 4.1), and 7 clozapine (M = 407 mg/day, SD = 67.2), and 6 ziprasidone (M= 120 mg/day, SD= 25.3) (11 patients received FGAs).

or on both quality of life measures (QLS, Q-LES-Q) compared with placebo treatment. Cohen's effect sizes for these variables also demonstrated no positive effect of DHEA versus placebo ($d = 0.06–0.13$). At the same time, we found a significant *period effect*, which means improvement in some symptoms, and extrapyramidal side effects during the trial period. In particular, when type and daily doses of antipsychotic agents, BMI and serum concentrations of DHEA(S) were controlled, patients receiving DHEA demonstrated improvement compared to baseline scores with respect to negative ($p = 0.019$), positive ($p = 0.033$), dysphoric mood ($p = 0.024$), and autistic preoccupation ($p = 0.016$) symptoms, and total PANSS scores ($p = 0.003$). These improvements from baseline, however, were not seen in the patients receiving placebo (all $ps > 0.05$). It should be noted, that while patients after DHEA augmentation showed significant improvement from baseline, *the difference between DHEA period and placebo period did not reach significance*.

During the trial no significant adverse events such as oily skin, acne, voice deepening, hirsutism, disinhibition or mania as noted by others with the use of DHEA under certain conditions were observed.

Overall, these observations indicate that DHEA administration is not associated with significant improvement of clinical symptoms, and side effects compared to placebo ("negative treatment trial"). Therefore, this study did not confirm conclusions from previous cross-sectional trials that DHEA augmentation of antipsychotic medication has value in the management of schizophrenia symptomatology[83,84] and extrapyramidal side effects.[85]

The lack of difference between the DHEA and placebo groups may have several possible explanations.

- There may be different study designs (crossover vs randomized single arm), different populations of patients (varying mix of early onset illness schizophrenia vs chronic schizophrenia illness), and different sample sizes (small vs large).
- For some schizophrenia patients DHEA augmentation may be effective, while for others, it may not. Indeed, it could be that DHEA is more efficacious in those with lower baseline DHEA plasma levels. For instance, basal blood DHEA levels of patients receiving DHEA in two "positive treatment trials" were much lower (21.2 ± 12.8 nmol/l[83]; and 15.7 ± 8.7 nmol/l[85]) compared to the "negative treatment trial" (48.2 ± 27.1 nmol/l[86]), and to the findings in most previous publications (41–90 nmol/l[93–95]).
- Conflicting reports underscore the fact that the neurobiology underlying the symptoms of schizophrenia is multifaceted.

Although there is evidence that neurosteroids are associated with feelings of "well-being" and enjoyment of "leisure" activities,[96] this study did not find a significant change in those two quality of life measures under DHEA treatment. However, taking into account a complexity of factor structure of health-related quality of life impairment in schizophrenia,[97] and short-term study time (6 weeks), additional analysis of health-related quality of life measures is warranted.

Limitations of this crossover trial include its relatively short length of period (6 weeks). The question of what might happen during longer time intervals is still standing. Therefore, further study is necessary in order to exclude any long-term effects including influence on the sex-hormonal axis.

17.4.3 Neurocognitive Impairment

Neurocognitive dysfunction, or impairment, constitutes a core or a hallmark feature of schizophrenia that includes problems in speed of processing, attention, working memory, learning, visual processing, executive functions, and social cognition.[98] Impairment in these domains is considered a reasonable treatment target in individuals with schizophrenia, partly because cognitive deficits contribute to poor quality of life and poor functional outcomes.

The beneficial effect of DHEA and DHEAS on the cognitive functioning animals was shown in several pre-clinical studies.[12–14] However, results from studies in humans were less promising. For instance, in healthy young subjects Wolf et al.[35] found no effect of a single dose of DHEA (300 mg) on memory. Alhaj et al.[99] have used DHEA (150 mg/day) or placebo in a double-blind, random, crossover study, in 24 healthy young men, over a 1-week period, and demonstrated the beneficial effects of on episodic memory and its neural correlates utilizing an event-related potential technique. Potential effects of DHEA on impaired cognition and memory were addressed by a double-blind randomized controlled trial including 58 patients with Alzheimer's disease receiving DHEA (100 mg/day) or placebo for 6 months.[100] This study showed transient minor improvements, narrowly missing statistical significance. However, DHEA supplementation may have adverse cognitive effects in elderly women, in particular. In a study by Parsons et al.,[101] DHEA administration increased negative associations between DHEA(S) levels and cognition. Increased cortisol does not explain the neurocognitive deficits associated with DHEA, suggesting a direct negative effect of exogenous DHEA on cognition.

We examined, repeatedly, the 55 schizophrenia patients participating in the DHEA crossover study[86] using the computerized Cambridge Automated Neuropsychological Test Battery (CANTAB; http://www.cantab.com/). These tests were run on an IBM-compatible personal computer with a touch-sensitive screen. The nonverbal nature of the CANTAB tests makes them largely language-independent and culture-free. Overall, 12 neuropsychological tests were grouped into five neurocognitive domains: visual and movement skills, attention and memory, learning, sustained attention, and executive function, which included Big/Little Circle (BLC), Reaction Time (RTI), Matching to Sample Visual Search (MTS), Delayed Matching to Sample (DMS), Pattern Recognition Memory (PRM), Spatial Recognition Memory (SRM), Spatial Span (SSP), Paired Associates Learning (PAL), Rapid visual information processing (RVP), Spatial working memory (SWM), Intra/Extra Dimensional Set Shift (IED), and Stockings of Cambridge (SOC).

Performance of neurocognitive tasks was presented using the standard z-score, which is given as the number of standard deviations (SD) from the mean performance, computed relative to an extensive database of raw scores for normal, healthy adult subjects matched by age and sex. Standardized *z*-scores were calculated by the CANTAB program on the basis of an extensive normative database included in CANTAB. A negative value of the *z*-score (≤ -1.0) indicates poorer than average performance. In addition, the following indices were evaluated[102] as average values of *z*-scores of neurocognitive tasks which were included in the same domain, namely: Visual and Movement Skills (VMS), Attention/Memory (AM), Executive Functions (EF), and Global Cognitive Index (GCI).

At baseline, neurocognitive performance of the patients show pronounced neurocognitive impairment in a range of domains, including visual and movement skills (VMS index = −2.4, SD = 3.0), attention and memory (AM index = −1.3, SD = 1), and executive function (EF index = −0.6, SD = 0.6) (Table 17.2). Following the study intervention, significant improvement with respect to *the visual sustained attention* (RVP test, $z = 2.0$, $p = 0.041$), and *visual and movement skills* (Visual/Movement Skills index, $z = 2.0$, $p = 0.044$; BLC test, $z = 2.1$, $p = 0.037$) were noted during the period that patients received DHEA as compared to the period during which they received placebo (Fig. 17.3). The difference between the DHEA and placebo periods in other CANTAB tasks and indices did not reach statistically significant levels (all *p*s > 0.05).

Visual processing deficits are an integral component of schizophrenia, and patients show such deficits in early-stage visual processing, potentially reflecting dysfunction of the magnocellular visual pathway mediated by glutamatergic receptors.[103] Consistent with glutamatergic theories, deficits in magnocellular processing may reflect more general impairments in neuronal system functioning, such as deficits in non-linear amplification and may thus represent an organizing principle for predicting neurocognitive dysfunction in schizophrenia.[104,105] It has further been hypothesized that such deficits may be associated with a more generalized disorganization syndrome in the illness that reflects a widespread deficit in the cognitive coordination of contextually related stimuli. This, in turn, would lead to dysfunctional organization of stimulus features in vision, thought and language.[106] While such findings may appear to be limited to an expression at the level of neurocognitive dysfunction, attentional control processes (as reflected by abnormalities in visuospatial tests and visual-perceptual function)[107] may be limiting factors for social functioning in the illness.[108]

Thus, this investigation, while preliminary, support some improvement noted in visual sustained attention, visual and motor skills due to DHEA administration. If confirmed, these findings should be taken given consideration in future trials.

17.4.4 Neurosteroids as Predictors of Neurocognitive Impairment

Current theories link efficacy in cognition to unique effects on glutamatergic neurotransmission in the prefrontal cortex, cortical dopaminergic and cholinergic function and improved patterns of connectivity in the brain during cognitive task

Table 17.2 Baseline neurocognitive functioning of 55 schizophrenia patients (CANTAB, *z*-scores, "crossover trial")

Domain	CANTAB task	Metrics		Mean[a]	SD	Min.	Max.
Motor skills	Big/little circle	Percent correct	BLC	−1.99	5.0	−3.4	0.2
	Reaction time	Five choice movement time[b]	RTI1	−0.59	1.8	−4.5	4.1
		Five choice reaction time[b]	RTI2	−2.80	3.4	−3.1	1.6
Attention and memory	Matching to sample visual search	Percent correct	MTS1	−0.67	2.0	−8.9	1.0
		Total number correct	MTS2	−0.80	1.3	−7.5	0.1
	Delayed matching to sample	Percent correct (all delays)	DMS1	−2.27	1.4	−6.0	1.2
		Percent correct (simultaneous)	DMS2	−0.46	1.4	−4.5	0.7
		Prob. error given error	DMS3	−1.63	1.1	−4.1	0.9
	Pattern recognition memory	Percent correct	PRM	−1.51	1.3	−4.6	1.4
	Spatial recognition memory	Percent correct	SRM	−1.56	1.4	−4.7	1.4
	Spatial span	Span length	SSP	−0.88	1.0	−2.6	1.8
Learning	Paired associates learning	Total errors, adjusted	PAL1	−2.16	2.0	−6.4	0.9
		Total errors, 6 shapes, adjusted	PAL2	−1.36	1.5	−3.7	0.7
Sustained attention	Rapid visual information processing	Sustained attention	RVP A'	−2.15	1.4	−7.5	0.4
Executive functions	Spatial working memory	Between errors	SWM1	−0.89	1.0	−3.8	1.5
		Strategy	SWM2	−0.87	0.7	−2.1	2.3
	Intra/extra dimensional set shift	Stages completed	IED1	−0.10	0.7	−1.7	0.4
		Total errors	IED2	−0.76	1.0	−2.9	0.9
		Total error (adjusted)	IED3	−0.17	0.6	−1.4	0.7
	Stockings of Cambridge	Mean initial thinking time[b]	SOC1	0.33	1.3	−6.7	1.4
		Subsequent thinking time[b]	SOC2	0.09	0.6	−2.1	0.7
		Problem solving in minimum moves	SOC3	−1.08	1.1	−2.5	1.9

[a]The standard *z*-score is given as the number of standard deviations from the mean performance computed relative to an extensive CANTAB database of raw score for normal, healthy adult subjects matched by age and sex. A negative value indicates performance worse than average in healthy control subjects.
[b]Milliseconds

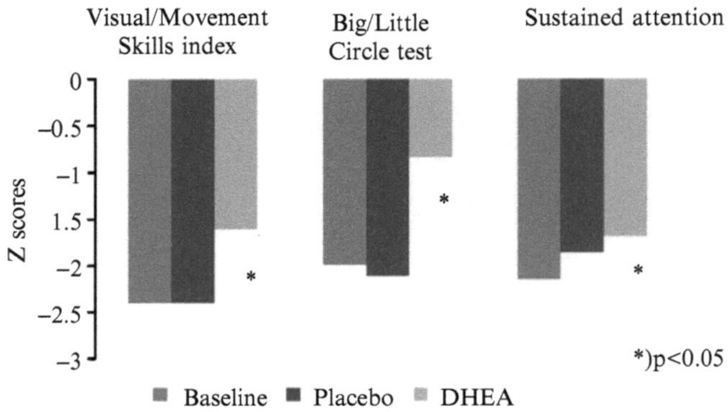

Fig. 17.3 The effect of add-on oral DHEA to ongoing antipsychotic treatment on neurocognitive schizophrenia patients (*z*-scores)

performance.[109,110] Generally the neurocognitive deficits, shown by patients with schizophrenia, can be attributed to pre-morbid intellectual function,[111] early age at onset,[112] severity of the negative and positive symptoms,[113–115] several antipsychotic agents,[116] and high-doses of antipsychotic agents.[117]

The factors underlying neurocognitive dysfunction in schizophrenia as well as the association between neurosteroids and neurocognitive function in general are not yet fully understood. More specifically, the relationship between DHEA and neurocognitive functions is intriguing, and much work is yet to be done in this field. Early DHEA studies investigating the neurocognitive functions in schizophrenia patients have shown conflicting results; neurocognitive impairment has been associated with low DHEAS levels,[118] high DHEAS levels[119] or high DHEA levels,[118] or no association was found altogether.[120] Controlling for age in 17 chronic, medicated schizophrenia or schizoaffective inpatients, higher DHEA levels were found to be significantly correlated with better performance on attention and working memory ($r=0.55$, $p<0.05$), and DHEAS levels were positively correlated with semantic memory ($r=0.51$, $p<0.05$).[121]

The relationship between DHEA administration and neurocognitive function in schizophrenia patients remains unclear. In order to study the association in DHEA administration trial between changes in circulating neurosteroids and the neurocognitive deficit in schizophrenia, we investigated the predictive role of neuroactive steroid concentrations. The twofold aim of this analysis was: applied using the NCSS-2000 PC program for predicting the 2-scores of fine.

- To delineate the relationship between blood levels of neurosteroids (DHEA, DHEAS, androstenedione, cortisol, and their molar ratios), and neurocognitive functions measured with CANTAB
- To identify, which neurosteroids, if any, contributed to predicting performance of attention and memory tasks, visual sustained attention, and executive functions as well as other relevant variables

Step-wise multiple regression analysis with a backward selection procedure was applied using the NCSS-2000 PC program[123] for predicting the *z*-scores of five

CANTAB tasks: MTS, PRM, RVP, IED and SWM. The following procedure was conducted for each neurocognitive test using an initial set of independent variables of the 55 schizophrenia patients who participated in the DHEA crossover trial at three time points: start of trial (i.e. baseline), after DHEA administration period, and after placebo administration period[86]:

- Circulating hormones (DHEA, DHEAS, androstenedione, cortisol), and molar ratios (cortisol to DHEA, cortisol to DHEAS*100, DHEAS to DHEA, and androstenedione to DHEA)
- Five PANSS factors (positive, negative, activation, dysphoric mood, autistic preoccupations)
- Daily doses of antipsychotic agents as CPZ equivalent (mg/day), and side effects (ESRS scores)
- Additional independent variables such as age, education, age of onset, and illness duration (years)

Table 17.3 presents summary of multiple regression analysis. Obtained regression models accounted for 20–39% of variance, and included neurosteroids, clinical and demographical predictors. In particular, blood neurosteroid levels and their molar ratios accounted for 16.5% of the total variance in sustained attention, 8–13% in visual memory tasks, and about 12% in executive functions. In addition, effects of symptoms (6–18%), illness duration (8–18%), daily doses of antipsychotic agents (4–6%), side effects (3.2%), education (7%), and age of onset (2.7%) accounted for variability in neurocognitive functioning in schizophrenia. Inherently, increasing blood DHEAS and androstenedione were associated with better performance on memory and visual sustained attention (PRM, and RVP) tests, while higher circulating DHEA, and some molar ratios (cortisol to DHEA(S), DHEAS to DHEA, androstenedione to DHEA) were associated with poorer neurocognitive functions.

In light of these findings we assert that neurocognitive dysfunctions in schizophrenia may be attributed, at least partially, to of variability in serum neurosteroids during DHEA treatment.

17.4.5 Hormonal Response on DHEA Administration

As expected, DHEA administration influences on circulating neuroactive steroids. For instance, Morales et al.[2] found that oral DHEA (100 mg/day) administered for 6 months resulted in elevation of circulating DHEA(S) concentrations and the DHEAS to cortisol molar ratio among healthy elderly men and women. Long-term DHEA administration (6–12 months, 25–50 mg/day) in postmenopausal women (PMW) was associated with significantly increased levels of the DHEA's metabolites: DHEAS, androstenedione, testosterone and dihydrotestosterone, estradiol and estrone.[51,53] At the end of 12-month study of 20 PMW receiving 25 mg/day DHEA serum concentrations of DHEAS had increased significantly, but those of testosterone had not changed.[123] A 12-month DHEA administration (25 mg/day) in a group of aging men with partial androgen deficiency resulted in great modification of the

Table 17.3 Predictors of neurocognitive performance ratings among schizophrenia patients across three examinations ("crossover trial")

Dependent variables	Domain	Positive predictors	Negative predictors
Matching to sample visual search $R^2 = 0.20$, adj. $R^2 = 0.16$, $F = 5.5$, df = 6, $p < 0.001$	Attention and memory	Dysphoric mood (PANSS) Education (year) Daily doses (CPZ, mg/day)	Autististic preoccupations (PANSS) Cortisol to DHEAS*100 ratio DHEAS to DHEA ratio
Pattern recognition memory $R^2 = 0.26$, adj. $R^2 = 0.23$, $F = 8.7$, df = 6, $p < 0.001$	Attention and memory	Dysphoric mood (PANSS) Androstenedione (nmol/l)	Autistic preoccupations (PANSS) Illness duration (year) DHEAS to DHEA ratio Androstenedione to DHEA ratio
Rapid visual information processing $R^2 = 0.25$, adj. $R^2 = 0.21$, $F = 7.3$, df = 6, $p < 0.001$	Visual sustained attention	Education (year) DHEAS (nmol/l) Androstenedione (nmol/l)	Negative symptoms (PANSS) DHEA (nmol/l) Androstenedione to DHEA ratio
Intra/extra dimensional set shift $R^2 = 0.39$, adj. $R^2 = 0.35$, $F = 10.7$, df = 8, $p < 0.001$	Executive functions	Dysphoric mood (PANSS) Daily doses (CPZ, mg/day)	Autistic preoccupations (PANSS) Side effects (ESRS) Age of onset (year) Illness duration (year) Cortisol to DHEAS*100 ratio DHEAS to DHEA ratio
Spatial working memory $R^2 = 0.31$, adj. $R^2 = 0.29$, $F = 12.7$, df = 5, $p < 0.001$	Working memory	Dysphoric mood (PANSS)	Autistic preoccupations (PANSS) Illness duration (year) DHEA (nmol/l) Androstenedione to DHEA ratio

Matching to Sample Visual Search tests ability to match visual samples and measures reaction and movement time (MTS, percent correct). Pattern Recognition Memory tests recognition memory for patterns (PRM, percent correct). Rapid visual information processing tests visual sustained attention (RVP). Intra/Extra Dimensional Set Shift tests rule acquisition and attentional set shifting (IED, stages completed). Spatial working memory tests working memory and strategy use (SWM, between errors).

endocrine profile: DHEA(S), androstenedione, testosterone, dihydrotestosterone, progesterone, 17-hydroxyprogesterone, estrone, and estradiol levels increased significantly with respect to baseline values.[49]

Assessments of hormonal response on DHEA administration among schizophrenia patients have revealed varying findings: administration of DHEA (100 mg/day) for *1 week* led to significant elevation in blood DHEAS compared to placebo.[85] After *6 weeks* between-group differences were found on both DHEA and DHEAS levels,[83] while *after 12 weeks* of DHEA treatment (150 mg/day) no significant between-group differences were detected.[84] In addition, several hormones were measured (TSH, prolactin, testosterone, insulin, and estradiol) in 27 chronic schizophrenia inpatients receiving DHEA (100 mg/day) or placebo augmentation for 6 weeks.[83,124] Complete data on these measurements were obtained for 16 subjects (8 men, 8 women). No significant changes in hormonal blood measures were noted.

In order to characterize the impact of DHEA administration on blood neuroactive steroids in schizophrenia, we analyzed circulating hormones in schizophrenia patients who participated in the aforementioned crossover study.[86,125] At baseline examination 55 schizophrenia patients exhibited *normal DHEA levels* ($F = 1.8$, $p = 0.19$), but significantly *decreased serum DHEAS levels* ($F = 8.3$, $p = 0.005$), and elevated levels of cortisol ($F = 30.3$, $p < 0.001$), prolactin ($F = 9.3$, $p < 0.001$), cortisol to DHEA ($F = 7.2$, $p = 0.009$) and cortisol to DHEAS ($F = 13.3$, $p < 0.001$) molar ratios when compared with 20 healthy subjects who were closely matched for gender, age, cigarette smoking, and body mass index ($df = 1.75$). As expected, DHEA administration (200 mg/day) *for a 6-week period* resulted in a significant increase in circulating DHEA ($p = 0.003$), DHEAS ($p < 0.001$), and androstenedione ($p = 0.024$), while cortisol to DHEA ($p = 0.004$) and cortisol to DHEAS ($p < 0.001$) molar ratios significantly decreased compared to the placebo period (Table 17.4). Changes in these hormones (with the exception of androstenedione) and molar ratios remain significant even after Bonferroni correction (0.05/7 = 0.007). No significant difference in impact of the DHEA-period and the placebo-period on DHEAS to DHEA ratio, and cortisol levels were observed.

These findings indicate that schizophrenia patients are characterized by significantly decreased basal serum DHEAS and by elevated cortisol to DHEA(S) molar ratios as compared to healthy subjects. While no definite proof exists for these baseline neurosteroids serum level observations, we witnessed elevations of DHEA(S) levels that were achieved by administration of DHEA.

17.4.6 Heterogeneity of Circulating DHEA(S)

It was hypothesized that schizophrenia patients may be quite heterogeneous regarding levels of circulating DHEA(S). In order to test this assumption, median serum concentrations of DHEA (32.2 nmol/l) and DHEAS (6,735 nmol/l) obtained from the distributions of healthy subjects were used as cut-off points to split DHEA and DHEAS into the two levels (high and low) among 55 schizophrenia patients.[86] This procedure identified 37 patients with high levels of serum DHEA, and 18 patients– with low levels, prior to the

Table 17.4 Circulating hormones among 55 schizophrenia patients ("crossover trial")

	Basal level		After DHEA		After Placebo		DHEA vs placebo effect[a]	
Serum hormones	Mean	SD	Mean	SD	Mean	SD	F	*p*
DHEA (nmol/l)	56.0	27.1	70.3	30.9	48.9	28.8	5.9	0.003*
DHEAS (nmol/l)	4,775	3,391	10,390	6,975	4,569	3,435	17.7	0.001*
Androstenedione (nmol/l)	8.7	2.9	10.5	4.3	8.5	3.3	3.8	0.024
Cortisol (nmol/l)	563	258	501	228	515	198	0.6	0.53
Cortisol to DHEA ratio	12.6	10.5	8.8	6.2	16.3	12.2	5.8	0.004*
Cortisol to DHEAS*100	19.1	15.3	8.5	9.1	20.0	15.6	8.4	0.001*
DHEAS to DHEA ratio	90.1	116	169	159	154	166	2.8	0.067

[a]Three-way ANCOVA for DHEA administration main effect (df = 2,165), gender as second factor (df = 1,165), and type of antipsychotic agents as third factor (FGA, SGA, and CT = FGA + SGA; df = 2,165) with controlling for body mass index (BMI), and CPZ equivalent (mg/day) of daily dose of antipsychotic agents

*Significance after the Bonferroni correction: 0.05/7 = 0.007

clinical trial, 47 with high levels and 8 with low levels during DHEA administration, and 33 patients with high levels and 22 with low levels during the placebo-period ($\chi^2 = 9.2$, df = 2, $p = 0.010$). A similar trend was found regarding DHEAS concentrations ($\chi^2 = 29.1$, df = 2, $p < 0.001$). Figure 17.4 shows that the proportion of patients with decreased DHEA and DHEAS levels is much lower when subjects receiving DHEA than when they receiving placebo (8 and 19 vs 22 and 44, respectively; $\chi^2 = 33.1$, df = 4, $p<0.001$). Nevertheless, as can be clearly seen, a significant portion of the patients receiving placebo maintained high physiological levels of either DHEA or DHEAS (28/55 = 50.1%), or both these neurosteroids (8/55 = 14.5%). The correlation coefficient between serum DHEA and DHEAS levels was slightly positive at the baseline examination ($r = 0.27$, $p = 0.047$) and after DHEA administration ($r = 0.29$, $p = 0.033$). This correlation was not seen following the placebo period ($r = 0.01$, $p = 0.94$).

Previous studies investigating DHEA(S) serum levels in schizophrenia demonstrated inconsistent findings that may be due to a wide clinical heterogeneity, variability of psychometric properties (distress, anxiety), prolactin levels, responsiveness of schizophrenia patients to antipsychotic treatment, small sample sizes, or differences in the age at examination, duration of the illness, tobacco and substance abuse as well as comorbid disorders (see more details in Chapter 14). In addition, two subgroups, one with low and one with high hormone levels, may be observed, and those indicate a definite heterogeneity in schizophrenia patients, and, consequently, in their responsiveness to DHEA administration.

17.4.7 *Circulating DHEA(S) and Outcome Measures*

Since treatment with DHEA resulted in an increase both DHEA and DHEAS plasma levels (see Table 17.4), and since evidence for the DHEA(S) heterogeneity of schizophrenia patients was already obtained, it seemed logical to examine

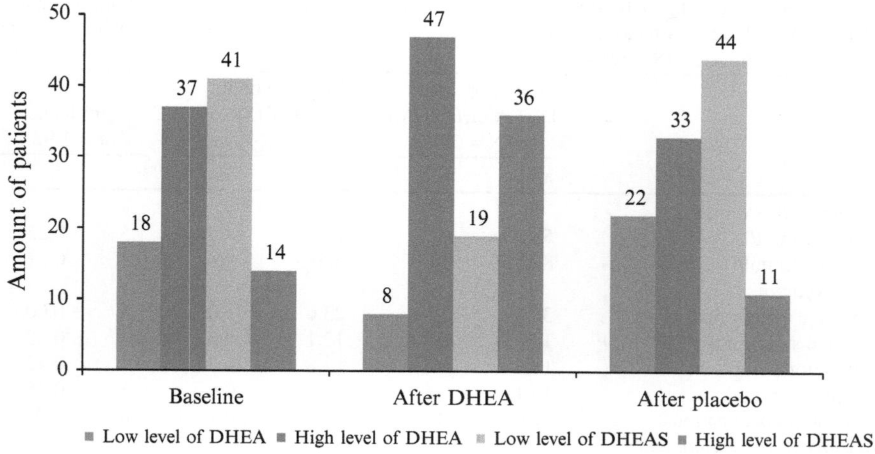

Fig. 17.4 Distribution of schizophrenia patients by serum DHEA(S) levels during crossover trial

whether it was possible to obtain outcome measures such as symptom severity, neurocognitive deficit, side effects, and poor quality of life associated with circulating DHEA, or DHEAS, or both the hormone levels? Therefore, a comparison was done between two groups of schizophrenia patients who participated in the crossover study[86] with regard to these outcome variables:

- DHEA or DHEAS levels increased in 22 patients during the DHEA-period of the treatment.
- Both DHEA and DHEAS concentrations increased in 24 patients during the DHEA-period of the treatment compared to baseline and placebo period. (Nine patients were excluded from this analysis since serum DHEA(S) either did not increase after DHEA treatment or did not decrease during placebo period, or both).

As can be seen in Table 17.5, a three-way ANCOVA model across DHEA augmentation and placebo periods indicate significant improvement in the subjects exhibiting increase in *both circulating DHEA and DHEAS* rather than just in one of them. Namely, improvement was seen in PANSS negative symptom scores (p = 0.008), global cognitive functioning (p = 0.023), executive performance (EF index, $p < 0.001$; initial thinking time, $p < 0.001$; and problem solving, p = 0.010) and observer-rated quality of life (QLS, p = 0.008) scores. However, after the Bonferroni correction (0.05/14 = 0.004) the only differences reaching a level of significance were in the global cognitive index and in the initial thinking time. No significant between-group differences were found in the serum DHEA(S) concentrations, and neither severity of side effects.

These preliminary findings indicate that elevations of both circulating DHEA and DHEAS may contribute to the improved state. If these effects are replicated,

Table 17.5 Circulating DHEA(S), the severity of symptoms, neurocognitive functioning, quality of life, and side effects measures among patients with elevated blood DHEA or DHEAS versus increasing both DHEA(S) levels

	Elevated single: DHEA or DHEAS (N = 22)		Elevated both: DHEA(S) (N = 24)		Significance (df = 1.92)	
Variables	Mean	SE	Mean	SE	F	p
Serum neurosteroids						
DHEA (nmol/l)	55.3	4.7	72.9	4.5	1.5	0.23
DHEAS (nmol/l)	8,569	1,007	6,583	964	0.4	0.52
PANSS factors						
Negative symptoms	32.4	1.0	23.6	1.0	7.5	0.008
Positive symptoms	14.6	0.7	12.1	0.6	1.4	0.23
Activation symptoms	14.8	0.8	12.8	0.8	0.6	0.42
Dysphoric mood	13.1	0.6	11.6	0.6	0.6	0.45
Autistic preoccupations	22.4	0.7	18.7	0.7	2.8	0.097
Neurocognitive functions (CANTAB tasks)[a]						
Executive function index	−1.7	0.1	−0.08	0.1	25.1	0.001*
Initial thinking time (SOC, milliseconds)[2]	−5.8	0.6	3.6	0.5	27.7	0.001*
Problem solving in minimum moves (SOC)[b]	−1.4	0.1	−0.25	0.1	6.9	0.010
Global cognitive index	−1.9	0.1	−0.93	0.1	5.4	0.023
Side effects						
AIMS (total score)	0.9	0.4	2.2	0.4	1.3	0.25
ESRS (total score)	6.3	0.9	5.0	0.8	0.2	0.64
Prolactin (nmol/l)	28.1	5.0	39.4	4.7	3.2	0.075
Quality of life						
Observer-rated (QLS)	57.7	1.8	73.1	1.8	7.3	0.008
Self-report (Q-LES-Q)	3.2	0.1	3.7	0.1	2.9	0.093

Mean and standard errors (SE) are shown

Three-way ANCOVA model across DHEA augmentation and placebo periods:

1. The first factor – serum DHEA effect across both stages of treatment (single hormone effect defined as serum elevation of either DHEA or DHEAS after DHEA treatment, while effect in both hormones means that both serum DHEA and DHEAS levels increased during DHEA treatment)
2. The second factor – type of antipsychotic agents (FGAs, SGAs, and CT)
3. Sex was entered as third factor (df = 1.92; data not shown, all $p > 0.05$)

BMI – Body mass index, CPZ – chlorpromazine equivalent (mg/day), and age are controlled as covariates across two time points

PANSS factors. Severity of psychopathology was assessed using 30 items of the Positive and Negative Syndromes Scale (PANSS),[126] which were analyzed by 5-factor models of White et al.[127]: negative factor (N_1, N_2, N_3, N_4, N_6, G_5, G_7, G_8, G_{13}, G_{14}), positive factor (P_1, P_3, P_5, G_1, G_9), activation (P_4, P_7, N_3, G_4, G_8, G_{14}), dysphoric mood (G_1, G2, G_3, G_4, G_6) and autistic preoccupations (P_3, N_5, N_7, G_{11}, G_{13}, G_{15})

AIMS – Abnormal Involuntary Movement Scale[128]

ESRS – Extrapyramidal Symptom Rating Scale[129,130]

QLS – Quality of Life Scale for rating the schizophrenic deficit syndrome[131]

Q-LES-Q – Quality of Life Enjoyment and Satisfaction Questionnaire[132]

*Significance after the Bonferroni correction: 0.05/14 = 0.004

[a]The standard z-scores are shown: a negative value indicates poorer than average performance of healthy subjects

[b]Stockings of Cambridge (SOC)

the underlying mechanisms by which DHEA(S) influence the outcome variables deserve further attention.

17.5 Concluding Remarks

DHEA is a fascinating molecule, which may have relevance to the pathophysiology of neuropsychiatric disorders. In this chapter we showed the following:

- Despite promising preclinical and human data the currently existing clinical studies provide only weak evidence, in favor of treating schizophrenia with DHEA. Cross-sectional trials[83–85] did not show significant improvements in clinical symptoms, side effects, and quality of life. That following 6 weeks of DHEA treatment in a crossover study of 55 schizophrenia patients.[86] This area requires further study. It is important to note that a total of only 100 schizophrenia patients have been studied only during short-term trials (7 days, 6–12 weeks). Further study is needed in order to investigate whether elevations of both circulating DHEA and DHEAS impact outcome variables more than each of them alone.
- While preliminary, our finding support some improvement in visual sustained attention, and visual and motor skills due to DHEA administration in schizophrenia patients,[86] and thus may offer an important means for further long-term clinical trials. In addition, multiple regression analysis suggests that alterations in blood neurosteroids may be present in the pathophysiological processes, and, at least partially, underlying neurocognitive dysfunctions in schizophrenia.
- We still do not know whether and how the neurosteroids are involved in the pathophysiology of schizophrenia. DHEA administration effect circulating neurosteroids mostly by increasing the circulating DHEA, DHEAS, and androstenedione, while progesterone, cortisol to DHEA and cortisol to DHEAS molar ratios show significantly decrease compared to the placebo arm. Changes in these hormones (with the exception of androstenedione) and molar ratios remain significant after the Bonferroni correction.
- The therapeutical efficacy of DHEA may depend highly on the DHEA(S) heterogeneity of schizophrenia patients, in particular, on the baseline hormonal status of patients. In this context, lack of efficacy of the DHEA augmentation may be observed among patients with already elevated baseline levels of DHEA(S), or their biological active metabolites. This issue deserves further attention. Future studies should consist of more homogeneous groups of schizophrenia patients than those used in previous studies. Baseline levels of DHEA(S) should be taken into account.
- The literature provides evidence that DHEA shows a relatively safe profile, albeit with some minimal effects, mostly acne and hirsutism, with unclear clinical relevance. However, it is important to note that in the clinical trials described above DHEA was administered using apparently supraphysiological doses (100–200 mg/day). Further investigation is recommended in order to verify these findings and confirm the safety of DHEA in administration to this subpopulation of patients.

- In summary, most clinical DHEA trials in schizophrenia have been limited, and results have been inconsistent due to the use of different dosage, and small samples, variation in length of period of DHEA administration, failure to adjust for baseline endogenous hormone level, and age and gender differences. From the inconsistent data in schizophrenia patients, future DHEA trials with these types of patients should be done under proper conditions. In view of the possibility that any neuroprotective effect of DHEA may only be evident in the long term, there is a need to undertake high quality trials in which the duration of DHEA treatment is longer than 1 year, and the number of participants is large enough to detect effects if they exist.[29] Indeed, there is still a lack of evidence from controlled long-term trials using validated measures of neurocognitive change. A long-term, large-scale study with a broader dosage range is warranted in order to further investigate DHEA's role in the management of schizophrenia.

Acknowledgements The author is deeply grateful the scientific contributions of his colleagues Drs. Anatoly Gibel, German Tsinovoy, Yael Ratner, Ms. Rachel Maayan, and Professor Abraham Weizman. Dr. Rael D. Strous is thanked for fruitful cooperation, as well as, for useful comments and suggestions, which improved the manuscript. The author would like to thank his clinical departmental staff, research assistant Ms. Moria Dayan, to Ms. Stella Lulinsky and Ms. Micky Gerchak for editorial services. He wishes to express his very special appreciation his wife Dr. Galina Ritsner for her generous support. He is also thankful of the Rappaport Faculty of Medicine, Technion – Israel Institute of Technology, and Sha'ar Menashe Mental Health Center for their support. The work was supported in part by the Stanley Foundation (USA).

References

1. Young J, Couzinet B, Nahoul K, et al. Panhypopituitarism as a model to study the metabolism of dehydroepiandrosterone (DHEA) in humans. J Clin Endocrinol Metab 1997; 82:2578–2585.
2. Morales AJ, Haubrich RH, Hwang JY, et al. The effect of six months treatment with a 100 mg daily dose of dehydroepiandrosterone (DHEA) on circulating sex steroids, body composition and muscle strength in age-advanced men and women. Clin Endocrinol (Oxf) 1998; 49:421–432.
3. Casson PR, Santoro N, Elkind-Hirsch K, et al. Postmenopausal dehydroepiandrosterone administration increases free insulin-like growth factor-I and decreases high-density lipoprotein: a six-month trial. Fertil Steril 1998; 70:107–110.
4. Arlt W, Justl HG, Callies F, et al. Oral dehydroepiandrosterone for adrenal androgen replacement: pharmacokinetics and peripheral conversion to androgens and estrogens in young healthy females after dexamethasone suppression. J Clin Endocrinol Metab 1998; 83:1928–1934.
5. Arlt W, Callies F, van Vlijmen JC, et al. Dehydroepiandrosterone replacement in women with adrenal insufficiency. N Engl J Med 1999; 341:1013–1020.
6. Mortola JF, Yen SS. The effects of oral dehydroepiandrosterone on endocrine-metabolic parameters in postmenopausal women. J Clin Endocrinol Metab 1990; 71:696–704.
7. Nestler JE, Barlascini CO, Clore JN, Blackard WG. Dehydroepiandrosterone reduces serum low density lipoprotein levels and body fat but does not alter insulin sensitivity in normal men. J Clin Endocrinol Metab 1988; 66:57–61.

8. Genazzani AD, Lanzoni C, Genazzani AR. Might DHEA be considered a beneficial replacement therapy in the elderly? Drugs Aging 2007; 24(3):173–185.
9. Obut TA, Ovsyukova MV, Cherkasova OP. Prolonged decrease in stress reactivity caused by dehydroepiandrosterone sulfate. Bull Exp Biol Med. 2006; 141:571–573.
10. Reddy DS, Kaur G, Kulkarni SK. Sigma (σ_1) receptor mediated antidepressant-like effects of neurosteroids in the Porsolt forced swim test. Neuroreport 1998; 9:3069–3073.
11. Urani A, Roman FJ, Phan V-L, et al. The antidepressant-like effect induced by σ_1-receptor agonists and neuroactive steroids in mice submitted to the forced swimming test. J Pharmacol Exp Ther 2001; 298:1269–1279.
12. Flood JF, Roberts E. Dehydroepiandrosterone sulphate proves memory in aging mice. Brain Res 1998; 448:178–181.
13. Fedotova J, Sapronov N. Behavioral effects of dehydroepiandrosterone in adult male rats. Prog Neuropsychopharmacol Biol Psychiatry 2004; 28:1023–7.
14. Juhasz-Vedres G, Rozsa E, Rakos G, et al. Dehydroepiandrosterone sulfate is neuroprotective when administered either before or after injury in a focal cortical cold lesion model. Endocrinology 2006; 147:683–686.
15. Tunez I, Munoz MC, et al. Treatment with dehydroepiandrosterone prevents oxidative stress induced by 3-nitropropionic acid in synaptosomes. Pharmacology 2005; 74:113–118.
16. Pelissier MA, Trap C, Malewiak MI, Morfin R. Antioxidant effects of dehydroepiandrosterone and 7alpha-hydroxy-dehydroepiandrosterone in the rat colon, intestine and liver. Steroids 2004; 69:137–144.
17. Patel MA, Katyare SS. Dehydroepiandrosterone (DHEA) treatment stimulates oxidative energy metabolism in the cerebral mitochondria from developing rats. Int J Dev Neurosci 2006; 24:327–334.
18. Patel MA, Katyare SS. Effect of dehydroepiandrosterone (DHEA) treatment on oxidative energy metabolism in rat liver and brain mitochondria. A dose-response study. Clin Biochem 2007; 40:57–65.
19. Patel MA, Katyare SS. Treatment with dehydroepiandrosterone (DHEA) stimulates oxidative energy metabolism in the cerebral mitochondria. A comparative study of effects in old and young adult rats. Neurosci Lett 2006; 402:131–136.
20. MacEwen EG, Kurzmann ID. Obesity in the dog: role of the adrenal steroid dehydroepiandrosterone. J Nutr 1991; 121:51–55.
21. Yen TT, Allan JV, Pearson DV. Prevention of obesity in Avy/A mice by dehydroepiandrosterone. Lipids 1997; 12:409–413.
22. Rontzsch A, Thoss K, Petrow PK, et al. Amelioration of murine antigen-induced arthritis by dehydroepiandrosterone (DHEA). Inflamm Res 2004; 53:189–198.
23. Wang L, Wang YD, Wang WJ, et al. Dehydroepiandrosterone improves murine osteoblast growth and bone tissue morphometry via mitogen-activated protein kinase signaling pathway independent of either androgen receptor or estrogen receptor. J Mol Endocrinol 2007; 38:467–479.
24. Schwartz AG, Pashko LL. Cancer chemoprevention with the adrenocortical steroid dehydroepiandrosterone and structural analogs. J Cell Biochem 1995; 22(Suppl):210–217.
25. Schwartz AG, Pashko LL. Dehydroepiandrosterone, glucose-6-phosphate dehydrogenase, and longevity. Ageing Res Rev 2004; 3:171–187.
26. Shilkaitis A, Green A, Punj V, et al. Dehydroepiandrosterone inhibits the progression phase of mammary carcinogenesis by inducing cellular senescence via a p16-dependent but p53-independent mechanism. Breast Cancer Res. 2005; 7:R1132–R1140.
27. Ravid D, Fridman L, Gispan-Herman I, Maayan R, et al. DHEA, a neurosteroid, decreases cocaine self-administration and reinstatement of cocaine-seeking behavior in rats. Neuropsychopharmacology. 2006; 31:2231–2236.
28. Kroboth PD, Salek FS, Pittenger AL, et al. DHEA and DHEA-S: a review. J Clin Pharmacol. 1999; 39:327–348.
29. Huppert F, Van Niekerk J. WITHDRAWN: Dehydroepiandrosterone (DHEA) supplementation for cognitive function. Cochrane Database Syst Rev 2007; 18:CD000304.

30. Olech E, Merrill JT. DHEA supplementation: the claims in perspective. Cleve Clin J Med. 2005; 72(11):965–966, 968, 970–971 passim.
31. Bovenberg SA, van Uum SH, Hermus AR. Dehydroepiandrosterone administration in humans: evidence based? Neth J Med. 2005; 63(8):300–304.
32. George O, Vallee M, Le Moal M, Mayo W. Neurosteroids and cholinergic systems: implications for sleep and cognitive processes and potential role of age-related changes. Psychopharmacology (Berl) 2006; 186:402–413.
33. Arlt W. Dehydroepiandrosterone replacement therapy. Curr Opin Endocrin Diab 2006; 13:291–305.
34. Nippoldt T B. Dehydroepiandrosterone supplements: bringing sense to sensational claims. Endocr Pract 1998; 4:106–111.
35. Wolf OT, Neumann O, Hellhammer DH, et al. Effects of a two-week physiological dehydroepiandrosterone substitution on neurocognitive performance and well-being in healthy elderly women and men. J Clin Endocrinol Metab 1997; 82:2363–2367.
36. Wolf OT, Naumann E, Hellhammer DH, Kirschbaum C. Effects of dehydroepiandrosterone (DHEA) replacement in elderly men on event related potentials (ERPs), memory and well-being. J Gerontol A Biol Sci Med Sci 1998; 53:385–390.
37. Kudielka BM, Hellhammer J, Hellhammer DH, et al. Sex differences in endocrine and psychological responses to psychosocial stress in healthy elderly subjects and the impact of a 2-week dehydroepiandrosterone treatment. J Clin Endocrinol Metab 1998; 83:1756–1761.
38. Morales AJ, Nolan JJ, Nelson JC, et al. Effects of replacement dose of dehydroepiandrosterone in men and women of advancing age. J Clin Endocrinol Metab 1994; 78:1360–1367.
39. Flynn MA, Weaver-Osterholtz D, Sharpe-Timms KL, et al. Dehydroepiandrosterone replacement in aging humans. J Clin Endocrinol Metab 1999; 84:1527–1533.
40. van Niekerk JK, Huppert FA, et al. Salivary cortisol and DHEA: association with measures of cognition and well-being in normal older men, and effects of three months of DHEA supplementation. Psychoneuroendocrinology 2001; 26:591–612.
41. Arlt W, Callies F, Koehler I, et al. Dehydroepiandrosterone supplementation in healthy men with an age-related decline of dehydroepiandrosterone secretion. J Clin Endocrinol Metab 2001; 86:4686–4692.
42. Gurnell EM, Chatterjee VK. Dehydroepiandrosterone replacement therapy. Eur J Endocrinol 2001; 145:103–106.
43. Allolio B, Arlt W. DHEA treatment: myth or reality? Trends Endocrinol Metab 2002; 13:288–294.
44. Kahn AJ, Haloran B. Dehydroepiandrosterone supplementation and bone turnover in middle-aged to elderly men. J Clin Endocrinol Metab 2002; 87:1544–1549.
45. Villareal DT, Holloszy JO. Effect of DHEA on abdominal fat and insulin action in elderly women and men: a randomized controlled trial. JAMA 2004; 292:2243–2248.
46. Baulieu EE, Thomas G, Legrain S, et al. Dehydroepiandrosterone (DHEA), DHEA sulfate, and aging: contribution of the DHEAge Study to a sociobiomedical issue. Proc Natl Acad Sci USA 2000; 97:4279–4284.
47. Villareal DT, Holloszy JO, Kohrt WM. Effects of DHEA replacement on bone mineral density and body composition in elderly women and men. Clin Endocrinol (Oxf) 2000; 53:561–568.
48. Jankowski CM, Gozansky WS, et al. Effects of dehydroepiandrosterone replacement therapy on bone mineral density in older adults: a randomized, controlled trial. J Clin Endocrinol Metab 2006; 91:2986–2993.
49. Genazzani AR, Inglese S, Lombardi I, et al. Long-term low-dose dehydroepiandrosterone replacement therapy in aging males with partial androgen deficiency. Aging Male 2004; 7:133–143.
50. Reiter WJ, Pycha A, Schatzl G, et al. Dehydroepiandrosterone in the treatment of erectile dysfunction: a prospective double-blind, randomized, placebo-controlled study. Urology 1999; 53:590–595.
51. Stomati M, Monteleone P, Casarosa E, et al. Six-month oral dehydroepiandrosterone supplementation in early and late postmenopause. Gynecol Endocrinol 2000; 14:342–363.

52. Finckh A, Berner IC, Aubry-Rozier B, et al. A randomized controlled trial of dehydroepiandrosterone in postmenopausal women with fibromyalgia. J Rheumatol 2005; 32:1336–1340.
53. Genazzani AD, Stomati M, Bernardi F, et al. Long-term low-dose dehydroepiandrosterone oral supplementation in early and late postmenopausal women modulates endocrine parameters and synthesis of neuroactive steroids. Fertil Steril 2003; 80:1495–1501.
54. Genazzani AD, Pluchino N, Begliuomini S, et al. Long-term low-dose oral administration of dehydroepiandrosterone modulates adrenal response to adrenocorticotropic hormone in early and late postmenopausal women. Gynecological Endocrinology 2006; 22:627–635.
55. Arlt W. Dehydroepiandrosterone replacement therapy. Seminars in Reproductive Medicine 2004; 22:379–388.
56. Saad F, Hoesl CE, et al. Dehydroepiandrosterone treatment in the aging male–what should the urologist know? Eur Urol 2005; 48:724–733.
57. Grimley Evans J, Malouf R, et al. Dehydroepiandrosterone (DHEA) supplementation for cognitive function in healthy elderly people. Cochrane Database Syst Rev 2006; Oct(4): CD006221.
58. Panjari M, Davis SR. DHEA therapy for women: effect on sexual function and wellbeing. Hum Reprod Update 2007; 13:239–248.
59. Sugino M, Ohsawa N, Ito T, et al. A pilot study of dehydroepiandrosterone sulfate in myotonic dystrophy. Neurology 1998; 51:586–589.
60. van Vollenhoven RF. Dehydroepiandrosterone for the treatment of systemic lupus erythematosus. Expert Opin Pharmacother 2002; 3:23–31.
61. Nordmark G, Bengtsson C, Larsson A, et al. Effects of dehydroepiandrosterone supplement on health-related quality of life in glucocorticoid treated female patients with systemic lupus erythematosus. Autoimmunity 2005; 38:531–540.
62. Sun Y, Mao M, Sun L, et al. Treatment of osteoporosis in men using dehydroepiandrosterone sulfate. Chin Med J 2002; 115:402–404.
63. Piketty C, Jayle D, Leplege A, et al. Double-blind placebo-controlled trial of oral dehydroepiandrosterone in patients with advanced HIV disease. Clin Endocrinol 2001; 55:325–330.
64. Brooke AM, Kalingag LA, et al. Dehydroepiandrosterone improves psychological well-being in male and female hypopituitary patients on maintenance growth hormone replacement. J Clin Endocrinol Metab 2006; 91:3773–3779.
65. Binello E, Gordon CM. Clinical uses and misuses of dehydroepiandrosterone. Curr Opin Pharmacol 2003; 3:635–641.
66. Arlt W, Haas J, Callies F. Biotransformation of oral dehydroepiandrosterone in elderly men: Significant increase in circulating estrogens. J Clin Endocrinol Metab 1999; 84:2170–2176.
67. Hunt PJ, Gurnell EM, Huppert FA, et al. Improvement in mood and fatigue after dehydroepiandrosterone replacement in Addison's disease in a randomized, double blind trial. J Clin Endocrinol Metab 2000; 85:4650–4656.
68. Arlt W, Callies F, Allolio B. Gynecol Six-month oral dehydroepiandrosterone supplementation in early and late postmenopause. Endocrinol 2000; 14:342–363.
69. Libe R, Barbetta L, et al. Effects of dehydroepiandrosterone (DHEA) supplementation on hormonal, metabolic and behavioral status in patients with hypoadrenalism. J Endocrinol Invest 2004; 27:736–741.
70. Stoffel-Wagner B. Neurosteroid metabolism in the human brain. Eur J Endocrinol 2001; 145:669–679.
71. Stoffel-Wagner B. Neurosteroid biosynthesis in the human brain and its clinical implications. Ann NY Acad Sci 2003; 1007:64–78.
72. Pisu MG, Serra M. Neurosteroids and neuroactive drugs in mental disorders. Life Sci 2004; 74:3181–3197.
73. Majewska MD, Demirgoren S, Spivak CE, London ED. The neurosteroid dehydroepiandrosterone sulfate is an allosteric antagonist of the GABAA receptor. Brain Res 1990; 526:143–146.
74. Demirgoren S, Majewska MD, Spivak CE, London ED. Receptor binding and electrophysiological effects of dehydroepiandrosterone sulfate, an antagonist of the GABAA receptor. Neuroscience 1991; 45:127–135.

75. Bergeron R, de Montigny C, Debonnel G. Potentiation of neuronal NMDA response induced by dehydroepiandrosterone and its suppression by progesterone: effects mediated via sigma receptors. J Neurosci 1996; 16:1193–1202.
76. Wolkowitz OM, Reus VI, et al. Dehydroepiandrosterone (DHEA) treatment of depression. Biol Psychiatry 1997; 41:311–318.
77. Wolkowitz OM, Reus VI, Keebler A, et al. Double-blind treatment of major depression with dehydroepiandrosterone. Am J Psychiatry 1999; 156:646–649.
78. Schmidt PJ, Daly RC, Bloch M, et al. Dehydroepiandrosterone monotherapy in midlife-onset major and minor depression. Arch Gen Psychiatry 2005; 62:154–162.
79. Hsiao CC. Difference in pre- and post-treatment plasma DHEA levels were significantly and positively correlated with difference in pre- and post-treatment Hamilton depression scores following successful therapy for major depression. Psychoneuroendocrinology 2006; 31:839–846.
80. Bloch M, Schmidt PJ, Danaceau MA, Adams LF, Rubinow DR. Dehydroepiandrosterone treatment of midlife dysthymia. Biol Psychiatry 1999; 45:1533–1541.
81. Rabkin JG, McElhiney MC, et al. Placebo-controlled trial of dehydroepiandrosterone (DHEA) for treatment of nonmajor depression in patients with HIV/AIDS. Am J Psychiatry 2006; 163:59–66.
82. Brebner K, Hayley S, Zacharko R, Merali Z, Anisman H. Synergistic effects of interleukin-1beta, interleukin-6 and tumor necrosis factor-alpha: central monoamine, corticosterone, and behavioral variations. Neuropsychopharmacology 2000; 22:566–580.
83. Strous RD, Maayan R, Lapidus R, et al. Value of Dehydroepiandrosterone (DHEA) Augmentation in the Management of Negative Symptoms in Schizophrenia. Arch Gen Psychiatry 2003; 60:133–141.
84. Strous RD, Stryjer R, Maayan R, et al. Analysis of clinical symptomatology, extrapyramidal symptoms and neurocognitive dysfunction following dehydroepiandrosterone (DHEA) administration in olanzapine treated schizophrenia patients: a randomized, double-blind placebo controlled trial. Psychoneuroendocrinology 2007; 32:96–105.
85. Nachshoni T, Ebert T, Abramovitch Y, et al. Improvement of extrapyramidal symptoms following dehydroepiandrosterone (DHEA) administration in antipsychotic treated schizophrenia patients: a randomized, double-blind placebo controlled trial. Schizophr Res 2005; 79:251–256.
86. Ritsner MS, Gibel A, Ratner Y, et al. Improvement of sustained attention and visual and movement skills, but not clinical symptoms, after dehydroepiandrosterone augmentation in schizophrenia: a randomized, double-blind, placebo-controlled, crossover trial. J Clin Psychopharmacol 2006; 26:495–499.
87. Welham J, Stedman T, Clair A. Choosing negative symptom instruments: issues of representation and redundancy. Psychiatry Res 1999; 87:47–56.
88. Norman RM, Malla AK, et al. A study of the interrelationship between and comparative interrater reliability of the SAPS, SANS and PANSS. Schizophr Res 1996; 19:73–85.
89. Spalletta G, Pasini A, et al. Patients with deficit, nondeficit, and negative symptom schizophrenia: do they differ during episodes of acute psychotic decompensation? Schizophr Res 1997; 24:341–348.
90. Galynker I, Cohen L, et al. Psychiatric symptom severity and length of stay on an intensive rehabilitation unit. Psychosomatics 2000; 41:114–120.
91. Smith RC, Infante M, et al. Effects of Cigarette Smoking on Psychopathology Scores in Patients With Schizophrenia: An Experimental Study. Subst Abus 2001; 22:175–186.
92. Feldman PD, Kaiser CJ, et al. Comparison of risperidone and olanzapine in the control of negative symptoms of chronic schizophrenia and related psychotic disorders in patients aged 50 to 65 years. J Clin Psychiatry 2003; 64:998–1004.
93. Ritsner M, Maayan R, Gibel A, et al. Elevation of the cortisol/dehydroepiandrosterone ratio in schizophrenia patients. Eur Neuropsychopharmacol 2004; 14:267–273.
94. Strous RD, Maayan R, Lapidus R, et al., Increased circulatory dehydroepiandrosterone and dehydroepiandrosterone-sulphate in first-episode schizophrenia: relationship to gender, aggression and symptomatology. Schizophr Res 2004; 71:427–434.
95. di Michele F, Caltagirone C, Bonaviri G, et al. Plasma dehydroepiandrosterone levels are strongly increased in schizophrenia. J Psychiatr Res 2005; 39:267–273.

96. Johnson M, Bebb R, Sirrs S. Uses of DHEA in aging and other disease states. Ageing Res Rev 2002; 1:29–41.
97. Ritsner M, Gibel A, Ratner Y. Determinants of changes in perceived quality of life in the course of schizophrenia. Qual Life Res 2006; 15:515–526.
98. Green MF. Cognitive impairment and functional outcome in schizophrenia and bipolar disorder. J Clin Psychiatry 2006; 67(Suppl 9):3–8; discussion 36–42.
99. Alhaj HA, Massey AE, McAllister-Williams RH. Effects of DHEA administration on episodic memory, cortisol and mood in healthy young men: a double-blind, placebo-controlled study. Psychopharmacology 2006; 188:541–551.
100. Wolkowitz OM, Kramer JH, Reus VI, et al. DHEA treatment of Alzheimer's disease: a randomized, double-blind, placebo-controlled study. Neurology 2003; 60:1071–1076.
101. Parsons TD, Kratz KM, et al. Dhea supplementation and cognition in postmenopausal women. Int J Neurosci 2006; 116:141–155.
102. Ritsner MS, Blumenkrantz H, Dubinsky T, Dwolatzky T. The detection of neurocognitive decline in schizophrenia using the Mindstreams Computerized Cognitive Test Battery. Schizophr Res 2006; 82:39–49.
103. Butler PD, Schechter I, Zemon V, et al. Dysfunction of early-stage visual processing in schizophrenia. Am J Psychiatry 2001; 158:1126–1133.
104. Butler PD, Zemon V, Schechter I, et al. Early-stage visual processing and cortical amplification deficits in schizophrenia. Arch Gen Psychiatry 2005; 62:495–504.
105. Butler PD, Martinez A, Foxe JJ, et al. Subcortical visual dysfunction in schizophrenia drives secondary cortical impairments. Brain 2007; 130:417–430.
106. Uhlhaas PJ, Phillips WA, Silverstein SM. The course and clinical correlates of dysfunctions in visual perceptual organization in schizophrenia during the remission of psychotic symptoms. Schizophr Res 2005; 75:183–192.
107. Lee TM, Cheung PP. The relationship between visual-perception and attention in Chinese with schizophrenia. Schizophr Res 2005; 72:185–193.
108. Prouteau A, Verdoux H, Briand C, et al. The crucial role of sustained attention in community functioning in outpatients with schizophrenia. Psychiatry Res 2004; 129:171–177.
109. Moghaddam B. Stress activation of glutamate neurotransmission in the prefrontal cortex: implications for dopamine-associated psychiatric disorders. Biol Psychiatry 2002; 51:775–787.
110. Reuter B, Kathmann N. Using saccade tasks as a tool to analyze executive dysfunctions in schizophrenia. Acta Psychol (Amst) 2004; 115:255–269.
111. Fiszdon JM, Choi J, Bryson GJ, Bell MD. Impact of intellectual status on response to cognitive task training in patients with schizophrenia. Schizophr Res 2006; 87:261–269.
112. Bellino S, Rocca P, Patria L, et al. Relationships of age at onset with clinical features and cognitive functions in a sample of schizophrenia patients. J Clin Psychiatry 2004; 65:908–914.
113. Bozikas VP, Kosmidis MH, Kioperlidou K, Karavatos A. Relationship between psychopathology and cognitive functioning in schizophrenia. Compr Psychiatry 2004; 45:392–400.
114. Brazo P, Delamillieure P, Morello R, et al. Impairments of executive/attentional functions in schizophrenia with primary and secondary negative symptoms. Psychiatry Res 2005; 133:45–55.
115. Villalta-Gil V, Vilaplana M, Ochoa S, et al. Neurocognitive performance and negative symptoms: are they equal in explaining disability in schizophrenia outpatients? Schizophr Res 2006; 87:246–253.
116. Weickert TW, Goldberg TE. First- and second-generation antipsychotic medication and cognitive processing in schizophrenia. Curr Psychiatry Rep 2005; 7:304–310.
117. Kawai N, Yamakawa Y, Baba A, et al. High-dose of multiple antipsychotics and cognitive function in schizophrenia: the effect of dose-reduction. Prog Neuropsychopharmacol Biol Psychiatry 2006; 30:1009–1114.
118. Yanase T, Fukahori M, Taniguchi S, et al. Serum dehydroepiandrosterone (DHEA) and DHEA-sulfate (DHEA-S) in Alzheimer's disease and in cerebrovascular dementia. Endocr J 1996; 43:119–123.

119. Morrison MF, Redei E, TenHave T, et al. Dehydroepiandrosterone sulfate and psychiatric measures in a frail, elderly residential care population. Biol Psychiatry 2000; 47:144–150.
120. Vallee M, Mayo W, Le Moal M. Role of pregnenolone, dehydroepiandrosterone and their sulfate esters on learning and memory in cognitive aging. Brain Res Brain Res Rev 2001; 37:301–312.
121. Harris DS, Wolkowitz OM, Reus VI. Movement disorder, memory, psychiatric symptoms and serum DHEA levels in schizophrenic and schizoaffective patients. World J Biol Psychiatry 2001; 2:99–102.
122. Hintze, JL NCSS 2007. Statistical System for Windows. User's Guide. Number Cruncher Statistical Systems, Kaysville, Utah, 2006 (http://www.ncss.com).
123. Lasco A, Frisina N, Morabito N, et al. Metabolic effects of dehydroepiandrosterone replacement therapy in postmenopausal women. Eur J Endocrinol 2001; 145:457–461.
124. Strous RD, Maayan R, Kotler M, Weizman A. Hormonal profile effects following dehydroepiandrosterone (DHEA) administration to schizophrenic patients. Clin Neuropharmacol 2005; 28:265–269.
125. Strous RD, Gibel A, Weizman A, Maayan R, Ritsner MS. Hormonal response on dehydroepiandrosterone administration in schizophrenia: findings from a randomized, double-blind, placebo-controlled, crossover study. J Clin Psychopharmacol 2007 (in press).
126. Kay SR, Fiszbein A, Opler LA. The Positive and Negative Syndrome Scale (PANSS) for schizophrenia. Schizophr Bull 1987; 13:261–276.
127. White L, Harvey PD, Opler L, Lindenmayer JP. Empirical assessment of the factorial structure of clinical symptoms in schizophrenia. A multisite, multimodel evaluation of the factorial structure of the Positive and Negative Syndrome Scale. The PANSS Study Group. Psychopathology 1997; 30:263–274.
128. Guy W. ECDEU Assessment Manual for Psychopharmacology, revised. Washington, DC: US Department of Health, Education and Welfare, 1976.
129. Chouinard G, Ross-Chouinard A, AnnAnnable L. Extrapyramidal Symptom Rating Scale. Can J Neurol Sci 1980; 7:233.
130. Chouinard G, Margolese HC. Manual for the Extrapyramidal Symptom Rating Scale (ESRS). Schizophr Res 2005; 76:247–265.
131. Heinrichs DW, Hanlon TE, Carpenter WT. The Quality of Life Scale: an instrument for rating the schizophrenic deficit scale. Schizophr Bull 1984; 10:388–398.
132. Endicott J, Nee J, Harrison W, Blumenthal R. Quality of Life Enjoyment and Satisfaction Questionnaire: a new measure. Psychopharmacol Bull 1993; 29:321–326.

Chapter 18
Allopregnanolone and Pregnenolone Alterations Following Pharmacological Agents in Rodents and Clinical Populations

Victoria M. Payne, Jason D. Kilts, Jennifer C. Naylor, Jennifer L. Strauss, Patrick S. Calhoun, Roger D. Madison, and Christine E. Marx

Abstract Recent research has revealed potential roles for neuroactive steroids in the pathophysiology and treatment of a number of psychiatric and neurological disorders. A wide array of preclinical investigations also reports neuroactive steroid actions in areas relevant to neuropsychiatric conditions, including neuroprotection, neuroplasticity, and neurogenesis. Levels of several neuroactive steroids are altered following the administration of various pharmacological agents, and it has been hypothesized that neuroactive steroid induction may contribute to the efficacy of selective serotonin reuptake inhibitors (SSRIs), atypical antipsychotics (also known as second-generation antipsychotics), and mood stabilizers. In addition, neuroactive steroids may have potential as agents for pharmacological intervention. The following chapter will focus primarily on changes in the neuroactive steroids allopregnanolone and pregnenolone in response to the administration of several classes of commonly prescribed psychiatric medications, although alterations in other neuroactive steroids will also be included. In some instances when appropriate, evidence of alterations in baseline neuroactive steroid levels in psychiatric disorders will be discussed for a more complete picture of clinically relevant findings. Given the multifaceted actions of neuroactive steroids in rodent models, including neuroprotective effects against a number of stressors, modulation of the hypothalamic–pituitary–adrenal (HPA) axis, enhancement of neurogenesis and myelination, and anti-inflammatory effects, it is perhaps not surprising that neuroactive steroids have been linked to clinically disparate psychiatric and neurological disorders that may be impacted by these diverse effects. The modulatory actions of neuroactive steroids at GABAergic and glutamatergic neurotransmitter systems, which constitute the major inhibitory and excitatory neurotransmitter systems in the mammalian brain, respectively, are also consistent with the relevance of neuroactive steroids to multiple neuropsychiatric conditions.

Keywords Neuroactive steroid, neurosteroid, antidepressant, antipsychotic, mood stabilizer, allopregnanolone, pregnenolone

---

Duke University Medical Center and Durham Veterans Affairs Medical Center, Durham, North Carolina, USA

M.S. Ritsner and A. Weizman (eds.), *Neuroactive Steroids in Brain Function, Behavior and Neuropsychiatric Disorders*

Abbreviations ACTH adrenocorticotropin hormone; CNS central nervous system; CSF cerebrospinal fluid; DHEA dehydroepiandrosterone; DHEAS dehydroepiandrosterone sulfate; GABA γ-aminobutyric acid; GC/MS gas chromatography/mass spectrometry; HPA hypothalamic–pituitary–adrenal; HPLC high-performance liquid chromatography; NMDA *N*-methyl-D-aspartate; PTSD post-traumatic stress disorder; RIA radioimmunoassay; SSRI selective serotonin reuptake inhibitor

18.1 Background

The term "neurosteroids" encompasses the group of steroids that are synthesized in the central nervous system (CNS) de novo from cholesterol or from peripheral steroid precursors that cross the blood–brain barrier readily. A number of neurosteroids are also neuroactive (designated "neuroactive steroids"), possessing actions at membrane-bound ligand-gated ion channel receptors such as inhibitory γ-aminobutyric acid type A ($GABA_A$) and excitatory *N*-methyl-D-aspartate (NMDA) receptors, among others. In addition to these receptor activities, a number of scientific reports have also used the term "neuroactive steroid" more generally to describe steroid actions on neuritic outgrowth and neurogenesis, among other CNS effects. In the current literature, the term "neuroactive steroid" therefore tends to encompass steroids that demonstrate some kind of activity within the CNS, regardless of whether they have been synthesized within the CNS or outside of the CNS in locales such as the adrenal glands or gonads. Throughout Chapter 18, the term "neuroactive steroid" will be predominantly used, as it is the more all-encompassing of the two.

Neuroactive steroids have garnered continually increased clinical interest based on the growing array of beneficial actions associated with them. The majority of these actions have been reported in the preclinical literature utilizing rodent models. For example, the GABAergic neuroactive steroid allopregnanolone (3α-hydroxy-5α-pregnan-20-one) demonstrates anxiolytic,[1] anticonvulsant,[2,3] and anti-aggressive[4] effects in rodents, and may also be relevant to antidepressant[5–7] and antipsychotic drug action.[8–11] Allopregnanolone has also been associated with the enhancement of myelination processes,[12,13] neurogenesis,[14] and neuroprotection in neurodegenerative disease models.[15] In addition to allopregnanolone, the neuroactive steroid pregnenolone (3β-hydroxy-5-pregnen-20-one) and/or its sulfated derivative demonstrate a number of properties in rodent models with potential clinical relevance. For example, pregnenolone and/or its sulfated derivative exhibit numerous beneficial effects in rodents, including the enhancement of learning and memory,[16–19] the amelioration of age-related cognitive deficits,[18] the enhancement of neuritic outgrowth,[20] and microtubule stabilization.[21] Rodent investigations suggest that pregnenolone may also contribute to the effects of both antidepressant[11,22] and antipsychotic drugs,[8,10,11,23] but may offset the anxiolytic effects of other compounds.[24] Furthermore, there is consistent animal model evidence of neuroactive steroid involvement in cellular functions that may be relevant to the treatment of psychiatric disorders, such as neuroprotection[15,25–27] and neurogenesis.[14,28,29]

Not all evidence for the potential therapeutic benefits of neuroactive steroids has come from animal studies, however, as a growing body of clinical literature also suggests an important role for neuroactive steroids in the evaluation and treatment of a variety of psychiatric disorders, including mood, anxiety, psychotic and cognitive disorders. For example, alterations in neuroactive steroid profiles have been reported in patients with major depression,[6,30] schizophrenia,[31,32] Alzheimer's Disease,[33–35] bipolar disorder,[31] post-traumatic stress disorder,[36,37] eating disorders,[38] and attention deficit hyperactivity disorder.[39] In addition, increased neuroactive steroid concentrations in patients with panic disorder have been hypothesized as a counterregulative mechanism against the occurrence of spontaneous panic attacks.[40]

Neuroactive steroid responses are mediated via several biochemical sites of action, discussed in detail in other chapters. Briefly, some neuroactive steroids such as allopregnanolone act as positive allosteric modulators of inhibitory $GABA_A$ receptors at locations separate from the benzodiazepine site.[41–43] Distinct binding sites for the potentiation of $GABA_A$ responses and for the direct activation of $GABA_A$ receptors by a neuroactive steroid have recently been identified within the $GABA_A$ receptor complex.[43,44] In the case of allopregnanolone, responses occur at nanomolar concentrations[45,46] and are very robust, potentiating $GABA_A$ receptor responses ten times more potently than the benzodiazepines diazepam and flurazepam, and approximately two hundred times more potently than the barbiturate pentobarbital.[45] Other neuroactive steroids, including pregnenolone sulfate and dehydroepiandrosterone sulfate (DHEAS), function in an opposite manner, acting as negative allosteric modulators of $GABA_A$ receptors.[42,47–49] Additionally, pregnenolone sulfate and DHEAS are both positive allosteric modulators of NMDA receptors,[50–52] and pregnenolone sulfate demonstrates modest negative modulatory activities at both kainate and AMPA receptors.[50] While less is known about other receptor activities, various neuroactive steroids have also been reported to interact with serotonin, sigma, glycine, nicotinic acetylcholine, and oxytocin receptors. Clearly, pharmacological interventions that produce marked changes in neuroactive steroid levels could thus affect numerous neurobiological processes and play a role in the pathophysiology of multiple psychiatric and neurological disorders.

Of note, direct comparison of neuroactive steroid levels among studies has remained somewhat challenging, based on the variability in reported levels depending upon different methodologies utilized for the examination of these compounds. Reported analytic strategies include gas chromatography/mass spectrometry (GC/MS) preceded by high-performance liquid chromatography (HPLC) purification, gas chromatography negative ion mass fragmentography (GC/NICI-MF), radioimmunoassay (RIA) preceded by HPLC, and RIA alone (reviewed in Purdy et al., 2005).[35,53] Thus, it is very important to consider individual methodologies employed in the literature when comparing and contrasting results across studies.

18.2 Neuroactive Steroid Alterations Following Antidepressants

18.2.1 Allopregnanolone

The modulation of serotonergic and/or catecholaminergic neurotransmission by antidepressants is generally thought to contribute to their mechanisms of action. However, considerable evidence suggests that antidepressant mechanisms of action may also include the normalization of the hypothalamic-pituitary-adrenal (HPA) system.[54,70,71] The neuroactive steroid allopregnanolone plays a modulatory role in the regulation of HPA axis activity,[55–58] and recent studies provide evidence that allopregnanolone may also be important to the actions of antidepressants. Several of these studies in rodent models report alterations in neuroactive steroid levels following administration of antidepressants, with some (but not all) antidepressants producing increases in allopregnanolone. For example, allopregnanolone levels increase dose-dependently in response to fluoxetine in whole rat brain, with significant increases specifically detected in frontal cortex, striatum, hippocampus, and olfactory bulb.[5] Consistent with this report, evidence suggests that hippocampal allopregnanolone levels triple in response to the administration of similar fluoxetine doses, and reach comparable hippocampal allopregnanolone levels to those achieved in the prior study utilizing a similar mass spectrometry-based technique.[11] Moreover, fluoxetine and norfluoxetine upregulate the brain content of allopregnanolone in socially isolated mice at doses insufficient to inhibit serotonin reuptake.[4,59] Remarkably, the inhibition of serotonin reuptake by fluoxetine and norfluoxetine requires doses 50 times greater than those necessary to increase brain allopregnanolone content, prompting the authors to suggest that allopregnanolone increases may be responsible for the observed behavioral results.[59]

Although these studies provide a consistent indication that allopregnanolone is elevated in response to selective serotonin reuptake inhibitors, two things should be noted. First, long-term downregulation of allopregnanolone may occur in the brief absence of selective serotonin reuptake inhibitor (SSRI) administration, as baseline allopregnanolone levels measured 2 days after chronic fluoxetine treatment for 2 weeks are significantly reduced (although subsequent injections of fluoxetine are able to overcome this effect and elevate allopregnanolone levels as expected).[22,60] Second, administration of some antidepressants, such as paroxetine and imipramine, produces far lesser or even no effect on allopregnanolone in rats compared to fluoxetine,[4,5] although long-term paroxetine administration appears to increase allopregnanolone in cortex, olfactory bulb, and hypothalamus in mice.[61]

As is commonly the case when animal models are immediately available for the study of a pharmacological agent, the number of animal studies examining alterations in allopregnanolone exceeds the current number of investigations in human subjects. The clinical data that is available for SSRIs, however, is generally consistent with neuroactive steroid findings in rodents. Two studies of extended fluoxetine administration in patients with major depression (fluoxetine for 4–7 weeks, or fluoxetine or fluvoxamine for 8–10 weeks) demonstrate significant increases in allopregnanolone in plasma[30] and cerebrospinal fluid[6] following SSRI treatment compared to baseline levels. Furthermore, several studies reveal decreased baseline allopregnanolone levels in depressed subjects relative to control subjects,[6,30,62] illuminating an obvious potential biochemical basis for SSRI therapeutic benefits secondary to neuroactive steroid induction. Supporting this possibility, increased allopregnanolone in cerebrospinal fluid in response to treatment with either fluoxetine or fluvoxamine correlates with reductions in depressive symptoms.[6] It should be noted, however, that allopregnanolone elevations in response to mirtazapine treatment in patients with depression do not appear to correlate with reductions in depressive symptoms, although mirtazapine does produce significant elevations in plasma allopregnanolone levels beginning at 3 weeks of treatment.[63] Interestingly, long-term mirtazapine administration in rodents affects stress-induced neuroactive steroid levels differently compared to long-term fluoxetine administration.[60] Finally, it is possible that paroxetine administration may not produce elevations in plasma allopregnanolone in subjects with panic disorder,[40,64] but the elucidation of potential paroxetine effects on neuroactive steroid levels may be challenging secondary to apparent baseline elevations in allopregnanolone levels in drug-free subjects with this disorder.[40]

18.2.2 Pregnenolone

Based upon its reported ability to enhance learning and memory, to ameliorate age-related cognitive decline, and to increase neurogenesis in rodent models, pregnenolone and/or its sulfated derivative may also be relevant to the mechanisms of action of antidepressant medications. Results from studies examining pregnenolone levels in rat brain and plasma have provided fairly consistent findings. Fluoxetine produces pregnenolone elevations in rat cerebral cortex and plasma,[22] and fluoxetine, olanzapine, or a combination of the two produces elevations of this neuroactive steroid in rat hippocampus and serum.[11] Also, long-term administration of fluoxetine may result in the downregulation of pregnenolone levels in both cerebral cortex and plasma following the absence of fluoxetine administration for 48 h; like allopregnanolone, fluoxetine challenge following this brief treatment cessation again results in elevated pregnenolone levels.[22]

In clinical populations, few data are currently available regarding antidepressant treatment and potential alterations in the neuroactive steroid pregnenolone. While baseline CSF pregnenolone is reduced in subjects with either unipolar or bipolar depression (with lowest levels seen during acute depressive episodes),[65] fluoxetine does not appear to alter pregnenolone levels in CSF[6] or plasma[30] in patients with major depression. A re-analysis of the Uzunova et al. study,[6] however, may suggest a possible antidepressant effect on pregnenolone levels.[66] More research will be required to fully characterize potential changes in pregnenolone following treatment with antidepressants.

18.2.3 Additional Indications of Neuroactive Steroids as Modulators of Antidepressant Activity

Evidence that neuroactive steroids also play roles in anxiety,[1] post-traumatic stress disorder (PTSD),[36,37] and panic disorder,[40] combined with the fact that many antidepressants are used for the amelioration of symptoms resulting from these conditions, further supports the possibility that neuroactive steroid activity is important to the function of this class of drugs. The introduction of a stressor[67] or the administration of exogenous corticotropin releasing hormone and adrenocorticotropin hormone (ACTH)[68,69] produces increases in allopregnanolone and/or pregnenolone in rat brain and plasma. Interestingly, subjects with depression whose Hamilton scores significantly improve after 6 weeks of fluoxetine administration also demonstrate significantly decreased plasma ACTH levels,[70] and administration of paroxetine for 12 months in females with PTSD results in reduced PTSD symptoms and reduced stress-induced cortisol levels.[71] As these SSRIs both reduce stress responses and normalize steroid profiles, the possibility of a role for neuroactive steroids in multiple SSRI actions becomes apparent, since several neuroactive steroids modulate the HPA axis and demonstrate anxiolytic effects. Given that the presently available data encompass only a small subset of commonly prescribed antidepressants, elucidation of the clinical importance of antidepressant-induced neuroactive steroid alterations will require further research.

18.3 Neuroactive Steroid Alterations Following Antipsychotics

18.3.1 Allopregnanolone

An increasing body of evidence suggests that neuroactive steroids may be relevant to the actions of atypical antipsychotics (also called second-generation antipsychotics). For example, atypical antipsychotics may modulate the stress response, producing changes in both behavioral and biochemical markers. Since a component of the therapeutic benefit of atypical antipsychotics may be derived from their stress modulatory and anxiolytic-like effects, neuroactive steroids possessing activity at inhibitory $GABA_A$ receptors may be implicated in their actions.[72] Examination of rodent neuroactive steroid levels in response to antipsychotic administration has resulted in the characterization of elevated neuroactive steroid levels in response to certain atypical antipsychotic medications with anxiolytic effects. Single administrations of both olanzapine[9,10] and clozapine[8,10] dose-dependently increase rat cerebral cortex levels of allopregnanolone, while neither risperidone[10] nor haloperidol[8,10] produces significant effects. Adrenalectomy prevents cerebral cortical elevations of allopregnanolone following the administration of both olanzapine and clozapine.[8,10] Similar dose-dependent increases in allopregnanolone, as well as elevations in other neuroactive steroids [progesterone and 3α,21-dihydroxy-5α-pregnan-20-one

(THDOC)] in response to clozapine but not haloperidol, have been reported in rat cerebral cortex and striatum.[8] The consistency of these antipsychotic-induced alterations in rodent models easily lead to the hypothesis that allopregnanolone is involved in their mechanisms of action, a conclusion supported by the requirement of allopregnanolone-induced modulation of $GABA_A$ receptors for the effects of olanzapine (but not risperidone or haloperidol) in two rodent behavioral tests.[73]

There is currently very limited information available on the potential effects of antipsychotics on allopregnanolone levels in humans, however. Long-term clozapine treatment in drug-resistant schizophrenia subjects in one study was reported to produce no changes in circulating allopregnanolone or THDOC levels.[74] It is possible, however, that methodological issues involving the assessment of neuroactive steroid levels may be impacting these findings. Specifically, a radioimmunoassay approach with potential antibody cross-reactivity issues was utilized in the prior study,[74] rather than a highly specific and sensitive mass spectrometry-based technique (reviewed in Purdy et al., 2005).[35,53] The elucidation of possible neuroactive steroid alterations in clinical populations following the administration of atypical antipsychotic agents will require additional investigation.

18.3.2 Pregnenolone and Other Neuroactive Steroids

Like allopregnanolone, concentrations of the neuroactive steroid pregnenolone in rat hippocampus also increase following clozapine and olanzapine administration, but pregnenolone is elevated to a greater degree in rat hippocampus and serum by clozapine at doses producing comparable striatal dopamine D_2 receptor occupancies consistent with clinically relevant ranges.[23] It has thus been hypothesized that these marked pregnenolone increases following clozapine administration may constitute a candidate mechanism contributing to its superior antipsychotic efficacy.[23] Other neuroactive steroids such as DHEA and DHEAS may also be altered following clozapine administration. For example, clozapine (but not haloperidol), decreases DHEA and DHEAS in rat cortex.[75]

18.4 Neuroactive Steroid Alterations Following Mood Stabilizers

A number of anticonvulsants with therapeutic mood stabilizing properties have been recognized to interact with peripheral-type benzodiazepine receptors.[76,77] The more recent demonstration of the role of these receptors in regulating neuroactive steroid synthesis[78,79] has led to hypotheses that neuroactive steroids modulate the actions of drugs effective in the treatment of bipolar disorder. Examination of this possibility has revealed that single and long-term administrations of the anticonvulsant carbamazepine induce dose-dependent elevations of several neuroactive steroids in rat cerebral cortex, including allopregnanolone and pregnenolone, as well as inducing steroidogenesis in the brains of adrenalectomized-orchidectomized rats.[80] Single administration of carbamazepine also induces dose-dependent elevations in plasma levels of neuroactive steroids.[80] Additionally, carbamazepine increases the synthesis of allopregnanolone in both rat frontal cortex and olfactory bulb,[81] prompting the authors of this finding to postulate a carbamazepine-induced enhancement of 3α-hydroxysteroid dehydrogenase activity, an enzyme previously shown to be stimulated by SSRIs.[82] However, a study using human 3α-hydroxysteroid dehydrogenase type III reports failure of carbamazepine and several antidepressants and antipsychotics to enhance the activity of this enzyme.[83] Long-term administration of lithium also appears to produce neuroactive steroid alterations, decreasing DHEA and DHEAS levels in rat frontal cortex and hippocampus.[84] Furthermore, chronic lithium administration significantly elevates rat frontal cortex allopregnanolone and also tends to produce higher pregnenolone levels, whereas valproate administration fails to alter the levels of either of these neuroactive steroids.[85]

Clinically, information on neuroactive steroid alterations in response to mood stabilizers provides a somewhat inconsistent picture. Carbamazepine treatment in patients with epilepsy decreases serum DHEAS levels.[86] Administration of carbamazepine to mood-disordered subjects did not consistently change CSF pregnenolone; these patients had exhibited significantly decreased CSF pregnenolone levels compared to control subjects prior to treatment with this medication.[65] Examination of postmortem brain tissue has revealed that pregnenolone and DHEA are significantly increased in both posterior cingulate and parietal cortex in patients with bipolar disorder compared to subjects with no psychiatric diagnosis.[31] Although little information in humans is available for additional agents demonstrated to possess anti-manic efficacy in humans, such as clozapine and olanzapine,[87,88] these antipsychotics have been found to increase rat central nervous system allopregnanolone and pregnenolone levels.[8,10,11,23] Both allopregnanolone and progesterone are increased in serum in female patients with bipolar I or bipolar II disorder in the luteal phase of the menstrual cycle compared to control subjects without psychiatric illness; however, specific drug regimens were not reported for individual subjects.[89] Although evidence suggests that neuroactive steroid levels are altered in bipolar disorder, the current clinical literature remains somewhat limited.

18.5 Neuroactive Steroids as Pharmacologic Agents

Since converging evidence in rodent models and clinical investigations suggests potential therapeutic benefits from increasing the levels of specific neuroactive steroids such as allopregnanolone and pregnenolone, the examination of clinical responses to the administration of these compounds is warranted. To date, however, controlled clinical trials are still very few. Pregnenolone has been reported to decrease fatigue[90] and improve psychomotor performance in response to stress and fatigue,[90] and produce improvement in multiple arthritic disorders.[91,92] In spite of these reported benefits, more recent examination of pregnenolone is limited to a single 4-week study in 17 healthy control subjects, reporting no effect of pregnenolone administration alone on mood, memory, self-rated sleep quality, or subjective well-being[66]; pregnenolone administration did, however, appear to antagonize the sedative effects of the benzodiazepine diazepam, but not its amnestic or anxiolytic effects.

Clinical studies investigating allopregnanolone administration are even fewer than those investigating pregnenolone. To our knowledge, three clinical studies utilizing allopregnanolone administration have been conducted. Increasing serum concentrations of allopregnanolone have been reported to correlate in healthy women with increases in subjective ratings of both sedation[93] and contentedness.[94] Conversely, increased allopregnanolone correlates with decreased contentedness in healthy men, and men also exhibit greater increases in serum allopregnanolone, both of which may indicate potential gender-specific differences in responses to allopregnanolone.[94] Finally, a clinical trial administering allopregnanolone vaginally to women undergoing estrogen replacement therapy examining the tolerability of allopregnanolone reported primarily only mild adverse events.[95]

While studies examining the clinical administration of these neuroactive steroids have only begun to be investigated, administration of several related compounds have been examined previously. To address rapid metabolism and poor water-solubility issues, several synthetic neuroactive steroids (including analogs of allopregnanolone) have been studied extensively. Two of these, alphaxolone and alphadolone, have generally been used in combination, producing potent anesthetic effects without the development of tolerance in rodent models.[96] It has been determined that a metabolite of alphadolone leads to the antinociceptive effects of this combination, involving spinal cord $GABA_A$ receptors,[97] but an adverse response to a vehicle component of the formulation has relegated this combination to use as a veterinary anesthetic.[97,98] Ganaxolone (3α-hydroxy-3β-methyl-5α-pregnan-20-one), a 3β-methylated synthetic analog of allopregnanolone, has received more extensive clinical examination with relatively promising results. Ganaxolone is a positive allosteric modulator of $GABA_A$ receptors with potency and efficacy similar to allopregnanolone,[99] demonstrating anticonvulsant properties in rodents[100–102] without development of tolerance.[103] Clinically, ganaxolone is well-tolerated and produces adverse events at the same rate as placebo,[102,104,105] and has demonstrated success in the reduction of epileptic symptoms in both children and adults.[102,105,106] Given the reported safety and tolerability of ganaxolone and its structural similarity to neuroactive steroids known to be involved in multiple psychiatric disorders, promising applications of this compound and related synthetic neuroactive steroids may be available in the future.

18.6 Summary

A rapidly expanding literature on the actions of neuroactive steroids reveals that they may play a role in the pathophysiology and therapeutics of a wide array of psychiatric and neurological disorders, a subset of which is discussed in this chapter. It is also possible that neuroactive steroid induction may contribute to the mechanisms of action of certain antidepressants, antipsychotics, and mood stabilizers, and that these molecules or their synthetic analogs may be extremely useful as pharmacological agents. Additional research efforts will be critical to the elucidation of these clinically relevant issues.

Acknowledgments Veterans Affairs Mid-Atlantic Mental Illness Research, Education, and Clinical Center, VA Advanced Research Career Development Award (CEM), National Alliance of Research on Schizophrenia and Affective Disorders (CEM), VA Research Enhancement Award Program (RDM). A provisional patent application has been filed by Duke University Medical Center and the Department of Veterans Affairs related to the content of this chapter (CEM).

References

1. Wieland S, Lan NC, Mirasedeghi S, et al. Anxiolytic activity of the progesterone metabolite 5α-pregnan-3α-ol-20-one. Brain Res 1991; 565:263–268.
2. Devaud LL, Purdy RH, Morrow AL. The neurosteroid, 3α-hydroxy-5α-pregnan-20-one, protects against bicuculline-induced seizures during ethanol withdrawal in rats. Alcohol Clin Exp Res 1995; 19:350–355.
3. Kokate TG, Cohen AL, Karp E, et al. Neuroactive steroids protect against pilocarpine- and kainic acid-induced limbic seizures and status epilepticus in mice. Neuropharmacology 1996; 35:1049–1056.
4. Pinna G, Dong E, Matsumoto K, et al. In socially isolated mice, the reversal of brain allopregnanolone down-regulation mediates the anti-aggressive action of fluoxetine. Proc Natl Acad Sci USA 2003; 100:2035–2040.
5. Uzunov DP, Cooper TB, Costa E, et al. Fluoxetine-elicited changes in brain neurosteroid content measured by negative ion mass fragmentography. Proc Natl Acad Sci USA 1996; 93:12599–12604.
6. Uzunova V, Sheline Y, Davis JM, et al. Increase in the cerebrospinal fluid content of neurosteroids in patients with unipolar major depression who are receiving fluoxetine or fluvoxamine. Proc Natl Acad Sci USA 1998; 95:3239–3244.
7. Khisti RT, Chopde CT, Jain SP. Antidepressant-like effect of the neurosteroid 3α-hydroxy-5α-pregnan-20-one in mice forced swim test. Pharmacol Biochem Behav 2000; 67:137–143.
8. Barbaccia ML, Affricano D, Purdy RH, et al. Clozapine, but not haloperidol, increases brain concentrations of neuroactive steroids in the rat. Neuropsychopharmacology 2001; 25:489–497.
9. Marx CE, Duncan GE, Gilmore JH, et al. Olanzapine increases allopregnanolone in the rat cerebral cortex. Biol Psychiatry 2000; 47:1000–1004.
10. Marx CE, VanDoren MJ, Duncan GE, et al. Olanzapine and clozapine increase the GABAergic neuroactive steroid allopregnanolone in rodents. Neuropsychopharmacology 2003; 28:1–13.
11. Marx CE, Shampine LJ, Khisti RT, et al. Olanzapine and fluoxetine administration and coadministration increase rat hippocampal pregnenolone, allopregnanolone and peripheral deoxycorticosterone: implications for therapeutic actions. Pharmacol Biochem Behav 2006; 84:609–617.
12. Azcoitia I, Leonelli E, Magnaghi V, et al. Progesterone and its derivatives dihydroprogesterone and tetrahydroprogesterone reduce myelin fiber morphological abnormalities and myelin fiber loss in the sciatic nerve of aged rats. Neurobiol Aging 2003; 24:853–860.
13. Ghoumari AM, Ibanez C, El-Etr M, et al. Progesterone and its metabolites increase myelin basic protein expression in organotypic slice cultures of rat cerebellum. J Neurochem 2003; 86:848–859.

14. Wang JM, Johnston PB, Ball BG, et al. The neurosteroid allopregnanolone promotes proliferation of rodent and human neural progenitor cells and regulates cell-cycle gene and protein expression. J Neurosci 2005; 25:4706–4718.
15. Griffin LD, Gong W, Verot L, et al. Niemann-Pick type C disease involves disrupted neurosteroidogenesis and responds to allopregnanolone. Nat Med 2004; 10:704–711.
16. Flood JF, Morley JE, Roberts E. Memory-enhancing effects in male mice of pregnenolone and steroids metabolically derived from it. Proc Natl Acad Sci USA 1992; 89:1567–1571.
17. Flood JF, Morley JE, Roberts E. Pregnenolone sulfate enhances post-training memory processes when injected in very low doses into limbic system structures: the amygdala is by far the most sensitive. Proc Natl Acad Sci USA 1995; 92:10806–10810.
18. Vallee M, Mayo W, Darnaudery M, et al. Neurosteroids: deficient cognitive performance in aged rats depends on low pregnenolone sulfate levels in the hippocampus. Proc Natl Acad Sci USA 1997; 94:14865–14870.
19. Akwa Y, Ladurelle N, Covey DF, et al. The synthetic enantiomer of pregnenolone sulfate is very active on memory in rats and mice, even more so than its physiological neurosteroid counterpart: distinct mechanisms? Proc Natl Acad Sci USA 2001; 98:14033–14037.
20. Fontaine-Lenoir V, Chambraud B, Fellous A, et al. Microtubule-associated protein 2 (MAP2) is a neurosteroid receptor. Proc Natl Acad Sci USA 2006; 103:4711–4716.
21. Hsu HJ, Liang MR, Chen CT, et al. Pregnenolone stabilizes microtubules and promotes zebrafish embryonic cell movement. Nature 2006; 439:480–483.
22. Serra M, Pisu MG, Muggironi M, et al. Opposite effects of short- versus long-term administration of fluoxetine on the concentrations of neuroactive steroids in rat plasma and brain. Psychopharmacology 2001; 158:48–54.
23. Marx CE, Shampine LJ, Duncan GE, et al. Clozapine markedly elevates pregnenolone in rat hippocampus, cerebral cortex, and serum: candidate mechanism for superior efficacy? Pharmacol Biochem Behav 2006; 84:598–608.
24. Melchior CL, Ritzmann RF. Pregnenolone and pregnenolone sulfate, alone and with ethanol, in mice on the plus-maze. Pharmacol Biochem Behav 1994; 48:893–897.
25. Lapchak PA, Chapman DF, Nunez SY, et al. Dehydroepiandrosterone sulfate is neuroprotective in a reversible spinal cord ischemia model: possible involvement of $GABA_A$ receptors. Stroke 2000; 31:1953–1956.
26. Labombarda F, Gonzalez SL, Gonzalez DM, et al. Cellular basis for progesterone neuroprotection in the injured spinal cord. J Neurotrauma 2002; 19:343–355.
27. Djebaili M, Guo Q, Pettus EH, et al. The neurosteroids progesterone and allopregnanolone reduce cell death, gliosis, and functional deficits after traumatic brain injury in rats. J Neurotrauma 2005; 22:106–118.
28. Mayo W, Lemaire V, Malaterre J, et al. Pregnenolone sulfate enhances neurogenesis and PSA-NCAM in young and aged hippocampus. Neurobiol Aging 2005; 26:103–114.
29. Brinton RD, Wang JM. Preclinical analyses of the therapeutic potential of allopregnanolone to promote neurogenesis in vitro and in vivo in transgenic mouse model of Alzheimer's disease. Curr Alzheimer Res 2006; 3:11–17.
30. Romeo E, Strohle A, Spalletta G, et al. Effects of antidepressant treatment on neuroactive steroids in major depression. Am J Psychiatry 1998; 155:910–913.
31. Marx CE, Stevens RD, Shampine LJ, et al. Neuroactive steroids are altered in schizophrenia and bipolar disorder: relevance to pathophysiology and therapeutics. Neuropsychopharmacology 2006; 31:1249–1263.
32. Ritsner M, Maayan R, Gibel A, et al. Differences in blood pregnenolone and dehydroepiandrosterone levels between schizophrenia patients and healthy subjects. Eur Neuropsychopharmacol 2007; 17:358–365.
33. Bernardi F, Lanzone A, Cento RM, et al. Allopregnanolone and dehydroepiandrosterone response to corticotropin-releasing factor in patients suffering from Alzheimer's disease and vascular dementia. Eur J Endocrinol 2000; 142:466–471.

34. Smith CD, Wekstein DR, Markesbery WR, et al. 3α,5α-THP: a potential plasma neurosteroid biomarker in Alzheimer's disease and perhaps non-Alzheimer's dementia. Psychopharmacology 2006; 186:481–485.
35. Marx CE, Trost WT, Shampine LJ, et al. The neurosteroid allopregnanolone is reduced in prefrontal cortex in Alzheimer's disease. Biol Psychiatry 2006; 60:1287–1294.
36. Spivak B, Maayan R, Kotler M, et al. Elevated circulatory level of $GABA_A$ – antagonistic neurosteroids in patients with combat-related post-traumatic stress disorder. Psychol Med 2000; 30:1227–1231.
37. Rasmusson AM, Pinna G, Paliwal P, et al. Decreased cerebrospinal fluid allopregnanolone levels in women with posttraumatic stress disorder. Biol Psychiatry 2006; 60:704–713.
38. Monteleone P, Luisi M, Colurcio B, et al. Plasma levels of neuroactive steroids are increased in untreated women with anorexia nervosa or bulimia nervosa. Psychosom Med 2001; 63:62–68.
39. Strous RD, Spivak B, Yoran-Hegesh R, et al. Analysis of neurosteroid levels in attention deficit hyperactivity disorder. Int J Neuropsychopharmacol 2001; 4:259–264.
40. Strohle A, Romeo E, di Michele F, et al. $GABA_A$ receptor-modulating neuroactive steroid composition in patients with panic disorder before and during paroxetine treatment. Am J Psychiatry 2002; 159:145–147.
41. Majewska MD, Harrison NL, Schwartz RD, et al. Steroid hormone metabolites are barbiturate-like modulators of the GABA receptor. Science 1986; 232:1004–1007.
42. Park-Chung M, Malayev A, Purdy RH, et al. Sulfated and unsulfated steroids modulate γ-aminobutyric $acid_A$ receptor function through distinct sites. Brain Res 1999; 830:72–87.
43. Hosie AM, Wilkins ME, da Silva HM, et al. Endogenous neurosteroids regulate $GABA_A$ receptors through two discrete transmembrane sites. Nature 2006; 444:486–489.
44. Hosie AM, Wilkins ME, Smart TG. Neurosteroid binding sites on $GABA_A$ receptors. Pharmacol Ther 2007; 116:7–19.
45. Morrow AL, Suzdak PD, Paul SM. Steroid hormone metabolites potentiate GABA receptor-mediated chloride ion flux with nanomolar potency. Eur J Pharmacol 1987; 142:483–485.
46. Shu HJ, Eisenman LN, Jinadasa D, et al. Slow actions of neuroactive steroids at $GABA_A$ receptors. J Neurosci 2004; 24:6667–6675.
47. Majewska MD, Mienville JM, Vicini S. Neurosteroid pregnenolone sulfate antagonizes electrophysiological responses to GABA in neurons. Neurosci Lett 1988; 90:279–284.
48. Majewska MD, Demirgoren S, Spivak CE, et al. The neurosteroid dehydroepiandrosterone sulfate is an allosteric antagonist of the $GABA_A$ receptor. Brain Res 1990; 526:143–146.
49. Imamura M, Prasad C. Modulation of gaba-gated chloride ion influx in the brain by dehydroepiandrosterone and its metabolites. Biochem Biophys Res Commun 1998; 243:771–775.
50. Wu FS, Gibbs TT, Farb DH. Pregnenolone sulfate: a positive allosteric modulator at the N-methyl-D-aspartate receptor. Mol Pharmacol 1991; 40:333–336.
51. Compagnone NA, Mellon SH. Dehydroepiandrosterone: a potential signalling molecule for neocortical organization during development. Proc Natl Acad Sci USA 1998; 95:4678–4683.
52. Johansson T, Le Greves P. The effect of dehydroepiandrosterone sulfate and allopregnanolone sulfate on the binding of [^{3}H]ifenprodil to the *N*-methyl-D-aspartate receptor in rat frontal cortex membrane. J Steroid Biochem Mol Biol 2005; 94:263–266.
53. Purdy RH, Fitzgerald R, Alomary A, et al. Practical neurochemistry: methods. In: Baker G (ed). Handbook of neurochemistry and molecular neurobiology, 3rd ed., Vol. 18. New York: Springer, 2006, pp. 1–15.
54. Holsboer F, Barden N. Antidepressants and hypothalamic-pituitary-adrenocortical regulation. Endocr Rev 1996; 17:187–205.
55. Patchev VK, Shoaib M, Holsboer F, et al. The neurosteroid tetrahydroprogesterone counteracts corticotropin-releasing hormone-induced anxiety and alters the release and gene expression of corticotropin-releasing hormone in the rat hypothalamus. Neuroscience 1994; 62:265–271.
56. Guo AL, Petraglia F, Criscuolo M, et al. Evidence for a role of neurosteroids in modulation of diurnal changes and acute stress-induced corticosterone secretion in rats. Gynecol Endocrinol 1995; 9:1–7.

57. Morrow AL, Devaud LL, Purdy RH, et al. Neuroactive steroid modulators of the stress response. Ann NY Acad Sci 1995; 771:257–272.
58. Patchev VK, Hassan AH, Holsboer DF, et al. The neurosteroid tetrahydroprogesterone attenuates the endocrine response to stress and exerts glucocorticoid-like effects on vasopressin gene transcription in the rat hypothalamus. Neuropsychopharmacology 1996; 15:533–540.
59. Pinna G, Costa E, Guidotti A. Fluoxetine and norfluoxetine stereospecifically and selectively increase brain neurosteroid content at doses that are inactive on 5-HT reuptake. Psychopharmacology 2006; 186:362–372.
60. Serra M, Pisul MG, Dazzi L, et al. Prevention of the stress-induced increase in the concentration of neuroactive steroids in rat brain by long-term administration of mirtazapine but not of fluoxetine. J Psychopharmacol 2002; 16:133–138.
61. Nechmad A, Maayan R, Spivak B, et al. Brain neurosteroid changes after paroxetine administration in mice. Eur Neuropsychopharmacol 2003; 13:327–332.
62. Strohle A, Romeo E, Hermann B, et al. Concentrations of 3α-reduced neuroactive steroids and their precursors in plasma of patients with major depression and after clinical recovery. Biol Psychiatry 1999; 45:274–277.
63. Schule C, Romeo E, Uzunov DP, et al. Influence of mirtazapine on plasma concentrations of neuroactive steroids in major depression and on 3α-hydroxysteroid dehydrogenase activity. Mol Psychiatry 2006; 11:261–272.
64. Brambilla F, Mellado C, Alciati A, et al. Plasma concentrations of anxiolytic neuroactive steroids in men with panic disorder. Psychiatry Res 2005; 135:185–190.
65. George MS, Guidotti A, Rubinow D, et al. CSF neuroactive steroids in affective disorders: pregnenolone, progesterone, and DBI. Biol Psychiatry 1994; 35:775–780.
66. Meieran SE, Reus VI, Webster R, et al. Chronic pregnenolone effects in normal humans: attenuation of benzodiazepine-induced sedation. Psychoneuroendocrinology 2004; 29:486–500.
67. Purdy RH, Morrow AL, Moore PH, Jr., et al. Stress-induced elevations of γ-aminobutyric acid type A receptor-active steroids in the rat brain. Proc Natl Acad Sci USA 1991; 88:4553–4557.
68. Torres JM, Ruiz E, Ortega E. Effects of CRH and ACTH administration on plasma and brain neurosteroid levels. Neurochem Res 2001; 26:555–558.
69. Torres JM, Ortega E. DHEA, PREG and their sulphate derivatives on plasma and brain after CRH and ACTH administration. Neurochem Res 2003; 28:1187–1191.
70. Inder WJ, Prickett TC, Mulder RT, et al. Reduction in basal afternoon plasma ACTH during early treatment of depression with fluoxetine. Psychopharmacology 2001; 156:73–78.
71. Vermetten E, Vythilingam M, Schmahl C, et al. Alterations in stress reactivity after long-term treatment with paroxetine in women with posttraumatic stress disorder. Ann NY Acad Sci 2006; 1071:184–202.
72. Marx CE, Grobin AC, Deutch AY, et al. Atypical antipsychotic drugs and stress. In: Steckler T, Kalin NH, Reul JMHM (eds). Handbook of stress and the brain, Vol. 15. Philadelphia, PA: Elsevier, 2005, pp. 301–313.
73. Ugale RR, Hirani K, Morelli M, et al. Role of neuroactive steroid allopregnanolone in antipsychotic-like action of olanzapine in rodents. Neuropsychopharmacology 2004; 29:1597–1609.
74. Monteleone P, Fabrazzo M, Serra M, et al. Long-term treatment with clozapine does not affect morning circulating levels of allopregnanolone and THDOC in patients with schizophrenia: a preliminary study. J Clin Psychopharmacol 2004; 24:437–440.
75. Nechmad A, Maayan R, Ramadan E, et al. Clozapine decreases rat brain dehydroepiandrosterone and dehydroepiandrosterone sulfate levels. Eur Neuropsychopharmacol 2003; 13:29–31.
76. Marangos PJ, Post RM, Patel J, et al. Specific and potent interactions of carbamazepine with brain adenosine receptors. Eur J Pharmacol 1983; 93:175–182.
77. Weiss SR, Post RM, Patel J, et al. Differential mediation of the anticonvulsant effects of carbamazepine and diazepam. Life Sci 1985; 36:2413–2419.
78. Mukhin AG, Papadopoulos V, Costa E, et al. Mitochondrial benzodiazepine receptors regulate steroid biosynthesis. Proc Natl Acad Sci USA 1989; 86:9813–9816.

79. Papadopoulos V, Mukhin AG, Costa E, et al. The peripheral-type benzodiazepine receptor is functionally linked to Leydig cell steroidogenesis. J Biol Chem 1990; 265:3772–3779.
80. Serra M, Littera M, Pisu MG, et al. Steroidogenesis in rat brain induced by short- and long-term administration of carbamazepine. Neuropharmacology 2000; 39:2448–2456.
81. Jaworska-Feil L, Budziszewska B, Leskiewicz M, et al. Effects of some centrally active drugs on the allopregnanolone synthesis in rat brain. Pol J Pharmacol 2000; 52:359–365.
82. Griffin LD, Mellon SH. Selective serotonin reuptake inhibitors directly alter activity of neurosteroidogenic enzymes. Proc Natl Acad Sci USA 1999; 96:13512–13517.
83. Trauger JW, Jiang A, Stearns BA, et al. Kinetics of allopregnanolone formation catalyzed by human 3α-hydroxysteroid dehydrogenase type III (AKR1C2). Biochemistry 2002; 41:13451–13459.
84. Maayan R, Shaltiel G, Poyurovsky M, et al. Chronic lithium treatment affects rat brain and serum dehydroepiandrosterone (DHEA) and DHEA-sulphate (DHEA-S) levels. Int J Neuropsychopharmacol 2004; 7:71–75.
85. Marx CE, Yuan P, Shampine LJ, et al. Neuroactive steroids, mood stabilizers, and neuroplasticity: alterations following lithium and changes in Bcl-2 knockout mice. Biol Psychiatry 2005; 57:205S.
86. Herzog AG, Drislane FW, Schomer DL, et al. Differential effects of antiepileptic drugs on neuroactive steroids in men with epilepsy. Epilepsia 2006; 47:1945–1948.
87. Suppes T, Webb A, Paul B, et al. Clinical outcome in a randomized 1-year trial of clozapine versus treatment as usual for patients with treatment-resistant illness and a history of mania. Am J Psychiatry 1999; 156:1164–1169.
88. Tohen M, Jacobs TG, Grundy SL, et al. Efficacy of olanzapine in acute bipolar mania: a double-blind, placebo-controlled study. Arch Gen Psychiatry 2000; 57:841–849.
89. Hardoy MC, Serra M, Carta MG, et al. Increased neuroactive steroid concentrations in women with bipolar disorder or major depressive disorder. J Clin Psychopharmacol 2006; 26:379–384.
90. Pincus G, Hoagland H. Effects of administered pregnenolone on fatiguing psychomotor performance. J Aviation Med 1944; 15:98–115.
91. Davison R, Koets P, Snow WG, et al. Effects of delta 5 pregnenolone in rheumatoid arthritis. Arch Intern Med 1950; 85:365–389.
92. Freeman H, Pincus G, Johnson CW, et al. Therapeutic efficacy of Δ^5-pregnenolone in rheumatoid arthritis. JAMA 1950; 142:1124–1127.
93. Timby E, Balgard M, Nyberg S, et al. Pharmacokinetic and behavioral effects of allopregnanolone in healthy women. Psychopharmacology 2006; 186:414–424.
94. van Broekhoven F, Backstrom T, van Luijtelaar G, et al. Effects of allopregnanolone on sedation in men, and in women on oral contraceptives. Psychoneuroendocrinology 2007; 32:555–564.
95. Navarro PA, Kaddouz D, de Ziegler D, et al. Vaginal administration of allopregnanolone to postmenopausal women undergoing estrogen replacement therapy: preliminary results. Maturitas 2003; 46:147–152.
96. Green CJ, Halsey MJ, Precious S, et al. Alphaxolone-alphadolone anaesthesia in laboratory animals. Lab Anim 1978; 12:85–89.
97. Nadeson R, Goodchild CS. Antinociceptive properties of neurosteroids III: experiments with alphadolone given intravenously, intraperitoneally, and intragastrically. Br J Anaesth 2001; 86:704–708.
98. Watkins J, Allen R, Ward AM. Adverse response to alphadolone/alphaxalone: possible role of IgD. Lancet 1978; 2:736.
99. Carter RB, Wood PL, Wieland S, et al. Characterization of the anticonvulsant properties of ganaxolone (CCD 1042; 3α-hydroxy-3β-methyl-5α-pregnan-20-one), a selective, high-affinity, steroid modulator of the γ-aminobutyric $acid_A$ receptor. J Pharmacol Exp Ther 1997; 280:1284–1295.
100. Gasior M, Ungard JT, Beekman M, et al. Acute and chronic effects of the synthetic neuroactive steroid, ganaxolone, against the convulsive and lethal effects of pentylenetetrazol in seizure-kindled mice: comparison with diazepam and valproate. Neuropharmacology 2000; 39:1184–1196.

101. Kaminski RM, Livingood MR, Rogawski MA. Allopregnanolone analogs that positively modulate $GABA_A$ receptors protect against partial seizures induced by 6-Hz electrical stimulation in mice. Epilepsia 2004; 45:864–867.
102. Bialer M, Johannessen SI, Kupferberg HJ, et al. Progress report on new antiepileptic drugs: a summary of the Eigth Eilat Conference (EILAT VIII). Epilepsy Res 2007; 73:1–52.
103. Reddy DS, Rogawski MA. Chronic treatment with the neuroactive steroid ganaxolone in the rat induces anticonvulsant tolerance to diazepam but not to itself. J Pharmacol Exp Ther 2000; 295:1241–1248.
104. Monaghan EP, Navalta LA, Shum L, et al. Initial human experience with ganaxolone, a neuroactive steroid with antiepileptic activity. Epilepsia 1997; 38:1026–1031.
105. Nohria V, Giller E. Ganaxolone. Neurotherapeutics 2007; 4:102–105.
106. Kerrigan JF, Shields WD, Nelson TY, et al. Ganaxolone for treating intractable infantile spasms: a multicenter, open-label, add-on trial. Epilepsy Res 2000; 42:133–139.

Chapter 19
Reconsidering Classifications of Depression Syndromes: Lessons from Neuroactive Steroids and Evolutionary Sciences

Bernardo Dubrovsky

Abstract Neurosteroids (NS) refers to steroids synthesized by the brain, neuroactive steroids (NAS) refers to steroids that, independent of their origin, are capable of modifying neural activities. NAS bind to and modulate different types of neurotransmitter receptors at the membrane level. The gamma-aminobutyric acid (GABA) and sigma receptor complexes have been the most extensively studied.

After intracellular oxidation, tetrahydroprogesterone (THP) to dihydroprogesterone (DHP), and tetrahydrodeoxycorticosterone (THDOC) to dihydrodeoxycorticosterone (DHDOC) bind to the progesterone intracellular receptor (PR), and in this way can also regulate membrane composition through gene expression.

Although NAS levels change in depression syndromes, in particular THP decreases while THDOC increases, levels return to normal baseline with recovery; however, normalization is not necessary for successful therapy.

Studies attempting to pinpoint the possible use of NAS levels as biological markers were unsuccessful. Data from studies on the evolution of nervous systems, evolutionary developmental biology, as well as anatomical and physiological findings, preclude that the disorders that result when the brain succumbs to psychiatric ailments can be categorically classified.

Persistence in maintaining essentialist classifications may help to explain why, until now, the search for biological markers in psychiatry has been an unrewarding effort. It is proposed that it would be more fruitful to focus on relationships between NAS and symptoms of psychiatric disorders, rather than with typologically defined disorders.

Keywords Neurosteroids, neuroactive steroids, Ring A reduced pregnanes, depression, biological markers, symptoms

---

McGill University

M.S. Ritsner and A. Weizman (eds.), *Neuroactive Steroids in Brain Function, Behavior and Neuropsychiatric Disorders*

Abbreviations ACTH adrenocorticotropic hormone; AD antidepressant; ADHD attention deficit hyperactivity disorder; ALDO aldosterone; AS androsterone sulfate; AVP arginine vasopressin; BDGF brain developmental growth factor; CFS chronic fatigue syndrome; CNS central nervous system; CPEP cytoplasmatic polyadezylation element binding protein; CRH corticotropic releasing hormone; CS corticosterone; CSF cerebrospinal fluid; DG dentate gyrus, DHEA dehydroepiandrosterone; DHEAS dehydroepiandrosterone sulfate; DHP dihydroprogesterone; DOC deoxycorticosterone; rECS repeated electroconvulsive shock; EPSP excitatory postsynaptic potential; ER evoked response; GABA gamma-aminobutyric acid; GC glucocorticoid; GFAP glial fibrillary acidic protein; GR glucocorticoid receptor; HPAA hypothalamic pituitary adrenal axis; 3α HSD 3α hydroxysteroid dehydrogenase; 17β HSD 17β hydroxysteroid dehydrogenase; IPSP inhibitory postsynaptic potential; LTD long-term depression, LTP long-term potentiation; MBP myelin basic protein; MC mineralocorticoid; MISS membrane-initiated steroid signaling; MR mineralocorticoid receptor; MRS magnetic resonance spectroscopy; NADPH nicotinamide adenine dinucleotide phosphate; NAS neuroactive steroid; NMDA *N*-methyl-D-aspartate; NS neurosteroid; 11β-OHD 11β-hydroxysteroid dehydrogenase; OT oxytocin; P progesterone; PBR peripheral benzodiazepine receptor; PKC phosphokinase C; PP perforant path; PPD postpartum depression; PS pregnenolone sulfate; PS population spike; PSD partial sleep deprivation; REM rapid eye movement; RF reticular formatio; SSRI selective serotonin reuptake inhibitor; SSNI selective serotonin noradrenaline inhibitor; TCA tricyclic antidepressant; THDOC allo-tetrahydrodeoxycorticosterone; THP allo-tetrahydroprogesterone

Classifying animals

Those that belong to the Emperor
Embalmed ones
Those that are trained
Suckling pigs
Mermaids
Fabulous ones
Stray dogs
Those that are included in this classification
Those that tremble as if they were mad
Innumerable ones
Those drawn with a very fine camel's-hair brush
Others
Those that have just broken a flower vase
Those that resemble flies from a distance
'The Celestial Emporium of Benevolent Knowledge'

Jorge Luis Borges

19.1 Introduction

The cornerstone of psychiatric practice stands on taxonomies constructed from clinical data. In Psychiatry, diagnosis relies almost exclusively on data obtained through the clinical method. The ideal working relationship between clinic and laboratory that was fundamental for the development of general medicine is rarely, if ever, used in Psychiatry.

The concept of disease is very controversial and difficult to define, probably a reason why neither the World Health Organization in the International Classification of Diseases, Injuries and Causes (ICD), nor the Diagnostic and Statistical Manual of Mental Disorders (DSM), have done so. In order to avoid the use of the word disease and the problems it raises, DSM IV uses the word disorder.

Psychiatric disorders comprise a large and heterogeneous population. Two methods of classification have been mainly used.[1] One is grouping members of the population into a number of subpopulations, each one showing some degree of homogeneity in one or various chosen aspects (Pace Rosh). In the other, the evaluator assigns a defined character into position in one or more axes or dimensions. These two classification methods are respectively referred to as categorical and dimensional.[1] In the ICD 10 and DSM IV classifications, psychiatric disorders are considered to be discrete, categorical units. These taxonomies are cast in a Platonic, essentialist mode, i.e., on the assumption that real things are copies of ideal types. For essentialists, all seemingly variable phenomena of nature can be sorted into classes, each one endowed with a unique essence that allows its identification.

For Kraepelin, a convinced Platonist, the task of psychiatrists is to discover the essential disorders and their basic forms as expressed through the symptoms of patients. "Such, in their essential features, are the points of view from which the clinical forms of insanity may be grouped today".[2] But notwithstanding the large amount of work done in search of essential features, i.e., biological and/or clinical markers of each psychiatric disorder, results have been unrewarding; neither clinical symptoms nor biological variable(s) can, so far, be specifically associated with a particular psychiatric disorder. Even in the two members of the Kraepelinian dichotomy, manic depression and schizophrenia, "there is hardly a symptom that we will not find at one time or another in both of the two major divisions of the Kraepelin dichotomy, and in no instance are the symptoms 'mutually exclusive and jointly exhaustive'".[1] The perennial search for biomarkers in Psychiatry brings to mind Gertude Stein's remark when looking for a nondescript American town: "When you get there, you discover there is no there, there."

The problem of finding single biomarkers is not restricted to psychiatry. In writing about the search for biomarkers to identify malignancy, Petricoin and Liotta[3] stated: "human diseases are a product of the tissue microenvironment. Interactions between the diseased or infected cells, the surrounding epithelial and stromal cells, the vascular channels, the extracellular matrix, and the immune system are the ultimate determinant of final pathology." Thus "a higher level of both specificity and sensitivity might be achieved by measuring the combination of markers emanating from both the diseased cells and the reactive cells in the environment."

19.2 NS and NAS as Biological Markers in Psychiatry: Basic Aspects

Compounds recently investigated as possible specific biological markers of psychiatric disorders include neurosteroids (NS), steroids synthesized by nervous systems, and neuroactive steroids (NAS), steroids acting on nervous systems irrespective of their origin.[4]

For most authors, the biosynthesis of adrenal and gonadal steroids, as well as NS, depend upon the tissue-specific and cell-specific array of steroidogenic enzymes. Neurosteroidogenic enzymes are classified as cytochrome P-450 and non-P450 enzymes. The steroidogenic P-450 enzymes are found as single genes in multiple species, but can mediate multiple enzymatic steps.[5,6] On the other hand, the non-P-450 enzymes, such as 3β-hydroxysteroid dehydrogenase (3β-HSD), 17β-hydroxysteroid dehydrogenase (17β-HSD), and 3α-hydroxysteroid dehydrogenase (3α-HSD), are found as multiple genes encoding different proteins. Each of these proteins mediates specific reactions.

Some differences between NS and adrenal steroid biosynthesis are known. Corticosteroid synthesis involves converting deoxycorticosterone (DOC) to either aldosterone by aldosterone synthase or to corticosterone by 11β-hydroxylase. In the adrenals the enzymes are never expressed in the same cell. But in the brain, the enzymes coexpress not only in the same region, but even within the same cell; therefore, aldosterone synthase and 11β-hydroxylase must compete for DOC.

In classical models of steroid action, pregnenes from the adrenal and gonads, e.g., mineralocorticoids (MC), glucocorticoids (GC), progesterone (P), and other hormones, bind to cytosolic receptors. Subsequently, these receptors change their conformation by dissociation from chaperone molecules, the heat shock proteins, and translocate to the nucleus where they bind as homodimers or heterodimers to the respective response elements of the genes, the promoters. Alternatively, these ligand-activated receptors can influence transactivation through protein–protein interaction with other transcription factors, thus acting as transcription factors in the regulation of gene expression.

NAS also modulate different kinds of membrane receptors. There are stringent structure–activity relationships for Ring A reduced pregnanes, an OH group in the α position at C3 is essential for them to act as positive allosteric modulators of the $GABA_A$ receptor complex. An OH group in β position at C3 converts THP into isopregnanolone which behaves as an antagonist at the $GABA_A$ receptor complex.[4]

Some molecules can exist in distinct mirror image forms, related to each other as a left hand is to a right. A number of characteristics of chiral molecules, as they are called, differ very much from each other. Some steroids have this property. In the case of the NAS pregnenolone sulfate (PS), enantiometers, the name given to the distinct handed forms of chiral molecules, differ in their potency. The synthetic negative enantiometer of PS is ten times more potent as a memory enhancer than the natural, positive enantiometer. Moreover, the memory enhancing effects of the (–) and (+) enantiometers involve different mechanisms. While the effects of the (+) form can be blocked by a selective NMDA receptor antagonist, its (–) enantiometer cannot.[7]

Do interactions between membrane responding sites to steroids and cytosolic receptors occur? Csaba[8] has argued that the "primordial receptor development for environment signals (e.g., hormones), took place on the surface of contact. Further, cytosol receptors (could) arise by internalization of membrane receptors." Allera and Wildt[9] as well as Daufeldt et al.[10] have identified a membrane initiated steroid signaling (MISS) protein that transfers steroids such as corticosterone, cortisol and particular gestagens and estrogens, into rat and human liver cells. In vitro and in vivo studies suggest that MISS mediates rapid, non-genomic responses to steroids.

It is believed that the biosynthesis of all steroids begins with the conversion of cholesterol to pregnenolone which is catalyzed by the cytochrome P-450 side-chain cleavage enzyme (P-450_{scc}). The rate-limiting step is the delivery of cholesterol to the inner mitochondrial membrane where P-450_{scc} is located. Transfer of cholesterol through the mitochondrial membrane is mediated by acute induction of steroidogenic acute regulatory protein.[11] The presence of translocator protein 18kDa is essential for this process. 18kDa is the new denomination for what was previously known as the peripheral-type benzodiazepine receptor (PBR).[12]

This version of steroid and NS biosynthesis has been reached by consensus. It is derived by piecing together the results obtained from several independent artificial in vitro incubation experiments. Extrapolation of such results from in vitro to in vivo conditions requires untested assumptions which introduce varying degrees of uncertainty. In vitro incubation experiments reveal only what is possible, not what actually prevails in situ. Lieberman and Kanshik[13] (2006) had advanced a hypothetical alternative rendition of steroidogenesis. This hypothesis suggests that some cytochromes P-450 catalyze the introduction of both oxygen atoms of dioxygen into an appropriate sterol precursor. The products are conceived as oxygen free radicals (peroxy or 1, 2-cyclic peroxy) which serve as the "reactive intermediates" (the precursors) for the hormones. The true intermediates are not stable hydroxylated compounds that can be isolated. Central to these new renditions, is the hypothesis that the appropriate P-450 introduces dioxygen into precursor yielding (a) 20peroxy sterol species, (b) species oxygenated at both C-17 and C-20 or (c) species oxygenated at both C-20 and C-21.

Sensitivity of $GABA_A$ receptors to NAS depends on their previous exposure to chemicals, e.g., alcohol, to other endocrines, or to hormonal withdrawal, e.g. P in termination of pseudo-pregnancy, Premenstrual Dysphoric Behavior Disorder (PMDD), or during the postpartum period.[14–16] P withdrawal results in a decrease in brain content of THP and a subsequent increase in the expression of the $GABA_A$ receptor α-4 subunit.[17] It has been proposed that the presence of the α-4 subunit is essential for the occurrence of P withdrawal effects, viz., increased anxiety and seizures, decreased sensitivity to positive allosteric $GABA_A$-receptor modulators THDOC and THP, benzodiazepines and ethanol, and increased sensitivity to negative allosteric $GABA_A$ receptor modulators like dehydroepiandrosterone sulfate (DEHA(S)).[14,18,19]

Based on their experimental results, Brussaard and Koksma[20] suggested an alternative explanation to account for sensitivity changes in the $GABA_A$ receptor complex. In addition to their allosteric interaction with $GABA_A$ receptors, neurosteroids play a role in translocation of particular protein-kinase C (PKC) isomers toward the membrane, which, in turn, would affect the neurosteroid sensitivity of $GABA_A$ receptors being expressed. For them, available evidence argues in favor of a simple model as being the most parsimonious: neurosteroid sensitivity affects $GABA_A$ receptors in a bidirectional manner, being dependent on, but also affecting, the metabotropic modulation of $GABA_A$ receptors. Brussaard and Koksma[20] question receptor subunit specificity and argue that their model presents a general concept that (a) also occurs in other brain areas, and (b) one which neurosteroids and other allosteric modulators might have in common. Their studies show that THP not only potentiates $GABA_A$ receptor function but also prevents its inhibition by PKC, and propose that neurosteroid sensitivity of the $GABA_A$ receptor complex itself is dependent on the balance between endogenous phosphatase and PKC activity and not, as previously suggested, on subunit composition changes of the $GABA_A$ receptor. These data imply that native $GABA_A$ receptors are only sensitive to 3α-DHP if there is endogenous phosphatase activity. In contrast, when due to endogenous release of oxytocin in the hypothalamus, the intracellular balance is shifted from high phosphatase activity toward a higher level of PKC-dependent phosphorylation, this leads to 3α-OH-DHP insensitivity of the $GABA_A$ receptor complex.[21]

However, the opposite is true for magnocellular neurons of the supraoptic nucleus in the hypothalamus in the late stages of pregnancy.[20] THP sensitivity of spontaneous inhibitory post-synaptic currents (sIPSC) recorded from magnocellular neurons prior to parturition can be prevented by stimulation of PKC whereas, conversely, the neurosteroid insensitivity of synaptic $GABA_A$ receptors post-parturition is restored by inhibiting PKC.[20] Interestingly, oxytocin which is released at parturition by magnocellular neurons, induces activation of PKC probably via an inositol triphosphate (IP_3) and diacylglycerol (DAG)-mediated mechanism.[22] Thus, PKC activation promoted by the timed-release of oxytocin may fulfill a physiological requirement to reduce the $GABA_A$ receptor-mediated inhibitory brake on the firing of magnocellular neurons, and consequently support lactation and uterine contraction.[21]

The isoforms of the kinases influencing neurosteroid sensitivity are not known. Recent findings implicate the PKCε subtype. Thus, mice with a targeted disruption of the PKCε gene display an enhanced behavioral sensitivity to a number of positive allosteric modulators of the $GABA_A$ receptor, including neurosteroids.[22] As NAS are part of the neuroendocrine response to acute stress, a PKCε-mediated phosphorylation process could potentially regulate, in a dynamic fashion, the physiological amplification of neuronal inhibition by neurosteroids following activation of the Hypothalamic-Pituitary-Adrenal (HPA) axis during the stress response.[23]

19.3 NS and NAS as Biological Markers for Depression Syndromes: Clinical Aspects

For the most part, and due mainly to practical reasons (being simple, fulfilling legal demands, habit), categorical classifications such as ICD-10 and DSM-IV are widely used in the clinic. However, examination of results of recent evo-devo studies, investigations on the evolution of the brain, recent neuroanatomical and neurophysiological data, and considerations of the structure of the brain, i.e., the set of all relations among its components, suggest that when human brains succumb to psychiatric ailments, they cannot give rise to disorders that can be classified as categorical, i.e., endowed with clinical pathognomonic and/or biological markers.[24–27]

The fact that categorical, typological classifications prevailed in psychiatry, and are still very much in use today, relates with the dominance of the platonic, essentialist underpinnings of western thought that prevailed until Darwin's introduction in 1859 of population thinking, the variational form of evolution.[28,29]

A brief review of recent results in fundamental neuroscience is necessary to substantiate the assertion that the structure and evolution of the human brain preclude the development of psychiatric disorders endowed with unique essences.

19.4 Evolutionary Developmental Biology and Brain Evolution Studies Relevant to Nosology in Psychiatry

At the end of the 20th century, evolutionary developmental biology (evo-devo) emerged as an independent discipline at the interface of embryology and evolutionary biology.[30–34] One of its first successes was the identification, isolation, and cloning of homeotic or Hox genes, so called because when mutated, they lead to the replacement of one part of the animal body by another.

Hox genes belong to the developmental family group of genes. Their physiology has characteristics that are similar to those described by Jacob and Monod[35] in their epoch-making paper "Genetic regulatory mechanisms in the synthesis of proteins". In this paper Jacob and Monod distinguished two types of genes: (a) structural genes, which signal the linear sequencing order of the amino acids of a protein (although not spatial orientation), and (b) regulatory genes, which lead to the synthesis of proteins that regulate the expression of structural or target genes, either directly or indirectly (by controlling the function of other regulatory genes).[30] The region in the DNA of a structural gene that contains the starting point of transcription and the site for binding the RNA polymerase is called promoter.

For Slack, the discovery of Hox genes is the Rosetta Stone of biology.[32] Hox genes are a highly conserved evolutionary set of genes that arose from a single ancestral cluster by duplication.[31,32,34,35] HOX genes, as Hox genes are termed in humans, have a well-characterized role in embryogenesis and anteroposterior axis specification. It has been proposed that HOX genes enable the retention of developmental plasticity in certain tissues that continue to undergo rapid developmental changes in the adult. This was shown to be the case in cells of adult tissues such as the female reproductive tract and hematopoietic system. Furthermore, available evidence reveals that endocrines (e.g., steroid hormones) regulate HOX genes during embryogenesis and in adult states.[33]

Hox genes have a well-characterized role in cell fate specification and regulate adult functional differentiation through similar mechanisms as used in development.[33]

It is plausible then that long term potentiation and longterm depression (LTP and LTD), phenomena associated with memory, could be regulated by Hox genes. For the same reasons, Hox genes may be of importance in the proliferation and regeneration of neurons, essential for antidepressant treatment.[4]

Hox genes bind to the promoter zone of structural genes and increase or decrease the expression of these genes. Selector genes are Hox genes expressed in the compartments of an animal's body plan and serve to distinguish each compartment. In fact, compartments are defined as the regions of the embryo in which selector genes are uniquely expressed.[36]

Neither selector nor target or slave genes are restricted to perform only one function.[30,33,34,36] Hox gene BF1, e.g., which regulates forebrain size, is closely related to genes controlling the formation of the gut.[37] Hox genes are extremely versatile and contribute to functions unrelated to development, e.g., intercellular matrix, metabolic activities. This is done through their molecular interactions with other genes and their products, a phenomenon referred to as promiscuity of interactions.[30] Genes controlling head and brain formation in fruit flies, gene empty spiracles, are closely related to genes that control the formation of the more anterior parts of the mammalian brain. Genes Emx1 and Emx2 that derive from genes empty spiracles regulate the formation of the cerebral cortex, an example of a phylogenetically recent part of the mammalian brain that is controlled by genes whose antecedents go back at least half a billion years.[31,37]

In mice, mutations of Emx1 can disrupt the formation of the corpus callosum, a structure present only in placental mammals. The homeobox is a short stretch of DNA with a specific sequence of building blocks (nitrogenous bases). A homeobox has about 180 of these blocks. More than one homeobox found in Hox genes have been linked with autism. Reports implicate the homeobox transcription factor EN2 on chromosome 7. $FOXP_2$ transcription factor has been found to affect praxis (vocal and sensory motor integration). Since defects in praxis occur in some autism cases, the suggestion has been made that $FOXP_2$ may underlie one of the symptoms associated with autism spectrum disorders. Fibroblast growth factor-8 (FGF8) is a protein critical in establishing neuronal apportionment in the developing brain. FGF8 controls Emx2, a gene that besides controlling cerebral cortex formation, also keeps in balance the growth of the hippocampus and the frontal cortex. Disabling Emx2 Hox gene increases frontal cortex size at hippocampus expense. Thus, both apportionment of new neural groups and evolution of the brain are regulated by genes which appeared at least half a billion years ago.

Following Bridges, Ohno[38] suggested that sets of banding proteins in the chromosome map of the fruit fly were probably sets of duplicate genes. He proposed that after duplication, identical genes could mutate separately, and then diversify their effects. While original existing genes keep structures and functions ongoing, their duplicates can engage in the development of new functions.[30] Co-option frequently follows gene duplication.[31] One of the functions of DNA molecules is to reproduce themselves. Hence, it is not surprising that gene duplication has been proposed as a major mechanism in the evolutionary process. Activities of replicate genes relate to increased diversity and expansion of behavioral repertoire of organisms, rather than with refinement of existing behaviors.[39,40]

In the early days of molecular biology it was thought that the origin of each protein could be traced back to a distinctive gene through a single chain of causation and construction, note Crick's central dogma of biology "DNA makes RNA, RNA makes protein". But genes alone do not make proteins directly. Genes replicate and serve as templates for the formation of RNA which, through a complex chain of biochemical processes, eventually assemble the vast array of proteins needed to construct the human body. Correct three-dimensional folding is critical for protein function. However, the signals for correct stereofolding are supplied by cell cofactors or chaperons, not genes. There are different DNA codes that correspond to the same amino acid, a phenomenon referred to as redundancy.

The vast majority of nucleotides in the genomes of complex organisms "appear" not to "make" anything of importance to bodies ("junk DNA"). This DNA presence is harmless to the organism, should it become harmful it would probably be disposed of as "garbage DNA" (see Gould).[28] Only about 1% of the human genome accounts for the (ca) 25,000 genes it contains. Genes are built of pieces of coding regions (exons) interspersed with other sequences of nucleotides that do not translate to RNA (introns). In assembling a gene, introns are left out and exons joined together to make an RNA sequence. If a gene consists of five exons, e.g., it can be envisioned that it can be involved in the synthesis of more than one protein, a process referred to as gene promiscuity. This can occur by combining the exons in different orders, or by leaving some of them out. The proteins that result can either function as enzymes in intermediary metabolism, or as transcription factors, proteins that bind to the DNA regulatory regions of genes, called promoters, where they switch on or off the expression of genes. These data explain how the (ca) 25,000 – 30,000 genes of the human genome can together codify the (ca) 150,000 messages originally expected for the human genome. The same generating set could also make many more messages that do not exist. Why these particular 150,000 and not others? For Gould,[28] the explanation lies principally on the contingency of molecular processes, for Conway Morris,[29] however, convergence during evolution provides a better explanation.

Results from evo-devo investigations overturned some tenets of classical genetics, e.g., the concept that each and every gene has acquired a fixed selective power specifically placing it in a hierarchy of commands during development is untenable today. Current data indicate that gene–gene interaction, along with the flexibility and promiscuity of such interactions, builds unique individuals. Moreover, evo-devo investigations revealed that animals, from flies to humans, share a large cohort of Hox genes, and furthermore, besides apportionment of new neural groups, the result of natural selection and adaptation interactions, brains also evolve by an expansion of the domain of some Hox genes independently of traditional adaptive interpretations.[37,41]

Transcription factors from master genes can affect activities of many target/slave genes. These include cells that form boundaries between regions that cause groups of cells to clump together. Consequently, selector genes exert effects which can extend in all directions.

Malfunctions of Hox genes can be one of the underlying factors of physio and psychological symptoms in psychiatric disorders. Due to their ubiquity and multiple actions in the brain, the effects of Hox genes malfunctions may be expressed in many psychiatric disorders, thus leading to the overlap of symptoms in those disorders. Plasma levels of neurotransmitters, hormones and NAS also overlap in different psychiatric disorders. These overlaps are so high that, together with the overlap of symptoms of psychiatric disorders likely produced by Hox genes malfunctions, they preclude categorical classification in psychiatry.

19.5 Normal and Pathological States

The ontological assumption underlying scientific medicine is the postulate of lawfulness, what Claude Bernard called "le principe du déterminisme". This principle maintains that abnormal states are just as lawful as normal states; "there are no nomological anomalies caused by supernatural agencies – but only departures from healthy states". This view embraces the normal and the pathological as a continuum and conceptualizes psychopathological states only as deviant from normal ones.

The view of disease as being immanent in the organization of healthy organisms is particularly entrenched in psychiatric and neurological disorders. H. Jackson[42] believed that "disease does not create, it sets free". More recently, Ey[43] postulated that "mental disease is implied in the organization of the psychic being, ambiguous symptoms are not caused, they are permitted". These formulations implicitly carry an erroneous interpretation of the evolution of nervous systems as solely "adding to the higher levels of integration and keeping down the lower levels",[44] hence the notion of disease as a release or regression to an ancestral state, a phenomenon for which Spencer coined the term "dissolution". The history of disease as a disorder of balance can be traced at least to Plato. The first medical writer who interpreted health as a state of equilibrium and disease as a disturbed harmony was Alcmacon.[44]

Reports of new putative neurotransmitters and the possibility that their deficits or excesses (imbalance) could be related to nervous disorders indicate that the doctrine of humours is very much alive. Although of heuristic value in some instances, models derived exclusively from the doctrine of harmony of humours are not sufficient to account for the variety of nervous disorders in clinical practice. An example from studies in patients suffering from specific phobias will illustrate the point.

The contingent negative variation (CNV) is a negative wave developing in the brain contingent to the association of two stimuli, the warning and the imperative. It appeared then, that a study of CNV in patients suffering from specific phobias while anticipating the phobogenic stimulus could serve as a valid behavioral paradigm to study phobic disorders. Anticipation is an important factor for both the physiological and the behavioral responses under study. CNV in patients suffering from specific phobias were abnormal, i.e., they showed significantly larger amplitude than reference controls. The resolution of the negative wave, taking more than 2 s, was also abnormal.[45] However, Knott et al.[46] had previously reported that in a group of normal subjects in which a similar paradigm was used (anticipation of a fearful stimulus, an electric shock to the finger), the CNV showed reduced rather than higher amplitude and fast rather than delayed resolution of the negativity, less than 2 s. An apparent inconsistency of results emerged from these two series of experiments. In one case, the anticipation of a phobogenic stimulus, the patient's brain generated abnormally large CNV, in the other case, anticipation of a fearful stimulus by normal subjects, produced CNV of significantly lower amplitude and normal duration. What appears to be a serious discrepancy in results can be reconciled if we consider discontinuity between normal and abnormal behavioral phenomena.

Clinical evaluation of patients suffering from specific phobias reveals that rather than an exacerbated fear of objects or situations, patients manifest a terrifying dread, different from an excessive fear. It is likely that in subjects anticipating a fearful stimulus and in phobic patients anticipating a phobogenic object or situation, we are dealing with different phenomena, the significant differences between their CNV suggest that their underlying neural processes may be different. The proposal that phobia is qualitatively different from excess fear is in keeping with two broad avenues of approach in psychophysiology. In the first, attempts are made to induce the abnormal phenomena in normal subjects by using various stimulation procedures. In the second, differences are sought between patients complaining of a disease and normal subjects. The first method is in the tradition of stimulus response psychology. The second has its antecedents in clinical medicine, where different laws may rule pathological processes. Different does not mean lawless. Pathological states may be ruled by different laws than those ruling healthy states.

Do hormones and neurotransmitters have the same effects in normal and in pathological organ structures? $GABA_A$ receptors may respond with excitation to GABA after seizures or injury. In the developing brain, GABA also elicits excitatory responses.[47] The effects of allopregnanolone on neuronal activity may shift from inhibitory to excitatory under similar circumstances. Some genes of the CNS involved in neuronal functions become sensitive to progesterone only after injury.[48]

Available results suggest that normal and injured or dysfunctional nervous systems may use different signaling mechanisms. Whether this is due to changes in receptor expression, their interaction with other co-operative signaling pathways, or to nuclear coregulators, or if steroids may use different signaling in normal and damaged neural tissue, is not yet known.[49,50] After being intracellularly oxidized THP and THDOC to DHP and DHDOC respectively bind to cytosolic receptors from where they influence the composition of $GABA_A$ receptor complex.[51] These results raise the issue of the validity of studying mechanisms of action of antidepressants in normal brains alone. Can the results be extended to brains of people suffering from depression syndromes, i.e., with affected brains?

19.6 Changing Views of Cortical Organization

Classical notions that the cerebral cortex is arranged into sensory, association and motor regions started to collapse in the 1960s.[52–54] Cortical zones that did not respond to peripheral sensory stimuli, or in which it was not possible at the time to demonstrate degenerated axons after thalamic lesions, silent cortices, were named associative cortices. Early in the 20th century, these regions were thought to receive afferent fibers exclusively from primary cortices. It was proposed that in these cortical regions association of sensations purportedly originating in primary cortices, gives rise to perceptions. But the introduction of new anatomical and physiological methods disclosed that association cortices do respond to stimuli from the periphery and show degenerated axons after thalamic lesions.[55] Subcortical fibers projecting to so-called association cortices signal specific attributes, different from those conveyed to primary receiving cortices, from the exterior and interior worlds of an individual[46,47,56]

All sensory systems are represented by various maps in the cerebral cortex. Presumably, each map is created by the duplication of existing genes. In each cortical map, different attributes of the external and internal world of an individual are processed. In the visual sensory system, e.g., form, color and velocity of an object are processed in different regions; cortical configurations of this type are referred to as mosaic.[57] The recognition that processing of sensory signals is carried out in parallel in primary and in so-called association cortices marked the end of the idea that sensory signals are processed in a sequential fashion.[57,58]

An important fact emerging from these studies is that within one transmission fiber group, we can see the "separation or factorization of different aspects of input into distinct processing channels, a process that appears to be a generalized one in the evolution of brains and a necessary condition to adapt to a varying environment".[59] Thus, the well-established principle of parallel processing of sensory signals in nervous systems results from the characteristics of the evolution of these systems.

The fact that sensory signals are processed in parallel also led to the demise of the notion that nervous systems are vertically and hierarchically organized, articulated early by H. Jackson and his followers. Inspired by Spencer's philosophical writings, Jackson[42] believed that nervous system evolution meant "adding on" higher levels of integration which, in turn, keeps lower levels under control. Jackson conceived nervous diseases as dissolution, a term coined by Spencer to indicate the reverse of evolution. His guiding plan was "The doctrine of evolution is not simply synonymous with Darwinism. ... For a long time it has seemed to me that our researches on the maladies of nervous system will be facilitated if we consider them as regressions – that is to say, as dissolutions".[42] His dictum: "destructive lesions never cause positive effects but induce a negative condition which permits positive symptoms to appear",[42] was a consequence of his principles of hierarchy and dissolution of function and structure. In Jackson's own words, "Disease is said to cause the symptoms of insanity. I submit that disease only produces negative mental symptoms answering to the dissolution, and that all elaborate positive mental symptoms (illusions, hallucinations, delusions, and extravagant conduct) are the outcome of activity of nervous elements untouched by any pathological process: but they arise during activity on the lower level of evolution remaining".[42] These notions, still part of many contemporary writings in psychiatry, are marred by what Armstrong[60] refers to as "the fatal flaw of elusive hierarchical views. [Its] unidirectional view, which does not take into account the multidimensional forms of life that have evolved". Thus clinical concepts like positive and negative symptoms, although descriptively useful, do not have heuristic value.

The question then emerges: how are different sensory components gathered into one global image to produce perceptual unity? The "homunculus fallacy" surfaces when a fundamental aspect of neural processing related to cognition is hypothesized to occur in unspecified, elusive, "higher centers" of the brain. These regions take on the aspect of "little men", or so called homunculus, a process that leads to infinite regression. Instead of the hidden homunculus hypothesis, the problem of perceptual unity has been approached by identifying the mechanisms by which different sensory components are gathered into a global image. This, "the binding problem", is a central problem in neuroscience. Proposed solutions revolve around the hypothesis that binding is the result of temporal conjunction, i.e., synchronicity of oscillatory activity of different brain regions, a mechanism by which different sensory components purportedly gather into one global image. As NAS play important roles in regulation of neural excitability, e.g., high doses of DHEA(S) can induce seizures while Ring A reduced metabolites of P and DOC, THP and THDOC have hypnotic properties, it is likely that NAS are involved in a broad spectrum of activities such as synchronization and desynchronization in the sleep–wake cycle, and in attention.

19.7 NAS and Salient Symptoms of Depression Syndromes

19.7.1 Memory

One of the most salient symptoms of depression syndromes is disturbance of memory systems. In fact, Lishman[61] considers that these disturbances play a primary role in the pathogenesis and maintenance of the syndromes. Consistent with this view, Weingartner and Silberman[62] showed that depressed patients use weak or incomplete encoding strategies to organize and transform events to be remembered. In patients suffering from depressive syndromes memory also undergoes qualitative changes, it turns towards negative, self-deprecating aspects of their lives. Moreover, with increased levels of depression, accessibility to negative memories increases, all these factors favor the presence of feelings of guilt so frequent in depression. Patients with depression remember unpleasant words with higher frequency than control (non-depressed) subjects, and have shorter latency in identifying negatively connotated words, they also show response biases as observed in a signal detection analysis study in which unpleasant material was handled in a preferential way to neutral or pleasant material.[63]

The fact that congruence between affective state and tone of material recalled interact during retrieval is well documented.[64,65] The internal milieu, of which steroids are a component, is one of the determinants of the complex set of autonomic, biochemical and emotional-cognitive factors that establishes mood at the moment of a particular experience. The reinstatement or persistence of this environment may then help to recreate or maintain one type of powerful source of context effect in memory: mood.[66,67] Memories with different connotations, i.e., positive or negative, may correspond with different biochemical mediators, different neuronal circuits modulated by different internal milieus or networks behaving in different modes.[68–70] These proposals, already made by von Monakow,[71] have received some experimental support.[72] For example, depressed and non-depressed individuals show different patterns of electrical brain activity during encoding of positive and negative information.[73]

Dysfunctions in associative type memories are also present in depression syndromes. A widely accepted electrophysiological model for associative-type memories is Long-Term Potentiation (LTP),[68] a prolonged enhancement of synaptic transmission first described in the rabbit's hippocampus (Fig. 19.1).[68] The hippocampus is essential for the establishment of declarative memories and for the temporary storage of long-term memories.[74]

Other reasons underlie the importance of LTP expression in the hippocampus. First, LTP can be elicited from entorhinal cortex, via the perforant path (PP) to the Dentate Gyrus (DG), from the mossy fibers, and from the Shaffer collaterals. By traveling through the hippocampal circuit, LTP allows this structure to modulate sensory signals that were first processed in cortical regions[68] (Fig. 19.1). Second, LTP can be rapidly induced, a single electrical presynaptic tetanus can double the strength of a synaptic connection, i.e., the amplitude of the response to the same stimulus can be twice the baseline.[75] A third reason is that once induced, LTP is stable for one or more hours, even days. The fourth one is that the hippocampus is the neural site from where LTP can be more easily elicited.[68] The hippocampus is also an important mediator for the behavioral effects of cortico-adrenal hormones and NAS, which also modulate amine receptor levels in the region.[76–79]

The hippocampus is also involved in attention processes. In turn, there is a great deal of interaction between memory and attention.[66] Attention determines the content of memory, and retrieved memories serve as the basis of expectations and direct attention. Thus, a change in memory would be expected to result in a change in attention and expectations, and vice versa. NAS modulate the LTP elicited in the DG by stimulation of the PP. DHEA(S) enhances LTP in a dose-related way.[80] The enhancement is probably caused by the negative modulatory effects that DHEA(S) exerts at $GABA_A$ receptor sites, GABA itself blocks the development of LTP.[81,82]

Harlan Meyer and Gruol[83] showed that DHEA(S) increased excitability in hippocampal slices. They explained the effect based on the facts that (a), DHEA(S) inhibits the fast inhibitory postsynaptic potential (IPSP) component of the evoked synaptic response and (b), DHEA(S) produces a direct enhancement of the excitatory postsynaptic potentials (EPSP). When simultaneously injected with corticosterone (CS), a depressogenic type hormone, DHEA(S) can counteract the decremental effects of high doses of CS on hippocampal LTP. These results suggested the potential use of DHEA(S) in the treatment of depression syndromes, in particular those presenting with hypercortisolemias.[76] The validity of this hypothesis was upheld in

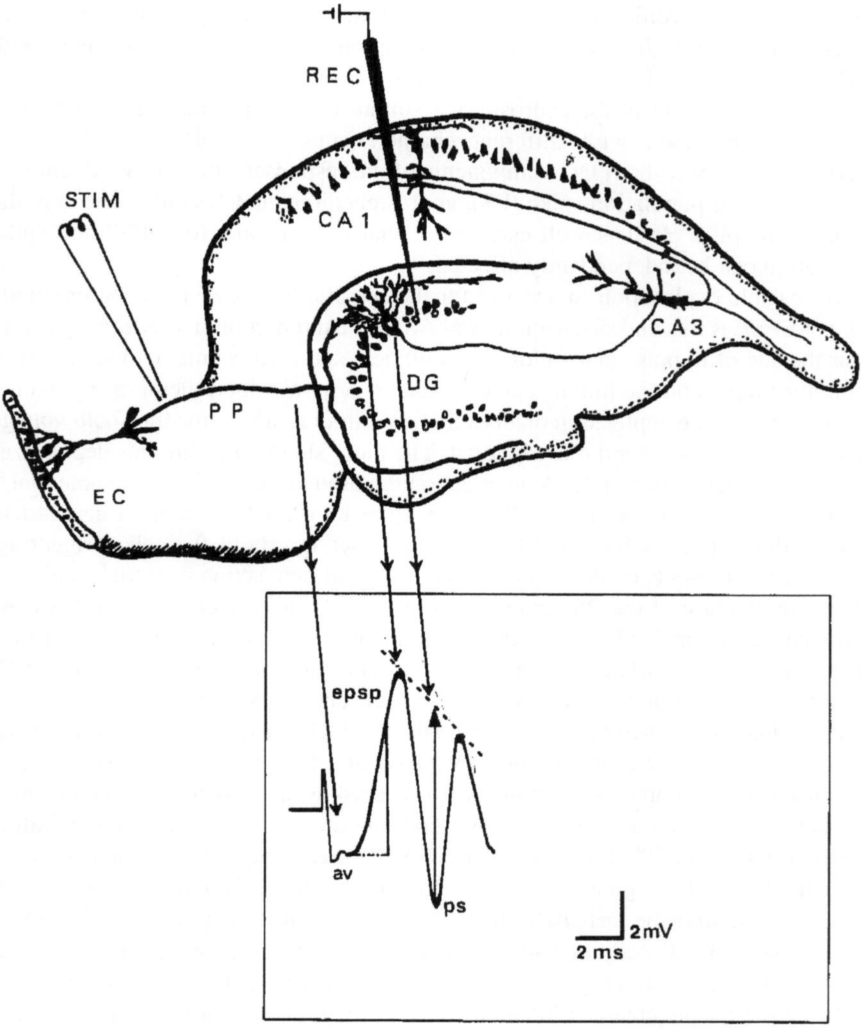

Fig. 19.1 Schematic diagram of mammalian hippocampus showing the stimulating electrode (STIM) in the perforant path (PP) and the recording electrode (REC) in the granular cells of the dentate gyrus (DG). EC, enthorhinal cortex. Inset, evoked potential generated in DG by stimulation of PP is illustrated, showing the afferent volley (av), excitatory postsynaptic potential (epsp) and population spike (ps) and with their respective field generators

the clinic. In open label as well as in double blind randomized placebo-controlled studies, oral administration of DHEA(S) decreased the intensity of the depression symptoms in patients with major forms of the syndrome, as well as with dysthymia.[84–86] The symptoms that improved most significantly with DHEA treatment were anhedonia, loss of energy, sadness, worry, emotional numbness, and lack of motivation. No specific effects on cognitive functions or sleep disturbances were noted. Thus, although it improves some symptoms of depression, treatment with DHEA(S) does not affect the entire spectrum.

Other androgenic NAS, androsterone sulfate (AS) and androstenedione, also enhance LTP, albeit with different characteristics than DHEA(S).[87,88] While DHEA(S) enhanced both DG components of the responses elicited by PP stimulation, EPSP and population spike (PS), androstenedione and AS enhanced only the population spike.[87,88] It is well established that mechanisms for EPSP and spike generation can be independently affected.[89]

A possible explanation to account for the specific increase in the PS amplitude but not EPSP, is that androstenedione and AS may exert part of their effects by acting near the site of impulse generation, axon hillock, and then counteract the effects of inhibitory inputs on the initial segment. The velocity at which membrane potential changes during the application of an excitatory stimulus affects the threshold voltage at which an action potential is triggered. The more slowly the stimulus depolarizes the membrane, the greater the depolarization required to initiate an action potential.[90] Even a minor increase in the EPSP slope may be translated into a faster depolarization of the initial segment and hence, in a lower threshold for spike triggering. Together these events could lead to a recruitment of new action potentials and, as a result, an increase of the PS amplitude with the NAS androstenedione and AS. As androstenedione and AS also modulate glutamate receptor and voltage gated Ca^{2+} channels, and Ca^{2+} influx is essential for LTP induction, the NAS could affect LTP generation through their effects on Ca^{2+} channels and glutamate receptors.[68,91,92]

Experimentally, memory enhancing effects of DHEA(S) in aging mice in an associative task were reported early by Flood and Roberts.[93] DHEA(S) was later shown to decrease impairments in memory produced by dizocilpine, a non-competitive NMDA receptor antagonist, and by scopolamine, a muscarinic acetylcholine receptor antagonist.[94,95] These effects were explained, in part, by the agonist effects of DHEA(S) on the sigma 1 receptor. Treatment with NE-100, a sigma 1 receptor antagonist, inhibits the DHEA(S) induced improvement in performance of memory tasks impaired by dizocilpine and scopolamine. A caveat to the generalization of the effects of DHEA(S) as a cognitive enhancer is that studies in humans did not always replicate the results obtained in animals.[96] Thus, e.g., treatment with DHEA(S) did not improve cognitive function in patients suffering from depression.[84,85]

On the other hand, Berr et al.[97] report positive effects on cognition and well-being with DHEA(S) treatment. At sigma receptor sites DHEA(S) acts as an agonist, while pregnenolone behaves as a sigma inverse agonist and then acts as an antagonist.[98] Enhancement of noradrenaline and serotonin neurotransmission has antidepressant effects, and in vitro data indicate that at sigma ligands, DHEA(S) inhibits noradrenaline presynaptic uptake in brain synaptosomes.[98,99] Furthermore, treatment with DHEA(S) increases the number of NMDA receptors and potentiates NMDA-evoked noradrenaline release via sigma receptors.[98,100]

Considering its benzodiazepine-like effects, Lancel et al.[101] expected that the effects of a $GABA_A$ positive allosteric modulator like THP would be detrimental to memory functions. Mayo et al.[102] reported that injection of 2 ng of THP in the nucleus basalis magnocellularis of rats, disrupted performance in a two-trial recognition task when injected before an acquisition trial. No effects were observed if injections were effected after acquisition trials, suggesting that THP interferes with learning processes during the acquisition phase. In contrast, pregnenolone sulfate infused into the ventricles enhanced memory performance in mice.[103] This NAS is believed to act, in part, by increasing hippocampal acetylcholine release, and as a positive modulator at NMDA receptor sites.[104]

In 2005, Spalletta et al.[105] reported that elevated THP plasma levels can also be associated with increased aggressiveness and hostility in a subpopulation of schizophrenic patients and can also induce aggressive behavior. Previously, Fish et al.[106] noticed a higher level of aggressiveness in mice under the effects of THP. Aggressive behavior can also be seen in women undergoing Hormonal Replacement Treatment (HRT). Andreen et al.,[107] e.g., reported a significant increase in negative mood symptoms in subjects showing high levels of THP serum concentrations during HRT. These observations at the behavioral level are in keeping with observations that, in contrast to THDOC which exerts uniform decremental effects on hippocampal LTP,[108] THP exerts dose-dependent effects.[109]

Experimental studies disclosed dose-dependent effects of THP in different components of the DG LTP, the EPSP and the PS. At a dose of 0.1 mg/kg, THP decreases the amplitude of the EPSP effect probably mediated by the positive allosteric modulation of THP on $GABA_A$ receptors. But at a higher dose of 0.5mg/kg, THP produced a significant enhancement of the population spike amplitude.

Data from Frye and Sturgis[110] revealed that, like DHEA, THP exerts dose-dependent effects. Spatial reference as well as working and long-term memory are enhanced with 6.4 mg/kg, and decreased with 3.2 mg/kg. Activation effects may relate to the fact that treatment with high doses of THP enhances dopamine release and dopaminergic responses to morphine in the rat nucleus accumbens.[111] Furthermore, THP potentiates clozapine and olanzapine actions on dopamine-mediated behaviors.[112]

Dopamine is important for maintaining long-term facilitation and modulating attention in spatial memory.[68] In 2003, Theis et al.,[113] reported that dopamine also regulates the amount of a prion-like protein, cytoplasmatic polyadezylation element binding protein (CPEP), which regulates local protein synthesis in neurons. Prions differ from other proteins in that they can fold into two configurations, one recessive and the other dominant. The shape of the latter is self perpetuating. The dominant shape causes the recessive configuration to become dominant, hence self-perpetuating, features that make the CPEP molecule ideal to behave as material for information retention.[68]

Relations between depression and the HPA axis were established early in the 20th century[114,115] and repeatedly confirmed afterwards.[116–119] Abnormalities of NAS levels in depression syndromes started to be revealed in the 1980s.[120,121] In patients suffering from major depression, THP concentrations in plasma and cerebrospinal fluid were found to be significantly lower than those measured in control subjects.[122] Simultaneously, there was an increase of an antagonist isomer of the NAS, 3β-5α-THP isopregnanolone.[123] In contrast to THP, THDOC was found at significantly higher concentrations.[124] Although with a different time course, concentration of both these NAS returned to normal baseline levels with successful treatment of the syndrome either with tricyclic antidepressants (TCA), or with selective serotonin reuptake inhibitor (SSRI) antidepressants (AD).[125] While THP levels normalize rapidly with successful treatment either with SSRI or TCA, THDOC returns to baseline levels only after (ca) 50 days of fluoxetine treatment.[117,118]

Neither the concentrations of P nor those of allopregnanediol, a metabolite of THP, vary in patients either before or after treatment with SSRI ADs.[126] These results suggest that the decreased THP concentrations in depressed patients are due to changes in the enzymes that are involved in the conversion of DHP into THP. Griffith and Mellon[126] found that the SSRI fluoxetine increased the affinity of 3α-HSOR, the enzyme that catalyses the reduction of the dihydrometabolites of P and DOC, for their substrates. This process, shifting conversion towards the tetrahydrometabolites of P and DOC, results in an increase in THP and THDOC, therefore in an enhanced GABAergic tone, a fact that may also account for the anticonvulsant effects of fluoxetine.[127,128] The SSRI sertraline decreases the conversion of THP to 5α-DHP by decreasing the affinity of 3α-HSOR for THP.

The Porsolt forced swim test has been proposed, and is used in rodents as an experimental model of depression-type behavior.[129] In the test, arrest or immobility in the water pool is interpreted as a state of despair in which rats or mice have learned that escape is impossible and resign themselves to the experimental conditions. Studies in which THP treatment was given to animals subjected to the test revealed that, like various AD medications THP and DHEA(S) extended the time of swimming prior to immobility in treated animals.[130] Serotonergic agents also modulate the AD-like actions of THP.[130] Recently, Uzunova et al.[131] showed that the significant decline of THP content in the prefrontal lobe after bilateral removal of the olfactory bulbs in rats, was reversed by chronic (3 weeks) treatment with three different classes of antidepressants (desipramine, fluoxetine, sertraline and venlafaxine).[132] As the effects of the four chronic antidepressant treatments on THP cerebrocortical levels are observed at a time which coincides with the drugs' ability to reverse the behavioral effects of the bulbectomy syndrome, Uzunova et al.[132] proposed that normalization of THP levels may contribute to the antidepressant-like effects of the drugs in the ofactory-bulbectomized rat by an upregulating of GABAergic tone.

Other preclinical studies are in keeping with these results. Pisu and Serra[133] reviewed animal experimental investigations and showed that acute (20 mg/kg ip) as well as chronic (10 mg/kg ip for 14 days) injections of fluoxetine or paroxetine, both increased THDOC and THP concentrations in the cerebral cortex. Taken together, these data lead to the suggestion that decrease of THP may be causally related with depression.[131,133]

However, not all clinical studies are in keeping with this proposal. Studies of women who became depressed during pregnancy or who developed postpartum depression (PPD), revealed that depression can be manifested even in the presence of low P concentrations, and higher, 14 nmol/l, than what are considered normal serum levels of THP, ~2.4 nmol/l.[134] What is the origin of the high THP and isopregnanolone levels when circulating levels of P are low? An abnormal functioning of 5α-reductase, the rate limiting enzyme for P? Other sources for THP are the adrenal glands and the ovaries.[135,136] A hypothetical source for circulating THP is the NS itself entering the bloodstream.

Nevertheless, the fact that depression recovery with non-pharmacological treatments, repetitive transcranial magnetic stimulation (rTMS), sleep deprivation or electroconvulsive therapy (ECT) occurs without changes in NAS levels indicates that NAS changes are not essential in recovery.[137] However, there is a caveat with the proposition. As adrenocorticotropic hormone (ACTH) can modulate NS biosynthesis as well as affect NAS levels, it is conceivable that after reaching specific CNS targets, e.g., hippocampus, amygdala, by volume transmission, an increase of ACTH with stress could disrupt the balance between excitatory and depressant NS biosynthesis, and affect core symptoms of depression such as anxiety, sleep disturbances, memory and sexual disfunctions.[138–140] The fact that peripheral levels do not always correlate with steroid dependent overt behaviors may relate to changes in NS concentration in specific regions of the CNS, independent of plasma steroid levels.

19.7.2 Sleep Disturbances and Fatigue in Depression

Sleep disturbances present with high incidence in depressive disorders. In a study of 35 patients with depression syndromes, 70% of them complained of sleep disturbances.[128] Pregnenes like DOC, P and their Ring A reduced metabolites decrease nervous systems excitability.[141–144] Lancel et al.[95] reported that THP affects sleeping in a benzodiazepine-like manner. While Mendelson et al.[145] showed that THP can also induce sleep in rats, Muller-Preuss et al.[146] reported the sleep inducing capacity by THDOC. An isomer of THDOC has been commercially used as an anesthetic in humans.[136] In normal subjects, approximately 20% of sleep is spent in stages 3 and 4, the latter also know as the delta stage.[147] In patients with Cushing's disease, as well as in those with depression syndromes, there is an absence of, or marked reduction in, stages 3 and 4.[148,149] Also, shorter rapid eye movement (REM) latency can be observed in both groups of patients.

The hypersomnia present in a subset of patients with depression may relate to high THDOC levels. Low THP levels may be associated with difficulties in sleep conciliation, insomnia and frequent awakening reported by many depressed patients. While not being a biological marker of depression syndromes, THP levels may relate to sleep disturbances in depression.

Should high blood concentration levels of THP and THDOC from the adrenals or of THP from the ovaries be necessary to modulate general CNS excitability, their circulating levels may reach those that can induce general anesthesia, interfering with appropriate behavioral responses. Some water beetles make use of the anesthetic effect of some steroids in their defense. When attacked by fish, these beetles release a milky fluid from their prothoracic glands that may include, among other steroids, DOC, 20-hydroxyprogesterone and 6-dehydroprogesterone. Added to water in concentrations of 2 mg/l, these steroids cause reversible paralysis in goldfish, observations that reveal the long phylogenetic history of the production of natural steroids with anesthetic properties.[135] Moreover, the enzymes for the metabolism of DOC and P are not equally distributed in the NS, a point in support of the notion that every behaviour has an anatomy.

"Endocrine evolution," Medawar[150] claimed, "is not an evolution of hormones, but an evolution of the uses to which they are put … is the evolution of new reactivities and tissue competences, not of chemical formulae." The tetrahydrometabolites of P and DOC exert powerful control of neural excitability. These actions are mainly the results of "new competences" appearing at hormonal targets, in this case nervous systems sites. The new competences are given by the presence of the metabolic chemical machinery to metabolize pregnane steroids to their tetrahydroderivatives in selective sites in the telencephalon. In this way, the effects of THP and THDOC, 30- to 40-fold more powerful than P and DOC, on CNS excitability,[140,142,144] can be restricted to specific regions of the brain, i.e., hippocampus, prefrontal cortex, brain stem, reticular formation, and can be limited in space.

Fatigue is an elusive medical symptom from which patients with depression syndromes frequently complain. Murphy et al.[151] found that isopregnanolone was significantly increased in the plasma of patients with chronic fatigue syndrome (CFS).

19.8 NS, NAS and Biological Amines: Their Role in Neuroprotection and Depression Recovery

Chronic stress is an important factor in the development of depression syndromes, a major contributor being chronic illness.[152] Each stress stimulus induces a particular hormonal profile that can be associated with distinct disorders.[153–155] Selye's (1932) monolithic concept of stress as consisting only of unspecific responses of the hypophyso-adrenal axis was criticized from the outset.[156]

How could the same pattern of hormone responses, Cannon asked, have adaptive utility in response to diverse stimuli which may pose diametrically opposite metabolic demands on the body as do, e.g., heat, cold, physical exhaustion, sickness, etc? Later, Mason[157] raised further criticisms. "I had long been troubled by the apparent incompatibility of the general adaptation syndrome with the concept of homeostasis." Mason challenged the concept of nonspecificity of the stress response and showed the opposite to be the case, i.e., that the responses of organisms to distinct stressors were selective and specific.[151] Further, he revealed that neuroendocrine responses to nocuous stimuli were not restricted to the HPA axis, but involved the entire neuroendocrine system. We know now that NAS also participate in stress responses.[158] Mason[151] reemphasized the importance of psychological factors in responses to stress, already noted by Cannon,[159] and that responses to stressful stimuli are mosaic-like.[63,160]

While the hormonal spectrum may be quite different for different stressors, the frequent overactivity of the HPA axis present in some patients with depression may have special significance. For Mason, “the stress concept should not be regarded primarily as a physiological but rather as a behavioural concept”.[161] The basic conclusion appears to be that the adrenal cortical responses occur in many different laboratory and life situations involving a wide variety of stimuli, because emotional reactions (which elicit adrenal cortical responses) occur commonly in a wide variety of situations in which animals or humans are subjected to “physical stress”. This “non-specific” response is of a primarily behavioural nature. The hormonal profile response elicited by different stress stimuli depends greatly on how individual subjects evaluate or appraise the stimuli. The “coping filter” is “the best understood, and best developed concept in the stress literature”, and “the psychological filtering mechanism of defence is only identified in humans”.[162] Moreover, choices of defences evolve throughout life and “may transmute irritating grains of sand into pearls”.[163]

Cell death, dendritic shrinkage, decreased levels of neurotrophins as well as of brain development growth factor (BDGF), and reduction of neurogenesis in hippocampal granule cells, occur in chronic stress.[164–168] They may be the neuropathological bases underlying, in part, some salient symptoms of depression, i.e., memory and attention deficits, sleep disruption, fatigue. Although with some similarities in their presentation, depressions may have different underlying endocrine physiopathologies and psychopathologies in different individuals. Different underlying physiopathologies follow from the established fact that the hormonal spectrum which makes up the responses to different stress stimuli are selective and specific for the stimulus. Furthermore, as separation of different aspects of input is a generalized process in the evolution of brains, a neurotransmitter for each stimulus can be specified even up to the final common path at neural and neuroendocrine levels.[155] In keeping with the specificity of hormonal responses to stressor agents, the nature of corticotropic releasing hormone (CRH) and other adrenocorticotropic hormone (ACTH) secretagogs, arginine vasopressine (AVP), oxytocin and catecholamines, varies depending on the stressor applied (hypoglycaemia, hypotension, hemorrhage, etc.). During chronic inflammation, e.g., there appears to be a shift from a primary CRH driven to an AVP driven HPA axis.[155]

Animals are endowed with mechanisms that can protect neural tissue and enhance neurongenesis, processes that play significant roles in recovery from depression syndromes. Some NAS and NS also impact these processes which converge, in their final stages, with processes of some biogenic amines. DHEA reduces neural and glial injury in acute models of spinal cord and cerebral ischemia.[169] When administered before or concurrent with the ischemic insult, DHEA decreases hippocampal injury. DHEA and DHEA(S) have also been shown to protect hippocampal neuronal cultures against glutamate-mediated neurotoxicity. Sulfated and unsulfated steroids modulate $GABA_A$ receptor function through distinct sites.[170,171]

The $GABA_A$ antagonist bicuculline abolished the neuroprotective effect of DHEA(S) in a model of reversible spinal cord ischemia, suggesting a role for GABA receptors in neuroprotection.[170] DHEA(S) also protects neurons from the toxic effects of high levels of corticosterone.[171] THP induces axonal regression in the developing hippocampus, suggesting that NAS can promote, guide and refine axonal growth as well as synaptic connections in the developing brain.[172] Allopregnanolone reduces glutamate-induced irreversible changes in intracellular Ca^{2+} concentrations in models of hippocampal neurotoxicity.[173] Progesterone, given to rats either chronically or acutely, decreases the infarct size in models of ischemia in both male and female rats, and also reduces cerebral edema following traumatic brain injury.[174] Allopregnanolone stimulates the proliferation of cerebral and hippocampal neuroprogenitor cells via $GABA_A$ receptor-activated, voltage-gated *L*-type calcium channels, and promotes neurogenesis.[174] Neuroactive steroids can also protect neural cells from apoptosis. Charalampopoulos et al.[175] reported that DHEA(S) and THP protect sympatho-adrenal medulla cells against apoptosis via anti-apoptotic Bcl-2 proteins. Decline of DHEA(S) during aging or stress makes the adrenal medulla more susceptible to proapoptotic challenges. Bcl-2 is physiologically involved in preventing cell death by decreasing the generation of reactive oxygen species (ROS) within mitochondria.[176]

Different stressors activate specific neuroendocrine and neurotransmitter circuits. On a short-term basis, these responses can benefit the organism. When chronically maintained, stressful stimuli can lead to pathological states, what Selye referred to as diseases of adaptation.[153] These states can be manifested in a qualitative shift from adrenal androgens and mineralocorticoids toward elevated secretion of GC, or in conditions of ACTH hypersecretion, such as pituitary-dependent Cushing's syndrome, where increased secretion of C19 steroids occurs due to an increase of 3β-hydroxy steroid dehydrogenase activity. In the case of depression, different types of syndromes are associated with different behaviours of the HPA axis: atypical depression with HPA axis hyporesponsiveness, melancholic depression with HPA axis hyperresponsiveness, and other types of depression disorders with disregulation of CRH. As organisms react to persistent but different life events with specific neuroendocrine responses, the totality of processes leading to disorders are particular for each individual, his/her genetic makeup and life experience. Presenting symptoms do not reveal the life history of the patient(s). A rational diagnosis and therapeutic approach should consider disorders as processes, not as final stages.[177]

Another factor that must be considered when extrapolating results from ancestral to recently derived species is that while animals and humans share many features, they do not share all of them. The domain with the least commonality involves intentions, values and behaviors associated with shame, guilt and pride. The animal evidence does little to inform these human states. These facts significantly restrict the use of animal models to study psychiatric disorders.[4] Environmental context must always be considered when evaluating plasma or CSF levels of steroids and NAS. For example, similar increases in plasma levels of cortisol or CS, its equivalent in rodents, can be observed under different, even opposite, behavioural circumstances, a pleasant one, watching a funny film or, on the contrary, a threatening, fearful stimulus.[178,179] Already in 1958, Nobelist, Alberto Houssay,[180] wrote: "in the whole organism one hormone never works alone. In every case, the action of one hormone is related to the balance of hormones present. If we study any function we find that it does not depend on one hormone, but on a balance between hormones acting together or in a consecutive way."

Glial activities are influenced by neurons; the existence of neurotransmitter-mediated signaling pathways in glial cells is well established.[181] Among neurotransmitters released in responses to stressful stimuli, isoproterenol and serotonin play important roles. Antidepressants (AD) initiate a cascade of processes that include differentiation of glial cells. They induce elevation of glial fibrillary acidic protein (GFAP) mRNA levels. The processes are dependent on the activation of $GABA_A$ receptors via Ring A reduced metabolites of pregnene steroids.[182] The expression of the GFAP gene is considered an index of glial cell differentiation. Progesterone (P), and its 5α-reduced metabolite dehydro P (DHP), increase GFAP gene mRNA levels in type 1 rat astrocytes. Since the effects are faster and more pronounced with DHP than with P, it is believed that DHP is the acting steroid.

Isoproterenol and serotonin elevate GFAP mRNA and protein levels in glial cells through the elevation of intracellular cyclic AMP levels.[183,184] The cyclic AMP-mediated intracellular signaling system, activated by indolamines-α and β-adrenergic stimulation, is involved in the regulation of glial cell metabolism and gene expression.[185] Adrenergic and serotonergic stimulation of glioma cells also enhances the expression of 5α-reductase genes in these cells.[177] Adrenergic activation increases 5α-reductase expression through the activation of cyclic AMP protein kinase A. Serotonin, in turn, induces 5α-reductase gene activation by enhancing the expression of transcription factor Egr-1.[182]

The increase of GFAB mRNA levels induced by isoproterenol and serotonin is suppressed by pretreatment of the cells with Finasteride, a 5α-reductase inhibitor which also abolishes the stimulatory effects of P and DHP on GFAP gene expression. Adrenergic and serotonergic increase of 5α-reductase gene expression in glial cells occurs prior to the activation of GFAP gene expression.[183,184] Isoproterenol and serotonin induced elevation of GFAP mRNA levels is also inhibited by pretreatment of the cells with bicuculline, a selective $GABA_A$ receptor antagonist.[185] Thus, the increase of GFAP mRNA levels produced by indol and catecholamines depends on DHP, a NAS that positively modulates GABA receptors directly, or likely after its conversion to THP.[186]

Regulation of astrocyte morphology and differentiation is partly mediated by GABA.[187] Adrenergic and serotonergic activation of GFAP gene expression depends on the effects of 5α-reduced steroid metabolites on the $GABA_A$ receptors of glioma cells.[183] Data indicate then that biological amines and NAS enhance glial growth differentiation, which enhances supporting neurogenesis recovery processes in depression.[186]

In the clinic, lower levels of GABA in CSF have been found in depressed patients.[188,189] Intravenous injections of the SSRI citalopram in control subjects, as well as following electroconvulsive seizure (ECS) treatment in depressed patients, increase GABA concentrations in the visual cortex, up to 50%, measured by magnetic resonance spectroscopy (MRS).[190,191] Long-term (more than 3 weeks) treatment with different classes of drugs used as antidepressants (AD), e.g., selective serotonin reuptake inhibitors (SSRI), selective serotonin noradrenaline reuptake inhibitors

(SNRI), tricyclic antidepressants (TCA), bupropion, an inhibitor of dopamine (DA) reuptake, and tianeptine, which enhances serotonin reuptake, all increase cell proliferation and neurogenesis in the hippocampus.[166–168] SSRI also have high affinities for σ1 receptors, while TCA have less. Activation of σ1 receptors can also enhance neurogenesis.[192] Actually, one way by which treatment with DHEA(S) improves some symptoms of depression is by acting as a sigma receptor agonist. Mirtazapine is an AD which acts as an antagonist of α2 $5HT_2$, $5HT_3$ and histamine H_1 receptors. Treatment with Mirtazapine for 5 weeks significantly increases 3α-5α-THP, 5α-5β-THP 5α-DHP and 5β-DHP concentrations.[137] ECS also increase hippocampal neurogenesis in the adult rat.[193] Simultaneously, there is an increase of endothelial cell proliferation which may act to support the increased neural proliferation and neuronal activity or vice versa, possibly leading to structural changes within the hippocampus of importance for the antidepressant effects of ECS.[194]

Chen et al.[195] reported that administration of DHEA(S) for 3 days after transient, incomplete forebrain ischemia prevented the severe impairments of LTP produced by the ischemia. Phosphorylation of NMDA receptor subunit 2B (NR2B) was significantly decreased after ischemia. Administration of DHEA(S) lessened the functional deficits as well as the reduction of tyrosine phosphorylation of NR2B produced by the ischemia. While co-administration of NE100, a specific σ1 receptor antagonist, completely abolished the protective effects of DHEA(S) on ischemia, simultaneous treatment with P attenuated DHEA(S) protection. For these reasons we are investigating the effects of THP, known to be superior to P as a neuroprotector, in conjunction with DHEA(S) to test if Ring A reduced metabolites of P, not only do not diminish the benefits of DHEA(S), but can also potentiate them.

When a neuron receives a signal related to a learning event, the intracellular concentration of cAMP increases. cAMP binds to and is phosphorylated by PKA. As the concentration of cAMP increases, PKA activates catalytic components of the kinase. The units move to the nucleus where they activate a molecule, cAMP-response binding protein (CREB). In turn, activated CREB binds to specific sites of DNA, where they switch on a number of other genes. CREB is a transcription factor originally identified as a contributor to the long term adaptations that underlie learning and memory.[68] Chronic antidepressant treatment increases CREB in the hippocampus and cerebral cortex in a time consistent with the time course necessary for the therapeutic action of AD.

In brief, increase of neurogenesis is a common effect of ECS, as well as of all AD presently in use. Experimental tests employed to look at the effects of AD in what are considered behaviors like depression include the Porsolt forced swim test, novelty suppressed feeding (also used in studies on anxiety), reduction of animal avoidance of new objects and increase exploration. However, the question remains: "Do any of these animal models have anything to do with human clinical depression?"

Be that as it may, available AD such as SSRI, TCA, SNRI, bupropion, an inhibitor of dopamine reuptake, tianeptine which enhances serotonin reuptake, all increase neurogenesis and activation of σ1 receptors. Moreover, Santarelli et al.[168] showed that AD require neurogenesis to modify behaviors. What we do not know yet, is if neurogenesis alone is sufficient for depression recovery. The time required to induce neurogenesis is a 3–4-week period, this coincides with the time necessary to see improvements in depression syndromes treated with AD in the clinic. Effects of AD on the dynamics of neurotransmitters at the synaptic cleft start after treatment, hence, they do not correlate in time with the beneficial effects of AD.

At least some of the delayed effects of AD acting on indol or catecholamines in glioma cells are mediated by the effects of 5α-reduced steroid metabolites on its $GABA_A$ receptors. Isoproterenol and serotonin-induced elevation of GFAP mRNA levels is inhibited by pre-treatment of the glioma cells with bicuculline, a selective $GABA_A$ receptor antagonist. Investigations on neurogenesis are not restricted to naturally existing molecules. In vivo studies showed that a compound, BC1540, which does not act upon serotonin, norepinephrine or dopamine, traditional pathways for AD, enhances neurogenesis, and does not have unwanted effects in humans.[196]

19.9 Potential Effects of NAS and Neurotransmitters in Changing the Mode of Neural Circuits

Activities of neural circuits depend on the influences that its components, membranes, synaptic and network, are being subjected to at the time of observation. Modulatory inputs: physical, e.g., electrical fields, chemical, e.g., hormones, environmental factors, i.e., social milieu, can throw the performance of an anatomical network into any one of several modes depending upon the particular combination of ongoing processes (polymorphic networks). Mode refers to the manner in which a network processes signals or generates an output pattern. Properties of neural microcircuit operations are emergent, i.e., they cannot be predicted from what is known of the action of single neurons.[54] Each mode represents the functional architecture of the network that gives rise to a particular function or task.

Afferent input to some neural networks may serve not only to activate, but also to configure them into one of several functional circuits. Thus, e.g., in the sea slug Tritonia the same neural network may mediate escape withdrawal or escape swimming, depending upon the pattern of activation that it receives. Such neural networks are referred to as polymorphic (Fig. 19.2).[197] An added factor to the array of functional possibilities of a specific network is revealed by studies that indicate that a given neuron, or even a given axon, may convey information at the same time in two or more distinct codes for different recipient cells. This high level of plasticity is manifested by all nervous systems, in particular human.[198] These neural network characteristics may account in part, for the overlap of brain regions showing functional abnormalities seen with MRI in patients with schizophrenia and with manic depression, as well as for the change of symptoms in time in these populations. The fact that qualitative neurophysiological abnormalities are not always associated with a unique symptom complex may also relate to the fact that neural circuits can function in different modes at different times.

NAS can change the mode or alter the operation of neural circuits, these changes may explain the presence of symptoms such as attention and memory deficits in depression. This view is supported by studies showing that steroids and NAS, as well

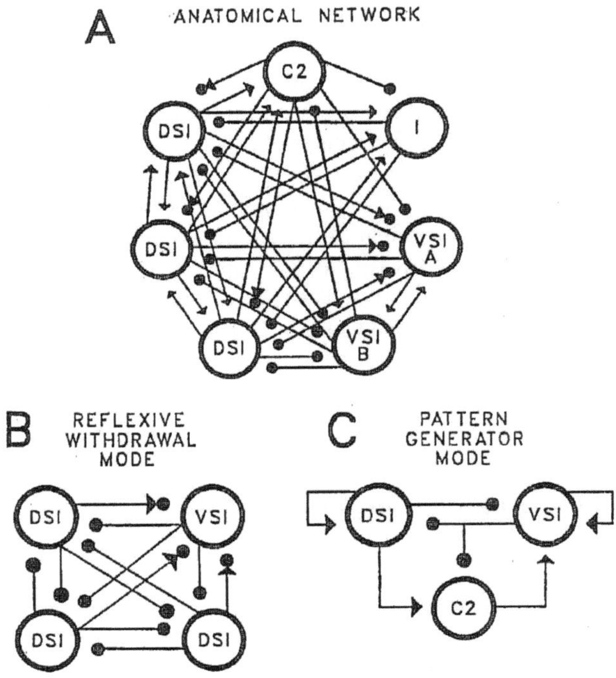

Fig. 19.2 Network diagram showing the monosynaptic connectivity between interneurons of the *Tritonia* escape swim system. **B** Network configuration reflecting the functional connectivity when **C2** is silent. In this configuration (mode), the network contributes to reflexive withdrawals. **C** Network configuration when **C2** is active. In this mode the network generates an alternating burst pattern between **DSI** and **VSI** which in turn activates motoneurons for each flexion movement. Pathways with more than one symbol indicate multicomponent synapses (Reprinted from the *Annual Review of Neuroscience*, Vol 12, 1989 by Annual Reviews www.AnnualReviews.org. With permission)

as neuropeptides, modulate LTP and long-term depression (LTD) in the hippocampal circuit.[4,76–78,120] Mode changes in hippocampal nets may account for impaired attention and memory functions in patients with depression syndromes.[70] Steroids and NAS are significantly affected in these disorders.[121,123,187] Contemporary views regard psychopathological symptoms such as fear and anxiety, as dysfunctions of specialized neural circuits rather than as the result of lesions in specific regions of the brain.[68]

19.10 NAS in Relation to Attention and Concentration Deficits as Symptoms of Depression

Patients with depression syndromes frequently report having problems concentrating. Neural processes associated with attention and memory have a great deal of interaction. As noted earlier, the hippocampal formation, a fundamental CNS region associated with these processes, is a main target for NAS, as well as for biosynthesis of NS. Disturbances and/or imbalances in these steroid compounds may then be expected to occur in patients with attention and concentration problems. Thus, plasma levels of DHEA(S) are inversely correlated with severity of symptoms in attention deficit disorder (ADD).[199]

Evidence is available supporting the thesis that attention is a lateralized function of the brain.[200] Geschwind and Galaburda[201] argued that the right hemisphere is dominant for neural mechanisms of attention. Data has also been presented indicating that right hemisphere disorders, e.g., seizures, lead more frequently to the development of affective disorders.[202] There is a group of right hemisphere functions that are closely linked precisely because they relate to several functions, monitoring the environment and making decisions as to shifting on the importance of the monitored stimuli for the survival of the self or species. The importance of the right hemisphere for attention and emotion, and for the configuration of space all appear to be possibly related to these functions.[192]

Dennemberg's work[203] on the effects of right hemisphere lesions in rats suggests that this form of cerebral lateralization of function appeared early in the animal kingdom. In rodents, NAS modulate behaviour requiring the use of spatial references, one can ask if there are brain asymmetries in the distribution and concentration of the enzymes that metabolize pregnanes that can be associated with this lateralization of functions.[104]

19.11 Commentary

Just as the conduction velocity of a nerve to transmit action potentials is constrained by physico-chemical and biological factors, among others the thickness of its myelin sheet and the diameter of its axons cross section, so is the brain constrained in its functions and dysfunctions, as evidenced by data from fundamental neuroscience. Characteristics of brain structure, the set of all data from evolutionary developmental biology and the evolutionary history of the human brain as reviewed previously, preclude the emergence of psychiatric disorders that can be categorically classified, i.e., endowed with essential features. Seen from this perspective, the lack of biological markers for psychiatric disorders is not surprising.

In recent years research efforts focused on measuring plasma levels of NAS as possible biological markers for psychiatry disorders. NAS changes and psychological symptoms are greatly overlapping in a variety of psychiatric disorders, an important factor for the lack of nosological specificity.[4,27,133,187]

However, nosological nonspecificity of a given biological variable does not mean it is not specific at all. It may be that the variable, e.g., NAS, relates not to disorders as defined in the DSM IV or ICD 10, but to specific components of the disorder, to particular psychological dysfunctions.[204] Strous et al.[205] (2003) presented data in keeping with this notion, they reported that treatment augmentation with DHEA(S) markedly improved negative, depressive and anxiety symptoms in schizophrenic patients, the same effects DHEA(S) has in depressed patients.[80]

It is proposed that a range of NAS plasma levels could contribute to the identification of some emotional and cognitive states. Neuroendocrine findings may be helpful in the validation of specific conditions, avoiding the usually lengthy wait for validation based exclusively on clinical data. For example, fear is a universal feeling present in all mammals that refers to real danger, while anxiety, which can be aroused by means of metaphorical associations, is uniquely human.[206] Do plasma levels of NAS show different profiles in these clinically close but distinct psychopathological states? If so, these profiles will be of help in differentiating between them. Moreover, precise identification of symptoms is a prerequisite for their proper pharmacological and psychotherapeutic treatment.

Acknowledgements Supported by the George Stairs Foundation. I would like to recognize the gracious and competent help of Ms. Marie-Elaine Huberdeau for the preparation of the manuscript.

References

1. Kendell RE. The role of diagnosis in psychiatry. Oxford: Blackwell, 1975.
2. Kraepelin E. Lectures on clinical psychiatry. Darien CT: Hafner, 1968.
3. Petricoin R, Liotta D. New developments in methods to study biological markers of tumors. Scientist 2006; 20:33–39.
4. Dubrovsky B. Neurosteroids, neuroactive steroids, and salient symptoms of affective disorders. Pharm Biochem Behav 2006; 84:644–655.
5. Mellon SH, Griffith LS. Neurosteroids: biochemical and clinical significance. Trends Endocrinol Metab 2002; 13:34–43.
6. Stoffel-Wagner B. Neurosteroid biosynthesis in the human brain and its critical implications. Ann NY Acad Sci 2003; 1007:64–67.
7. Akwa Y, Ladurelle N, Covey DF, Baulieu EE. The synthetic enantiometer of pregnenolone sulfate is very active on memory in rats and mice, even more so than its physiological neurosteroid counterpart: distinct mechanisms. Proc Natl Acad Sci USA 2001; 98:14033–14037.
8. Csaba G. Phylogeny and ontogeny of hormone receptors: the selection theory if receptor formation and hormonal imprinting. Bio Rev 1980; 55:47–56.
9. Allera A, Wildt L. Glucocorticoid-recognizing and glucocorticoid-effector sites in rat liver plasma membrane. Kinestics of corticosterone uptake by isolated membrane vehicules: 2. Comparative influx and efflux. J Steroid Biochem 1992; 42(7):575–771.
10. Danfeldt S, Lanz R, Allera A. Membrane-initiated signaling system (MISS): genomic steroid action starts at the plasma membrane. J Steroid Biochem 2003; 85:9–23.
11. Stocco DM. The role of Star protein in steroidogenesis: challenges for the future. J Endocrinol 2000; 164:247–253.
12. Papadopoulos V, Baraldi M, Guilarte TR, Knudsen TB, Lacapère JJ, Lindemann P, Norenberg MD, Nutt D, Weizman A, Zhang MR, Gavish M. Translocator protein (18kDA): new nomenclature for the peripheral-type benzodiazepine receptor based on its structure and molecular function. Trends Pharmacol Sci 2006; 27:402–409.
13. Lieberman S, Kaushik G. Other conceivable renditions of some oxidative processes used in the biosynthesis of steroid hormones. J Steroid Biochem Mol Biol 2006; 101.
14. Finn DA Ford MM, Wiren KM, Roselli CE, Crabbe JC. The role of pregnane neurosteroids in ethanol withdrawal: behavioral genetic approaches. Pharmacol Ther 2004; 101:91–112.
15. Biggio G, Follesa P, Sanna E, Purdy RH, Concas A. GABAA related plasticity during long-term exposure to and withdrawal from progesterone. Int Rev Neurobiol 2001; 46:207–241.

16. Bitran D, Smith SS. Termination of pseudopregnancy in the rat produces an anxiogenic-like response that is associated with an increase in benzodiazepine receptor binding densitity and a decrease in GABAA-stimulated chloride influx in the hippocampus. Brain Res Bull 2005; 64:511–518.
17. Smith SS. Withdrawal properties of a neuroactive steroid: implications for GABAA receptor gene regulation in the brain and anxiety behaviour. Steroids 2002; 67(6):519–528.
18. Khisti RT, Penland SN, VanDoren MJ, Grobin AC, Morrow AL. GABAenergic neurosteroid modulation of ethanol actions. World J Biol Psychiatry 2002; 3:87–95.
19. Smith SS, Ruderman Y, Frye C, Homanics G, Yuan M. Steroid withdrawal in the mouse results in anxiogenic effects of 3alpha, 5beta-THP; a possible model for premenstrual dysphoric disorder. Psychopharmacology (Berl) 2005:1–11.
20. Brussard AB, Koksma JJ. Conditional regulation of neurosteroid sensitivity of GABAA receptors. Ann NY Acad Sci 2003; 1007:29–36.
21. Koksma JJ, van Kesteren RE, Rosahl TW, Kits KS, Brussard AB. Oxytocin regulates neurosteroid modulation of GABAA receptors in supraoptic nucleus around parturition. J Neurosci 2003; 23:788–797.
22. Belelli D, Herd MB, Mitchell EA, Peden DR, Vardy AW, Gentet L, Lambert. Neuroactive steroids and inhibitory neurotransmission: mechanisms of action and physiological relevance. J Neurosci 2006; 138:821–829.
23. Purdy RH, Morrow AL, Moore PH, Jr., Paul SM. Stress-induced elevations of gamma-aminobutyric acid type A receptor-active steroids in the rat brain. Proc Natl Acad Sci USA 1991; 88:4553–4557.
24. Dubrovsky B. Un confronto fra visione adattamentistica e visione pluralistica dell'evoluzione in psichiatria. In: Canali S, Corbellini (eds) Medicina Darwiniana. L'approcio evoluzionista all malattia. Aperion; 2004a, pp. 91–143.
25. Dubrovsky B. Fundamental neuroscience and the classification of psychiatric disorders. Neurosci Biobehav Rev 1995; 19:511–518.
26. Dubrovsky B. Evolutionary psychiatry. Adaptationists and non-adaptationists conceptualization. Prog Neuro-psychopharmacol Biol Psychiatry 2002; 27:1–19.
27. Dubrovsky B. Evolution and nervous systems and psychiatry: consequences of the vertical and horizontal duality of the evolutionary process. J Psychiatry Neurosci. 1993; 18:245–259.
28. Gould SJ. The architecture of evolutionary theory. Cambridge, MA: Harvard University Press, 2002.
29. Conway Morris S. Life's solution. Inevitable humans in a lonely universe. Cambridge, MA: Cambridge University Press, 2003.
30. Arthur W. Biased Embryos and Evolution. Cambridge: Cambridge University Press, 2004.
31. Arthur W. Creatures of accident: the rise of the animal kingdom. Farran, Strauss and Giroux, 2006.
32. Carroll S-B. The new science of evo-devo. Endless forms most beautiful: the new science of evo-devo and the making of the animal kingdom. New York: W. W. Norton, 2005.
33. Daftary GS, Taylor HS. Endocrine regulation of HOX genes. Endocrine Reviews 2006; 27:331–355.
34. Nusslem-Volhard C. Coming to life. Carlsbad, CA: Kales, 2006.
35. Jacob F, Monod J. Genetic regulatory mechanisms in the synthesis of proteins. J Mol Biol 1961; 2:318–356.
36. Dover G, Darwin DM. Letters on the evolution of life and human nature. London: Weinfeld & Nicolson, 2000.
37. Allman JM. Evolving brains. New York: Scientific American Library, 1999.
38. Ohno S. Evolution by gene duplication. New York: Springer, 1970.
39. Kirschner MW, Gerhart JC. The plausibility of life. New Haven, CT: Yale University Press, 2005.
40. Dubrovsky B, Davelaar E, Garcia-Rill E. The role of dorsal columns in serial order acts. Exp Neurol 1971; 33:93–102.

41. Linden DJ. The accidental mind. Cambridge, MA: The Belknap Press of Harvard University Press, 2007.
42. Jackson H. Selected writings. In: Taylor J (ed). Evolution and dissolution of the nervous system. Vol II. New York: Basic Books, 1955.
43. Ey H. The organo-dynamic concept of psychiatry. In: Natanson M, Ey H (eds). Psychiatry and philosophy. Berlin, Germany: Springer, 1969.
44. Riese H. The concept of disease: its history, its versions and its nature. New York: Philosophical Library, 1953.
45. Dubrovsky B, Barbas H, Solyom L. Characteristics of CNV in patients suffering from specific phobias. Biol Psychiatry 1978; 13N5:531–540.
46. Knott JR, Irwin A. Anxiety, stress and the contigent negative variation. Arch Gen Psychiatry 1973; 29:538–541.
47. Ben-Ari Y. Excitatory actions of GABA during the development: the nature of the nurture. Nature Rev Neurosci 2002; 3(9):728–739.
48. Schumacher M, Guennon R, Robert F, Carelli C, Gago N, Ghoumari A. Local synthesis and dual actions of progesterone in the nervous system: neuroprotection and myelination. Growth Horm IGF Res 2004; 14(Suppl A):S18–S33.
49. Fonte C, Grenier J, Trousson A, Chauchereau A, Lahuna O, Baulieu EE et al. Involment of (beta)-catenin and unusual behavior of CBP and p300 in glucocorticosteroid signaling in Schwann cells. Proc Natl Acad Sci USA 2005; 102:14260–14265.
50. Grenier J, Trousson A, Chauchereau A, Cartaud J, Shumacher M, Massaad C. Differential recruitment of p160 coactivators by glucocorticoid receptor between Schwann cells and astrocytes. Molceular Endocrinol 2006; 20:254–267.
51. Rupprecht R, Reul JMM, Trapp T, van Steezsel B, Wetzl C, Damm K, Zieglzansberger W, Holsboer F. Progesterone receptormediated effects of neuroactive steroids. Neuron 1993; 11:523.
52. Diamond LT, Hall WC. Evolution of neocortex. Science 1969; 164:251–262.
53. Diamond IT. Parallel pathways in the auditory, visual and somatic systems. In: Macchi G, Rustioni A, Spreafico R (eds). Somatosensory integration in the thalamus. New York: Elsevier Science 1983; 251–272.
54. Mountcastle VB. Perceptual neuroscience. The cerebral cortex. Cambridge, MA: Cambridge University Press, 1998.
55. Sherman SM, Guillery RW. The role of the thalamus in the flow of information to the cortex. Philos Trans R Soc 2002; 357:1695–1708.
56. Butler AB, Hodos W. Comparative vertebrate neuroanatomy. Evolution and adaptation. New York: Wiley-Liss, 2004.
57. Zeki SA. A vision of the brain. Oxford: Blackwell, 1993.
58. Dubrovsky B, Garcia-Rill E. Convergence of tectal and visual cortex input in cat motosensory cortex. Exp Neurol 1971; 33:471–484.
59. McCormack M, Dubrovsky B. Impairment of limb actions after dorsal funiculi section in cats. Exp Brain Res 1979; 37:31–40.
60. Armstrong E. The limbic system and culture. Hum Nat 1991; 2:117–136.
61. Lishman WA. Selective factors in memory: Part 2. Affective disorder. Psychol Med 1972; 2:248–253.
62. Weingartner H, Silberman E. Models of cognitive impairment: cognitive changes in depression. Psychopharmacol Bull 1982; 18:27–42.
63. Dunbar GC, Lishman WA. Depression, recognition-memory and hedonic tone. A signal detection analysis. Br J Psychiatry 1984; 144:365–382.
64. Schacter DL. The seven sins of memory. How the mind forgets and remembers. New York: Houghton Mifflin, 2001.
65. Teasdale JC, Howard RJ, Cox SG, Ha Y, Brammer MJ, Williams SCR, et al. Functional MRI study of the cognitive generation of affect. Am J Psychiatry 1999; 156:209–215.
66. Bower GH. Affect and cognition. Philos Trans R Soc Lond, B Biol Sci 1983; 302:387–402.
67. Teasdale JD. Affect and accessibility. Philos Trans R Soc Lond, B Biol Sci 1983; 302:403–412.

68. Kandel ER. In search of memory. New York: W. W. Norton, 2006.
69. Dubrovsky B. The specificity of stress responses to different nocuous stimuli: neurosteroids and depression. Brain Res Bull 2000; 51:443–455.
70. Dubrovsky B. Dynamics of neural networks. A proposed mechanism(s) to account for changes in clinical symptomatology through time in patients with psychotic diseases. Med Hypotheses 2001; 57:439–445.
71. von Monakow C. The emotions, morality and the brain. New York: Nervous and Mental Disease Publishing, 1925.
72. Schacter DL. The seven sins of memory. How the mind forgets and remembers. New York: Houghton Mifflin, 2001.
73. Deldin PJ, Deveney CM, Kim AS, Casas BR, Best JL. A slow wave investigation of working memory biases in mood disorders. J Abnorm Psychology 2001; 110(2):267–281.
74. Squire LR, Kandel ER. Memory from mind to molecules. New York: Scientific American Library, 1999.
75. Dubrovsky B, Filipini D, Gijsbers K, Birmingham MK. Early and late effects of steroids hormones on the central nervous system. In: Simmonds, MA (ed). Steroids and neuronal activity: Ciba Foundation Symposium. Chichester: Wiley, 1990, Vol 153, pp. 240–260.
76. Dubrovsky B, Gijsbers K, Filipini DL, Birmingham MK. Effects of adrenocortical steroids on long-term potentiation in the limbic system. Basis mechanisms and behavioural consequences. Cell Mol Neurobiol 1993; 13:399–414.
77. Filipini D, Birmingham MK, Dubrovsky B. Modulatory actions of adrenal steroids on limbic structures. J Steroid Biochem Mol Biol 1991a; 39:245–253.
78. Filipini D, Birmingham MK, Dubrovsky B. Effects of adrenal steroids and their reduced metabolites on hippocampal long-term potentiation. J Steroid Biochem Mol Biol 1991b; 40:87–92.
79. Joels M. Corticosteroid actions in the hippocampus. J Neuroendocrinol 2001; 13:657–669.
80. Yoo A, Harris J, Dubrovsky B. Dose response study of dehydropiandrosterone sulfate on dentaye gyrus in LTP. Exp Neurol 1996; 137:151–156.
81. Majewska MD. Neurosteroids; endogenous bimodal modulators of the GABAA receptor. Mechanism of action and physiological significance. Prog Neurobiol 1992; 38:379–395.
82. Scharfman H, Sarvey JM. Postsynaptic firing during repetitive stimulation is required for long term potentiation in hippocampus. Brain Res 1985; 331:267–274.
83. Harlan Meyer J, Gruol DL. Dehydroepiandrosterone sulfate alters synaptic potentials in area CA1 of the hippocampal slice. Brain 1994; 633:253.
84. Wolkowitz OM, Reus M, Keebler A, Nelson N, Friedland M, Brizendine L, Roberts E. Double blind treatment of major depression with dehydroepiandrosterone. Am J Psychiatry 1999; 156:646–649.
85. Schmidt PJ, Daly RC, Bloch M, Smith MJ, Danaceau MA, St Clair LS, et al. Dehydroepiandrosterone monotherapy in midlife-onset major and minor depression. Arch Gen Psychiatry 2005; 62:154–162.
86. Bloch M, Schmidt PS, Danaceau MA, Adams LF, Rubinow DR. Dehydroepiandrosterone treatment of midlife dysthymia. Biol Psychiatry 1999; 45:1533–1541.
87. Urbanoski K, Harris J, Gijsbers K, Dubrovsky B. Effects of androsterone sulphate on the population spike of the rat dentate gyrus following tetanisation. Physiol Behav 2000; 71:435–440.
88. Schwartz M, Harris J, Chu L, Gijsbers K, Dubrovsky B. Effects of androstenedione on long term potentiation in the rat dentate gyrus. Relevance for affective and degenerative diseases. Brain Res Bull 2002; 58:207–211.
89. Taube JS, Schwartkroin PS. Mechanisms of Long-Term Potentiation; EPSP/spike dissociation, intradendritic recordings and glutamate sensitivity. J Neurosci 1988; 8:1632–1644.
90. Frankenhaeuser B, Vallbo AB. Accomodation in myelinated nerve fibres of Xenopus laevis as computed on the basis of voltage clamp data. Acta Physiol Scand 1965; 63:1–12.
91. Irwin RP, Maragakis NJ, Rogawski MA, Purdy RH, Farb H, Paul SM. Pregnenolone sulfate augments NMDA receptor mediated increases in intracellular Ca^{2+} in cultured rat hippocampal neurons. Neurosci Lett 1992; 141:30–34.

92. Irwin RP, Lin SZ, Rogawski A, Purdy RH, Paul SM. Steroid potentiation and inhibition of *N*-methyl-D-aspartate receptor-mediated intracellular Ca^{++} responses: structure activity studies. J Pharmacol Exp Ther 1994; 271:677–682.
93. Flood JF, Roberts E. Dehydroepiandrosterone sulfate improves memory in aging mice. Brain Res 1998; 448:178–181.
94. Urani A, Privat A, Maurice T. The modulation by neurosteroids of the scopolamine-induced learning impairment in mice involves an interaction with sigma I (sigma 1) receptors. Brain Res 1998; 799:64–77.
95. Zou LD, Yamada K, Sasa M, Nakata Y, Nabeshima T. Effects of sigma (1) receptor agonist SA4503 and neuroactive steroids on performance in a radical arm maze task in rat. Neuropharmacology 2000; 39:1617–1627.
96. Huppert FA, Van Niekerk JK. Dehydroepiandrosterone (DHEA) supplementation for cognitive function (Cochrane Review). The Cocharane Library, issue 2. Oxford: Update Software 2001.
97. Berr C, Lafont S, Debuire B, Dartigues JF, Baulieu EE. Relationship of dehydroepiandrosterone sulfate in the eldery with functional, psychological, and mental status, and short-term mentality; a French community-based study. Proc Natl Acad Sci USA 1996; 93:13410–13415.
98. Monnet FP, Mahe V, Robel P, Baulieu EE. Neurosteroids via sigma receptors modulate the (3H) norepinephrine release evoked by *N*-methyl-D-aspartate in the rat hippocampus. Proc Natl Acad Sci USA 1995; 92:3774–3778.
99. Kinouchi K, Maeda S, Saito K, Inoki R, Fukumsita K, Yoshiya I. Effects of D and L-pentazocine on the release and uptake of norepinephrine in rat brain cortex. Res Commun Chem Pathol Pharmacol 1989; 63:201–213.
100. Wen S, Dong K, Onolfo JP, Vincens M. Treatment with dehydroepiandrosterone sulphate increases NMDA receptors in hippocampal and cortex. Eur J Pharmacol 2001; 430:373–374.
101. Lancel M, Faulhaber J, Schiffelholz T, Romeo E, di Michele F, Holsboer F, Rupprecht R. Allopregnanolone affects sleep in a benzodiazepine-like fashion. J Pharmacol Exp Ther 1997; 282:1213–1218.
102. Mayo W, Dellu F, Robel P, Cherkaoui J, Le Moal M, Baulieu EE. Infusion of neurosteroids into the nucleus basalis magnocellelaris affects cognitive processes in the rat. Brain Res 1993; 607:324–328.
103. Ladurelle N, Exchenne B, Denton D, Blair West J, Schumacher M, Baulieu. Prolonged intracerebro ventricular infusion of neurosteroids affects cognitive performances in the mouse. Brain Res 2000; 858:371–379.
104. Griffith LD, Mellon SH. Selective serotonin reuptake inhibitors directly alter activity of neurosteroidogenic enzymes. Proc Natl Acad Sci USA 199; 96:13512–13517.
105. Spalletta G, Romeo E, Bonaviri G, Bernardi G, Caltagirone C, di Michele F. Preliminary evidence for an association between aggressive and hostile behaviors and 3alpha, 5alpha tetrahydroprogesterone plasma levels in schizophrenia. J Psychiatry Neurosci 2005; 30:49–52.
106. Fish EW, Faccidomo S, DeBold JF, Micek KA. Alcohol allopregnanolone and aggression in mice. Psychopharmacology 2001; 153:173–183.
107. Andreen I, Sundstrom-Poromaaa I, Bixo M, Andersson A, Nyberg S, Backstrom T. Relationship between allopregnanolone and negative mood in postmenopausal women taking sequential hormonal replacement therapy with vaginal progesterone. Psychoneuroendocrinology 2005; 30:212–224.
108. Dubrovsky B, Harris J, Tatarinov-Levin A. Effects of the active neurosteroid THDOC on long term potentiation. Implications for depression. Prog neuron-psychopharmacol Biol Psychiatry 2004; 28:1029–1034.
109. Gijsbers K, Dubrovsky B, Harris J, Tatarinov-Levin A. Dose dependent effects of THP on LTP in the hippocampus of the rat. In press.
110. Frye CA, Sturgis JD. Neurosteroids affect spatial reference, working, and long-term memory of female rats. Neurobiol Learn Mem 1995; 64:83–96.

111. Rouge-Pont F, Mayo W, Marinelli M. The neurosteroid allopregnanolone increases dopamine release and dopaminergic response to morphine in the rat nucleus accumbens. Eur J Neurosci 2002; 16:169–173.
112. Ugale RR, Hirani K, Morelli M, Chopde CT. Role of the NAS THP in the antipsychotic-like action of olanzapine in rodents. Neruopsychopharmacol 2004; 29:1597–1609.
113. Theis M, Si K, Kandel E. Two previously undescribed members of the mouse CPEB family of genes and their inducible expression in the principal layers of the hippocampus. Proc Natl Acad Sci USA 2003; 100:9602–9607.
114. Cushing H. Psychiatric disturbances associated with disorders of the ductless glands. Am J Insan 1913; 69:965–990.
115. Cushing H. The basophil adenomas of the pituitary body and their clinical manifestations. Bull John Hopkins Hosp 1932; 50:137–195.
116. De Kloet ER. Stress in the brain. Eur J Pharmacol 2000; 405:187–198.
117. Gold PW, Goodwin FK, Chrousos GP. Clinical and biochemical manifestations of depression: relation to the neurobiology of stress. N Engl J Med 1988; 319:348–353.
118. Gold PW, Chrousos GP. Organization of the stress system and its dysregulation in melancholic and atypical depression: high vs low CRH/NE states. Mol Psychiatry 2002; 7:254–275.
119. Holsboer F. The corticosteroid receptor hypothesis of depression. Neuropsychopharmacol 2002a; 23:477–501.
120. Broekhoven van F, Verkes RJ. Neurosteroids in depression: a review. Psychopharmacology (Berl) 2003; 165:97–110.
121. Holsboer F. Stress, hypercortisolism and corticosteroid receptors in depression: implications for therapy. J Affect Disord 2001; 62:77–91.
122. Uzunova V, Sheline Y, Davis JM, Rasmusson A, Uzunov DP, Costa E, Guidotti A. Increase in the cerebrospinal fluid content of neurosteroids in patients with unipolar major depression who are receiving fluoxetine or fluvoxamine. Proc Natl Acad Sci USA 1998; 95:3239–3244.
123. Romeo E, Ströhle A, Spaelleta G, di Michele F, Herman B, Holsboer F. Effects of antidepressant treatment on neuractive steroids in major depression. Am J Psychiatry 1998; 155:910–913.
124. Ströhle A, Romeo E, Hermann B, Pasinin A, Spalletta G, di Michele F, Holsboer F, Rupprecht R. Concentrations of 3 a reduced neuroactive steroids and their precursors in plasma of patients with major depression and after clinical recovery. Biol Psychiatry 1999; 45:274–277.
125. Ströhle A, Pasini A, Romeo E, Hermann B, Spalletta G, Michele F, Holsboer F, Rupprecht R. Fluoxetina decreases concentration of 3alpha 5 alpha tetra-hydrodeoxycorticosterone (THDOC) in major depression. J Psychiatr Res 2000; 34:183–186.
126. Griffith LD, Mellon SH. Selective serotonin reuptake inhibitors directly alter activity of neurosteroidogenic enzymes. Proc Natl Acad Sci USA 1999; 96:13512–13517.
127. Gigli GL, Diomedi M, Troisi A, Marciani MG, Pasini A. Fluoxetine and seizures. Neurology 1996; 47:303–307.
128. Pasini A, Tortorella A, Gale K. The anticonvulsant action of fluoxetine in substantia nigra is dependent upon endogenous serotonin. Brain Res 1996; 724:84–88.
129. Porsolt RD, Le Pechon M, Jalfre M. Depression: a new animal model sensitive to antidepressant treatment. Nature 1978; 266:730–732.
130. Khisti RT, Chopde CT, Jain SP. Antidepressant-like effect of the neurosteroid 3aplha-hyrdroxy-5alpha-pregnan-20-one in mice forced swim test. Pharmacol Biochem Behav 2000; 67:137–143.
131. Uzunova V, Wrynn AS, Kinnunen A, Ceci M, Kohler C, Uzunov DP. Chronic antidepressants reverse cerebrocortical allopregnanolone decline in the olfactory-bulbectomized rat. Eur J Pharmacol 2004; 486:31–34.
132. Uzunova V, Ceci M, Kohler C, Uzunov DP, Wrynn AS. Region-specific dysregulation of allopregnanolone brain content in the olfactory bulbectomized rat model of depression. Brain Res 2003; 976:1–8.

133. Pisu MG, Serra M. Neurosteroids and neuroactive drugs in mental disorders. Life Sci 2004; 74:3181–3197.
134. Murphy BEP, Steinberg SI, Hu FY, Allison CM. Neuroactive Ring A-reduced metabolites of progestrone in human plasma during pregnancy: elevated levels of 5a-dihydroprogesterone in depressed patients during the latter half of pregnancy. J Clin Endocrinol Metab 2001; 86:5981–5987.
135. Holzbauer M. Physiological aspects of steroids with anaesthetic properties. Med Biol 1976; 54:227–242.
136. Backström T, Andersson A, Baird D, Selstam G. The human corpus luteum secretes 5 a pregnane-3, 20-dione. Acta Endocrinol 1986b; 111:116–121.
137. Eser D, Schüle C, Baghai TC, Romeo E, Vzunov DP, Rupprecht R. Neuroactive steroids and affective disorders. Pharmac Biochem Behav 2006; 84:656–666.
138. Torres JM, Ruiz E, Ortega E. Effects of CRH and ACTH administration on plasma and brain neurosteroid levels. Neurochem Rev 2001; 26:555–558.
139. Torres JM, Ortega E. DHEA, PREG and their sulphate derivatives on plasma and brain after CRH and ACTH administration. Neurochem Res 2003; 28:1187–1191.
140. Agnati LF, Fuxe C, Nicholson C, Sykova E. Volume transmission revisited. Progress in Brain Research, Chap. 1, Vol 125. Amsterdam: Elsevier Neurosteroids 2000
141. Dubrovsky B, Williams D, Kraulis I. Effects of deoxycorticosterone and its Ring A-reduced derivates on the nervous system. Exp Neurol 1982; 78:728–739.
142. Kraulis I, Foldes G, Dubrovsky B, Traikov H, Birmingham MK. Distribution metabolism and biological activity of deoxycorticosterone in the central nervous system. Brain Res 1975; 88:1–14.
143. Backström T, Dubrovsky B, Landgren S, Brain excitability, steroid hormone and menstrual cycle. In: Dennerstein S, Fraser I (eds). Hormone and behavior. Elsevier, 1986a, pp. 137–142.
144. Landgren S, Backstrom T, Dubrovsky B. The effect of progesterone and its metabolites on the interictal epileptiform discharge in the cat's cerebral cortex. Acta Physiol Scand 1987; 131:33–42.
145. Mendelson WB, Martin JV, Perlis M, Wagner R, Majewska MD, Paul SM. Sleep induction by an adrenal steroid in the rat. Psychopharmacol 1987; 93:226–229.
146. Muller-Preuss P, Rupprecht R, Lancel M. The effects of the neuroactive steroid 3 alpha,5alpha-THDOC on sleep in the rat. Neuroreport 2002; 13:487–490.
147. Hobson AJ. Sleep. Scientific American Library. New York: W. H. Freeman, 1989, pp. 213.
148. Krieger DT. The central nervous system and Cushing's disease. Med Clin North Am 1978; 62:264–268.
149. Kupfer Dj, Foster FG, Detre TP. Sleep continuity changes in depression. Dis Nerv Syst 1973; 34:192–195.
150. Medawar PB. Some immunogogical and endocrinological problems raised by the evolution of viviparity. SSEB N VI Evolution. Cambridge: Cambridge University Press, 1953, pp. 320–338.
151. Murphy BEP, Abbot FV, Allison CM, Watts C, Ghadirian AM. Elevated levels of some neuroactive progesterone metabolites, particularly isopregnanolone, in women with chronic fatigue syndrome. Psychoneuroendocrinology 2004; 29:245–268.
152. Modell S, Holsboer F. Depression and effects of antidepresssant drugs on the stress systems. In: Steckler T, Kalin NH, Reul JM (eds). Handbook of stress and the brain. Amsterdam: Elsevier, 2005, pp. 273–286.
153. Dubrovsky B. Steroids, neuroactive steroids and neurosteroids in psychopathology. Prog Neuro-Psychopharm Biol Psychiatry 2005; 29:169–192.
154. Dubrovsky B, Yoo A, Harris J. Electrophysiological effects and steroid hormones on nervous systems. In: Stone T (ed). CNS transmitters and neuromodulators. Neuroactive steroids. Boca Raton, FL: CRC, 1996, pp. 89–102.
155. Reincke M, Schmann R, Karl M, Magiakon A, Chrousos GP, Allolio B. Severe illness neurodocrinology. Ann NY Acad Sci 1995; 771:556–569.
156. Cannon WB. Stresses and strains of homeostasis. Am J Med Sci 1935; 189:1–14.

157. Mason JW. Specificity in the organization of neuroendocrine response profiles. In: Seeman P, Brown G (eds). Frontiers in neurology and neuroscience research. Toronto ON: University of Toronto Press, 1974; pp. 68.
158. Barbaccia ML, Roscetti G, Trabucchi M, Mostallino MC, Concas A, Purdy RH. Time-dependent changes in rat brain neuroactive steroid concentrations and $GABA_A$ receptor function after acute stress. Neuroendocrinology 1996; 63:166–172.
159. Cannon WB. The mechanism of emotional disturbance of bodily functions. N Engl J Med 1928; 198:87–84.
160. Pakak K, Palkovits, M. Stressor specificity of central neuroendocrine responses; implication fro stress related disorders. Endocr Rev 2001; 22:425–450.
161. Mason JW. A reevaluation of the concept of non-specificity in stress theory. J Psychiatr Res 1971; 8:323–333.
162. Ursin H. The psychology in psychoneuroendocrinology. Psychoneuroendocrinology 1998; 23:550–570.
163. Vaillant G. The wisdom of the ego. Cambridge, MA: Harvard University Press, 1993.
164. Castren, E. Neurotropic effects of antidepressant drugs. Curr Opin Pharmacol 2004; 4:58–64.
165. Czeh B, Michaelis T, Watanabe T, McEwen B. Stress induced changes in cerebral metabolites, hippocampal volume and cell proliferation are prevented by antidepressant treatment with tianeptine. Proc Natl Acad Sci USA 2001; 98:12796–12801.
166. Mahlberg JE. Implications of adult hippocampal neurogenesis in antidepressant action. Rev Psychiatry Neurosci 2004; 29:196–205.
167. Nestler EJ, Barrot M, DiLeone RJ, Eisch AJ, Gold, SJ, Monteggia, LM. Neurobiology of depression. Neuron 2002; 34:13–25.
168. Santarelli L, Saxe M, Gross C, Surget A, Battaglia F, Dulawa S, Weisstaub N, Lee J, Duman R, Arancio O, Belzung C, Hen R. Requirement of hippocampal neurogenesis for the behavioural effects of antidepressants. Sci 2003; 301:805–809.
169. Malik AS, Narayan RK, Wendling WW. A novel dehydroepiandrosterone analog improves functional recovery in a rat traumatic brain injury model. J Neurotrauma 2003; 20:463–476.
170. Lapchak PA, Araujo DM. Preclinical development of neurosteroids as neuron-protective agents for the treatment of neurodegeneratice diseases. Int Rev Neurobiol 2001; 46:379–397.
171. Kimonides VG, Spillantini MG, Sofroniew MV, Fawcett JW, Herbert J, Dehydroepiandrosterone (DHEA) antagonises the neurotoxic effects of corticosterone, translocation of SAPK 3 in hippocampal primary cultures. Neuro Sci 1999; 89:429–436.
172. Stein DG. The case for progesterone. Ann NY Acad Sci 2005; 1052:152–169.
173. Frank C, Sagratella S. Neuroprotective effects of allopregnenolone on hippocampal irreversible neurotoxicity in vitro. Prog Neuropsychopharmacol Biol Psychiatry 2000; 24:1117–1126.
174. Brinton RD, Wang JM. Preclinical analyses of the therapeutic potential of allopregnanolone to promote neurogenesis in vitro and in vivo in transgenic mouse model of Alzheimer's disease. Curr Alzheimer Res 2006; 3:11–17.
175. Charalampous I, Tsatsanis C, Demitzaki E, Alexaki V-I, Castana E, Margioris AN et al. Dehydroepiandrosterone and allopregnanolone protect sympathoadrenal medulla cells against apoptosis via antiapoptotic Bel-2 protein. PNAS 2004; 101:8209–8214.
176. Jayanthi S, Deng X, Bordelon M, McCoy MI, Cadet JL. Methamphetamine causes differential regulation of pro-death and anti-death Bel-2 genes in the mouse neocortex. FASEB J 2004; 29911:15L1745052.
177. Kubie LS. Biological bases of psychiatry. The viewpoint of the psychoanalyst. In: Wortis J (ed). Recent advances in biological psychiatry, Vol VII. New York: Plenum, 1964.
178. Hubert W, Moller M, de-Jong-Meyer R. Film-induced amusement change in saliva cortisol levels. Psychoneuroendocrinology 1993; 18:265–272.
179. Kim JJ, Diamond DM. The stressed hippocampus, synaptic plasticity and lost memories. Natl Rev Neurosci 2002; 3:453.
180. Houssay A (ed). Fisiologia humana. Buenosaires: El Ateneo, 1958.

181. Morita K, Arimochi H, Her A. Serotonergic 5-HT2A receptor stimulation induces steroid 5a-reductase gene expression in rat C6 glioma cells via transcription factor Egr-1. Mol Brain Res 2005; 139:193–200.
182. Morita K, Hideki A, Hiroyuki I, Song H, et al. Possible involvement of 5α-reduced neurosteroids in adrenergic and serotonergic stimulations of GFAP gene expression in rat C6 glioma cells. Brain Res 2006; 1085:49–56.
183. Papadopoulos V, Guarneri P. Regulation of C6 glioma cell steroidogenesis by adenosine 3 ,5 -cyclic monophosphate. Glia 1994; 10:75–78.
184. Segovia J, Lawless GM, Tillakaratne NJ, Brenner M, Tobin AJ. Cyclic AMP decreases the expression of a neuronal marker (GAD67) and increases the expression of an astroglia marker (GFAP) in C6 cells. J Neurochem 1994; 63:1218–1225.
185. Chui SY, Kriegler S. Neurotransmitter-mediated signaling between axons and glial cells. Glia 1994; 11:191–200.
186. Mong JA, Nunez JL, McCarthy MM. GABA mediates steroid-induced astrocyte differentiation in the neonatal rat hypothalamus. J Neuroendocrinol 2002; 14:45–55.
187. Dubrovsky B. Potential use of neurosteroids and neuroactive steroids as modulators of symptoms of depression, anxiety, and psychotic disorders. Drug Dev Res 2005; 65:318–334.
188. Gerner RH, Hare TA. CSF GABA in normal subjects and patients with depression, schizophrenia, mania and anorexia nervosa. Am J Psychiatry 1981; 138:1098–1101.
189. Gold BI, Bowers Jr MB, Roth RH, Sweeney W. GABA levels in CSF of patients with psychiatric disorders. Am J Psychiatry 1980; 137:362–364.
190. Bhagwagar Z, Wylezinska M, Taylor M, Jezzard P, Matthews P, Cowen J. Increased brain GABA concentrations following acute administration of a selective serotonin reuptake inhibitor. Am J Psychiatry 2004; 161:368–370.
191. Sanacora G, Mason GF, Rothman DL, Hyder F, Ciarcia JJ, Ostroff RB, Berman RM, Krystal JH. Increased cortical GABA concentrations in depressed patients receiving ECT. Am J Psychiatry 2003; 160:577–579.
192. Takebayashi M, Hayashi T, Su TP. A perspective on the new mechanism of antidepressants: neuritogenesis through sigma-1 receptor. Pharmacopsychiatry 2004; 37:S208–S213.
193. Hellsten J, Wennström M, Mohapel B, Ekdahl CT, Bengzon J, Tingström A. Electroconvulsive seizures increase hippocampal neurogenesis after chronic corticosterone treatment. Eur J Neurosci 2002; 16:283–290.
194. Hellsten J, Wennström M, Bengzon J, Mohapel B, Tingström A. Electroconvulsive seizures induce endothelial cell proliferation in adult rat hippocampus. Biol Psychiatry 2004; 55:20–27.
195. Chen L, Kis B, Hashimoto H, Busija DW, Takei Y, Yamashita H, Ueta Y. Adrenomedullin 2 protects rat cerebral endothelial cells from oxidative damage in vitro. Brain Res 2006; 1086(1):42–49.
196. Greens K. Growing a new antidepressant. The Scientist 2007; 21:40–46.
197. Getting PA. Emerging principles governing the operation of neural networks. Ann Rev Neurosci 1989; 12:185–204.
198. Kaas JH. The reorganization of sensory and motor maps in adult mammals. In: Gazzaniga M (ed). The cognitive neurosciences. Cambridge MA: Bardford Books/MIT, 1995, pp. 51–71.
199. Strous RD, Spivak B, Yran-Hegesh R, Maayan R, Averbuch E, Kotler M, Mester R, Weizman A. Analysis of neurosteroid levels in attention deficit hyperactivity disorder. Int J Neuropsychopharmacol 2001; 4:259–264.
200. Mesulman MM. A cortical network for directed attention and unilateral neglect. Ann Neurol 1981; 10:309–325.
201. Geschwind N, Galaburda A. Cerebral lateralization. Biological mechanisms, associations and pathology. Cambridge, MA: MIT, 1987.
202. Gruzelier J, Flor-Henry P. Hemispheric asymmetries of function in psychopathology. Amsterdam: Elsevier/North-Holland, 1979.
203. Dennemberg V. Hemispheric laterality in animals and the effects of early experience. Behav Brain Sci 1981; 4:1–49.

204. Van Praag HM, Kahn R, Asnis GM. Denosologisation of biological psychiatry. On the specificity of 5-HT disturbances in psychiatric disorders. J Affect Disord 1987; 13:1–8.
205. Strous RD, Maayan R, Lapidus R, Stryjer R, Lustig M, Kotler M, Weizman A. Dehydroepiandrostrone augmentation in the management of negative, depressive, and anxiety symptoms in schizophrenia. Arch Gen Psychiatry 2003; 60:133–140.
206. Modell AH. Imagination and the meaningful brain. Cambridge, MA: MIT, 2006.

Chapter 20
Neuroactive Steroids in Brain and Relevance to Mood

Torbjörn Bäckström[1], Lotta Andréen[1], Marie Bixo[1], Inger Björn[1], Guillén Fernández[2], Inga-Maj Johansson[1], Per Lundgren[1], Magnus Löfgren[1], Sigrid Nyberg[1], Gianna Ragagnin[1], Inger Poromaa-Sundström[1], Jessica Strömberg[1], Frank van Broekhoven[3], Guido van Wingen[2,3], and Ming-De Wang[1]

Abstract Depression and anxiety often affect women in relation to reproductive events like menarche, premenstrual periods, post-partum and perimenopause. A prominent example of the interaction between mood, neuroactive-steroids and the GABA system is premenstrual dysphoric disorder (PMDD). Severe premenstrual negative mood symptoms occur in 3–8% of women. Sex and stress hormones are metabolized to neuroactive steroids with effects on brain function as positive modulators of the $GABA_A$ receptor (called GABA-steroids) similar to benzodiazepines, barbiturates and alcohol. One example of a neuroactive sex steroid is allopregnanolone, and other GABA-steroids, are produced within the brain, by the adrenals at stress and from the ovary during the menstrual cycle. Animal and human studies show that benzodiazepines, barbiturates, alcohol and allopregnanolone have a bimodal effect on behavior. In high dosages or concentrations the positive $GABA_A$ receptor modulators are CNS depressants, anesthetic, and anxiolytic, whereas in certain sensitive individuals low concentrations instead of being anxiolytic cause severe anxiety, irritability, aggressiveness and depressive mood in 3–6% of individuals, and moderate symptoms in up to 30%. Low concentrations of GABA-steroids are found endogenously during the luteal phase and induce adverse emotional reactions. In women with PMDD/PMS this paradoxical effect of neuroactive steroids seems to provoke negative mood symptoms as tension, irritability and depression. The mechanism behind the effect is called disinhibition that acts together with tolerance development by $GABA_A$ receptor active substances. Effective treatments are inhibition of ovarian steroid production or changing the CNS response to neuroactive steroids.

Keywords $GABA_A$ receptor, premenstrual dysphoric disorder, negative mood, menstrual cycle, allopregnanolone, progesterone, paradoxical effect, gaba-steroids

[1] Umeå Neurosteroid Research Center, Department of Clinical Sciences, University of Umeå Sweden

[2] F.C. Donders Centre for Cognitive Neuroimaging, Radboud University Nijmegen, Kapittelweg 29, 6525 EN, The Netherlands

[3] Department of Psychiatry, Radboud University Nijmegen Medical Center, Reinier Postlaan 10, 6525 GC, Nijmegen, The Netherlands

M.S. Ritsner and A. Weizman (eds.), *Neuroactive Steroids in Brain Function, Behavior and Neuropsychiatric Disorders*

Abbreviations CNS central nervous system; fMRI functional magnetic resonance imaging; GABA gamma-aminobutyric acid; $GABA_A$ one receptor for GABA; GABA-steroids positive modulators of $GABA_A$ receptor; HT hormone therapy; MPA medroxyprogesterone; MRI magnetic resonance imaging; PMDD premenstrual dysphoric disorder (PMDD); PMS premenstrual syndrome; RU486 mifepristone

20.1 Introduction

Mood disorders are common health problems affecting women, especially during the reproductive years. Women are about two times as likely as men to report a lifetime history of major depression or anxiety disorder and the sex difference begins in the early adolescence and persists through the mid-1950s.[1,2] It has been suggested that periods of hormonal variability, i.e., menarche,[3] premenstrual periods,[4] postpartum,[5,6] and perimenopause[7] increase the risk of mood disorders in certain women. Therefore it seems likely that sex steroid hormones can provide one possible explanation to the differences in mood disorders seen between genders. The central nervous system (CNS) is both a producer and target of sex steroids and three obvious examples where there seem to be evidence for the interaction between mood, steroids and CNS are the premenstrual dysphoric disorder (PMDD), side effects of oral contraceptives and negative mood symptoms encountered during sequential progestagen addition to estrogen treatment in postmenopausal women. Neuroendocrine factors such as neuroactive steroids are likely to contribute to the overall increased risk of developing mood disorders in women and in order to approach this vast problem a deeper understanding of the underlying mechanisms is most important.

One very obvious relation between sex steroids and mood symptoms are the symptoms related to the menstrual cycle. The sex hormones estradiol and progesterone display regular predictable changes during the menstrual cycle. In parallel with the progesterone increase an increase also occur in serum neuroactive steroids allopregnanolone (3α-OH-5α-pregnan-20-one) and pregnanolone (3α-OH-5β-pregnan-20-one).[8] Allopregnanolone, can be synthesized in the central nervous system but the major contributor to the concentration in the brain is the corpus luteum of the ovary.[9,10] Allopregnanolone and progesterone are produced in parallel during the luteal phase of the menstrual cycle.[8,11] In fertile women plasma levels of allopregnanolone are approximately 0.2–0.5 nmol/l in the follicular phase and up to 4 nmol/l in the luteal phase. In the third trimester of a pregnancy these levels increase up to more than 100 nmol/l.[12,13] Pregnanolone displays a similar luteal phase increase.[8,14] Allopregnanolon, pregnanolone and several other metabolites of steroid hormones are modulators of the Gamma Amino Butyric Acid-A ($GABA_A$) receptor and therefore called GABA-steroids.

20.2 Menstrual Cycle-Linked Mood Changes

The relation between the luteal phase of the menstrual cycle and symptom development in PMDD/PMS is obvious. The symptoms start at the time of ovulation and increase in parallel with the rise in serum progesterone during the luteal phase. The symptom severity reaches a peak during the last 5 premenstrual days or the first day of menstruation. Thereafter the symptoms decline and disappear 3–4 days after the onset of menstrual bleeding. During the postmenstrual phase there is a period of well-being closely related to the estradiol peak.[15] This suggests that there is a symptom-provoking factor produced by the corpus luteum of the ovary. In anovulatory cycles, spontaneous or induced, when a corpus luteum is not formed and progesterone or allopregnanolone is not produced the symptom cyclicity disappears.[16–18] There is, however, a lag time of 4–5 days between peak of luteal steroids and peak of symptoms indicating that the symptom development takes some time to develop. This may be related to secondary gene transcriptional activity and subsequent protein synthesis due to enhanced $GABA_A$ receptor activation. This would eventually modulate the $GABA_A$ receptor composition and hence leading to the changes in GABA sensitivity occurring during the luteal phase.[8] The same mechanism occurs during tolerance development with change in the $GABA_A$ receptor subunit composition, which has been shown to occur during allopregnanolone treatment (see below).

The fact that progesterone and progestagens induce negative mood symptoms similar as in PMDD/PMS is shown in several studies, e.g., in postmenopausal women receiving estrogen/progesterone hormone replacement therapy[19–22] and in PMDD patients with induced anovulation.[23] Several types of progestagens are investigated and all seem to induce negative mood in certain sensitive individuals.[20,21,23]

20.3 Pathogenesis of Symptom Induction

The endocrine progesterone receptor is expressed in the brain and it is conceivable that some of progesterone's effects could be mediated via the classical endocrine progesterone receptor, resulting in changes in the $GABA_A$ receptor composition. However, the classical hormonal receptor for progesterone seems not to be involved in PMS/PMDD pathophysiology. Treatment with the endocrine progesterone receptor antagonist mifepristone (RU486) failed to reduce the physical or behavioral manifestations of PMS.[24]

Another possibility of importance for the pathogenesis of PMDD/PMS is related to the metabolism of progesterone, which is rapidly and to a high degree metabolized to allopregnanolone and pregnanolone in the liver, brain and other parts of the body.[25,26] Both steroids are acting as agonists on the $GABA_A$ receptor complex in the brain.[27] The GABA transmitter system is the major inhibitory system in CNS. When GABA binds to the $GABA_A$ receptor, the influx of chloride ions increases, leading to a hyperpolarizing of the post-synaptic membrane, thus rendering the postsynaptic cell less prone to excitation. Allopregnanolone and pregnanolone are $GABA_A$ receptor positive modulators and enhances the effect of GABA on the receptor similar to ethanol, barbiturates, and benzodiazepins. Neurosteroids, benzodiazepines, barbiturates, alcohol and most anaesthetic agents bind to the $GABA_A$ receptor and increase the GABA-induced chloride ion influx by interacting with allosteric binding sites.[28]

20.4 Involvement of Positive $GABA_A$ Receptor Modulators in Mood Symptoms

Studies in animals and humans have reported typical $GABA_A$ receptor agonistic effects of high doses allopregnanolone and pregnanolone such as sedation/anaesthesia,[29,30] anti-epileptic effects,[31] and anxiolytic effects in animals.[32] However, reports from human and animal studies indicate that in certain individuals all $GABA_A$ receptor agonists also can induce negative symptoms with anxiety, irritability/aggressiveness. Strong irritability/aggression is induced in 3–6% of individuals and moderate symptoms are induced in 20–30%.[33,34] Interestingly, the prevalence of PMDD among women in reproductive age is in the similar range, 3–8%, of women in reproductive age and 25–35% have milder symptom severity as in PMS.[35]

Why an increase in allopregnanolone during the menstrual cycle is related to development of negative mood is puzzling as allopregnanolone should be anxiolytic like benzodiazepines. The answer seems to be the fact that all $GABA_A$ receptor agonists like benzodiazepines, barbiturates, alcohol and allopregnanolone have paradoxical anxiogenic effects in certain individuals. As mentioned above low concentrations or doses give severe adverse emotional reactions in a subset of individuals (3–6%) and moderate reactions in up to 20–30%. This paradoxical effect is induced by allopregnanolone[36,37] benzodiazepines,[38,39] barbiturates,[34,40,41] and ethanol.[37,42,43] The symptoms induced by these $GABA_A$ receptor active drugs are depressive mood, irritability, aggression and other symptoms known to occur during the luteal phase in women with PMS/PMDD. A bimodal effect has also been noted of different dosages of medroxyprogesterone (MPA) and natural progesterone in postmenopausal women taking hormone therapy (HT). These women feel worse on a lower dosage of MPA or progesterone than on a higher dosage or placebo.[22,44,45]

Thus allopregnanolone seem to have a bimodal effect on mood with an inverted U-shaped relationship between concentration and effect. In postmenopausal women receiving progesterone a biphasic relation between the negative mood symptoms and the allopregnanolone concentrations in blood is noted. The negative mood increases with the increase in serum concentration of allopregnanolone up to the maximum concentration seen during the luteal phase but with further increase in allopregnanolone concentration there is a decrease in symptom severity.[22,45] An inverted U-shaped relation between allopregnanolone dosage and irritability/aggression has also been noted in rats.[37]

Benzodiazepines also induce paradoxical reactions in certain individuals with irritability, aggression, depression, confusion, violent behavior and loss of impulse control compared to placebo.[38,46–48] The paradoxical effects of midazolam in patients who underwent surgery were effectively treated with flumazenil, a benzodiazepine receptor antagonist that effectively reversed the midazolam-induced paradoxical behaviours.[49] In rats the benzodiazepine-heightened aggressive behaviour induced by midazolam or triazolam was also antagonised by flumazenil and the $GABA_A$ receptor antagonists β-CCt and 3-PBC.[50,51]

20.5 FMRI and MRI Studies

The GABA turnover has been studied in PMDD patients and controls using magnetic resonance imaging (MRI) studies of occipital cortex and they indicate that the GABAergic system is substantially modulated during the menstrual cycle. PMDD patients and controls show different patterns in brain GABA concentration changes during the menstrual cycle suggesting that PMDD patients have a dysfunction in the GABA system.[52] During the menstrual cycle in control women there are significant changes in fMRI responses in adult human brain related to hormone variations but there are also region and task specific effects. (Fernández et al.[53]). FMRI studies can investigate the activity in defined brain areas during, e.g., emotional stimulation and under drug treatments. The amygdala is one part of the brain that is related to emotional experiences. Therefore it is interesting to study the responses to emotional stimulations during the menstrual cycle and when progesterone is given. Healthy women given progesterone show a modulation of amygdala reactivity to emotional stimuli. Progesterone administration increases the neural response to angry and fearful faces selectively in the amygdala compared to placebo at moderate progesterone and allopregnanolone plasma concentrations.[54] These results therefore show a neural mechanism by which progesterone could induce adverse effects on anxiety and mood. Because the acute effects of progesterone are likely mediated by allopregnanolone, this paradoxical amygdala activity increase might reflect the disinhibition of the principal neurons of the amygdala via inhibition of inhibitory interneurons. However, higher progesterone and allopregnanolone concentrations are associated with a decrease in amygdala reactivity during the intentional encoding of neutral and happy faces into memory.[54] Also benzodiazepines giving an anxiolytic response have been reported to decrease amygdala fMRI responses to angry and fearful face stimuli.[55] These fMRI results support the observation that allopregnanolone seems to induce negative mood changes in a nonlinear inverted U-shaped curve.[37,45,56] The increased amygdala response in the fMRI studies was observed when allopregnanolone levels had increased from the follicular phase range to the luteal phase or early pregnancy range.[8,57] The increased amygdala response might therefore be specific for this concentration change, and as shown in our study.[54] Supra-physiological concentrations seem to give another response as high doses of allopregnanolone injected into amygdala gives anxiolysis.[58]

20.6 PMDD/PMS Patients Have Different Sensitivity in the GABA-system

It seems thus as a subset of individuals are very sensitive to low doses or concentrations of allopregnanolone and respond with severe adverse emotional reactions when provoked. There is evidence that the sensitivity in the brain for steroids differs between PMS/PMDD patients and controls. Negative effects of oral contraceptives on mood were found mainly in women with PMS/PMDD and it was the lowest progestagen concentration that provoked symptoms.[59] Ad-back of estradiol or progesterone to women with PMS/PMDD and medically inhibited ovarian hormone production resulted in recurrence of symptoms. This relapse of symptoms did not occur in normal women or in PMS/PMDD women during placebo treatment.[23] Postmenopausal women with a history of PMS/PMDD responded with more negative symptoms on progestagens compared to women without a PMS/PMDD history.[21]

In PMS/PMDD patients but not in healthy controls the sedative response to intravenous pregnanolone, diazepam and alcohol is reduced in the luteal phase compared to follicular phase.[14,60–62] In addition, patients with severe symptoms were less sensitive to the given pregnanolone or benzodiazepines compared to patients with more moderate symptoms. These findings suggest that patients with PMS/PMDD develop tolerance for $GABA_A$ receptor allosteric agonists during the luteal phase. In an animal model of PMS/PMDD the allopregnanolone effect occurs in parallel with an up-regulation of the hippocampal alpha4 subunit of the $GABA_A$ receptor and decreased benzodiazepine sensitivity.[63] This is in line with the decreased benzodiazepine sensitivity in women with PMDD.[60] Especially animals with a high risk taking behavior develop withdrawal symptoms on progesterone treatment.[64]

Most $GABA_A$ receptor active substances will induce tolerance when used long term and this is also true for allopregnanolone. It has been shown that chronic administration of allopregnanolone will down regulate the $GABA_A$ receptor with a decreased GABA sensitivity of the $GABA_A$ receptor.[65] A tolerance to allopregnanolone is in rats noted already after 90 min of anesthesia.[66] Also, there is a relation between tolerance development and changes in the $GABA_A$ receptor subunit alpha4 in thalamus.[67] Tolerance to GABA-active neurosteroids might contribute to the symptoms in women with mood disorders and explain the decreased sensitivity to pregnanolone[14] benzodiazepines[60] and alcohol[62] seen in PMDD. A change in the $GABA_A$ receptor subunit composition has also been shown at allopregnanolone.[68] A withdrawal effect has also been suggested to induce negative mood when high levels of allopregnanolone in the luteal phase of the menstrual cycle suddenly decline.[68] The exact mechanisms and relation between tolerance developments, sensitivity to GABA-steroids and mood induction is not known but intense research is going on in the area to understand the pathogenesis of PMDD/PMS.

20.7 Role of Estradiol in Progesterone-Induced Mood Symptoms

Estradiol concentration is also of importance for the mood inducing effect of progesterone. Estradiol alone seems to be related to well-being in women with PMDD/PMS.[15] However, together with progesterone or progestagen the effect seems different. Higher estradiol dosage in postmenopausal hormone therapy (HT) during the progestagen period gave more severe symptoms compared to lower estradiol dosage in the same women but only during the period when the progestagen was given. During the period of unopposed estrogen no difference in mood severity was noted related to the estrogen dosage.[69] Similar results were seen in women with PMS/PMDD but not in controls with interrupted ovarian function as both estradiol and progesterone induced symptoms.[23] Increased estradiol and progesterone plasma levels during the luteal phase in patients with PMS are related to more severe symptoms compared to cycles in the same individuals with lower levels.[70] Moreover, estradiol treatment during the luteal phase induced more negative symptoms than placebo in PMS/PMDD patients.[71] Estradiol and progesterone acting together seem to induce another response in the central nervous system than when they act separately.

20.8 Conclusion

Ovarian steroid sex hormones and their metabolites are of fundamental importance for inducing negative mood in PMS/PMDD. We are starting to understand the mechanism in that the $GABA_A$ receptor sensitivity seems to be changed in women with PMS/PMDD and sensitive persons seem to react upon $GABA_A$ receptor agonists in a bimodal inverted U-shaped manner.

Acknowledgements This work is supported by EU structural fund objective 1, Swedish research council Medicine (proj. 4X-11198), Västerbottens county, Umeå Municipal, Northern Sweden health region, Norrlands University Hospital, Umeå University foundations.

References

1. Kessler RC, McGonagle KA, Zhao S, et al. Lifetime and 12-month prevalence of DSM-III-R psychiatric disorders in the United States. Results from the National Comorbidity Survey. Arch Gen Psychiatry 1994; 51:8–19.
2. Weissman MM, Klerman GL. Sex differences and the epidemiology of depression. Arch Gen Psychiatry 1977; 34:98–111.
3. Angold A, Costello EJ, Erkanli A, et al. Pubertal changes in hormone levels and depression in girls. Psychol Med 1999; 29:1043–1053.

4. Soares CN, Cohen LS, Otto MW, et al. Characteristics of women with premenstrual dysphoric disorder (PMDD) who did or did not report history of depression: a preliminary report from the Harvard Study of Moods and Cycles. J Womens Health Gend Based Med 2001; 10:873–878.
5. Kendell RE, McGuire RJ, Connor Y, et al. Mood changes in the first three weeks after childbirth. J Affect Disord 1981; 3:317–326.
6. Chaudron LH, Klein MH, Remington P, et al. Predictors, prodromes and incidence of postpartum depression. J Psychosom Obstet Gynaecol 2001; 22:103–112.
7. Freeman EW, Sammel MD, Liu L, et al. Hormones and menopausal status as predictors of depression in women in transition to menopause. Arch Gen Psychiatry 2004; 61:62–70.
8. Wang M, Seippel L, Purdy RH, et al. Relationship between symptom severity and steroid variation in women with premenstrual syndrome: study on serum pregnenolone, pregnenolone sulfate, 5 alpha-pregnane-3,20-dione and 3 alpha-hydroxy-5 alpha-pregnan-20-one. J Clin Endocrinol Metab 1996; 81:1076–1082.
9. Bixo M, Andersson A, Winblad B, et al. Progesterone, 5alpha-pregnane-3,20-dione and 3alpha-hydroxy-5alpha-pregnane-20-one in specific regions of the human female brain in different endocrine states. Brain Res 1997; 764:173–178.
10. Ottander U, Poromaa IS, Bjurulf E, et al. Allopregnanolone and pregnanolone are produced by the human corpus luteum. Mol Cell Endocrinol 2005; 239:37–44.
11. Genazzani AR, Petraglia F, Bernardi F, et al. Circulating levels of allopregnanolone in humans: gender, age, and endocrine influences. J Clin Endocrinol Metab 1998; 83:2099–20103.
12. Bicikova M, Lapcik O, Hampl R, et al. A novel radioimmunoassay of allopregnanolone. Steroids 1995; 60:210–213.
13. Luisi S, Petraglia F, Benedetto C, et al. Serum allopregnanolone levels in pregnant women: changes during pregnancy, at delivery, and in hypertensive patients. J Clin Endocrinol Metab 2000; 85:2429–2433.
14. Sundström I, Andersson A, Nyberg S, et al. Patients with premenstrual syndrome have a different sensitivity to a neuroactive steroid during the menstrual cycle compared to control subjects. Neuroendocrinology 1998; 67:126–138.
15. Bäckström T, Sanders D, Leask RM, et al. Mood, sexuality, hormones and the menstrual cycle. II. Hormone levels and their relationship to premenstrual syndrome. Psychsom Med 1983; 45:503–507.
16. Hamarbäck S, Bäckström T. Induced anovulation as treatment of premenstrual tension syndrome. A double-blind cross-over study with GnRH-agonist versus placebo. Acta Obstet Gynecol Scand 1988; 67:159–166.
17. Hammarbäck, S, Ekholm UB, Bäckström, T. Spontaneous anovulation causing disappearance of cyclical symptoms in women with the premenstrual syndrome. Acta Endocrinol 1991; 125:132–137.
18. Mortola JF, Girton L, Fischer U. Successful treatment of severe premenstrual syndrome by combined use of gonadotropin-releasing hormone agonist and estrogen/progestin. J Clin Endocrinol Metab 1991; 72:252A–252F.
19. Hammarbäck S, Bäckström T, Holst J, et al. Cyclical mood changes as in the premenstrual tension syndrome during sequential estrogen-progestagen postmenopausal replacement treatment. Acta Obstet Gynecol Scand 1985; 64:393–397.
20. Magos AL, Brewster E, Sing R, et al. The effect of norethisterone in postmenopausal women on oestrogen therapy: a model for the premenstrual syndrome. Br J Obstet Gynaecol 1986; 93:1290–1296.
21. Björn I, Bixo M, Strandberg-Nöjd K, et al. Negative mood changes during hormone replacement therapy: a comparison between two progestogens. Am J Obstet Gynecol 2000; 183:1419–1426.
22. Andréen L, Sundström-Poromaa I, Bixo M, et al. Relationship between allopregnanolone and negative mood in postmenopausal women taking sequential hormone replacement therapy with vaginal progesterone. Psychoneuroendocrinology 2005; 30:212–224.

23. Schmidt PJ, Nieman LK, Danaceau MA, et al. Differential behavioral effects of gonadal steroids in women with and in those without premenstrual syndrome. N Engl J Med 1998; 338:209–216.
24. Chan AF, Mortola JF, Wood SH, et al. Persistence of premenstrual syndrome during low-dose administration of the progesterone antagonist RU 486. Obstet Gynecol 1994; 84:1001–1005.
25. Baulieu EE. Neurosteroids: a new function in the brain. Biol Cell 1991; 71:3–10.
26. Mellon SH. Neurosteroids: biochemistry, modes of action, and clinical relevance. J Clin Endocrinol Metab 1994; 78:1003–1008.
27. Majewska MD, Harrison NL, Schwartz RD, et al. Steroid hormone metabolites are barbiturate-like modulators of the GABA receptor. Science 1986; 232:1004–1007.
28. Sieghart W. Structure and pharmacology of gamma-aminobutyric acidA receptor subtypes. Pharmacol Rev 1995; 47:181–234.
29. Carl P, Hogskilde S, Nielsen JW, et al. Pregnanolone emulsion. A preliminary pharmacokinetic and pharmacodynamic study of a new intravenous anaesthetic agent. Anaesthesia 1990; 45:189–197.
30. Timby E, Balgård M, Nyberg S, et al. Pharmacokinetic and behavioral effects of allopregnanolone in healthy women. Psychopharmacology (Berl) 2006; 186:414–424.
31. Landgren S, Wang MD, Backstrom T, et al. Interaction between 3 alpha-hydroxy-5 alpha-pregnan-20-one and carbachol in the control of neuronal excitability in hippocampal slices of female rats in defined phases of the oestrus. Acta Physiol Scand 1998; 162:77–88.
32. Wieland S, Lan NC, Mirasedeghi S, et al. Anxiolytic activity of the progesterone metabolite 5 alpha-pregnan-3 alpha-o1-20-one. Brain Res 1991; 565:263–268.
33. Masia SL, Perrine K, Westbrook L, et al. Emotional outbursts and post-traumatic stress disorder during intracarotid amobarbital procedure. Neurology 2000; 54:1691–1693.
34. Weinbroum AA, Szold O, Ogorek D, et al. The midazolam-induced paradox phenomenon is reversible by flumazenil. Epidemiology, patient characteristics and review of the literature. Eur J Anaesthesiol 2001; 18:789–797.
35. Sveindottir H, Backstrom T. Prevalence of menstrual cycle symptom cyclicity and premenstrual dysphoric disorder in a random sample of women using and not using oral contraceptives. Acta Obstet Gynecol Scand 2000; 79:405–413.
36. Beauchamp MH, Ormerod BK, Jhamandas K, et al. Neurosteroids and reward: allopregnanolone produces a conditioned place aversion in rats. Pharmacol Biochem Behav 2000; 67:29–35.
37. Miczek KA, Fish EW, De Bold JF. Neurosteroids, GABAA receptors, and escalated aggressive behavior. Horm Behav 2003; 44:242–257.
38. Ben-Porath DD, Taylor SP. The effects of diazepam (valium) and aggressive disposition on human aggression: an experimental investigation. Addict Behav 2002; 27:167–177.
39. Weinbroum AA, Szold O, Ogorek D, Flaishon R. The midazolam-induced paradox phenomenon is reversible by flumazenil. Epidemiology, patient characteristics and review of the literature. Eur J Anaesthesiol, 2001; 18(12):789–797.
40. Kurthen M, Linke DB, Reuter BM, et al. Severe negative emotional reactions in intracarotid sodium amytal procedures: further evidence for hemispheric asymmetries? Cortex 1991; 27:333–337.
41. Lee GP, Loring DW, Meador KJ, et al. Severe behavioral complications following intracarotid sodium amobarbital injection: implications for hemispheric asymmetry of emotion. Neurology 1988; 38:1233–1236.
42. Cherek DR, Spiga R, Egli M. Effects of response requirement and alcohol on human aggressive responding. J Exp Anal Behav 1992; 58:577–587.
43. Dougherty DM, Cherek DR, Bennett RH. The effects of alcohol on the aggressive responding of women. J Stud Alcohol 1996; 57:178–186.
44. Björn I, Bixo M, Nöjd K, et al. The impact of different doses of medroxyprogesterone acetate on mood symptoms in sequential hormonal therapy. Gynecol Endocrinol 2002; 16:1–8.
45. Andréen L, Sundström-Poromaa I, Bixo M, et al. Allopregnanolone concentration and mood-a bimodal association in postmenopausal women treated with oral progesterone. Psychopharmacology (Berl) 2006; 187:209–221.

46. Hall RC, Zisook S. Paradoxical reactions to benzodiazepines. Br J Clin Pharmacol 1981; 11(Suppl 1):99S–104S.
47. Honan VJ. Paradoxical reaction to midazolam and control with flumazenil. Gastrointest Endosc 1994; 40:86–88.
48. Wenzel RR, Bartel T, Eggebrecht H, et al. Central-nervous side effects of midazolam during transesophageal echocardiography. J Am Soc Echocardiogr 2002; 15:1297–1300.
49. Weinbroum AA, Szold O, Ogorek D, et al. The midazolam-induced paradox phenomenon is reversible by flumazenil. Epidemiology, patient characteristics and review of the literature. Eur J Anaesthesiol 2001; 18:789–797.
50. Gourley SL, Debold JF, Yin W, et al. Benzodiazepines and heightened aggressive behavior in rats: reduction by GABA(A)/alpha(1) receptor antagonists. Psychopharmacology (Berl) 2005; 178:232–240.
51. Weerts EM, Tornatzky W, Miczek KA. "Anxiolytic" and "anxiogenic" benzodiazepines and beta-carbolines: effects on aggressive and social behavior in rats and squirrel monkeys. Psychopharmacology (Berl) 1993; 110:451–459.
52. Epperson CN, Haga K, Mason GF, et al. Cortical gamma-aminobutyric acid levels across the menstrual cycle in healthy women and those with premenstrual dysphoric disorder: a proton magnetic resonance spectroscopy study. Arch Gen Psychiatry 2002; 59:851–858.
53. Fernández G, Weis S, Stoffel-Wagner B, et al. Menstrual cycle-dependent neural plasticity in the adult human brain is hormone, task, and region specific. J Neurosci 2003; 23:3790–3795.
54. van Wingen G, van Broekhoven F, Verkes RJ, et al. Progesterone selectively increases amygdala reactivity in women. Molecular Psychiatry 2007; June 19 [epub ahead of print].
55. Paulus MP, Feinstein JS, Castillo G, et al. Dose-dependent decrease of activation in bilateral amygdala and insula by lorazepam during emotion processing. Arch Gen Psychiatry 2005; 62:282–288.
56. N-Wihlbäck AC, Sundström-Poromaa I, Bäckström T. Action by and sensitivity to neuroactive steroids in menstrual cycle related CNS disorders. Psychopharmacology (Berl) 2006; 186:388–401.
57. Parizek A, Hill M, Kancheva R, et al. Neuroactive pregnanolone isomers during pregnancy. J Clin Endocrinol Metab 2005; 90:395–403.
58. Akwa Y, Purdy RH, Koob GF, et al. The amygdala mediates the anxiolytic-like effect of the neurosteroid allopregnanolone in rat. Behav Brain Res 1999; 106:119–125.
59. Cullberg J. Mood changes and menstrual symptoms with different gestagen/estrogen combinations. A double blind comparison with placebo. Acta Psychiat Scand 1972; 236 (Suppl):1–84.
60. Sundström I, Ashbrook D, Bäckström T. Reduced benzodiazepine sensitivity in patients with premenstrual syndrome: a pilot study. Psychoneuroendocrinology 1997; 22:25–38.
61. Sundström I, Nyberg S, Bäckström T. Patients with premenstrual syndrome have reduced sensitivity to midazolam compared to control subjects. Neuropsychopharmacology 1997; 17:370–381.
62. Nyberg S, Wahlström G, Bäckström T, et al. Altered sensitivity to alcohol in the late luteal phase among patients with premenstrual dysphoric disorder. Psychoneuroendocrinology 2004; 29:767–777.
63. Gulinello M, Gong QH, Li X, et al. Short-term exposure to a neuroactive steroid increases alpha4 GABA(A) receptor subunit levels in association with increased anxiety in the female rat. Brain Res 2001; 910:55–66.
64. Löfgren M, Johansson IM, Meyerson B, et al. Progesterone withdrawal effects in the open field test can be predicted by elevated plus maze performance. Horm Behav 2006; 50:208–215.
65. Yu R, Follesa P, Ticku MK. Down-regulation of the GABA receptor subunits mRNA levels in mammalian cultured cortical neurons following chronic neurosteroid treatment. Brain Res Mol Brain Res 1996; 41:163–168.

66. Zhu D, Birzniece V, Bäckström T, et al. Dynamic aspects of acute tolerance to allopregnanolone evaluated using anaesthesia threshold in male rats. Br J Anaesth 2004; 93:560–567.
67. Birzniece V, Türkmen S, Lindblad C, et al. GABA(A) receptor changes in acute allopregnanolone tolerance. Eur J Pharmacol 2006; 535:125–134.
68. Smith SS, Gong QH, Hsu FC, et al. GABA(A) receptor alpha4 subunit suppression prevents withdrawal properties of an endogenous steroid. Nature 1998; 392:926–930.
69. Björn I, Sundström-Poromaa I Bixo M, et al. Increase of estrogen dose deteriorates mood during progestin phase in sequential hormonal therapy. J Clin Endocrinol Metab 2003; 88:2026–2030.
70. Hammarbäck S, Damber JE, Bäckström T. Relationship between symptom severity and hormone changes in women with premenstrual syndrome. J Clin Endocrinol Metab 1989; 68(1):125–130.
71. Dhar V, Murphy BE. Double-blind randomized crossover trial of luteal phase estrogens (Premarin) in the premenstrual syndrome (PMS). Psychoneuroendocrinology 1990; 15:489–493.

Chapter 21
The Role of Neuroactive Steroids in Anxiety Disorders

Erin M. MacKenzie, Glen B. Baker, and Jean-Michel Le Mellédo

Abstract Given that neuroactive steroids modulate the activity of a number of neurotransmitter receptors in the brain thought to be involved in the pathophysiology of mood and anxiety disorders (including $GABA_A$ and NMDA receptors), it is not surprising that increasing evidence suggests that levels of some neuroactive steroids may be altered in individuals suffering from these disorders. This chapter will provide an in-depth summary of the findings regarding neuroactive steroid abnormalities in panic disorder, generalized anxiety disorder, generalized social phobia and post-traumatic stress disorder. How these abnormalities may be related to illness symptoms or whether they may reflect compensatory mechanisms for maintaining homeostasis will be explored. Furthermore, the regulation of steroidal abnormalities by drugs used to treat the above-mentioned disorders will be discussed.

Keywords Neuroactive steroids, neurosteroids, anxiety, post-traumatic stress disorder, panic disorder, social phobia, hypothalamic-pituitary-adrenal axis

Abbreviations 3β,5α-THP (3β-hydroxy-5α-pregnan-20-one; 3β,5α-tetrahydroprogesterone; epiallopregnanolone); ACTH (adrenocorticotrophic hormone); ALLO (3α-hydroxy-5α-pregnan-20-one; 3α,5α-tetrahydroprogesterone; allopregnanolone); CCK-4 (cholecystokinin tetrapeptide); CRH (corticotropin releasing hormone); CSF (cerebral spinal fluid); DHEA (3β-hydroxy-androst-5-en-17-one; dehydroepiandrosterone); DHEAS (3β-hydroxy-androst-5-en-17-one sulfate; dehydroepiandrosterone sulfate); GABA (gamma-aminobutyric acid); HPA (hypothalamic pituitary adrenal); NMDA (*N*-methyl-D-aspartate); P (progesterone); PD (panic disorder); PregS (3β-hydroxypregn-5-en-20-one sulfate; pregnenolone sulfate); PTSD (post-traumatic stress disorder); SP (social phobia); SSRI (selective serotonin reuptake inhibitor); THDOC (3α,20-dihydroxy-5α-pregnan-20-one; 3α,5α- tetrahydrodeoxycorticosterone)

Neurochemical Research Unit, Department of Psychiatry, University of Alberta, Edmonton, Alberta, Canada

M.S. Ritsner and A. Weizman (eds.), *Neuroactive Steroids in Brain Function, Behavior and Neuropsychiatric Disorders*

21.1 Introduction

Neuroactive steroids are steroids that produce rapid, non-genomic actions in the brain, generally through actions on ligand-gated ion channels. The role of these steroids in anxiety disorders such as post-traumatic stress disorder (PTSD), panic disorder (PD) and social phobia (SP) has not been studied as extensively as their role in other psychiatric disorders, although a growing number of studies are investigating these relationships. Results to date are often contradictory, indicating that further studies are needed. In this chapter, we will summarize the findings pertaining to neuroactive steroid function in individuals suffering from anxiety disorders.

While neuroactive steroids exert their effects on a variety of ion channel subtypes, their modulatory effects at the $GABA_A$ receptor have been most extensively studied. The 3α-reduced metabolites of progesterone (P), i.e., 3α-hydroxy-5α-pregnan-20-one (3α,5α-tetrahydroprogesterone; allopregnanolone; ALLO) and 3α,20-dihydroxy-5α-pregnan-20-one (tetrahydrodeoxycorticosterone; THDOC), are among the most potent known positive allosteric modulators of the $GABA_A$ receptor, whereas 3β-hydroxy-androst-5-en-17-one (dehydroepiandrosterone; DHEA), DHEA sulfate (DHEAS) and 3β-hydroxypregn-5-en-20-one sulfate (pregnenolone sulfate; PregS) are negative modulators at this receptor complex.[1,2] Given the strong evidence for GABAergic dysregulation in anxiety disorders, it is not surprising that neuroactive steroids affecting the activity of the $GABA_A$ receptor may also be involved in the pathophysiology of anxiety disorders.

Altered levels of neuroactive steroids may be related to the pathophysiology of anxiety disorders in many ways. Firstly, altered neuroactive steroid levels may play a direct role in the behaviours observed in anxiety disorders. Evidence that positive modulation of the $GABA_A$ receptor by neuroactive steroids is associated with anxiolytic activity is extensive. For example, administration of ALLO and THDOC to rodents induces anxiolytic activity in behavioural measures of anxiety, such as the elevated plus maze, the Geller-Seifter test, the lick suppression test, the open-field test, the light/dark transition test and the burying behaviour paradigm.[3–6] Changes in anxiety-like behaviour have been observed with natural steroid fluctuations in rodents (such as over the course of the ovarian cycle,[7] or during pregnancy[8]), with higher brain levels of 3α-reduced steroids corresponding to decreased anxiety-like behaviour. Treatment of rodents with ALLO prevents anxiety-like behaviour induced by caffeine administration[9] and predator odour exposure,[10] and the anxiolytic effects of ethanol have been postulated to be the result of ethanol-induced increases in brain ALLO.[11] Although ALLO also antagonizes nicotinic acetylcholine receptor activity, the anxiolytic effect of this steroid is thought to be mediated solely via the $GABA_A$ receptor given that other neuroactive steroids that antagonize the nicotinic acetylcholine receptor are behaviourally inert.[12] From these examples, it is logical to suggest that alterations in the endogenous levels of these steroids could alter anxiety-related behaviours in anxiety disorders.

Secondly, neuroactive steroids interact with the hypothalamic–pituitary–adrenal (HPA) axis, which might be dysfunctional in patients with anxiety disorders. The HPA axis is under tight GABAergic control, and therefore could be indirectly regulated by neuroactive steroids affecting GABAergic transmission. However, neuroactive steroids have also been shown to directly affect HPA axis function.

The physiological and behavioural responses to stress are achieved upon activation of the HPA axis. Briefly, corticotropin releasing hormone (CRH) and arginine-vasopressin (AVP) are released from the paraventricular nucleus of the hypothalamus in response to stress, which stimulates the release of adrenocorticotrophic hormone (ACTH) from the pituitary gland. ACTH stimulates the release of glucocorticoids from the adrenal cortex. Negative feedback mechanisms exist at various sites on the HPA axis to regulate its activity, and evidence suggests that stress-induced changes in neuroactive steroid levels may be part of this feedback mechanism. For example, acute stress markedly increases brain and plasma levels of P, ALLO and THDOC,[13–15] likely counteracting the decrease in $GABA_A$ receptor activity observed in response to acute stress. This stress-induced increase in neuroactive steroids may also directly contribute to regulation of HPA axis activity, as ALLO was shown to suppress methoxamine-stimulated CRH release in vitro,[16] administration of ALLO to rodents was shown to reduce the ACTH and corticosterone response to emotional stress,[17] and central administration of corticosterone antiserum was shown to increase the corticosterone response to cold swimming stress.[18] THDOC was also shown to reduce stress-induced increases in plasma corticosterone.[19] Interestingly, adrenalectomized and gonadectomized animals do not exhibit stress-induced increases in neuroactive steroid concentrations,[13,15] suggesting that activation of the HPA axis by stress may stimulate neuroactive steroid release of adrenal origin. During chronic stress, GABAergic function is also reduced, but brain and plasma levels of P, ALLO and THDOC are also significantly decreased. Studies have shown that the HPA axis is hyperactive during chronic stress,[20–22] and the reduced GABAergic transmission and reduced levels of steroids potentiating GABAergic transmission may be related to this effect.

It should be noted that DHEA may also affect HPA axis activity in a mechanism independent of GABAergic transmission, given that this neuroactive steroid has been shown to possess antiglucocorticoid activity in the brain.[23] In line with this, it was shown in postmenopausal women that the ACTH response to psychological stress was increased after DHEA treatment.[24] In this way, abnormal concentrations of these steroids could also influence the reactivity of the HPA axis, leading to the dysfunctional stress responses observed in individuals with anxiety disorders.

Further support for a role of neuroactive steroids in anxiety disorders lies in the fact that many drugs used to treat anxiety disorders increase GABAergic transmission. The selective serotonin reuptake inhibitiors (SSRIs) exhibit anxiolytic effects in rodents[25] and humans,[26,27] and are most often prescribed for the treatment of anxiety disorders. Acute treatment with the SSRI citalopram was shown to significantly increase GABA levels in the occipital cortex of healthy volunteers,[28] although to our knowledge this finding has not been extended to chronic SSRI treatment. Furthermore, the SSRIs paroxetine, sertraline and fluoxetine were shown to alter the activity of neurosteroidogenic enzymes in the rodent brain.[29] This finding supports the results from other studies reporting that paroxetine and fluoxetine significantly increased brain levels of ALLO[30,31] and THDOC,[31] whereas brain levels of P and Preg were either increased[31] or not changed,[30] and that fluoxetine and norfluoxetine reversed the chronic stress-related decrease in brain ALLO observed in mice.[32,33] Although chronic fluoxetine administration was shown to decrease brain ALLO levels in mice (which is in contrast to chronic paroxetine treatment, which reportedly increased brain ALLO[34]), significant increases in brain ALLO were still achieved after acute fluoxetine challenge in these animals, indicating that while chronic drug treatment may decrease basal brain levels, each administration would still evoke increases in brain ALLO, and a role of this increase in the therapeutic effects of the drug cannot be discounted. SSRIs increase brain levels of ALLO at doses that are ineffective in blocking serotonergic reuptake, possibly explaining why SSRIs are anxiolytic at doses that are below those required for antidepressant activity. In fact, Pinna et al.[35] have suggested that SSRIs should be redefined as "selective brain steroidogenic stimulants," given their potential to exert therapeutic effects resulting from alterations in brain neuroactive steroid levels.

Other classes of drugs that are used for the treatment of anxiety disorders include benzodiazepines and anticonvulsants. Benzodiazepines bind to their active site on the $GABA_A$ receptor, facilitating the effects of GABA. They are often used for the rapid alleviation of the symptoms of some (but not all) anxiety disorders, although tolerance may occur. Anticonvulsant drugs, such as vigabatrin, gabapentin, topiramate and valproate, have also been shown to be effective in the treatment of anxiety in both rodents and humans. The mechanisms of action of these drugs have in common the ability to increase GABAergic transmission, probably contributing to their anxiolytic effects. The fact that GABAergic drugs are effective in treating the symptoms of anxiety disorders suggests that endogenous modulators of the $GABA_A$ receptor (i.e., neuroactive steroids) may be involved in the expression and/or reduction of anxiety symptoms.

In the following sections, findings of neuroactive steroid levels in patients with PTSD, PD and SP will be described. These results will be discussed with respect to how these alterations may affect behaviour and HPA axis functioning, and also how these alterations may be related to the pathophysiology of the illnesses. While studies are often conflicting, patterns of dysregulation are beginning to emerge; however, it is evident that further studies are warranted in these populations.

21.2 Post-Traumatic Stress Disorder (PTSD)

PTSD is an often-severe disorder characterized by symptoms of re-experiencing, avoidance and hyperarousal that arise following exposure to a traumatic event. While a number of neurobiological and neuroendocrine systems have been suggested to be altered in PTSD compared to healthy volunteers, studies suggest that altered GABAergic function may be a contributing factor to the development and/or pathophysiology of PTSD. Plasma GABA levels were significantly lower following motor vehicle accident trauma in individuals who went on to develop acute PTSD compared to those that did not,[36] and higher GABA levels were shown to be protective against the development of chronic PTSD.[37] These results must be interpreted with caution, however, as conflicting results have also been reported; decreased benzodiazepine receptor binding was demonstrated in the frontal lobes of male Vietnam war veterans with chronic PTSD,[38] although this result was not confirmed in Gulf War veterans with a shorter history of disease.[39] Furthermore, drugs that increase GABAergic transmission have been shown to be effective in reducing some symptoms of PTSD in some, but not all, studies.

Given the role of GABA in the regulation of HPA axis activity, it is not surprising that disruption of the HPA axis in these patients is well-established; studies have shown that individuals with PTSD exhibit decreased circulating cortisol levels,[40,41] decreased urinary cortisol excretion,[42] and decreased cortisol response to stress and dexamethasone challenge (although the latter finding may be related to trauma exposure rather than to PTSD). Enhanced sensitivity of pituitary glucocorticoid receptors may be responsible for these effects, in that hypersensitivity of the negative feedback mechanism in the HPA axis of PTSD patients results in blunted cortisol levels and responses.

Studies have also investigated $GABA_A$ receptor-modulating neuroactive steroids in individuals suffering from PTSD. ALLO was shown to be remarkably decreased in the CSF of female patients with PTSD compared to non-traumatized controls, and ALLO levels were lower in patients with co-morbid major depression compared to PTSD patients without depression.[12] Additionally, the ALLO to DHEA ratio was inversely correlated with re-experiencing symptoms (such as intrusive thoughts, nightmares, flashbacks, and emotional and physiological reactivity at trauma cue exposure). CSF levels of 5α-DHP and progesterone were not significantly different between PTSD patients and healthy controls.[12]

Levels of DHEA and its sulfated ester DHEAS (negative modulators of the $GABA_A$ receptor and positive modulators of the glutamate NMDA receptor[1]) have also been investigated in PTSD. Increased plasma DHEA and/or DHEAS levels and lower cortisol to DHEA ratios were observed in males with combat-related PTSD[43,44] and in refugees with PTSD[45] compared to healthy controls (although no changes in plasma DHEA and DHEAS levels in patients with PTSD compared to healthy controls have also been reported[12,40]). It has also been shown that females with PTSD have an increased capacity for DHEA release upon adrenal stimulation (via intravenous ACTH administration) compared to healthy controls, and that peak DHEA levels in response to ACTH are significantly higher in PTSD patients compared to healthy controls and traumatized non-PTSD controls.[46] Furthermore, significantly greater plasma DHEA levels were observed in male veterans with PTSD who had attempted suicide compared to those who had not.[47]

Dysregulation of excitatory/inhibitory transmission in the brain may contribute to the pathophysiology of PTSD, by affecting both anxious behaviour and HPA axis activity. For example, a decrease in overall GABAergic function may exist in patients with PTSD, given the decrease in positively modulating GABAergic neuroactive steroids and an increase in the negatively modulating steroids, resulting in increased anxiety. Furthermore, the hyperactivity observed in the HPA axis of

PTSD patients may arise from this excitatory/inhibitory imbalance; decreased concentrations of ALLO, a positive modulator of $GABA_A$ receptors, together with increased circulating levels of DHEA and/or DHEAS, negative modulators of the $GABA_A$ receptor and positive modulators of NMDA receptors, could lead to a decrease in inhibitory control over HPA axis activity.[12,46] Furthermore, DHEA exerts antiglucocorticoid activity in the brain, which has been shown to alter HPA axis activity; treatment with an antiglucocorticoid ligand was shown to increase the reactivity of the HPA axis in both animals and humans in a similar manner to that observed in patients with PTSD.[48,49]

Interestingly, however, it has also been shown in females with PTSD that DHEA reactivity was inversely correlated with symptom severity, suggesting that DHEA may be protective against PTSD.[46] This is supported by a report in which PTSD symptoms in five women with chronic, treatment-resistant PTSD were significantly reduced following treatment with 3-acetyl-7-keto-DHEA, a metabolite of DHEA with antiglucocorticoid properties.[50]

Drugs that are frequently used for the treatment of PTSD include anticonvulsants[51] and SSRIs.[52] It was shown that after 12 months of treatment with the SSRI paroxetine, there was a trend for decreased basal salivary cortisol levels, and cognitive stress-induced increases in salivary cortisol were significantly reduced compared to pretreatment stress-induced levels, even though the perception of stress was similar.[53] It is possible that the therapeutic effects of the drugs used to treat the symptoms of PTSD do so by regulating HPA axis activity, either via drug-induced changes in neuroactive steroid levels, or resulting in changes in neuroactive steroid levels. The use of benzodiazepines alone in PTSD has not been very successful, and is generally not recommended.[54]

21.3 Panic Disorder

PD is an anxiety disorder characterized by unexpected and repeated panic attacks and persistent anxiety regarding these attacks, leading to considerable social and psychological consequences. As with other anxiety disorders, evidence suggests that GABAergic transmission may be abnormal in individuals suffering from PD. Although patients with PD were shown to have normal plasma[55] and CSF[56] GABA levels, imaging studies have shown reduced brain GABA levels in individuals with PD compared to healthy controls.[57,58] Furthermore, benzodiazepine binding has been shown to be decreased in the brains of PD patients compared to healthy controls,[59,60] and the GABAergic response to benzodiazepines was shown to be blunted in PD compared to controls.[61]

HPA axis irregularities have also been demonstrated in PD in some (but not all) studies, and dysregulation of GABAergic transmission could contribute to HPA axis-related dysfunction. Some studies have shown that the HPA axis is hyperresponsive in PD,[62] and it has been proposed that HPA hyperactivity may occur to a greater extent in PD compared to controls in situations that are novel, threatening or uncontrollable.[63] Conflicting results regarding HPA axis reactivity in PD have been suggested by Abelson and colleagues[63] to result from differences in "how basal" the basal states are before experimental testing; for example, a short accommodation period before HPA axis testing would result in HPA axis hyperactivity, whereas a prolonged period of rest before testing would remove the factors contributing to this HPA axis hyperactivity, leading to normal HPA axis function.

Neuroactive steroids affecting GABAergic activity have also been reported to be altered in individuals with PD, and dysregulation of these steroids may play a role in the pathophysiology of the illness. Plasma P was reported to be increased in unmedicated male patients with PD,[64] as well as in female patients during the mid-luteal phase of the menstrual cycle,[65] compared to healthy controls (although P has also been reported to be normal in PD[66]). Plasma ALLO was reported to be elevated in unmedicated patients with PD[66] (and in unmedicated females with PD during the follicular phase[65]) compared to controls, while other studies observed normal circulating ALLO levels in unmedicated PD patients.[64] Plasma 5α-DHP[66] levels were also similar in PD patients and healthy controls, and THDOC levels were similar[64] or higher[65] in PD compared to controls. Interestingly, epiallopregnanolone (3β,5α-THP), a functional antagonist of the $GABA_A$ receptor, was shown to be significantly decreased in PD patients compared to controls.[66] Hypersecretion of the 3α-reduced neuroactive steroids, together with reduced circulating levels of 3β,5α-THP may be part of a homeostatic mechanism compensating for low GABAergic activity that may be inherent in PD. This might serve not only to reduce anxiety and the incidence of spontaneous panic attacks,[65] but also to reduce hyperactivity of the HPA axis.

Studies have also reported that plasma DHEA levels were significantly increased in males with PD compared to healthy controls.[64] While DHEA itself is a negative modulator of the $GABA_A$ receptor, it has also been shown to exert antiglucocorticoid effects in the brain; increased DHEA could be a compensatory effect to counteract HPA axis hyperactivity in PD. Interestingly, studies in mice have reported that frontal cortex DHEA levels were positively correlated with anti-anxiety behaviour in the elevated-plus maze,[67] further supporting a compensatory role for increased DHEA in PD.

Experimental induction of panic is frequently used to mimic the spontaneous panic attacks occurring in PD. Panicogenic agents such as sodium lactate, pentagastrin and cholecystokinin tetrapeptide (CCK-4) induce short-lived panic-like behaviour in patients with PD and to a lesser extent in healthy subjects, and neuroactive steroid concentrations have been measured during these pharmacologically induced panic attacks. Panic induction in patients with PD using sodium lactate and CCK-4 was shown to cause a marked reduction in circulating ALLO and 3α,5β-THP and a marked increase in plasma 3β,5α-THP compared to baseline values. Interestingly, plasma ALLO concentrations were not altered in healthy controls after administration of panicogenic agents, regardless of the intensity of the panic response,[68,69] although plasma THDOC levels were significantly increased after panic induction with CCK-4.[70] Unfortunately THDOC has not, to our knowledge, been measured in PD patients after panic induction. These results suggest that the neuroactive steroid response to experimentally induced panic attacks in patients with PD is different than that observed in healthy controls, and may shed some light on how patients with PD are more prone to panic attacks (due to decreased GABAergic transmission). Interestingly, it has been reported that increases in circulating cortisol resulted from anticipation of panic attacks rather than the actual experience of panic[71]; investigations such as this with neuroactive steroids should shed some light as to whether these changes in steroids are related to HPA axis activity or occur independently of the stress response.

Interestingly, PD is often treated with SSRIs, which have been shown in rodents to increase brain levels of ALLO. Studies in PD have shown that circulating levels of neuroactive steroids do not change from baseline after chronic treatment with paroxetine,[66] although the possibility of drug-induced changes in brain levels of anxiolytic neuroactive steroids cannot be ruled out, particularly given the finding in mice that chronic paroxetine treatment increased brain, but not plasma, ALLO concentrations.[34] Anticonvulsant drugs such as vigabatrin and tiagabine have also been shown to be effective in treating PD, and SSRIs, anticonvulsants, and benzodiazepines have shown some utility in reducing pharmacologically induced panic.[68]

21.4 Social Phobia

SP, or social anxiety disorder, is characterized by pervasive fear and anxiety regarding social and/or performance situations that can lead to panic attacks, and is associated with significant distress. Although GABA agonists show promising results in the treatment of PD, data regarding GABAergic function in patients with SP is lacking, and conflicting results regarding HPA axis functioning have been reported. While basal cortisol and ACTH levels appear to be normal in patients with SP,[72–75] it has been reported that patients with SP exhibited significantly greater increases in salivary cortisol following psychological stress (a speech task) compared to healthy controls, (although cortisol responses to a physical stressor were similar between the two groups).[72] Further, the maximal cortisol response to mental stress in patients with SP was greater than that observed in controls.[75] These results might indicate that patients with SP exhibit hyperactivity of the HPA axis, although these results are in contrast to other studies showing no differences in cortisol responses to stress between patients with SP and controls.[71,76] Normal cortisol responses in response to dexamethasone administration were also observed in patients with SP.[74]

Few studies have investigated the levels of neuroactive steroids in patients with SP. Decreased plasma concentrations of PregS were observed in patients with SP compared to healthy controls,[77] while observed normal plasma levels of ALLO, DHEA, DHEAS and Preg were in individuals with SP.[77,78] The decreased concentration of PregS was suggested to reflect a homeostatic decrease in negative modulation of the $GABA_A$ receptor to reduce anxiety,[77] although the plasma levels of other GABA-modulating neuroactive steroids were not significantly different from controls. Interestingly, no studies on SP have investigated the concentrations of neuroactive steroids during a stressful event or task; given the findings that HPA axis functioning is also normal during baseline conditions, further studies investigating the differential effects on neuroactive steroid concentrations in patients with SP and controls undergoing stress are warranted.

GABAergic drugs such as the monoamine oxidase inhibitor phenelzine[79] (which significantly elevates brain GABA levels in addition to elevating biogenic amines[80]), benzodiazepines,[81] SSRIs[82,83] and GABAergic anticonvulsants[84,85] have been shown to be effective in treating SP. The anticonvulsant pregabalin[86] have also been shown to be useful, and although there is still some discussion about its mechanism of action, it has been reported to bind to calcium channels, modulate calcium influx and influence GABAergic neurotransmission.[87] Thus, although a role of neuroactive steroids in abnormal behaviour or physiological functioning in patients with SP is not clear, they certainly warrant further research given their interaction with GABA and the inhibitory/excitatory balance in the brain.

21.5 Conclusions

The role of neuroactive steroids in the development, physiology and presentation of anxiety disorders is far from clear. The results of studies investigating neuroactive steroid levels in patients with anxiety disorders are conflicting, and it appears that there are no trends in neuroactive steroid dysregulation that can be generalized to anxiety disorders as a group; instead, each anxiety disorder appears to have a specific pattern of steroid dysregulation. The alterations in neuroactive steroid concentrations could lead to behaviours associated with anxiety disorders, and could also lead to the physiological changes (such as HPA axis functioning) observed in anxiety disorders. Compensatory changes in neuroactive steroid levels appear to occur in some forms of anxiety, such as PD, while changes in steroid concentrations may contribute to the symptomatology and/or pathophysiology of other forms of anxiety, such as PTSD.

Further studies investigating neuroactive steroid levels in anxiety disorders are warranted, not only to determine altered levels of particular steroids, but also to ascertain the ratios of steroids relative to each other in order to determine the net effect, given that steroids often have opposite effects on a given system or receptor complex. In addition, further studies are needed to determine the precise role of neuroactive steroids in the treatment of anxiety symptoms; pharmacological agents used to treat the symptoms of anxiety disorders often alter brain steroid levels, and understanding the role of these steroid changes in the therapeutic effects of these drugs may in the future lead to more specific and effective drug treatments.

Acknowledgements The authors are grateful to the following for ongoing funding: the Canadian Institutes for Health Research, the Alberta Heritage Foundation for Medical Research, the Canada Research Chairs and Canada Foundation for Innovation programs, the Davey Endowment and the Berger Endowment.

References

1. Baulieu EE, Robel P. Dehydroepiandrosterone (DHEA) and dehydroepiandrosterone sulfate (DHEAS) as neuroactive neurosteroids. Proc Natl Acad Sci USA 1998; 95:4089–4091.
2. Rupprecht R. Neuroactive steroids: mechanisms of action and neuropsychopharmacological properties. Psychoneuroendocrinology 2003; 28:139–168.
3. Bitran D, Hilvers RJ, Kellogg CK. Anxiolytic effects of 3 alpha-hydroxy-5 alpha[beta]-pregnan-20-one: endogenous metabolites of progesterone that are active at the GABAA receptor. Brain Res 1991; 561:157–161.
4. Mayo W, Vallee M, Darnaudry M, et al. Neurosteroids. Behavioral studies. In: Baulieu EE, Robel P, Schumacher M (eds). Contemporary endocrinology: neurosteroids. A new regulatory function in the nervous system. Totowa, NJ: Humana, 1999, pp. 317–335.
5. Crawley JN, Glowa JR, Majewska MD, et al. Anxiolytic activity of an endogenous adrenal steroid. Brain Res 1986; 398:382–385.
6. Wieland S, Lan NC, Mirasedeghi S, et al. Anxiolytic activity of the progesterone metabolite 5 alpha-pregnan-3 alpha-o1–20-one. Brain Res 1991; 565:263–268.
7. Frye CA,Walf AA. Changes in progesterone metabolites in the hippocampus can modulate open field and forced swim test behavior of proestrous rats. Horm Behav 2002; 41:306–315.
8. de Brito Faturi C, Teixeira-Silva F,Leite JR. The anxiolytic effect of pregnancy in rats is reversed by finasteride. Pharmacol Biochem Behav 2006; 85:569–574.
9. Jain NS, Hirani K,Chopde CT. Reversal of caffeine-induced anxiety by neurosteroid 3-alpha-hydroxy-5-alpha-pregnane-20-one in rats. Neuropharmacology 2005; 48:627–638.
10. Kavaliers M, Wiebe JP,Galea LA. Reduction of predator odor-induced anxiety in mice by the neurosteroid 3 alpha-hydroxy-4-pregnen-20-one (3 alpha HP). Brain Res 1994; 645:325–329.
11. Hirani K, Sharma AN, Jain NS, et al. Evaluation of GABAergic neuroactive steroid 3alpha-hydroxy-5alpha-pregnane-20-one as a neurobiological substrate for the anti-anxiety effect of ethanol in rats. Psychopharmacology (Berl) 2005; 180:267–278.
12. Rasmusson AM, Pinna G, Paliwal P, et al. Decreased cerebrospinal fluid allopregnanolone levels in women with posttraumatic stress disorder. Biol Psychiatry 2006; 60:704–713.
13. Purdy RH, Morrow AL, Moore PH, Jr., et al. Stress-induced elevations of gamma-aminobutyric acid type A receptor-active steroids in the rat brain. Proc Natl Acad Sci USA 1991; 88:4553–4557.
14. Barbaccia ML, Roscetti G, Bolacchi F, et al. Stress-induced increase in brain neuroactive steroids: antagonism by abecarnil. Pharmacol Biochem Behav 1996; 54:205–210.
15. Barbaccia ML, Roscetti G, Trabucchi M, et al. The effects of inhibitors of GABAergic transmission and stress on brain and plasma allopregnanolone concentrations. Br J Pharmacol 1997; 120:1582–1588.
16. Patchev VK, Shoaib M, Holsboer F, et al. The neurosteroid tetrahydroprogesterone counteracts corticotropin-releasing hormone-induced anxiety and alters the release and gene expression of corticotropin-releasing hormone in the rat hypothalamus. Neuroscience 1994; 62:265–271.
17. Patchev VK, Hassan AH, Holsboer DF, et al. The neurosteroid tetrahydroprogesterone attenuates the endocrine response to stress and exerts glucocorticoid-like effects on vasopressin gene transcription in the rat hypothalamus. Neuropsychopharmacology 1996; 15:533–540.
18. Guo AL, Petraglia F, Criscuolo M, et al. Evidence for a role of neurosteroids in modulation of diurnal changes and acute stress-induced corticosterone secretion in rats. Gynecol Endocrinol 1995; 9:1–7.
19. Owens MJ, Ritchie JC,Nemeroff CB. 5 alpha-pregnane-3 alpha, 21-diol-20-one (THDOC) attenuates mild stress-induced increases in plasma corticosterone via a non-glucocorticoid mechanism: comparison with alprazolam. Brain Res 1992; 573:353–355.
20. Aguilera G,Rabadan-Diehl C. Vasopressinergic regulation of the hypothalamic-pituitary-adrenal axis: implications for stress adaptation. Regul Pept 2000; 96:23–29.

21. Bartanusz V, Jezova D, Bertini LT, et al. Stress-induced increase in vasopressin and corticotropin-releasing factor expression in hypophysiotrophic paraventricular neurons. Endocrinology 1993; 132:895–902.
22. Herman JP, Adams D,Prewitt C. Regulatory changes in neuroendocrine stress-integrative circuitry produced by a variable stress paradigm. Neuroendocrinology 1995; 61:180–190.
23. Kalimi M, Shafagoj Y, Loria R, et al. Anti-glucocorticoid effects of dehydroepiandrosterone (DHEA). Mol Cell Biochem 1994; 131:99–104.
24. Kudielka BM, Hellhammer J, Hellhammer DH, et al. Sex differences in endocrine and psychological responses to psychosocial stress in healthy elderly subjects and the impact of a 2-week dehydroepiandrosterone treatment. J Clin Endocrinol Metab 1998; 83:1756–1761.
25. Silva MT, Alves CR,Santarem EM. Anxiogenic-like effect of acute and chronic fluoxetine on rats tested on the elevated plus-maze. Braz J Med Biol Res 1999; 32:333–339.
26. Ballenger JC. Remission rates in patients with anxiety disorders treated with paroxetine. J Clin Psychiatry 2004; 65:1696–1707.
27. Hedges DW, Brown BL, Shwalb DA, et al. The efficacy of selective serotonin reuptake inhibitors in adult social anxiety disorder: a meta-analysis of double-blind, placebo-controlled trials. J Psychopharmacol 2007; 21:102–111.
28. Bhagwagar Z, Wylezinska M, Taylor M, et al. Increased brain GABA concentrations following acute administration of a selective serotonin reuptake inhibitor. Am J Psychiatry 2004; 161:368–370.
29. Griffin LD,Mellon SH. Selective serotonin reuptake inhibitors directly alter activity of neurosteroidogenic enzymes. Proc Natl Acad Sci USA 1999; 96:13512–13517.
30. Uzunov DP, Cooper TB, Costa E, et al. Fluoxetine-elicited changes in brain neurosteroid content measured by negative ion mass fragmentography. Proc Natl Acad Sci USA 1996; 93:12599–12604.
31. Serra M, Pisu MG, Muggironi M, et al. Opposite effects of short- versus long-term administration of fluoxetine on the concentrations of neuroactive steroids in rat plasma and brain. Psychopharmacology (Berl) 2001; 158:48–54.
32. Pinna G, Costa E,Guidotti A. Fluoxetine and norfluoxetine stereospecifically facilitate pentobarbital sedation by increasing neurosteroids. Proc Natl Acad Sci USA 2004; 101:6222–6225.
33. Matsumoto K, Puia G, Dong E, et al. GABA(A) receptor neurotransmission dysfunction in a mouse model of social isolation-induced stress: possible insights into a non-serotonergic mechanism of action of SSRIs in mood and anxiety disorders. Stress 2007; 10:3–12.
34. Nechmad A, Maayan R, Spivak B, et al. Brain neurosteroid changes after paroxetine administration in mice. Eur Neuropsychopharmacol 2003; 13:327–332.
35. Pinna G, Costa E,Guidotti A. Fluoxetine and norfluoxetine stereospecifically and selectively increase brain neurosteroid content at doses that are inactive on 5-HT reuptake. Psychopharmacology (Berl) 2006; 186:362–372.
36. Vaiva G, Thomas P, Ducrocq F, et al. Low posttrauma GABA plasma levels as a predictive factor in the development of acute posttraumatic stress disorder. Biol Psychiatry 2004; 55:250–254.
37. Vaiva G, Boss V, Ducrocq F, et al. Relationship between posttrauma GABA plasma levels and PTSD at 1-year follow-up. Am J Psychiatry 2006; 163:1446–1448.
38. Bremner JD, Innis RB, Southwick SM, et al. Decreased benzodiazepine receptor binding in prefrontal cortex in combat-related posttraumatic stress disorder. Am J Psychiatry 2000; 157:1120–1126.
39. Fujita M, Southwick SM, Denucci CC, et al. Central type benzodiazepine receptors in Gulf War veterans with posttraumatic stress disorder. Biol Psychiatry 2004; 56:95–100.
40. Olff M, Guzelcan Y, de Vries GJ, et al. HPA- and HPT-axis alterations in chronic posttraumatic stress disorder. Psychoneuroendocrino 2006; 31:1220–1230.
41. Yehuda R, Teicher MH, Trestman RL, et al. Cortisol regulation in posttraumatic stress disorder and major depression: a chronobiological analysis. Biol Psychiatry 1996; 40:79–88.
42. Yehuda R, Kahana B, Binder-Brynes K, et al. Low urinary cortisol excretion in Holocaust survivors with posttraumatic stress disorder. Am J Psychiatry 1995; 152:982–986.

43. Spivak B, Maayan R, Kotler M, et al. Elevated circulatory level of GABA(A)-antagonistic neurosteroids in patients with combat-related post-traumatic stress disorder. Psychol Med 2000; 30:1227–1231.
44. Yehuda R, Brand SR, Golier JA, et al. Clinical correlates of DHEA associated with post-traumatic stress disorder. Acta Psychiatr Scand 2006; 114:187–193.
45. Sondergaard HP, Hansson LO,Theorell T. Elevated blood levels of dehydroepiandrosterone sulphate vary with symptom load in posttraumatic stress disorder: findings from a longitudinal study of refugees in Sweden. Psychother Psychosom 2002; 71:298–303.
46. Rasmusson AM, Vasek J, Lipschitz DS, et al. An increased capacity for adrenal DHEA release is associated with decreased avoidance and negative mood symptoms in women with PTSD. Neuropsychopharmacol 2004; 29:1546–1557.
47. Butterfield MI, Stechuchak KM, Connor KM, et al. Neuroactive steroids and suicidality in posttraumatic stress disorder. Am J Psychiatry 2005; 162:380–382.
48. Bertagna X, Escourolle H, Pinquier JL, et al. Administration of RU 486 for 8 days in normal volunteers: antiglucocorticoid effect with no evidence of peripheral cortisol deprivation. J Clin Endocrinol Metab 1994; 78:375–380.
49. van Haarst AD, Oitzl MS, Workel JO, et al. Chronic brain glucocorticoid receptor blockade enhances the rise in circadian and stress-induced pituitary-adrenal activity. Endocrinology 1996; 137:4935–4943.
50. Sageman S,Brown RP. 3-Acetyl-7-oxo-dehydroepiandrosterone for healing treatment-resistant posttraumatic stress disorder in women: 5 case reports. J Clin Psychiatry 2006; 67:493–496.
51. Berigan T. Treatment of posttraumatic stress disorder with tiagabine. Can J Psychiatry 2002; 47:788.
52. van der Kolk BA, Dreyfuss D, Michaels M, et al. Fluoxetine in posttraumatic stress disorder. J Clin Psychiatry 1994; 55:517–522.
53. Vermetten E, Vythilingam M, Schmahl C, et al. Alterations in stress reactivity after long-term treatment with paroxetine in women with posttraumatic stress disorder. Ann NY Acad Sci 2006; 1071:184–202.
54. Davidson JR. Use of benzodiazepines in social anxiety disorder, generalized anxiety disorder, and posttraumatic stress disorder. J Clin Psychiatry 2004; 65(Suppl 5):29–33.
55. Goddard AW, Narayan M, Woods SW, et al. Plasma levels of gamma-aminobutyric acid and panic disorder. Psychiatry Res 1996; 63:223–225.
56. Rimon R, Lepola U, Jolkkonen J, et al. Cerebrospinal fluid gamma-aminobutyric acid in patients with panic disorder. Biol Psychiatry 1995; 38:737–741.
57. Goddard AW, Mason GF, Almai A, et al. Reductions in occipital cortex GABA levels in panic disorder detected with 1H-magnetic resonance spectroscopy. Arch Gen Psychiatry 2001; 58:556–561.
58. Ham BJ, Sung Y, Kim N, et al. Decreased GABA levels in anterior cingulate and basal ganglia in medicated subjects with panic disorder: a proton magnetic resonance spectroscopy (1H-MRS) study. Prog Neuropsychopharmacol Biol Psychiatry 2007; 31:403–411.
59. Bremner JD, Innis RB, White T, et al. SPECT [I-123]iomazenil measurement of the benzodiazepine receptor in panic disorder. Biol Psychiatry 2000; 47:96–106.
60. Malizia AL, Cunningham VJ, Bell CJ, et al. Decreased brain GABA(A)-benzodiazepine receptor binding in panic disorder: preliminary results from a quantitative PET study. Arch Gen Psychiatry 1998; 55:715–720.
61. Goddard AW, Mason GF, Appel M, et al. Impaired GABA neuronal response to acute benzodiazepine administration in panic disorder. Am J Psychiatry 2004; 161:2186–2193.
62. Erhardt A, Ising M, Unschuld PG, et al. Regulation of the hypothalamic-pituitary-adrenocortical system in patients with panic disorder. Neuropsychopharmacol 2006; 31:2515–2522.
63. Abelson JL, Khan S, Liberzon I, et al. HPA axis activity in patients with panic disorder: review and synthesis of four studies. Depress Anxiety 2007; 24:66–76.
64. Brambilla F, Mellado C, Alciati A, et al. Plasma concentrations of anxiolytic neuroactive steroids in men with panic disorder. Psychiatry Res 2005; 135:185–190.

65. Brambilla F, Biggio G, Pisu MG, et al. Neurosteroid secretion in panic disorder. Psychiatry Res 2003; 118:107–116.
66. Strohle A, Romeo E, di Michele F, et al. GABAA receptor-modulating neuroactive steroid composition in patients with panic disorder before and during paroxetine treatment. Am J Psychiatry 2002; 159:145–147.
67. Maayan R, Touati-Werner D, Ram E, et al. The protective effect of frontal cortex dehydroepiandrosterone in anxiety and depressive models in mice. Pharmacol Biochem Behav 2006; 85:415–421.
68. Zwanzger P, Eser D, Padberg F, et al. Neuroactive steroids are not affected by panic induction with 50 microg cholecystokinin-tetrapeptide (CCK-4) in healthy volunteers. J Psychiatr Res 2004; 38:215–217.
69. Strohle A, Romeo E, di Michele F, et al. Induced panic attacks shift gamma-aminobutyric acid type A receptor modulatory neuroactive steroid composition in patients with panic disorder: preliminary results. Arch Gen Psychiatry 2003; 60:161–168.
70. Eser D, di Michele F, Zwanzger P, et al. Panic induction with cholecystokinin-tetrapeptide (CCK-4) Increases plasma concentrations of the neuroactive steroid 3alpha,5alpha-tetrahydrodeoxycorticosterone (3alpha,5alpha-THDOC) in healthy volunteers. Neuropsychopharmacol 2005; 30:192–195.
71. Martel FL, Hayward C, Lyons DM, et al. Salivary cortisol levels in socially phobic adolescent girls. Depress Anxiety 1999; 10:25–27.
72. Furlan PM, DeMartinis N, Schweizer E, et al. Abnormal salivary cortisol levels in social phobic patients in response to acute psychological but not physical stress. Biol Psychiatry 2001; 50:254–259.
73. Potts NL, Davidson JR, Krishnan KR, et al. Levels of urinary free cortisol in social phobia. J Clin Psychiatry 1991; 52(Suppl):41–42.
74. Uhde TW, Tancer ME, Gelernter CS, et al. Normal urinary free cortisol and postdexamethasone cortisol in social phobia: comparison to normal volunteers. J Affect Disord 1994; 30:155–161.
75. Condren RM, O'Neill A, Ryan MC, et al. HPA axis response to a psychological stressor in generalised social phobia. Psychoneuroendocrino 2002; 27:693–703.
76. Levin AP, Saoud JB, Strauman T, et al. Responses of "generalized" and "discrete" social phobics during public speaking. J Anxiety Disord 1993; 7:207–221.
77. Heydari B, Le Melledo JM. Low pregnenolone sulphate plasma concentrations in patients with generalized social phobia. Psychol Med 2002; 32:929–933.
78. Laufer N, Maayan R, Hermesh H, et al. Involvement of GABAA receptor modulating neuroactive steroids in patients with social phobia. Psychiatry Res 2005; 137:131–136.
79. Heimberg RG, Liebowitz MR, Hope DA, et al. Cognitive behavioral group therapy vs phenelzine therapy for social phobia: 12-week outcome. Arch Gen Psychiatry 1998; 55:1133–1141.
80. Baker GB, Wong JT, Yeung JM, et al. Effects of the antidepressant phenelzine on brain levels of gamma-aminobutyric acid (GABA). J Affect Disord 1991; 21:207–211.
81. Davidson JR, Potts N, Richichi E, et al. Treatment of social phobia with clonazepam and placebo. J Clin Psychopharmacol 1993; 13:423–428.
82. van Vliet IM, den Boer JA, Westenberg HG. Psychopharmacological treatment of social phobia; a double blind placebo controlled study with fluvoxamine. Psychopharmacology (Berl) 1994; 115:128–134.
83. Lepola U, Bergtholdt B, St Lambert J, et al. Controlled-release paroxetine in the treatment of patients with social anxiety disorder. J Clin Psychiatry 2004; 65:222–229.
84. Van Ameringen M, Mancini C, Pipe B, et al. An open trial of topiramate in the treatment of generalized social phobia. J Clin Psychiatry 2004; 65:1674–1678.
85. Pande AC, Davidson JR, Jefferson JW, et al. Treatment of social phobia with gabapentin: a placebo-controlled study. J Clin Psychopharmacol 1999; 19:341–348.
86. Pande AC, Feltner DE, Jefferson JW, et al. Efficacy of the novel anxiolytic pregabalin in social anxiety disorder: a placebo-controlled, multicenter study. J Clin Psychopharmacol 2004; 24:141–149.
87. Lyrica (Pregabalin) – New gabapentinoid with wide clinical application. Pfizer. Available at: http://www.drugdevelopment-technology.com/projects/pregabalin. Accessed on June 2, 2007.

Chapter 22
The Role of Midbrain 3α,5α-THP in Mediating Exploration, Anxiety, Social, and Reproductive Behavior

Cheryl A. Frye[1] **and Madeline E. Rhodes**[2]

Abstract Ovarian steroids, including 17β-estradiol (E_2) and progesterone (P_4) influence the onset and duration of reproductive behavior of female rodents. Separate lines of investigation also suggest that E_2 and P_4 influence exploratory, anxiety, and social behaviors. In addition to ovarian secretion, E_2 and P_4 are also neurosteroids produced *de novo* in brain. The midbrain ventral tegmental area (VTA), which is involved in motivation and reward, is an important brain area for mediating steroids' effects on reproductive behavior. As such, this chapter discusses research from our laboratory on the role, sources, and substrates of steroid hormones' modulation of exploratory, anxiety, social, and reproductive behaviors. The approach that we have used is manipulating, in the VTA, E_2 and P_4 and its metabolites, dihydroprogesterone (DHP) and 5α-pregnan-3α-ol-20-one (3α,5α-THP), their *de novo* production, and subsequent effects on behavior via traditional actions at intracellular progestin receptors (PRs) and non-traditional substrates, such as $GABA_A$, NMDA, and dopamine receptors. Endpoints examined include behavior in the open field, elevated plus maze, social choice, social interaction, and paced mating tasks and levels of E_2 and progestins in serum, midbrain, hippocampus, striatum, and cortex. Manipulating 3α,5α-THP in the VTA influences exploration, anxiety, social and reproductive behavior, as well as neurosteroidogenesis in the VTA, hippocampus, and cortex.

Keywords Neurosteroids, non-genomic, lordosis, reproductive behavior, learning, anxiety

Abbreviations E_2 17β-estradiol; P_4 progesterone; VTA midbrain ventral tegmental area; DHP dihydroprogesterone; 3α,5α-THP 5α-pregnan-3α-ol-20-one; PRs intracellular progestin receptors; $GABA_A$ receptors $GABA_A$/benzodiazepine receptors

[1] The Psychology and Biology Departments, Centers for Life Science and Neuroscience Research, The University at Albany, Albany, New York, USA

[2] The Department of Psychology Smith College, Northampton, Massachusetts, USA

M.S. Ritsner and A. Weizman (eds.), *Neuroactive Steroids in Brain Function, Behavior and Neuropsychiatric Disorders*

Sequential actions of ovarian hormones, such as 17β-estradiol (E_2) and progesterone (P_4) have actions in the brain to mediate the onset and duration of sexual receptivity.[1] Our research has focused on the central mechanisms by which P_4 mediates lordosis. Findings from localization, lesion, and hormone implant studies reveal that P_4's actions in the hypothalamus *and* the midbrain ventral tegmental area (VTA) are important for mediating lordosis of E_2-primed rodents. There are higher concentrations of P_4 in the midbrain VTA than the hypothalamus (cortex or hippocampus) of female rodents.[2] Lesions to either the midbrain VTA or hypothalamus, or knife cuts between these regions, attenuate P_4-facilitated lordosis of rats and hamsters.[3–5] Furthermore, applying P_4 to the midbrain VTA and/or hypothalamus of ovariectomized, E_2-primed rodents consistently enhances receptivity of rats,[6–8] hamsters[7,9,10] and mice.[11–14] Given that P_4's actions in the VTA for facilitating lordosis are conserved across rodent species, this implies that progestins actions in this region are particularly important in mediating reproductive behavior.

It is important to understand further the role and mechanisms of P_4's actions in the midbrain VTA to mediate lordosis. This review focuses on research from our laboratory, regarding the role of progestins' in the VTA for lordosis, as well as, the source and substrates for progestins' actions in these sites. This review will be accomplished in the following manner. First, we discuss results from our laboratory in which the actions of progestins in the midbrain VTA are manipulated and effects on lordosis behavior are examined. Second, the role of these actions of progestins in other systems, and putative relevance for the etiology and/or therapeutic basis of neuropsychiatric disorders, are also discussed. Third, we review findings from our laboratory which demonstrate that the role of progestins' actions in the VTA extend beyond effects on lordosis, and influence exploration, anxiety, social and reproductive behavior, as well as endocrine responses in other brain regions. As such, the latter research approach may be particularly relevant as a model for elucidating progestins' actions related to stress-associated neuropsychiatric disorders.

22.1 Effects on Lordosis of Manipulating Progestins in the Midbrain VTA

Our research has investigated the mechanisms by which progestins' have their actions in the midbrain VTA to mediate the onset and duration of sexual behavior of female rodents. One approach has been to use a model system in which progestins' mechanisms are manipulated in the midbrain VTA of rodents and subsequent effects on lordosis, the stereotypic posture, which female rodents assume to enable mating to occur (given appropriate hormonal and environmental stimulation), are examined. In effect, our research has used lordosis behavior as a bioassay, or an index, which is sensitive to, and will reveal effects of, manipulations to P_4's mechanisms that are essential to the normative expression of this behavior. Using this paradigm, we have found that P_4 has classic actions through cognate intracellular progestin receptors (PRs – "genomic" actions) in the hypothalamus to initiate lordosis, whereas in the midbrain VTA, the 5α-reduced, 3-hydroxylated metabolite of P_4, 3α-hydroxy-5α-pregnan-20-one (3α,5α-THP) has actions independent of PRs ("non-genomic" actions) to modulate lordosis. 3α,5α-THP's effects in the VTA to modulate lordosis of rodents occur via its actions at γ-aminobutyric acid ($GABA_A$)/benzodiazepine receptor complexes ($GABA_A$), glutamate type *N*-methyl-D-asparate receptors (NMDARs), and dopamine (DA) type 1 receptors (D_1) and their downstream signal transduction processes. The findings supporting this are discussed below.

22.1.1 Are P_4's Facilitatory Effects on Lordosis Mediated by 3α,5α-THP?

Evidence suggest that some of P_4's actions to mediate lordosis may involve formation of 3α,5α-THP. P_4 is secreted by the ovaries and is metabolized in the brain through actions of the enzyme 5α-reductase, which forms 5α-pregnane-3,20-dione (DHP). DHP is then reduced by the 3α-hydroxysteroid dehydrogenase (3α-HSD) enzyme to form 3α,5α-THP, which can also oxidize 3α,5α-THP back to DHP. Given the ready conversion of P_4 to 3α,5α-THP, whether P_4's effects to facilitate lordosis involve its metabolism to 3α,5α-THP, is of interest.

Increases in both circulating P_4 and 3α,5α-THP are associated with sexual receptivity of rodents. Gonadally intact female rodents typically display lordosis in response to appropriate stimulation from males during the proestrous phase of the estrous cycle, when plasma levels of E_2, P_4, and 3α,5α-THP levels are elevated compared to the diestrous phase, when female rodents are typically not receptive.[14–18] Furthermore, among proestrous mice, those with higher levels of 3α,5α-THP exhibit greater lordosis responses.[13] Removal of the primary source of progestins, the ovaries, eliminates natural increases in steroid hormones and subsequent facilitation of sexual receptivity. Systemic replacement of E_2 followed by P_4 or 3α,5α-THP to ovariectomized rodents reinstates lordosis.[12,19] Rapid facilitation of sexual receptivity (within 2–3 min) occurs when ovariectomized, E_2-primed, rats or mice are intravenously infused with P_4 or 3α,5α-THP.[12,20–25] These data suggest P_4's facilitation of lordosis may involve central actions of 3α,5α-THP in the hypothalamus, midbrain VTA, and/or other sites.

Effects of various progestins to facilitate lordosis, when applied to the hypothalamus and midbrain VTA, have been compared. In the hypothalamus, implants of P_4 are more effective than its 5α-reduced metabolite, DHP, at facilitating lordosis within 30 min. However, application of P_4, DHP and/or 3α,5α-THP to the hypothalamus significantly (and similarly) increase lordosis compared to vehicle infusions 2 h later.[26] Facilitative effects of P_4 to the hypothalamus can be blocked by co-administration of the PR and glucocorticoid antagonist, RU38486.[27] In the midbrain VTA, implants or infusions of 3α,5α-THP produce higher levels of lordosis than does P_4.[6,12,28] Further, 3α,5α-THP to the VTA significantly increases lordosis within 5 min; whereas, P_4 can require a longer latency to produce its facilitatory effects, which implies that P_4's actions in the midbrain may require its metabolism to 3α,5α-THP. Notably, the two enzymes, 5α-reductase and 3α-HSD, which convert P_4 to 3α,5α-THP have been identified in the midbrain VTA[11] as well as other brain regions (olfactory bulb, cortex, thalamus, hypothalamus, hippocampus, basal ganglia, cerebellum, pons, medulla).[29,30] Activity of the 5α-reductase enzyme (the limiting enzyme in the conversion of P_4 to 3α,5α-THP) is greater in the midbrain tegmentum than the hypothalamus, hippocampus and/or cortex in mouse and rat brain.[31,32] Thus, there is the capacity for P_4 in the midbrain VTA to be readily converted to 3α,5α-THP.

In order to address whether P_4's metabolism to 3α,5α-THP is necessary for progestin-facilitated lordosis, we have manipulated the 5α-reductase or 3α-HSD enzymes and examined subsequent effects on lordosis. Systemic or intra-VTA administration of 5α-reductase or 3α-HSD inhibitors attenuates lordosis and midbrain 3α,5α-THP levels of naturally receptive or ovariectomized, hormone-primed rodents.[11,13,19,33] Conversely, infusions to the VTA of fluoxetine, which shifts 3α-HSD activity to favor reduction over oxidation,[34] facilitates lordosis and increases midbrain 3α,5α-THP levels.[35] These findings suggest that P_4's metabolism to 3α,5α-THP in the VTA is important for lordosis and that the VTA may be a primary site where P_4 is metabolized to 3α,5α-THP for mediating lordosis.

3α,5α-THP is produced in peripheral endocrine glands and brain. In the adrenals, 3α,5α-THP is derived from P_4 as a result of ACTH and cAMP-induced steroidogenesis.[36] Notably, central concentrations of 3α,5α-THP are independent of circulating levels. In support, brain levels of 3α,5α-THP are higher than plasma, and detectable in adrenalectomized animals.[14,17,18,37–40] The brain expresses diazepam binding inhibitor, which is an endogenous ligand for central and peripheral mitochondrial benzodiazepine receptors (MBRs), and has been implicated in adrenal biosynthesis of 3α,5α-THP.[41] In glial cells, MBRs facilitate the translocation of cholesterol across the mitochondrial membranes to make it available to the P450 side chain cleavage enzyme (P450scc), which converts it to pregnenolone.[42] Pregnenolone is subsequently metabolized by 3β-hydroxysteroid dehydrogenase (3β-HSD) to P_4, which is converted by 5α-reductase and 3α-HSD to 3α,5α-THP. Each of the enzymes that are necessary for neurosteroidogenesis of 3α,5α-THP (P450scc, 3β-HSD, 5α-reductase, 3α-HSD) have been localized to midbrain of rats.[11,43] Thus, we have investigated whether *de novo* synthesis of 3α,5α-THP in the VTA is essential in the mediation of lordosis.

We have investigated effects on lordosis of manipulating *de novo* synthesis of 3α,5α-THP in the VTA. Infusions of MBR antagonists to the VTA decreases lordosis and midbrain 3α,5α-THP levels of naturally receptive or hormone-primed rats and hamsters.[44,45] Infusions of P450scc or 3β-HSD inhibitors also attenuate lordosis of proestrous rats.[46] Conversely, application of MBR agonists, or other neurosteroidogenic agents, such as fluoxetine or olanzapine, to the VTA increase lordosis and midbrain 3α,5α-THP concentrations of naturally receptive or hormone-primed rodents.[36,45,47] Together these data support the notion that induction of female sexual behavior is, in part, dependent upon biosynthesis of neurosteroids in the VTA.

22.2 Effects on Lordosis of Manipulating Substrates in the Midbrain VTA Through Which Progestins May Have Actions

22.2.1 Progestins' Actions at PRs in the VTA Are Not Required to Facilitate Sexual Receptivity

Intracellular progestin receptors (PRs) belong to a family of ligand-activated transcription factors, which include receptors for steroid hormones, thyroid hormones, retinoic acid, and vitamin D. The regulation of gene expression by E_2 and/or P_4 through their interaction with PRs occurs as follows. First, these steroids interact with PRs that in the unliganded form are associated with proteins, such as heat shock proteins, HSP90 and HSP70, FKBP51, FKBP52 and/or p23. After hormone binding, PR dissociates from these chaperone proteins, and dimerizes with another PR molecule (PR to PR). Next, in the liganded form PR is phosphorylated, and forms complexes with cofactors, such as the co-activator SRC-1 or the co-repressor SMRT that modify chromatin organization. PR interacts with short specific DNA sequences, known as hormone responsive elements, mainly located in the promoter region of progestin target genes. This interaction also induces PR phosphorylation and PR recruits the basic machinery of transcription involved in gene expression,[48,49] which results in subsequent changes in protein synthesis, and ultimately cellular function.[50,51]

One approach that has been used to begin to address whether progestins have actions via PRs is to examine the latency for their effects. Changes in protein synthesis initiated by P_4-PR may require a minimum of 10–15 min to occur.[52] As such, faster effects of progestins are typically attributed to mechanisms that may be independent of PRs. P increases cell firing in the VTA 60 s after intravenous (IV) infusion to E_2-primed rats; within 120 s lordosis is enhanced and cell firing in the VTA is increased further.[21] These findings imply that the rapid effects of P_4 in the VTA to facilitate lordosis may be independent of actions at intracellular PRs.

Evidence suggests that P_4's facilitatory effects on lordosis involve classic actions at PRs in the hypothalamus[53,54] and are independent of PRs in the VTA. First, E_2 and P_4 regulate the expression of PRs both at transcriptional and translational levels in a tissue and hormone-dependent manner. For example, in hypothalamus, like most tissues, there are many PRs which are up-regulated by E_2 and down-regulated by P_4.[11,55–57] Whereas there are few PRs in the midbrain VTA, and those identified in this region are not induced by E_2.[11,12,56–58] Second, blocking the function of PRs in the hypothalamus with antagonists or antisense oligonucleotides produces decrements in lordosis; whereas these same infusions to the VTA do not alter lordosis of rodents.[27,59–61] Third, some progestins that facilitate lordosis have low relative binding affinities at PRs. Although P_4 and DHP bind with high affinity for PRs, 3α,5α-THP, is devoid of affinity for PRs in physiological concentrations.[62] Fourth, P_4 and 3α,5α-THP can facilitate lordosis of mice that are bred specifically to have mutations in PRs (PRKOs).[12] Fifth, P_4-conjugated to bovine serum album (P/BSA), which does not readily cross cell membranes, has effects similar to free P_4 when applied to the VTA to facilitate lordosis of E_2-primed rodents or PRKO mice.[12,63] In the hypothalamus, only free P_4, not P:BSA, is effective at facilitating lordosis,[6,64] indicating that P/BSA does not become unbound to produce its behavioral effects. Together these data suggest that while P_4's actions in the hypothalamus can be PR-mediated, in the VTA, P_4's actions for facilitation of lordosis can occur independent of cognate PRs. This implies that there must be other actions of P_4 in the VTA to mediate some of these effects on lordosis.

22.2.2 Progestins' Actions at $GABA_A$ Receptors in the VTA to Facilitate Sexual Receptivity

Progestins are positive, allosteric modulators of $GABA_A$/benzodiazepine receptors ($GABA_A$ receptors) and can increase binding of GABA to $GABA_A$ receptors.[65,66] Progestins differ in their efficacy at enhancing the function of $GABA_A$ receptors. The two structural requirements associated with progestins that are the most effective positive modulators of GBRs are a 5α- or 5β-reduced steroid A-ring and a 3α-OH group,[66] which are both characteristics of 3α,5α-THP. Of all the steroids that have been examined and occur endogenously, 3α,5α-THP is the most potent modulator of $GABA_A$ receptors.[66–68] In nanomolar concentrations, 3α,5α-THP can increase GABA-stimulated chloride influx in rat brain cortical synaptoneurosomes[68] and enhance noncompetitive binding of $GABA_A$ receptors.[67,69] Given that 3α,5α-THP, when in nanomolar to low micromolar range, has positive modulatory effects on $GABA_A$ receptors, this implies the physiological significance of these interactions. Evidence that progestin-facilitated lordosis may involve actions at $GABA_A$ receptors include that during proestrous when 3α,5α-THP levels are high, the number, density, and affinity of $GABA_A$ receptors are enhanced compared to other phases of the estrous cycle, when progestin levels are low.[70,71]

There is evidence that GABA, acting at $GABA_A$ receptors in the hypothalamus, facilitates lordosis behavior of rodents. Within the hypothalamus, E_2 and P_4 enhance muscimol binding.[72,73] P_4 when administered in the micromolar range increases GABA-induced chloride currents in the hypothalamus.[74] In proestrous female rats, endogenous levels of GABA are increased in the hypothalamus coincident with sexual receptivity,[75] which may be due to increased concomitant activity of glutamic acid decarboxylase (GAD) enzymes, which synthesize GABA from glutamate. GAD 67, an enzyme isoform that is important for the synthesis of tonic, synaptically released GABA, is increased on the afternoon of proestrous within the hypothalamus.[76] These findings suggest that in addition to progestins actions via PRs in the hypothalamus (discussed above), there may also be actions of progestin at $GABA_A$ receptors in the hypothalamus to facilitate sexual receptivity.

In the VTA, there are changes in GABAergic function associated with endogenous variations in 3α,5α-THP and sexual receptivity of rodents. There are GABAergic neurons in the VTA[77] and midbrain GABA and $GABA_A$ receptor sensitivity change over the estrous cycle and with mating.[11,78] Midbrain GABA levels are higher during proestrus, than diestrus, and increased after mating.[11,78] As well, during proestrus, in midbrain tissues of rodents less GABA is needed to displace muscimol, from $GABA_A$ receptors.[11,78] Thus, during proestrus, when progestins levels are high, and rodents are sexually receptive, GABA and $GABA_A$ receptor function in the midbrain VTA are greater, than that observed during diestrus.

There are also effects of manipulating GABA and/or $GABA_A$ receptor function in the VTA to alter progestin-facilitated lordosis. Infusions of GAD antisense to the VTA knocks down GAD immunoreactivity and disrupts lordosis of naturally receptive or hormone-primed rodents compared to that seen following vehicle or sense infusions.[12,21] Infusions to the VTA of a GABA-transaminase inhibitor, which increases GABA by preventing its degradation, facilitates lordosis of hamsters.[79] Infusions of $GABA_A$ receptor agonists and antagonists to the VTA respectively enhance and inhibit P-facilitated lordosis of rodents.[6,11,21,79] Similarly, progestins' efficacy to facilitate lordosis when applied to the VTA corresponds with their ability to activate $GABA_A$ receptors in the midbrain VTA.[12] Thus, altering GABA levels or $GABA_A$ receptor function in the VTA influences progestin-facilitated sexual receptivity of rodents, which suggests that progestins may facilitate lordosis in the VTA, in part, by enhancing GABAergic function via actions at $GABA_A$ receptors.

22.2.3 Can P_4 Have Actions via NMDARs to Mediate Lordosis?

P_4 may also have actions to modulate sexual receptivity by interacting with NMDA receptors. First, there are NMDA receptors (NMDARs) in the hypothalamus and VTA,[80] which when manipulated pharmacologically, produce both electrophysiological and behavioral changes.[81] Second, progestins can modulate NMDARs in a rapid fashion by altering NMDA-activated calcium influx and blocking formation of 3α,5α-THP attenuates this effect.[82,83] Indeed, 3α,5α-THP in micromolar concentrations exerts antagonist activity at NMDA receptors. Third, P_4 decreases NMDA receptor binding in the frontal cortex of ovariectomized rats.[84] Fourth, P_4 attenuates excitatory responses in cerebellar Purkinje cells produces by NMDA receptor agonists and persistent inflammatory hyperalgesia.[83] Fifth, in the hypothalamus, progestins can modulate the release of glutamate and thereby alter NMDA receptor function.[85] Sixth, blocking NMDA receptors decreases sexual receptivity of rodents. Systemic administration of NMDA receptor blockers attenuates lordosis of hormone-primed rodents.[86,87] Thus, progestins can alter NMDA receptor function and influence sexual behavior of rodents; however, whether these effects are due to actions in the hypothalamus and/or midbrain are of interest.

Manipulating NMDARs in the hypothalamus influences the expression of lordosis behavior of rodents. Infusions of a noncompetitive NMDAR antagonist, to the hypothalamus, decreases lordosis of hormone-primed rat or hamsters.[11,86] Lordosis of hormone-primed rats is also inhibited with intrahypothalamic infusions of competitive NMDAR antagonists.[88] Thus, blocking NMDA receptors in the hypothalamus can inhibit lordosis of rodents.

Manipulating NMDARs in the midbrain VTA alters lordosis behavior of rodents. Infusions of a noncompetitive NMDAR antagonist to the VTA increase lordosis of hormone-primed rats or hamsters, above those seen with vehicle infusions.[11,86] These functional effects of NMDAR antagonists to the VTA are similar to that produced by intra-VTA infusions of $GABA_A$ agonists.[6,11] Indeed, progestins' actions at NMDARs and GBRs in the VTA may mediate female sexual behavior, in part, through NMDAR's and GBR's subsequent effects at dopaminergic neurons. Increasing $GABA_A$ function and decreasing NMDARs in the VTA can elevate and reduce respectively extracellular DA concentrations, which are associated with changes in sexual receptivity.[11,89,90] For example, mesolimbic DA function is increased during proestrus, attenuated by ovariectomy, and reinstated by E_2 + P_4, but neither E_2 nor P_4 alone.[91–94] Mating, particularly when females can pace their sexual contacts, further increases DA concentrations in the VTA.[11] Increasing the functioning of DA neurons in the VTA enhances sexual receptivity.[90,95–97] Thus, in the VTA, it may be that NMDA neurons excite, and GABAergic neurons inhibit, dopaminergic cells to regulate DA levels and thereby lordosis.

22.2.4 Can P_4 Have Actions via D_1 to Mediate Lordosis?

Progestins and D_1 ligands can have actions in the hypothalamus via PRs to facilitate lordosis. Intracerebroventricular infusions of P_4 or the D_1 agonist, SKF38393, facilitate lordosis of E_2-primed rats,[98,99] but their facilitative effects are attenuated when D_1 or PRs are knocked down in the hypothalamus.[98,99] Furthermore, among PR knockout mice, infusions of SKF38393 are ineffective at facilitating lordosis and P_4 has marginal facilitative effects compared to those seen among wild-type mice.[98] Thus, in the hypothalamus, enhancing effects of P_4 and D_1 ligands on lordosis may involve PRs.

In the midbrain VTA, P_4 and D_1 ligands may enhance lordosis independent of actions at PRs. In support, lordosis of rats and hamsters is attenuated when actions at D_1 receptors in the midbrain VTA are blocked. Infusions of anti-D_1 mRNA or of the D_1 antagonist, SCH23390VTA, significantly reduce lordosis of cycling rats and hamsters in behavioral estrus, compared to vehicle-infused animals.[12] D_1 activity in the midbrain VTA is also necessary for P_4-facilitated lordosis. Ovariectomized, E_2-primed rats infused with the D_1 antagonist SCH23390 to the midbrain VTA and P_4 30 min later showed significantly less lordosis than rats that received vehicle infusions. In contrast, infusions of the D_1 agonist SKF3839 (but not vehicle) to the midbrain VTA, followed by P_4 30 min later, produces enhanced lordosis responses in ovariectomized, E_2-primed rats. These data suggest that progestin-facilitated lordosis may involve actions at membrane-bound D_1 receptors.

22.2.5 Is cAMP a Substrate for Progestin-facilitated Lordosis?

Activation of adenosine 3′,5′-monophosphate (cAMP) may be involved in P_4's facilitation of lordosis. *In vitro*, P_4 rapidly (within 15 min) augments forskolin-induced cAMP in neocortical slices from ovariectomized rats. P_4's ability to enhance forskolin-induced increases in cAMP is attenuated by presence of bicuculline or pictotoxin, which suggests that this process is GBR-dependent. Further, concentrations of 3α,5α-THP, a potent positive modulator of GBRs, were positively correlated increases in cAMP levels. Hence, P_4-induced increases in cAMP may be mediated by actions of 3α,5α-THP at GBRs.[100]

Increasing cAMP *in vivo* also enhances lordosis. Systemic administration of P_4, but not vehicle, to ovariectomized, E_2-primed rats significantly increases central cAMP concentrations and lordosis.[101] Ovariectomized, E_2-primed rats administered SC injections of the cAMP agonist dibutyryl cAMP have increased lordosis responses. Notably, dibutyryl cAMP's facilitative effects are attenuated when administered with the anti-progestin RU38486. Thus, cAMP may be a substrate for progestin-facilitated lordosis, which may be dependent, in part, on PRs in the hypothalamus.

In the hypothalamus, cAMP is likely an important substrate for facilitating lordosis. Ovariectomized, E_2-primed rats administered systemic P_4 have significantly higher levels of cAMP in the hypothalamus following mating than do vehicle-administered controls.[101] Augmenting cAMP directly in the hypothalamus also produces increases in lordosis. Lordosis of E_2-primed rats is facilitated by ICV administration of the cAMP analogue, 8-bromo-cAMP.[102] Activation of P_4 or D_1 with SKF38393 increases levels of cAMP and cAMP-dependent protein kinase A (PKA) in the hypothalamus.[102] ICV infusions of the Rp isomer of 8-bromo-cAMP (Rp-cAMP) concomitantly inhibits PKA and attenuates lordosis of E_2- and P_4-primed rats and prevents SKF38393- or 8-bromo-cAMP-facilitated lordosis.[102] Together, these data suggest that camp is an important substrate through which lordosis is facilitated by P_4 and/or D_1 activation.

In the midbrain VTA, progestin-facilitated lordosis may also be dependent upon cAMP. Levels of cAMP in the midbrain are higher in hormone-primed or proestrous compared to ovariectomized rats. Pharmacological manipulations of cAMP in the midbrainVTA produce alterations in lordosis. Infusions of VTA infusions of 8-bromo-cAMP, but not vehicle, to the midbrain VTA increase lordosis of rats late in behavioral estrus, when sexual behavior is naturally waning. Similarly, lordosis of ovariectomized, E_2- and P_4-primed rats is enhanced by infusions of 8-bromo-cAMP to the midbrain VTA. These findings suggest that, in the midbrain VTA, intracellular actions of camp may be involved in P_4-facilitated lordosis.

In the hypothalamus, direct actions at membrane DA receptors and subsequent initiation of signal transduction and phosphorylation-dependent mechanisms may be involved in the mediation of progestin-facilitated lordosis.[102] In the hypothalamus, PRs are required for facilitation of lordosis. Activation of PRs can occur through traditional ligand-dependent activation by P_4 or via "ligand-independent" actions. D_1 agonists can increase cAMP, PKA, and DA- and cAMP-regulated phosphoprotein of

32 kDa (DARPP-32) phosphorylation.[103] This process inhibits protein phosphatase 1 (PP-1), which increases phosphorylation of many substrate proteins,[103] including PRs, steroid receptor coactivator proteins (SRCs), or other proteins that may enhance lordosis.[102] Further, ICV infusions of P_4 or SKF38393, but not vehicle, similarly increase concentrations of cAMP, PKA, and the phosphorylated form of DARPP-32 in the hypothalamus of ovariectomized, E_2-primed rats.[102] Finally, SKF38393-facilitated lordosis is blocked by co-administration of the PR antagonist RU38486,[98] which suggests that, in the hypothalamus, DA may require PRs to facilitate lordosis.

Progestins' actions in the midbrain VTA to enhance lordosis may involve D_1, hence, we have begun to investigate the extent to which P_4's actions in the midbrain VTA include second messengers. In the midbrain VTA, it is possible that progestins facilitate lordosis by activating the classical D_1 signaling pathway, which initiates a signaling cascade that results in increases in cAMP and PKA, and the facilitation of phosphorylation of DARPP-32.[102,103] In its phosporylated state, DARPP-32 inhibits PP-1 and thereby reduces protein dephosphorylation,[102,103] which may result in a prolonged phosphorylation of proteins important for mediating lordosis. In support, activating and inhibiting D_1 receptors in the midbrain VTA increases and decreases, respectively, P_4's ability to enhance lordosis. Increasing cAMP in the midbrain VTA enhances lordosis of ovariectomized, E_2- or E_2- and P_4-primed rats. Further, inhibiting DARPP-32 in the midbrain VTA with infusions of antisense oligonucleotides attenuates P_4-facilitated lordosis of ovariectomized, E_2-primed rats. These data, together with those reviewed above, suggest that progestins, in the midbrain VTA, may enhance lordosis via a D_1-initiated intracellular pathway that includes cAMP, PKA, and/or DARPP-32. Ongoing experiments in our laboratory are being conducted to elucidate the precise mechanisms involved.

22.3 Summary of Research on Progestins' Mechanisms for Lordosis

Using lordosis as a bioassay to identify actions of progestins has revealed multiple mechanisms involved in this behavioral response. In the hypothalamus, P_4 and dopamine can enhance lordosis through requisite actions at PRs. In constrast, in the midbrain VTA, P_4 has actions that are independent of PRs. P_4, in the midbrain VTA is converted to 3α,5α-THP, which has local actions via $GABA_A$ receptors, NMDARs, and D_1 to enhance ion flux, and activity of downstream second messengers. An important goal of our research is to ascertain the extent to which progestins actions that have been identified using this reductionist approach are meaningful in a broader context. As such, we discuss the relevance of these findings for the etiology and/or treatment of neuropsychiatric disorders. As well, we summarize findings from another aspect of our research that considers the mechanisms of progestins actions in a more ethologically relevant context.

22.4 The Role of Progestins in Stress-Sensitive Neuropsychiatric Disorders

Actions of 3α,5α-THP may influence the incidence, expression, and/or therapeutic treatment of stress-related, neuropsychiatric disorders. Summarized below are some of the findings which indicate that 3α,5α-THP has a putative role in the diathesis of stress-related, neuropsychiatric disorders. This is followed by information on sex/gender differences in stress, 3α,5α-THP levels, and incidence/expression of neuropsychiatric disorders. Moreover, evidence that some therapeutics used in the treatment of stress-related, neuropsychiatric disorders, can alter 3α,5α-THP levels is reviewed. Together, these findings demonstrate that 3α,5α-THP may play an important role in the pathophysiology of stress-related, neuropsychiatric disorders.

3α,5α-THP may be an important neuroendocrine mediator that may serve in part to buffer an individual's response to stressful stimuli in their environment. First, biosynthesis of 3α,5α-THP is readily increased following exposure to extreme stress or "fight-or-flight" stimuli. 3α,5α-THP biosynthesis occurs among gonadectomized and adrenalectomized rodents indicating that such effects are independent of peripheral gland secretion.[11] Second, 3α,5α-THP can attenuate hypothalamic pituitary adrenal (HPA) axis activation in response to stress.[104] For example, administration of 3α,5α-THP to female rats reduces the elevation of plasma adrenocorticotropin (ACTH) and serum corticosterone following emotional stress.[104] Third, as discussed above, 3α,5α-THP has potent agonist-like actions at $GABA_A$ receptors. 3α,5α-THP is the most potent endogenous modulator of $GABA_A$ receptors.[66,67] Fourth, these effects of 3α,5α-THP are conserved across species. Biosynthesis of levels of 3α,5α-THP that are sufficient to modulate $GABA_A$ receptors and HPA function have been reported for various avian, amphibian, and mammalian species.[11,105,106] Fifth, these effects of 3α,5α-THP have been demonstrated in mammals across development. Indeed, 3α,5α-THP has been detected in rodents as early as embryonic day 17.[107] 3α,5α-THP concentrations are increased in response to maternal separation as early as post-natal day 6.[108,109] These effects persist, such that rats which were handled neonatally demonstrate high basal 3α,5α-THP levels in adulthood. Together, these findings have contributed to the notion that biosynthesis of 3α,5α-THP acts as a rapid neuroendocrine mediator of an organism's response to environmental stressors.

There are sex differences in 3α,5α-THP-mediated responses to stress. Although extreme stressors, such as cold-water swim, reliably increase biosynthesis of 3α,5α-THP in the brains of gonadectomized and adrenalectomized rats,[110,111] females may be more stress-responsive in this capacity. For example, female, compared to male rats, demonstrate greater increases in 3α,5α-THP biosynthesis in response to more modest stressors.[112] Neonatal handling, such as social isolation for 1 h on post-natal days 2–9, has a greater effect among female compared to male rats pups to increase biosynthesis of 3α,5α-THP.[113] These effects may be pervasive, given that in adulthood female rats that were neonatally handled do not adapt to chronic restraint and loose more weight than do their male counterparts.[112] Thus, there are sex differences in stress-induced 3α,5α-THP biosynthesis.

22.4.1 Sex Differences in 3α,5α-THP

In addition to stress-induced biosynthesis of 3α,5α-THP, females and males differ in their basal levels of 3α,5α-THP. In general, females, compared to males, have higher progestin levels that are more labile. Although progestin levels of men and women during the follicular phase are similarly low, during the luteal phase and pregnancy, womens' progestin levels become much higher than are mens'.[114,115] This pattern is also seen in female and male rodents.[16,17]

22.4.2 Gender Differences in Depression, Schizophrenia, and 3α,5α-THP

Men and women differ in the prevalence, onset, and symptomology of stress-related neuropsychiatric disorders, such as depression and schizophrenia. Women are more likely to be diagnosed with depression[116] and they are uniquely vulnerable to depressive disorders associated with changes in hormonal status, such as premenstrual syndrome, premenstrual dysphoric disorder, and postpartum depression. Moreover, there is increasing recognition that depression and anxiety can increase around the climacteric period. In contrast, gender differences in schizophrenia favor women. Women, compared to men, with schizophrenia experience later age of onset, less debilitating psychiatric symptoms, fewer psychiatric hospitalizations, better pre- and post-functioning, and more rapid and greater response to antipsychotics.[117] Moreover, some gender differences in stress-related neuropsychiatric disorders may be manifest as more subtle alterations in patterns of secretion. For example, we (in collaboration with Dr. Ed Blanchard) have examined differences in stress responsiveness of men and women with post-traumatic stress disorder (PTSD), due to a motor vehicle accident. There were no robust gender differences in physiological measures (heart rate, blood pressure, galvanic skin responses, or absolute levels of stress hormones). However, among women with PTSD, cortisol levels were more suppressed than were mens', albeit the normal diurnal decline in salivary cortisol levels was observed among women, but not men (Fig. 22.1).

In addition to gender differences, there are hormonal influences in the vulnerability to stress-related neuropsychiatric disorders. For example, women are more likely to be hospitalized during the perimenstrum.[118] Further, during pregnancy, circulating concentrations of 3α,5α-THP become higher than at any other point in the lifespan and these effects are sustained for a long period of time. Within minutes of parturition, 3α,5α-THP levels drop to nadir. Precipitous decline in 3α,5α-THP levels produces anxiety and depressive behavior in animal models[119,120] and this paradigm is used as a model of post-partum depression.

Depression and schizophrenia are neuropsychiatric disorders that are characterized by aberrant stress responses.[116] Notably, people with depression and schizophrenia

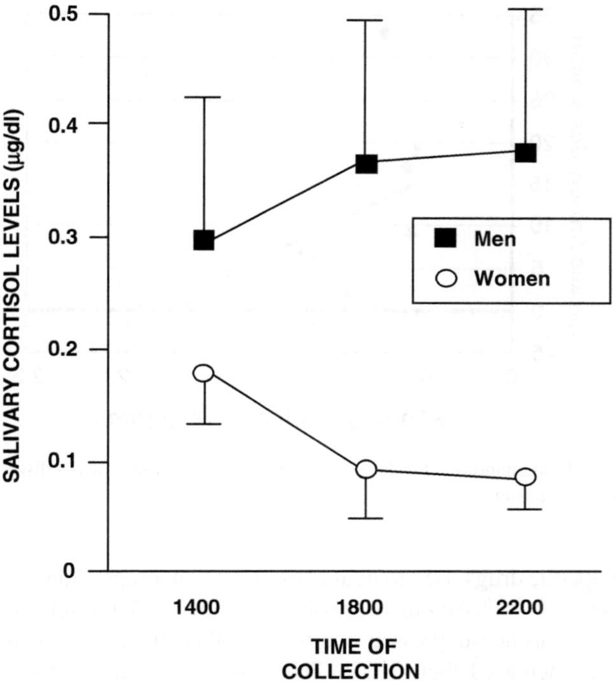

Fig. 22.1 Mean basal cortisol levels (±SEM) collected at 1400h, 1800h and 2200h from men (filled squares) and women (open circles) with post-traumatic stress disorder

have demonstrated atypical stress responsiveness compared to those without these psychopathologies.[116,121] Genetic deficits, such as a novel polymorphism and genetic mutations in the sequence encoding the gene for the MBR, which is necessary for 3α,5α-THP biosynthesis in glial cells, has been found among schizophrenics, and may create a predisposition to oversensitivity to stress.[122] Further, there is a negative relationship between 3α,5α-THP levels and anxiety reports among individuals with stress-related disorders. For example, in collaboration with Drs. Roache and Casada, we have found that plasma 3α,5α-THP levels of people with post-traumatic stress disorder that are acutely exposed to trauma-relevant stimuli are lower among those that report greater levels of anxiety (Fig. 22.2).[123] In animal models of depression, such as social isolation, levels of 3α,5α-THP are decreased.[124,125] Findings, such as these, suggest that stress-related neuropsychiatric disorders may be characterized by altered capacity to synthesize 3α,5α-THP in the brain, which may thereby impact sensitivity to stress. Given gender differences in response to stress, and hormonal influences on stress-responsiveness, changes in 3α,5α-THP concentrations may account in part for some of gender differences in the expression of depression and schizophrenia. Research on this topic is lacking. More information is needed about how gender and hormones influence the expression of stress-related neuropsychiatric disorders.

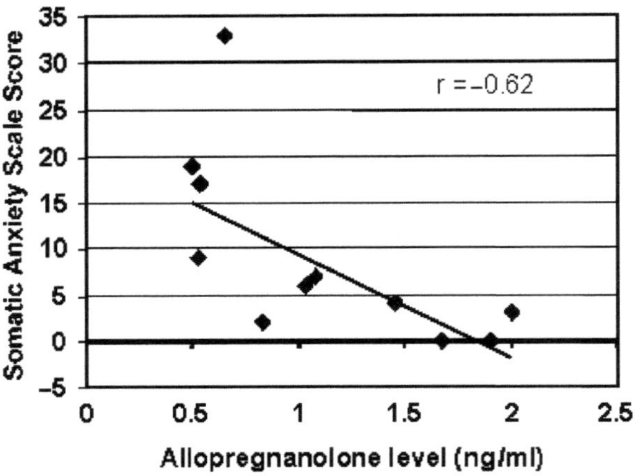

Fig. 22.2 Mean allopregnanolone levels of people with post-traumatic stress disorder as a function of subjective anxiety rating

Some therapeutic drugs used to treat stress-related neuropsychiatric disorders alter 3α,5α-THP levels.[47,126–128] Among depressed men, 3α,5α-THP levels in cerebral spinal fluid are lower than non-depressed controls. Further, treatment of their depression with fluoxetine increased their 3α,5α-THP levels concomitant with alleviation of their depressive symptomology.[128] Notably, fluoxetine increases the activity of the 3α-hydroxysteroid dehydrogenase to favor 3α,5α-THP production.[34] Similarly, some women with premenstrual dysphoric disorder had significant improvement of symptoms following stabilization of plasma 3α,5α-THP levels with anti-depressants.[129] Administration of the atypical antipsychotic, olanzapine, increases 3α,5α-THP.[47,127] Together, these findings suggest that 3α,5α-THP may be involved in the pathophysiology and/or treatment of stress-related neuropsychiatric disorders.

Based upon these findings that therapeutics can alter 3α,5α-THP levels, we have examined their effects on behavior and biosynthesis. Olanzapine, when administered systemically, or when applied to the midbrain VTA enhances social and anti-anxiety behavior of rodents and increases midbrain 3α,5α-THP levels.[47] Indeed, some of the weight gain associated with olanzapine may be related to the hyperphagic effects of 3α,5α-THP,[130,131] although this has not been systematically investigated. We have also examined effects of fluoxetine on behavior and 3α,5α-THP biosynthesis. When administered systemically, fluoxetine inhibits sexual receptivity and decreases 3α,5α-THP levels[132] and administration of 3α,5α-THP can reinstate these effects.[132] However, when fluoxetine is infused into the midbrain VTA, lordosis is enhanced and midbrain 3α,5α-THP levels are increased.[35] These findings demonstrate that in animal models these therapeutics can alter 3α,5α-THP biosynthesis and behavior. Furthermore, they indicate a relationship between therapeutics, 3α,5α-THP levels, and sexual behavior. Indeed, sexual side effects of these therapeutics can hinder compliance and screening of compounds in the midbrain VTA may be a sensitive model.

In addition to manipulating 3α,5α-THP levels in the midbrain VTA and examining effects on behavior, we have also examined effects of behavior on 3α,5α-THP biosynthesis. Exposing rodents to stressors such as cold water swim, ether, predator odor, shock and restraint are known to increase 3α,5α-THP biosynthesis. Exposure of male rats to mating increases 3α,5α-THP biosynthesis. We have examined effects of mating on 3α,5α-THP levels. Among female proestrous rats, plasma, whole brain and midbrain levels of 3α,5α-THP are increased with mating. Midbrain levels of 3α,5α-THP are also increased with mating among ovariectomized, adrenalectomized E_2-primed rats. These findings indicate that sexual experience can alter 3α,5α-THP biosynthesis and that the midbrain may be particularly sensitive to these effects.

22.5 Summary of Progestins' Role in Stress-Sensitive Neuropsychiatric Disorders

3α,5α-THP may play an important role in the etiology, expression and/or treatment of stress related neuropsychiatric disorders. There are sex/gender and hormonal differences in stress responses, 3α,5α-THP levels, and neuropsychiatric disorders. Moreover, some therapeutics can alter 3α,5α-THP levels used in the treatment of stress-related, neuropsychiatric disorders. Although these findings imply that biosynthesis of 3α,5α-THP and basal patterns of secretion may interplay to mitigate the pathophysiology of stress-related, neuropsychiatric disorders, it is not feasible to fully characterize these effects in people given limitations of measuring biosynthesis. As such, it is imperative to have good animal models to ascertain interactions between circulating steroid hormones, 3α,5α-THP biosynthesis and how they influence stress responses. The midbrain VTA is exquisitely sensitive to manipulations in 3α,5α-THP and behavioral-induced biosynthesis. Further, the midbrain VTA plays an important role in mediating motivation, reward and stress responses. For these reasons, we have moved beyond our reductionist model of examining effects of midbrain 3α,5α-THP on lordosis. Recently, we have begun to investigate effects of manipulating 3α,5α-THP, pharmacologically or behaviorally, in a broader social context to ascertain effects on stress-related measures as an animal model of hormonally influenced neuropsychiatric disorders.

22.6 Progestins' Actions in the Midbrain VTA Extend Beyond Effects on Lordosis

Results of the research discussed above demonstrate that progestins have actions in the midbrain VTA to modulate lordosis. Mating among rodents requires that females are receptive to, and respond appropriately to (i.e., display lordosis), male mounting. However, mating is also characterized by exploration and affiliations, social behaviors that bring individuals together. Successful reproduction requires that female rodents increase exploratory behavior to find mates and suppress fearful and/or aggressive responses to potential mates.[134] In the laboratory, the multiple components of mating can be examined using various models of anxiety and social behavior, in addition to reproductive indices. We have utilized a battery of behavioral tasks to examine effects of progestins on anxiety (open field, elevated plus maze), social (social choice, social interaction), and sex behavior (lordosis, pacing). The following pages describe our findings using these indices.

There are behavioral changes and fluctuations in progestins across the estrous cycle. First, rats in proestrus have high circulating levels of 3α,5α-THP that are sufficient to produce agonist-like actions at $GABA_A$ receptors, and increase anti-anxiety behavior compared to that of diestrous rats, with lower levels of 3α,5α-THP.[135–137] Second, removal of the primary source of endogenous hormones, the ovaries, increases anxiety behavior and replacement with systemic or central administration of P_4 or 3α,5α-THP reverses this effect.[138,139] Third, blocking formation of 3α,5α-THP attenuates P_4's anti-anxiety effects.[140,141] Thus, P_4's effects on anxiety behavior are mediated in part by 3α,5α-THP; however, the brain areas involved for these effects have not been elucidated.

To begin to elucidate brain areas that might be important for 3α,5α-THP's effects on anti-anxiety, pro-social, and sex behaviors, we investigated effects of 3α,5α-THP infusions to various brain areas of proestrous (high E_2 and P_4) and diestrous (low E_2 and P_4) rats. Rats were surgically implanted with bilateral guide cannulae aimed at the VTA, or nearby control sites, such as the substantia nigra (SN) or central grey (CG), which are progestin-sensitive brain areas for effects on seizures[142] and catatonia,[143] respectively. Rats were randomly assigned to receive infusions of 3α,5α-THP (100 ng) or β-cyclodextran vehicle while in diestrus or proestrus. Ten minutes following infusions, rats were tested sequentially through a test battery that examined exploratory, anxiety, social, and reproductive behaviors. Exploratory and anxiety behavior were examined using the 5 min open field and elevated plus maze tasks. Rats were then returned to the open field so that social choice (time spent in non-physical close proximity to a stimulus male or female) could be evaluated in a 5 min test. Social interaction was assessed by recording the amount of time rats spent engaging an ovariectomized conspecific in social behavior in a 5 min test. Immediately after this, rats were placed in a pacing chamber opposite a sexually responsive stimulus male. For one ejaculatory series, or 15 min, the percent occurrence (lordosis quotient) and intensity (lordosis rating) of the experimental female's lordosis response to each sexual contact was evaluated. The percentage of sexual contacts that were preceded by solicitation/proceptive (hopping, darting, ear wiggling) behaviors were also recorded and quantitated (proceptivity quotients). The percentage of sexual contacts for which the female entered and exited the side of the pacing chamber with the male was also recorded (percent exits). Immediately following the pacing test, trunk blood and whole brain were collected from each rat. Radioimmunoassay was used to measure E_2, P_4, DHP, and 3α,5α-THP in plasma, midbrain, hippocampus, diencephalon, and cortex.

As observed previously in separate reports, proestrous rats exhibited increased exploratory, anti-anxiety, social, and reproductive behaviors than did diestrous rats.

Proestrous rats entered more central squares in the open field, spent more time on the open arms of the elevated plus maze, spent more time in close proximity to a male and interacted for longer with a female (Table 22.1). Proestrous rats also engaged in more sex behavior than did diestrous rats. Rats in proestrous had significantly higher lordosis quotients and ratings, and proceptivity quotients, and paced their sexual contacts more than did diestrous rats (Table 22.1).

Infusions of 3α,5α-THP to the VTA enhanced exploratory, anti-anxiety, social, and reproductive behaviors of diestrous and proestrous rats. 3α,5α-THP infusions to the VTA significantly increased central entries in the open field, time on the open arms of the elevated plus maze, time in proximity to a male, and social interaction with a female conspecific, as compared to effects of vehicle infusions (Table 22.1). Infusions of 3α,5α-THP to the VTA, compared to vehicle, also significantly increased lordosis quotients, lordosis ratings, proceptivity quotients, and percent exits (Table 22.1). Notably, infusions of 3α,5α-THP to diestrous rats produced levels of reproductive and non-reproductive behaviors that were similar to those observed for rats in proestrous that received vehicle infusions. Infusions of 3α,5α-THP to proestrous rats further increased exploratory and anti-anxiety behaviors in the open field and plus maze and pacing of sexual contacts over that of vehicle-infused proestrous rats. Infusions of 3α,5α-THP to the SN or CG did not enhance exploratory, anti-anxiety, social, or reproductive behaviors of either diestrous or proestrous rats. Proestrous, compared to diestrous, rats had higher E_2 and P_4, DHP, and 3α,5α-THP concentrations in plasma, midbrain, hippocampus, diencephalon, and cortex (Table 22.1). Notably, infusions of 3α,5α-THP increased DHP and 3α,5α-THP levels in midbrain, hippocampus, diencephalon, and cortex (Table 22.1).

In summary, results of this study revealed that actions of 3α,5α-THP in the VTA modulate exploratory, anti-anxiety, social, and reproductive behaviors and that manipulations of 3α,5α-THP in the VTA can increase DHP and/or 3α,5α-THP levels in midbrain, hippocampus, diencephalon, and cortex. In support, infusions of 3α,5α-THP to the VTA, but not SN or CG, enhanced exploratory, anti-anxiety, social, and reproductive behaviors of diestrous rats and increased DHP and 3α,5α-THP levels in midbrain, hippocampus, diencephalon, and cortex similar to that of proestrous rats. In sum, infusions of 3α,5α-THP facilitated reproductive and non-reproductive behavior and concomitantly increased DHP and 3α,5α-THP levels in midbrain, hippocampus, and diencephalon, brain areas that may be involved in these behavioral processes.

Findings from the above study suggest that DHP and 3α,5α-THP levels in midbrain, hippocampus, diencephalon, and/or cortex may mediate reproductive and non-reproductive behavior examined. However, E_2 also influences many of the behaviors examined. Estrogen has modulatory effects on mating and courtship behaviors, in part through actions in the VTA.[144–146] E_2 can also enhance biosynthesis of progestins by increasing activity of the 5α-reductase enzyme.[147] In ovariectomized and/or adrenalectomized rats, E_2-priming increases progestin levels.[148] Whether priming with and/or biosynthesis of E_2 and/or progestins in the VTA influence reproductive behavior needs to be established. In the aforementioned study, E_2 was monitored but not manipulated, as in the study described below.

Table 22.1 Behavioral and endocrine measures of diestrous and proestrous rats infused with vehicle or 3α,5α-THP to the midbrain VTA, substantia nigra (SN), or central grey (CG)

Task	Diestrus + vehicle	Diestrus + 3α, 5α-THP VTA	Diestrus + 3α, 5α-THP SN	Diestrus + 3α, 5α-THP CG	Proestrus + vehicle	Proestrus + 3α, 5α-THP VTA	Proestrus + 3α, 5α-THP SN	Proestrus + 3α, 5α-THP CG
Behavioral measures								
Central entries	27 ± 3	47 + 3	34 ± 4	26 ± 3	42 ± 4	57 + 2[a]	43 ± 5[a]	39 ± 3[a]
Total entries	267 ± 31	258 + 12	248 ± 25	252 ± 32	229 ± 8	218 + 18	203 ± 21	229 ± 25
Open arm time	17 ± 3	49 + 13	18 ± 6	41 ± 10	67 ± 15	61 + 7	81 ± 8[a]	66 ± 13[a]
Total open arm entries	10 ± 1	13 + 5	11 ± 1	7 ± 1	15 ± 5	10 + 1	12 ± 1	9 ± 2
Time/w male	96 ± 9	173 + 23	119 ± 20	99 ± 9	164 ± 21	150 + 6	144 ± 12[a]	194 ± 9[a]
Social interaction	44 ± 6	76 + 19	28 ± 7	29 ± 7	90 ± 21	100 + 5	91 ± 6[a]	91 ± 6[a]
Lordosis quotient	0 ± 0	39 + 10	12 ± 12	12 ± 9	86 ± 14	89 + 7	68 ± 13[a]	88 ± 9[a]
Lordosis rating	0 ± 0	1.4 + 0.3	0.2 ± 0.2	0.1 ± 0.1	2.0 ± 0.3	2.1 + 0.2	1.2 ± 0.4[a]	2.3 ± 0.3[a]
Perceptivity quotient	0 ± 0	12 + 11	0 ± 0	0 ± 0	66 ± 16	71 + 3	25 ± 16[a]	47 ± 17[a]
Percent exits	3 ± 1	12 + 4	10 ± 10	0 ± 0	25 ± 4	35 + 3	33 ± 6	43 ± 7[a]
Neuroendocrine measures								
Midbrain E_2	1.6 ± 0.2	1.4 ± 0.2	1.6 ± 0.7	1.6 ± 0.7	2.3 ± 0.4[a]	2.9 + 0.1	2.1 ± 0.1[a]	2.4 ± 0.3[a]
Hippo E2	1.5 ± 0.1	1.7 ± 0.2	1.4 ± 0.1	1.4 ± 0.09	2.5 ± 0.2[a]	3.0 + 0.2	2.5 ± 0.2[a]	2.5 ± 0.2[a]
Diencephalon E2	1.1 ± 0.1	1.0 ± 0.1	1.9 ± 0.5	1.2 ± 0.1	2.1 ± 0.5[a]	1.8 + 0.1	2.1 ± 0.3[a]	2.2 ± 0.3[a]
Cortex E2	1.6 ± 0.2	1.3 ± 0.1	1.5 ± 0.5	1.8 ± 0.1	2.7 ± 0.3[a]	1.2 + 0.1	2.0 ± 0.5[a]	3.0 ± 0.2[a]
Midbrain P4	1.4 ± 0.2	1.2 ± 0.1	1.3 ± 0.3	1.5 ± 0.2	2.3 ± 0.2	2.1 ± 0.2	2.1 ± 0.2[a]	2.2 ± 0.2
Hippo P4	1.8 ± 0.2	1.7 ± 0.3	1.6 ± 0.2	1.6 ± 0.3	2.1 ± 0.3[a]	2.1 ± 0.3[a]	2.6 ± 0.3[a]	3.4 ± 0.7[a]
Diencephalon P4	2.1 ± 0.3	1.9 ± 0.2	1.6 ± 0.3	2.1 ± 0.3	3.2 ± 0.2[a]	2.9 ± 0.1[a]	2.5 ± 0.4[a]	3.0 ± 0.2[a]
Cortex P4	2.0 ± 0.2	1.7 ± 0.1	2.3 ± 0.4	1.8 ± 0.4	3.1 ± 0.7[a]	3.3 ± 0.2[a]	3.2 ± 0.2[a]	2.5 ± 0.3[a]
Midbrain DHP	2.1 ± 0.2	4.6 ± 0.2	2.5 ± 0.3	2.6 ± 0.3	9.3 ± 0.7[a]	11.2 ± 0.4[a]	6.3 ± 0.5[a]	8.6 ± 1.0[a]
Hippo DHP	5.1 ± 0.5	4.9 ± 0.4	3.5 ± 0.4	4.2 ± 0.5	21.0 ± 1.9[a]	23.0 ± 1.3[a]	23.0 ± 0.8[a]	21.0 ± 1.2[a]
Diencephalon DHP	5.8 ± 0.7	5.4 ± 0.3	5.9 ± 0.7	6.1 ± 0.6	10.0 ± 1.1[a]	9.4 ± 1.9[a]	13.0 ± 0.5[a]	12.0 ± 1.2[a]
Cortex DHP	4.7 ± 0.7	4.3 ± 0.9	4.5 ± 0.2	5.8 ± 1.0	11.0 ± 1.7[a]	10.2 ± 1.5[a]	14.0 ± 1.2[a]	13.0 ± 0.8[a]
Midbrain 3α, 5α-THP	2.0 ± 0.8	79 ± 0.2	1.7 ± 0.3	2.7 ± 0.3	8.5 ± 1.5[a]	10.2 + 1.2	12 ± 1.6[a]	9.6 ± 1.2[a]
Hippo 3α, 5α-THP	3.4 ± 0.6	3.0 ± 0.2	4.7 ± 0.7	3.3 ± 0.5	19.0 ± 4.8[a]	18.9 + 1.7	16.0 ± 0.8[a]	22.0 ± 3.1[a]
Diencephalon 3α, 5α-THP	2.0 ± 0.7	1.7 ± 0.4	1.4 ± 0.1	2.0 ± 0.4	4.0 ± 0.7[a]	4.1 + 1.0	4.2 ± 0.5[a]	4.2 ± 0.7[a]
Cortex 3α, 5α-THP	1.3 ± 0.3	1.0 ± 0.1	1.3 ± 0.1	1.8 ± 0.3	2.4 ± 0.9[a]	5.9 + 0.4	2.5 ± 0.2[a]	2.5 ± 0.3[a]

[a] indicates significant difference from respective control group.

To elucidate the putative role of E_2 in the VTA in modulating effects of progestins on exploratory, anti-anxiety, social, and sex behaviors, we compared effects of 3α,5α-THP or vehicle infusions to the VTA, SN, or CG of ovariectomized, E_2-primed rats to that of ovariectomized controls. Ovariectomized rats were primed with 10 μg E_2 or sesame oil vehicle 48 h prior to testing. Rats were then infused with 3α,5α-THP or vehicle to the VTA, SN, or CG. Ten minutes following infusions, rats were tested sequentially in the open field, elevated plus maze, social choice, social interaction, and pacing tests. Immediately after testing, tissues were collected for later measurement of E_2, P_4, DHP, and 3α,5α-THP.

3α,5α-THP infusions to the VTA, but neither the SN nor CG, enhanced non-reproductive and reproductive behaviors. Infusions of 3α,5α-THP to the VTA of ovariectomized, E_2-primed rats significantly increased the number of central square entries in the open field, open arm time on the elevated plus maze, time spent in close proximity to a stimulus male, social interaction with a conspecific, lordosis quotients, lordosis ratings, proceptivity quotients, and percentage of exits following mating contacts compared to of E_2-primed ovx rats receiving vehicle infusions, which were increased compared to ovariectomized controls (Table 22.2). Infusions

Table 22.2 Behavioral and endocrine measures of ovariectomized rats primed with vehicle or E_2 and infused with 3α,5α-THP to the midbrain VTA, substantia nigra (SN), or central grey (CG)

	Vehicle control	E_2 control	VTA 3α, 5α-THP	SN 3α, 5α-THP	CG 3α, 5α-THP
Central entries	9 ± 2	15 ± 2	51 ± 9	18 ± 4	20 ± 3
Total entries	248 ± 16	276 ± 25	219 ± 12	236 ± 16	228 ± 10
Open arm time	6 ± 2	15 ± 3	73 ± 10	13 ± 4	18 ± 6
Total arm entries	6 ± 1	9 ± 1	12 ± 4	9 ± 1	9 ± 1
Time/w male	44 ± 11	34 ± 9	181 ± 24	47 ± 12	42 ± 7
Social interaction	25 ± 3	34 ± 5	99 ± 19	46 ± 8	58 ± 15
Lordosis quotient	0 ± 0	12 ± 12	91 ± 12	19 ± 7	24 ± 6
Lordosis rating	0 ± 0	0.2 ± 0.1	2.3 ± 0.1	0.3 ± 0.2	0.4 ± 0.2
Perceptivity quotient	0 ± 0	0 ± 0	75 ± 14	0 ± 0	0 ± 0
% exit	0 ± 0	0 ± 0	29 ± 2	9 ± 3	7 ± 4
Midbrain E_2	1.2 ± 0.1	2.6 ± 0.3[a]	2.0 ± 0.2	2.6 ± 0.3[a]	2.7 ± 0.5[a]
Hippo E_2	1.6 ± 0.3	3.2 ± 0.5[a]	3.0 ± 0.4	3.7 ± 0.4[a]	3.3 ± 0.3[a]
Diencephalon E_2	1.0 ± 0.2	2.1 ± 0.2[a]	2.3 ± 0.2	2.5 ± 0.3[a]	2.3 ± 0.2[a]
Cortex E_2	1.2 ± 0.2	1.5 ± 0.1	2.4 ± 0.2	1.5 ± 0.1	1.7 ± 0.2
Midbrain P_4	1.5 ± 0.1	1.6 ± 0.1	2.2 ± 0.1	1.8 ± 0.5	1.8 ± 0.2
Hippo P_4	1.9 ± 0.2	1.8 ± 0.2	1.6 ± 0.1	1.8 ± 0.3	1.8 ± 0.3
Diencephalon P_4	1.7 ± 0.6	1.6 ± 0.1	1.8 ± 0.2	2.0 ± 0.4	1.8 ± 0.5
Cortex P_4	1.9 ± 0.3	1.9 ± 0.2	1.6 ± 0.2	1.7 ± 0.2	1.9 ± 0.2
Midbrain DHP	1.5 ± 0.1	1.7 ± 0.1	10.9 ± 0.2	1.6 ± 0.1	1.7 ± 0.1
Hippo DHP	1.5 ± 0.1	1.8 ± 0.3	20.0 ± 1.1	1.5 ± 0.1	1.6 ± 0.2
Diencephalon DHP	1.5 ± 0.1	1.7 ± 0.1	9.7 ± 1.2	1.7 ± 0.1	1.8 ± 0.2
Cortex DHP	1.3 ± 0.4	2.1 ± 0.6	9.9 ± 1.1	1.8 ± 0.3	1.7 ± 0.4
Midbrain 3α, 5α-THP	1.1 ± 0.2	1.0 ± 0.1	10.3 + 1.5	1.3 ± 0.1	1.4 ± 0.1
Hippo 3α, 5α-THP	2.0 ± 0.33	1.7 ± 0.2	18.0 + 1.2	1.7 ± 0.2	2.1 ± 0.3
Diencephalon 3α, 5α-THP	1.6 ± 0.3	1.8 ± 0.2	4.2 + 1.6	2.0 ± 0.2	2.2 ± 0.2
Cortex 3α,5α-THP	2.8 ± 0.5	3.3 ± 0.3	5.2 + 0.9	2.8 ± 0.3	2.2 ± 0.4

[a] indicates significant difference from vehicle control group.

of 3α,5α-THP to the SN or CG did not enhance behavior of E_2-primed rats. Notably, E_2-priming alone was insufficient to produce behavior akin to that of proestrous rats seen above or in our prior experiments.

E_2-priming and infusions of 3α,5α-THP to the VTA, but neither the SN nor CG, also increased the endocrine measures examined. E_2, but not vehicle, administration increased E_2, but not progestin, concentrations in serum, midbrain, hippocampus, diencephalon. 3α,5α-THP infusions to the VTA, but not the SN or CG, increased concentrations of DHP and 3α,5α-THP in serum, midbrain, hippocampus, diencephalon, and cortex (Table 22.2). Given that increases in DHP and 3α,5α-THP were not seen following infusions to the SN or CG, it is unlikely that increases observed following infusions to the midbrain were due to diffusion of infusate.

Results of this study, together with those discussed above, suggest that actions of progestins may be more important than E_2 in mediating the factors examined. Ovariectomized rats with infusions of 3α,5α-THP to the VTA, but not the SN or CG, entered more central squares in the open field and spent more time on the open arms of the elevated plus maze and in close proximity to a male, irrespective of E_2-priming. Social interaction with a conspecific and mating behaviors were enhanced most among ovariectomized rats that were E_2-primed and received infusions of 3α,5α-THP to the VTA. 3α,5α-THP infusions to the VTA, but not SN or CG, increased 3α,5α-THP in midbrain, hippocampus, diencephalon, and cortex. Thus, 3α,5α-THP to the VTA enhances exploratory and anti-anxiety behavior, and social interactions of E_2-primed rats, concomitant with increased 3α,5α-THP in midbrain, hippocampus, diencephalon, and cortex.

The studies discussed above suggest that manipulating progestins, and to some extent E_2 levels, alters reproductive and non-reproductive behaviors; however, whether engaging in these behaviors alters neuroendocrine outcomes was not revealed. Biosynthesis of E_2 and P_4 occurs in glial cells, independent of peripheral secretion. In oligodendrocytes and astrocytes, mitochondrial benzodiazepine receptors translocate cholesterol across the mitochondrial membrane, whereupon the steroidogenic regulatory protein facilitates its transport to the inner membrane (a rate-limiting step in steroid biosynthesis),[41,42,149] where it is converted to pregnenolone by the P450 side chain cleavage (P450scc) enzyme.[150] Pregnenolone is then converted to P_4 by 3β-hydroxysteroid dehydrogenase. The 5α-reductase and 3α-hydroxysteroid dehydrogenase enzymes then metabolize P_4 to DHP and 3α,5α-THP, respectively.[151] Pregnenolone, via a number of enzymatic steps, can also be converted to testosterone, which is aromatized to E_2.[152] Circulating and/or central E_2 and progestin levels are maintained following gonadectomy and/or adrenalectomy.[11, 152–154] Many behavioral and environmental factors can produce rapid increases in E_2 and progestins, independent of peripheral gland secretion.[40,154] Thus, it is important to consider the role of progestins and E_2 produced de novo in brain for mediating reproductive behaviors.

We have investigated the extent to which engaging in reproductive and non-reproductive behaviors alters endocrine measures of rats in proestrus or diestrus. Rats were individually tested in the battery indicated above, with, or without, paced mating. As previously observed, proestrous rats showed enhanced non-reproductive and reproductive behaviors over diestrous rats. Rats in proestrus made more central entries in the open field, spent more time on the open arms of the elevated plus maze, spent more time in proximity to a male than a female rat, and interacted more with a female conspecific than did diestrous experimental rats. Rats in proestrus also had higher lordosis quotients, lordosis ratings, proceptivity quotients, and paced more than did diestrous rats.

Rats that were tested in all tasks, including mating, had higher DHP and/or 3α,5α-THP levels in some regions of the brain than did rats that were tested in all tasks, except mating. Proestrous rats that were mated had higher levels of DHP in midbrain, hippocampus, diencephalon, and cortex, but not serum. Among diestrous rats that were mated, only levels of DHP in midbrain, but not other tissues examined, were increased. 3α,5α-THP in hippocampus and midbrain were increased by mating in proestrous and diestrous rats, respectively compared to rats that were not mated.

This investigation revealed that engaging in reproductive behaviors can alter endocrine outcomes. Proestrous, compared to diestrous, rats had concomitant increases in exploration, anti-anxiety, social, approach, anti-conflict, and sexual behavior and had increased circulating and central levels of E_2, P_4, DHP, and 3α,5α-THP. Mating increased levels of 3α,5α-THP in midbrain and hippocampus of rats in proestrus. In proestrous rats, 3α,5α-THP's immediate precursor, DHP, was also increased in midbrain, hippocampus, diencephalon, and cortex following paced mating. Diestrous rats that were behaviorally tested with paced mating, had 3α,5α-THP and DHP concentrations that were increased only in midbrain. Together, these data suggest that higher levels of 3α,5α-THP in hippocampus, diencephalon, and cortex are associated with enhanced exploratory, anti-anxiety, social, and reproductive behaviors. Further, exposure to paced mating has site-specific effects to increase 3α,5α-THP and/or its pre-cursor, DHP, in midbrain, hippocampus, diencephalon, and cortex of proestrous rats and midbrain of diestrous, rats.

The study above supports previous reports that stress-responsiveness varies across the estrous cycle. Effects of paced mating on increased progestin levels were more salient in proestrous, compared to diestrous, rats. In proestrous rats, DHP and 3α,5α-THP were increased following mating in the midbrain, hippocampus, diencephalon, and cortex. However, in diestrous rats, increases in DHP and 3α,5α-THP were only observed in the midbrain. Basal and stress-induced corticosterone levels are typically higher among proestrous, compared to diestrous,[155–157] albeit, not all investigations report these differences.[158] Differential responsiveness to stressors of proestrous and diestrous rats may explain the findings discussed above. First, mating alone may differentially activate stress responses in proestrous, compared to diestrous, rats to enhance 3α,5α-THP biosynthesis in brain areas other than the midbrain. Second, 3α,5α-THP biosynthesis may have been invoked in areas other than the midbrain of proestrous rats in response to the behavioral tasks that preceded mating in the above studies. These tasks, which are mediated in part by the hippocampus, diencephalon, and/or cortex, are likely less-stressful events.[159–161] Proestrous rats may have a lower threshold for stress-induced biosynthesis than diestrous rats, which may underlie the differences in progestin biosynthesis observed above. Hence, these data suggest that mating is a salient stimulus that can invoke biosynthesis in the midbrain of rodents. However, increased progestin biosynthesis in other brain areas may be mediated by threshold for stress-responsiveness and/or additional exposure to other reproductively relevant behaviors that lead up to mating and are mediated by other brain regions. Given that there may have been additive effects of stressors, in the above study, associated with engaging in a test battery, we have also examined effects on endocrine measures of engaging in a single behavioral task.

Whether engaging in single behavioral tasks is sufficient to produce increases in progestin biosynthesis was investigated next. Proestrous rats were randomly assigned to be tested in individual tasks including the open field, elevated plus maze, partner preference, social interaction, or no test control, in conjunction with paced mating or no mating. All rats had E_2 and P_4 levels that were within the physiological range previously reported for proestrous rats[16] and there were no effects of behavioral testing on these endocrine measures. Paced mating, but not any other behavioral experience, increased DHP and 3α,5α-THP concentrations in midbrain, hippocampus, diencephalon, and cortex compared to non-tested proestrous controls (Table 22.3). Notably, rats that actively engaged in paced mating, compared to rats that were exposed to paced mating, but did not engage in pacing, had higher levels of DHP and 3α,5α-THP concentrations in midbrain, hippocampus, diencephalon, or cortex.[162] Thus, engaging in paced mating, but not other individual tasks, significantly increased DHP and 3α,5α-THP levels in midbrain, hippocampus, diencephalon, and cortex. These findings suggest that paced mating is a more salient stimulus capable of increasing progestin concentrations; however, it is possible that there may be additive effects of exploratory and/or social experiences that could increase progestin concentrations, analogous to that of paced mated rats.

To begin to address the extent to which there might be additive effects of behavioral tasks in our battery altered E_2 and/or progestin concentrations in brain, we have investigated the effects of engaging in multiple exploratory (open and elevated plus maze) or social (partner preference and social interaction) on progestin concentrations. Proestrous rats were tested in combinations of the above tasks (exploration: open field and elevated plus maze or social behavior: partner preference and social interaction) with or without paced mating. Neither E_2 nor P_4 concentrations varied as a function of behavioral testing in any tissues examined. However, both DHP and 3α,5α-THP levels in midbrain, hippocampus, diencephalon, and cortex were increased by mating, but not exploratory or social behaviors alone (Table 22.4).

These latter studies revealed that engaging in paced mating, irrespective of prior exposure to other tasks, either individually or in combination, increases DHP and 3α,5α-THP concentrations in midbrain, hippocampus, diencephalon, and cortex. There were no differences in E_2 or P_4 concentrations in any brain areas examined as a result of engaging in various exploratory (open field, elevated plus maze), social (partner preference, social interaction) or sexual behaviors. Together, these data suggest that engaging in paced mating behavior produces increases in DHP and 3α,5α-THP in areas associated with reproduction (midbrain, diencephalon), as well as exploration/anxiety behavior (hippocampus) and social choice (cortex).

Table 22.3 Endocrine data of proestrous rats exposed to the open field, elevated plus maze, social choice, social interaction, or no task with or without paced mating

	Open field		Elevated plus maze		Social choice		Social interaction		Control	
	No paced mating	Paced mating	No paced mating	Paced mating	No paced mating	Paced mating	No paced mating	Paced mating	No paced mating	Paced mating
Midbrain E_2	2.2 ± 0.3	2.0 ± 0.2	3.0 ± 0.4	2.8 ±0.2	2.1 ± 0.1	2.3 ± 0.2	2.4 ± 0.2	3.0 ± 0.5	2.9 ± 0.2	2.7 ± 0.2
Hippo E_2	3.0 ± 0.4	3.0 ± 0.4	2.8 ± 0.2	2.1 ± 0.1	2.3 ± 0.2	2.4 ± 0.2	3.0 ± 0.5	2.9 ± 0.2	2.7 ± 0.2	2.4 ± 0.2
Diencephalon E_2	1.9 ± 0.2	1.7 ± 0.2	1.6 ± 0.2	1.5 ± 0.2	1.8 ± 0.2	1.5 ± 0.1	2.1 ± 0.1	2.3 ± 0.2	2.2 ± 0.3	2.0 ± 0.2
Cortex E_2	2.8 ± 0.2	2.1 ± 0.1	2.3 ± 0.2	2.4 ± 0.2	2.8 ± 0.2	2.1 ± 0.1	2.3 ± 0.2	2.4 ± 0.2	2.1 ± 0.1	2.3 ± 0.2
Midbrain P_4	1.9 ± 0.2	2.2 ± 0.1	2.2 ± 0.1	2.2 ± 0.6	1.6 ± 0.1	1.5 ± 0.1	1.8 ± 0.2	1.6 ± 0.2	1.9 ± 0.1	1.9 ± 0.2
Hippo P_4	2.5 ± 1.2	2.0 ± 0.1	2.3 ± 0.2	2.3 ± 0.1	2.4 ± 0.2	2.6 ± 0.1	2.9 ± 0.2	2.5 ± 0.2	2.2 ± 0.2	2.2 ± 0.2
Diencephalon P_4	2.2 ± 0.6	1.6 ± 0.1	1.5 ± 0.1	1.8 ± 0.2	2.3 ± 0.1	2.4 ± 0.2	2.6 ± 0.1	2.9 ± 0.2	2.1 ± 0.1	2.3 ± 0.2
Cortex P_4	2.1 ± 0.1	2.3 ± 0.2	1.9 ± 0.2	1.7 ± 0.2	1.6 ± 0.2	1.5 ± 0.2	2.1 ± 0.2	2.2 ± 0.2	2.1 ± 0.1	2.3 ± 0.2
Midbrain DHP	9.3 ± 0.7	11.9 ± 0.2	10.9 ± 0.3	14.2 ± 0.5	8.7 ± 0.2	13.4 ± 0.2	8.2 ± 0.4	12.7 ± 0.1	7.3 ± 0.2	12.7 ± 0.1
Hippo DHP	21.1 ± 1.9	32.0 ± 1.2	20.0 ± 1.1	31.3 ± 0.9	19.9 ± 0.9	28.7 ± 0.2	18.0 ± 1.3	32.1 ± 1.1	17.2 ± 1.1	29.2.1 ± 0.1
Diencephalon DHP	10.0 ± 1.1	12.1 ± 0.9	9.7 ± 1.2	13.1 ± 1.0	9.6 ± 1.1	12.9 ± 0.8	9.4 ± 1.9	11.3 ± 1.6	8.1 ± 1.3	12.4 ± 1.2
Cortex DHP	11.0 ± 1.7	13.2 ± 1.1	9.9 ± 1.1	11.9 ± 1.0	10.9 ± 1.1	12.1 ± 0.1	10.2 ± 1.5	11.9 ± 1.2	8.9 ± 1.4	13.0 ± 1.4
Midbrain 3α, 5α-THP	8.5 ± 1.5	14.2 ± 1.0	10.3 + 1.5	16.6 + 1.0	9.5 ± 1.0	19.1 ± 1.3	10.2 + 1.2	17.8 + 1.0	8.8 + 1.4	17.8 + 1.0
Hippo 3α, 5α-THP	19.0 ± 4.8	29.2 ± 3.1	18.0 + 1.2	31.1 + 2.3	17.3 ± 4.0	28.3 ± 2.6	18.9 + 1.7	27.7 + 1.2	17.6 + 1.4	28.1 + 1.0
Diencephalon 3α, 5α-THP	4.0 ± 0.7	8.1 ± 1.2	4.2 + 1.6	7.5 + 1.3	4.0 ± 0.9	9.3 ± 0.7	4.1 + 1.0	8.6 + 1.0	3.2 + 1.2	7.8 + 1.5
Cortex 3α,5α-THP	2.4 ± 0.9	5.1 ± 1.2	2.6 + 0.9	5.2 + 0.4	2.3 ± 0.5	4.3 ± 1.5	2.2 ± 0.4	6.1 ± 1.5	1.9 ± 0.6	5.7 ± 1.3

Table 22.4 Endocrine data of proestrous rats exposed to exploratory (open field, and elevated plus maze), social (social choice and social interaction), or no task with or without paced mating

	Open field + elevated plus maze		Social choice + social interaction		Control	
	No paced mating	Paced mating	No paced mating	Paced mating	No paced mating	Paced mating
Midbrain E_2	2.2 ± 0.1	2.2 ± 0.2	2.1 ± 0.1	2.0 ± 0.2	2.1 ± 0.1	2.0 ± 0.2
Hippo E_2	2.3 ± 0.5	2.4 ± 0.3	2.7 ± 0.6	2.5 ± 0.3	1.8 ± 0.1	2.0 ± 0.1
Diencephalon E_2	1.9 ± 0.7	2.0 ± 0.5	1.7 ± 0.6	1.7 ± 0.3	1.5 ± 0.2	2.0 ± 0.7
Cortex E_2	1.8 ± 0.2	1.5 ± 0.1	1.5 ± 0.4	1.8 ± 0.6	1.3 ± 0.4	2.2 ± 0.7
Midbrain P_4	2.0 ± 0.5	2.2 ± 0.1	2.1 ± 0.1	2.0 ± 0.2	2.1 ± 0.1	2.0 ± 0.2
Hippo P_4	2.5 ± 0.7	2.0 ± 0.5	2.3 ± 0.6	2.5 ± 0.3	2.4 ± 0.2	2.3 ± 0.7
Diencephalon P_4	1.6 ± 0.2	1.5 ± 0.1	1.4 ± 0.4	1.7 ± 0.6	1.5 ± 0.4	1.2 ± 0.7
Cortex P_4	1.7 ± 0.1	1.5 ± 0.3	1.4 ± 0.2	1.5 ± 0.3	1.3 ± 0.1	1.5 ± 0.6
Midbrain DHP	8.9 ± 0.1	15.3 ± 0.5	7.8 ± 0.4	12.9± 0.7	9.1 ± 0.9	11.9 ± 0.2
Hippo DHP	19.3 ± 1.0	27.9 ± 1.5	18.4 ± 1.5	30.0 ± 1.3	21.1 ± 1.9	32.0 ± 1.2
Diencephalon DHP	9.0 ± 1.1	13.9 ± 1.0	9.2 ± 1.5	11.1 ± 1.1	10.0 ± 1.1	12.1 ± 0.9
Cortex DHP	10.1 ± 1.1	12.5 ± 0.9	9.8 ± 1.2	11.9 ± 1.5	11.0 ± 1.7	13.2 ± 1.1
Midbrain 3α, 5α-THP	10.3 + 1.5	16.8 + 1.5	10.6 + 0.9	16.7 + 1.3	8.5 ± 1.5	14.2 ± 1.0
Hippo 3α, 5α-THP	152.4 + 1.1	30.6 + 1.4	16.4 + 1.2	28.4 + 1.6	19.0 ± 4.8	29.2 ± 3.1
Diencephalon 3α, 5α-THP	3.8 + 1.4	6.9 + 1.0	4.0 + 1.2	7.8 + 1.3	4.0 ± 0.7	8.1 ± 1.2
Cortex 3α, 5α-THP	2.5 + 0.9	4.9 + 0.7	2.1 ± 0.9	5.4 ± 1.1	2.4 ± 0.9	5.1 ± 1.2

Our findings to date using an integrated behavioral approach to examine effects of manipulating 3α,5α-THP in the VTA on concomitant changes in reproductive and non-reproductive behavior and endocrine levels suggest the following. Increasing levels of 3α,5α-THP in the midbrain VTA enhances exploratory, anxiety, social, and reproductive behaviors, and only modest levels of E_2 are necessary for such changes to occur. Further, manipulations of 3α,5α-THP in the VTA elicit biosynthesis of DHP and 3α,5α-THP, but not E_2 or P_4, in midbrain, hippocampus, diencephalon and/or cortex of diestrous rats. Findings from ovariectomized E_2-primed rats revealed that E_2 is not necessary for progestin-enhanced exploratory and anti-anxiety behaviors, while social and reproductive behaviors are facilitated by progestins only in the presence of E_2. There may be differences in sensitivity for effects of 3α,5α-THP in the VTA to modulate exploratory, anxiety, and social behaviors compared to reproductive behavior, as demonstrated by ceiling effects of 3α,5α-THP infusions to the VTA of proestrous rats on reproductive behavior. Together, these findings suggest that 3α,5α-THP manipulations in the VTA enhance exploratory, anxiety, social, and reproductive behaviors and elicit progestin biosynthesis in areas other than the VTA, including hippocampus, diencephalon, and cortex, which may underlie some of the other functional changes observed.

Our findings to date examining how engaging in reproductive and/or non-reproductive behavior alter endocrine levels in the brain are as follows. Paced mating, in conjunction with other behavioral tasks, increases DHP levels in midbrain, hippocampus, diencephalon, and cortex and 3α,5α-THP levels in midbrain and hippocampus of proestrous rats and DHP in midbrain of diestrous rats.[163] Findings from studies in which increases in DHP and 3α,5α-THP in midbrain, hippocampus, diencephalon, and cortex of proestrous rats were observed following paced mating, independent of exposure to other behavioral tasks, suggest that engaging in paced mating is sufficient to produce progestin biosynthesis in midbrain, hippocampus, diencephalon, and cortex, areas which are important for the modulation of reproductive, anxiety, exploratory, anxiety, and social behaviors.[162]

22.7 Conclusions

Using this model of investigating progestins' actions in the midbrain VTA for lordosis has enabled us to elucidate progestins' source, mechanisms, and function. This approach has revealed that "neurosteroids", such as 3α,5α-THP, are synthesized in the brain and act locally at membrane to produce their effects. Some membrane effects include rapid actions through changes in non-classical substrates, such as GBRs, NMDARS, and D_1, as well as second messengers. These actions of 3α,5α-THP on socio-sexual behaviors are also relevant for neuropsychiatric disorders. Indeed, we have found that therapeutic drugs for depression and schizophrenia demonstrate classic neurosteroidogenic effects and that these neurosteroidogenic effects improve social functioning.

What is the broader functional significance of these results beyond importance for reproduction? Our data to date suggest that modulation of exploratory, anxiety, social, and reproductive behaviors and/or increased progestin biosynthesis in areas may be a normative functional process. Notably, among people with schizophrenia and depression, 3α,5α-THP levels are altered.[128,164] Congruent with these clinical reports, disruptions in stress-responsiveness early in development alter proximate and later progestin biosynthesis of rats.[162] Together, these data suggest that disruptions in progestin biosynthesis may underlie some of the functional deficits observed among people with various stress-related neuropsychiatric disorders. This may also have implications for drug abuse, given that the VTA is involved in motivation and reward and there are sex differences and effects of progestins in response to drugs of abuse.[165] Further, from a basic research perspective, it seems that 3α,5α-THP may be a very sensitive and important neuroendocrine mediator of stress-related functional and endocrine responses. 3α,5α-THP may help buffer response to stressors so that social approach can be maintained and given the increases in 3α,5α-THP in the hippocampus, diencephalon, and cortex, biosynthesis may play a role in consolidation of help consolidate information about challenging situations that will later need to be recalled.

Our research model that reveals that paced mating has salient effects on progestins may be particularly relevance for neuropsychiatric disorders, such as schizophrenia. Pacing behavior involves approach, avoidance, social choice, and complex social interactions, which are typically disrupted in schizophrenia. Schizophrenia is characterized by dysregulation in stress responses.[166] Negative symptoms (avolution, alogia) of schizophrenia correlate with loss of social function, and plasma levels of the stress hormone, cortisol.[166,167] Evidence that 3α,5α-THP may underlie pathophysiology of schizophrenia and/or antipsychotic drug action are as follows. First, stress-reactivity may be involved in the etiology and/or manifestation of schizophrenia. People with schizophrenia have elevated cortisol levels and stress can precipitate psychiatric episodes.[166,167] Second, stress-induced 3α,5α-THP production may be disrupted in schizophrenia. A novel polymorphism and genetic mutation in the sequence encoding the gene for the mitochondrial benzodiazepine receptor, which is necessary for stress-induced 3α,5α-THP biosynthesis, has been identified among some schizophrenics.[168] As well, plasma levels of 3α,5α-THP precursors are lower among schizophrenics.[167] Third, 3α,5α-THP metabolized in the brain from peripheral progesterone may reduce the incidence and/or expression of schizophrenia. Women vs men, have higher 3α,5α-THP levels from metabolism of ovarian progesterone, and women, compared to men, with schizophrenia experience later age of onset, less debilitating psychiatric symptoms, fewer hospitalizations, and more rapid and better response to antipsychotics than do men.[169] Third, effective pharmacotherapies for schizophrenia can alter 3α,5α-THP levels. The atypical antipsychotic drug, olanzapine, enhances social functioning and increases 3α,5α-THP levels.[47,127] Because of the putative clinical significance of 3α,5α-THP for schizophrenia, and other neuropsychiatric disorders, further investigation of 3α,5α-THP's functional effects and how behavior can mediate 3α,5α-THP secretion, is essential.

Acknowledgements This research was supported by a grant from the National Institute of Mental Health (MH06769801).

References

1. Feder HH. Hormones and sexual behavior. Ann Rev Psychol 1984; 35:165–200.
2. Whalen RE, Luttge WG. Differential localization of progesterone uptake in brain. Role of sex, estrogen pretreatment and adrenalectomy. Brain Res 1971; 33:147–155.
3. Malsbury CW, Kow LM, Pfaff DW. Effects of medial hypothalamic lesions on the lordosis response and other behaviors in remale golden hamsters. Physiol Behav 1977; 19:223–237.
4. Muntz JA, Rose JD, Shults RC. Disruption of lordosis by dorsal midbrain lesions in the golden hamster. Brain Res Bull 1980; 5:359–364.
5. Rose JD. Changes in hypothalamic neuronal function related to hormonal induction of lordosis in behaving hamsters. Physiol Behav 1990; 47:1201–1212.
6. Frye CA, Gardiner SG. Progestins can have a membrane-mediated action in rat midbrain for facilitation of sexual receptivity. Horm Behav 1996; 30:682–691.
7. Pleim ET, Lisciotto CA, DeBold JF. Facilitation of sexual receptivity in hamsters by simultaneous progesterone implants into the VMH and ventral mesencephalon. Horm Behav 1990; 24:139–151.
8. Rubin BS, Barfield RJ. Progesterone in the ventromedial hypothalamus facilitates estrous behavior in ovariectomized, estrogen-primed rats. Endocrinology 1983; 113(2):797–804.
9. DeBold JF, Malsbury CW. Facilitation of sexual receptivity by hypothalamic and midbrain implants of progesterone in female hamsters Physiol Behav 1989; 46:655–660.

10. Takahashi LK, Lisk RD Dual progesterone action in diencephalon facilitates the induction of sexual receptivity in estrogen-primed golden hamsters Physiol Behav 1988; 44:741–747.
11. Frye CA. The role of neurosteroids and non-genomic effects of progestins and androgens in mediating sexual receptivity of rodents. Brain Res Brain Res Rev 2001; 37(1–3):201–222.
12. Frye CA, Vongher JM. Progestins' rapid facilitation of lordosis when applied to the ventral tegmentum corresponds to efficacy at enhancing GABA(A) receptor activity. J Neuroendocrinol 1999; 11:829–837.
13. Frye CA, Vongher JM. Ventral tegmental area infusions of inhibitors of the biosynthesis and metabolism of 3α,5α-THP attenuate lordosis of hormone-primed and behavioural oestrous rats and hamsters. J Neuroendocrinol 2001; 13:1076–1086.
14. Corpechot C, Young J, Calvel M, Wehrey C, Veltz JN, Touyer G, Mouren M, Prasad VV, Banner C, Sjovall J, et al. Neurosteroids: 3α-hydroxy-α-pregnan-20-one and its precursors in the brain, plasma, and steroidogenic glands of male and female rats. Endocrinology 1993; 133:1003–1009.
15. Freeman MC, Dupke KC, Croteau CM. Extinction of the estrogen-induced daily signal for LH release in the rat: a role for the proestrous surge of progesterone. Endocrinol 1976; 99:223–229.
16. Frye CA, Bayon LE. Mating stimuli influence endogenous variations in the neurosteroids 3α,5α-THP and 3α-Diol. J Neuroendocrinol 1999; 11:839–847.
17. Holzbauer M. Physiological variations in the ovarian production of 5α-pregnane derivatives with sedative properties in the rat. J Steroid Biochem 1975; 6:1307–1310.
18. Purdy RH, Morrow AL, Blinn JR, Paul SM. Synthesis, metabolism, and pharmacological activity of 3α-hydroxy steroids which potentiate GABA-receptor-mediated chloride ion uptake in rat cerebral cortical synaptoneurosomes. J Med Chem 1990; 33:1572–1581.
19. Frye CA, Bayon LE, Pursnani NK, Purdy RH. The neurosteroids, progesterone and 3α,5α-THP, enhance sexual motivation, receptivity, and proceptivity in female rats. Brain Res 1998; 808:72–83.
20. Beyer C, Gonzalez-Flores O, Gonzalez-Mariscal G. Ring A reduced progestins potently stimulate estrous behavior in rats: paradoxical effect through the progesterone receptor. Physiol Behav 1995; 58:985–993.
21. Frye CA, Bayon LE, Vongher JM. Intravenous progesterone elicits a more rapid induction of lordosis in rats than does SKF38393. Psychobiol 2000; 28:99–109.
22. Lisk RD. A comparison of the effectiveness of intravenous, as opposed to subcutaneous, injection of progesterone for the induction of estrous behavior in the rat. Can J Biochem Physiol 1960; 38:1381–1383.
23. Meyerson BJ. Relationship between the anesthetic and gestagenic action and estrous behavior-inducing activity of different progestins. Endocrinology 1967; 81:369–374.
24. Meyerson B. Latency between intravenous injection of progestins and the appearance of estrous behavior in estrogen-treated ovariectomized rats. Horm Behav 1972; 3:1–9.
25. Kubli-Garfias C, Whalen RE. Induction of lordosis behavior in female rats by intravenous administration of progestins. Horm Behav 1977; 9:380–386.
26. Beyer C, Gonzalez-Mariscal G, Eguibar JR, Gomora P. Lordosis facilitation in estrogen primed rats by intrabrain injection of pregnanes. Pharmacol Biochem Behav 1988; 31:919–926.
27. Gonzalez-Mariscal G, Gonzalez-Flores O, Beyer C. Intrahypothalamic injection of RU486 antagonizes the lordosis induced by ring A-reduced progestins. Physiol Behav 1989; 46:435–438.
28. Frye CA, DeBold JF. P-3-BSA, but not P-11-BSA, implants in the VTA rapidly facilitate receptivity in hamsters after progesterone priming to the VMH. Behav Brain Res 1993; 53:167–175.
29. Mellon SH, Griffin LD, Compagnone NA. Biosynthesis and action of neurosteroids. Brain Res Brain Res Rev 2001; 37:3–12.
30. Robel P, Young J, Corpechot C, Mayo W, Perche F, Haug M, Simon H, Baulieu EE. Biosynthesis and assay of neurosteroids in rats and mice: functional correlates. J Steroid Biochem Mol Biol 1995; 53:355–360.

31. Li X, Bertics PJ, Karavolas HJ. Regional distribution of cytosolic and particulate 5α-dihydroprogesterone 3α-hydroxysteroid oxidoreductases in female rat brain. J Steroid Biochem Mol Biol 1997; 60:311–318.
32. Roselli CE, Snipes CA. Progesterone 5α-reductase in mouse brain. Brain Res 1984; 305:197–202.
33. Frye CA, Leadbetter EA. 5α-reduced progesterone metabolites are essential in hamster VTA for sexual receptivity. Life Sci 1994; 54:653–659.
34. Griffin LD, Mellon SH. Selective serotonin reuptake inhibitors directly alter activity of neurosteroidogenic enzymes. Proc Natl Acad Sci USA 1999; 96:13512–13517.
35. Frye CA, Petralia SM, Rhodes ME, Stein B. Fluoxetine may influence lordosis of rats through effects on midbrain 3α,5α-THP concentrations. Ann NY Acad Sci 2003; 1007:37–41.
36. Dubrovsky B. A paracrine component of salient symptoms of depression in Cushing's of diencephalic origin, and in perimenstrual syndromes: a hypothesis. Med Hypotheses 2006; 66:936–938.
37. Cheney DL, Uzunov D, Guidotti A. Pregnenolone sulfate antagonizes dizocilpine amnesia: role for allopregnanolone. Neuroreport 1995; 6:1697–1700.
38. Ichikawa S, Sawada T, Nakamura Y, Morioka H. Ovarian secretion of pregnane compounds during the estrous cycle and pregnancy in rats. Endocrinology 1974; 94:1615–1620.
39. Paul SM, Purdy RH. Neuroactive steroids. FASEB J 1992; 6:2311–2322.
40. Purdy RH, Morrow AL, Moore PH, Jr., Paul SM. Stress-induced elevations of gamma-aminobutyric acid type A receptor-active steroids in the rat brain. Proc Natl Acad Sci USA 1991; 88:4553–4557.
41. Krueger KE, Papadopoulos V. Peripheral-type benzodiazepine receptors mediate translocation of cholesterol from outer to inner mitochondrial membranes in adrenocortical cells. J Biol Chem 1990; 265:15015–15022.
42. Papadopoulos V, Amri H, Boujrad N, Cascio C, Culty M, Garnier M, Hardwick M, Li H, Vidic B, Brown AS, Reversa JL, Bernassau JM, Drieu K. Peripheral benzodiazepine receptor in cholesterol transport and steroidogenesis. Steroids 1997; 62:21–28.
43. Stoffel-Wagner B. Neurosteroid metabolism in the human brain. Eur J Endocrinol 2001; 145:669–679.
44. Frye CA, Petralia SM. Mitochondrial benzodiazepine receptors in the ventral tegmental area modulate sexual behaviour of cycling or hormone-primed hamsters. J Neuroendocrinol 2003; 15:677–686.
45. Frye CA, Petralia SM. Lordosis of rats is modified by neurosteroidogenic effects of membrane benzodiazepine receptors in the ventral tegmental area. Neuroendocrinology 2003; 77:71–82.
46. Petralia SM, Jahagirdar V, Frye CA. Inhibiting biosynthesis and/or metabolism of progestins in the ventral tegmental area attenuates lordosis of rats in behavioural oestrus. J Neuroendocrinol 2005; 17:545–552.
47. Frye C, Seliga A. Effects of olanzapine infusions to the ventral tegmental area on lordosis and midbrain 3α,5α-THP concentrations in rats. Psychopharmacology (Berl) 2003; 170:132–139.
48. Gronemeyer H. Transcription activation by estrogen and progesterone receptors. Annu Rev Genet 1991; 25:89–123.
49. Hernandez MP, Chadli A, Toft DO. HSP40 binding is the first step in the HSP90 chaperoning pathway for the progesterone receptor. J Biol Chem 2002; 277:11873–11881.
50. Etgen AM. Progestin receptors and the activation of female reproductive behavior: a critical review. Horm Behav 1984; 18:411–430.
51. McGinnis MY, Krey LC, MacLusky NJ, McEwen BS. Steroid receptor levels in intact and ovariectomized estrogen-treated rats: an examination of quantitative, temporal and endocrine factors influencing the efficacy of an estradiol stimulus. Neuroendocrinology 1981; 33:158–165.
52. Pfaff DW, McEwen BS. Actions of estrogens and progestins on nerve cells. Science 1983; 219:808–814.
53. Acosta-Martinez M, Gonzalez-Flores O, Etgen AM. The role of progestin receptors and the mitogen-activated protein kinase pathway in delta opioid receptor facilitation of female reproductive behaviors. Horm Behav 2006; 49:458–462.

54. Pollio G, Xue P, Zanisi M, Nicolin A, Maggi A. Antisense oligonucleotide blocks progesterone-induced lordosis behavior in ovariectomized rats. Brain Res Mol Brain Res 1993; 19:135–139.
55. Brown TJ, MacLusky NJ, Shanabrough M, Naftolin F. Comparison of age- and sex-related changes in cell nuclear estrogen-binding capacity and progestin receptor induction in the rat brain. Endocrinology 1990; 126:2965–2972.
56. MacLusky NJ, McEwen BS. Progestin receptors in rat brain: distribution and properties of cytoplasmic progestin-binding sites. Endocrinology 1980; 106:192–202.
57. Parsons B, MacLusky NJ, Krey L, Pfaff DW, McEwen BS. The temporal relationship between estrogen-inducible progestin receptors in the female rat brain and the time course of estrogen activation of mating behavior. Endocrinology 1980; 107:774–779.
58. Munn AR 3rd, Sar M, Stumpf WE. Topographic distribution of progestin target cells in hamster brain and pituitary after injection of [3H]R5020. Brain Res 1983; 274:1–10.
59. Ogawa S, Pfaff DW. Application of antisense DNA method for the study of molecular bases of brain function and behavior. Behav Genet 1996; 26:279–292.
60. Rainbow TC, Parsons B, McEwen BS. Sex differences in rat brain oestrogen and progestin receptors. Nature 1982; 300:648–649.
61. Vathy IU, Etgen AM, Barfield RJ. Actions of progestins on estrous behaviour in female rats. Physiol Behav 1987; 40:591–595.
62. Frye CA. The role of neurosteroids and nongenomic effects of progestins in the ventral tegmental area in mediating sexual receptivity of rodents. Horm Behav 2001; 40:226–233.
63. Krebs CJ, Jarvis ED, Chan J, Lydon JP, Ogawa S, Pfaff DW. A membrane-associated progesterone-binding protein, 25-Dx, is regulated by progesterone in brain regions involved in female reproductive behaviors. Proc Natl Acad Sci USA 2000; 97:12816–12821.
64. Frye CA, Mermelstein PG, DeBold JF. Evidence for a non-genomic action of progestins on sexual receptivity in hamster ventral tegmental area but not hypothalamus. Brain Res 1992; 578:87–93.
65. Harrison NL, Majewska MD, Harrington JW, Barker JL. Structure-activity relationships for steroid interaction with the gamma-aminobutyric acidA receptor complex. J Pharmacol Exp Ther 1987; 241:346–353.
66. Majewska MD, Harrison NL, Schwartz RD, Barker JL, Paul SM. Steroid hormone metabolites are barbiturate-like modulators of the GABA receptor. Science 1986; 232:1004–1007.
67. Belelli D, McCauley L, Gee KW. Heterotropic cooperativity between putative recognition sites for progesterone metabolites and the atypical benzodiazepine Ro 5-4864. J Neurochem 1990; 55:83–87.
68. Morrow AL, Suzdak PD, Paul SM. Steroid hormone metabolites potentiate GABA receptor-mediated chloride ion flux with nanomolar potency. Eur J Pharmacol 1987; 142:483–485.
69. Gee KW, Bolger MB, Brinton RE, Coirini H, McEwen BS. Steroid modulation of the chloride ionophore in rat brain: structure-activity requirements, regional dependence and mechanism of action. J Pharmacol Exp Ther 1988; 246:803–812.
70. Masco D, Weigel R, Carrer HF. Gamma aminobutyric acid mediates ventromedial hypothalamic mechanisms controlling the execution of lordotic responses in the female rat. Behav Brain Res 1986; 19:153–162.
71. Westerling P, Lindgren S, Meyerson B. Functional changes in $GABA_A$ receptor stimulation during the oestrous cycle of the rat. Br J Pharmacol 1991; 103:1580–1584.
72. O'Connor LH, Nock B, McEwen BS. Regional specificity of gamma-aminobutyric acid receptor regulation by estradiol. Neuroendocrinology 1988; 47:473–481.
73. Lopez-Colome AM, McCarthy M, Beyer C. Enhancement of [3H]muscimol binding to brain synaptic membranes by progesterone and related pregnanes. Eur J Pharmacol 1990; 176:297–303.
74. Wu FS, Gibbs TT, Farb DH. Inverse modulation of gamma-aminobutyric acid- and glycine-induced currents by progesterone. Mol Pharmacol 1990; 37:597–602.
75. McCarthy MM, Coirini H, Schumacher M, Pfaff DW, McEwen BS, Schwartz-Giblin S. Ovarian steroid modulation of [3H]muscimol binding in the spinal cord of the rat. Brain Res 1991; 556:321–323.

76. McCarthy MM, Felzenberg E, Robbins A, Pfaff DW, Schwartz-Giblin S. Infusions of diazepam and allopregnanolone into the midbrain central gray facilitate open-field behavior and sexual receptivity in female rats. Horm Behav 1995; 29:279–295.
77. Bayer VE, Pickel VM. GABA-labeled terminals form proportionally more synapses with dopaminergic neurons containing low densities of tyrosine hydroxylase-immunoreactivity in rat ventral tegmental area. Brain Res 1991; 559:44–55.
78. DeBold JF, Frye CA. Genomic and non-genomic actions of progesterone in the control of female hamster sexual behavior. Horm Behav 1994; 28:445–453.
79. Frye CA, Mermelstein PG, DeBold JF. Bicuculline infused into the hamster ventral tegmentum inhibits, while sodium valproate facilitates, sexual receptivity. Pharmacol Biochem Behav 1993; 46:1–8.
80. Kole MH, Fuchs E, Ziemann U, Paulus W, Ebert U. Changes in 5-HT1A and NMDA binding sites by a single rapid transcranial magnetic stimulation procedure in rats. Brain Res 1999; 826:309–312.
81. Chu HP, Etgen AM. A potential role of cyclic GMP in the regulation of lordosis behavior of female rats. Horm Behav 1997; 32(2):125–132.
82. Gibbs RB. Treatment with estrogen and progesterone affects relative levels of brain-derived neurotrophic factor mRNA and protein in different regions of the adult rat brain. Brain Res 1999; 844:20–27.
83. Smith SS. Progesterone administration attenuates excitatory amino acid responses of cerebellar Purkinje cells. Neuroscience 1991; 42:309–320.
84. Cyr M, Ghribi O, Di Paolo T. Regional and selective effects of oestradiol and progesterone on NMDA and AMPA receptors in the rat brain. J Neuroendocrinol 2000; 12(5):445–452.
85. Carbone S, Szwarcfarb B, Losada M, Moguilevsky JA. Effect of ovarian hormones on the hypothalamic excitatory amino acids system during sexual maturation in female rats. Neuroendocrinology 1995; 61:235–242.
86. Petralia SM, Debold JF, Frye CA. MK-801 infusions to the ventral tegmental area and ventromedial hypothalamus produce opposite effects on lordosis of hormone-primed rats. Pharmacol Biochem Behav 2007; 86(2):377–385.
87. Fleischmann A, Vincent PA, Etgen AM. Effects of non-competitive NMDA receptor antagonists on reproductive and motor behaviors in female rats. Brain Res 1991; 568:138–146.
88. Kow LM, Harlan RE, Shivers BD, Pfaff DW. Inhibition of the lordosis reflex in rats by intrahypothalamic infusion of neural excitatory agents: evidence that the hypothalamus contains separate inhibitory and facilitatory elements. Brain Res 1985; 341:26–34.
89. Westerink BH, de Boer P, de Vries JB, Kruse CG, Long SK. Antipsychotic drugs induce similar effects on the release of dopamine and noradrenaline in the medial prefrontal cortex of the rat brain. Eur J Pharmacol 1998; 361:27–33.
90. Caggiula AR, Herndon JG, Jr., Scanlon R, Greenstone D, Bradshaw W, Sharp D. Dissociation of active from immobility components of sexual behavior in female rats by central 6-hydroxydopamine: implications for CA involvement in sexual behavior and sensorimotor responsiveness. Brain Res 1979; 172:505–520.
91. Becker JB, Rudick CN, Jenkins WJ. The role of dopamine in the nucleus accumbens and striatum during sexual behavior in the female rat. J Neurosci 2001; 21:3236–3241.
92. Becker JB, Ramirez VD. Experimental studies on the development of sex differences in the release of dopamine from striatal tissue fragments in vitro. Neuroendocrinology 1981; 32:168–173.
93. Fernandez-Ruiz JJ, Hernandez ML, de Miguel R, Ramos JA. Nigrostriatal and mesolimbic dopaminergic activities were modified throughout the ovarian cycle of female rats. J Neural Transm Gen Sect 1991; 85:223–229.
94. Morissette M, Di Paolo T. Sex and estrous cycle variations of rat striatal dopamine uptake sites. Neuroendocrinology 1993; 58(1):16–22.
95. Harden DG, King D, Finlay JM, Grace AA. Depletion of dopamine in the prefrontal cortex decreases the basal electrophysiological activity of mesolimbic dopamine neurons. Brain Res 1998; 794:96–102.

96. Herndon JG, Jr., Caggiula AR, Sharp D, Ellis D, Redgate E. Selective enhancement of the lordotic component of female sexual behavior in rats following destruction of central catecholamine-containing systems. Brain Res 1978; 141:137–151.
97. Thomas DN, Post RM, Pert A. Central and systemic corticosterone differentially affect dopamine and norepinephrine in the frontal cortex of the awake freely moving rat. Ann NY Acad Sci 1994; 746:467–469.
98. Mani SK, Blaustein JD, Allen JM, Law SW, O'Malley BW, Clark JH. Inhibition of rat sexual behavior by antisense oligonucleotides to the progesterone receptor. Endocrinology 1994; 135:1409–414.
99. Apostolakis EM, Garai J, Fox C, Smith CL, Watson SJ, Clark JH, O'Malley BW. Dopaminergic regulation of progesterone receptors: brain D5 dopamine receptors mediate induction of lordosis by D1-like agonists in rats. J Neurosci 1996; 16:4823–4834.
100. Mani SK, Reyna AM, Chen JZ, Mulac-Jericevic B, Conneely OM. Differential response of progesterone receptor isoforms in hormone-dependent and -independent facilitation of female sexual receptivity. Mol Endocrinol 2006; 20:1322–1332.
101. Collado ML, Rodriguez-Manzo G, Cruz ML. Effect of progesterone upon adenylate cyclase activity and cAMP levels on brain areas. Pharmacol Biochem Behav 1985; 23:501–504.
102. Mani SK, Fienberg AA, O'Callaghan JP, Snyder GL, Allen PB, Dash PK, Moore AN, Mitchell AJ, Bibb J, Greengard P, O'Malley BW. Requirement for DARPP-32 in progesterone-facilitated sexual receptivity in female rats and mice. Science 2000; 287:1053–1056.
103. Greengard P, Nairn AC, Girault JA, Ouimet CC, Snyder GL, Fisone G, Allen PB, Fienberg A, Nishi A. The DARPP-32/protein phosphatase-1 cascade: a model for signal integration. Brain Res Brain Res Rev 1998; 26:274–284.
104. Patchev VK, Hassan AH, Holsboer DF, Almeida OF. The neurosteroid tetrahydroprogesterone attenuates the endocrine response to stress and exerts glucocorticoid-like effects on vasopressin gene transcription in the rat hypothalamus. Neuropsychopharmacology 1996; 15:533–540.
105. Mensah-Nyagan AG, Do-Rego JL, Beaujean D, Luu-The V, Pelletier G, Vaudry H. Regulation of neurosteroid biosynthesis in the frog diencephalon by GABA and endozepines. Horm Behav 2001; 40:218–225.
106. Schlinger BA, Soma KK, London SE. Neurosteroids and brain sexual differentiation. Trends Neurosci 2001; 24:429–431.
107. Kellogg CK, Frye CA. Endogenous levels of 5 alpha-reduced progestins and androgens in fetal vs. adult rat brains. Brain Res Dev Brain Res 1999; 24:429–431.
108. Kehoe P, Mallinson K, McCormick CM, Frye CA. Central allopregnanolone is increased in rat pups in response to repeated, short episodes of neonatal isolation. Brain Res Dev Brain Res 2000; 124:133–136.
109. McCormick CM, Kehoe P, Mallinson K, Cecchi L, Frye CA. Neonatal isolation alters stress hormone and mesolimbic dopamine release in juvenile rats. Pharmacol Biochem Behav 2002; 73(1):77–85.
110. Drugan RC, Paul SM, Crawley JN. Decreased forebrain [35S]TBPS binding and increased [3H]muscimol binding in rats that do not develop stress-induced behavioral depression. Brain Res 1993; 631:270–276.
111. Serra M, Pisu MG, Floris I, Floris S, Cannas E, Mossa A, Trapani G, Latrofa A, Purdy RH, Biggio G. Social isolation increases the response of peripheral benzodiazepine receptors in the rat. Neurochem Int 2004; 45:141–148.
112. Papaioannou A, Dafni U, Alikaridis F, Bolaris S, Stylianopoulou F. Effects of neonatal handling on basal and stress-induced monoamine levels in the male and female rat brain. Neuroscience 2002; 114:195–206.
113. McCormick CM, Linkroum W, Sallinen BJ, Miller NW. Peripheral and central sex steroids have differential effects on the HPA axis of male and female rats. Stress 2002; 5:235–247.
114. Pearson Murphy BE, Allison CM. Determination of progesterone and some of its neuroactive ring A-reduced metabolites in human serum. J Steroid Biochem Bol Biol 2000; 74:137–142

115. Holzbauer M. Physiological variations in the ovarian production of 5alpha-pregnane derivatives with sedative properties in the rat. J Steroid Biochem 1975; 6:1307–1310.
116. Young EA. Sex differences and the HPA axis: implications for psychiatric disease. J Gend Specif Med 1998; 1:21–27.
117. Leung A, Chue P. Sex differences in schizophrenia, a review of the literature. Acta Psychiatr Scand Suppl 2000; 401:3–38.
118. Bergemann N, Parzer P, Nagl I, Salbach B, Runnebaum B, Mundt Ch, Resch F. Acute psychiatric admission and menstrual cycle phase in women with schizophrenia. Arch Womens Ment Health 2002; 5:119–126.
119. Gallo MA, Smith SS. Progesterone withdrawal decreases latency to and increases duration of electrified prod burial: a possible rat model of PMS anxiety. Pharmacol Biochem Behav 1993; 46:897–904.
120. Frye CA, Walf AA. Hippocampal 3α,5α-THP may alter depressive behavior of pregnant and lactating rats. Pharmacol Biochem Behav 2004; 78:531–540.
121. Norman RM, Malla AK. Stressful life events and schizophrenia. I: a review of the research. Br J Psychiatry 1993; 162:161–166.
122. Kurumaji A, Nomoto H, Yoshikawa T, Okubo Y, Toru M. An association study between two missense variations of the benzodiazepine receptor (peripheral) gene and schizophrenia in a Japanese sample. J Neural Transm 2000; 107:491–500.
123. Casada, JH, Roache, JD. Neurosteriods in PTSD. International Society for Traumatic Stress Studies Annual Meeting, 2004.
124. Dong E, Matsumoto K, Uzunova V, Sugaya I, Takahata H, Nomura H, Watanabe H, Costa E, Guidotti A. Brain 5α-dihydroprogesterone and allopregnanolone synthesis in a mouse model of protracted social isolation. Proc Natl Acad Sci USA 2001; 98:2849–2854.
125. Pinna G, Costa E, Guidotti A. Changes in brain testosterone and allopregnanolone biosynthesis elicit aggressive behavior. Proc Natl Acad Sci USA 2005; 102:2135–2140.
126. Marx CE, VanDoren MJ, Duncan GE, Lieberman JA, Morrow AL. Olanzapine and clozapine increase the GABAergic neuroactive steroid allopregnanolone in rodents. Neuropsychopharmacology 2003; 28:1–13.
127. Marx SG, Wentz MJ, Mackay LB, Schlembach D, Maul H, Fittkow C, Given R, Vedernikov Y, Saade GR, Garfield RE Effects of progesterone on iNOS, COX-2, and collagen expression in the cervix. J Histochem Cytochem 2006; 54:623–639.
128. Uzunova V, Sheline Y, Davis JM, Rasmusson A, Uzunov DP, Costa E, Guidotti A. Increase in the cerebrospinal fluid content of neurosteroids in patients with unipolar major depression who are receiving fluoxetine or fluvoxamine. Proc Natl Acad Sci USA 1998; 95:3239–3244.
129. Freeman EW, Jabara S, Sondheimer SJ, Auletto R. Citalopram in PMS patients with prior SSRI treatment failure: a preliminary study. J Womens Health Gend Based Med 2002; 11:459–464.
130. Assuncao SS, Ruschel SI, Rosa Lde C, Campos JA, Alves MJ, Bracco OL, de Lima MS. Weight gain management in patients with schizophrenia during treatment with olanzapine in association with nizatidine. Rev Bras Psiquiatr 2006; 28:270–276.
131. Monteleone P, Bencivenga R, Longobardi N, Serritella C, Maj M. Differential responses of circulating ghrelin to high-fat or high-carbohydrate meal in healthy women. J Clin Endocrinol Metab 2003; 88:5510–5514.
132. Torres H, Rhodes ME, Frye CA. Fluoxetine-induced deficits in sexual behavior of female hamsters are attenuated with 3α,5α-THP-replacement. McNair Program Research Symposium, 2003.
133. Frye CA, Rhodes ME, Petralia SM, Walf AA, Sumida K, Edinger KL. 3α-hydroxy-5α-pregnan-20-one in the midbrain ventral tegmental area mediates social, sexual, and affective behaviors. Neuroscience 2006; 138:1007–1014.
134. Carter CS. Neuroendocrine perspectives on social attachment and love. Psychoneuroendocrinology 1998; 23:779–818.
135. Frye CA, Petralia SM, Rhodes ME. Estrous sycle and sex differences in performance on anxiety tasks coincide with increases in hippocampal progesterone and 3α,5α-THP. Pharmacol Biochem Behav 2000; 67:587–596.

136. Mora S, Dussaubat N, Diaz-Veliz G. Effects of the estrous vyvle and ovarian hormones on behavioral indices of anxiety in female rats. Psychoneuroendocrinology 1996; 21:609–620.
137. Vinogradova OS, Kitchigina VF, Kudina TA, Zenchenko KI. Spontaneous activity and sensory responses of hippocampal neurons during persistent theta-rhythm evoked by median raphe nucleus blockade in rabbit. Neuroscience 1999; 94:745–753.
138. Galeeva A, Tuohimaa P. Analysis of mouse plus-maze behavior modulated by ovarian steroids. Behav Brain Res 2001; 119:41–47.
139. Laconi MR, Casteller G, Gargiulo PA, Bregonzio C, Cabrera RJ. The anxiolytic effect of allopregnanolone is associated with gonadal hormonal status in female rats. Eur J Pharmacol 2001; 417:111–116.
140. Rhodes ME, Frye CA. Inhibiting progesterone metabolism in the hippocampus of rats in behavioral estrus decreases anxiolytic behaviors and enhances exploratory and antinociceptive behaviors. Cogn Affect Behav Neurosci 2001; 1:287–296.
141. Walf AA, Rhodes ME, Frye CA. Ovarian steroids enhance object recognition in naturally cycling and ovariectomized, hormone-primed rats. Neurobiol Learn Mem 2006; 86:35–46.
142. Schindler CJ, Slamberova R, Vathy I. Bicuculline seizure susceptibility and nigral $GABA_A$ α1 receptor mRNA is altered in adult prenatally morphine-exposed females. Psychoneuroendocrinology 2003; 28:348–363.
143. McCarthy MM, Felzenberg E, Robbins A, Pfaff DW, Schwartz-Giblin S. Infusions of diazepam and allopregnanolone into the midbrain central gray facilitate open-field behavior and sexual receptivity in female rats. Horm Behav 1995; 29:279–295.
144. Fernandez-Guasti A, Vega-Matuszczyk J, Larsson K. Synergistic action of estradiol, progesterone and testosterone on rat proceptive behavior. Physiol Behav 1991; 50:1007–1011.
145. Steel E. Female sexual behaviour: roles of gonadal hormones in the Syrian hamster. Physiol Behav 1983 Oct; 31(4):453–459.
146. Pfeifle JK, Edwards DA. Midbrain lesions eliminate sexual receptivity but spare sexual motivation in female rats. Physiol Behav 1983; 31:385–389.
147. Cheng YJ, Karavolas HJ. Conversion of progesterone to 5α-pregnane-3,20-dione and 3α-hydroxy-5-α-pregnan-20-one by rat medical basal hypothalami and the effects of estradiol and stage of estrous cycle on the conversion. Endocrinology 1973; 93:1157–1162.
148. Sinchak K, Mills RH, Tao L, LaPolt P, Lu JK, Micevych P. Estrogen induces de novo progesterone synthesis in astrocytes. Dev Neurosci 2003; 25:343–348.
149. Jung-Testas I, Hu ZY, Baulieu EE, Robel P. Neurosteroids: biosynthesis of pregnenolone and progesterone in primary cultures of rat glial cells. Endocrinology 1989; 125:2083–2091.
150. Schumacher M, Guennoun R, Robert F, Carelli C, Gago N, Ghoumari A, Gonzalez Deniselle MC, Gonzalez SL, Ibanez C, Labombarda F, Coirini H, Baulieu EE, De Nicola AF. Local synthesis and dual actions of progesterone in the nervous system: neuroprotection and myelination. Growth Horm IGF Res 2004; 14(Suppl A):S18–S33.
151. Zwain IH, Yen SS, Cheng CY. Astrocytes cultured in vitro produce estradiol-17β and express aromatase cytochrome P-450 (P-450 AROM) mRNA. Biochim Biophys Acta 1997; 1334:338–348.
152. Robel P, Young J, Corpechot C, Mayo W, Perche F, Haug M, Simon H, Baulieu EE. Biosynthesis and assay of neurosteroids in rats and mice: functional correlates. J Steroid Biochem Mol Biol 1995; 53:355–360.
153. Balthazart J, Foidart A. Brain aromatase and the control of male sexual behavior. J Steroid Biochem Mol Biol 1993; 44:521–540.
154. Wood GE, Beylin AV, Shors TJ. The contribution of adrenal and reproductive hormones to the opposing effects of stress on trace conditioning in males versus females. Behav Neurosci 2001; 115:175–187.
155. Cavigelli SA, Monfort SL, Whitney TK, Mechref YS, Novotny M, McClintock MK. Frequent serial fecal corticoid measures from rats reflect circadian and ovarian corticosterone rhythms. J Endocrinol 2005; 184:153–163.
156. Rhodes RD, Harrison DW, Demaree HA. Hostility as a moderator of physical reactivity and recovery to stress. Int J Neurosci 2002; 112:1671–1686.

157. Viau V, Meaney MJ. Variations in the hypothalamic-pituitary-adrenal response to stress during the estrous cycle in the rat. Endocrinology 1991; 129:2503–2511.
158. Figueiredo HF, Dolgas CM, Herman JP. Stress activation of cortex and hippocampus is modulated by sex and stage of estrus. Endocrinology 2002; 143:2534–2540.
159. Bitran D, Dugan M, Renda P, Ellis R, Foley M. Anxiolytic effects of the neuroactive steroid pregnanolone (3α-OH-5β-pregnan-20-one) after microinjection in the dorsal hippocampus and lateral septum. Brain Res 1999; 850:217–224.
160. Godsil BP, Blackmore MA, Fanselow MS. Modulation of an activity response with associative and nonassociative fear in the rat: a lighting differential influences the form of defensive behavior evoked after fear conditioning. Learn Behav 2005; 33:454–463.
161. McHugh EM, Zhu W, Milgram S, Mager S. The GABA transporter GAT1 and the MAGUK protein Pals1: interaction, uptake modulation, and coexpression in the brain. Mol Cell Neurosci 2004; 26:406–417.
162. Frye CA, Walf AA, Petralia SM. In the ventral tegmental area, progestins have actions at D1 receptors for lordosis of hamsters and rats that involve $GABA_A$ receptors. Horm Behav 2006; 50:332–337.
163. Frye CA, Rhodes ME. Infusions of 5α-pregnan-3α-ol-20-one (3α,5α-THP) to the ventral tegmental area, but not the substantia nigra, enhance exploratory, anti-anxiety, social and sexual behaviours and concomitantly increase 3α,5α-THP concentrations in the hippocampus, diencephalon and cortex of ovariectomised oestrogen-primed rats. J Neuroendocrinol 2006; 18:960–975.
164. Shirayama Y, Hashimoto K, Suzuki Y, Higuchi T. Correlation of plasma neurosteroid levels to the severity of negative symptoms in male patients with schizophrenia. Schizophr Res 2002; 58:69–74.
165. Frye CA. Progestins influence motivation, reward, conditioning, stress, and/or response to drugs of abuse. Pharmacol Biochem Behav 2007; 86:209–219.
166. Al-Amin HA, Weinberger DR, Lipska BK. Exaggerated MK-801-induced motor hyperactivity in rats with the neonatal lesion of the ventral hippocampus. Behav Pharmacol 2000; 11:269–278.
167. Andreasen PA. Changes in net charge of glucocorticoid receptors by activation, and evidence for a biphasic activation kinetics. Mol Cell Endocrinol 1982; 28:563–586.
168. Bachevalier J, Beauregard M, Alvarado MC. Long-term effects of neonatal damage to the hippocampal formation and amygdaloid complex on object discrimination and object recognition in rhesus monkeys (Macaca mulatta). Behav Neurosci 1999; 113:1127–1151.
169. Barbazanges A, Piazza PV, Le Moal M, Maccari S. Maternal glucocorticoid secretion mediates long-term effects of prenatal stress. J Neurosci 1996; 16:3943–3949.

Chapter 23
The Role of Progesterone and its Metabolites in Premenstrual Disorders of Affect

Akiko Dohi, Glenn H. Dillon, and Meharvan Singh

Abstract It is estimated that between 70% and 90% of women in their reproductive years experience some level of discomfort and experience affective and/or somatic symptoms during the premenstrual (luteal) phase of their menstrual cycle. These symptoms range from mild discomfort to severe emotional and somatic symptoms, with the latter seen particularly in women diagnosed with Premenstrual Dysphoric Disorder (PMDD). Although the neurobiological basis of the Premenstrual syndrome (PMS) and PMDD is still not completely understood, there appears to be a relationship between the levels of certain gonadal steroids and their metabolites on the development or exacerbation of these premenstrual disorders of affect. Here we review our current understanding of how progesterone, and its metabolite, allopregnanolone, may relate to such mood disturbances, focusing particularly on both conventional and more recently described, alternative mechanisms by which these neurosteroids exert their effects on the GABA-A receptor. Based on published reports and recent data from our laboratory, we propose that progesterone and its 5α-reduced metabolite, allopregnanolone, may have distinct effects on GABA-A receptor function, resulting from distinct mechanisms of action. This difference, we assert, may help explain the discrepant association between the levels of progesterone and/or its metabolites and the development of premenstrual disorders of affect, including PMDD.

Keywords Progesterone, neurosteroid, cell signaling, GABA-A receptor, premenstrual syndrome, premenstrual dysphoric disorder

Abbreviations CaMK calcium–calmodulin-dependent kinase; CNS central nervous system; ERK extracellular signal-regulated kinase; GABA γ-aminobutyric acid; MAPK mitogen-activated protein kinase; PI3-K phosphotidyl inositol-3 kinase; PKA protein kinase A (cAMP-dependent kinase); PKC protein kinase C; PKG protein kinase G; PMDD premenstrual dysphoric disorder; PMS premenstrual

---

Department of Pharmacology and Neuroscience, Institute for Aging and Alzheimer's disease Research, and Center FOR HER, University of North Texas Health Science Center, Fort Worth, Texas, USA

M.S. Ritsner and A. Weizman (eds.), *Neuroactive Steroids in Brain Function, Behavior and Neuropsychiatric Disorders*

syndrome; PR progesterone receptor; SSRI selective serotonin reuptake inhibitor; TM transmembrane

23.1 Introduction

The premenstrual period, defined here as the period after ovulation but prior to the onset of menses, has been associated with numerous physiological, emotional, and behavioral changes in women. Depending on the severity of the symptoms, and along with specific diagnostic criteria, women may be diagnosed as having Premenstrual Syndrome (PMS) or Premenstrual Dysphoric Disorder (PMDD). The cyclical appearance of these symptoms supports the potential involvement of gonadal steroids and/or their metabolites in the etiology of these disorders. For the purposes of this review, we point out that the symptoms of PMS and PMDD generally occur during a period of elevated progesterone levels (coincident with the luteal phase of the menstrual cycle). Among the neurotransmitter systems affected by progesterone, and whose perturbation may be relevant to the symptoms of PMS and PMDD, is the GABA-A receptor. The dogmatic view of the relationship between progesterone and the GABA-A receptor is that allopregnanolone, the 5α-reduced metabolite of progesterone, is the principle effector of progesterone's actions, where it binds to, and allosterically modulates, the GABA-A receptor. However, in this review, we implicate alternative mediators of progesterone on cellular systems that impact mood, including specific cell signaling pathways that impact GABA-A receptor function, and suggest the potential importance of other mechanisms other than the direct binding of the metabolite of progesterone to the GABA-A receptor as the primary influence on mood.

23.2 Premenstrual Syndrome and Premenstrual Dysphoric Disorder

Since the original descriptions of "premenstrual tension", as described by Frank[1] and then by Israel,[2] much information has been gathered about the characteristics and prevalence of PMS and PMDD. It is estimated that as high as 90% of women in their reproductive years experience some level of discomfort, and experience affective and/or somatic symptoms, during the premenstrual phase of their menstrual cycle. While approximately 40% of women experience relatively mild symptoms that may simply be interpreted as predicting the onset of menses, another 40% of women experience symptoms that can be classified as PMS where medical/pharmacological intervention may be sought.[3] The prevalence of PMDD, on the other hand, is much lower, where reports estimate its occurrence to be between 3% and 8% of women in their reproductive years.[4]

To date, more than 200 symptoms have been described in women who experience PMS and/or PMDD, the most common of which are tension, irritability and dysphoria.[5] A major characteristic and criteria for diagnosing PMS and PMDD is that the symptoms, which generally appear during the mid to late luteal phase of the menstrual cycle, disappear in the week immediately following the menstrual period (*Diagnostic and Statistical Manual of Mental Disorders*, 4th ed., pp. 717–718), and thus, helps distinguish between the possible confound of an existing and underlying affective disorder, whose symptoms may be exacerbated during the luteal phase of the menstrual cycle.

23.3 Progesterone, Allopregnanolone and PMS/PMDD

Levels of progesterone and its metabolites in PMS/PMDD patients have been assessed in numerous studies. However, the correlation between absolute levels of those steroids and the susceptibility to PMS/PMDD is not clear. For example, several studies have described that the levels of progesterone and its metabolites (5α-dihydroprogesterone (5α-DHP) and allopregnanolone) are lower in women diagnosed with PMS.[6–8] although some of these studies report reductions only during specific phases of the menstrual cycle.[6,7] In contrast, some studies have failed to identify a correlation between the levels of these neuroactive steroids and clinical symptoms of PMS,[9,10] and in one recent instance, associated the *relief* of severe PMS symptoms, in response to a GnRH agonist, with a *reduction* in both progesterone and allopregnanolone levels.[11]

While absolute levels of progesterone (or its metabolites) do not appear to provide clear insight into the relationship between progesterone and the symptoms of PMS, it may be necessary to consider other factors relating to progesterone biology as well. For example, the pattern of pulsatile progesterone release may be relevant as suggested by Fachinetti et al., who determined that women with PMS have a higher frequency and lower amplitude pattern of release compared to their respective controls.[12]

Despite the apparent lack of association between the absolute levels of progesterone or its metabolites and expression of clinical symptoms of PMS/PMDD, it is noteworthy that in several studies, treatment with progesterone has been shown to alleviate certain symptoms of PMS. This strategy is based on the premise that progesterone, through its metabolite, allopregnanolone, would allosterically enhance the function of $GABA_A$ receptor, the major inhibitory neurotransmitter receptor, resulting in anxiolysis, thereby alleviating at least some of the affective symptoms associated with PMS and/or PMDD.[13] And while selective serotonin reuptake inhibitors (SSRIs), which are currently considered as the primary treatment for PMDD, generally work through the serotonin receptor system, it is noteworthy that certain SSRIs (such as fluoxetine and norfluoxetine) can attenuate such anxiety associated behaviors such as aggression, in a manner that is distinct from its actions on serotonin uptake mechanisms.[14] Instead, Pinna and colleagues found that fluoxetine, at doses that were insufficient to inhibit serotonin reuptake, enhanced brain allopregnanolone content.[14] Thus, the effectiveness of even SSRIs may be linked to regulation of brain neurosteroid levels.

Several studies that assessed the effects of progesterone (or its metabolite, allopregnanolone) on symptoms of PMS have yielded positive results. For example, PMS patients treated with progesterone, via different routes of administration (vaginal suppositories, orally) and different durations of treatment (ranging from 2 to 7 months), resulted in significant improvement in symptoms relating to anxiety and irritability.[15–17] However, as was the case when associating the levels of circulating hormones with the manifestation of clinical symptoms, the efficacy of progesterone has been equivocal as other groups have found progesterone to be ineffective at relieving symptoms of PMS.[18,19]

These conflicting reports underscore the fact that the neurobiology underlying the symptoms of PMS and/or PMDD is likely multifaceted. Further, these data may suggest that there is a subpopulation of women for which progesterone therapy may be effective, while for others, it may not. In the following sections, we review some basic biological aspects of the relationship between progesterone and the GABA-A receptor and outline alternative means by which progesterone may elicit its effects on GABA-A function, and as such, have different outcomes with respect to relieving some of the affective symptoms associated with the premenstrual syndrome.

23.4 GABA, the GABA-A Receptor and PMS/PMDD

Gamma-aminobutyric acid (GABA) binds to the $GABA_A$ receptor, resulting in opening of a Cl^- channel. The direction of movement of Cl^- ions through this channel depends upon the relative concentration of intra- to extracellular Cl^-. In the adult CNS, the extracellular concentration of Cl^- ($[Cl^-]_o$) is typically higher than the intracellular concentration ($[Cl^-]_i$), and thus Cl^- moves into the cell. This causes hyperpolarization of the neuron, and results in reduced neuronal activity.

Alterations in GABAergic neurotransmission have been implicated in a variety of CNS disorders. For example, work in recent years has established a strong link between GABAergic signaling and schizophrenia (see[20,21]). Moreover, and pertinent to the symptoms of premenstrual disorders, GABA-A receptors have also been implicated in anxiety.[22] In fact, disturbances in cortical GABAergic neuronal function may be a relevant factor in the development of PMDD[23] and could be contributed by the influence of neurosteroids,[23] or alterations in levels of GABA itself.[24]

And while the current treatment of choice for symptoms of PMS/PMDD are generally selective serotonin reuptake inhibitors (SSRI), targeting the GABA-A receptor has also been found to be effective. For example, the benzodiazepine, Alprazolam, has been shown to have reasonable efficacy in the treatment of severe PMS patients,[25] further underscoring the value of considering the GABA-A receptor in the pathogenesis of premenstrual disorders of affect as well as a biological substrate/target that may be exploited in therapeutic strategies.

23.5 Progesterone-Induced Regulation of the GABA-A Receptor

When considering the influence of progesterone on the GABA-A receptor, the dogmatic view is that progesterone's effects are mediated by its metabolite, allopregnanolone. Progesterone itself is produced by the gonads but can also be synthesized de novo in the central nervous system (CNS).[26] Progesterone's metabolites can also be synthesized de novo in the brain[27,28] and include 5α-dihydroprogesterone (5α-DHP) and 3α, 5α-tetrahydroprogesterone (3α, 5α-THP, or allopregnanolone). Allopregnanolone was among the first steroids that were shown to modulate neuronal excitability by binding to and regulating the GABA-A receptor.

While direct binding of allopregnanolone to the $GABA_A$ receptor is certainly an important mechanism of regulating GABA-gated currents, and thus, neuronal excitability, it has become clear that these receptors are also targets of signal transduction pathways such that phosphorylation of specific subunits can influence the function of the $GABA_A$ receptor. In fact, a number of protein kinases such as protein kinase C (PKC), protein kinase A (PKA), protein kinase G (PKG), tyrosine kinase and Ca^{2+}/calmodin-dependent kinase II (CAMKII), have been reported to modulate $GABA_A$Rs by direct phosphorylation of residues at the large intracellular loop between TM3 and TM4 (for review, see[29–33]). Further, recent studies have shown that, in addition to the modulation of GABA receptor gating, protein kinases also influence surface receptor trafficking of GABA-A receptor, which could also be an important mechanism for regulating neuronal excitability.

An example of this is PKC-mediated receptor endocytosis. Down-regulation of $GABA_A$R function by PKC activation has been well-documented in the *Xenopus* oocyte expression system[34–37] and may be mediated by direct phosphorylation and/or receptor endocytosis.[35–37] However, depending on the cellular model used, activation of PKC may enhance,[38] inhibit[39,40] or have no effect[41] on GABA-gated currents. Interestingly, PKC-mediated endocytosis does not appear to be dependent on direct receptor phosphorylation because mutation of the known PKC sites within GABA-A receptor subunits (S409 in β1/3 and S410 in β2) does not alter the effect of PKC on endocytosis.[35,40] It is possible, however, that the downstream effectors of PKC, such as ERK, may serve as the kinase that phosphorylate the GABA-A receptor directly. Support for the latter comes from the recent description of putative ERK phosphorylation sites on the alpha subunits of the GABA-A receptors.[42]

As introduced above, other signaling proteins and pathways may modulate GABA-A receptor function through phosphorylation of its subunits. These include calcium–calmodulin-dependent protein kinase (type II) and protein kinase G (PKG), which can phosphorylate the β1, β3 and γ2 subunits of the GABA-A receptor.[43–45] In addition, cyclic AMP-dependent protein kinase (PKA) and Src have also been shown to phosphorylate purified preparations of neuronal $GABA_A$ receptors in vitro.[46–48] But as suggested, depending on the identity and location of the sites phosphorylated on the $GABA_A$ receptor, diverse functional effects may result, ranging from enhanced chloride conductance to the inhibition of such function.[29,38,49,50]

While the "classical" mechanism of progesterone action involves binding to and activation of the intracellular progesterone receptor (PR), leading to transcriptional regulation, progesterone has also been found to elicit its effects via the recruitment of second messenger/signal transduction systems, including cAMP/PKA,[51] the MAPK[52,53] and PI-3K/Akt.[52] Some of these effects may be mediated by novel progesterone receptors, including the recently cloned membrane progesterone receptors.[54,55] Thus, with respect to the influence of progesterone on the GABA-A receptor, it is important to recognize that the effects may be mediated not only through the neuroactive metabolite, allopregnanolone, but may also be mediated through rapid, "non-genomic" mechanisms elicited by the parent compound, progesterone. The significance of the latter is also underscored by the fact that activation of certain signaling pathways, such as ERK, may have the opposite effect on the GABA-A receptor, relative to the effect of progesterone's metabolites. For example, our laboratory has recently shown that the ERK/MAPK pathway exerts a negative influence on GABA-gated currents.

Since inhibition of the GABA-A receptor may contribute to some of the symptoms of PMS and PMDD, the relative abundance of allopregnanolone (which enhances GABA-A receptor function) and its precursor, progesterone (which through signaling, may dampen synaptic inhibition mediated by the GABA-A receptor), may drive whether or not progesterone is effective in treatment of at least certain symptoms of PMS and/or PMDD. By inference, the capacity to convert progesterone to allopregnanolone (i.e., the levels of 5α-reductase) may be critical.

23.6 Conclusion

Our current understanding of the mechanistic basis of the symptoms of PMS and/or PMDD certainly appear to implicate the GABA-A receptor, making it a relevant target for study and treatment. With respect to the interrelationship with progesterone, some of the positive effects of this neurosteroid on symptoms of PMS and PMDD may be mediated by its metabolite, allopregnanolone, while in numerous studies that describe no effect or even negative consequences of progesterone, this may be a result of a cell signaling-mediated inhibition of GABA-A receptor function. With respect to the latter, progesterone may be the culprit rather than the therapeutic in terms of its relationship to the symptoms of PMS and/or PMDD. And conversely, lower progesterone levels may be associated with reduced symptom severity, or alternatively, a reduced propensity to develop symptoms of PMS and/or PMDD. In this regard, it is interesting that symptom relief in a population of women suffering from extreme PMS that had been treated with the GnRH agonist, buserelin, was associated with a decrease in serum progesterone levels.[11]

The difference between whether progesterone may be effective as a therapeutic or alternatively, having a role in the pathogenesis of PMS and/or PMDD may depend on such factors as the level (and function) of the converting enzyme, 5α-reductase, or even subunit make-up of GABA-A receptors (where the same neurosteroid may have different consequences on the GABA-A receptor depending on the subunit make-up of the GABA-A receptor pentamer). As such, it is important that steps be taken to define the population of women experiencing symptoms of PMS and/or PMDD. A mere understanding or cataloging the symptoms may be useful in arriving at a diagnosis, but is unlikely to be adequate in defining what type of therapy may be most effective at reducing or eliminating life-disruptive symptoms of severe PMS and/or PMDD. Future studies that aim to understand the neurobiology of gonadal hormones, their neurosteroid derivatives and mood influencing neuronal systems (including the GABA-A receptor and serotonin receptor systems) will be critical in helping define the mechanistic basis of the disorder, which can then lead to better and more effective therapeutic strategies to treat these disorders.

Acknowledgments Some of the work presented herein was supported, in part, by NIH grants AG022550 and AG023330, and a NARSAD (National Alliance for Research on Schizophrenia and Depression) Young Investigator Award.

References

1. Frank RT. The hormonal causes of premenstrual tension. Arch Neurol Psych 1931; 26:1053–1057.
2. Israel SL. Premenstrual tension. JAMA 1938; 110:1721–1723.
3. Dell DL. Premenstrual syndrome, premenstrual dysphoric disorder, and premenstrual exacerbation of another disorder. Clin Obstet Gynecol 2004; 47:568–575.
4. Steiner M. Premenstrual syndrome and premenstrual dysphoric disorder: guidelines for management. J Psychiatry Neurosci 2000; 25:459–468.
5. Steiner M, Born L. Diagnosis and treatment of premenstrual dysphoric disorder: an update. Int Clin Psychopharmacol 2000; 15(Suppl 3):S5–S17.
6. Monteleone P, Luisi S, Tonetti A, et al. Allopregnanolone concentrations and premenstrual syndrome. Eur J Endocrinol 2000; 142:269–273.
7. Bicikova M, Putz Z, Hill M, et al. Serum levels of neurosteroid allopregnanolone in patients with premenstrual syndrome and patients after thyroidectomy. Endocr Regul 1998; 32:87–92.
8. Wang M, Seippel L, Purdy RH, et al. Relationship between symptom severity and steroid variation in women with premenstrual syndrome: study on serum pregnenolone, pregnenolone sulfate, 5 alpha-pregnane-3,20-dione and 3 alpha-hydroxy-5 alpha-pregnan-20-one. J Clin Endocr Metab 1996; 81:1076–1082.
9. Hsiao CC, Liu CY, Hsiao MC. No correlation of depression and anxiety to plasma estrogen and progesterone levels in patients with premenstrual dysphoric disorder. Psychiatry Clin Neurosci 2004; 58:593–599.
10. Schmidt PJ, Purdy RH, Moore PH, Jr., et al. Circulating levels of anxiolytic steroids in the luteal phase in women with premenstrual syndrome and in control subjects. J Clin Endocr Metab 1994; 79:1256–1260.
11. Nyberg S, Backstrom T, Zingmark E, et al. Allopregnanolone decrease with symptom improvement during placebo and gonadotropin-releasing hormone agonist treatment in women with severe premenstrual syndrome. Gynecol Endocrinol 2007; 23:257–266.
12. Facchinetti F, Genazzani AD, Martignoni E, et al. Neuroendocrine changes in luteal function in patients with premenstrual syndrome. J Clin Endocr Metab 1993; 76:1123–1127.
13. Majewska MD, Harrison NL, Schwartz RD, et al. Steroid hormone metabolites are barbiturate-like modulators of the GABA receptor. Science 1986; 232:1004–1007.

14. Pinna G, Costa E, Guidotti A. Fluoxetine and norfluoxetine stereospecifically and selectively increase brain neurosteroid content at doses that are inactive on 5-HT reuptake. Psychopharmacology (Berl) 2006; 186:362–372.
15. Baker ER, Best RG, Manfredi RL, et al. Efficacy of progesterone vaginal suppositories in alleviation of nervous symptoms in patients with premenstrual syndrome. J Assist Reprod Genet 1995; 12:205–209.
16. Dennerstein L, Spencer-Gardner C, Gotts G, et al. Progesterone and the premenstrual syndrome: a double blind crossover trial. Brit Med J 1985; 290:1617–1621.
17. Magill PJ. Investigation of the efficacy of progesterone pessaries in the relief of symptoms of premenstrual syndrome. progesterone Study Group. Br J Gen Pract 1995; 45:589–593.
18. Freeman E, Rickels K, Sondheimer SJ, et al. Ineffectiveness of progesterone suppository treatment for premenstrual syndrome. JAMA 1990; 264:349–353.
19. Vanselow W, Dennerstein L, Greenwood KM, et al. Effect of progesterone and its 5 alpha and 5 beta metabolites on symptoms of premenstrual syndrome according to route of administration. J Psychosom Obstet Gynaecol 1996; 17:29–38.
20. Caruncho HJ, Dopeso-Reyes IG, Loza MI, et al. A GABA, reelin, and the neurodevelopmental hypothesis of schizophrenia. Crit Rev Neurobiol 2004; 16:25–32.
21. Costa E, Davis JM, Dong E, et al. A GABAergic cortical deficit dominates schizophrenia pathophysiology. Crit Rev Neurobiol 2004; 16:1–23.
22. Nemeroff CB. The role of GABA in the pathophysiology and treatment of anxiety disorders. Psychopharmacol Bull 2003; 37:133–146.
23. Epperson CN, Haga K, Mason GF, et al. Cortical gamma-aminobutyric acid levels across the menstrual cycle in healthy women and those with premenstrual dysphoric disorder: a proton magnetic resonance spectroscopy study. Arch Gen Psychiatry 2002; 59:851–858.
24. Halbreich U, Petty F, Yonkers K, et al. Low plasma gamma-aminobutyric acid levels during the late luteal phase of women with premenstrual dysphoric disorder. Am J Psychiatry 1996; 153:718–720.
25. Diegoli MS, da Fonseca AM, Diegoli CA, et al. A double-blind trial of four medications to treat severe premenstrual syndrome. Int J Gynaecol Obstet 1998; 62:63–67.
26. Sinchak K, Mills RH, Tao L, et al. Estrogen induces de novo progesterone synthesis in astrocytes. Dev Neurosci 2003; 25:343–348.
27. Baulieu EE. Neurosteroids: a new function in the brain. Biol Cell 1991; 71:3–10.
28. Baulieu EE. Neurosteroids: a novel function of the brain. Psychoneuroendocrinology 1998; 23:963–987.
29. Moss SJ, Smart TG. Modulation of amino acid-gated ion channels by protein phosphorylation. Int Rev Neurobiol 1996; 39:1–52.
30. Mody I, Pearce RA. Diversity of inhibitory neurotransmission through GABA(A) receptors. Trends Neurosci 2004; 27:569–575.
31. Kittler JT, Moss SJ. Modulation of GABAA receptor activity by phosphorylation and receptor trafficking: implications for the efficacy of synaptic inhibition. Curr Opin Neurobiol 2003; 13:341–347.
32. Brandon N, Jovanovic J, Moss S. Multiple roles of protein kinases in the modulation of gamma-aminobutyric acid(A) receptor function and cell surface expression. Pharmacol Ther 2002; 94:113–122.
33. Smart TG. Regulation of excitatory and inhibitory neurotransmitter-gated ion channels by protein phosphorylation. Curr Opin Neurobiol 1997; 7:358–367.
34. Leidenheimer NJ, McQuilkin SJ, Hahner LD, et al. Activation of protein kinase C selectively inhibits the gamma-aminobutyric acidA receptor: role of desensitization. Mol Pharmacol 1992; 41:1116–1123.
35. Chapell R, Bueno OF, Alvarez-Hernandez X, et al. Activation of protein kinase C induces gamma-aminobutyric acid type A receptor internalization in Xenopus oocytes. J Biol Chem 1998; 273:32595–32601.
36. Filippova N, Sedelnikova A, Zong Y, et al. Regulation of recombinant gamma-aminobutyric acid (GABA)(A) and GABA(C) receptors by protein kinase C. Mol Pharmacol 2000; 57:847–856.

37. Kellenberger S, Malherbe P, Sigel E. Function of the alpha 1 beta 2 gamma 2S gamma-aminobutyric acid type A receptor is modulated by protein kinase C via multiple phosphorylation sites. J Biol Chem 1992; 267:25660–25663.
38. Lin YF, Angelotti TP, Dudek EM, et al. Enhancement of recombinant alpha 1 beta 1 gamma 2L gamma-aminobutyric acid A receptor whole-cell currents by protein kinase C is mediated through phosphorylation of both beta 1 and gamma 2L subunits. Mol Pharmacol 1996; 50:185–195.
39. Cinar H, Barnes EM Jr. Clathrin-independent endocytosis of GABA(A) receptors in HEK 293 cells. Biochemistry 2001; 40:14030–14036.
40. Connolly CN, Kittler JT, Thomas P, et al. Cell surface stability of gamma-aminobutyric acid type A receptors. Dependence on protein kinase C activity and subunit composition. J Biol Chem 1999; 274:36565–36572.
41. Ticku MK, Mehta AK. Gamma-aminobutyric acid-A receptor desensitization in mice spinal cord cultured neurons: lack of involvement of protein kinases A and C. Mol Pharmacol 1990; 38:719–724.
42. Bell-Horner CL, Dohi A, Nguyen Q, et al. ERK/MAPK pathway regulates GABAA receptors. J Neurobiol 2006; 66:1467–1474.
43. McDonald BJ, Moss SJ. Conserved phosphorylation of the intracellular domains of GABA(A) receptor beta2 and beta3 subunits by cAMP-dependent protein kinase, cGMP-dependent protein kinase protein kinase C and $Ca2^{+}$/calmodulin type II-dependent protein kinase. Neuropharmacology 1997; 36:1377–1385.
44. McDonald BJ, Moss SJ. Differential phosphorylation of intracellular domains of gamma-aminobutyric acid type A receptor subunits by calcium/calmodulin type 2-dependent protein kinase and cGMP-dependent protein kinase. J Biol Chem 1994; 269:18111–18117.
45. Moss SJ, Doherty CA, Huganir RL. Identification of the cAMP-dependent protein kinase and protein kinase C phosphorylation sites within the major intracellular domains of the beta 1, gamma 2S, and gamma 2L subunits of the gamma-aminobutyric acid type A receptor. J Biol Chem 1992; 267:14470–14476.
46. Kirkness EF, Bovenkerk CF, Ueda T, et al. Phosphorylation of gamma-aminobutyrate (GABA)/benzodiazepine receptors by cyclic AMP-dependent protein kinase. Biochem J 1989; 259:613–616.
47. Browning MD, Bureau M, Dudek EM, et al. Protein kinase C and cAMP-dependent protein kinase phosphorylate the beta subunit of the purified gamma-aminobutyric acid A receptor. Proc Natl Acad Sci USA 1990; 87:1315–1318.
48. Valenzuela CF, Machu TK, McKernan RM, et al. Tyrosine kinase phosphorylation of GABAA receptors. Brain Res Mol Brain Res 1995; 31:165–172.
49. Kapur J, Macdonald RL. Cyclic AMP-dependent protein kinase enhances hippocampal dentate granule cell GABAA receptor currents. J Neurophysiol 1996; 76:2626–2634.
50. McDonald BJ, Amato A, Connolly CN, et al. Adjacent phosphorylation sites on GABAA receptor beta subunits determine regulation by cAMP-dependent protein kinase. Nat Neurosci 1998; 1:23–28.
51. Collado ML, Rodriguez-Manzo G, Cruz ML. Effect of progesterone upon adenylate cyclase activity and cAMP levels on brain areas. Pharmacol Biochem Behav 1985; 23:501–504.
52. Singh M. Ovarian hormones elicit phosphorylation of Akt and extracellular-signal regulated kinase in explants of the cerebral cortex. Endocrine 2001; 14:407–415.
53. Migliaccio A, Piccolo D, Castoria G, et al. Activation of the Src/p21ras/Erk pathway by progesterone receptor via cross-talk with estrogen receptor. EMBO J 1998; 17:2008–2018.
54. Zhu Y, Bond J, Thomas P. Identification, classification, and partial characterization of genes in humans and other vertebrates homologous to a fish membrane progestin receptor. Proc Natl Acad Sci USA 2003; 100:2237–2242.
55. Zhu Y, Rice CD, Pang Y, et al. Cloning, expression, and characterization of a membrane progestin receptor and evidence it is an intermediary in meiotic maturation of fish oocytes. Proc Natl Acad Sci USA 2003; 100:2231–2236.

Chapter 24
Neurosteroid Derangement in Women Diagnosed with Eating Disorders

Daniel Stein[1], Rachel Maayan[2], Ron Loewenthal[3], and Abraham Weizman[2,4]

Abstract *Background*: The neurosteroids allopregnanolone, dehydroepiandrosterone (DHEA), and dehydroepiandrosterone-sulphate (DHEA-S) may have a role in the pathophysiology and control of feeding behaviors. This influence is related to neurosteroid modulation of neurotransmitters that affect feeding, the influence that stress might have on their activity, and the changes found in eating disorders (ED) patients in the activity of the hypothalamic–pituitary–adrenal cortex (HPA) axis, which regulates neurosteroid secretion. The aim of this review is to summarize the findings of studies assessing neurosteroids in EDs.
Method: The review is based on an updated comprehensive systematic literature search of the Cochrane, PUBMED, PSYCHLIT, PSYCHINFO, and ERIC databases.
Results: We found 12 studies assessing neurosteroids in EDs, most of them in anorexia nervosa (AN), and a few in bulimia nervosa and binge eating disorder. Significant inconstancies were found among these studies. Whereas cortisol was usually elevated in underweight AN patients, DHEA and DHEA-S were decreased in some studies, not different from controls in other, and elevated in still other studies. Similar discrepancies in DHEA and DHEA-S were found weight-restored AN patients, whereas cortisol levels usually decreased.
Conclusions: Despite these inconsistencies, several conclusions can be drawn with respect to the role of neurosteroids in EDs, particularly AN. The studies showing elevated levels of DHEA and DHEA-S in underweight AN patients relate this derangement to HPA hyperactivity, associated with malnutrition. The studies

---

[1]Pediatric Psychosomatic Department, The Safra Children's Hospital, The Chaim Sheba Medical Center, Tel Hashomer

[2]Laboratory of Biological Psychiatry, Felsenstein Medical Research Center, Rabin Medical Center, Beilinson Campus, Petah Tiqva

[3]Tissue-Typing Laboratory and Bone Marrow Registry, The Safra Children's Hospital, The Chaim Sheba Medical Center, Tel Hashomer

[4]Geha Mental Health Center, Rabin Medical Center, Beilinson Campus, Petah Tiqva
All affiliated with the Sackler Faculty of Medicine, Tel Aviv University, Israel

M.S. Ritsner and A. Weizman (eds.), *Neuroactive Steroids in Brain Function, Behavior and Neuropsychiatric Disorders*

showing reduced DHEA and DHEA-S but normal or high cortisol levels in malnourished AN patients, suggest that this dissociation represents a regression to prepubertal HPA system functioning. The reduction in cortisol but persistence of elevated DHEA and DHEA-S in weight restored AN patients may account for the feeding problems these patients encounter; this because cortisol increases food intake whereas DHEA and DHEA-S decrease it.

Keywords Neurosteroids, neuroactive steroids, allopregnanolone, dehydroepiandrosterone, dehydroepiandrosterone-sulphate, anorexia nervosa, bulimia nervosa, binge eating disorder, eating disorder

Abbreviations DHEA dehydroepiandrosterone; DHEA-S dehydroepiandrosterone-sulphate; DHEA (-S) DHEA and DHEA-S; ALP allopregnanolone; HPA hypothalamic–pituitary–adrenal cortex axis; ACTH adrenocorticotropic hormone; CRF corticotrophin-releasing factor; 5HT serotonin; AN anorexia nervosa; AN-R restricting AN; AN-B/P bingeing/purging AN; BN bulimia nervosa; BED binge eating disorder; ED eating disorder; BMI body mass index; W/H ratio weight to height ratio; IBW ideal body weight; WR weight restoration; RIA radio immunoassay

24.1 Introduction

Neurosteroids, also referred to as neuroactive steroids, are steroids synthesized de novo in the brain from peripheral steroid precursors that cross the blood–brain barrier readily, and that have definite effects on the brain.[1] Some suggest that steroids synthesized in the periphery that affect the brain as a result of their lypophylic nature, can be also conceptualized as neurosteroids.[2] The term neurosteroids applies to: (1) *allopregnanolone* [$3\alpha,5\alpha$ (tetrahydroprogesterone)-THP], synthesized within the CNS from pregnanolone, that can be converted within the glial cells to progesterone, which in turn is sequentially metabolized to 5α-THP and $3\alpha,5\alpha$ THP or allopregnanolone[1]; (2) *dehydroepiandrosterone* (DHEA); and (3) the sulfated metabolite of DHEA – *dehydroepiandrosterone-sulphate* (DHEA-S).

Neurosteroids can alter neuronal functions by modulating synaptic transmission,[3] as a result of their binding to ligand-gated ion channel-coupled neurotransmitter receptors at the cell membrane. They may act at inhibitory γ-amino-butyric acid type-A (GABA-A) and glycine receptors, at excitatory glutamate receptors, and at serotonin (5HT), N-Methyl-D aspartate, sigma type 1, and nicotinic acetylcholine receptors.[1,4]

Converging evidence suggests that neurosteroids may be relevant to the pathophysiology of psychiatric disorders such as depression,[5] schizophrenia,[6] post-traumatic stress disorder,[7] panic disorder[8]; attention deficit hyperactivity disorder,[9]

conduct disorder,[10] and autism.[11] Specifically, across different psychiatric diagnoses, allopregnanolone, DHEA, and DHEA-S may have a modulatory effect on such traits as mood, anxiety, sexuality, aggression, cognition, general well-being, and activity.[1,2,5,10,12,13]

Neurosteroids may have a putative role of in the pathophysiology and control of feeding behaviors. Experimental data have found that allopregnanolone produces hyperphagic, and DHEA and DHEA-S hypophagic effects in hungry mice.[14] Allopregnanolone, being a positive allosteric modulator (agonist) of inhibitory GABA-A and glycine receptors and a negative modulator (antagonist) of $5HT_3$ receptors may increase feeding behavior and weight. Conversely, DHEA and DHEA-S, being negative modulators (antagonists) of GABA-A and glycine receptors and positive modulators (agonists) of excitatory glutamate receptors and of hypothalamic 5HT activity, may decrease food intake and weight.[4,15,16]

Several mechanisms have been postulated to account for the association between neurosteroid abnormalities and eating disorders (EDs). The modulation of 5HT by allopregnanolone, DHEA and DHEA-S, may be of relevance in disordered eating, as hypothalamic 5HT is involved in the regulation of feeding behavior.[17] Furthermore, down-regulation of serotonin has been reported to occur in individuals ill with anorexia nervosa (AN) and bulimia nervosa (BN), whereas elevated 5HT activity may be found in long-term remitted AN and BN patients.[18] Additionally, findings from studies in long-term recovered AN[19] and BN[20] patients suggest that elevated serotonin activity may increase eating disorders (ED)-related personality traits indicative of behavioral constraint and dysphoria, including obsessionality, rigidity, perfectionism, harm-avoidance, depression and anxiety.

Stress-related neurosteroid changes may be also of importance in EDs, as malnutrition represents a chronic stress condition.[2] Neurosteroid activity is closely related to stress. Allopregnanolone is known to increase with stress,[21] its stress-induced elevations ultimately suppressing the hypothalamic–pituitary–adrenal cortex (HPA) axis. These stress-induced changes in the activity of the HPA axis might have a considerable influence on the feeding behavior of AN and BN patients.[2]

An interaction between neurosteroid activity and feeding behaviors might be postulated also from the changes in the HPA axis that occur with malnutrition. Underweight individuals with AN usually,[20] although not always,[22] show elevated plasma cortisol secretion with attenuated diurnal variation, reflecting hypersecretion and hyperactivity of endogenous corticotrophin-releasing factor (CRF) and an overdrive of the CRF/ACTH (adrenocorticotrophic hormone) system. The secretion of CRF and ACTH in AN normalizes after weight restoration, leading to normalization of cortisol levels.[20]

In this review will analyze the results of the studies evaluating neurosteroid changes in EDs. In the case that similar results are found with respect to both DHEA and DHEA-S, we will use the term DHEA (-S) (see also Table 24.1).

Table 24.1 The main findings in studies investigating neurosteroids in female eating disorders patients

	Diagnosis, number of participants	Age (years)	W/H ratio (% of IBW)/ BMI (kg/m^2)	Duration of illness (months)	Method of neurosteroid measurement	Findings
Zumoff et al.[25]	a. 14 AN patients b. 6 patients 5–35 months after partial WR c. 10 controls	a. 19–29 c. 21–30	a. 60–90% (m = 72%) b. 77–95% (m = 83%) c. 87–104%		24 h mean plasma concentrations Cortisol–protein binding assay DHEA-(s)-RIA	Cortisol ↑ (a vs c) DHEA ↓ (a vs c) DHEA-S = N (a vs c) DHEA/cortisol ↓ (a vs c) Cortisol ↓ (b vs a) DHEA = N (b vs a) DHEA/cortisol ↑ (b vs a)
Winterer et al.[22]	a. 15 AN patients b. 6 patients 2–4 weeks after WR c. 6 patients >6 months after WR d. 23 controls	d. 20–23	a. 48–69% (m = 58%) b. 77–81% (m = 81%) c. 82–97% (m = 86%) b- and c- menstruating d. 107 ± 3%		7 measurements during 6 h (only ACTH-stimulated) Cortisol and DHEA (-S) – RIA	Cortisol = N (a, b, c vs d) DHEA (-S) ↓ (a, b vs d) DHEA (-S)/cortisol ↓ (a, b vs d) ↑ (normalization) of DHEA (-S) and DHEA (-S)/cortisol >6 months of weight ↑ (c = d)
Devesa et al.[26]	a. 7 AN patients b. 7 controls	a. 14–23 b. 17–23	a. 55–80%		Baseline and ACTH-stimulated (two measurements)	Cortisol = N (a vs b) DHEA-S ↓ (a vs b)
Sokoya et al.[27]	a. 19 AN outpatients b. 19 controls	a. 12.8–18.5 b. 13.1–17.8	a. BMI = 25–90th percentile	2–72 (m =19 ± 5)	24 h free cortisol DEHA-S- one measure ment RIA	Cortisol, DHEA-S = N (a vs b)

Table 24.1 (continued)

	Diagnosis, number of participants	Age (years)	W/H ratio (% of IBW)/ BMI (kg/m²)	Duration of illness (months)	Method of neurosteroid measurement	Findings
Gordon et al.[30]	a. 15 AN patients, no controls b. 3 months follow-up	a. 17.3 ± 2.7		3–99. (m = 29.1 ± 27.6)	a. Single measurement Double-antibody-RIA b. measurement after 1, 2, 3 months	DHEA (-S) ↓ (a vs mean for age) b. – all patients received DHEA treatment
Gordon et al.[31,32]	a. 61 AN out-patients, no controls	a. 17.8 ±2.9 Tanner-5, post-menarche	a. BMI = 14–21 (m = 17.7 ± 1.4)		a. Single measurement Double-antibody-RIA	DHEA (-S) ↓ (a vs mean for age)
	b. 51 of these patients – 12 months follow up				b. measurements after 3,6, 9, 12 months	No change in DHEA –S with long-term (b vs a) weight ↑ in AN patients not receiving DHEA treatment
Monteleone et al.[2]	a. 20 AN-R, 10 AN-B/P b. 32 BN outpatients c. 30 controls	a. 23.4 ± 5.6 b. 23.9 ± 3.0 c. 24.1 ± 3.9	a. BMI = 16.1 ± 1.6 b. BMI = 21.0 ± 3.4 c. BMI = 21.8 ± 2.0	a. 50.4 ± 54	Single measurement. Double-antibody RIA for cortisol and DHEA (-S); ALP-special RIA	Cortisol, ALP DHEA (-S) ↑ (a, b vs c) DHEA (-S)/ cortisol = N (a, b vs c) (no differences between AN-R and AN-B/P)
Galderisi et al[24]	a. 12 AN-R, 2AN-B/P b. 31 BN c. 11 controls	a,b. 23 ± 4.5 c. 24.2 ± 4.8	a. BMI = 15.5 ± 2.2 b. BMI = 21.6 ± 2.3 c. BMI = 22.0 ± 1.7		Single measurement. Assays similar to Monteleone et al.[2]	Cortisol = N (a, b vs c) ALP ↑ (a, b vs c; a vs b) DHEA = N (a vs c) ↑ (b vs c) DHEA-S = N (a vs c) ↑ (b vs a,c)
Monteleone et al.[24]	a. 9 non-obese BED, b. 16 obese BED c. 12 obese non-BED out patients, d. 31 controls	a. 24.8 ± 3.3 b. 30.1 ± 10.8 c. 37.4 ± 15.4 d. 23.8 ± 3.7	a. BMI = 24.9 ± 2.1 b. BMI = 39.9 ± 6.8 c. BMI = 37.0 ± 2.8 d. BMI = 22.6 ± 3.7	a. 3.4 ± 2.5 b. 8.6 ± 7.0	Single measurement. Assays similar to Monteleone et al.[2]	Cortisol = N (a, b, c, d) ALP, DHEA (-S.), DHEA (-S)/ cortisol – BED (a, b) ↑ than non-BED (c, d) ALP, DHEA (-S.), DHEA (-S)/ cortisol ↑ (a vs d) ALP ↑, DHEA ↓, DHEA-S = N DHEA/ cortisol ↓ DHEA-S/cortisol = N (c vs d)

(continued)

Table 24.1 (continued)

	Diagnosis, number of participants	Age (years)	W/H ratio (% of IBW)/ BMI (kg/m^2)	Duration of illness (months)	Method of neurosteroid measurement	Findings
Stein et al.[28]	a. 19 AN-R inpatients b_1. 1 month follow-up b_2. 4 months follow-up c. 16 controls	15.5 ± 1.6	a. BMI = 15.6 ± 4.3 b_1. BMI = 18.0 ± 1.5 b_2. BMI = 20.1 ± 1.7 c. BMI = 20.2 ± 2.1	a. 26.4 ±30	Single baseline measurement; 1 and 4 months follow up (single measurement) Cortisol and DHEA (-S) – RIA	Cortisol, DHEA (-S), DHEA/cortisol = N (a vs c) DHEA-S/cortisol ↑ (a vs c) Cortisol ↓ ($b_{1,2}$ vs a); DHEA (-S), DHEA (-S) /cortisol = N ($b_{1,2}$ vs a)
Monteleone et al,[29]	a. 16 AN outpatients b. 10 controls	a. 21.3 ± 3.6 b. 32.6 ± 2.5	a. BMI = 15.4 ± 2.1 b. BMI = 21.9 ± 1.7		Baseline and four measurements within 24h following DST. Assays similar to Monteleone et al.[2]	Cortisol ↑ (baseline) DHEA ↑ (baseline and during 24h after DST) (a vs b) ALP = N (baseline and during 24h after DST) (a vs b)

AN – anorexia nervosa; AN-R – restricting AN; AN-B/P – bingeing/purging AN; BN – bulimia nervosa; BED – binge eating disorder; BMI – body mass index (kg/m2); W/H ratio – weight to height ratio, compared with ideal body weight (IBW) for age and gender; ALP – Allopregnanolone; DHEA – dehydroepiandrosterone; DHEA-S – dehydroepiandrosterone-sulphate; DHEA (-S) – refers to both DHEA and DHEA-S; ACTH – adrenocorticotropic hormone; WR – weight restoration; m = mean; N – no difference between patients and controls; ↑/↓ results elevated/decreased in patients compared to controls; RIA-radio immunoassay; DST – dexamethasone suppression test

24.2 Analysis of the Studies Assessing Neurosteroids in Eds

An updated comprehensive systematic literature search of the Cochrane, PUBMED, PSYCHLIT, PSYCHINFO, and ERIC databases revealed 12 studies investigating neurosteroid changes in EDs. The results of these studies are summarized in Table 24.1. These studies mainly assessed AN patients, with two studies[2,23] investigating BN and one study[24] BED patients. All findings relate to females.

In the malnourished state of AN, when compared to control women, DHEA was found to be decreased in three studies,[22,25,26] within the range of control participants in three studies,[23,27,28] and elevated in two studies.[2,29] Similarly, DHEA-S was decreased in two studies,[25,26] not different from controls in another three studies,[22,23,28] and elevated in one study.[2] Gordon et al.[30,31] who did not include control samples, found decreased values of DHEA (-S) in their malnourished AN patients compared to mean values of normal female populations matched for age.

Similar inconsistencies were found with respect to the DHEA (-S) to cortisol ratios in malnourished AN patients in comparison to controls. The DHEA to cortisol ratio was, thus, decreased in two studies,[22,25] and not different from controls in two studies.[2,28] Similarly, the DHEA-S to cortisol ratio was decreased in one study,[22] not different from controls in two studies,[2,25] and elevated in one study.[28]

Only one group[2,23,29] assessed allopregnanolone in emaciated AN patients. Despite similar procedures and biochemical analyses in these studies, allopregnanolone was elevated in two studies,[2,23] and not different from controls in one study.[29]

A few studies attempted to investigate neurosteroid changes in AN following short-term (a few weeks) medium (a few months) and long-term (12 months) weight restoration. In the only study assessing short-term changes, Winterer et al.[22] found no changes in cortisol, DHEA (-S), and DHEA(-S) to cortisol ratios in six AN patients following 2–4 weeks of refeeding (the mean weight to height [W/H] ratio of these patients at follow-up was 81% of ideal body weight [IBW], which was still in the range of AN, although much improved from the mean of 58% of IBW at baseline). This group also studied another six AN patients after more than 6 months of weight restoration (mean W/H ratio 86% of IBW, i.e., just slightly over the minimal cutoff point of 85% of IBW that defines return to normal weight), and found a significant increase of DHEA (-S) and DHEA (-S) to cortisol ratios in comparison to the acute stage; the levels of these neurosteroids at follow-up were no longer different from those of controls. Stein et al.,[28] assessing 19 restricting adolescent AN (AN-R) patients at baseline (BMI = $15.6 \pm 4.3\,kg/m^2$) and following 1 month (BMI = $18.0 \pm 1.5\,kg/m^2$) and 4 months (BMI = $20.1 \pm 1.7\,kg/m^2$) of refeeding, found a reduction in cortisol, but no changes in DHEA (-S) and DHEA (-S) to cortisol ratios at both follow-up points, although there was a trend for a decrease in these ratios after 4 months compared to baseline. Zumoff et al.,[25] who studied six AN patients after 5–35 months of refeeding (mean W/H ratio at follow-up was 83% of IBW, which was still in the range of AN, compared to 72% of IBW in malnourished patients), found a reduction in cortisol, no change in DHEA, and an elevation of the DHEA to cortisol ratio at follow-up in comparison to baseline. In the only study yet assessing the effect of long-term weight restoration, Gordon et al.,[32] who studied 51 malnourished AN patients treated for bone loss with either DHEA or estrogen replacement for 12 months in addition to refeeding, found no DHEA-S change in those patients who were not receiving DHEA, despite a significant increase in their weight.

Two studies conducted by Monteleone and his associates assessed neurosteroids in BN. In the earlier of these studies, Monteleone et al.[2] found elevated levels of allopregnanolone, cortisol and DHEA (-S) in 30 actively ill BN females in comparison to control women, no between-group differences in DHEA (-S) to cortisol ratios, and no differences in any of these neurosteroids between AN and BN patients. Similar results were found in a more recent study of this group,[23] although DHEA-S was reduced, and allopregnanolone elevated, in AN vs BN patients.

In another study, Monteleone et al.[24] assessed neurosteroid concentrations in 9 non-obese BED women (BMI = 24.9 ± 2.1 kg/m^2), 16 obese BED women (BMI = 39.9 ± 6.8 kg/m^2), 12 non-BED obese women (BMI = 37.0 ± 2.8 kg/m^2), and 31 controls (BMI = 23.8 ± 3.7 kg/m^2). No between-group differences were found for cortisol, whereas allopregnanolone, DHEA (-S), and DHEA (-S) to cortisol ratios were significantly greater in two BED groups in comparison to the two non-BED populations. Furthermore, significant differences were shown when comparing the neurosteriod findings in non-obese BED patients vs non-obese non-BED controls, and in non-BED obese patients vs non-BED non-obese controls.

24.3 Discussion

Our in-depth analysis demonstrated inconsistent findings and considerable differences in the results of the studies assessing neurosteroids in AN. Most groups used relatively small samples, with the exception of the studies of Monteleone et al.[2] and Gordon et al.[31,32] However, Gordon and her associates were primarily interested in the investigation of bone loss and its correction in AN, with the study of neurosteroids being only a part of this larger project.

The methodological inconsistencies in the studies assessing neurosteroids in AN do not enable a valid comprehensive meta-analysis of the findings. These inconsistencies are described in Table 24.1.

24.4 Sample Inconsistencies

Inpatients vs outpatients, with some studies failing to describe the exact source of their patients. *Age differences*, e.g., adolescents vs adults. This difference is of relevance, as DHEA (-S) secretion is accelerated through puberty, and these neurosteroids reach their peak levels between the ages of 20 and 30 years, and decrease thereafter.[33] *Differences in AN type*, i.e., AN-R in some studies, a mixture of AN-R and AN-B/P patients in other studies, and no description of the type of AN in several samples. Although Monteleone et al.[2] found no differences in neurosteroid concentrations between restricting and bingeing/purging AN patients, the repeated fluctuations between reduced food intake (malnutrition) and over-eating (bingeing) found in AN-B/P but not in AN-R, might be associated with dysregulation between neurosteroids that decrease feeding [DHEA (-S)] and those putatively increasing feeding (cortisol, allopregnanolone).[4] This, however, to the best of our knowledge, has not been studied yet. *Differences in nutritional condition or in duration of AN* may influence the extent of neurosteroid derangement, as reciprocal interrelationships likely exist between these steroids and body weight and food intake.[2] In this respect, whereas the older studies define the nutritional status of their patients with the weight to height ratio in relation to the IBW for age, the more recent studies rely on the BMI method. *Comorbidity*: AN patients are often diagnosed with comorbid psychiatric disorders, primarily affective disorders and anxiety disorders.[34] This is of relevance, as the severity of depression and anxiety traits in AN patients can be potentially enhanced by cortisol and reduced by DHEA and DHEA-S.[2] Most studies assessed by us have not related to the psychiatric comorbidity of their ED patients.

24.5 Inconsistencies in Neurosteroid Measurement

Some studies have used 24 h analyses, others several daily measurements, and still others a single daily measurement (usually in the early morning hours, with or with ACTH stimulation). Additionally, whereas patients in all studies were not taking medications for an adequate period before neurosteroid analysis, and neurosteroids were analyzed in controls and menstruating patients primarily in the follicular phase, controlling for physical activity, caffeine consumption and smoking, all potentially affecting neurosteroid concentrations,[6] was not always the case.

Despite these methodological inconsistencies, several important conclusions can be drawn with respect to the role of neurosteroids in EDs. The finding that changes in neurosteroid concentrations appear not only in emaciated AN patients, but also in normal-weight BN patients and in obese and normal-weight BED patients suggests that neurosteroid derangement is associated not only with changes in weight, but also with the presence of maladaptive consummatory patterns of restricting and bingeing/purging.

24.5.1 DHEA (-S) in Malnourished AN Patients

Monteleone et al.[2] and Galdarisi et al.[23] who have found elevated levels of DHEA (-S) in malnourished AN patients (in the study of Galderisi et al.,[23] only DHEA-S concentrations are found elevated), suggest that as the major source of blood DHEA (-S) is the adrenal gland, and as their secretion is regulated by the CRF/ACTH axis that is often hyperactive in AN, the increased concentrations of these neurosteroids in AN are likely the result of CRF/ACTH axis overdrive. Nevertheless, following dexamethasone suppression test, cortisol concentrations decrease in malnourished AN to values similar to those found in healthy controls, whereas DHEA levels remain significantly higher than in controls[29]; this suggests that factors other than CRF/ACTH hyperactivity, that are not influenced by exogenous corticosteroids, may contribute to the enhanced production of DHEA in AN.

The pattern of HPA overactivity in AN resembles that seen in chronic stress conditions.[2,23] In the same token, those studies showing reduced DHEA (-S) to cortisol ratios in malnourished AN patients in comparison to controls,[22,25] suggest this pattern to represent impaired stress processing related to dysregulated HPA activity,[6] as is the case in chronic schizophrenia[6] and major depressive disorder.[5]

The studies showing reduced DHEA (-S) but normal or high cortisol levels in malnourished AN patients[22,25,26,30,31] propose that a dissociation in the secretion of cortisol and DHEA (-S) may occur in AN. As the secretion of DHEA (-S), but not of cortisol, normally accelerates during puberty,[33] the occurrence of elevated cortisol but reduced DHEA (-S) in malnourished AN women may signify a regression to prepubertal HPA system functioning.[25] In the same token, since prepubertal girls show significantly lower DHEA (-S) to cortisol ratios than adult women,[25] the reduced DHEA (-S) to cortisol ratios in acutely ill postpubertal AN women in comparison to healthy controls[22,25] supports the notion that that HPA regression can occur with malnutrition. However, as other studies have found no difference[2] or even an increase[28] in DHEA (-S) to cortisol ratios in AN patients in comparison to controls, the regression of the HPA system in AN to prepubertal functioning is apparently not always the case.

From a different perspective, as DHEA (-S) may be involved in the modulation of depression and anxiety,[1,2,35] changes in these neurosteroids in AN patients can putatively be associated with the elevated depression and anxiety often found in these patients.[34] However, two studies,[2,28] although not assessing anxiety, have found no associations between neurosteroid concentrations and depression,[2,28] aggression,[2] obsessionality and core-ED personality traits,[28] suggesting that these traits are not associated with neurosteroid derangement in AN patents.

One study attempted to assess the influence of neurosteroid derangement on the neurocognitive functioning of acutely ill AN and BN patients.[23] Both patient groups were slower than controls on incidental learning tasks, but more accurate on a spatial executive task. DHEA (-S) concentrations, being elevated in the ED patients in comparison to controls (DHEA was elevated only in the BN patients), were positively correlated with accuracy on the executive task. These findings support the notion that DHEA (-S) may have neurocognitive enhancing effects, and might explain, at least in part, why in many ED patients, despite malnutrition, metabolic alterations and brain pseudoatrophy, cognitive dysfunctions are relatively subtle and often compatible with high achievements at school and at work.[23]

24.5.2 Changes in DHEA (-S) Following Weight Restoration in AN Patients

Following weight restoration, cortisol levels usually decrease to normal levels, suggesting that the elevated HPA activity in acutely ill AN patients is likely secondary to malnutrition.[18] By contrast, although one study has found an increase in DHEA (-S) to normal levels with restored weight,[22] most studies report no changes in these neurosteroids with refeeding.[25,28,32]

This dissociation in the normalization of cortisol but not of DHEA (-S) in weight- restored AN patients, suggests the changes in DHEA (-S) secretion following refeeding might be regulated by systems other than the CRF/ACTH axis. The serotonergic system might be of relevance in this respect, as DHEA (-S) are positive modulators of post-synaptic ion channel-coupled serotonin receptors; thus they might modulate feeding behavior by increasing the amount or activity of hypothalamic 5HT.[15]

The persistence of elevated DHEA (-S) following refeeding suggests the likelihood of elevated 5HT activity in weight-restored patients. Indeed, recent neuroimaging studies have shown elevated 5HT activity in long-term recovered AN patients.[36] Elevated serotonin activity can be associated with the decreased food intake often found in weight-restored AN patients by virtue of its satiety enhancing activity.[18] Additionally, it might be associated with the persistence of personality traits indicative of behavioral constraint and dysphoria (e.g., obsesionality and rigidity) in weight-restored patients,[37] that by themselves contribute to decreased feeding.[18]

Alternatively, as DHEA-(S) tend to decrease food consumption and cortisol to increase it, the decrease of cortisol but the persistence of elevated DHEA (-S) in weight-restored AN patients, lend another explanation for the feeding problems encountered by these patients. Stein et al.[28] have additionally found that a greater increase in DHEA-S to cortisol ratio at 4 months of refeeding compared to baseline, is correlated with less increase in BMI, a finding lending further support for the influence of the derangement in DHEA-S secretion on the decrease in daily caloric intake during refeeding.[32]

24.5.3 Allopregnanolone in AN

Monteloene et al.[2] and Galderisi et al.[23] (although not Monteleone et al.[29]), found elevated allopregnanolone levels in malnourished AN patients compared to controls. These authors suggest that as allopregnanolone reduces the anxiety evoked by CRF[38] and attenuates the response of the HPA axis to stress,[39] the elevation of allopregnanolone in acutely ill AN might represent an endogenous regulatory mechanism to counteract HPA axis hyperactivity. Furthermore, as allopregnanolone increases feeding and DHEA (-S) reduce it, elevated allopregnanolone might also act to balance the reduction in feeding in malnourished AN patients due to elevated DHEA (-S).[2]

24.5.4 DHEA Treatment in AN

Several studies found that the administration of DHEA can improve depressive symptoms in patients with major depressive disorder,[40] as well as the overall psychological well-being of elderly individuals.[41] Only one study[32] has yet treated AN patients with DHEA, in an attempt to correct the profound bone loss often found in these patients. Gordon and her associates,[31] in a study of physiologic regulators of bone turnover in female adolescent AN patients, found that DHEA-S levels (which were significantly lower in comparison to normal female populations matched for age) were negatively correlated with physiologic markers of bone resorption, supporting a putatively antiresorptive property of DHEA-S. Short-term (3 months) DHEA replacement therapy was found to reduce bone resorption markers in AN patients.[30] However, a long-term (12 months) prospective randomized control study of this group,[32] in which oral DHEA was compared to standard treatment (estrogens) for the correction of bone loss in AN, failed to show any beneficiary effect of both treatments after controlling for the effect of weight gain. Whereas both treatments significantly reduced the levels of bone resorption markers, DHEA, but not estrogens, exhibited anabolic properties, including enhanced levels of bone formation markers. Both treatments were shown to be safe, with no evidence of significant adverse effects. Furthermore, patients treated with DHEA, but not with the standard treatment, showed significant decrease from baseline to 12-month follow-up in anxiety, maladaptive eating-related preoccupations and behaviors, and body image concerns.

24.5.5 Nerosteroids in BN and BED

The two studies assessing neurosteroids in acutely ill BN patients,[2,23] found that similar to malnourished underweight AN patients, normal-weight BN patients show elevated concentrations of cortisol (only in Monteleone et al.[2]), DHEA (-S), and allopregnanolone, compared to control women. The authors suggest that similar to AN, the increased concentrations of cortisol and DHEA (-S) in BN are the result of CRF/ACTH axis overdrive, likely associated in BN patients with the repeated fluctuations between reduced food intake (malnutrition) and over-eating (bingeing). Interestingly, whereas in one study,[2] BN and AN patients were similar in all neurosteroids assessed, in the other study,[23] DHEA-S was reduced, and allopregnanolone elevated, in AN compared to BN patients.

In another study, this group[24] found elevated levels of DHEA (-S), and allopregnanolone, and significantly higher DHEA (-S) to cortisol ratios in BED women, whether obese or not, in comparison to obese and normal-weight controls. Cortisol levels were not different in BED and non-BED women.

The authors suggest that the elevation of DEHA (-S) in BED is apparently not the result of CRF/ACTH axis overdrive, since the production of cortisol is not enhanced in BED patients. It is also not related to adiposity per se. As increased adiposity is associated with conversion of androgens to estrogens, DHEA (-S), being precursors of androgens and estrogens, may be converted to estrogens in obese individuals, likely leading to reduction in their circulating levels. Indeed, Monteleone et al.[24] have found lower DHEA levels in non-BED obese women in comparison to normal-weight controls. The authors speculate that chronic BED patients (mean duration of illness in obese and non-obese BED patients 8.6 ± 7.0 and 3.4 ± 2.5 years, respectively) may develop insulin resistance, which might lead to an increased production and/or reduced clearance of DHEA (-S). The mechanism(s) underlying the elevation of allopregnanolone in BED women are still unclear.

Monteleone et al.[24] conclude that if malnutrition is involved in the determination of increased plasma levels of neuroactive steroids in individuals with AN and BN, then different mechanisms may induce similar effects in BED patients. Alternatively, common, as yet unknown, factors might account for the similar neurosteroid changes in all EDs. It has been suggested that whereas individuals with AN, BN and BED differ in symptom and personality presentation during the active phase of the disorder,[42] the resemblance in enduring traits following remission in AN and BN suggests the possibility of some shared vulnerability for these disorders[43] (there are yet no studies of long-term remitted BED patients). Further support for a shared vulnerability comes from twin studies which have found increased rates of all EDs among co-twins in whom the affected twin had either one of these disorders.[44] The similarity in neurosteroid profile in AN, BN, and BED patients, despite considerable differences in weight and consummatory patterns in acutely ill patients, lends further support for some shared vulnerability for these EDs.

24.6 Suggestions for Future Research

This review demonstrates significant methodological inconsistencies in the studies investigating neurosteroids in EDs, likely associated with the discrepant and inconsistent findings of these studies. Accordingly, our suggestions to future research are: (1) inclusion of larger, more homogenous, samples; (2) relating to the issue of psychiatric comorbidity, e.g., the influence of the depression or anxiety of ED patients on neurosteroid derangement; (3) comparison of the findings in adolescent and adult ED patients; (4) prospective longitudinal assessment of neurosteroids in acutely ill ED patients, upon achieving symptomatic improvement, and in recovered vs chronic patients; (5) studies in families of ED patients; (6) the association of neurosteroid derangement in EDs with other relevant aspects, e.g., serotonin alterations, neurocognitive dysfunction, and ED-relevant personality temperaments.

References

1. Rupprect R. Neuroactive steroids: mechanism of action and neuropharmacological properties. Psychoneuroendocrinology 2002; 28:139–168.
2. Monteleone P, Luisi M, Colurcio B, et al. Plasma levels of neuroactive steroids are increased in untreated women with anorexia nervosa or bulimia nervosa. Psychosom Med 2001; 63:52–68.

3. McEwen BS. Steroid hormones are multifunctional messengers to the brain. Trends Endocrinol Metab 1991; 2:62–67.
4. Engel SR, Grant KA. Neurosteroids and behavior. Int Rev Neurobiol 2001; 46:321–348.
5. Michael A, Jenaway A, Paykel ES, et al. Altered salivary dehydroepiandrosterone levels in major depression in adults. Biol Psychiatry 2000; 48:989–995.
6. Ritsner M, Maayan R, Gibel A, et al. Elevation of the cortisol/dehydroepiandrosterone ratio in schizophrenia patients. Eur Neuropsychopharmacol 2004; 14:267–273.
7. Spivak B, Maayan R, Kotler M, et al. Elevated circulatory levels of GABA (A)-antagonistic neurosteroids in patients with combat-related post-traumatic stress disorder. Psychol Med 2000; 30:1227–1231.
8. Strohle A, Romeo E, DiMichele F, et al. GABA-A receptor modulatory neuroactive steroid composition in panic disorder and during paroxetine treatment. Am J Psychiatry 2002; 159:145–147.
9. Strous RD, Spivak B, Yoran-Hegesh R, et al. Analysis of neurosteroid levels in attention deficit hyperactivity disorder. Int J Neuropsychopharmacol 2001; 4:259–264.
10. Van Goozen SHM, Matthys W, Cohen-Kettenis PT, et al. Adrenal androgens and aggression in conduct disorder prepubertal boys and normal controls. Biol Psychiatry 1998; 43:156–158.
11. Strous RD, Golobchik P, Maayan R, et al. Lowered DHEA-S plasma levels in adult individuals with autistic disorder. Eur Neuropsychopharmacol 2003; 15:305–309.
12. Wolf OT, Kirschbaum C. Actions of dehydroepiandrosterone and its sulphate in the central nervous system: effects on cognition and emotion in animals and humans. Brain Res Rev 1999; 30:264–288.
13. Cawood EH, Bancroft J. Steroid hormones, the menopause, sexuality and well-being of women. Psychol Med 1996; 26:925–936.
14. Pham J, Porter J, Svec D, et al. The effect of dehydroepiandrosterone on Zucker rats selected for fat food preference. Physiol Behav 2000; 70:431–441
15. Abadie JM, Wright B, Correa G, et al. Effect of dehydroepiandrosterone on neurotransmitter levels and appetite regulation of the obese Zucker rat. Diabetes 1993; 42:662–669.
16. Kaur G, Kulkarni SK. Subchronic studies on modulation of feeding behavior and body weight by neurosteroids in female mice. Methods Find Exp Clin Pharmacol 2001; 23:115–119.
17. Blundel J. Pharmacological approaches to appetite suppression. Trends Pharmacol Sci 1991; 12:147–157.
18. Kaye WH, Strober M, Jimerson D. The neurobiology of eating disorders. In: Charney DS, Nestler EJ (eds). The neurobiology of mental illness. New York: Oxford, 2004, pp. 1112–1128.
19. Srinivasagam NM, Kaye WH, Plotnicov KH, et al. Persistent perfectionism, symmetry, and exactness in anorexia nervosa after long time recovery. Am J Psychiatry 1995; 152:1630–1634.
20. Kaye WH, Greeno CG, Moss H, et al. Alterations in serotonin activity and psychiatric symptomatology after recovery from bulimia nervosa. Arch Gen Psychiatry 1998; 55:927–935.
21. Purdy RH, Morrow AL, Moore PH, et al. Stress-induced elevations of γ-amino-butyric acid type-A receptor-active steroids in the rat brain. Proc Natl Acad Sci USA 1991; 88:4553–4557.
22. Winterer J, Gwirtsman HE, George DT, et al. Adrenocorticotropin-stimulated adrenal androgen secretion in anorexia nervosa: impaired secretion at low weight with normalization after long term recovery. J Clin Endocrinol Metab 1985; 61:693–697.
23. Galderisi S, Mucci A, Monteleone P, et al. Neurocognitive functioning in subjects with eating disorders: the influence of neuroactive steroids. Biol Psychiatry 2003; 53:921–927.
24. Monteleone P, Luisi M, De Filippis G, et al. Circulating levels of neuroactive steroids in patients with binge eating disorder: a comparison with nonobese healthy controls and non-binge eating obese subjects. Int J Eat Disord 2003; 34:432–440.
25. Zumoff B, Walsh BT, Katz JL, et al. Subnormal plasma dehydroepiandrosterone to cortisol ratio in anorexia nervosa: a second hormonal parameter of ontogenic regression. J Clin Endocrinol Metab 1983; 56:668–672.
26. Devesa J, Perez-Fernandez R, Bokser L, et al. Adrenal androgen secretion and dopaminergic activity in anorexia nervosa. Horm Metab Res 1987; 20:57–60.

27. Soyka LA, Grinspoon S, Levitsky L, et al. The effects of anorexia nervosa on bone metabolism in female adolescents. J Clin Endocrinol Metab 1999; 84:4489–4496.
28. Stein D, Maayan R, Ram A, et al. Circulatory neurosteroid levels in underweight anorexia nervosa patients and following weight restoration. Eur Neuropsychopharmacol 2005; 15:647–653.
29. Monteleone P, Luisi M, Martiadis V, et al. Impaired reduction of enhanced levels of dehydroepiandrosterone by oral dexamethasone in anorexia nervosa. Psychoendocrinology 2006; 31:537–542.
30. Gordon CM, Grace E, Emans SJ, et al. Effects of oral DHEA on bone turnover markers in anorexia nervosa. J Bone Miner Res 1999; 14:136–145.
31. Gordon CM, Goodman E, Emans SJ, et al. Physiologic regulators of bone turnover in young women with anorexia nervosa. J Pediatr 2002; 141:64–70.
32. Gordon CM, Grace E, Emans SJ, et al. Effects of oral dehydroepiandrosterone on bone density in young women with anorexia nervosa: a randomized trial. J Clin Endocrinol Metab 2002; 87:4935–4941.
33. Baulieu EE. Dehydroepiandrosterone (DHEA): a fountain of youth? J Clin Endocrinol Metab 1996; 81:3147–3151.
34. Lilenfeld LR., Kaye WH, Greeno CG, et al. A controlled family study of anorexia nervosa and bulimia nervosa: psychiatric disorders in first-degree relatives and effects of proband comorbidity. Arch Gen Psychiatry 1998; 55:603–610.
35. Strohle A, Romeo E, Hermann B, et al. Concentrations of 3α-reduced neuroactive steroids and precursors in plasma of patients with major depression and after clinical recovery. Biol Psychiatry 1999; 45:274–277.
36. Kaye WH, Frank GK, Bailer UF, et al. Serotonin alterations in anorexia and bulimia nervosa: new insights from imaging studies. Physiol Behav 2005; 85:73–81.
37. Pollice C, Kaye WH, Greeno CG, et al. Relationship of depression, anxiety, and obsessionality to state of illness in anorexia nervosa. Int J Eat Disord 1997; 21:367–376.
38. Patchev VK, Shoaib M, Holsboer DF, et al. The neurosteroid tetrahydroprogesterone counteracts corticotropin-releasing hormone-induced anxiety and alters the release and gene expression of corticotropin-releasing hormone in the rat hypothalamus. Neuroscience 1994; 62:265–271.
39. Patchev VK, Holsboer DF, Almeida OFX. The neurosteroid tetrahydroprogesterone attenuates the endocrine response to stress and exerts glucocorticoid-like effects on vasopressin gene transcription in the rat hypothalamus. Neuropsychopharmacology 1996; 15:533–541.
40. Wolkowitz OM, Reus VI, Keebler A, et al. Double-blind treatment of major depression with dehydroepiandrosterone. Am J Psychiatry 1999; 156:646–649.
41. Morales AJ, Haubrich RH, Hwang JY, et al. The effect of 6 months treatment with a 100 mg daily dose of dehydroepiandrosterone (DHEA) on circulating sex hormones, body composition and muscle strength in age-advanced men and women. Clin Endocrinol 1998; 49:421–432.
42. Vitousek K, Manke F. Personality variables and disorders in anorexia nervosa and bulimia nervosa. J Abnorm Psychol 1994; 103:137–147.
43. Lilenfeld LR, Stein D, Bulik CM, et al. Personality traits among the first-degree female relatives of women with bulimia nervosa. Psychol Med 2000; 30:1399–1410.
44. Lilenfeld LR, Kaye WH. Genetic studies of anorexia and bulimia nervosa. In: Hoek HW, Treasure JL, Katzman MA (eds). Neurobiology in the treatment of eating disorders. New York: Wiley, 1998, pp. 169–194.

Chapter 25
Neurosteroids in Alcohol and Substance Use

Brett C. Ginsburg[1,3], Lisa R. Gerak[2,3], Lance R. McMahon[2,3], and John D. Roache[1,3]

Abstract Allopregnanolone and related neurosteroids share effects with ethanol at several critical targets. Though data are equivocal in humans, acute administration of ethanol (and several other drugs of abuse) increases brain and plasma levels of allopregnanolone in animals. Evidence exists suggesting that this effect of ethanol is mediated via the peripheral or mitochondrial benzodiazepine receptor (PBR). PBR activation can stimulate neurosteroid synthesis and results in anxiolytic effects similar to ethanol, benzodiazepines, and neurosteroids. Further, finasteride, an inhibitor of neurosteroidogenesis, blunts some effects of ethanol in animals and humans. Thus, neurosteroids may modulate or mediate effects of ethanol, benzodiazepines, and barbiturates. Like ethanol and allopregnanolone, benzodiazepines and barbiturates are positive GABA-A receptor modulators. Accordingly, neurosteroids share discriminative stimulus properties with ethanol, benzodiazepines, and barbiturates. Allopregnanolone and related neurosteroids are self-administered by animals, similar to ethanol and other abused drugs. A role for neurosteroids in ethanol tolerance and withdrawal has also emerged, as rodents undergoing ethanol withdrawal are sensitized to neurosteroid effects and alcoholics in withdrawal display reduced allopregnanolone levels. While symmetrical cross tolerance develops between ethanol and benzodiazepines, asymmetrical cross tolerance develops between neurosteroids and benzodiazepines or ethanol, which could be exploited to help treat ethanol, benzodiazepine or barbiturate dependence and withdrawal.

Keywords Ethanol, benzodiazepine, barbiturate, cocaine, cannabinoid, GABA, NMDA, drug abuse, tolerance, withdrawal

---

[1] Division of Alcohol and Drug Addiction, Department of Psychiatry, The University of Texas Health Science Center at San Antonio, Texas, USA

[2] Department of Pharmacology, The University of Texas Health Science Center at San Antonio, Texas, USA

[3] Behavioral Pharmacology Group, The University of Texas Health Science Center at San Antonio, Texas, USA

M.S. Ritsner and A. Weizman (eds.), *Neuroactive Steroids in Brain Function, Behavior and Neuropsychiatric Disorders*

Abbreviations GABA-A γ-aminobutyric acid type A; DHEA dehydroepiandrosterone; AMPA alpha-amino-3-hydroxy-5-methyl-4-isoxazolepropionic acid; NMDA *N*-methyl-D-aspartic acid; 5-HT 5-hydroxytryptamine (serotonin); HPA hypothalamic–pituitary–adrenal; DBI diazepam binding inhibitor; PBR peripheral (mitochondrial) benzodiazepine receptor; CRH corticotropin releasing hormone; ACTH adrenocorticotropic hormone; StAR steroidogenic acute regulatory protein; CBR central benzodiazepine receptor

25.1 Introduction

The discovery that some steroid molecules can exert effects on the same cell-surface receptors that several common drugs of abuse target has led to a surge of interest in neurosteroids among drug abuse researchers. As shown in Fig. 25.1, the number of publications related to neurosteroids and ethanol or benzodiazepines has tended to increase each year over the last 15 years. However, the field is still developing, and the available data remain limited and in some cases contradictory. Neuroactive steroids, particularly those with a 3α-OH conformation, have pharmacological actions similar to those of ethanol and benzodiazepines. Administration of ethanol, as well as several other drugs of abuse, increases plasma and brain levels of 3α-OH neurosteroids. Neurosteroids with a 3α-OH group also share behavioral effects with ethanol and benzodiazepines, including anxiolytic, anesthetic, motor depressant, and response rate disruptive effects. Studies of drug discrimination, which provide a pharmacologically selective assay of systemic drug effects, consistently reveal similarities between 3α-OH neurosteroids and ethanol or benzodiazepines. The reinforcing effects of 3α-OH neurosteroids remain unclear; however, administration of these compounds before ethanol self-administration may enhance the reinforcing effectiveness of ethanol. In contrast, 3β-OH neurosteroids have pharmacological actions that oppose those of 3α-OH neurosteroids, ethanol and benzodiazepines, do not share behavioral effects, and may blunt the reinforcing effects of ethanol. These lines of evidence have led some to propose that the behavioral effects of ethanol are mediated by increased levels of 3α-OH neurosteroids. While this notion is not fully supported by the available data, a strong case can be made for the modulation of ethanol effects by changes in levels of endogenous neurosteroids. Neurosteroids could also be involved in the development of tolerance to ethanol, as brain-region specific changes in γ-aminobutyric acid type A (GABA-A) receptor subunits that result in ethanol tolerance and withdrawal are influenced by neuroactive steroids. Indeed, cross-tolerance develops between neuroactive steroids and ethanol. However, cross tolerance does not appear to develop between neurosteroids and benzodiazepines, suggesting that neurosteroids could be used as an adjunct treatment in order to reduce the therapeutic dose of benzodiazepines, thereby limiting their abuse and dependence liability. Further, the lack of cross

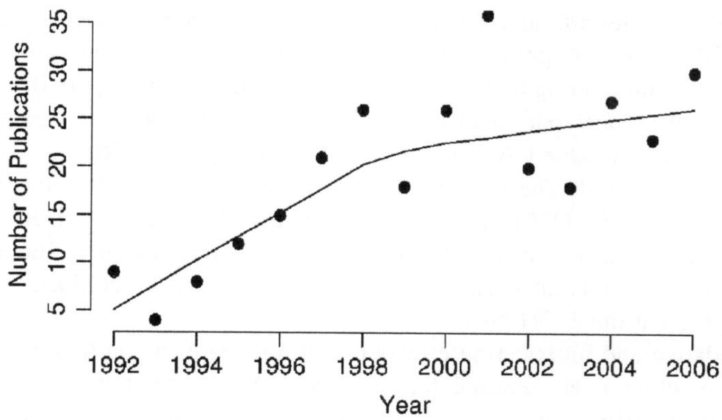

Fig. 25.1 The number of publications involving neuroactive steroids and ethanol or benzodiazepines, by year. The total number of publications registered in the National Library of Medicine satisfying this search criteria was compiled for each year. The line represents a Lowess curve through the points

tolerance between benzodiazepines and neurosteroids could provide a means by which dependence on benzodiazepines could be treated.

25.2 Cell Surface Receptor Pharmacology

Neurosteroids produce rapid, non-genomic effects in the brain.[1–4] Steroids with a 3α-OH reduction share effects on several cell-surface receptors and behaviors with ethanol and sedatives. Steroids with a 3β-OH reduction act as competitive antagonists of 3α-OH neurosteroids[5] and may reverse effects of chronic ethanol.[6] Neurosteroid action at the GABA-A receptor complex is most widely studied, but other sites of action apparently include *N*-methyl-D-aspartate (NMDA), serotonin-3 (5-HT3), and other ligand-gated ion channels. Neurosteroids also exert indirect effects on dopamine (and presumably other monoamine) receptors via alterations in catecholamine production and release.

25.2.1 GABA-A Receptor

Modulation of the GABA-A receptor is the most widely understood cell-surface receptor effect of neurosteroids. Research suggests that a 3α-OH group confers positive modulatory activity to neurosteroids at GABA-A receptors, while a 3β-OH group confers negative or antagonistic activity, assessed using GABA facilitation of [^{3}H]flunitrazepam binding, Cl^- flux, or inhibition of a population spike in the hippocampus.[5,7,8] In fact, 3β-OH neurosteroids may act as competitive antagonists to 3α-OH neurosteroids at GABA-A receptors.[9,10] Prince and Simmonds[9] report that the 3β isomer of pregnanolone competitively inhibits the potentiation of [^{3}H]flunitrazepam binding by GABA and pregnanolone with a K_i of 10.5 micromolar. Wang et al.[5] report that 3α-OH neurosteroid inhibition of an evoked population spike in neurons from the CA1 region of rat hippocampus is selectively blocked by a 3β-OH neurosteroid. These results are consistent with the notion that GABA-A receptor effects of 3α-OH neurosteroids are competitively antagonized by 3β-OH neurosteroids. Thus, neurosteroids can act as positive or negative modulators of GABA-A receptor-mediated chloride flux, depending on the conformation of the hydroxy group at the 3-OH position.

The sedative and intoxicating effects of benzodiazepines, barbiturates, and to a large extent ethanol are exerted through the GABA-mediated Cl^- ion channel, known as the GABA-A receptor.[11–14] GABA binding opens the channel to inhibitory Cl^- ion currents. Sedative-hypnotic drugs including 3α-OH neurosteroids, ethanol, benzodiazepines, barbiturates, and volatile anesthetics bind to distinct sites on the α and β subunits of the receptor, allosterically promoting chloride ion flux.[15] Although primarily considered positive modulators of the GABA-A receptor complex, neuroactive steroids appear to be more similar to barbiturates than to benzodiazepines or ethanol in that, at large concentrations, these drugs can directly activate the channel to increase Cl^- flux.[1] In fact, a recent report indicates that GABA-A receptors have two distinct sites to which neuroactive steroids can bind; actions at one site result in potentiation of the effects of GABA and actions at the other site result in direct activation of the channel.[16]

The GABA-A receptor is a transmembrane oligomeric protein composed of heterogenous combinations of α1–α6, β1–β3, γ1–γ3, δ, ε, and π polypeptide subunits.[17,18] Different brain areas show regional variation in the specific subunit compositions of the GABA-A receptor, but the α1β2γ2 combination is the most common. The potentiation site for neurosteroids is located in the transmembrane domains of α subunits whereas the direct activation site is located between α and β subunits.[16] These regions of the subunits are highly conserved among GABA-A receptors; consequently, they likely are present on all GABA-A receptors.[16] In contrast, benzodiazepine modulatory sites are formed when a γ2 subunit is coexpressed with any α and any β subunit[19] with the α subunit conferring selectivity to benzodiazepine ligands.[20] Thus, benzodiazepine modulatory sites are present on only a subset of GABA-A receptors while neurosteroid sites are present within most GABA-A complexes.

25.2.2 NMDA Receptor

Another site of neurosteroid action is at *N*-methyl D-aspartate (NMDA) receptors, wherein neurosteroids with a 3α-OH group appear to negatively modulate NMDA receptors.[21,22] These neurosteroids reduce the NMDA receptor-mediated current in cultured cells.[21] Allopregnanolone reduced striatal dopamine release induced by 10 min striatal microinfusions of NMDA as measured with *in vivo* microdialysis in rats.[22] This effect was similar to the effect observed following co-administration of NMDA and the NMDA antagonist D-AP5.[22] Again, the 3β-OH neurosteroids appear to act in opposition to the 3α-OH neurosteroids as pregnenolone sulfate acts as a positive allostatic modulator of NMDA receptors.[23] However, positive or negative modulation of the NMDA receptor by neurosteroids is mediated by distinct binding sites for 3α-OH vs 3β-OH enantiomers.[24] It should be noted that, in one study, DHEA-sulfate (a 3α-OH neurosteroid) potentiated NMDA receptor-mediated release of [^{3}H]norepinephrine in hippocampal slices, while pregnenolone sulfate acted as an inverse agonist.[25] Nonetheless, the majority of evidence suggests that 3α-OH neurosteroids are negative modulators of NMDA receptors while 3β-OH neurosteroids are positive modulators.

25.2.3 5-HT3 Receptor

Neurosteroids with a 3α-OH group also inhibit ion flux through 5-HT3 receptors. Unlike the GABA-A receptor, the 5-HT3 ion channel is homomeric, although some evidence of a second subunit has been reported.[26,27] Neurosteroids such as allopregnanolone can act as functional antagonists of the 5-HT3 ion channel.[27] Neurosteroids bind to membranes of HEK 293 cells expressing 5-HT3 receptors, but not native HEK 293 cells, indicative of a specific binding site on the 5-HT3 receptor. However, the binding affinity of the specific 5-HT3 receptor antagonist [^{3}H]GR65630 was unaltered by neurosteroids, suggesting that serotonin and neursteroids have distinct binding sites. Neurosteroids do not dose-dependently displace steroid binding to membranes of cells expressing 5-HT3 receptors, prompting some to conclude that steroids insert into the membrane and allosterically alter channel conformation, thus blocking ion flux through the 5-HT3 receptor.[27]

25.2.4 Other Receptors

Other ligand-gated ion channels are also sensitive to modulation by neuroactive steroids. Nicotinic acetylcholine receptors appear to be allosterically inhibited by progesterone and perhaps by other related steroids.[28,29] Kainate (AMPA)-type glutamate receptors are sensitive to neurosteroids, and evidence is also available that neurosteroids can bind to sigma receptors.[27] The G-protein coupled oxytocin receptor can bind neurosteroids, as well.[27] Despite some evidence that these cell-surface receptors are sensitive to neuroactive steroids, most of the work on the implications of neuroactive steroids on drug abuse have focused on GABA-A, NMDA and 5-HT3 receptors. This is due to the known effects of ethanol and benzodiazepines on these receptor systems. Future research may implicate neuroactive steroids in components of the behavioral effects of nicotine and other drugs of abuse, due to less studied actions of neuractive steroids at these receptor types.

25.2.5 *Effects on Extracellular Dopamine in the Mesolimbic System*

Increased dopamine levels in the medial forebrain, particularly in the nucleus accumbens, is a hallmark of drugs of abuse.[30] Some studies indicate that the 3α-OH neurosteroid allopregnanolone can enhance dopamine release and potentiate extracellular dopamine overflow following ethanol administration.[31,32] Allopregnanolone (12.5, 50, and 100 pmol icv) increases extracellular dopamine levels in the nucleus accumbens of rats and enhances the dopaminergic response to 1 mg/kg (sc) morphine in the same region.[31] However, in a study from the same laboratory, the 3β-OH neurosteroid pregnenolone sulfate also increases the accumbens dopamine levels as well as the dopaminergic response to morphine.[33] Others have shown data that agrees with the report by Barrot et al.[33]: 3α-OH neurosteroids depress extracellular dopamine levels or produce behavioral effects similar to those produced by dopamine antagonists, and 3β-OH neurosteroids increase dopamine release.[35–38] This is surprising in light of the fact that allopregnanolone and pregnenolone have opposing actions at GABA-A and NMDA receptors.[34] These results might be reconciled by a report suggesting that both 3α-OH and 3β-OH neurosteroids can stimulate catecholamine production by acting directly upon tyrosine hydroxylase, although in this report the effect requires 24 h.[39] Further studies on the dopaminergic effects of other neurosteroids with both 3α-OH and 3β-OH conformations are required to explain these results.

25.2.6 *Summary of Cell-Surface Pharmacology*

Drugs of abuse share cell-surface effects with neurosteroids. As previously described, sedative-hypnotic drugs including ethanol, benzodiazepines, barbiturates, and volatile anesthetics act as positive modulators of the GABA-A receptor complex; 3α-OH neurosteroids share this effect. Ethanol also acts as a negative modulator of NMDA receptors and inhibits 5-HT3 receptors, actions shared by 3α-OH neurosteroids.[12,27,40] In contrast, 3β-OH neurosteroids exert actions that oppose ethanol and other sedative-hypnotic drugs and act as competitive antagonists of 3α-OH neurosteroids at GABA-A and NMDA receptors.

25.3 Endogenous Neurosteroid Response to Exogenous Administration of Drugs of Abuse

Ethanol administration increases levels of 3α-OH neurosteroids in the brain and plasma of rats.[41,42] The doses of ethanol used in these studies were behaviorally relevant (1.3–4.0 g/kg). Ethanol-induced increases in brain allopregnanolone and

THDOC levels were greater in rats selectively bred to prefer ethanol compared with non-preferring counterparts.[43] Adrenalectomy–orchiectomy abolishes this effect, suggesting that it is peripherally mediated.[44] Further evidence that ethanol-induced increases in neurosteroids mediate the effects of ethanol comes from studies with finasteride. Finasteride, an inhibitor of the 5α-reductase enzyme that converts progesterone into allopregnanolone, blocks many of the cellular and behavioral effects of ethanol in rats.[41,45] However, levels of 3β-OH neurosteroids, which have opposing actions to those of ethanol at GABA-A, NMDA, and 5-HT3 receptors, are unaffected by ethanol administration.[45]

In humans, there have been reports suggesting that levels of the 3α-OH neurosteroid allopregnanolone increase in the plasma of ethanol-intoxicated adolescent males[46] and females, regardless of menstrual cycle.[47] However a more recent study reports that ethanol (0.7 g/kg) increases progesterone levels, but not allopregnanolone levels, in women during the luteal, but not during the follicular, phase of the menstrual cycle.[48] Another study[49] reports reduced allopregnanolone levels after moderate ethanol doses given to 17 male, but not 15 female, subjects. This same group has attempted to block the neurosteroidal effects of ethanol with finasteride. Finasteride reduces the subjective effects of ethanol on ratings from the Alcohol Sensation Scale in only the seven subjects homozygous for the A-allele of the GABA-α2 subunit polymorphism,[50] suggesting the possibility that ethanol effects may be mediated by 3α-OH neurosteriods in genetically sensitive subgroups. Therefore, despite consistent findings in rodents, it remains unclear whether allopregnanolone levels should be expected to rise in humans exposed to ethanol.

Administration of other drugs of abuse also results in increased levels of circulating neurosteroids. A dose of 5 mg/kg Δ9-tetrahydrocannabinol, the primary active constituent of cannabis, is as effective at increasing allopregnanolone in the cortex and plasma of rats as 2 g/kg ethanol.[41,51] Nicotine administration (0.5–2 mg/kg) results in dose-dependent increases in 3α-OH neurosteroids (allopregnanolone and 3α, 5β-THDOC), a 3β-OH neurosteroid (pregnenolone) and the neurosteroid precursor progesterone in the cerebral cortex of rats 40–60 min after administration.[52] The abused club drug gamma-hydroxybutyrate (75–1,000 mg/kg) increases cortical levels of the same neurosteroids 30–60 min after injection.[44,53] This effect is abolished in adrenalectomized-orchiectomized rats, suggesting a peripheral mechanism.[44] Morphine administration (15 mg/kg) also significantly increases cortical allopregnanolone in this study, but the relevance of the rather small increase achieved is questioned by the authors.[51]

While it is tempting to link drug-induced increases in these neurosteroids to abuse liability, not all drugs that increase levels of allopregnanolone are prone to abuse. Indeed, treatment with acute fluoxetine or baclofen results in increased allopregnanolone levels in the brain and plasma of rats, although chronic fluoxetine treatment results in decreased levels of neurosteroids.[54,56] These drugs, one a selective serotonin reuptake inhibitor and the other a GABA-B receptor agonist, do not have reinforcing effects and have even been proposed as anti-addiction medications.[57–59] Furthermore, administration of the highly addictive drug cocaine had no effect on cortical allopregnanolone levels in rats.[51] In addition, there are some notable gaps in the literature with regard to the effects of other abused drugs on neurosteroid levels. For example, the effects of benzodiazepines on circulating levels of neurosteroids have not been well characterized. Overall, while ethanol-induced increases in neurosteroid levels might mediate the reinforcing, discriminative stimulus, or motor activating effects of ethanol,[42] it is hard to make the same case for nicotine, Δ9-tetrahydrocannabinol, and cocaine, and clearly it is not the case for fluoxetine or baclofen, which have no abuse liability.

25.4 HPA Axis and Peripheral Benzodiazepine Receptor

Based on evidence from several lines of research, we are beginning to understand how ethanol administration could promote neurosteroidogenesis. Activation of the Hypothalamic-Pituitary-Adrenal (HPA) axis stimulates neurosteroidogenesis via adrenocorticotropic hormone (ACTH) stimulation of diazepam binding inhibitor (DBI) release which, in turn, acts directly on the peripheral (also known as the mitochondrial) benzodiazepine receptors (PBR). Activation of the PBR promotes cholesterol transport across the mitochondrial membrane where enzymes convert cholesterol into neurosteroids. The PBRs are found in adrenal cortex as well as in glial cells; thus, HPA axis activation could promote neurosteroidogenesis in both the adrenal cortex and brain, leading to the increases reported in brain and plasma following drug administration.

One line of evidence that HPA activation is involved in ethanol-induced neurosteroidogenesis comes from studies in socially isolated rats. Imposition of social isolation in rats immediately after weaning is considered to be stressful and results in hypersensitivity of the HPA axis, as assessed by isoniazid administration.[60] Such social isolation increases plasma corticosterone and decreases brain and plasma levels of some neuroactive steroids. Similar results have been reported in socially isolated mice.[61] Consistent with the notion that HPA axis activation is involved in ethanol-induced neurosteroidogenesis, ethanol administration results in greater percentage increases in allopregnanolone, 3α-5β-THDOC, pregnenolone, and progesterone in socially isolated rats compared with group-housed subjects.[60] Similarly, the percentage increase in allopregnanolone following acute fluoxetine administration is greater in socially isolated mice compared with group-housed controls, suggesting the mechanism by which fluoxetine exerts effects on neurosteroid levels is similar to that of ethanol.[61]

Ethanol administration activates the HPA axis and increases corticosterone levels.[62,63] Studies of intraventricular ethanol injection have suggested this is due to a direct activation by ethanol of corticotropin releasing hormone (CRH) expression and release from the hypothalamus.[64] In humans, ethanol abuse or intoxication activates the HPA axis. Compared to non-drinkers, heavy drinkers show increased basal levels of plasma cortisol.[65] Compared to control subjects coming into an emergency room for non-ethanol related trauma, young adults coming into an emergency room with acute ethanol intoxication show elevations in serum ACTH and cortisol.[66] In "light", but not "heavy" drinkers without abuse or dependence, cortisol is elevated after drinking an intoxicating (0.8 g/kg) but not moderate (0.4 g/kg) dose of ethanol.[67] In moderate to heavy drinkers without abuse or dependence, 1.0 g/kg ethanol, but not 0.5 g/kg ethanol, increases cortisol,[68] and in social drinkers given a low ethanol dose cortisol is decreased.[69]

Activation of the HPA axis results in release of CRH from the hypothalamus. Circulating CRH, in turn, promotes release of ACTH from the pituitary which acts within the adrenal cortex to stimulate production of neurosteroids, among other steroids. The mechanism by which ACTH promotes neurosteroidogenesis is becoming more clear. ACTH stimulates production of diazepam binding inhibitor (DBI), perhaps through a transcriptional mechanism. DBI is an endogenous polypeptide that has been classified as an endozepine because it binds to benzodiazepine-sensitive binding sites. In particular, DBI binds with high affinity to the peripheral benzodiazepine receptor located on the outer membrane of mitochondria. Activation of the PBR by DBI or exogenous benzodiazepines enhances the delivery of cholesterol from the cytoplasm of the cell to the inner mitochondrial membrane.[70] This is the rate-limiting step in steroid synthesis. Once transported, cholesterol is subject to conversion to pregnenolone by P450scc within the mitochondria. Steroidogenic acute regulatory protein (StAR) is also clearly involved in this process; however the interaction between the PBR and StAR is still being delineated.[71] PBR is certainly necessary but not sufficient to promote neurosteroidogenesis. Activation of PBR can occur in the brain and in the periphery.[72,73]

In addition to binding the central benzodiazepine receptor (CBR), associated with GABA-A receptors, diazepam also binds to the PBR, though only with micromolar affinity, as do several other benzodiazepines but with even lower affinity. However, ligands have been developed which bind the PBR with substantially higher affinity than the CBR. The prototypical PBR agonist Ro5-4864 and antagonist PK11195 bind to the PBR with nanomolar affinity but do not bind to the CBR with such high affinity.[74] Consistent with the notion that activation of the PBR results in neurosteroidogenesis, diazepam and Ro5–4894, but not chlordiazepoxide, stimulate steroidogenesis in mitochondria derived from bovine adrenal cortex.[75] Similarly, PBR agonists increase pregnenolone formation in glioma cells, suggesting that the high levels of PBR found in normal glial cells are responsible for neurosteroidogenesis in the brain.[73]

Thus, we have a model by which neurosteroids act as modulators of HPA activation. HPA activation increases circulating DBI which stimulates PBR in the mitochondria in both brain and adrenal cortex. PBR activation promotes cholesterol transport and conversion to pregnenolone, which can then be converted into other neuroactive steroids such as allopregnanolone. These products are then able to function as a negative feedback mechanism for HPA activation (specifically CRF and ACTH production) which allows the system to return to homeostasis once the HPA response has reached effective levels. Additionally these neurosteroids act at specific sites on cell-surface receptors and exert anxiolytic effects which influences the organism's behavioral response to HPA activation.

Taken together, inhibition of the HPA axis by benzodiazepines, ethanol, and neurosteroids might appear to be involved in the reinforcing effects of these drugs.

Activation of PBR by hormones released by the HPA following administration of benzodiazepines and ethanol could elicit neurosteroid production, which could mediate the reinforcing effects of these drugs through activity at cell-surface GABA and glutamate receptors. However, the story is clearly more complex as evidenced by studies with fluoxetine and cocaine. Fluoxetine is not reinforcing under any reported conditions, and yet it increases brain levels of neuroactive steroids, apparently by a mechanism that involves HPA activation.[54,55] Cocaine is a highly reinforcing drug under many conditions, activates the HPA, but does not increase brain level of GABA-A active neurosteroids.[51] Thus, HPA activation does not necessarily leads to neurosteroid production, and drug-induced neurosteroid production via HPA activation is not necessarily reinforcing.

25.5 Behavioral Effects of Neurosteroids

25.5.1 Anxiolytic Effects

One behavioral effect that is common to drugs that increase neurosteroid production as well as neurosteroids themselves is an anxiolytic effect. The anxiolytic effects of benzodiazepines have been well described, and the GABA-A receptor remains a target for anxiolytic drug development.[77] Allopregnanolone exerts anxiolytic effects in several different paradigms, including the elevated plus-maze and the Geller–Seifter conflict test, and following administration of anxiogenic doses of caffeine.[77–80] This effect appears to be mediated through the amygdala, as evidenced by anxiolytic effects of local injections of allopregnanolone (but not pregnenolone or DHEA sulfate) into this brain region.[81] Other 3α-OH neurosteroids also exert anxiolytic effects.[82] Similar anxiolytic effects have been reported following systemic administration of selective agonists of the PBR which increase brain and plasma neurosteroid levels.[72,83] Acute administration of ethanol also results in an anxiolytic effect as assessed by the elevated plus maze in rats and social anxiety-related stimuli in normal humans.[84,85] Fluoxetine, but not desipramine, produces anxiolytic effects in the elevated plus maze[86] and similar drugs have been suggested as treatments for anxiety in humans.[87] Therefore, neurosteroids could have some utility as anxiolytics, and the anxiolytic effects of some drugs may involve neurosteroidogenesis.

25.5.2 Antinociception and Anesthesia

Neurosteroids (especially those with a 3β-OH group) are effective antinociceptive agents, similar to ethanol, opioids, cannabinoids, and cocaine.[88] This is likely due to neurosteroid action at peripheral T-type calcium channels. 3α-OH Neuroactive steroids are also effective anesthetics, similar to ethanol, but not benzodiazepines, opioids, cannabinoids, fluoxetine, or cocaine. The anesthetic effect of neurosteroids has been linked to a 3α-OH conformation, specifically allopregnanolone.[8,89] Both neurosteroids and benzodiazepines increase the duration of loss of the righting reflex due to ethanol.[90] While such common effects could suggest overlapping mechanisms of action, other behavioral studies are more informative.

25.5.3 *Motor and Response Rate Disruptive Effects*

Ethanol, benzodiazepines, and 3α-OH neurosteroids reduce motor function in mice and rats.[90] The motor-depressant effect of neurosteroids is unaltered by ethanol in rats, and slightly enhanced by ethanol in mice, though the motor effects of benzodiazepines are greatly enhanced in both species.[90] Neurosteroids with a 3α-OH group can reverse locomotor depression due to pentylenetetrazole, bicuculline, picrotoxin, and yohimbine but are without effect on locomotor stimulation due to cocaine, methamphetamine, dizocilpine, or phencyclidine.[91]

Both ethanol and neuroactive steroids disrupt operant responding under a variety of different schedules. For example, in six rats responding under a fixed ratio 20 schedule of food presentation, pregnanolone (Fig. 25.2) produces dose- and time-dependent decreases in response rates with a dose of 17.8 mg/kg of pregnanolone decreasing rates to <20% of control. This disruption in responding occurs within

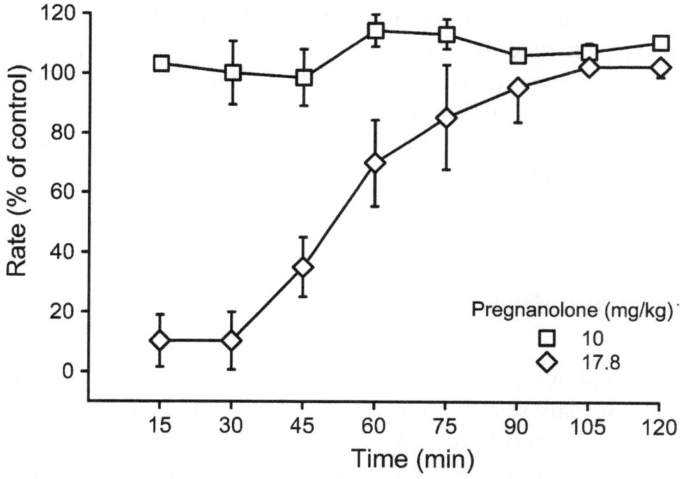

Fig. 25.2 Rate-decreasing effects of pregnanolone in six rats responding under a fixed-ratio 20 schedule of food presentation. Pregnanolone was administered immediately before 2 h sessions which were divided into 15 min cycles; response rate was measured during the last 5 min of each cycle. A dose of 10 mg/kg does not markedly alter responding whereas a larger dose (17.8 mg/kg) decreases rates to <20% of control. Maximum effects are evident for 30 min and rate-decreasing effects are no longer evident 90 min after administration. Abscissa: time after pregnanolone administration in min. Ordinate: average rate expressed as a percentage of control response rate ±1 SEM

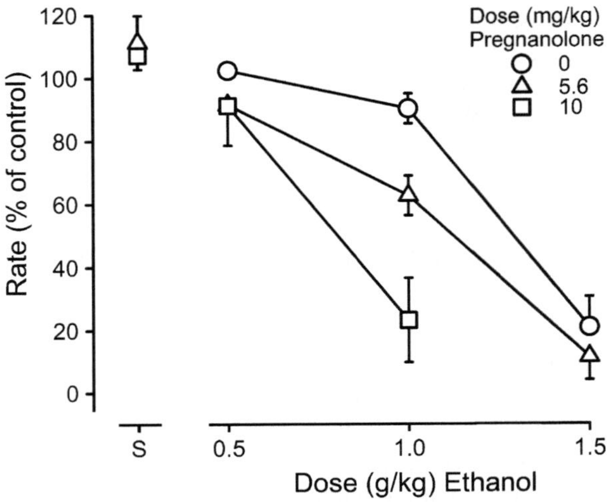

Fig. 25.3 Rate-decreasing effects of ethanol determined in the absence and presence of pregnanolone (n = 6). Ethanol, administered alone or concurrently with pregnanolone, was given immediately before 2 h sessions which were divided into 15 min cycles; response rate was measured during the last 5 min of each cycle. Dose–effect curves were constructed using data from the first cycle of sessions. Points above S represent the effects of pregnanolone administered alone. Doses of pregnanolone that did not alter response rates shifted the ethanol dose-effect curve to the left. Abscissa: dose of ethanol in g/kg. Ordinate: average rate expressed as a percentage of control response rate ±1 SEM

15 min of administration and is no longer evident 90 min after pregnanolone administration. When an ineffective dose of pregnanolone (10 mg/kg; Fig. 25.2) is administered concurrently with ethanol, the potency of ethanol is increased (Fig. 25.3). Moreover, the time course of the rate-decreasing effects of combinations of ethanol and pregnanolone is similar to the time course of pregnanolone alone (data not shown), suggesting that the increased potency of ethanol in the presence of pregnanolone is not due to a kinetic interaction.

25.6 Discriminative Stimulus Effects

Among behavioral assays, drug discrimination is a highly sensitive, quantitative, and pharmacologically selective procedure for evaluating mechanisms responsible for the behavioral effects of drugs including drugs increasing GABAergic transmission at GABA-A receptors. One common drug discrimination procedure involves training an animal to make one operant response (i.e., press a left lever for food) after administration of a dose of a drug and another operant response (i.e., press a right lever for food) after administration of a vehicle solution. The drug is considered to have been successfully established as a discriminative stimulus when the animal responds reliably and almost exclusively on the correct lever (i.e., left lever in the presence of the drug and right lever in the absence of the drug). Under test conditions and in an animal that has been trained to reliably discriminate a dose of a drug, other doses of the training drug and other drugs sharing a mechanism of action with the training drug can produce responding on the lever paired with the training drug (i.e., drug-appropriate responding). Administration of pharmacologically unrelated drugs, on the other hand, does not produce drug-appropriate responding, but instead, animals will respond on the lever paired with vehicle. Drugs that act at a pharmacologically related site, but that have additional sites of action, may produce only partial substitution.

In animals trained to discriminate neuroactive steroids from vehicle, ethanol administration results in 3α-OH neurosteroid-appropriate responding, although sex- and strain-dependent differences have been reported. Ethanol administration (greater than 2 g/kg, ip) results in 100% pregnanolone-appropriate responding in female DBA/2J mice trained to discriminate 10 mg/kg pregnanolone from saline.[93] However, ethanol administration only results in 50–60% pregnanolone-appropriate responding in male mice of the same strain. Interestingly, ethanol also produces only about 60% pregnanolone-appropriate responding in female C57BL/6 mice, suggesting that strain-dependent differences are also present.[92] When trained to discriminate a lower pregnanolone dose (5.6 mg/kg) from saline, 2.5 g/kg ethanol results in only about 60% pregnanolone-appropriate responding in male DBA/2J mice.[93] In male C57BL/6 mice, ethanol produces predominantly saline-appropriate responding at doses that do not suppress response rate.[93] In rats trained to discriminate 5 mg/kg pregnanolone from saline, 1 g/kg ethanol results in almost 100% pregnanolone-appropriate responding.[94] For the most part, these results suggest that 3α-OH neurosteroids and ethanol, especially low doses of ethanol, share discriminative stimulus effects. Incomplete substitution of ethanol for neurosteroids could be due to variations in endogenous steroid levels as seen in monkeys over the estrous cycle, or it could reflect differences in the pharmacology of ethanol and neurosteroids. Indeed, McCauley and Gee[95] report that pregnanolone allosterically inhibits [35S]TBPS binding to GABA-A receptors in cortical homogenates from estrus rats, but not diestrus rats. Inhibition of [35S]TBPS binding in brain from overectomized rats is not different from inhibition of binding in male brain homogenates.[95] This may indicate changes in neurosteroid pharmacology that depend on the menstrual cycle.

Engel et al.[94] have performed a study in which rats were trained to discriminate 5 mg/kg pregnanolone from saline. While all of the GABA-A positive modulators tested dose-dependently produced pregnanolone-appropriate responding, the direct GABA-site agonists 4,5,6,7-tetrahydroisoxazolo[4,5-c]pyridin-3-ol and muscimol failed to produce responding for pregnanolone. Both ethanol and the σ1 receptor agonist SKF 10047 produced responding that did not differ from the training dose of pregnanolone, while the NMDA antagonist MK-801 was only able to maintain 75% pregnanolone-lever responding. The 5-HT3 agonist SR 57227A produced pregnanolone-appropriate responding, whereas the other 5-HT3 agonist tested, m-chlorophenylbiguanide, only produced about 60% pregnanolone-appropriate responding. These results suggest that pregnanolone discriminative stimulus effects are due to positive GABA-A modulation, but not direct agonism; however, muscimol does not readily cross the blood brain barrier and studies with other direct-acting GABA-A receptor agonists might help to clarify a role for GABA-A receptor agonism in the behavioral effects of pregnanolone and other positive GABA-A receptor modulators. Additionally, antagonism of NMDA receptors and activation of 5-HT3 and σ1 receptors can, to some extent, mimic the discriminative stimulus effects of pregnanolone.

In animals trained to discriminate ethanol from water, a variety of positive GABA-A receptor modulators, including barbiturates, benzodiazepines, and neuroactive steroids, substitute for the discriminative stimulus effects of ethanol across a range of conditions.[96] Therefore, based upon the results of drug discrimination studies, it appears that facilitation of GABA-A receptor-mediated Cl^- flux is related to the mechanism by which ethanol produces its behavioral effects. Drug discrimination has further shown that training dose of ethanol is an important determinant of the extent to which positive GABA-A modulators and other drugs share discriminative stimulus effects with ethanol. Thus, whereas barbiturates, benzodiazepines, and neuroactive steroids substitute for the discriminative stimulus effects of ethanol at a relatively small training dose (1 g/kg), these drugs are less likely to substitute for ethanol at a larger training dose (2 g/kg) in rats.[96] Thus, the involvement of GABA-A receptors in the behavioral effects of ethanol appears to vary as a function of ethanol training dose, with GABA-A receptors playing a greater role at relatively small doses of ethanol, and other mechanisms (glutamate receptor) playing a greater role at relatively large doses of ethanol.

In animals trained to discriminate ethanol from water, 3α-OH neurosteroids result in ethanol-appropriate responding in mice, rats, and monkeys. Pregnanolone administration results in over 80% ethanol-appropriate lever responding in DBA/2J and C57BL/6 strains of mice trained to discriminate 1.5 g/kg ethanol from vehicle.[97] Systemic administration of the endogenous 3α-OH neurosteroid THDOC results in over 80% ethanol-appropriate responding in rats trained to discriminate 1.0 g/kg ethanol from saline.[98] Similarly, the synthetic 3α-OH neurosteroid alphaxalone results in 65% and 80% ethanol-appropriate lever responding in rats trained to discriminate 0.8 and 1.2 g/kg ethanol (respectively) from saline.[99] Both systemic administration and microinjection of allopregnanolone into the nucleus accumbens results in over 80% ethanol-appropriate responding in rats trained to discriminate 1 g/kg ethanol from saline; however, microinjection of allopregnanolone into the hippocampus results in only approximately 50% ethanol-appropriate responding.[100] In female cynomologous monkeys, allopregnanolone administration results in ethanol-appropriate responding.[101] It appears that the phase of the menstrual cycle can alter monkeys' sensitivity to discriminative stimulus properties of ethanol and ethanol-like discriminative stimulus properties of allopregnanolone. In each monkey, ethanol always results in ethanol-appropriate responding, although the potency varies depending on menstrual-cycle phase. Yet, in some subjects allopregnanolone administration does not result in ethanol-appropriate responding during the follicular phase.[102]

While there is considerable overlap in the discriminative stimulus effects of ethanol and neuroactive steroids, failure of these drug classes to share discriminative stimulus effects under some conditions is consistent with some differences in their mechanism of action. Differences in the mechanism of action of ethanol and neuroactive steroids are likely to include not only differences at NMDA, serotonin 5-HT3, and other receptor-gated ion channels, but also at GABA-A receptors. These might include differences in the mechanism by which these drugs increase GABA-A receptor transmission (i.e., facilitation of GABA-A receptor-mediated Cl^- flux or direct activation of GABA-A receptors), or they might include differences in selectivity for the 20 or more GABA-A receptor subtypes found in mammalian brain.

Further evidence of the complexity of the discriminative stimulus effects of 3α-OH neurosteroids comes from a study in which rats were trained to discriminate 200 mg/kg γ-hydroxybutyrate (GHB). This chemical has generated interest as a drug of abuse used extensively as a "club drug." GHB appears to have a complex pharmacology, with effects at GABA-A, GABA-B, and GHB-specific receptors. Administration of the GABA-B receptor agonist baclofen results in approximately 80% GHB-appropriate responding, while diazepam generates only about 60% GHB-lever responding.[103] In contrast, pregnanolone administration generates less than 20% GHB-appropriate responding up to doses that significantly reduce response rates.[103] These results highlight differences in the behavioral pharmacology of 3α-OH neurosteroids and other GABAergic drugs of abuse. Taken together, results from studies of the discriminative stimulus effects of positive modulators of GABA-A receptors suggest that the discriminative stimulus effects of neurosteroids appear to be primarily determined by activity at the GABA-A receptor versus effects at other targets such as the NMDA or 5-HT3 receptors.

In opposition to positive GABA-A receptor modulators, negative GABA-A receptor modulators can decrease or inhibit GABA-A-mediated Cl^- flux, and negative modulators at one site of action on the GABA-A receptor complex (benzodiazepine receptors) have the potential for producing functional (as opposed to competitive) antagonism of positive modulators acting at another site (neuroactive steroids). Negative GABA-A receptor modulators, therefore, could be used to confirm a role for GABA-A receptors in the behavioral effects of neuroactive steroids.

As might be expected based on similar GABAergic effects, alphaxalone or pregnanolone administration results in 100% midazolam-appropriate responding in rhesus monkeys trained to discriminate 0.32 mg/kg midazolam from saline.[104,105] Negative GABA-A receptor modulators acting at benzodiazepine sites antagonized the discriminative stimulus of pregnanolone or midazolam in these same monkeys, suggesting the discriminative stimulus of pregnanolone is primarily mediated by positive modulation of the GABA-A receptor.[106] Further evidence for this is provided by a study in which rats were trained on a three lever discrimination of ethanol (1 g/kg, ig plus saline ip), midazolam (3 mg/kg, ip plus water ig), and water (ig plus saline ip). In this study, allopregnanolone administration resulted in 75% midazolam-appropriate responses and only about 10% ethanol-appropriate responses, suggesting that the discriminative stimulus of allopregnanolone is more like a benzodiazepine than a moderate dose of ethanol.[107] Nevertheless, ethanol appears to have a mechanism of action that is more similar to barbiturates and neuroactive steroids than to benzodiazepines inasmuch as ethanol is more likely to substitute for the discriminative stimulus effects of barbiturates and neuroactive steroids than benzodiazepines.[108]

Because 3β-OH neurosteroids act as antagonists in cell or slice preparations, some have attempted to block 3α-OH neurosteroid or ethanol effects with 3β-OH enantiomers. Consistent with in vitro studies demonstrating opposing cell-surface effects, 3β-OH neurosteroids do not produce ethanol-appropriate responding in rats or 3α-OH neurosteroid-appropriate responding in mice.[98,99,92] In contrast with in vitro work, the discriminative stimulus effects of ethanol or 3α-OH neurosteroids are not antagonized by 3β-OH neurosteroids.[98,99] However, it is important to note that 3β-OH neurosteroids do not compete with ethanol or 3α-OH neurosteroids for GABA-A receptor binding sites.[15,24] Further, picrotoxin (a GABA-A receptor antagonist) and NMDA or D-cycloserine (an agonist and positive modulator at the NMDA receptor) do not antagonize the discriminative stimulus effects of ethanol, so it remains unclear whether 3β-OH neurosteroids should be expected to antagonize the discriminative stimulus effects of ethanol.[109] It must also be noted that in one study the 3β-OH neurosteroid epipregnanolone potentiates the discriminative stimulus effects of ethanol in rats.[110] Interpretation of this study is complicated by the fact that a 3α-OH neurosteroid has no effect on the ethanol discriminative stimulus in this study, a result that contrasts with several other similar studies.[98,99,111] Thus, there is almost no evidence that 3β-OH neurosteroids can inhibit the discriminative stimulus effects of ethanol or 3α-OH neurosteroids; however it remains unclear whether such inhibition should be expected.

Taken together, the results of drug discrimination studies in animals show that ethanol and neuroactive steroids have strikingly similar discriminative stimulus effects. These studies suggest that there is considerable overlap in the receptor mechanisms responsible for the behavioral effects of neuroactive steroids and ethanol, especially at GABA-A receptors, but also at NMDA and 5-HT3 receptors.

25.7 Reinforcing Effects

Based on similarities in the cell-surface pharmacology and discriminative stimulus effects of 3α-OH neurosteroids and ethanol or other sedatives, one might expect 3α-OH neurosteroids to be positively reinforcing. There is some evidence that this is the case. Finn et al.[112,113] demonstrated dose-dependent increases in the time spent in an environment paired with allopregnanolone versus vehicle under a conditioned place preference procedure in mice, suggesting that allopregnanolone is preferred. These results are consistent with a later study in which rats were given unlimited access to two bottles, one containing water and the other containing 50 μg/ml allopregnanolone. These rats accelerated their consumption of the allopregnanolone solution and developed a clear preference for this solution over water over 10 consecutive days.[114] However, the same concentration of allopregnanolone does not maintain operant responding unless sucrose is included in the solution.[114] Other 3α-OH neurosteroids do maintain operant self-administration. The androgenic 3α-OH neurosteroid 3α-DIOL is consumed at higher levels than vehicle in a two-bottle choice procedure by male hamsters,[115] produces modest conditioned place preference in overiectomized female rats, and maintains operant responding for intravenous injections.[116] In rhesus monkeys, iv delivery of pregnanolone or the 3α-OH neurosteroid prodrug Co 8-7071 each maintains responding at levels above saline in some subjects.[117] All of these studies utilize peripheral administration of the neurosteroid (ip injections and iv or ig self-administration). It remains unknown whether central delivery of allopregnanolone would maintain self-administration behavior; however, in one study, central (icv) administration of 5 μg or 25 μg (but not 0 or 30 μg) allopregnanolone produced conditioned place *aversion*.[118] Thus, a weak case can be made that 3α-OH neurosteroids are reinforcing in the same manner as ethanol or other drugs of abuse, though this may depend on the dose and the route of administration.

Evidence for neurosteroids modulating the reinforcing effects of ethanol is similarly equivocal. Some early studies report that allopregnanolone alters the reinforcing effects of ethanol. Janak et al.[119] report that allopregnanolone (20 min pretreatment, ip) produces biphasic changes in the number of responses performed by rats for presentations of a 10% ethanol solution. A dose of 3 mg/kg allopregnanolone significantly increases the number of responses during a 30 min session, while 10 mg/kg decreases the number of responses, though not significantly. When ethanol (10%) and sucrose (1%) are available under independent concurrent fixed-ratio schedules, 3 and 5.6 mg/kg allopregnanolone increase ethanol responding without affecting sucrose responding.[120] These results contrast with effects of diazepam, muscimol, and baclofen in the same subjects where only non-significant decreases are observed across a dose range of 1–10 mg/kg.[119,120]

Others have also reported on the effect of neurosteroids on ethanol self-administration. Allopregnanolone does not alter ethanol consumption in female mice when the solution is 5% or 10% (v/v) ethanol.[121] Though in male mice, allopregnanolone (3.2–24 mg/kg) increases ethanol consumption when administered immediately before limited access to a 10% (v/v) ethanol solution during the first hour of access.[121,122] When the ethanol solution concentration is 5% (v/v), 10 mg/kg allopregnanolone is without effect, though 17 mg/kg significantly increases consumption.[121] However, saccharin consumption (0.033%, w/v) is also significantly increased during the first hour of access in male mice following 10 mg/kg allopregnanolone, while quinine consumption (0.015 nM) is not. It should also be noted that 10 mg/kg allopregnanolone results in significant decreases in ethanol consumption in the second hour of ethanol access in male mice. This is likely due to an interaction between consumed ethanol and allopregnanolone, as saccharin consumption (in the absence of ethanol) is unaltered in male mice during the second hour of access.[121]

Consistent with opposing actions of 3α-OH and 3β-OH enantiomers, the 3β-OH enantiomer of pregnanolone, epipregnanolone (10 and 20 mg/kg 45 min before ethanol access), significantly reduces the number of ethanol deliveries earned.[123] In another study, epipregnanolone has no effect on ethanol consumption in male C57BL6 mice, while allopregnanolone increases drinking.[122]

Some studies have failed to confirm allopregnanolone effects on ethanol reinforcement. Neither allopregnanolone (3.2, 10, or 17 mg/kg) nor the 5α-reductase inhibitor finasteride (50 or 100 mg/kg) alters conditioned place preference associated with ethanol in mice.[124] However, ethanol administration increases whole brain allopregnanolone levels threefold as expected, and allopregnanolone administration results in dose-dependent increases in whole brain allopregnanolone levels, suggesting effective injections.[124] A similar study utilizing finasteride to produce a 99% reduction in whole brain allopregnanolone levels also reports no effect on either ethanol (2 g/kg)-induced conditioned place preference or on c-fos protein expression in response to ethanol in several discrete brain regions thought to be involved in ethanol reinforcement.[125] These studies suggest that the reinforcing effect of ethanol is not altered by exogenous or endogenous allopregnanolone, in contradiction with the notion that the behavioral effects of ethanol (including reinforcement) are mediated via allopregnanolone.[46] The limitations (and strengths) inherent in the conditioned place preference procedure have been noted elsewhere,[126] and it is difficult to justify discounting the earlier experiments utilizing voluntary self-administration [121,122] based on the results of these conditioned place preference experiments.

Just as central administration results in opposite effects compared to peripherally administered allopregnanolone under conditioned place preference,[113,118] administration of allopregnanolone directly into the hippocampus of rats reduces ethanol intake compared with saline-injected controls on the day of the injection as well as on the day that followed.[127] Pregnanolone has a similar effect on the injection day, but not the following day. As a comparison, intra-hippocampal allopregnanolone also reduces glucose intake on the injection day and the day that followed in a second group.[127] Together, these data underscore potential differences in the effects of neurosteroids on ethanol reinforcement that depend on the route of administration. Thus, conclusive evidence demonstrating that neurosteroids can modulate ethanol reinforcement remains elusive.

Results of studies investigating modulatory effects of neurosteroids on reinforcing properties of other drugs of abuse are also equivocal. Romieu et al.[128] demonstrates that co-administration of the 3β-OH neurosteroids DHEA or pregnenolone with cocaine increases cocaine-paired place preference, suggesting an enhancement of the rewarding properties of cocaine. In contrast, co-administration with progesterone or finasteride reduces cocaine conditioned place preference.[128] Doron et al.[129] also report that DHEA reduces cocaine self-administration. Further studies will be necessary to sort out the effects of neurosteroids on cocaine reinforcement.

25.8 Reinstatement

The reinstatement procedure has gained acceptance as a model of relapse to drug use. In this procedure, animals are trained to self-administer a drug or ethanol, then access is removed, functionally reducing the self-administration behavior. After some time passes, the experimenter administers the drug or ethanol then the animals are placed into the experimental chamber and the amount of responding on the lever associated with drug delivery increases. Treatments that reduce this reinstatement of responding are thought to have utility against relapse, while treatments that increase reinstatement responding are thought to precipitate relapse.[130,131] Allopregnanolone administration results in reinstatement of extinguished ethanol responding but not extinguished sucrose responding.[132] This allpregnanolone-induced reinstatement of ethanol responding is not surprising because allopregnanolone and ethanol share discriminative stimulus effects, and low doses of ethanol also increase reinstatement of ethanol responding. Reinstatement research with other neurosteroids or drugs of abuse is lacking, but DHEA reduces cocaine reinstatement.[129] Overall, these studies suggest that neurosteroids could be involved in relapse to drug use.

25.9 Tolerance and Withdrawal

Many studies in rodents implicate neurosteroids in tolerance and dependence to ethanol.[45,133] In particular, allopregnanolone influences the brain region-specific changes in the expression of the GABA-A receptor subunits that mediate tolerance to ethanol and sensitivity to withdrawal.[133–135] Generally, the neuroadaptive changes in the GABA-A receptor that reduce its sensitivity to positive allosteric modulation also increase its sensitivity to negative modulators[34] and increased binding of DHEA and DHEA-S to the [^{35}S]TBPS binding site of GABA-A receptors has been observed in the cerebellum and cerebral cortex of male rats chronically exposed to ethanol.[136] Interestingly, subunit changes in the GABA-A receptor that confer tolerance to ethanol preserve sensitivity to allopregnanolone. During ethanol withdrawal, rats show enhanced anticonvulsant effects of allopregnanolone which are associated with increased expression of $\beta 2$, $\beta 3$, and $\gamma 1$ subunits in the cerebral cortex and also with increased sensitivity of GABA-A receptors to allopregnanolone-mediated chloride conductance in synaptoneurosomes.[137] Thus, tolerance to ethanol is associated with brain region-specific alterations in GABA-A receptor subunit expression which account for cross tolerance to benzodiazepines and barbiturates[138] without necessarily conferring cross tolerance to neuroactive steriods.

Chronic, intermittent doses of ethanol sensitize rats to the anticonvulsant and anxiolytic effects of alphaxalone; at the same time, levels of allopregnanolone are reduced as are the messenger RNAs for both the 3α- and 5α- reductase enzymes involved in its synthesis.[139] Other studies show that chronic ethanol reduces baseline allopregnanolone levels and ethanol-stimulated allopregnanolone release.[45,140] Thus, chronic ethanol exposure may cause sensitization to exogenous neurosteroid administration, while simultaneously reducing endogenous circulating levels.

In humans, long-term, heavy ethanol consumption is known to inhibit progesterone levels and disrupt women's menstrual cycles. Compared to 17 healthy controls, 20 women undergoing ethanol detoxification exhibit decreased levels of progesterone and reduced ratios of allopregnanolone to pregnenolone sulfate.[141] In a group of heavy ethanol "abusers" undergoing uncomplicated withdrawal, days 1–5 of ethanol withdrawal are associated with lower allopregnanolone levels and higher anxiety and depression ratings in comparison to 15 days after withdrawal.[140,142] Furthermore, randomized treatment with either fluoxetine or indomethacin (two drugs known to increase neurosteroid synthesis) increases allopregnanolone levels and lowers the anxiety and depression symptoms in these patients relative to patients who received no such treatment.[142] In comparison to healthy controls, currently abstinent ethanol-dependent patients show increased DHEA responses to CRH challenge and an increased DHEA to ACTH ratio.[143]

In chronic ethanol intoxication or dependence, alterations in HPA function indicate a suppressed HPA axis.[63,144] In abstinent alcoholics, ethanol challenge does not increase cortisol[145] but intoxicated alcoholics in detoxification treatment have elevated cortisol in saliva samples.[146] Abstinence-related withdrawal increases cortisol release[147,148] which tends to normalize over time with continued abstinence. Compared to healthy controls, abstinent alcoholics usually,[149,150] but not always,[151] demonstrate a blunted cortisol response to social stress. The nature of alcoholic disturbances of the HPA have been demonstrated best by neuroendocrine challenge studies where ethanol dependence is associated with suppressed HPA activity. Compared to normal control subjects, ethanol-dependent subjects generally have shown reduced ACTH and cortisol responses to CRH challenge.[152,153] Though Adinoff et al.[154] do not observe reduced ACTH release following CRH challenge, cortisol is reduced, and a cosyntropin challenge test shows that adrenocortical activation of adrenal release of cortisol[155] is reduced in ethanol-dependent male subjects compared to healthy controls.

Tolerance develops to positive modulators that act at the benzodiazepine[156–159] or barbiturate[158,160] binding sites; however, tolerance does not appear to develop to neuroactive steroids under a variety of conditions.[161–163] Clinically, the development of tolerance is an undesirable effect because it necessitates increases in dose to maintain a particular effect. Moreover, if increasing doses are needed, dependence will be more likely to develop and the withdrawal syndrome will be more robust when chronic treatment is terminated.

Cross tolerance also seems to develop differentially among positive GABA-A modulators. Cross tolerance clearly develops among positive modulators that act at benzodiazepine sites as well as among positive modulators that act at barbiturate sites on the GABA-A receptor complex.[159,164,165] In contrast, cross tolerance may not develop to drugs acting at modulatory sites that are different from the site of action of the drug administered chronically.[156,163,165] In studies where cross tolerance developed, it appears to be asymmetrical. For example, there is no change in sensitivity to barbiturates following benzodiazepine treatment whereas sensitivity to benzodiazepines decreases following barbiturate treatment.[164] In other studies, cross tolerance, but not tolerance, develops; during chronic treatment with the neuroactive steroid ganaxolone, there is no change in sensitivity to ganaxolone but a decreased sensitivity to diazepam.[162] If cross tolerance develops in the absence of primary tolerance, then the use of neuroactive steroids for the treatment of benzodiazepine or ethanol abuse would provide a great clinical advantage. Because tolerance would not develop to the neuroactive steroid, the dose of the treatment drug that would be needed to prevent the emergence of withdrawal would not change over a long treatment period. However, cross tolerance would develop during chronic treatment with the neuroactive steroid, thereby decreasing sensitivity to benzodiazepines or ethanol. Thus, treating benzodiazepine or ethanol-abusing patients with neurosteroids could increase the quantity of the benzodiazepine or ethanol required to achieve the desired effect. This could reduce the incentive to abuse these drugs.

Just as benzodiazepines replaced barbiturates resulting in improved treatment 40 years ago, other drugs are being investigated to provide similar therapeutic effects to benzodiazepines with reduced dependence liability. Although novel treatments with drugs affecting different neurotransmitter systems (e.g., buspirone and ramelteon) have been developed, GABA-A receptors remain the primary target for treatment of many forms of anxiety, insomnia, and ethanol withdrawal. Advances in our understanding of the GABA-A receptor complex have resulted in renewed interest in neuroactive steroids. Initial characterization suggests that neuroactive steroids would produce therapeutic effects comparable to those of benzodiazepines, including anxiolytic, sedative, and anticonvulsant effects. Despite these similarities in the acute effects of benzodiazepines and neuroactive steroids, the clinical use of neuroactive steroids has been limited. Recently, differences between the effects of benzodiazepines and those of neuroactive steroids have been observed during chronic diazepam treatment.[163,164] Although some similarities between neuroactive steroids and benzodiazepines, particularly discriminative stimulus and reinforcing effects in nonhuman subjects, suggest that neuroactive steroids will have abuse liability, differences observed during chronic treatment imply that neuroactive steroids will offer distinct clinical advantages over benzodiazepines.

25.10 Summary and Conclusion

In summary, 3α-OH neurosteroids share cell-surface receptor and behavioral effects with several drugs of abuse, particularly ethanol. Neuroactive steroids with a 3α-OH configuration are positive modulators of GABA-A receptors, and negative modulators of NMDA and 5-HT3 receptors. Neuroactive steroids with 3β-OH configuration have opposing actions, at least at GABA-A and probably NMDA receptors. Much of the research related to neuroactive steroids and drugs of abuse has been directed at GABA-A receptor effects, as neuroactive steroids, ethanol, benzodiazepines, and barbiturates all facilitate and/or enhance GABA-mediated chloride flux through this ion channel. Sedative, depressant, and discriminative stimulus effects of neuroactive steroids appear to be mediated in large part by action at this receptor. Effects of neuroactive steroids on the NMDA and 5-HT3 receptors have indicated that symmetry between the discriminative stimulus effects of neuroactive steroids and ethanol are likely due to similar effects at these receptors, in addition to the well-characterized effects at the GABA-A receptor. Administration of ethanol, in non-human and in some human studies, might increase circulating levels of neuroactive steroids. Based on the similar pharmacological effects of neuroactive steroids and ethanol, some have proposed that ethanol-induced increases in levels of neuroactive steroids mediate the effects of ethanol. Several lines of evidence suggest that neuroactive steroids more likely modulate the effects of ethanol, rather than mediating ethanol effects. The increase in circulating neuroactive steroids following drug administration is not unique to ethanol and appears to be due to activation of the HPA axis. Activation of the HPA axis results in a classic endocrine response that includes neurosteroidogenesis. Increased levels of neurosteroids may in turn act as negative feedback modulators of the HPA axis. Chronic exposure to neuroactive steroids does not appear to lead to the development of tolerance or dependence on neuroactive steroids but does lead to cross-tolerance to ethanol or benzodiazepines. This asymmetrical development of tolerance and dependence could be exploited to help treat ethanol or benzodiazepine dependence. The lack of tolerance or dependence development during chronic neurosteroid treatment could help reduce the abuse liability of benzodiazepines in clinical use.

To conclude, a compelling role for neurosteroids in studies of drug abuse continues to emerge. Much more research is required in order to complete this picture; however neuroactive steroids are clearly involved in the effects of ethanol and likely the effects of benzodiazepines. Future research may begin to answer critical questions about the importance and impact of the menstrual cycle, different neurosteroid isomers, and regulatory mechanisms that maintain circulating levels of the different isomers. As such results are reported, the potential therapeutic use of neurosteroids in addiction and as an adjunct in order to reduce the abuse liability of other sedatives will likely gain more interest.

References

1. Lambert JJ, Belelli D, Hill-Venning C, Peters JA. Neurosteroids and GABAA receptor function. Trends Pharmacol Sci 1995; 16(9):295–303.
2. McEwen BS. What makes a steroid a neurosteroid? Neurosteroids and brain function, sponsored by the Fidia research foundation, New Orleans, LA, USA, November 8–9, 1991. New Biol 1992; 4(3):212–216.
3. McNeil C. Potential drug DHEA hits snags on way to clinic. J Natl Cancer Inst 1997; 89(10):681–683.
4. Paul SM, Purdy RH. Neuroactive steroids. FASEB J 1992; 6(6):2311–2322.
5. Wang MD, Backstrom T, Landgren S. The inhibitory effects of allopregnanolone and pregnanolone on the population spike, evoked in the rat hippocampal CA1 stratum pyramidale in vitro, can be blocked selectively by epiallopregnanolone. Acta Physiol Scand 2000; 169(4):333–341.
6. Barbosa AD, Morato GS. Effect of epipregnanolone and pregnenolone sulfate on chronic tolerance to ethanol. Pharmacol Biochem Behav 2000; 67(3):459–464.
7. Lan NC, Bolger MB, Gee KW. Identification and characterization of a pregnane steroid recognition site that is functionally coupled to an expressed GABAA receptor. Neurochem Res 1991; 16(3):347–356.
8. Covey DF, Nathan D, Kalkbrenner M, Nilsson KR, Hu Y, Zorumski CF, Evers AS. Enantioselectivity of pregnanolone-induced gamma-aminobutyric acid(A) receptor modulation and anesthesia. J Pharmacol Exp Ther 2000; 293(3):1009–1016.

9. Prince RJ, Simmonds MA. Differential antagonism by epipregnanolone of alphaxalone and pregnanolone potentiation of [3H]flunitrazepam binding suggests more than one class of binding site for steroids at GABAA receptors. Neuropharmacology 1993; 32(1):59–63.
10. Maitra R, Reynolds JN. Modulation of GABA(A) receptor function by neuroactive steroids: evidence for heterogeneity of steroid sensitivity of recombinant GABA(A) receptor isoforms. Can J Physiol Pharmacol 1998; 76(9):909–920.
11. Mehta AK, Ticku MK. An update on GABAA receptors. Brain Res Brain Res Rev 1999; 29(2–3):196–217.
12. Faingold CL, N'Gouemo P, Riaz A. Ethanol and neurotransmitter interactions – from molecular to integrative effects. Prog Neurobiol 1998; 55(5):509–535.
13. Grobin AC, Matthews DB, Devaud LL, Morrow AL. The role of GABA(A) receptors in the acute and chronic effects of ethanol. Psychopharmacology 1998; 139(1–2):2–19.
14. Kumar S, Fleming RL, Morrow AL. Ethanol regulation of gamma-aminobutyric acid A receptors: genomic and nongenomic mechanisms. Pharmacol Ther 2004; 101(3):211–226.
15. Rick CE, Ye Q, Finn SE, Harrison NL. Neurosteroids act on the GABA(A) receptor at sites on the N-terminal side of the middle of TM2. Neuroreport 1998; 9(3):379–383.
16. Hosie AM, Wilkins ME, da Silva HM, Smart TG. Endogenous neurosteroids regulate GABAA receptors through two discrete transmembrane sites. Nature 2006; 444(7118):486–489.
17. Sieghart W. Structure and pharmacology of gamma-aminobutyric acid A receptor subtypes. Pharmacol Rev 1995; 47(2):181–234.
18. Whiting PJ. The GABA-A receptor gene family: new targets for therapeutic intervention. Neurochem Int 1999; 34(5):387–390.
19. Luddens H, Korpi ER, Seeburg PH. GABAA/benzodiazepine receptor heterogeneity: neurophysiological implications. Neuropharmacology 1995; 34(3):245–254.
20. Doble A, Martin IL. Multiple benzodiazepine receptors: no reason for anxiety. Trends Pharmacol Sci 1992; 13(2):76–81.
21. Park-Chung M, Wu FS, Farb DH. 3 alpha-Hydroxy-5 beta-pregnan-20-one sulfate: a negative modulator of the NMDA-induced current in cultured neurons. Mol Pharmacol 1994; 46(1):146–150.
22. Sadri-Vakili G, Johnson DW, Janis GC, Gibbs TT, Pierce RC, Farb DH. Inhibition of NMDA-induced striatal dopamine release and behavioral activation by the neuroactive steroid 3alpha-hydroxy-5beta-pregnan-20-one hemisuccinate. J Neurochem 2003; 86(1):92–101.
23. Wu FS, Gibbs TT, Farb DH. Pregnenolone sulfate: a positive allosteric modulator at the *N*-methyl-D-aspartate receptor. Mol Pharmacol 1991; 40(3):333–336.
24. Park-Chung M, Wu FS, Purdy RH, Malayev AA, Gibbs TT, Farb DH. Distinct sites for inverse modulation of *N*-methyl-D-aspartate receptors by sulfated steroids. Mol Pharmacol 1997; 52(6):1113–1123.
25. Monnet FP, Mahe V, Robel P, Baulieu EE. Neurosteroids, via sigma receptors, modulate the [3H]norepinephrine release evoked by *N*-methyl-D-aspartate in the rat hippocampus. Proc Natl Acad Sci USA 1995; 92(9):3774–3778.
26. Davies PA, Pistis M, Hanna MC, Peters JA, Lambert JJ, Hales TG, Kirkness EF. The 5-HT3B subunit is a major determinant of serotonin-receptor function. Nature 1999; 397(6717):359–363.
27. Rupprecht R. Neuroactive steroids: mechanisms of action and neuropsychopharmacological properties. Psychoneuroendocrinology 2003; 28(2):139–168.
28. Valera S, Ballivet M, Bertrand D. Progesterone modulates a neuronal nicotinic acetylcholine receptor. Proc Natl Acad Sci USA 1992; 89(20):9949–9953.
29. Bullock AE, Clark AL, Grady SR, Robinson SF, Slobe BS, Marks MJ, Collins AC. Neurosteroids modulate nicotinic receptor function in mouse striatal and thalamic synaptosomes. J Neurochem 1997; 68(6):2412–2423.
30. Kuhar MJ, Ritz MC, Boja JW. The dopamine hypothesis of the reinforcing properties of cocaine. Trends Neurosci 1991; 14(7):299–302.
31. Rouge-Pont F, Mayo W, Marinelli M, Gingras M, Le Moal M, Piazza PV. The neurosteroid allopregnanolone increases dopamine release and dopaminergic response to morphine in the rat nucleus accumbens. Eur J Neurosci 2002; 16(1):169–173.

32. Dazzi L, Serra M, Seu E, Cherchi G, Pisu MG, Purdy RH, Biggio G. Progesterone enhances ethanol-induced modulation of mesocortical dopamine neurons: antagonism by finasteride. J Neurochem 2002; 83(5):1103–1109.
33. Barrot M, Vallee M, Gingras MA, Le Moal M, Mayo W, Piazza PV. The neurosteroid pregnenolone sulphate increases dopamine release and the dopaminergic response to morphine in the rat nucleus accumbens. Eur J Neurosci 1999; 11(10):3757–3760.
34. Baulieu EE. Neurosteroids: a novel function of the brain. Psychoneuroendocrinology 1998; 23(8):963–987.
35. Murray HE, Gillies GE. Differential effects of neuroactive steroids on somatostatin and dopamine secretion from primary hypothalamic cell cultures. J Neuroendocrinol 1997; 9(4):287–295.
36. Khisti RT, Deshpande LS, Chopde CT. The neurosteroid 3 alpha-hydroxy-5 alpha-pregnan-20-one affects dopamine-mediated behavior in rodents. Psychopharmacology 2002; 161(2):120–128.
37. Jaworska-Feil L, Budziszewska B, Leskiewicz M, Lason W. Opposite effects of inhibitory and excitatory neurosteroids on [3H]dopamine release from rat nucleus accumbens. Pol J Pharmacol 1998; 50(6):449–452.
38. Matsunaga M, Ukena K, Baulieu EE, Tsutsui K. 7alpha-Hydroxypregnenolone acts as a neuronal activator to stimulate locomotor activity of breeding newts by means of the dopaminergic system. Proc Natl Acad Sci USA 2004; 101(49):17282–17287.
39. Charalampopoulos I, Dermitzaki E, Vardouli L, Tsatsanis C, Stournaras C, Margioris AN, Gravanis A. Dehydroepiandrosterone sulfate and allopregnanolone directly stimulate catecholamine production via induction of tyrosine hydroxylase and secretion by affecting actin polymerization. Endocrinology 2005; 146(8):3309–3318.
40. Kostowski W. 5-HT3 receptors and central effects of ethanol. Pol J Pharmacol 1996; 48(3):243–254.
41. VanDoren MJ, Matthews DB, Janis GC, Grobin AC, Devaud LL, Morrow AL. Neuroactive steroid 3alpha-hydroxy-5alpha-pregnan-20-one modulates electrophysiological and behavioral actions of ethanol. J Neurosci 2000; 20(5):1982–1989.
42. Morrow AL, Janis GC, VanDoren MJ, Matthews DB, Samson HH, Janak PH, Grant KA. Neurosteroids mediate pharmacological effects of ethanol: a new mechanism of ethanol action? Alcohol Clin Exp Res 1999; 23(12):1933–1940.
43. Barbaccia ML, Affricano D, Trabucchi M, Purdy RH, Colombo G, Agabio R, Gessa GL. Ethanol markedly increases "GABAergic" neurosteroids in alcohol-preferring rats. Eur J Pharmacol 1999; 384(2–3):R1–R2.
44. Porcu P, Sogliano C, Ibba C, Piredda M, Tocco S, Marra C, Purdy RH, Biggio G, Concas A. Failure of gamma-hydroxybutyric acid both to increase neuroactive steroid concentrations in adrenalectomized-orchiectomized rats and to induce tolerance to its steroidogenic effect in intact animals. Brain Res 2004; 1012(1–2):160–168.
45. Morrow AL, VanDoren MJ, Penland SN, Matthews DB. GABAergic neuroactive steroids in ethanol action, tolerance and dependence. Brain Res Brain Res Rev 2001; 37(1–3):98–109.
46. Torres JM, Ortega E. Alcohol intoxication increases allopregnanolone levels in male adolescent humans. Psychopharmacology 2004; 172(3):352–355.
47. Torres JM, Ortega E. Differential regulation of steroid 5alpha-reductase isozymes expression by androgens in the adult rat brain. FASEB J 2003; 17(11):1428–1433.
48. Holdstock L, Penland SN, Morrow AL, de Wit H. Moderate doses of ethanol fail to increase plasma levels of neurosteroid 3alpha-hydroxy-5alpha-pregnan-20-one-like immunoreactivity in healthy men and women. Psychopharmacology 2006; 186(3):442–450.
49. Pierucci-Lagha A, Covault J, Feinn R, Khisti RT, Morrow AL, Marx CE, Shampine LJ, Kranzler HR. Subjective effects and changes in steroid hormone concentrations in humans following acute consumption of alcohol. Psychopharmacology 2006; 186(3):451–461.
50. Pierucci-Lagha A, Covault J, Feinn R, Nellissery M, Hernandez-Avila C, Oncken C, Morrow AL, Kranzler HR. GABRA2 alleles moderate the subjective effects of alcohol, which are attenuated by finasteride. Neuropsychopharmacology 2005; 30(6):1193–1203.

51. Grobin AC, VanDoren MJ, Porrino LJ, Morrow AL. Cortical 3 alpha-hydroxy-5 alpha-pregnan-20-one levels after acute administration of Delta 9-tetrahydrocannabinol, cocaine and morphine. Psychopharmacology 2005; 179(3):544–550.
52. Porcu P, Sogliano C, Cinus M, Purdy RH, Biggio G, Concas A. Nicotine-induced changes in cerebrocortical neuroactive steroids and plasma corticosterone concentrations in the rat. Pharmacol Biochem Behav 2003; 74(3):683–690.
53. Sanna E, Talani G, Busonero F, Pisu MG, Purdy RH, Serra M, Biggio G. Brain steroidogenesis mediates ethanol modulation of GABAA receptor activity in rat hippocampus. J Neurosci 2004; 24(29):6521–6530.
54. Uzunov DP, Cooper TB, Costa E, Guidotti A. Fluoxetine-elicited changes in brain neurosteroid content measured by negative ion mass fragmentography. Proc Natl Acad Sci USA 1996; 93(22):12599–12604.
55. Serra M, Pisu MG, Muggironi M, Parodo V, Papi G, Sari R, Dazzi L, Spiga F, Purdy RH, Biggio G. Opposite effects of short- versus long-term administration of fluoxetine on the concentrations of neuroactive steroids in rat plasma and brain. Psychopharmacology 2001; 158(1):48–54.
56. Barbaccia ML, Colombo G, Affricano D, Carai MA, Vacca G, Melis S, Purdy RH, Gessa GL. GABA(B) receptor-mediated increase of neurosteroids by gamma-hydroxybutyric acid. Neuropharmacology 2002; 42(6):782–791.
57. Pettinati HM. The use of selective serotonin reuptake inhibitors in treating alcoholic subtypes. J Clin Psychiatry 2001; 62(Suppl 20):26–31.
58. Colombo G, Addolorato G, Agabio R, Carai MA, Pibiri F, Serra S, Vacca G, Gessa GL. Role of GABA(B) receptor in alcohol dependence: reducing effect of baclofen on alcohol intake and alcohol motivational properties in rats and amelioration of alcohol withdrawal syndrome and alcohol craving in human alcoholics. Neurotox Res 2004; 6(5):403–414.
59. Brebner K, Childress AR, Roberts DC. A potential role for GABA(B) agonists in the treatment of psychostimulant addiction. Alcohol Alcohol 2002; 37(5):478–484.
60. Serra M, Pisu MG, Floris I, Cara V, Purdy RH, Biggio G. Social isolation-induced increase in the sensitivity of rats to the steroidogenic effect of ethanol. J Neurochem 2003; 85(1):257–263.
61. Matsumoto K, Pinna G, Puia G, Guidotti A, Costa E. Social isolation stress-induced aggression in mice: a model to study the pharmacology of neurosteroidogenesis. Stress 2005; 8(2):85–93.
62. Gianoulakis C. Alcohol-seeking behavior: the roles of the hypothalamic-pituitary-adrenal axis and the endogenous opioid system. Alcohol Health Res World 1998; 22(3):202–210.
63. Haddad JJ. Alcoholism and neuro-immune-endocrine interactions: physiochemical aspects. Biochem Biophys Res Commun 2004; 323(2):361–371.
64. Lee S, Selvage D, Hansen K, Rivier C. Site of action of acute alcohol administration in stimulating the rat hypothalamic-pituitary-adrenal axis: comparison between the effect of systemic and intracerebroventricular injection of this drug on pituitary and hypothalamic responses. Endocrinology 2004; 145(10):4470–4479.
65. Gianoulakis C, Dai X, Brown T. Effect of chronic alcohol consumption on the activity of the hypothalamic-pituitary-adrenal axis and pituitary beta-endorphin as a function of alcohol intake, age, and gender. Alcohol Clin Exp Res 2003; 27(3):410–423.
66. Frias J, Torres JM, Miranda MT, Ruiz E, Ortega E. Effects of acute alcohol intoxication on pituitary-gonadal axis hormones, pituitary-adrenal axis hormones, beta-endorphin and prolactin in human adults of both sexes. Alcohol Alcohol 2002; 37(2):169–173.
67. King AC, Houle T, de Wit H, Holdstock L, Schuster A. Biphasic alcohol response differs in heavy versus light drinkers. Alcohol Clin Exp Res 2002; 26(6):827–835.
68. McCaul ME, Wand GS, Stauffer R, Lee SM, Rohde CA. Naltrexone dampens ethanol-induced cardiovascular and hypothalamic- pituitary-adrenal axis activation. Neuropsychopharmacology 2001; 25(4):537–547.
69. Kokavec A, Crowe SF. The effect of a moderate level of white wine consumption on the hypothalamic-pituitary-adrenal axis before and after a meal. Pharmacol Biochem Behav 2001; 70(2–3):243–250.

70. Papadopoulos V, Berkovich A, Krueger KE, Costa E, Guidotti A. Diazepam binding inhibitor and its processing products stimulate mitochondrial steroid biosynthesis via an interaction with mitochondrial benzodiazepine receptors. Endocrinology 1991; 129(3):1481–1488.
71. Miller WL. Steroidogenic acute regulatory protein (StAR), a novel mitochondrial cholesterol transporter. Biochim Biophys Acta 2007 March 6 (epub ahead of print).
72. Bitran D, Foley M, Audette D, Leslie N, Frye CA. Activation of peripheral mitochondrial benzodiazepine receptors in the hippocampus stimulates allopregnanolone synthesis and produces anxiolytic-like effects in the rat. Psychopharmacology 2000; 151(1):64–71.
73. Papadopoulos V, Lecanu L, Brown RC, Han Z, Yao ZX. Peripheral-type benzodiazepine receptor in neurosteroid biosynthesis, neuropathology and neurological disorders. Neuroscience 2006; 138(3):749–756.
74. Casellas P, Galiegue S, Basile AS. Peripheral benzodiazepine receptors and mitochondrial function. Neurochem Int 2002; 40(6):475–486.
75. Yanagibashi K, Ohno Y, Nakamichi N, Matsui T, Hayashida K, Takamura M, Yamada K, Tou S, Kawamura M. Peripheral-type benzodiazepine receptors are involved in the regulation of cholesterol side chain cleavage in adrenocortical mitochondria. J Biochem 1989; 106(6):1026–1029.
76. Whiting PJ. The GABAA receptor gene family: new opportunities for drug development. Curr Opin Drug Discov Devel 2003; 6(5):648–657.
77. Bitran D, Hilvers RJ, Kellogg CK. Anxiolytic effects of 3 alpha-hydroxy-5 alpha[beta]-pregnan-20-one: endogenous metabolites of progesterone that are active at the GABAA receptor. Brain Res 1991; 561(1):157–161.
78. Bitran D, Purdy RH, Kellogg CK. Anxiolytic effect of progesterone is associated with increases in cortical allopregnanolone and GABAA receptor function. Pharmacol Biochem Behav 1993; 45(2):423–428.
79. Brot MD, Akwa Y, Purdy RH, Koob GF, Britton KT. The anxiolytic-like effects of the neurosteroid allopregnanolone: interactions with GABA(A) receptors. Eur J Pharmacol 1997; 325(1):1–7.
80. Jain NS, Hirani K, Chopde CT. Reversal of caffeine-induced anxiety by neurosteroid 3-alpha-hydroxy-5-alpha-pregnane-20-one in rats. Neuropharmacology 2005; 48(5):627–638.
81. Akwa Y, Purdy RH, Koob GF, Britton KT. The amygdala mediates the anxiolytic-like effect of the neurosteroid allopregnanolone in rat. Behav Brain Res 1999; 106(1–2):119–125.
82. Vanover KE, Hogenkamp DJ, Lan NC, Gee KW, Carter RB. Behavioral characterization of Co 134444 (3alpha-hydroxy-21-(1'-imidazolyl)-3beta-methoxymethyl-5alpha- pregnan-20-one), a novel sedative-hypnotic neuroactive steroid. Psychopharmacology 2001; 155(3):285–291.
83. Kita A, Kohayakawa H, Kinoshita T, Ochi Y, Nakamichi K, Kurumiya S, Furukawa K, Oka M. Antianxiety and antidepressant-like effects of AC-5216, a novel mitochondrial benzodiazepine receptor ligand. Br J Pharmacol 2004; 142(7):1059–1072.
84. Langen B, Dietze S, Fink H. Acute effect of ethanol on anxiety and 5-HT in the prefrontal cortex of rats. Alcohol 2002; 27(2):135–141.
85. Gerlach AL, Schiller A, Wild C, Rist F. Effects of alcohol on the processing of social threat-related stimuli in socially phobic women. Br J Clin Psychol 2006; 45(Pt 3):279–295.
86. Drapier D, Bentue-Ferrer D, Laviolle B, Millet B, Allain H, Bourin M, Reymann JM. Effects of acute fluoxetine, paroxetine and desipramine on rats tested on the elevated plus-maze. Behav Brain Res 2007; 176(2):202–209.
87. Argyropoulos SV, Sandford JJ, Nutt DJ. The psychobiology of anxiolytic drug. Part 2: Pharmacological treatments of anxiety. Pharmacol Ther 2000; 88(3):213–227.
88. Todorovic SM, Pathirathna S, Brimelow BC, Jagodic MM, Ko SH, Jiang X, Nilsson KR, Zorumski CF, Covey DF, Jevtovic-Todorovic V. 5beta-reduced neuroactive steroids are novel voltage-dependent blockers of T-type Ca2 + channels in rat sensory neurons in vitro and potent peripheral analgesics in vivo. Mol Pharmacol 2004; 66(5):1223–1235.
89. Mok WM, Bukusoglu C, Krieger NR. 3 alpha-hydroxy-5 alpha-pregnan-20-one is the only active anesthetic steroid in anesthetized mouse brain. Steroids 1993; 58(3):112–114.
90. Vanover KE, Suruki M, Robledo S, Huber M, Wieland S, Lan NC, Gee KW, Wood PL, Carter RB. Positive allosteric modulators of the GABA(A) receptor: differential interaction

of benzodiazepines and neuroactive steroids with ethanol. Psychopharmacology 1999; 141(1):77–82.
91. Ungard JT, Beekman M, Gasior M, Carter RB, Dijkstra D, Witkin JM. Modification of behavioral effects of drugs in mice by neuroactive steroids. Psychopharmacology 2000; 148(4):336–343.
92. Shannon EE, Porcu P, Purdy RH, Grant KA. Characterization of the discriminative stimulus effects of the neuroactive steroid pregnanolone in DBA/2J and C57BL/6J inbred mice. J Pharmacol Exp Ther 2005; 314(2):675–685.
93. Shannon EE, Purdy RH, Grant KA. Discriminative stimulus effects of 5.6 mg/kg pregnanolone in DBA/2J and C57BL/6J inbred mice. Alcohol 2005; 37(1):35–45.
94. Engel SR, Purdy RH, Grant KA. Characterization of discriminative stimulus effects of the neuroactive steroid pregnanolone. J Pharmacol Exp Ther 2001; 297(2):489–495.
95. McCauley LD, Gee KW. Influence of the estrus cycle on the discrimination of apparent neuroactive steroid site subtypes on the gamma-aminobutyric acidA receptor complex in the rat. J Pharmacol Exp Ther 1995; 275(3):1412–1417.
96. Grant KA. Strategies for understanding the pharmacological effects of ethanol with drug discrimination procedures. Pharmacol Biochem Behav 1999; 64(2):261–267.
97. Shelton KL, Grant KA. Discriminative stimulus effects of ethanol in C57BL/6J and DBA/2J inbred mice. Alcohol Clin Exp Res 2002; 26(6):747–757.
98. Bienkowski P, Kostowski W. Discriminative stimulus properties of ethanol in the rat: effects of neurosteroids and picrotoxin. Brain Res 1997; 753(2):348–352.
99. Ginsburg BC, Lamb RJ. Alphaxalone and epiallopregnanolone in rats trained to discriminate ethanol. Alcohol Clin Exp Res 2005; 29(9):1621–1629.
100. Hodge CW, Nannini MA, Olive MF, Kelley SP, Mehmert KK. Allopregnanolone and pentobarbital infused into the nucleus accumbens substitute for the discriminative stimulus effects of ethanol. Alcohol Clin Exp Res 2001; 25(10):1441–1447.
101. Grant KA, Azarov A, Bowen CA, Mirkis S, Purdy RH. Ethanol-like discriminative stimulus effects of the neurosteroid 3 alpha-hydroxy-5 alpha-pregnan-20-one in female Macaca fascicularis monkeys. Psychopharmacology 1996; 124(4):340–346.
102. Grant KA, Azarov A, Shively CA, Purdy RH. Discriminative stimulus effects of ethanol and 3 alpha-hydroxy-5 alpha-pregnan-20-one in relation to menstrual cycle phase in cynomolgus monkeys (Macaca fascicularis). Psychopharmacology 1997; 130(1):59–68.
103. Carter LP, Flores LR, Wu H, Chen W, Unzeitig AW, Coop A, France CP. The role of GABAB receptors in the discriminative stimulus effects of gamma-hydroxybutyrate in rats: time course and antagonism studies. J Pharmacol Exp Ther 2003; 305(2):668–674.
104. McMahon LR, France CP. Combined discriminative stimulus effects of midazolam with other positive GABAA modulators and GABAA receptor agonists in rhesus monkeys. Psychopharmacology 2005; 178(4):400–409.
105. McMahon LR, France CP. Differential behavioral effects of low efficacy positive GABAA modulators in combination with benzodiazepines and a neuroactive steroid in rhesus monkeys. Br J Pharmacol 2006; 147(3):260–268.
106. McMahon LR, France CP. Negative GABA(A) modulators attenuate the discriminative stimulus effects of benzodiazepines and the neuroactive steroid pregnanolone in rhesus monkeys. Psychopharmacology 2005; 181(4):697–705.
107. Porcu P, Grant KA. Discriminative stimulus effects of ethanol in rats using a three-choice ethanol-midazolam-water discrimination. Behav Pharmacol 2004; 15(8):555–567.
108. Massey BW, Woolverton WL. Discriminative stimulus effects of combinations of pentobarbital and ethanol in rhesus monkeys. Drug Alcohol Depend 1994; 35(1):37–43.
109. Bienkowski P, Stefanski R, Kostowski W. Discriminative stimulus effects of ethanol: lack of antagonism with *N*-methyl-D-aspartate and D-cycloserine. Alcohol 1997; 14(4):345–350.
110. Bowen CA, Purdy RH, Grant KA. An investigation of endogenous neuroactive steroid-induced modulation of ethanol's discriminative stimulus effects. Behav Pharmacol 1999; 10(3):297–311.
111. Ator NA, Grant KA, Purdy RH, Paul SM, Griffiths RR. Drug discrimination analysis of endogenous neuroactive steroids in rats. Eur J Pharmacol 1993; 241(2–3):237–243.

112. Finn DA, Phillips TJ, Okorn DM, Chester JA, Cunningham CL. Rewarding effect of the neuroactive steroid 3 alpha-hydroxy-5 alpha-pregnan-20-one in mice. Pharmacol Biochem Behav 1997; 56(2):261–264.
113. Sinnott RS, Mark GP, Finn DA. Reinforcing effects of the neurosteroid allopregnanolone in rats. Pharmacol Biochem Behav 2002; 72(4):923–929.
114. Frye CA. Some rewarding effects of androgens may be mediated by actions of its 5alpha-reduced metabolite 3alpha-androstanediol. Pharmacol Biochem Behav 2007; 86(2):354–367.
115. Jorge JC, Velazquez KT, Ramos-Ortolaza DL, Lorenzini I, Marrero J, Maldonado-Vlaar CS. A testosterone metabolite is rewarding to ovariectomized female rats. Behav Neurosci 2005; 119(5):1222–1226.
116. Rowlett JK, Winger G, Carter RB, Wood PL, Woods JH, Woolverton WL. Reinforcing and discriminative stimulus effects of the neuroactive steroids pregnanolone and Co 8–7071 in rhesus monkeys. Psychopharmacology 1999; 145(2):205–212.
117. Beauchamp MH, Ormerod BK, Jhamandas K, Boegman RJ, Beninger RJ. Neurosteroids and reward: allopregnanolone produces a conditioned place aversion in rats. Pharmacol Biochem Behav 2000; 67(1):29–35.
118. Janak PH, Redfern JE, Samson HH. The reinforcing effects of ethanol are altered by the endogenous neurosteroid, allopregnanolone. Alcohol Clin Exp Res 1998; 22(5):1106–1112.
119. Janak PH, Michael Gill T. Comparison of the effects of allopregnanolone with direct GABAergic agonists on ethanol self-administration with and without concurrently available sucrose. Alcohol 2003; 30(1):1–7.
120. Sinnott RS, Phillips TJ, Finn DA. Alteration of voluntary ethanol and saccharin consumption by the neurosteroid allopregnanolone in mice. Psychopharmacology 2002; 162(4):438–447.
121. Ford MM, Nickel JD, Phillips TJ, Finn DA. Neurosteroid modulators of GABA(A) receptors differentially modulate Ethanol intake patterns in male C57BL/6J mice. Alcohol Clin Exp Res 2005; 29(9):1630–1640.
122. O'Dell LE, Purdy RH, Covey DF, Richardson HN, Roberto M, Koob GF. Epipregnanolone and a novel synthetic neuroactive steroid reduce alcohol self-administration in rats. Pharmacol Biochem Behav 2005; 81(3):543–550.
123. Gabriel KI, Cunningham CL, Finn DA. Allopregnanolone does not influence ethanol-induced conditioned place preference in DBA/2J mice. Psychopharmacology 2004; 176(1):50–56.
124. Murphy NP, Sakoori K, Okabe C. Lack of evidence of a role for the neurosteroid allopregnanolone in ethanol-induced reward and c-fos expression in DBA/2 mice. Brain Res 2006; 1094(1):107–118.
125. Bardo MT, Bevins RA. Conditioned place preference: what does it add to our preclinical understanding of drug reward? Psychopharmacology 2000; 153(1):31–43.
126. Martin-Garcia E, Darbra S, Pallares M. Intrahippocampal allopregnanolone decreases voluntary chronic alcohol consumption in non-selected rats. Prog Neuropsychopharmacol Biol Psychiatry 2007; 31(4):823–831.
127. Romieu P, Martin-Fardon R, Bowen WD, Maurice T. Sigma 1 receptor-related neuroactive steroids modulate cocaine-induced reward. J Neurosci 2003; 23(9):3572–3576.
128. Doron R, Fridman L, Gispan-Herman I, Maayan R, Weizman A, Yadid G. DHEA, a neurosteroid, decreases cocaine self-administration and reinstatement of cocaine-seeking behavior in rats. Neuropsychopharmacology 2006; 31(10):2231–2236.
129. Epstein DH, Preston KL, Stewart J, Shaham Y. Toward a model of drug relapse: an assessment of the validity of the reinstatement procedure. Psychopharmacology 2006; 189(1):1–16.
130. Katz JL, Higgins ST. The validity of the reinstatement model of craving and relapse to drug use. Psychopharmacology 2003; 168(1–2):21–30.
131. Nie H, Janak PH. Comparison of reinstatement of ethanol- and sucrose-seeking by conditioned stimuli and priming injections of allopregnanolone after extinction in rats. Psychopharmacology 2003; 168(1–2):222–228.
132. Finn DA, Ford MM, Wiren KM, Roselli CE, Crabbe JC. The role of pregnane neurosteroids in ethanol withdrawal: behavioral genetic approaches. Pharmacol Ther 2004; 101(2):91–112.

133. Follesa P, Biggio F, Caria S, Gorini G, Biggio G. Modulation of GABA(A) receptor gene expression by allopregnanolone and ethanol. Eur J Pharmacol 2004; 500(1–3):413–425.
134. Sanna E, Mostallino MC, Busonero F, Talani G, Tranquilli S, Mameli M, Spiga S, Follesa P, Biggio G. Changes in GABA(A) receptor gene expression associated with selective alterations in receptor function and pharmacology after ethanol withdrawal. J Neurosci 2003; 23(37):11711–11724.
135. Mehta AK, Ticku MK. Unsulfated and sulfated neurosteroids differentially modulate the binding characteristics of various radioligands of GABA(A) receptors following chronic ethanol administration. Neuropharmacology 2001; 40(5):668–675.
136. Devaud LL, Purdy RH, Finn DA, Morrow AL. Sensitization of gamma-aminobutyric acidA receptors to neuroactive steroids in rats during ethanol withdrawal. J Pharmacol Exp Ther 1996; 278(2):510–517.
137. Kumar S, Fleming RL, Morrow AL. Ethanol regulation of gamma-aminobutyric acid A receptors: genomic and nongenomic mechanisms. Pharmacol Ther 2004; 101(3):211–226.
138. Cagetti E, Pinna G, Guidotti A, Baicy K, Olsen RW. Chronic intermittent ethanol (CIE) administration in rats decreases levels of neurosteroids in hippocampus, accompanied by altered behavioral responses to neurosteroids and memory function. Neuropharmacology 2004; 46(4):570–579.
139. Romeo E, Brancati A, De Lorenzo A, Fucci P, Furnari C, Pompili E, Sasso GF, Spalletta G, Troisi A, Pasini A. Marked decrease of plasma neuroactive steroids during alcohol withdrawal. Clin Neuropharmacol 1996; 19(4):366–369.
140. Hill M, Popov P, Havlikova H, Kancheva L, Vrbikova J, Kancheva R, Pouzar V, Cerny I, Starka L. Altered profiles of serum neuroactive steroids in premenopausal women treated for alcohol addiction. Steroids 2005; 70(8):515–524.
141. Romeo E, Pompili E, di Michele F, Pace M, Rupprecht R, Bernardi G, Pasinib A. Effects of fluoxetine, indomethacine and placebo on 3 alpha, 5 alpha tetrahydroprogesterone (THP) plasma levels in uncomplicated alcohol withdrawal. World J Biol Psychiatry 2000; 1(2):101–104.
142. Adinoff B, Kiser JM, Martin PR, Linnoila M. Response of dehydroepiandrosterone to corticotropin-releasing hormone stimulation in alcohol-dependent subjects. Biol Psychiatry 1996; 40(12):1305–1307.
143. Adinoff B, Krebaum SR, Chandler PA, Ye W, Brown MB, Williams MJ. Dissection of hypothalamic-pituitary-adrenal axis pathology in 1-month-abstinent alcohol-dependent men, part 1: adrenocortical and pituitary glucocorticoid responsiveness. Alcohol Clin Exp Res 2005; 29(4):517–527.
144. O'Malley SS, Krishnan-Sarin S, Farren C, Sinha R, Kreek MJ. Naltrexone decreases craving and alcohol self-administration in alcohol-dependent subjects and activates the hypothalamo-pituitary-adrenocortical axis. Psychopharmacology 2002; 160(1):19–29.
145. Adinoff B, Ruether K, Krebaum S, Iranmanesh A, Williams MJ. Increased salivary cortisol concentrations during chronic alcohol intoxication in a naturalistic clinical sample of men. Alcohol Clin Exp Res 2003; 27(9):1420–1427.
146. Adinoff B, Iranmanesh A, Veldhuis J, Fisher L. Disturbances of the stress response: the role of the HPA axis during alcohol withdrawal and abstinence. Alcohol Health Res World 1998; 22(1):67–72.
147. Zimmermann U, Hundt W, Spring K, Grabner A, Holsboer F. Hypothalamic-pituitary-adrenal system adaptation to detoxification in alcohol-dependent patients is affected by family history of alcoholism. Biol Psychiatry 2003; 53(1):75–84.
148. Lovallo WR, Dickensheets SL, Myers DA, Thomas TL, Nixon SJ. Blunted stress cortisol response in abstinent alcoholic and polysubstance-abusing men. Alcohol Clin Exp Res 2000; 24(5):651–658.
149. Junghanns K, Backhaus J, Tietz U, Lange W, Bernzen J, Wetterling T, Rink L, Driessen M. Impaired serum cortisol stress response is a predictor of early relapse. Alcohol Alcohol 2003; 38(2):189–193.
150. Munro CA, Oswald LM, Weerts EM, McCaul ME, Wand GS. Hormone responses to social stress in abstinent alcohol-dependent subjects and social drinkers with no history of alcohol dependence. Alcohol Clin Exp Res 2005; 29(7):1133–1138.

151. von Bardeleben U, Holsboer F. Human corticotropin releasing hormone: clinical studies in patients with affective disorders, alcoholism, panic disorder and in normal controls. Prog Neuropsychopharmacol Biol Psychiatry 1988; 12(Suppl):S165–S187.
152. Costa A, Bono G, Martignoni E, Merlo P, Sances G, Nappi G. An assessment of hypothalamo-pituitary-adrenal axis functioning in non-depressed, early abstinent alcoholics. Psychoneuroendocrinology 1996; 21(3):263–275.
153. Adinoff B, Junghanns K, Kiefer F, Krishnan-Sarin S. Suppression of the HPA axis stress-response: implications for relapse. Alcohol Clin Exp Res 2005; 29(7):1351–1355. Erratum in: Alcohol Clin Exp Res 2006; 30(3):585.
154. Adinoff B, Krebaum SR, Chandler PA, Ye W, Brown MB, Williams MJ. Dissection of hypothalamic-pituitary-adrenal axis pathology in 1-month-abstinent alcohol-dependent men, part 2: response to ovine corticotropin-releasing factor and naloxone. Alcohol Clin Exp Res 2005; 29(4):528–537.
155. Cesare DA, McKearney JW. Tolerance to suppressive effects of chlordiazepoxide on operant behavior: lack of cross tolerance to pentobarbital. Pharmacol Biochem Behav 1980; 13(4):545–548.
156. McMillan DE. Effects of drugs on behavior before and during chronic diazepam administration. Eur J Pharmacol 1992; 215(2–3):145–152.
157. McMillan DE, Leander JD. Chronic chlordiazepoxide and pentobarbital interactions on punished and unpunished behavior. J Pharmacol Exp Ther 1978; 207(2):515–520.
158. Pugh SL, Boone MS, Emmett-Oglesby MW. Tolerance, cross-tolerance and withdrawal in rats made dependent on diazepam. J Pharmacol Exp Ther 1992; 262(2):751–758.
159. Kim CK, Kalynchuk LE, Pinel JP, Kippin TE. Tolerance to the anticonvulsant and ataxic effects of pentobarbital: effect of an ascending-dose regimen. Pharmacol Biochem Behav 1995; 52(4):825–829.
160. Kokate TG, Yamaguchi S, Pannell LK, Rajamani U, Carroll DM, Grossman AB, Rogawski MA. Lack of anticonvulsant tolerance to the neuroactive steroid pregnanolone in mice. J Pharmacol Exp Ther 1998; 287(2):553–558.
161. Reddy DS, Rogawski MA. Chronic treatment with the neuroactive steroid ganaxolone in the rat induces anticonvulsant tolerance to diazepam but not to itself. J Pharmacol Exp Ther 2000; 295(3):1241–1248.
162. McMahon LR, France CP. Acute and chronic effects of the neuroactive steroid pregnanolone on schedule-controlled responding in rhesus monkeys. Behav Pharmacol 2002; 13(7):545–555.
163. Khanna JM, Kalant H, Chau A, Shah G. Rapid tolerance and crosstolerance to motor impairment effects of benzodiazepines, barbiturates, and ethanol. Pharmacol Biochem Behav 1998; 59(2):511–519.
164. McMahon LR, France CP. Daily treatment with diazepam differentially modifies sensitivity to the effects of gamma-aminobutyric acid(A) modulators on schedule-controlled responding in rhesus monkeys. J Pharmacol Exp Ther 2002; 300(3):1017–1025.
165. McMahon LR, Javors MA, France CP. Changes in relative potency among positive GABA(A) receptor modulators upon discontinuation of chronic benzodiazepine treatment in rhesus monkeys. Psychopharmacology 2007; 192(1):135–145.

Chapter 26
The Role of Neurosteroids in Development of Pediatric Psychopathology

Pavel Golubchik[1,2] **and Abraham Weizman**[2–4]

Abstract Neurosteroids play a significant role in neurodevelopment and are involved in a wide variety of psychopathological processes. There is accumulating evidence on their role in adult psychopathology, little is known, however, about the possible role of neurosteroids in child and adolescent psychopathology although there is increasing evidence for their critical role from the early stages of brain development until adolescence. In this chapter we focus on the involvement of neurosteroids in neurodevelopment and mental disorders in children and adolescents. Adequate physiological levels protect the developing neural system from insult and contribute to the regulation of brain organization and function. Neurosteroids may be involved in the pathophysiology and pharmacotherapy of a variety of disorders in children and adolescents, including schizophrenia, autism, depression, eating disorders, aggressive behavior and attention deficit. The complex interaction between neurosteroids, neurodevelopment, life-events, genetics and mental disorders in children and adolescents merits further investigation.

Keywords Neurosteroid, neurodevelopment, child, adolescent, psychopathology

Abbreviations AN anorexia nervosa; ADHD attention deficit hyperactivity disorder; CD conduct disorder; DHEA dehydroepiandrosterone; DHEA-S dehydroepiandrosterone-sulfate; ECT electro-convulsive therapy; ED eating disorders; GABA gamma-aminobutyric acid; $GABA_A$ gamma-aminobutyric acid type A; MDD major depressive disorder; ODD oppositional defiant disorder; PTSD post-traumatic stress disorder

---

[1]Child and Adolescent Outpatient Unit, Geha Mental Health Center, Petah Tiqva

[2]Sackler Faculty of Medicine, Tel Aviv University, Tel Aviv

[3]Research Unit, Geha Mental Health Center, Petah Tiqva

[4]Laboratory of Biological Psychiatry, Felsenstein Medical Research Center, Rabin, Israel

M.S. Ritsner and A. Weizman (eds.), *Neuroactive Steroids in Brain Function, Behavior and Neuropsychiatric Disorders*

26.1 Introduction

Neurosteroids contribute to shaping behavioral function during early development and may modulate the expression of psychopathology.[1] Human studies on DHEA and DHEA-S have shown age-related changes in neurosteroid serum levels.[2] From the first few months of life until the age of 5 years their peripheral concentrations decline. During puberty, serum DHEA levels increase progressively with age in both boys and girls.[3] This elevation seems to be adrenarche-independent, since gonadotropins and estrogen have no effect on DHEA levels and the alterations in neurosteroids and sex hormones are not linked across time.[1] Peak levels are reached between the ages of 20 and 30 years.[4]

Neurosteroids play an important role in various psychological processes. They have been found to be involved in the response to stress, depression, anxiety, sleep, aggression, epilepsy, memory formation and learning. Some studies have indicated their neurotrophic and neuroprotective properties, their role in the development of the nervous system and underlying neurodevelopmental pathological processes.[5,6] There is accumulating data on their role in adults, including Alzheimer's disease,[7] schizophrenia,[8–11] mood disorders,[1,12] anxiety disorders[13] and post-traumatic stress disorder (PTSD).[14]

Despite the significant role of neurosteroids in the regulation of neurotransmission and their influence on the developing neural system, little is known about their possible role in child and adolescent psychopathology. Golubchik et al. (2007)[15] reviewed some of the relevant literature regarding the role of neurosteroids in pediatric neuropsychiatric disorders. The current chapter is aimed to analyze in more details the accumulating data on the possible involvement of neurosteroids in the pathophysiology and pharmacotherapy of psychopathological conditions in children and adolescents. Table 26.1 summarized the updated data on neurosteroid blood levels in a variety of psychiatric disorders among pediatric population.

26.1.1 Neurosteroids: Relevance of their Neuroprotective and Neurotrophic Function to Brain Development

The relevance of protective and trophic effects of neurosteroids in neurodevelopmental disorder is as yet unclear. Neurosteroids are synthesized during the maturation of the nervous system by glial cells and during embryonic life by some populations of neurons. Cell culture experiments suggest that neurosteroids may promote the survival and differentiation of both neurons and glial cells.[5] The impact of neurosteroids on neuronal cell is summarized in Table 26.2.

Table 26.1 Neurosteroid blood levels in a variety of pediatric psychiatric disorders

Disorder	Neurosteroid	Change	Reference
Autism	DHEA-S	No difference	[38]
		↑	[39[a]]
ADHD	DHEA, DHEA-S,	↓	[49]
	Pregnenolone	No difference	[49]
	DHEA, DHEA-S,	↓	[50]
Anorexia Nervosa	DHEA, DHEA-S	No difference	[71]
	Cortisol/DHEA-S ratio	↓	[71]
Depression	DHEA-S	↓	[80]
	DHEA	↓	[80]
	Cortisol/DHEA ratio	↑	[91]
Aggressive behavior, CD, ODD	DHEA, DHEA-S	No difference	[59]
	DHEA-S	↑	[60]
	DHEA-S	↑	[62]

ADHD – attention deficit hyperactivity disorder; CD – conduct disorder; DHEA – dehydroepiandrosterone; DHEA-S – dehydroepiandrosterone-sulphate; ODD – oppositional defiant disorder
[a]Adults

Neurosteroids also regulate the synthesis of myelin protein. Blocking the local synthesis or action of progesterone impairs remyelination of the regenerating axons, whereas administration of progesterone to lesion sites promotes the formation of new myelin sheaths. Thus it was proposed that progesterone may improve and facilitate myelinization.[16–18]

It was suggested by Sakamoto et al.[19] that progesterone promotes both dendritic outgrowth and synaptogenesis in Purkinje cells during cerebellar development. This promotion occurs via intranuclear receptor-mediated mechanisms. Such actions may contribute to the formation of the cerebellar neuronal circuit. Schumacher et al.[20] demonstrated that progesterone promotes the formation of new myelin sheaths after neuronal lesion, suggesting that it plays an important role during regeneration of the nervous system.

DHEA and DHEA-S are found in the brain of many species of animals. Furthermore enzymes crucial for their synthesis, particularly 17α-hydroxylase/c17,20 lyase (P450c17), are expressed in a developmentally regulated, region-specific fashion in the developing rodent brain. One region of embryonic expression of P450c17, namely the neocortical subplate, has been postulated to play a role in guiding cortical projections to their appropriate targets. In primary cultures of mouse embryonic neocortical neurons, DHEA increased the length of neuritis. These findings show that DHEA and DHEA-S exert neurobiological actions that are relevant to neurodevelopment, especially to the neocortex neural organization. Thus, DHEA and DHEA-S have been implicated

Table 26.2 Neuroprotective and neurotrophic function of neurosteroids relevant to neurodevelopmental disorders

Neurosteroid	Trophic function	Protective function	Reference
Progesterone	Improve, facilitate myelinization and formation of new myelinl sheaths and promotes both dendritic outgrowth and synaptogenesis Promotes both dendritic outgrowth and in synaptogenesis in Purkinje cells during cerebellar development Promotes the formation of new myelin sheaths	Neuroprotection of of rat cerebellar Purkinje cells following oxygen-glucose deprivation Mediates progestins' effects to against adrenal-ectomyinduced cell death the dentate female and male rats	[16–20, 92, 93]
DHEA and DHEA-S	Signaling molecules for neocortical organization Enhanced neuronal and glial survival and/or differentiation Increased neuronal survival and differentiation and reduced astroglial proliferation rates	Protection of the hippocampal neurons against glutamate-induced toxicity Protection of the hippocampal neurons against oxidative stress of hydrogen peroxide and against glutamate Possible protection against cytokine toxicity in the gestational period Increase neuron survival following anoxia	[5, 21–23, 25–27, 86]
Allopregnanolone	Modest effects on neuron survival that did not attain statistical significance		[5]

as potential signaling molecules for neocortical organization during neuronal development, suggesting that they have neurotrophic or neurotopic activity.[6,21,22]

It was also reported that low concentrations of DHEA and DHEA-S enhanced neuronal and glial survival and/or differentiation in dissociated cultures of 14-day mouse embryo brains.[23] A supplement of 10^{-8} M of DHEA or DHEA-S was sufficient to increase neuronal survival and differentiation and reduced astroglial proliferation rates in mouse brain cells in cultures. Bologa et al.[21] proposed that DHEA and DHEA-S are needed for normal function and that their presence may prevent neuronal loss and/or damage.

Neurosteroids affect survival of nervous cells in culture that are subjected to toxic insult. DHEA-S showed marked neuroprotective ability, while pregnenolone-sulfate facilitated the excitotoxic damage.[24] It was reported that DHEA-S, but not DHEA[22] protected hippocampal neurons against glutamate-induced toxicity.[25] The neuroprotective ability of DHEA-S, was also reported in primary cultures of rat hippocampal neurons subjected to an oxidative stress, induced by direct application of hydrogen peroxide.[26]

DHEA-S protected hippocampal neurons against glutamate but little protection was observed with equivalent doses of DHEA.

Neuroprotective effects have also been demonstrated in the gestational period. A significant negative correlation was found between DHEA and cytokine levels (which are thought to be toxic to the developing brain tissue). Serum DHEA levels are increased in the first and second trimesters of gestation. This may suppress the immune reaction during pregnancy. Such DHEA-induced suppression of cytokine production may be involved in modifying the maternal immune responses during and after pregnancy.[27]

It was also shown that neurosteroids increase embryonic neuronal survival following anoxia in rat embryonic day 18 cerebral cortical cultures. DHEA and DHEA-S significantly increased neuron survival following anoxia, by 85–87% and by 74%, respectively.[5] In contrast, allopregnanolone had modest and nonsignificant effects on neuron survival. These data indicate that pretreatment with DHEA and DHEA-S at physiologically relevant concentrations promotes neuronal survival following anoxia in embryonic rat cerebral cortical cultures.[5] The authors proposed that due to their modulating effect on anoxia in embryonic neurons, DHEA and DHEA-S may be relevant to neurodevelopmental disorders involving such insult.

These studies demonstrate a trophic activity of DHEA-S and indicate neuroprotective activity against in vitro toxicity models and suggests that DHEA-S in the brain may have an important role in the viability of neurons in pathological conditions.

26.2 Neurodevelopmental Disorders

Evidence suggests that it is likely that brain insult in utero or at birth plays a role in the development of some childhood mental disorders.[28,29] Many studies have been performed in search of early developmental factors that may cause neurodevelopmental disorders like autism, early-onset schizophrenia and ADHD.[30,31] Various environmental determinants, including gestational malnutrition, prenatal viral infection, obstetric complications, as well as pre- and peri-natal stress have been proposed as risk factors.[32]

The data are consistent with a neurodevelopmental model in which an early brain insult interacts with brain maturational events. It is proposed that maturation of specific brain areas is affected by early developmental pathology, particularly the prefrontal cortex and hippocampus.[33] Neonatal insult may lead to the emergence of abnormalities in a number of dopamine-related behaviors in young adulthood.[33]

It is of note that neurosteroids may influence glutamatergic, GABA-ergic dopaminergic and serotonergic activity.[6] All of these neurotransmitters are implicated in the pathophysiology of neurodevelopmental disorders.

The complex interaction between neurosteroids, neurotrophic factors, neuroprotective compounds and neurotransmitters mediating brain maturation and organization, as well as the relevance of such interactions to the emergence of neurodevelopmental disorders, and its possible preventive and therapeutic role at a critical periods of brain development merit further investigation.

Despite the accumulating data on the neuroprotective properties of various neurosteroids relevant to schizophrenia, there are no studies concerning the evaluation of the role of neurosteroids in early-onset schizophrenia.

In a laboratory animal model, neonatal allopregnanolone administration to rats disrupted the normal development of the prefrontal cortex and medial dorsal halamus, indicating that allopregnanolone affect the development of thalamo-cortical systems and may play a role in neurodevelopmental disorders such as schizophrenia.[34]

26.2.1 Childhood Autism

Autism was suggested to associated with disturbances of several systems of neurotransmitters, such as dopaminergic, adrenergic and serotonergic systems.[35–37]

Only one study has examined levels of DHEA-S in children and adolescent with autism. Tordjman et al.[38] found no difference in DHEA-S levels between young males with autism and controls. Strous et al.[39] demonstrated that un-medicated adult autistic patients have lower levels of plasma DHEA-S compared to normal controls. These findings, may provide evidence for reduced DHEA-S levels in autism. The apparently contradictory findings in those two studies may stem from the age difference between the two populations.

Since DHEA-S is a GABA antagonist and glutamate enhancer, the lower DHEA-S level may result in excessive GABA-ergic tone and a hypoglutamatergic state which may be relevant to pathophysiology of autism.[40–2]

Further investigation of autistic patients is required to clarify the role of DHEA-S and other neurosteroids in autism.

26.2.2 Attention Deficit

Results from studies in rodents have demonstrated that DHEA(S) administration can enhance attention and memory.[43–45] These effects may be mediated by the antagonistic activity of DHEA-S on the $GABA_A$ receptor[46] and agonistic action on the sigma-1 receptor.[47]

Wolf et al.[48] examined the effects of DHEA replacement on memory and selective attention before and after stress exposure in the elderly patients. DHEA replacement increased DHEA-S levels into ranges found in young subjects.

DHEA-substituted subjects showed a trend towards a larger cortisol stress response. In the attention test subjects under DHEA treatment performed better than subjects from the placebo group after stress.

Plasma levels of DHEA, DHEA-S and pregnenolone have been measured in attention deficit hyperactivity disorder (ADHD) patients.[49] They demonstrated inverse correlation between ADHD severity and levels of DHEA and pregnenolone. Higher blood levels of DHEA and DHEA-S were associated with lower ADHD scores, particular hyperactivity symptomatology. The researchers suggested a possible beneficial effect of various neurosteroids on the clinical manifestation of ADHD symptomatology.

In another study[50] the effect of a 3-month course of methylphenidate treatment on circulatory DHEA, DHEA-S and cortisol was assessed in ADHD children (15 males with ADHD combined type). Methylphenidate treatment led to significant clinical improvement. Parallel to a significant increase in serum levels of DHEA and DHEA-S, without increase in circulatory levels of cortisol. These findings suggest that DHEA and DHEA-S may play a role in the therapeutic effect of methylphenidate in ADHD. The modulatory effect of those neurosteroids on serotonergic, GABA-ergic, glutamatergic, sigma-1, noradrenergic and dopaminergic pathways may explain their possible role in the pharmacotherapy of ADHD.[51,52]

26.3 Other Disorders

26.3.1 Aggressive Behavior

An animals study has shown that DHEA administration inhibited the attack behavior of castrated mice.[53,54] In contrast, administration of DHEA-S displayed a dose-dependent increase in aggressive behavior.[55] The $GABA_A$ positive modulator allopregnenolone exhibited a bimodal effect on aggression similar to those for alcohol and benzodiazepines, suggesting that the $GABA_A$ receptor complex may be relevant to the modulatory effect of neurosteroids in aggressive behavior.[56] Although direct stimulation of $GABA_A$ receptors generally suppresses aggression, a number of studies have found that positive allosteric modulation of $GABA_A$ receptors can increase aggressive behavior. Thus some neurosteroids, similar to alcohol and benzodiazepines, are positive modulators of the $GABA_A$ receptor and can increase aggressive behavior. These effects are dose-dependent and higher doses of the compounds generally shift from heightening aggressive behavior to being sedative and anti-aggressive.[57]

While high doses of allopregnenolone effectively decrease aggressive behavior it in animals and produce sedation, at moderate doses this neurosteroid enhances aggression behavior.[58] No significant differences in DHEA and DHEA-S levels were found between aggressive and nonaggressive hospitalized children.[59] In contrast, a significant difference was observed in adolescent violent patients. The relationship between androgens and aggression has been examined in pre-pubertal boys who were diagnosed as suffering from severe aggressive behavior.[60] Boys with conduct disorder had significantly higher levels of DHEA-S than controls. There was a significant positive correlation between DHEA-S levels and the severity of aggression and delinquency (as rated by both parents and teachers). Additionally, children with oppositional defiant disorder (ODD) had higher DHEA-S levels than either the psychiatric control or healthy control groups. The authors were able to classify children as having either ODD or ADHD on the basis of their DHEA-S levels.[61]

Dmitrieva et al.[62] also reported higher DHEA-S levels in children with conduct disorder. They also reported a positive correlation between DHEA-S levels and disruptive behavior. Higher DHEA-S levels were found also in adults with cocaine addiction with history of conduct disorder in childhood.[63]

26.3.2 Anorexia Nervosa

The relevance of neurosteroids to mood and anxiety disorders was studied extensively, but there is little data regarding their role in eating disorders (ED). Zumoff et al.[64] found decreased mean daily values of plasma DHEA and DHEA-S in 14 anorexia nervosa (AN) patients and a concomitant increase in plasma cortisol concentrations. Winterer et al.[65] reported reduced baseline and ACTH-stimulated concentrations of DHEA and DHEA-S in six underweight AN women. Other studies have also found reduced levels of both DHEA and DHEA-S in acute 1 AN patients.[66,67] Additionally, Monteleon et al.[68] found increased plasma levels of allopregnanolone, DHEA, DHEA-S, and cortisol in 30 female AN patients, compared with 30 age-matched healthy controls.

Kaur and Kulkarni[69] found that treatment with progesterone increased body weight and elicited hyperphagia in female mice. On the other hand, treatment with DHEA-S produced a significant decrease in food intake and body weight and significantly suppressed progesterone-induced weight gain and hyperphagia.

Galderisi et al.[70] studied the influence of neuroactive steroids on neurocognitive functioning in subjects with eating disorders. There was a positive correlation between DHEA and DHEA-S levels and accuracy on executive tasks. There was also a positive correlation between cortisol and the speed of non-effortful learning. The researchers proposed that changes in neuroactive steroids may influence neurocognitive performance in ED patients.

A recent study by Stein et al.[71] did not detect differences in the plasma levels of cortisol, DHEA, and DHEA-S between female AN-restrictive type (AN-R) adolescent inpatients and non-ED control females. The patient group had significantly lower Cortisol to DHEA-S ratios compared to the controls. Since DHEA and cortisol modulate feeding behavior, it was proposed that lower cortisol/DHEA-S ratios in the AN-R group of female adolescent patients may contribute to the decrease in food intake in the acute phase of AN. Further investigation is needed to elucidate the role of neurosteroids in eating disorders in adolescents.

26.3.3 Adolescent Depression

DHEA has demonstrated beneficial effects in adults in mood, energy, interest, confidence and activity.[12,72,73] Clinical studies have demonstrated that neurosteroids are efficacious in the treatment of both dysthymia and major depression as well as depressive features of schizophrenia.[10,12,74,75]

Recent studies indicate that the pharmacological profiles of fluoxetine and fluvoxamine are correlated with their ability to increase the brain and cerebrospinal fluid content of allopregnanolone, a potent positive allosteric modulator of $GABA_A$ receptor activity.[76] Antidepressants elevate 3α-reduced neurosteroid levels in rodent brains, and their clinically effective antidepressive activity is associated with normalization of plasma and cerebrospinal fluid concentrations of endogenous neuroactive steroids in adult depressed patients, revealing a possible involvement of neuroactive steroids in the mode of action of antidepressants.[77,78] Moreover, blood DHEA-S levels may serve as a predictor of response or resistance of major depressed adults to electro-convulsive therapy (ECT).[79]

Accumulating data suggest the influence of neurosteroids on mood and its possible role in affective disorders in children and adolescents. Hyposecretion of salivary DHEA was significantly associated with MDD in young patients.[80] In 31% of MDD patients aged 8–16 years salivary levels of DHEA-S were below the 20th percentile of the control subjects.

A prospective study in the community demonstrated that in adolescents at high risk of psychopathology there is a significant association between evening DHEA peaks and subsequent MDD. This finding predicted MDD independently of recent life events, long-term difficulties or level of pre-morbid depressive symptoms.[81] The authors also suggested that peak rather than basal salivary levels may reflect an additional peripheral index of hypothalamic–pituitary–adrenal axis dysregulation. Furthermore, the variations in peak levels of neurosteroids may be related to adverse environmental or genetic influences.

It has also been shown that only the morning salivary cortisol/DHEA ratio predicted persistence of major depression in MDD adolescents.[82] It is as yet unclear whether these pre-morbid psychoendocrine properties are specific for MDD.

26.3.4 Stress and Anxiety

It has become clear that the neurosteroids play an important role in stress and anxiety. The assumption that neurosteroids have anxiolytic effects is based on various animal models of anxiety.[83–86] The role of neurosteroids in anxiety disorders among adults, such as social phobia, generalized anxiety disorder, panic disorder and PTSD is well studied.[13,14,87,88] Involvement of neurosteroids in stress and anxiety disorders is probably based mainly on their interaction with GABA-ergic neurotransmission.

Despite a large amount of evidence for involvement of neurosteroids in stress and anxiety in adults, little is known about their involvement in children and adolescents.

Recent studies support the hypothesis that exposure to stress or trauma during early childhood may disturb the formation of functional brain pathways, in particular, of the limbic circuits, known as relevant to stress response and emergence of anxiety disorders. The effects of exposure to early life trauma (juvenile stress) on emotional and cognitive aspects of behavior in adulthood was examined in animals by Avital et al.[89] It was found that an exposure to juvenile stress increased DHEA-S but not DHEA concentrations both in the hypothalamus and the entorhinal cortex. These findings indicated that an exposure to juvenile stress has long-lasting effects on behavior and DHEA-S levels in the hypothalamus and the entorhinal cortex.

These effects may be of relevance to our understanding of early life stress-related disorders such as PTSD, anxiety disorders and major depression. Shen et al.[90] recently showed that allopregnanolone, in contrast to its anxiety-reducing effect in adults, increase anxiety in pubertal female mice. It was reported that allopregnanolone reduces the tonic inhibition of pyramidal cells in hippocampus, leading to increased excitability.

This paradoxical effect of allopregnanolone results from inhibition of $GABA_A$ receptors, normally expressed at very low levels, but at puberty, their expression is increased and therefore, inhibition of $GABA_A$ receptors provides a mechanism for the generation of anxiety at puberty.

Consistent with the in vitro findings from Shen et al.[90] study, the onset of puberty reversed the behavioral effect of allopregnanolone from decreasing anxiety, as normally observed, to increasing anxiety. The anxiety-promoting effect of allopregnanolone at the onset of puberty might contribute to the aversive effects of stress that emerge at puberty in humans.

Further investigation is needed to elucidate the role of neurosteroids in the pathophysiology of stress and anxiety in children and adolescents.

26.3.5 Future Directions and the Limitations of Studies in Pediatric Population

As demonstrated in laboratory animal models of neurodevelopmental disorders, neurosteroids can regulate the synthesis of myelin and synaptogenesis. They are implicated as potential signaling molecules during neuronal development, enhanced neuronal and glial survival, and as such possess trophic and neuroprotective activity. Such neuromodulatory activities indicate important role of neurosteroids in the maintenance of the viability of neurons, brain development.[94]

Although there is increasing evidence for critical role of neurosteroids from the early stages of brain development until adolescence, the relevance of their protective and trophic effects in neurodevelopmental disorders is as yet unclear, and it may be the focus of future studies.

Neonatal insult may lead to the emergence of abnormalities in a number of neurotransmitter systems, as glutamatergic, GABA-ergic, serotonergic and dopaminergic that are relevant in pathogenesis of early psychopathology. All these neurotransmitter systems are regulated by neurosteroids indicating their additional role at the neurochemical level. Thus, evaluation of neurosteroid levels in a wide variety of stages of human brain development, including pre- and peri-natal critical period are essential for better understanding of the function of neurosteroids in normal and abnormal brain organization and pediatric psychiatric disorders.

Studying additional risk factors and possible neuroprotective properties of neurosteroids is needed and long-term follow-up studies from neonatal period up to adolescence may provide additional information.

Little is known about the changes in neurosteroid levels during infancy and childhood. Assessment of neurosteroid levels in there periods may help to understand better the continuity of neurosteroids availability across life cycle. It is especially important to evaluate neurosteroid levels in critical periods of vulnerability across childhood and adolescent development that seems to be relevant to neurodevelopmental disorders, such as schizophrenia and autism, in an attempt to clarify the interaction between neurosteroidogenesis and pathological processes, as well as the role of neurosteroids in the early prevention and pharmacotherapy of neurodevelopmental disorders and stress-related mental disorders.

26.4 Conclusion

Neurosteroids play a significant role in neurodevelopment and stress and mood regulation. The presence of appropriate brain levels protect the developing neural system from insult and contribute to adequate brain development.

Neurosteroids have a significant role in pediatric psychopathology, including depression, eating disorders, anxiety disorders, autistic disorders, aggressive behavior and ADHD. The complex interaction between neurosteroids, neurodevelopment, response to meaningful stressful life-events, genetic background and psychiatric disorders in children and adolescents, merits further investigation.

References

1. Goodyer IM, Park RJ, Netherton CM, et al. Possible role of cortisol and dehydroepiandrosterone in human development and psychopathology. Br J Psychiatry 2001; 79:243–249.
2. Kroboth FS, Salek AL, Pittenger TJ, et al. DHEA and DHEA-S: a review, J Clin Pharmacol 1999; 39:327–348.
3. Fadalti M, Petraglia F, Luisi S, et al. Changes of serum allopregnanolone levels in the first 2 years of life and during pubertal development, Pediatr Res 1999; 46:323–327.
4. Labrie F, Belander A, Cusan L. Marked decline in serum concentrations of adrenal C19 sex steroid precursors and conjugated androgen metabolites during aging. J Clin Endocrinol Metab 1997; 82:2396–2402.

5. Marx CE, Jarskog LF, Lauder JM, et al. Neurosteroid modulation of embryonic neuronal survival in vitro following anoxia. Brain Res 2000; 871:104–112.
6. Compagnone NA, Mellon SH. Dehydroepiandrosterone: a potential signalling molecule for neocortical organization during development. Proc Natl Acad Sci USA 1998; 95:4678–4683.
7. Schumacher M, Weill-Engerer S, Liere P, et al. Steroid hormones and neurosteroids in normal and pathological aging of the nervous system. Prog Neurobiol 2003; 71:3–29.
8. Harris DS, Wolkowitz OM, Reus VI. Movement disorder, memory, psychiatric symptoms and serum DHEA levels in schizophrenic and schizoaffective patients. World J Biol Psychiatry 2001; 2:99–102.
9. Shirayama Y, Hashimoto K, Suzuki Y., et al. Correlation of plasma neurosteroid level to the severity of negative symptoms in male patients with schizophrenia. Schizophr Res 2002; 58:9–74.
10. Ritsner M, Gibel A, Ram E, et al. Alterations in DHEA metabolism in schizophrenia: two-month case-control study. Eur Neuropsychopharmacol 2006; 16:137–146.
11. Ritsner MS, Gibel A, Ratner Y, et al. Improvement of sustained attention and visual and movement skills, but not clinical symptoms, after dehydroepiandrosterone augmentation in schizophrenia: a randomized, double-blind, placebo-controlled, crossover trial. J Clin Psychopharmacol 2006; 26: 495–499.
12. Wolkowitz OM, Reus VI. Neurotransmitters, neurosteroids and neurotrophins: new models of the pathophysiology and treatment of depression. World J Biol Psychiatry 2003; 4:8–102.
13. Brambilla F, Biggio G, Pisu MG, et al. Neurosteroid secretion in panic disorder. Psychiatry Res 2003; 118:107–116.
14. Rasmusson AM, Vythilingam M, Morgan CA, III. The neuroendocrinology of post-traumatic stress disorder: new directions. CNS Spectr 2003; 8:651–667.
15. Golubchik P, Lewis M, Maayan R, et al. Neurosteroids in child and adolescent psychopathology. Eur Neuropsychopharmacol 2007; 17:157–164.
16. Baulieu E, Schumacher M. Progesterone as a neuroactive neurosteroid, with special reference to the effect of progesterone on myelination. Steroids 2000; 65:605–612.
17. Chan JR, Phillips LJ, Glaser M. Glucocorticoids and progestins signal the initiation and enhance the rate of myelin formation. Proc Natl Acad Sci USA 1998; 95:10459–10464.
18. Koenig HL, Schumacher M, Ferzaz B, et al. Progesterone synthesis and myelin formation by Schwann cells. Science 1995; 268:1500–1503.
19. Sakamoto H, Ukena K, Tsutsui K. Effects of progesterone synthesized de novo in the developing Purkinje cell on its dendritic growth and synaptogenesis. J Neurosci 2001; 21:6221–6232.
20. Schumacher M, Robel P, Baulieu EE. Development and regeneration of the nervous system: a role for neurosteroids. Dev Neurosci 1996; 18:6–21.
21. Bologa L, Sharma J, Roberts E. Dehydroepiandrosterone and its sulfated derivative reduce neuronal death and enhance astrocytic differentiation in brain cell cultures. J Neurosci Res 1987; 17:225–234.
22. Mao X, Barger SW. Neuroprotection by dehydroepiandrosterone-sulfate: role of an NFkappaB-like factor. NeuroReport 1998; 9:59–63.
23. Roberts E, Bologa L, Flood JF, et al. Effects of dehydroepiandrosterone and its sulfate on brain tissue in culture and on memory in mice. Brain Res 1987; 406:357–362.
24. Weaver CE, Jr., Wu FS, Gibbs TT, et al. Pregnenolone sulfate exacerbates NMDA-induced death of hippocampal neurons. Brain Res 1998; 803:129–136.
25. Kimonides VG, Khatibi NH, Svendsen CN, et al. Dehydroepiandrosterone (DHEA) and DHEA-sulfate (DHEA-S) protect hippocampal neurons against excitatory amino acid-induced neurotoxicity. Proc Natl Acad Sci USA1998; 95:1852–1857.
26. Bastianetto S, Ramassamy C, Poirier J, et al. Dehydroepiandrosterone (DHEA) protects hippocampal cells from oxidative stress-induced damage. Mol Brain Res 1999; 66:35–41.
27. Tagawa N, Hidaka Y, Takano T, et al. Serum concentrations of dehydroepiandrosterone and dehydroepiandrosterone sulfata and their relation to cytokine production during and after normal pregnancy. Clin Chim Acta 2004; 340:187–193.

28. Marenco S, Weinberger DR. The neurodevelopmental hypothesis of schizophrenia: following a trail of evidence from cradle to grave. Dev Psychopathol 2000; 12:501–527.
29. Weinberger DR, Lipska BK. Cortical maldevelopment, antipsychotic drugs, and schizophrenia: in search of common ground. Schizophr Res 1995; 16:87–110.
30. Maimburg RD, Vaeth M. Perinatal risk factors and infantile autism. Acta Psychiatr Scand 2006; 114:257–264.
31. Milberger S, Biederman J, Faraone SV, et al. Pregnancy, delivery and infancy complications and attention deficit hyperactivity disorder: issues of gene-environment interaction. Biol Psychiatry 1997; 41:65–75.
32. Maki P, Veijola J, Jones PB, et al. Predictors of schizophrenia – a review. Br Med Bull 2005; 73:1–15.
33. Lipska BK. Using animal models to test a neurodevelopmental hypothesis of schizophrenia. J Psychiatry Neurosci 2004; 29:282–286.
34. Gizerian SS, Morrow AL, Lieberman JA, et al. Neonatal neurosteroid administration alters parvalbumin expression and neuron number in medial dorsal thalamus of adult rats. Brain Res 2004; 1012:66–74.
35. Anderson GM. Monoamines in autism: an update of neurochemical research on a pervasive developmental disorder. Med Biol 1987; 65:67–74.
36. Buitelaar JK, Willemsen-Swinkels SH. Autism: current theories regarding its pathogenesis and implications for rational pharmacotherapy. Paediatr Drugs 2000; 2:67–81.
37. Chugani DC. Role of altered brain serotonin mechanisms in autism. Mol Psychiatry 2002; 7: S16–S17.
38. Tordjman S, Anderson GM, McBride PA, et al. Plasma androgens in autism. J Autism Dev Disord 1995; 25:295–304.
39. Strous RD, Golubchik P, Maayan R, et al. Lowered DHEA-S plasma levels in adult individuals with autistic disorder. Eur Neuropsychopharmacol 2003; 15:305–309.
40. Cohen BI. GABA-transaminase, the liver and infantile autism. Med Hypotheses 2001; 57:673–674.
41. Dhossche D, Applegate H, Abraham A, et al. Elevated plasma gamma-aminobutyric acid (GABA) levels in autistic youngsters: stimulus for a GABA hypothesis of autism. Med Sci Monit 2002; 8:1–6.
42. Carlsson ML. Hypothesis: is infantile autism a hypoglutamatergic disorder? Relevance of glutamate – serotonin interactions for pharmacotherapy. J Neural Transm 1998; 105:525–535.
43. Flood JF, Roberts E. Dehydroepiandrosterone sulfate improves memory in aging mice. Brain Res 1988; 448:178–181.
44. Melchior CL, Ritzmann RF. Neurosteroids block the memory-impairing effects of ethanol in mice. Pharmacol Biochem Behav 1996; 53:51–56.
45. Frye CA, Sturgis JD. Neurosteroids affect spatial/reference, working, and long-term memory of female rats. Neurobiol Learn Mem 1995; 64:83–96.
46. Majewska MD. Neurosteroids: endogenous bimodal modulators of the GABA-A receptor. Mechanism of action and physiological significance. Prog Neurobiol 1992; 38:379–395.
47. Monnet FP, Mahe V, Robel P, Baulieu EE. Neurosteroids, via sigma receptors, modulate the [^{3}H]norepinephrine release evoked by *N*-methyl – aspartate in the rat hippocampus. Proc Natl Acad Sci USA 1995; 92:3774–3778.
48. Wolf OT, Kudielka BM, Hellhammer DH, et al. Opposing effects of DHEA replacement in elderly subjects on declarative memory and attention after exposure to a laboratory stressor. Psychoneuroendocrinology 1998; 23:617–629.
49. Strous RD, Spivak B, Yoran-Hegesh R, et al. Analysis of neurosteroid levels in attention deficit hyperactivity disorder. Int J Neuropsychopharmacol 2001; 4:259–264.
50. Maayan R, Yoran-Hegesh R, Strous R, et al. Three-month treatment course of methylphenidate increases plasma levels of dehydroepiandrosterone (DHEA) and dehydroepiandrosterone-sulfate (DHEA-S) in attention deficit hyperactivity disotder. Neuropsychobiology 2003; 48:111–115.

51. Maurice T, Gregoire C, Espallergues J. Neuro(active)steroids actions at the neuromodulatory sigma1 (sigma1) receptor: biochemical and physiological evidences, consequences in neuroprotection. Pharmacol Biochem Behav 2006; 84:581–597.
52. Gibbs TT, Russek SJ, Farb DH. Sulfated steroids as endogenous neuromodulators. Pharmacol Biochem Behav 2006; 84:555–567.
53. Haug M, Spetz JF, Schlegel ML, et al. Dehydroepiandrosterone inhibits aggressive behavior in castrated male mice. Comptes Rendus Seances Acad Sci 1983; 296:975–977.
54. Schlegel ML, Spetz JF, Robel P, et al. Studies on the affects of dehydroepiandrosterone and its metabolites on attack by castrated mice on lactating intruders. Physiol Behav 1985; 34:867–870.
55. Nicolas LB, Pinoteau W, Papot S, et al. Aggressive behavior induced by the sulfatase inhibitor COUMATE and by DHEA-S in CBA/H mice. Brain Res 2001; 922:216–222.
56. Engel SR, Grant KA. Neurosteroids and behavior. Int Rev Neurobiol 2001;46:321–348.
57. Miczek KA, Fish EW, De Bold JF. Neurosteroids, $CABA_A$ receptors, and escalated aggressive behavior. Horm Behav 2003; 44:242–257.
58. Fish EW, Faccidomo S, DeBolt JF, et al. Alcohol, allopregnanolone and aggression in mice. Psychopharmacology 2001; 153:391–399.
59. Constantino JN, Grosz D, Saenger P, et al. Testosterone and aggression in children. J Am Acad Child Adolesc Psych 1993; 32:1212–1222.
60. Van Goozen SHM, Matthys W, Cohen-Kettenis PT, et al. Adrenal androgens and aggression in conduct disorder prepubertal boys and normal controls. Biol Psychiatry 1998; 43:156–158.
61. Van Goozen SH, van den Ban E, Matthys W, Cohen-Kettenis PT, et al. Increased adrenal androgen functioning in children with oppositional defiant disorder: a comparison with psychiatric and normal controls. J Am Acad Child Adolesc Psych 2000; 39:1446–1451.
62. Dmitrieva TN, Oades RD, Hauffa BP, et al. Dehydroepiandrosterone sulphate and corticotropin levels are high in young male patients with conduct disorder: comparison for growth factors, thyroid and gonadal hormones. Neuropsychobiology 2001; 43:134–140.
63. Buydens-Branchey L, Branchey M. Cocaine addicts with conduct disorder are typified by decreased cortisol responsivity and high plasma levels of DHEA-S. Neuropsychobiology 2004; 50:161–166.
64. Zumoff B, Walsh BT, Katz JL, et al. Subnormal plasma dehydroisoandrosterone to cortisol ratio in anorexia nervosa: a second hormonal parameter of ontogenic regression. J Clin Endocrinol Metab 1983; 56:668–672.
65. Winterer J, Gwirtsman HE, George DT, et al. Adrenocorticotropin-stimulated adrenal androgen secretion in anorexia nervosa: impaired secretion at low weight with normalization after long term recovery. J Clin Endocrinol Metab 1985; 61:693–697.
66. Devesa J, Perz-Fernandes R, Bokser L, et al. Adrenal androgen secretion and dopaminergic activity in anorexia nervosa. Horm Metab Res 1987; 20:57–60.
67. Gordon CM, Goodman E, Emans SJ, et al. Physiologic regulation of bone turnover in adolescent girls with anorexia nervosa. J Pediatr 2000; 141:64–70.
68. Monteleon P, Luisi M, Colurcio B, et al. Plasma levels of neuroactive steroids are increased in untreated women with anorexia nervosa or bulimia nervosa. Psychosom Med 2001; 63:52–68.
69. Kaur G, Kulkarni SK. Subchronic studies on modulation of feeding behavior and body weight by neurosteroids in female mice. Methods Find Exp Clin Pharmacol 2001; 23:115–119.
70. Galderisi S, Mucci A, Monteleone P, et al. Neurocognitive functioning in subjects with eating disorders: the influence of neuroactive steroids. Biol Psychiatry 2003; 53:921–927.
71. Stein D, Maayan R, Ram A, et al. Circulatory neurosteroid levels in underweight female adolescent anorexia nervosa inpatients and following weight restoration. Eur Neuropsychopharmacol 2005; 15:647–653.
72. Wolkowitz OM, Reus VI, Keebler A, et al. Double-blind treatment of major depression with dehydroepiandrosterone. Am J Psychiatry 1999; 156:646–649.
73. Herbert J. Neurosteroids, brain damage, and mental illness. Exp Gerontol 1998; 33:713–727.
74. Bloch M, Rubinow DR, Adams LF, et al. Dehydroepiandrosterone treatment of midlife dysthymia. Biol Psychiatry 1999; 45:1531–1532.

75. Schmidt PJ, Daly RC, Bloch M, et al. Dehydroepiandrosterone monotherapy in midlife-onset major and minor depression. Arch Gen Psychiatry 2005; 62:154–162.
76. Pinna G, Costa E, Guidotti A. Fluoxetine and norfluoxetine stereospecifically and selectively increase brain neurosteroid content at doses that are inactive on 5-HT reuptake. Psychopharmacology 2006; 86:362–372.
77. Uzunova V, Sheline Y, Davis JM, et al. Increase in the cerebrospinal fluid content of neurosteroids in patients with unipolar major depression who are receiving fluoxetine or fluvoxamine. Proc Natl Acad Sci USA 1998; 95:3239–3244.
78. Uzunova V, Sampson L, Uzunov DP. Relevance of endogenous 3alpha-reduced neurosteroids to depression and antidepressant action. Psychopharmacology 2006; 186:351–361.
79. Maayan R, Yagorowski Y, Grupper D, et al. Basal plasma dehydroepiandrosterone sulfate level: a possible predictor for response to electroconvulsive therapy in depressed psychotic inpatients. Biol Psychiatry 2000; 48:693–701.
80. Goodyer IM, Herbert J, Altman PM, et al. Adrenal secretion during major depression in 8–16-year-olds: I. Altered diurnal rhythms in salivary cortisol and dehydroepiamdrosterone (DHEA) presentation. Psychol Med 1996; 26:245–256.
81. Goodyer IM, Herbert J, Tamplin A, et al. Recent life events, cortisol, dehydroepiandrosterone and the onset of major depression in high-risk adolescents. Br J Psychiatry 2000; 177:499–504.
82. Goodyer IM, Herbert J, Tamplin A. Psychoendocrine antecedents of persistent first-episode major depression in adolescents: a community-based longitudinal enquiry. Psychol Med 2003; 33:601–610.
83. Serra M, Pisu MG, Littera M, et al. Social isolation-induced decreases in both the abundance of neuroactive steroids and GABA(A) receptor function in rat brain. J Neurochem 2000; 75:732–740.
84. Ovsiukova MV, Amikishieva AV, Kudriavtseva NN, et al. [Anxiolytic effect of dehydroepiandrosterone sulfate in male mice with high anxiety level]. Zh Vyssh Nerv Deiat Im I P Pavlova 2003; 53:789–793.
85. Reddy DS, O'Malley BW, Rogawski MA. Anxiolytic activity of progesterone in progesterone receptor knockout mice. Neuropharmacology 2005; 48:14–24.
86. Cohen H, Maayan R, Touati-Werner D, et al. Decreased circulatory levels of neuroactive steroids in behaviourally more extremely affected rats subsequent to exposure to a potentially traumatic experience. Int J Neuropsychopharmacol 2006; 4:1–7.
87. Heydari B, Le Melledo JM. Low pregnenolone sulphate plasma concentrations in patients with generalized social phobia. Psychol Med 2002; 32:929–933.
88. Semeniuk T, Jhangri GS, Le Melledo JM. Neuroactive steroid levels in patients with generalized anxiety disorder. J Neuropsychiatry Clin Neurosci 2001; 13: 396–398.
89. Avital A, Ram E, Maayan R, et al. Effects of early-life stress on behavior and neurosteroid levels in the rat hypothalamus and entorhinal cortex. Brain Res Bull 2006; 15;68: 419–424.
90. Shen H, Gong QH, Aoki C, et al. Reversal of neurosteroid effects at $GABA_A$ receptors triggers anxiety at puberty. Nat Neurosci 2007; 10:469–477.
91. Goodyer IM, Herbert J, Altham PM. Adrenal steroid secretion and major depression in 8- to 16-year-olds, III. Influence of cortisol/DHEA ratio at presentation on subsequent rates of disappointing life events and persistent major depression. Psychol Med 1998; 28:265–273.
92. Ardeshiri A, Kelley MH, Korner IP, et al. Mechanism of progesterone neuroprotection of rat cerebellar Purkinje cells following oxygen-glucose deprivation. Eur J Neurosci 2006; 24:2567–2574.
93. Rhodes ME, McCormick CM, Frye CA. 3alpha,5alpha-THP mediates progestins' effects to protect against adrenalectomy-induced cell death in the dentate gyrus of female and male rats. Pharmacol Biochem Behav 2004; 78:505–512.
94. Mellon SH. Neurosteroid regulation of central nervous system development. Pharmacol Ther 2007;16.

Index

MIX
Papier aus verantwortungsvollen Quellen
Paper from responsible sources
FSC® C105338

If you have any concerns about our products,
you can contact us on
ProductSafety@springernature.com

In case Publisher is established outside the EU,
the EU authorized representative is:
Springer Nature Customer Service Center GmbH
Europaplatz 3, 69115 Heidelberg, Germany

Printed by Libri Plureos GmbH
in Hamburg, Germany